Benchmark Papers in Acoustics

Series Editor: R. Bruce Lindsay
Brown University

Other volumes in the series

Acoustics: Its Historical and Philosophical Development, R. Bruce Lindsay
Musical Acoustics: Volume I, The Violin Family, Carleen Hutchins
Psychological Acoustics, Arnold M. Small, Jr.
Physical Acoustics, R. Bruce Lindsay
Noise and Noise Control, Lewis S. Goodfriend
Architectural Acoustics, Thomas D. Northwood
Light and Sound Interaction, Osman K. Mawardi
Physiological Acoustics, Arnold M. Small and Joel S. Wernick
Vibration Problems, Arturs Kalnins

Benchmark Papers in Acoustics

UNDERWATER SOUND

Edited by
VERNON M. ALBERS
Pennsylvania State University

Dowden, Hutchinson & Ross, Inc.
Stroudsburg, Pennsylvania

Library of Congress Catalog Card Number: 72–79141
ISBN: 0–87933–006–6

Manufactured in the United States of America.

Exclusive distributor outside the United States and Canada:
JOHN WILEY & SONS, INC.

Acknowledgements and Permissions

ACKNOWLEDGEMENTS

The Acoustical Institute at the Academy of Sciences of USSR—SOVIET PHYSICS AND ACOUSTICS
"Acoustic Nondirectional Ceramic Receiving Transducers"

PERMISSIONS

The following papers have been reprinted with the permission of the authors, publishers, and the present copyright owners.

The American Physical Society—THE PHYSICAL REVIEW
"Propagation of Radiation in a Medium with Random Inhomogeneities"
"Nucleation by Cosmic Rays in Ultrasonic Cavitation"

The Acoustical Society of America—JOURNAL OF THE ACOUSTICAL SOCIETY OF AMERICA
"Wave Propagation in a Randomly Inhomogeneous Medium: I"
"Range Dependence of Acoustic Fluctuations in a Randomly Inhomogeneous Medium"
"Normal-Mode Theory Applied to Short Range Propagation in an Underwater Acoustic Surface Duct"
"Axial Focusing of Sound in the SOFAR Channel"
"Fluctuation in Horizontal Acoustic Propagation Over Short Depth Increments"
"Refraction Correction in Constant-Gradient Media"
Abstract from "Sound Absorption in Aqueous Solutions of Magnesium Sulfate and in Sea Water"
"Equation for the Speed of Sound in Sea Water"
"Sound Speed in Pure Water and Sea Water"
"A Decade of Experience with Velocimeters"
"Ambient Noise under Arctic Sea Ice"
"Origin of the Knudsen Spectra"
"Acoustical Ambient Noise in the Ocean—Spectra and Sources"
"Sound Scattering by Solid Cylinders and Spheres"
"Scattering of Acoustic Energy by Solid and Air Filled Cylinders in Water"
"The Scattering of Sound from the Sea Surface"
"Reflection of a Plane Sound Wave from a Sinusoidal Surface"
"Acoustic Characteristics of Underwater Bottoms"
"Sound Reflection from a Low Velocity Bottom"
"The Electroacoustic Sensitivity of Cylindrical Ceramic Tubes"
"Origin of Mechanical Bias for Transducers"
"Acoustic Focusing with Spherical Structures"
"Propagation of Displacement Waves in the Sea"
"Absolute Measurement of Sound without a Primary Standard"
"Laboratory Calibrator for Gradient Hydrophones"
"Underwater-Sound-Transducer Calibration from Nearfield Data"
"Sonar Transducer Calibration in a High-Pressure Tube"
Abstract from "Parametric End-Fire Array"
Abstract from "Experimental Investigation of a Parametric End Array"
"Parametric Acoustic Array"

continued

Acknowledgements and Permissions, continued

"Connection between the Fay and Fubini Solutions for Plane Sound Waves of Finite Amplitude"
"Experiments on the Acoustic Modulation of Large Amplitude Waves"
"A Novel Technique for Measuring the Strength of Liquids"

The American Institute of Physics—THE REVIEW OF SCIENTIFIC INSTRUMENTS
"Sing-Around Ultrasonic Velocimeter for Liquids"

The American Institute of Physics—THE JOURNAL OF APPLIED PHYSICS
"Reflection of Sound from Randomly Rough Surfaces"

The Sears Foundation for Marine Research, Yale University—THE JOURNAL OF MARINE RESEARCH
"Underwater Ambient Noise"

The Royal Society, London—PROCEEDINGS
"On Sound Generated Aerodynamically: I—General Theory"
"On Sound Generated Aerodynamically: II—Turbulence as a Source of Sound"

University of Cambridge—THE JOURNAL OF FLUID MECHANICS
"Resolution and Structure of the Wall Pressure Field Beneath a Turbulent Boundary Layer"

S. Hirzel-Verlag, Stuttgart—ACUSTICA
"A Mechanism of Acoustic Echo Formation"
"The High Frequency Echo Structure of Some Simple Body Shapes"

Institute of Physics, London—PROCEEDINGS OF THE PHYSICAL SOCIETY
"Underwater Explosions as Acoustic Sources"

Series Editor's Preface

This volume is one of a series devoted to the reproduction of seminal articles in various branches of acoustics, that is, papers which significantly influenced the development of the science in a certain direction and introduced concepts and methods which possess basic utility in modern acoustics. Each volume is prefaced by an introduction summarizing the technical significance of the field being covered. Each article is accompanied by editorial commentary on the scientific status of the author or authors and the relation of their work to the field as a whole, as well as explanatory notes where necessary. Articles in languages other than English are either translated or abstracted in English. An adequate subject index completes each volume.

It is the hope of the publisher and editor that these volumes will constitute a working library of the most important technical literature of acoustics of value to students and research workers alike.

R. Bruce Lindsay

Contents

III. AMBIENT NOISE IN THE SEA

IV. FLOW NOISE

V. SOUND SCATTERING AND REVERBERATION

VI. UNDERWATER TRANSDUCERS AND CALIBRATION

VII. FINITE AMPLITUDE EFFECTS

VIII. ACOUSTIC CAVITATION IN LIQUIDS

Author List

Introduction

Relatively little research was done in the field of underwater acoustics prior to World War II. The speed of sound in fresh water was measured in Lake Geneva in 1826 by Colladon and Sturm. Although the measurements were made with primitive apparatus, the results were surprisingly accurate.

Before the turn of the century, it was recognized that sound waves were propagated in water with less attenuation than any other form of energy.

As early as 1902, there was a considerable industry, in England and the United States of America, devoted to producing underwater bells as shoal warnings and for avoidance of ship collisions.

The loss of the Titanic in 1912 gave rise to proposals for ultrasonic echo ranging as a means of detecting obstacles such as icebergs. The efforts in the U.S.A. culminated in the Fessenden oscillator which operated in the audio frequency region. The first echo from an iceberg was received in 1914 at a range of about two miles at a frequency of about 1100 Hz. The first echo at ultrasonic frequency was obtained by Langevin in France in 1916, and the first echoes from submerged submarines were obtained nearly simultaneously, in 1917, by Langevin in France, and by Boyle in England.

During World War I, much work was done under the sponsorship of the Board of Inventions of the Royal Naval Scientific Service on the development and application of hydrophones for use in combating submarines. Until 1917, vacuum tube amplifiers were not available and the sound was measured in terms of the currents generated in the hydrophones. When amplifiers became available, it was possible to utilize piezoelectric materials as the sensitive elements in hydrophones.

When the seriousness of the submarine menace became apparent during the first World War, the Secretary of the Navy, Josephus Daniels, asked Thomas A. Edison to form a technical advisory group to serve the Navy in matters of inventions and scientific endeavors.[1] This group was called the Naval Consulting Board of the United States. They proposed the immediate establishment of a laboratory, but the Naval Research Laboratory was not established until 1923.

[1] E. Klein, Underwater sound and naval acoustical research and applications before 1939, *J. Acoust. Soc. Amer.* **43**, 931–947 (1968).

One device developed was the C tube which consisted of a beam with a rubber air-filled bulb at each end. The air in each bulb was connected to one side of a stethoscope and, as the beam was turned so that the sound appeared to come from directly ahead of the observer, the direction to the submarine could be determined.

The scientific attachés in England, France, and Italy kept the American scientists informed about the developments in Europe.

The ultrasonic transducers developed in Europe were based on the quartz–steel sandwich type but the Americans worked on and developed synthetic Rochelle salt crystal transducers.[2]

The Naval Research Laboratory, in cooperation with the B. F. Goodrich Company, developed RHO—C rubber and were able to produce an acoustically transparent housing for the transducers.

Although piezoelectric crystal transducers appeared promising, research was undertaken with magnetostrictive materials. Magnetostrictive transducers promised to produce more powerful signals in the water than piezoelectric transducers and were, therefore, better adapted to echo-ranging systems. Thin-walled nickel tubes were used at 25 kHz and were used for echo-ranging transducers and in depth finding systems.

Because of the importance of underwater sound in combating the submarine menace, a high level of effort was undertaken immediately prior to and during World War II under the U.S. Navy and Division 6 of the National Defense Research Committee (NDRC). In this program, there was a large research effort in studying propagation of sound in the ocean, physical properties of the ocean medium which influence the propagation of sound, and ambient noise in the ocean. There was also a major effort in the development of transducers and other instrumentation because of their importance in the development of more effective echo/ranging systems to which the name sonar was ultimately given.

Because of the importance of refraction on sound propagation in the ocean, means of measuring the temperature structure were developed and the deep scattering layers were discovered and studied by means of fathometers.

The research conducted under the NDRC was mostly classified during the war, but the results of much of the underwater sound research was published in volumes 7 and 8 of the National Defense Research Committee Summary Technical Report. These reports have been reprinted by the Department of the Navy, Naval Material Command.[3,4] Because these two NDRC Report volumes are readily available, none of the individual NDRC reports are reproduced in this volume.

Although the research in underwater acoustics during World War II was done for the purpose of solving military problems such as detecting and tracking enemy submarines and the development of acoustic homing torpedoes, there was a great

---

[2]W. G. Cady, Piezoelectricity and ultrasonics, *Sound, Its Uses and Control* **2**, 46–52 (1963).

[3]*Principles and Applications of Sound in the Sea,* U.S. Govt. Printing Office, Washington, D.C. 20204

[4]*Physics of Sound in the Sea,* U.S. Govt. Printing Office, Washington, D.C. 20204

deal of development of technology and instrumentation useful for other applications of underwater sound. At the same time, much was learned about the physical properties of the ocean medium which affects the propagation of sound in the ocean.

Since World War II, the military applications of underwater sound have been pursued at a relatively high level of effort. Modern submarines operate at higher speeds and at greater depths than those used during World War II, and nuclear propulsion eliminates the necessity for a submarine to surface. These characteristics of submarines impose the requirement of greater range for sonar and homing torpedoes than were necessary during World War II. Refraction of sound in the ocean medium thus becomes of increasing importance; this has resulted in a great deal of research, both theoretical and experimental, on propagation of sound in the ocean. Although this research is motivated by its military importance, it is yielding much information about the physical properties of the ocean medium.

Fortunately, this work, except for its specific military applications, is unclassified and the results are published in the unclassified journals. This work is supported, to a large extent, by the navies in the various countries in government laboratories and in university laboratories under contract to the navies. This is an important example of research conducted for its military importance which has contributed greatly to our understanding of the physical properties of the ocean medium. The technology developed in this research has been applied in oceanography, marine geology, marine biology, and navigation.

The publications reproduced in this volume are grouped in the following categories: I. Propagation, Including Fluctuations and Attenuation; II. Velocity of Sound; III. Ambient Noise in the Sea; IV. Flow Noise; V. Sound Scattering and Reverberation; VI. Underwater Transducers and Calibration; VII. Finite Amplitude Effects; and VIII. Acoustic Cavitation in Liquids.

Propagation Including Fluctuations and Attenuation

I

In the many measurements of sound propagation made during the war, there was much indication of large fluctuations in the level of signal received when the level of signal transmitted was held constant.

Since such experiments were necessarily made in the open ocean using transducers suspended from or attached to ships, it was difficult to determine if these fluctuations were caused by the properties of the medium or if they were due to the motions of the transducers as the ships pitched and rolled.

Dr. Bergmann was Assistant Director of the Sonar Analysis Group, Division of War Research, Columbia University, and the Oceanographic Institution at Woods Hole, Massachusetts. He has done research in the fields of relativistic field theories, wave propagation and scattering, electron optics, tactosols, stochastic problems, and irreversible processes. He is currently Professor of Physics at Syracuse University, Syracuse, New York.

This is a theoretical study, based on ray theory, to predict the effect of the inhomogeneities of the medium in producing the observed fluctuations.

This paper is reprinted with the permission of the author and the Editors of the *Physical Review*.

Physical Review, 1946, V. 70, p. 486–492.

Propagation of Radiation in a Medium with Random Inhomogeneities*

1

PETER G. BERGMANN
Sonar Analysis Group, 64th Floor, 350 Fifth Avenue, New York 1, New York

(Received May 13, 1946)

By means of the methods of geometrical optics, approximate formulae are being derived which correlate the statistical properties of the inhomogeneities of the transmitting medium with the fluctuations to be expected in the signal level of radiative energy. Through a further simplification of the formulae obtained, it is possible to predict the dependence of signal fluctuation on range without detailed knowledge of the statistical parameters of the "microstructure" of the transmitting medium.

INTRODUCTION

WHEN radiative energy is propagated over considerable distances, the transmitting medium is rarely the homogeneous expanse it is assumed to be in elementary theory. The medium is bounded, and it may have an internal structure, such as a density gradient. In addition, the most important non-solid media of propagation, the atmosphere and the ocean, are known to possess a rapidly changeable random structure of comparatively small dimensions, which is caused by local heating, convective currents, and similar factors. In this paper, the modification of the radiative field which is caused by random structure will be derived for the case of small changes and on the assumption of a wave-length so short that the formulae of ray optics are valid. The results obtained may find application in the propagation of either electromagnetic or sound waves of high frequencies in the atmosphere or in a similarly extended medium.

THE PROBLEM

We shall restrict ourselves at once to the application of ray optical methods. Treatment of the same problem by means of wave optics is planned for the future. The basic equations of ray optics in the stationary case may be written in the form

$$\left.\begin{aligned}(\nabla S)^2 &= n^2,\\ \nabla\cdot\left(\frac{I}{n}\nabla S\right) &= 0.\end{aligned}\right\} \qquad (1)$$

In these equations, the function $S(\mathbf{r})$ is the optical path length counted from the radiative source to the point $\mathbf{r}$ along the connecting ray path. Each constant value of $S>0$ corresponds to a wave front in the ray optical approximation. $n(\mathbf{r})$ is the local index of refraction, and $I(\mathbf{r})$ is the intensity of the radiative field, measured in units of energy passing per unit time through a unit area cross section perpendicular to the rays. In electromagnetic theory, I is the averaged magnitude of Poynting's vector, while in acoustics, I is the mean square pressure divided by ρc. The first of the Eqs. (1) expresses Huygens' principle that the distance between consecutive wave fronts is inversely proportional to the local index of refraction; while the second equation is the law of conservation of energy. Absorption and scattering are disregarded. For the following discussion, it is convenient to introduce the level L, defined as the logarithm (on some base) of I/n.[1] We shall measure the level in nepers, that is on the base e. In terms of level, Eqs. (1) may be rewritten in the form

$$\begin{aligned}(\nabla S)^2 &= n^2,\\ \nabla S\cdot\nabla L+\nabla^2 S &= 0.\end{aligned} \qquad (2)$$

It will now be assumed that n is very nearly equal to unity. We shall set

$$n = 1+\lambda n', \qquad (3)$$

with λ a constant parameter,

$$\lambda \ll 1. \qquad (4)$$

* This work represents in part results of research carried out by the Sonar Analysis Group under contracts between Columbia University and the Office of Scientific Research and Development and between Woods Hole Oceanographic Institution and the Bureau of Ships, Navy Department.

[1] The quantity I/n has no direct physical significance and is chosen merely for convenience. However, because n very nearly equals unity, according to the assumptions made, I/n does not differ significantly from I. In most practical cases, n differs from unity by an amount less than 10^{-2}.

An approximate solution of Eqs. (1) will be obtained in terms of definite integrals.

APPROXIMATE SOLUTION OF THE BASIC EQUATIONS

If n' vanishes, the solution of Eq. (1) for a point source is

$$\left.\begin{aligned} S_0&=r,\\ L_0&=b(\Omega)-2\ln r,\end{aligned}\right\} \tag{5}$$

where r is the distance from the source and b is a constant or, in the case of a directional source, a function of the angle Ω. That the pair S_0, L_0, is a solution can be verified directly by substituting the expressions (5) into Eqs. (2) with n equal to unity. We shall call S_0, L_0 the zero approximation.

The first approximation is obtained if Eq. (2) is expanded into a power series with respect to the parameter λ. Denoting all first-order variables by primes, we obtain the following conditions for the first approximation:

$$\left.\begin{aligned} \mathbf{r}_0\cdot\nabla S'&=n',\\ \mathbf{r}_0\cdot\nabla L'&=-\nabla^2S'-\nabla L_0\cdot\nabla S',\end{aligned}\right\} \tag{6}$$

where $\mathbf{r}_0$ is the unit vector pointing directly away from the source.

Fortunately, the two Eqs. (6) need not be solved simultaneously. They can be solved with the further condition that the variables S' and L' are to vanish at the location of the source.

The solution of the first Eq. (6) is provided by the integral

$$S'(\mathbf{r})={}^*\!\!\int_{\rho=0}^{\mathbf{r}} n'(\rho)d\rho. \tag{7}$$

The symbol ${}^*\!\!\int_0^{\mathbf{r}}$ refers to integration along a straight line from the location of the source to the point characterized by the radius vector $\mathbf{r}$. ρ is the variable of integration. For what follows, we shall require the gradient and the Laplacian of S'. Figure 1 shows how the path of integration must be transformed if a differentiation of an integral of the type (7) is to be carried out. Considering the unspecified integral

$$J(\mathbf{r})={}^*\!\!\int_0^{\mathbf{r}} A(\rho)d\rho, \tag{8}$$

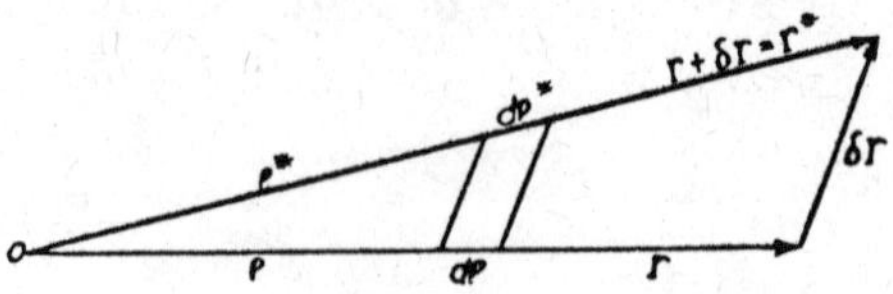

FIG. 1. Variation of the line integral.

we find that the variation of this integral may be written as

$$\delta J(\mathbf{r})={}^*\!\!\int_0^{\mathbf{r}} \delta[A(\rho)d\rho]$$

$$={}^*\!\!\int_0^{\mathbf{r}} \delta A(\rho)d\rho+{}^*\!\!\int_0^{\mathbf{r}} A(\rho)\delta d\rho. \tag{9}$$

The local variation $\delta A(\rho)$ is, according to the figure,

$$\delta A(\rho)=\frac{\rho}{r}\nabla_\rho A(\rho)\cdot\delta\mathbf{r} \tag{10}$$

while the variation of $d\rho$ equals

$$\delta d\rho=\frac{\mathbf{r}_0\cdot\delta\mathbf{r}}{r}d\rho. \tag{11}$$

We find, therefore,

$$\begin{aligned}\delta J(\mathbf{r})&=\frac{1}{r}\delta\mathbf{r}\cdot{}^*\!\!\int_0^{\mathbf{r}}[\rho\nabla_\rho A(\rho)+\mathbf{r}_0A(\rho)]d\rho\\ &=\frac{1}{r}\delta\mathbf{r}\cdot{}^*\!\!\int_0^{\mathbf{r}}\nabla_\rho[\rho A(\rho)]d\rho.\end{aligned} \tag{12}$$

The gradient of S', Eq. (7), is thus

$$\nabla S'(\mathbf{r})=\frac{1}{r}{}^*\!\!\int_0^{\mathbf{r}}\nabla_\rho(\rho n')d\rho. \tag{13}$$

Iteration of this process of differentiation leads to the result

$$\nabla^2S'(\mathbf{r})=\nabla\left(\frac{1}{r}\right)\cdot{}^*\!\!\int_0^{\mathbf{r}}\nabla_\rho(\rho n')d\rho$$

$$+\frac{1}{r^2}{}^*\!\!\int_0^{\mathbf{r}}\nabla_\rho\cdot[\rho\nabla_\rho\,\rho n')]d\rho. \tag{14}$$

This expression can be further simplified with the help of the identity

$$\mathbf{r}_0\cdot\overset{*}{\int_0^{r}}\nabla_\rho Q(\rho)d\rho\equiv Q(\mathbf{r})-Q(0). \qquad (15)$$

We obtain, after a straightforward computation,

$$\nabla^2 S'(\mathbf{r})=\frac{2}{r}n'(\mathbf{r})+\frac{1}{r^2}\overset{*}{\int_0^{r}}\rho^2\nabla_\rho{}^2 n'(\rho)d\rho. \qquad (16)$$

We shall now write out the explicit expressions occurring on the right-hand side of the second Eq. (6). We have for ∇L_0 the expression

$$\nabla L_0=\nabla b-\frac{2}{r}\mathbf{r}_0, \qquad (17)$$

with the further condition

$$\mathbf{r}_0\cdot\nabla b=0. \qquad (18)$$

The second Eq. (6) therefore assumes the form

$$\mathbf{r}_0\cdot\nabla L'=-\frac{1}{r^2}\overset{*}{\int_0^{r}}\rho^2\nabla_\rho{}^2 n'd\rho$$
$$-\frac{1}{r}\nabla b\cdot\overset{*}{\int_0^{r}}\rho\nabla_\rho n'd\rho\equiv B(\mathbf{r}). \qquad (19)$$

The two terms on the right-hand side of Eq. (19) can be interpreted quite simply. The first term corresponds to the "lens action" of local inhomogeneities, while the second term, which depends on the directivity of the source, represents the change in local intensity caused by the lateral displacement of the "beam." In what follows we shall consider a non-directional source and drop the second term.

The solution of Eq. (19) is the integral

$$\left.\begin{aligned} L'&=\overset{*}{\int_0^{r}}B(\rho)d\rho\\ &=-\overset{*}{\int_0^{r}}\left(\frac{1}{\rho}-\frac{1}{r}\right)\rho^2\nabla_\rho{}^2 n'd\rho,\end{aligned}\right\} \qquad (20)$$

by a simple transformation of the double integrals.

Equations (7) and (20) are solutions of the differential Eq. (6). If conditions at the source are disregarded, these solutions are not unique; it is possible to write down the general solution of Eq. (6). The difference between two solutions of the first Eq. (6) is a function of the angle only. The most general solution is, therefore,

$$\begin{aligned} S_g'&=S_0'+\phi(\Omega),\\ \mathbf{r}_0\cdot\nabla\phi&=0.\end{aligned} \qquad (21)$$

S_0' is the particular solution (7), and $\phi(\Omega)$ is an arbitrary function of the solid angle Ω. The addition $\phi(\Omega)$ produces a discontinuous addition to the variable S at the source, and is, thus, inconsistent with the assumption that for vanishing r, S' is to vanish. We find that the expressions (7) and (20) represent the only solution which satisfies the conditions at the source.

As for the higher approximations, the expansion described here leads, at each stage, to equations for $S^{(n)}$ and $L^{(n)}$, which have the same type left-hand sides as Eq. (6). The right-hand sides, however, get progressively more involved. At any rate, each successive approximation has a unique solution, which can be expressed in the form of explicit line integrals. These higher approximations will not be considered in the remaining sections of this paper.

THE SELF-CORRELATION FUNCTION OF THE INDEX OF REFRACTION

In the following sections, we shall derive the mean and the standard deviation of S, the optical path length from the source, and of L, the level at a fixed point $\mathbf{r}$. First, however, the microstructure must be characterized by certain statistical properties. If we consider a configuration of microstructure patterns, or a time series of distributions of n' in the same medium, then we shall assume that there is a possibility of averaging all quantities which depend on the spatial distribution of n' and its derivatives. The averaged quantities shall be enclosed in angular brackets, $\langle\,\rangle$. Obviously, it is reasonable to define the standard wave velocity so that

$$\langle n'(x,y,z)\rangle=0. \qquad (22)$$

The significance of this normalization is that the deviation of n' from unity is equally likely to be positive or negative. The second assumption is that the spatial correlation function,

$$\langle n'(x_1, y_1, z_1)n'(x_2, y_2, z_2)\rangle,$$

exists and is a function of the coordinate differences only,

$$\left.\begin{aligned}\langle n'(x_1, y_1, z_1)n'(x_2, y_2, z_2)\rangle \\ \equiv N(x_1-x_2, y_1-y_2, z_1-z_2) \\ \equiv N(x_2-x_1, y_2-y_1, z_2-z_1).\end{aligned}\right\} \quad (23)$$

In other words, it is assumed that the statistical characteristics of the microstructure, while not necessarily isotropic, are homogeneous.

The function N satisfies a number of inequalities. A few shall be listed here. From the inequality

$$[n'(0)\pm n'(\mathbf{r})]^2 \geqslant 0 \quad (24)$$

we have

$$0\leqslant \langle [n'(0)\pm n'(\mathbf{r})]^2\rangle = 2[N(0)\pm N(\mathbf{r})] \quad (25)$$

or, for all coordinate differences $\mathbf{r}$,

$$\begin{gathered} N(0)\geqslant 0, \\ |N(\mathbf{r})|\leqslant N(0). \end{gathered} \quad (26)$$

An inequality between the N-values belonging to two arguments $\mathbf{r}_1$ and $\mathbf{r}_2$ is obtained by means of another positive definite expression. Introduce a parameter α (or β or γ) which serves to number the cases which constitute the "population" for purposes of averaging and then consider the square of the determinant

$$D=\begin{vmatrix} n_\alpha'(0), & n_\alpha'(\mathbf{r}_1), & n_\alpha'(\mathbf{r}_2) \\ n_\beta'(0), & n_\beta'(\mathbf{r}_1), & n_\beta'(\mathbf{r}_2) \\ n_\gamma'(0), & n_\gamma'(\mathbf{r}_1), & n_\gamma'(\mathbf{r}_2)\end{vmatrix}, \quad (27)$$

averaged three times, over α, β, and γ. This triple mean square is

$$\left.\begin{aligned}\langle D^2\rangle = 6\{N^3(0)+2N(\mathbf{r}_1)N(\mathbf{r}_2)N(\mathbf{r}_1\pm\mathbf{r}_2) \\ -N(0)[N^2(\mathbf{r}_1)+N^2(\mathbf{r}_2)+N^2(\mathbf{r}_1\pm\mathbf{r}_2)]\}.\end{aligned}\right\} \quad (28)$$

The resulting inequality takes the form

$$\left.\begin{aligned}N^3(0)+2N(\mathbf{r}_1)N(\mathbf{r}_2)N(\mathbf{r}_1\pm\mathbf{r}_2) \\ -N(0)[N^2(\mathbf{r}_1)+N^2(\mathbf{r}_2) \\ +N^2(\mathbf{r}_1\pm\mathbf{r}_2)]\geqslant 0.\end{aligned}\right\} \quad (29)$$

This inequality can be transformed if we solve it with respect to $N(\mathbf{r}_1\pm\mathbf{r}_2)$. We find the inequality

$$\begin{aligned}\frac{1}{N(0)}[N(\mathbf{r}_1)N(\mathbf{r}_2)-R]\leqslant N(\mathbf{r}_1\pm\mathbf{r}_2) \\ \leqslant\frac{1}{N(0)}[N(\mathbf{r}_1)N(\mathbf{r}_2)+R],\end{aligned} \quad (30)$$

$$R\equiv\{[N^2(0)-N^2(\mathbf{r}_1)][N^2(0)-N^2(\mathbf{r}_2)]\}^{\frac{1}{2}}.$$

It follows that N will be continuous everywhere if it is continuous at the point 0. For if $\mathbf{r}_2$ is a very small coordinate difference $\boldsymbol{\delta}$, and if

$$N(\boldsymbol{\delta})=N(0)(1-\epsilon), \quad (31)$$

where ϵ is again a small quantity, then (30) goes over into

$$\left.\begin{aligned}-\epsilon N(\mathbf{r})-[\epsilon(2-\epsilon)]^{\frac{1}{2}}[N^2(0)-N^2(\mathbf{r})]^{\frac{1}{2}} \\ \leqslant N(\mathbf{r}\pm\boldsymbol{\delta})-N(\mathbf{r})\leqslant -\epsilon N(\mathbf{r}) \\ +[\epsilon(2+\epsilon)]^{\frac{1}{2}}[N^2(0)-N^2(\mathbf{r})]^{\frac{1}{2}}.\end{aligned}\right\} \quad (32)$$

Both the lower and the upper bound converge toward zero with ϵ. It follows further from the inequality (32) that the rate of change of N, defined as

$$\lim_{|\delta|\to 0}\left\{\frac{1}{|\boldsymbol{\delta}|}[N(\mathbf{r}+\boldsymbol{\delta})-N(\mathbf{r})]\right\}, \quad (33)$$

is everywhere bounded if the following limit exists

$$\lim_{|\delta|\to 0}\left\{\frac{1}{|\boldsymbol{\delta}|}\left[1-\frac{N(\boldsymbol{\delta})}{N(0)}\right]^{\frac{1}{2}}\right\}, \quad (34)$$

for fixed direction of $\boldsymbol{\delta}$. If that limit vanishes, $N(\mathbf{r})$ is constant everywhere and equal to $N(0)$.

Another set of inequalities can be obtained by considering positive definite expressions of the type

$$Q={}^*\!\!\int_0^r \phi(\rho)n'(\rho)d\rho\cdot{}^*\!\!\int_0^r \phi^*(\sigma)n'(\sigma)d\sigma\geqslant 0, \quad (35)$$

where the asterisk indicates transition to the conjugate complex. If we choose in particular for the arbitrary function $\phi(\rho)$ the function $e^{ik\rho}$, then we obtain the following

$$\begin{aligned}Q&={}^*\!\!\iint_0^r e^{ik(\rho-\sigma)}N(\rho-\sigma)d\rho d\sigma \\ &=2\,{}^*\!\!\int_0^r (r-\rho)\cos k\rho\cdot N(\rho)d\rho\geqslant 0\end{aligned} \quad (36)$$

or

$$\int_0^r \left(1-\frac{\rho}{r}\right) \cos k\rho \cdot N(\rho)d\rho \geqslant 0. \tag{37}$$

THE MEAN SQUARE DEVIATION OF S'

From Eq. (7), we shall now compute an expression for the mean square deviation of the optical path length from the geometrical path length. We have

$$\left.\begin{aligned}\sigma_S^2 &= \langle (S_0+\lambda S'+\lambda^2 S'')^2\rangle - \langle S_0+\lambda S'+\lambda^2 S''\rangle^2 \\ &= \lambda^2[\langle S'^2\rangle - \langle S'\rangle^2] + \text{higher terms.}\end{aligned}\right\} \tag{38}$$

In this expression, $\langle S'\rangle$, the mean deviation of the optical from the geometrical path length, vanishes,

$$\langle S'\rangle = {}^*\!\int_0^r \langle n'(\rho)\rangle d\rho = 0. \tag{39}$$

For $\langle S'^2\rangle$ we find

$$\langle S'^2\rangle = {}^*\!\int\!\!\int_{\rho,\sigma=0}^{r} N(\rho-\sigma)d\sigma d\rho. \tag{40}$$

This double integral can be transformed as follows:

$$\left.\begin{aligned}\langle S'^2\rangle &= \int_{\sigma=0}^{r}\int_{\tau=-\sigma}^{r-\sigma} N(\tau)d\tau d\sigma \\ &= \int_{\tau=0}^{r}\int_{\sigma=0}^{r-\tau} N(\tau)d\sigma d\tau + \int_{\tau=-r}^{0}\int_{\sigma=-\tau}^{r} N(\tau)d\sigma d\tau,\end{aligned}\right\} \tag{41}$$

and finally

$$\langle S'^2\rangle = 2\,{}^*\!\int_0^r (r-\rho)N(\rho)d\rho. \tag{42}$$

We find, thus, that the mean optical path length equals the actual distance r, while the r.m.s. deviation is given in this approximation by the expression

$$\sigma_S = \left[2\,{}^*\!\int_0^r (r-\rho)N(\rho)d\rho\right]^{\frac{1}{2}}. \tag{43}$$

The integral is to be extended over the straight line connecting the source with the point at which the observer is located.

It may be assumed that in most practical cases the function $N(\mathbf{r})$ decreases rapidly for large values of the argument. As a result, expression (42) may be replaced, in fair approximation, by

$$\sigma_S^2 \sim 2r\,{}^*\!\int_0^\infty N(\rho)d\rho. \tag{44}$$

Without any knowledge of the details of the function N, it can be predicted that, for sufficiently large distances, the r.m.s. fluctuation of the optical path length will increase with the square root of the distance.

THE SIGNAL LEVEL FLUCTUATION

From Eqs. (19) and (20), it is possible to obtain an expression for signal level fluctuation. By an argument analogous to that leading to the mean square deviation of the optical path length, we find first that

$$\sigma_L^2 = \langle L'^2\rangle. \tag{45}$$

This expression is, in turn, equal to

$$\langle L'^2\rangle = {}^*\!\int\!\!\int_{\rho,\sigma=0}^{r} \left(\frac{1}{\rho}-\frac{1}{r}\right)\left(\frac{1}{\sigma}-\frac{1}{r}\right) \times \rho^2\sigma^2\langle \nabla_\rho^2 n'(\rho)\nabla_\sigma^2 n'(\sigma)\rangle d\sigma d\rho. \tag{46}$$

The integrand can be simplified, first of all, by the introduction of the iterated Laplacean of $N(\mathbf{r})$, in accordance with the identity

$$\langle \nabla^2 n'(\rho)\nabla^2 n'(\sigma)\rangle \equiv \nabla^2\nabla^2 N(\rho-\sigma). \tag{47}$$

Equation (46) then assumes the form

$$\langle L'^2\rangle = {}^*\!\int\!\!\int_{\rho,\sigma=0}^{r} \left(\frac{1}{\rho}-\frac{1}{r}\right)\left(\frac{1}{\sigma}-\frac{1}{r}\right) \times \rho^2\sigma^2\nabla^2\nabla^2 N(\rho-\sigma)d\sigma d\rho. \tag{48}$$

It is now possible to convert the double integral into a single integral by means of transformations similar to those leading to Eq. (42) above. We obtain the expression

$$\langle L'^2\rangle = \frac{1}{15}\frac{1}{r^2}\,{}^*\!\int_{\rho=0}^{r} (r^2+3\rho r+\rho^2) \times (r-\rho)^3\nabla^2\nabla^2 N(\rho)d\rho. \tag{49}$$

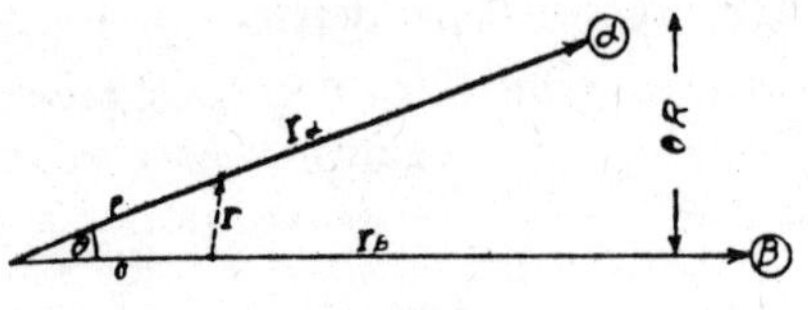

FIG. 2. Two receiving stations.

If $N(\rho)$ is small for large values of the argument, then the expression (49) can be approximated by the much cruder approximation

$$\langle L'^2\rangle \sim \frac{1}{15} r^3 \overset{*}{\int_{\rho=0}^{\infty}} \nabla^2\nabla^2 N(\rho)d\rho. \quad (50)$$

New observations, made recently at the Laboratory of the University of California, Division of War Research, at San Diego, appear to show that the fluctuation of supersonic sound signals over paths in the deep ocean, well removed from both surface and bottom, shows an increase with increasing distance, but the rate of increase did not agree quantitatively with Eq. (50).

CORRELATION AT TWO RECEIVING STATIONS

In the case of two stations receiving the same signals (Fig. 2), the levels received at either station are given by the expression (20), so that we have

$$\left.\begin{aligned} L_\alpha' &= -\overset{*}{\int_0^{r_\alpha}} \left(\frac{1}{\rho}-\frac{1}{r_\alpha}\right)\rho^2\nabla_\rho^2 n'(\rho)d\rho, \\ L_\beta' &= -\overset{*}{\int_0^{r_\beta}} \left(\frac{1}{\sigma}-\frac{1}{r_\beta}\right)\sigma^2\nabla_\sigma^2 n'(\sigma)d\sigma. \end{aligned}\right\} \quad (51)$$

The average product of L_α' and L_β' is given by the expression

$$\langle L_\alpha' L_\beta'\rangle = \overset{*}{\int_{\rho=0}^{r_\alpha}}\overset{*}{\int_{\sigma=0}^{r_\beta}} \left(\frac{1}{\rho}-\frac{1}{r_\alpha}\right)\left(\frac{1}{\sigma}-\frac{1}{r_\beta}\right) \times \rho^2\sigma^2\nabla^2\nabla^2 N(\mathbf{r})d\sigma d\rho, \quad (52)$$

where the two integrals are to be extended over two different paths. $\mathbf{r}$ is a vector which connects the two points (σ) and (ρ), as indicated in Fig. 2. This double integral cannot be converted into a single integral, because the vector $\mathbf{r}$ assumes values depending on two parameters. However, it is possible to replace the expression (52) by an approximate expression which, like (50), gives

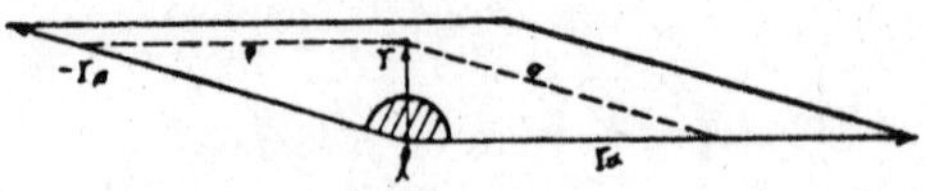

FIG. 3. Area of integration in the r-plane.

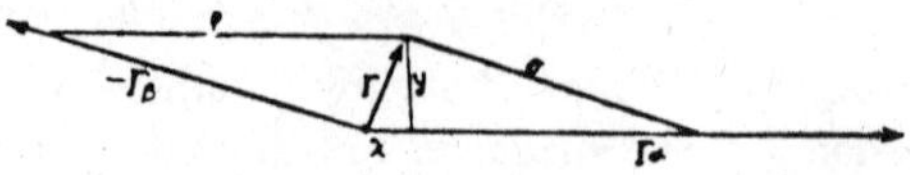

FIG. 4. Coordinate transformation in the r-plane.

information on the dependence on distance for large distances. First of all, we shall assume that the argument $N(\mathbf{r})$ is significantly different from zero only for small values of the argument $\mathbf{r}$, let us say for

$$|\mathbf{r}| < l, \quad (53)$$

and further, that l itself is small compared with the lateral separation of the stations α and β, that is,

$$l \ll \tfrac{1}{2}\theta(r_\alpha + r_\beta). \quad (54)$$

It will further be assumed that θ is a small angle and that

$$\frac{|r_\alpha - r_\beta|}{r_\alpha + r_\beta} \ll 1. \quad (55)$$

The average vector from source to receivers will be called $\mathbf{R}$,

$$\mathbf{R} \equiv \tfrac{1}{2}(\mathbf{r}_\alpha + \mathbf{r}_\beta). \quad (56)$$

In that case, the only portion of the integral which contributes significantly is the obtuse circular section in the $\mathbf{r}$-plane indicated in Fig. 3. In this region, the expression (52) may be replaced, in fair approximation, by

$$\langle L_\alpha' L_\beta'\rangle \sim \int\int \rho\sigma\nabla^2\nabla^2 N(\mathbf{r})d\rho d\sigma. \quad (57)$$

This integral, in turn, can be further transformed. If we describe the area of integration by rectangular coordinates, x and y, Fig. 4,

$$\left.\begin{aligned} x &= \rho - \sigma\cos\theta, \\ y &= \sigma\sin\theta, \end{aligned}\right\} \quad (58)$$

or

$$\left.\begin{aligned} \rho &= x + y\cot\theta, \\ \sigma &= y\operatorname{cosec}\theta, \end{aligned}\right\} \quad (59)$$

with the Jacobian

$$\frac{\partial\rho}{\partial x}\frac{\partial\sigma}{\partial y} - \frac{\partial\rho}{\partial y}\frac{\partial\sigma}{\partial x} = \operatorname{cosec}\theta, \quad (60)$$

Eq. (53) assumes the form

$$\langle L_\alpha' L_\beta' \rangle \sim \operatorname{cosec}^2 \theta \iint_{x,y} y(x+y\cot\theta) \times \nabla^2\nabla^2 N(x,y,z)dxdy. \quad (61)$$

It can be shown that the expression which remains under the integral sign is changed but little if the circular sector is expanded into a semicircle. Thus, the expression (57) can be further simplified into

$$\langle L_\alpha' L_\beta' \rangle \sim \frac{1}{2\theta^2} \iint_{x,y=-\infty}^{\infty} xy\nabla^2\nabla^2 N dxdy + \frac{1}{2\theta^3} \iint_{x,y=-\infty}^{\infty} y^2\nabla^2\nabla^2 N dxdy. \quad (62)$$

Of these two terms, the second one is large compared with the first, so that the final expression is

$$\langle L_\alpha' L_\beta' \rangle \sim \frac{1}{2\theta^3} \iint_{x,y=-\infty}^{\infty} y^2\nabla^2\nabla^2 N dxdy, \quad (63)$$

where y is the direction perpendicular to **R** in the plane of source and receivers.

To obtain the usual correlation coefficient, the expression (63) has to be divided through by (50). It is found that the correlation coefficient is independent of the distance between source and receivers and inversely proportional to the cube of the distance between the two receiving stations perpendicular to the line connecting the source with the location of the receivers, δ, as indicated in Fig. 2.

In the first of the two following papers, Mintzer develops the theory of fluctuation of sound propagation in the medium on the basis of the Born approximation to the wave equation. This theory is in better agreement with experimental results published by Sheehy than the theory developed in the previous paper by Bergmann based on ray optical theory.

The second paper describes a model experiment where the fluctuation in propagation was measured in a tank in which the inhomogeneities were generated by means of a heater at the bottom of the tank. There is good agreement between the results obtained in the experiment with the predictions of the theory in the first paper.

The first paper was published while Dr. Mintzer was at Harvard University, and the second paper was published while he was at Yale University and it is based on Dr. Stone's doctoral thesis. Dr. Mintzer is now Acting Dean of the Technological Institute, Northwestern University.

These papers are reprinted with the permission of the authors and the Acoustical Society of America.

Reprinted from THE JOURNAL OF THE ACOUSTICAL SOCIETY OF AMERICA, Vol. 25, No. 5, 922–927, September, 1953

2

Wave Propagation in a Randomly Inhomogeneous Medium. I*

DAVID MINTZER
Department of Physics, Brown University, Providence, Rhode Island
(Received June 5, 1953)

The propagation of sound pulses from a point source in a medium where the index of refraction varies randomly is studied by means of the Born approximation to the wave equation. The coefficient of variation (standard deviation of the amplitude of a series of pulses, expressed as a percentage of the mean amplitude of the series) is evaluated for pulse lengths short compared with the time in which the refractive index varies significantly, and for ranges large compared with the wavelength of the sound. The results are in agreement with the experiments of Sheehy.

I. INTRODUCTION

WHEN a series of sound pulses are transmitted between two points in the ocean, it is found that the pressure amplitude fluctuates in a random manner about a mean value. It is believed that these fluctuations are caused by the presence of small, random variations in the temperature of the medium, which cause random variations in its refractive index. It is the purpose of this paper to derive some of the properties of the pressure amplitude fluctuations using an approximate method of solution of the wave equation.

Data have been published[1] by M. J. Sheehy which show that the amount of fluctuation for deep-water transmission increases with increasing range for directly received signals. In his experiment a series of sound pulses are sent from a source at a fixed range; the maximum amplitudes of the received pulses are determined, and the coefficient of variation (standard deviation of the series of pulses, expressed as a percentage of the mean amplitude) is found. A straight line fitted to the data indicates that the coefficient of variation increases as the square root of the range, though the scatter of the data seems large (see Fig. 1, V_{exp}).

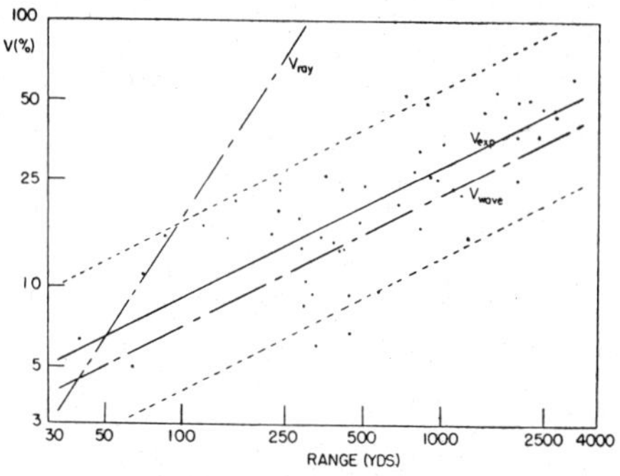

FIG. 1. Coefficient of variation of the amplitudes of direct signals *vs* range from source to receiver. Points are from M. J. Sheehy's data. V_{exp}: straight line fitted to data by Sheehy. V_{ray}: coefficient of variation derived from ray theory. V_{wave}: coefficient of variation derived from wave thoery. Short-dashed lines enclose 90 percent of the data.

A theoretical investigation of these fluctuations, based upon ray theory, has been made by P. G. Bergmann.[2] He found that the coefficient of variation (or, rather, a close approximation to it) varies as the three-halves power of the range, but does not explain the discrepancy. Liebermann,[3] who has experimentally determined some of the properties of the temperature variations, uses his data in Bergmann's theory in an attempt to obtain Sheehy's results. This is unsuccessful, however, as can be seen in Fig. 1, (V_{ray}).

Since the medium (the ocean) is one in which the refractive index varies continuously with position, interference effects should be of importance, so it may be expected that a ray theory will not give satisfactory results, since it cannot take account of phase effects. The method of attack used in this paper is the Born approximation to the wave equation.[4] A first-order approximation is derived for the pressure in an acoustic pulse traveling in a medium with small variations in the refractive index. Instead of the maximum value of the pressure amplitude, which was used in the experiment, the time average (over the pulse length) of the pressure amplitude is then determined, assuming that the refractive index at a point in space does not change significantly during the time the acoustic pulse passes through the point. The coefficient of variation for a series of pulses is then found, and the resulting expression is evaluated in a form which can be applied to Sheehy's data.

II. BORN APPROXIMATION FOR THE PRESSURE

Let us consider a medium whose index of refraction has a mean value of unity, but which has small random variations about this mean:

$$\mu(x, y, z) = 1 + \alpha n(x, y, z), \quad \alpha \ll 1 \qquad (1)$$

where μ is the refractive index, $n(x, y, z)$ is the refractive index variation (whose rms value has been

* Supported by U. S. Office of Naval Research, and presented at the Forty-fifth Meeting of the Acoustical Society of America, Philadelphia, May 7, 1953.
[1] M. J. Sheehy, J. Acoust. Soc. Am. 22, 24 (1950).
[2] P. G. Bergmann, Phys. Rev. 70, 486 (1946).
[3] L. Liebermann, J. Acoust. Soc. Am. 23, 563 (1951).
[4] C. L. Pekeris, Phys. Rev. 71, 268 (1942); R. H. Ellison, J. Atm. and Terrest. Phys. 2, 14 (1951).

normalized to unity), and α is the rms value of the refractive index variations.

Since α is very small, we shall neglect terms in α^2, so that the wave equation for the pressure can be written as

$$\nabla^2 p + k_0^2 p = -2k_0^2 \alpha n p, \tag{2}$$

where k_0 is the wave number with respect to the mean sound velocity, c_0. The solution of this equation in integral form is (for an initial wave of unit amplitude)

$$p(x,y,z) = \frac{e^{ik_0 r}}{r} - \frac{2k_0^2\alpha}{4\pi}\int_V n(x',y',z') \times \frac{e^{ik_0|\mathbf{r}-\mathbf{r}'|}}{|\mathbf{r}-\mathbf{r}'|} p(x',y',z')dv', \tag{3}$$

where the first term is the required solution to the homogeneous wave equation [right-hand side of Eq. (2) equals zero]; $\mathbf{r}$ is the vector distance from the origin to the observation point (x, y, z); $\mathbf{r}'$ is the vector distance from the origin to the scattering volume dv' at (x', y', z'); and the integration is over all space irradiated by the pressure pulse. We may take the approximation for single-scattering, i.e., the Born approximation, by using the incident wave as a first approximation to the pressure in the integral. We therefore obtain

$$p(x,y,z) = \frac{e^{ik_0 r}}{r} - \frac{k_0^2\alpha}{2\pi}\int_V n(x',y',z') \times \frac{e^{ik_0|\mathbf{r}-\mathbf{r}'|}}{|\mathbf{r}-\mathbf{r}'|} \frac{e^{ik_0 r'}}{r'} dv'. \tag{4}$$

Here the term $\exp(ik_0 r')/r'$ gives the magnitude and phase of the incident wave at the scattering center dv', and $\exp(ik_0|\mathbf{r}-\mathbf{r}'|)/|\mathbf{r}-\mathbf{r}'|$ gives the magnitude and phase of the scattered wave at the observation point. This single scattering approximation should be valid for values of range depending upon the size of the variations of refractive index and will be investigated in a later paper..

Since we are only interested in the variations of the pulse amplitude at the observation point during the time that the pulse passes through the observation point, we must integrate over only that region of space, the scattered waves of which reach the observation point between times t_0 and t_0+T, where $t_0 = r/c_0$ is the time for the leading edge of the pulse from the source to reach the receiver, and T is the pulse length. (The scattered waves coming after t_0+T are part of the reverberation due to volume scattering.) The scattered waves coming from the volume element at (x', y', z') have traveled a distance $[r' + |\mathbf{r}-\mathbf{r}'|]$; all the scattered waves which reach the observation point at a given time $t(\geq t_0)$ have come from volume elements whose distance from the origin satisfies the inequality $c_0(t-T) \leq [r' + |\mathbf{r}-\mathbf{r}'|] \leq c_0 t$, where the upper limit is for those elements which have been irradiated by only the beginning of the pulse, and the lower limit for those elements for which the irradiation is just ending. In order to simplify the introduction of these limits of integration, we change from the rectangular coordinates (x', y', z') to the prolate spheroidal coordinates[5] (ξ', η', ϕ'):

$$\begin{aligned} x' &= (r/2)\xi'\eta'; \quad y' = (r/2)[(\xi'^2-1)(1-\eta'^2)]^{\frac{1}{2}} \cos\phi'; \\ z' &= (r/2)[(\xi'^2-1)(1-\eta'^2)]^{\frac{1}{2}} \sin\phi'; \\ \xi' &= (1/r)[r' + |\mathbf{r}-\mathbf{r}'|]; \quad \eta' = (1/r)[r' - |\mathbf{r}-\mathbf{r}'|]; \\ dv' &= dx'dy'dz' = (r^3/8)(\xi'^2-\eta'^2)d\xi' d\eta' d\phi', \end{aligned} \tag{5}$$

where, in general, $1 \leq \xi' \leq \infty$; $-1 \leq \eta' \leq 1$; $0 \leq \phi' \leq 2\pi$.

The second term in Eq. (4) becomes for $t > r/c_0$:

$$T_2 = -\frac{k_0^2\alpha r}{4\pi}\int_0^{2\pi} d\phi' \int_{-1}^{1} d\eta' \int_{(c_0/r)(t-T)}^{c_0 t/r} d\xi' \times e^{ik_0 r\xi'} n(\xi', \eta', \phi'), \tag{6}$$

where the lower ξ' limit is 1 until $t = t_0 + T$.

In taking the time average over the pulse length $(r/c_0) \leq t \leq (r/c_0) + T$, the assumption is made that the index of refraction at a point does not change during the passage of the pulse past the point, but that random changes occur between successive pulses. Since the refractive index depends upon the scattering coordinates, and therefore upon time, this assumption allows for part of the fluctuation within a pulse, but suppresses the fluctuation due to refractive index changes (in time) during the passage of the pulse past the scattering point; it is these fluctuations between the pulses (so that each pulse traverses a "new" medium) which cause the variations in amplitude from one pulse to another. We get for the average pulse amplitude

$$\bar{p} = \frac{e^{ik_0 r}}{r} - \frac{k_0^2\alpha r}{4\pi}\int_0^{2\pi} d\phi' \int_{-1}^{1} d\eta' \int_1^{1+(c_0T/r)} d\xi' \times \left(1 + \frac{r}{c_0T} - \frac{r}{c_0T}\xi'\right) e^{ik_0 r\xi'} n(\xi', \eta', \phi'). \tag{7}$$

III. THE COEFFICIENT OF VARIATION

Equation (7) is an expression for the time-averaged amplitude of a pressure pulse as it passes by the observation point located in a slightly inhomogeneous medium. Each pulse passes through a medium in which the refractive index changes between pulses in a random manner; we may now average over a large number of these pulses. This will involve averaging over many different distributions of refractive index;[6] this ensemble

[5] J. A. Stratton, *Electromagnetic Theory* (McGraw-Hill Book Company, Inc., New York, 1941), p. 56.

[6] This is related to the time average, taken long enough so that many pulses have been received; the correlation function to be defined in Eq. (9) is similar to the time-correlation function. See H. Staras, J. Appl. Phys. 23, 1152 (1952).

average will be denoted by angular brackets $\langle\ \rangle$:

$$\langle\bar{p}\rangle=\frac{e^{ik_0r}}{r}\frac{k_0{}^2\alpha r}{4\pi}\int_0^{2\pi}d\phi'\int_{-1}^{1}d\eta'\int_1^{1+(c_0T/r)}d\xi'\times\left(1+\frac{r}{c_0T}-\frac{r}{c_0T}\xi'\right)e^{ik_0r\xi'}\langle n(\xi',\eta',\phi')\rangle. \quad (8)$$

Since the refractive index is assumed to vary randomly about a mean value of unity, the average of the "excess" refractive index, n, is zero, so that the ensemble average of the pressure amplitude is just $\exp(ik_0r)/r$. The average of the product of the refractive indexes at points separated in space is just the autocorrelation function:

$$\langle n(x',y',z')n(x'',y'',z'')\rangle=N(x'-x'',y'-y'',z'-z''). \quad (9)$$

The correlation function is taken as a function of the coordinate differences only, assuming that the statistical character of the refractive index is homogeneous, but perhaps anisotropic. (Here, homogeneity means that the dependance is upon coordinate differences only, and not on the values of the coordinates themselves; anisotropy means that there is a different functional dependance for each of the coordinate differences.)

The coefficient of variation V is defined as the standard deviation of the series expressed as a percentage of the mean amplitude (we shall use fractional values, but plot percentages). It is given by

$$V^2=\frac{\langle|\bar{p}|^2\rangle-|\langle\bar{p}\rangle|^2}{|\langle\bar{p}\rangle|^2}, \quad (10)$$

where we have used absolute values to find the magnitude of the mean pressure amplitude. Using Eqs. (8) and (9), we get

$$|\langle\bar{p}\rangle|^2=1/r^2,\quad \langle|\bar{p}|^2\rangle=(1/r^2)+I(x,y,z), \quad (11)$$

where

$$I(x,y,z)=\frac{\alpha^2k_0{}^4r^2}{16\pi^2}\int_0^{2\pi}d\phi'\int_0^{2\pi}d\phi''\int_{-1}^{1}d\eta'\int_{-1}^{1}d\eta''\times\int_1^{1+(c_0T/r)}d\xi'\int_1^{1+(c_0T/r)}d\xi''\left(1+\frac{r}{c_0T}-\frac{r}{c_0T}\xi'\right)\times\left(1+\frac{r}{c_0T}-\frac{r}{c_0T}\xi''\right)\exp[ik_0r(\xi'-\xi'')]\times N(|\boldsymbol{\varrho}'-\boldsymbol{\varrho}''|), \quad (12)$$

where $N(|\boldsymbol{\varrho}'-\boldsymbol{\varrho}''|)$ is given by Eq. (9) with the coordinate differences expressed in terms of the spheroidal coordinates (the $\boldsymbol{\varrho}$'s being used symbolically).

Let us now consider the three-dimensional Fourier transform of the autocorrelation function:

$$\mathfrak{N}(k_1,k_2,k_3)=\frac{1}{(2\pi)^{\frac{3}{2}}}\int\int\int_{-\infty}^{\infty}N(x,y,z)\times e^{-i(k_1x+k_2y+k_3z)}dxdydz, \quad (13)$$

where $x=x'-x''$, $y=y'-y''$, $z=z'-z''$. We therefore have

$$N(x,y,z)=\frac{1}{(2\pi)^{\frac{3}{2}}}\int\int\int_{-\infty}^{\infty}\mathfrak{N}(k_1,k_2,k_3)\times e^{i(k_1x+k_2y+k_3z)}dk_1dk_2dk_3 \quad (13')$$

To use Eq. (13′) in (12) we replace the rectangular coordinates by spheroidal coordinates, using Eq. (5); we also change from rectangular coordinates in the k's to spherical coordinates,

$$k_1=k\cos\theta;\quad k_2=k\sin\theta\cos\phi;\quad k_3=k\sin\theta\sin\phi. \quad (14)$$

We therefore get for Eq. (12)

$$I=\frac{\alpha^2k_0{}^4r^2}{32\pi^3(2\pi)^{\frac{3}{2}}}\int_0^{2\pi}d\phi\int_0^{\pi}\sin\theta d\theta\int_0^{\infty}k^2dk\mathfrak{N}(k,\theta,\phi)\times\Bigg\{\Bigg(\int_0^{2\pi}d\phi'\int_{-1}^{1}d\eta'\int_1^{1+(c_0T/r)}d\xi'\times\left(1+\frac{r}{c_0T}-\frac{r}{c_0T}\xi'\right)e^{ik_0r\xi'}e^{\frac{1}{2}ikr\xi'\eta'\cos\theta}\times\exp\{\tfrac{1}{2}ikr[(\xi'^2-1)(1-\eta'^2)]^{\frac{1}{2}}\sin\theta\cos(\phi'-\phi)\}\Bigg)\times(\text{complex conjugate})\Bigg\}, \quad (15)$$

where the $\mathfrak{N}(k,\theta,\phi)$ is from Eq. (13) using (14).

IV. EVALUATION OF THE INTEGRALS

Equation (15) is partially evaluated in Appendix A to give for the first bracketed part:

$$T_1=(\ )=\frac{4\pi^{\frac{3}{2}}}{(kr)^{\frac{1}{2}}}\int_1^{1+(c_0T/r)}d\xi'\left(1+\frac{r}{c_0T}-\frac{r}{c_0T}\xi'\right)\times e^{ik_0r\xi'}\frac{J_{\frac{1}{2}}[\frac{1}{2}kr(\xi'^2-\sin^2\theta)^{\frac{1}{2}}]}{(\xi'^2-\sin^2\theta)^{\frac{1}{4}}}, \quad (16)$$

where $J_{\frac{1}{2}}$ is the Bessel function of order $\frac{1}{2}$. By repeated integrations by parts, and approximating the Hankel function of argument k_0r by the first term of its asymptotic expansion (good for ranges large compared with the wavelength), Eq. (16) is evaluated in Appendix B to give

$$T_1 = 2\pi \frac{e^{ik_0 r}}{(-ik_0 r)} e^{-i(k^2 r/8k_0)} \int_0^\pi e^{\frac{1}{2}ikr\cos\theta\cos\theta'} \times e^{i(k^2 r/8k_0)\cos^2\theta'} \sin\theta' d\theta'. \quad (17)$$

Using this in Eq. (15), we obtain our basic equation, good for $k_0 r \gg 1$:

$$I = \frac{\alpha^2 k_0^2}{8\pi(2\pi)^{\frac{1}{2}}} \int_0^{2\pi} d\phi \int_0^\pi \sin\theta d\theta \int_0^\infty k^2 dk \mathfrak{N}(k, \theta, \phi) \times \int_0^\pi \sin\theta' d\theta' \int_0^\pi \sin\theta'' d\theta'' \times \exp[\tfrac{1}{2}ikr\cos\theta(\cos\theta' - \cos\theta'')] \times \exp\left[i\frac{k^2 r}{8k_0}(\cos^2\theta' - \cos^2\theta'')\right]. \quad (18)$$

Returning to rectangular coordinates in the k's [see Eq. (14)], and letting $r\cos\theta' = x'$, $r\cos\theta'' = x''$, we may write the above equation as

$$I = \frac{\alpha^2 k_0^2}{8\pi(2\pi)^{\frac{1}{2}} r^2} \iiint_{-\infty}^{\infty} dk_1 dk_2 dk_3 \mathfrak{N}(k_1, k_2, k_3) \times \int_{-r}^{r} dx' \int_{-r}^{r} dx'' \exp[\tfrac{1}{2}ik_1(x' - x'')] \times \exp\left[i\frac{(k_1^2 + k_2^2 + k_3^2)}{8k_0 r}(x'^2 - x''^2)\right]. \quad (19)$$

If the second exponential term in the integral,

$$\left\{\exp\left[i\frac{(k_1^2 + k_2^2 + k_3^2)}{8k_0 r}(x'^2 - x''^2)\right]\right\},$$

is expanded, it gives rise to a series of terms, each of order $1/(k_0 a)^n$, where a is the mean size of the inhomogeneities; only the first term is of importance.

$$\therefore I \simeq \frac{\alpha^2 k_0^2}{8\pi(2\pi)^{\frac{1}{2}} r^2} \int_{-r}^{r} dx' \int_{-r}^{r} dx'' \times \iiint_{-\infty}^{\infty} dk_1 dk_2 dk_3 \mathfrak{N}(k_1, k_2, k_3) e^{\frac{1}{2}ik_1(x' - x'')}$$

$$= \frac{\alpha^2 k_0^2}{4r^2} \int_{-r}^{r} dx' \int_{-r}^{r} dx'' N\left(\frac{x' - x''}{2}, 0, 0\right), \quad (20)$$

where $N(x, 0, 0)$ is the autocorrelation function of the refractive index variations *taken along the line between source and receiver* [see Eq. (9)]. If we let $\rho = x' - x''$, and

$$N(\rho) \equiv N(x' - x'', 0, 0), \quad (21)$$

we may change variables, using $N(-\rho) = N(\rho)$, and perform one of the integrations to get for the coefficient of variation by Eq. (10), (11), and (20):

$$V^2 = 2\alpha^2 k_0^2 \int_0^r (r - \rho) N(\rho) d\rho. \quad (22)$$

For ranges large compared with the correlation distance (r large compared with the distance in which $N(r)$ becomes small), we have

$$V^2 = 2k_0^2 \alpha^2 r \int_0^\infty N(\rho) d\rho, \quad (23)$$

where, as before, k_0 is the wave number of the sound, r is the range from source to receiver, and α is the rms refractive index variation.

It is of interest to note that the above coefficient of variation of the amplitude, Eq. (22), is the same as the coefficient of variation of the phase as derived by Bergmann[7] using ray theory. This seems to bear out the idea that it is the fluctuations in the phase of the wave which are of importance for the pressure amplitude fluctuations.

V. COMPARISON WITH EXPERIMENTS

To compare quantitatively Eq. (23) with the experiments of Sheehy, a knowledge of the autocorrelation function and the rms refractive index variation is necessary; it is to be noted, however, that the result has the same range dependance as found by Sheehy ($V \sim r^{\frac{1}{2}}$).

The correlation function defined by Eq. (9) is found by averaging over time, i.e., taking an average over a large number of consecutive pulses is equivalent to taking an average over the time during which these pulses were emitted. We may assume, however, that the large-time average is the same as averaging over a large region of space; the correlation function defined in terms of the space average has been measured by Liebermann,[3] who has also measured the rms variations of refractive index. We will use here

$$\alpha^2 = 5 \times 10^{-9} \quad N(x) = \exp(-x^2/a^2), \quad a = 60 \text{ cm}. \quad (24)$$

The coefficient of variation for the above Gaussian correlation function is

$$V = (\pi^{\frac{1}{2}} k_0^2 \alpha^2 a)^{\frac{1}{2}} r^{\frac{1}{2}}. \quad (25)$$

Using the numerical values of Eq. (24), we obtain for 24-kc sound pulses (the frequency used by Sheehy) the line shown in Fig. 1 (V_{wave}). It is to be noted that 90

[7] Bergmann, see reference 2, p. 490, Eq. (43).

percent of the data (between the two short-dashed lines of Fig. 1) can be fitted with values of $\alpha^2 a$ varying between about 1.5×10^{-6} to 1.5×10^{-7}; Sheehy's line of best fit to the data gives a value of $\alpha^2 a$ of 5×10^{-7}. It is to be noted that the theoretically derived coefficient of variation is for values of the pressure amplitude averaged over the pulse length; the measurements are from the maximum values of pressure amplitude in each pulse. A study by members of the U. S. Navy Electronics Laboratory[8] indicate that the maximum-value coefficient of variation should be slightly larger than the mean-value coefficient of variation.

It is to be noted that Liebermann measured the refractive index variations in the horizontal plane; there is no reason to suppose, however, that the correlation length of 60 cm is also characteristic of the vertical plane. One would therefore expect that the correlation length, a, used in Eq. (25) should be an average over both the horizontal and vertical correlation lengths. In fact, if a Gaussian correlation function is assumed in both horizontal and vertical directions, with correlation lengths a and b, respectively, we should replace a in Eq. (25) by $ab/[a^2\sin^2\theta+b^2\cos^2\theta]^{\frac{1}{2}}$, where θ is the angle from the horizontal. It is therefore quite interesting that such a small range of values of $\alpha^2 a$ is needed to fit all of Sheehy's data.

ACKNOWLEDGMENT

Part of the above work was done during the author's visit to the Naval Electronics Laboratory; he wishes to thank M. J. Sheehy for discussing with him the experimental data.

APPENDIX A

Starting from the bracketed expression in Eq. (15), we may evaluate the ϕ' and ϕ'' integrals:[9]

$$\int_0^{2\pi} d\phi' \exp\{\tfrac{1}{2}ikr[(\xi'^2-1)(1-\eta'^2)]^{\frac{1}{2}}\sin\theta\cos(\phi'-\phi)\} = 2\pi J_0\{\tfrac{1}{2}kr\sin\theta[(\xi'^2-1)(1-\eta'^2)]^{\frac{1}{2}}\}. \quad \text{(A1)}$$

If we now let $\eta'=\cos v'$ and $\eta''=\cos v''(0\leq v',\ v''\leq\pi)$, we obtain terms of the form [10]

$$\int_0^{\pi} \sin v' dv' \exp[\tfrac{1}{2}ikr\xi'\cos\theta\cos v'] \times J_0[\tfrac{1}{2}kr(\xi'^2-1)^{\frac{1}{2}}\sin\theta\sin v'] = \frac{2\pi^{\frac{1}{2}}}{(kr)^{\frac{1}{2}}}\frac{J_{\frac{1}{2}}[\frac{1}{2}kr(\xi'^2-\sin^2\theta)^{\frac{1}{2}}]}{(\xi'^2-\sin^2\theta)^{\frac{1}{4}}}. \quad \text{(A2)}$$

The resulting expression is given in Eq. (16).

[8] E. C. Westerfield and W. E. Batzler, "Interval-maximum method of measurement for underwater reverberation and noise," Report 239, Navy Electronics Laboratory (October 1, 1952).

[9] G. N. Watson, *Bessel Functions* (The Cambridge University Press, Cambridge, 1948), p. 48, Eq. (6).

[10] Watson; p. 379, Eq. (1); we have here $\nu=\frac{1}{2}$; $r=0$; $C_0^{\frac{1}{2}}(z)=P_0(z)=1$; $z\cos\psi=\frac{1}{2}kr\xi'\cos\theta$; $z\sin\psi=\frac{1}{2}kr(\xi'^2-1)^{\frac{1}{2}}\sin\theta$.

APPENDIX B

Consider the integral:

$$I_1=\int_1^{1+(c_0T/r)} dx e^{ik_0rx}\frac{J_{\frac{1}{2}}[\frac{1}{2}kr(x^2-\sin^2\theta)^{\frac{1}{2}}]}{(x^2-\sin^2\theta)^{\frac{1}{4}}}. \quad \text{(B1)}$$

This may be written as

$$I_1=\left(\frac{\pi}{2}\right)^{\frac{1}{2}}\int_1^{1+(c_0T/r)} dx(k_0rx)^{\frac{1}{2}}H_{-\frac{1}{2}}(k_0rx) \times\frac{J_{\frac{1}{2}}[\frac{1}{2}kr(x^2-\sin^2\theta)^{\frac{1}{2}}]}{(x^2-\sin^2\theta)^{\frac{1}{4}}}, \quad \text{(B2)}$$

where we have used the definition[11] of the Hankel function of the first kind of order $-\frac{1}{2}$. Using the following formulas:[12]

$$\frac{d}{dx}\left\{\frac{J_{n+\frac{1}{2}}[\frac{1}{2}kr(x^2-\sin^2\theta)^{\frac{1}{2}}]}{(x^2-\sin^2\theta)^{\frac{1}{2}(n+\frac{1}{2})}}\right\} = -\frac{krx}{2}\frac{J_{n+\frac{3}{2}}[\frac{1}{2}kr(x^2-\sin^2\theta)^{\frac{1}{2}}]}{(x^2-\sin^2\theta)^{\frac{1}{2}(n+\frac{3}{2})}}$$

$$\int(k_0rx)^{n+\frac{1}{2}}H_{n-\frac{1}{2}}(k_0rx)dx = \frac{1}{k_0r}(k_0rx)^{n+\frac{1}{2}}H_{n+\frac{1}{2}}(k_0rx), \quad \text{(B3)}$$

we may repeatedly integrate by parts, obtaining the following infinite series:

$$I_1=\left(\frac{\pi}{2}\right)^{\frac{1}{2}}\sum_{m=0}^{\infty}\left(\frac{k}{2k_0}\right)^m\frac{J_{m+\frac{1}{2}}[\frac{1}{2}kr(x^2-\sin^2\theta)^{\frac{1}{2}}]}{(x^2-\sin^2\theta)^{\frac{1}{2}(m+\frac{1}{2})}} \times\frac{(k_0rx)^{m+\frac{1}{2}}}{(k_0r)^{m+1}}H_{m+\frac{1}{2}}(k_0rx)\Bigg|_1^{1+(c_0T/r)} \quad \text{(B4)}$$

For the case of $k_0r\gg1$, we have $k_0rx\gg1$ for all x in the range of integration. We may therefore use the first term of the asymptotic representation of the Hankel function[13] to give

$$I_1\simeq\frac{e^{ik_0rx}}{ik_0r}\sum_{m=0}^{\infty}e^{-\frac{1}{2}m\pi i}\left(\frac{k}{2k_0}\right)^m x^m \times\frac{J_{m+\frac{1}{2}}[\frac{1}{2}kr(x^2-\sin^2\theta)^{\frac{1}{2}}]}{(x^2-\sin^2\theta)^{\frac{1}{2}(m+\frac{1}{2})}}\Bigg|_1^{1+(c_0T/r)} \quad \text{(B5)}$$

[11] Jahnke-Emde, *Tables of Functions* (Dover Publications, New York, 1943), p. 136.

[12] Reference 12, p. 145.

[13] See reference 9, p. 198.

If we now consider the integral

$$I_2=\int_1^{1+(c_0T/r)} xdxe^{ik_0rx}\frac{J_{\frac{1}{2}}[\frac{1}{2}kr(x^2-\sin^2\theta)^{\frac{1}{2}}]}{(x^2-\sin^2\theta)^{\frac{1}{4}}}, \quad \text{(B6)}$$

we may proceed as before (starting with the Hankel function of order $+\frac{1}{2}$), obtaining

$$I_2=\left(\frac{\pi}{2}\right)^{\frac{1}{2}}\frac{i}{k_0r}\sum_{m=0}^{\infty}\left(\frac{k}{2k_0}\right)^m\frac{J_{m+\frac{1}{2}}[\frac{1}{2}kr(x^2-\sin^2\theta)^{\frac{1}{2}}]}{(x^2-\sin^2\theta)^{\frac{1}{2}(m+\frac{1}{2})}}$$

$$\times\frac{(k_0rx)^{m+\frac{3}{2}}}{(k_0r)^{m+1}}H_{m+\frac{3}{2}}(k_0rx)\Bigg|_1^{1+(c_0T/r)}$$

$$\simeq\frac{e^{ik_0rx}}{ik_0r}\sum_{m=0}^{\infty}e^{-\frac{1}{2}m\pi i}\left(\frac{k}{2k_0}\right)^m x^{m+1}$$

$$\times\frac{J_{m+\frac{1}{2}}[\frac{1}{2}kr(x^2-\sin^2\theta)^{\frac{1}{2}}]}{(x^2-\sin^2\theta)^{\frac{1}{2}(m+\frac{1}{2})}}\Bigg|_1^{1+(c_0T/r)} \quad \text{(B7)}$$

Applying Eqs. (B5) and (B7) to Eq. (16), we obtain

$$T_1=\frac{4\pi^{\frac{1}{2}}}{(kr)^{\frac{1}{2}}}\frac{e^{ik_0r}}{(-ik_0r)}\sum_{m=0}^{\infty}\left(\frac{-ik}{2k_0}\right)^m\frac{J_{m+\frac{1}{2}}(\frac{1}{2}kr\cos\theta)}{(\cos\theta)^{m+\frac{1}{2}}}. \quad \text{(B8)}$$

Using an integral representation for the Bessel function,[14] we may sum the series:

$$T_1=2\pi\frac{e^{ik_0r}}{(-ik_0r)}\sum_{m=0}^{\infty}\left(\frac{-ik^2r}{8k_0}\right)^m\frac{1}{m!}$$

$$\times\int_0^{\pi}e^{\frac{1}{2}ikr\cos\theta\cos\theta'}(\sin\theta')^{2m+1}d\theta'$$

$$=2\pi\frac{e^{ik_0r}}{(-ik_0r)}\int_0^{\pi}e^{\frac{1}{2}ikr\cos\theta\cos\theta'}$$

$$\times e^{-i(k^2r/8k_0)\sin^2\theta'}\sin\theta'd\theta'. \quad \text{(B9)}$$

This leads directly to Eq. (17).

[14] See reference 9, p. 48, Eq. (6).

Reprinted from THE JOURNAL OF THE ACOUSTICAL SOCIETY OF AMERICA, Vol. 34, No. 5, 647–653, May, 1962

3

Range Dependence of Acoustic Fluctuations in a Randomly Inhomogeneous Medium*†

ROBERT G. STONE AND DAVID MINTZER
Laboratory of Marine Physics, Yale University, New Haven 11, Connecticut
(Received October 27, 1961)

Measurements have been made to relate the acoustic fluctuations, resulting when a series of uniform sound pulses is transmitted through a medium in which there are random variations in space and time of the index of refraction, to the range from source to receiver, the acoustic frequency, and the parameters of the scattering medium. This paper considers the range dependence of the coefficient of variation (which is a measure of the acoustic fluctuations). A tank of water is heated from below causing the warmer, less dense, layers of water to rise by convection. This produces a medium of almost homogeneous turbulence with time and space variations in the temperature microstructure. Experimental results are in good agreement with a theory developed by Mintzer [J. Acoust. Soc. Am. **25**, 922 (1953)] which predicts that the coefficient of variation of the acoustic pulses varies as the square root of range from source to receiver. There is some evidence that this dependence may be correct under broader conditions than are required by the theory.

I. INTRODUCTION

RANDOM variations about the mean amplitude are observed when acoustic pulses are transmitted through the ocean. These "acoustic fluctuations" are caused by random variations, in space and time, of the water temperature; this is known as the "temperature microstructure." The variations in temperature cause a corresponding change in sound velocity; the sound wave is scattered as it traverses the medium, and the amplitude of the transmitted signal fluctuates as the microstructure changes in time. Similar phenomena are observed for propagation of sound waves in air, and for radio waves in the ionosphere and troposphere. In the latter cases scattering effects may be altered slightly by the polarization of the waves.

Particularly in connection with sonar transmission, acoustic scattering due to temperature inhomogeneities has been observed in the ocean.[1–3] In this case the temperature microstructure is, at least in part, produced by the heating of the water by the sun and by the surface cooling; the latter causes the upper layers to become denser than the lower layers so that turbulent mixing occurs. This process, however, greatly depends on daily and seasonal variations. Moreover, warm-water currents may underlie cooler surface waters if there is a sufficient salinity difference; this source of heat can also cause turbulent mixing.

In the most familiar example of classical scattering, that is, Rayleigh scattering, the scatters are small compared to the wavelength; these small scatterers radiate equally well in all directions. Much of the scattering of the radio and acoustic waves, on the other hand, is caused primarily by scatterers which are larger or at least comparable in size to the wavelength of the radiation. Such large scatterers behave like large radiation sources, and have directionality characteristics; the scattering in their proximity can, as well, be thought of as the focusing and defocusing effect due to a lens action. The change in time of the distribution of scatterers causes a change in time of the scattered radiation received at each point.

Sufficiently far from the scatterers, the scattering pattern is mainly caused by interference effects. The waves refracted from different parts of the scatterers will eventually acquire sufficient phase difference to cause interference among themselves. In this latter region the laws of geometrical optics cannot be applied, and one must consider the wave nature of the radiation. Here the amplitude and phase become random variables in time and space, governed by appropriate probability laws.

The investigation of the radiation propagated through a medium with a random distribution of inhomogeneities which is changing in time is well suited for a labora-

* Presented at the sixty-first meeting of the Acoustical Society of America, May 11, 1961, Philadelphia, Pennsylvania.
† Based upon a Ph.D. thesis submitted by the first author to Yale University, 1961.
[1] M. J. Sheehy, J. Acoust. Soc. Am. **22**, 22 (1950).
[2] F. H. Sagar, J. Acoust. Soc. Am. **27**, 1092 (1955).
[3] F. H. Sagar, J. Acoust. Soc. Am. **32**, 112 (1960).

tory experiment. Not only is the cost of such an experiment done in the ocean very great, but also there is no way of controlling the microstructure parameters of interest. However, in a model experiment performed in the laboratory, the microstructure parameters can be controlled and measured. Furthermore, the relative motion of source and receiver in the ocean may introduce an intensity modulation; this modulation will result in additional fluctuations of the received signal.[2] This does not occur in the laboratory, so that the fluctuations due only to the temperature inhomogeneities will be measured.

In this paper we are concerned with the effects of the temperature microstructure on the propagation of an acoustic wave. The theory indicates that the acoustic fluctuations are related to certain parameters of the temperature microstructure. There is then the interesting possibility of determining the parameters of the temperature microstructure by means of acoustic waves. (In a more general sense, a great deal of statistical information concerning any scattering medium may be determined from the fluctuations of radiation propagated through the medium.)

The experiment discussed in this and in subsequent papers is performed by transmitting a series of uniform acoustic pulses through a tank of water heated from below. The turbulent mixing of the warmer (more buoyant) water gives rise to the temperature and, hence, refractive index inhomogeneities. The received acoustic signals are converted to an electrical output which is recorded; the fluctuations can then be measured. The acoustic fluctuations are determined as a function of range between source and receiver, acoustic frequency, and microstructure parameters. The significant parameters of the microstructure are separately measured with thermistors; provision is made to determine the temperature fluctuation at a single point or the difference in fluctuations at two points. The resultant output voltage is recorded, and analyzed to determine the microstructure parameters. The relations between the acoustic and microstructure statistics are then determined.

II. SUMMARY OF THEORY

From the point of view of the scattering theory, the microstructure of the medium is characterized by the rms index of refraction variation from unity α and the correlation distance a.

If $\mu(\mathbf{r})$ is the index of refraction, α the rms index of refraction deviation from unity, and $n(\mathbf{r})$ is the normalized index of refraction variation from unity, then, by their definitions:

$$\mu(\mathbf{r})=1+\alpha n(\mathbf{r}),$$
$$\langle\mu(\mathbf{r})-1\rangle=\alpha\langle n(\mathbf{r})\rangle=0, \tag{1}$$
$$\langle[\mu(\mathbf{r})-1]^2\rangle=\alpha^2,$$

where the angular brackets refer to the ensemble average. The time dependence of the refractive index variation is left out in the notation since all quantities are taken at the same time.

The correlation function $N(\boldsymbol{\rho})$ is defined, for the case of isotropic homogeneous statistics, as

$$N(\rho)=\langle n(\mathbf{r})n(\mathbf{r}+\boldsymbol{\rho})\rangle, \tag{2}$$

where ρ is the separation between the two points $(\mathbf{r})$ and $(\mathbf{r}+\boldsymbol{\rho})$. The correlation function may be utilized to find the characteristic length a from

$$a=\int_0^\infty N(\rho)d\rho. \tag{3}$$

For example, the Gaussian correlation function

$$N(\rho)=\exp(-\rho^2/a^2) \tag{4}$$

has been used frequently in the theory. Although many of the available experimental data fit the exponential function,

$$N(\rho)=\exp(-\rho/a), \tag{5}$$

the correlation function should be even and flat at the origin.

Usually the theories of scattering consider propagation in a medium whose properties deviate only slightly from their average value. Then the wave equation for the pressure can be written as

$$\nabla^2P+k_0^2P=-2k_0^2\alpha nP, \tag{6}$$

where k_0 is the wave number for a homogeneous medium in which the sound velocity is c_0. Terms in α^2 have been neglected because α is assumed small.

Mintzer[4] has applied the single scattering approximation to the wave equation for a point source of sound (which gives rise to spherical waves). An expression is obtained which relates the statistics of the scattering medium to the statistics of the acoustic pulse fluctuations. In terms of the coefficient of variation V (fractional standard deviation of a series of pulses) as used by Mintzer, the following result is obtained,

$$V^2=2\alpha^2k_0^2r\int_0^\infty N(\rho)d\rho, \tag{7}$$

where $N(\rho)$ is the autocorrelation function of the refractive index variations taken along the line between source and receiver. This equation is subject to the conditions $k_0r\gg1$ and $k_0a\gg1$. Furthermore, the single scattering approximation is shown to be valid provided $k_0^2\alpha^2ar\ll1$.

Potter and Murphy[5] have shown that the pressure term used in Mintzer's original calculation of V is not the observable since it contains a phase term. These

[4] D. Mintzer, Part I, J. Acoust. Soc. Am. **25**, 922 (1953); Part II, **25**, 1107 (1953); Part III, **26**, 186 (1954).

[5] D. S. Potter and S. R. Murphy, J. Acoust. Soc. Am. **29**, 197 (1957).

authors have shown that the proper definition of V is

$$V^2=\frac{\langle|P|^2\rangle-\langle|P|\rangle^2}{\langle|P|\rangle^2}. \tag{8}$$

This introduces a factor of $\frac{1}{2}$ into Mintzer's results; the corrected value is used in the subsequent discussion:

$$V^2=\alpha^2 k_0^2 r\int_0^\infty N(\rho)d\rho. \tag{7'}$$

For a Gaussian correlation function,

$$V^2=\tfrac{1}{2}(\pi)^{\frac{1}{2}}k_0^2\alpha^2 ar, \tag{9}$$

and for an exponential correlation function (since only the area under the correlation function is necessary here, the exponential can be used as an approximation),

$$V^2=k_0^2\alpha^2 ar. \tag{10}$$

Mintzer has also shown that, when r/ka^2 is much less than unity, his wave solution is equivalent to the ray theory solution developed by Bergmann.[6] Several theories have been developed which are valid for the entire frequency range. Potter and Murphy[5] evaluated the exact fluctuation integral (for a Gaussian correlation function) in Mintzer's theory by a series development and obtained a solution identical with Eq. (9) in the low-frequency limit and with Bergmann's solution in the high-frequency limit. Obukhov[7] as well as Karavainikov,[8] obtained similar results by using the single scattering approximation but, in addition, employed only a first approximation to the wave equation.

III. EXPERIMENTAL APPARATUS AND PROCEDURE

The experimental investigation involves the propagation of a series of uniform acoustic pulses through a medium (water) in which there are random time and space variations of the index of refraction. An appropriate electrical signal is transformed into an acoustic wave. The acoustic wave is transmitted to a receiver through a region of "thermal turbulence" in which the index of refraction varies. The fluctuations of the received acoustic signal as well as the temperature fluctuations along the path of propagation are measured and analyzed. The acoustic wavelength may be varied over a range of values; the distance from the sound source to receiver may be varied; the variations in the index of refraction are related to the applied heat input.

The experimental apparatus to make these observations may be divided into three categories: (a) the electrical and acoustical elements which produce, transmit, receive, and measure the acoustic signal; (b) the temperature sensors and associated system to measure the temperature microstructure; (c) the tank which contains the water, the heaters, and associated mechanical devices.

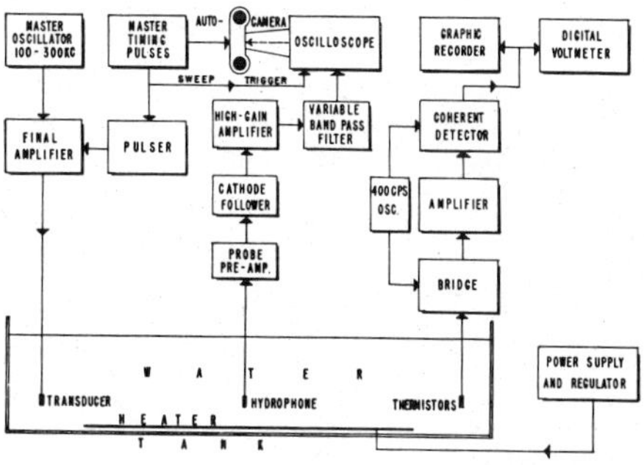

FIG. 1. Block diagram of experimental apparatus.

In the over-all design an attempt was made to keep the apparatus as versatile as possible, thus commercially built equipment was used wherever feasible. The block diagram, Fig. 1, shows schematically the general arrangement of the experimental units.

A master oscillator (which can be crystal controlled or can be operated as a variable-frequency oscillator) supplies driving voltage to the final amplifier which in turn supplies electrical power to the transducer. A pulser unit turns the final amplifier on for a predetermined time to produce sharp pulses. The master timing pulses determine the rate at which the pulser operates as well as triggering the horizontal sweep of the oscilloscope at the same rate. The electrical pulses are converted to acoustic pulses by the transducer and are received by hydrophones; the transducer is a 6-in.-diameter barium titanate mosaic, and the hydrophones are 0.062 ×0.062-in. Gulton 1C1 barium titanate cylinders. Here the acoustic signal is converted to an electrical pulse which is amplified in the probe preamplifier. The signal is further amplified in a high-gain amplifier and sent through a bandpass filter, where much of the noise and ac hum are eliminated. The received signal is then displayed on an oscilloscope, where it is photographed.

Temperature variations are measured with thermistors which form one arm of a 400-cps Wheatstone bridge. The unbalanced bridge voltage is amplified and detected. The output voltage, which is proportional to the temperature variations, is presented either on a graphic recorder or on a digital voltmeter.

The water in the tank is heated from below to produce turbulent convection. Since the Mach number of the water flow is of the order of 10^{-5}, while the rms refractive index variation due to the temperature microstructure is of the order of 10^{-4}, it is only the latter which is effective in producing the acoustic scattering. Heat input is controlled by means of powerstats and regulators. The tank is also equipped with a supporting

[6] P. G. Bergmann, Phys. Rev. **70**, 486–02 (1946).
[7] V. A. Krasilnikov and A. M. Obukhov, Soviet Phys.—Acoustics **2**, 103 (1956).
[8] V. N. Karavainikov, Soviet Phys.—Acoustics **2**, 175 (1957).

structure for the positioning of the various probes. The heater grid is supported 1 ft above the bottom of the tank. This structure simulates the grid structure in a wind tunnel to a certain extent. The local gradients between wires also facilitate the formation of turbulence. Measurements of the microstructure, and the acoustic measurements, are usually made after power has been supplied to the heater for at least 1 hr. This allows ample time for the turbulent field to become homogeneous.

If power is applied to the heaters until the ambient water temperature reaches a constant value, acoustic and temperature measurements indicate that statistical homogeneity does not exist. On the other hand, measurements taken during the first few hours after power is applied to the heaters ($\bar{T}$, the average temperature at a point, rises approximately linearly during this period) indicate that statistical homogeneity exists during this time.

Measurements of α and a taken during this period are roughly constant over horizontal planes of sufficient thickness to contain the acoustic beam. α and a are, however, slowly varying functions of height above the heater. This is an advantage because different microstructure parameter values can be obtained by measurements at several different heights above the heater. The variations of the parameters about any fixed height are small, so that a definite a and α can be defined for each height. The lateral extent of homogeneity is sufficient to contain the main acoustic beam.

Homogeneity was checked by measuring $\langle(\Delta T)^2\rangle^{\frac{1}{2}}$ at various points in the medium, where ΔT is the deviation from the mean temperature at a point. The region over which $\langle(\Delta T)^2\rangle^{\frac{1}{2}}$ is constant defines the region of homogeneity. The details of techniques and measurements to obtain values of α and a will be given in a subsequent paper.

The acoustic signal which has traveled through the medium, and has been received at the hydrophone, is converted to an electrical output as indicated in the block diagram, Fig. 1. Because of the extreme writing speed required in displaying the pulse on an oscilloscope, a peak detector is used so that only the envelope of the signal need be photographed. This makes the photography simpler and has a further advantage: The high-frequency noise components are filtered out and this makes the amplitude measurement simpler. The received pulse is displayed on the oscilloscope and is photographed. The camera film speed is adjusted to provide 30 exposures per ft of film. At a pulse rate of one every 2 sec, the temperature microstructure has sufficiently changed so that the acoustic signal "sees" a new medium, as required by the theory.

The film is developed and is read on a modified 35-mm film viewer which is equipped with a screen ruled in millimeter spacings. The viewer is also equipped with a slow film drive to facilitate positioning of the pulses on the screen for reading. The theory calls for the mean amplitude of the pulses, but the maximum pulse amplitude can be read far more easily. However, the coefficient of variation of the maximum values should be only slightly larger than the mean-value coefficient of variation,[9] and has been used in the experiments which have been carried out in the ocean. The sweep speed is also adjusted so that only a small pulse width is displayed on the oscilloscope; the actual pulse width was about 0.2 msec. The pulse width (length) must be sufficiently long to allow equilibrium of response of the receiving system, and must be sufficiently short so that the index of refraction of the medium at a point does not change during the time that the pulse passes the point. This condition is easily satisfied for the present experiment since the microstructure changes slowly.

Determinations of V usually involve measuring and analyzing a sequence of 500 to 1000 pulses. Although the absolute pressure amplitude is not measured, a quantity proportional to it is measured, so that when V is determined this proportionality factor is normalized out. The values of pulse amplitude are read on the film viewer and from this one computes the coefficient of variation, V.

Before any data run is taken with the heater power on, an ambient temperature run is taken. The coefficient of variation obtained from the "quiet medium" not only offers a check on the proper operation of the equipment but also gives a measure of the combined errors due to system and background noise.

The range over which the experiment can be performed is limited to a 180-cm span between 90 and 270 cm from the transducer. The minimum range from the source is fixed by the requirement for the far-field approximation of the sound field, treating the source as a simple piston. The maximum range is limited by the extent of the heated surface. All range measurements were therefore made over the central region of the heater.

Five hydrophones are spaced at fixed distances from the transducer. The outputs of all five hydrophones can be displayed on the same oscilloscope trace. The time required for the acoustic signal to traverse the medium is much less than the time in which the medium changes significantly. Then the coefficient of variation for each of the five ranges will have been determined from data taken over the same time interval, and secular variations tend to be averaged out. Furthermore, the time dependence of the microstructure will not influence range dependence measurements. Each run is analyzed and the resulting points plotted on log-log paper, from which the slope can be conveniently determined.

Departures of the transmitting medium from homogeneous statistics will introduce an error which is difficult to evaluate because the parameters of the temperature microstructure are determined at a single point, whereas the acoustic fluctuations are produced by

[9] E. C. Westerfield and W. E. Batzler, *Interval-Minimum Method of Measurement for Underwater Reverberation and Noise.* (Rept. 239, U. S. Navy Electronics Laboratory, October 1952).

scatterers along the path from source to receiver. Consequently, variations from homogeneity in one region may not be apparent from the microstructure data but will cause a change in the coefficient of variation. This point will be discussed further in a subsequent paper. Furthermore, when measurements are made in the region less than a foot above the heater, local gradients exist which change slowly with time. These cause some refraction of the sound beam; this usually causes a slow drift in the average pulse amplitude. The latter difficulty can be minimized by computing V for groups of 50 to 100 pulses and averaging the 10 or 20 groups so obtained; this assumes that drift is of a frequency much lower than the fluctuations of the pulse. The signal-to-noise level could be increased by using large hydrophones or transmitting stronger pulses. The former method cannot be used if the hydrophones are to be small compared to the acoustic wavelength and the latter method requires the construction of a larger driver which will deliver more power.

IV. RESULTS AND CONCLUSIONS

In this paper, the measurements of the coefficient of variation V as a function of range from source to receiver are discussed. During these measurements, a, α, and k_0 are held constant; subsequent papers will discuss the dependence of the coefficient of variation on the acoustic frequency and on the parameters of the temperature microstructure. The conditions discussed earlier will be shown to be satisfied for the range of parameters encountered in the experiment, so that Mintzer's theory should be applicable.

We shall now examine more closely the values of the several parameters encountered in the experimental observations in order to see how well the experiment satisfies these conditions. Experimental measurements to determine the correlation length a indicate that $2.5 \leq a \leq 4.0$ cm under the present range of experimental conditions. At a frequency of 200 kc, $20.8 \leq k_0 a \leq 33.2$. Since the range r varies between 90 and 240 cm, the minimum value of $k_0 r$ is about 360. Consequently $k_0 a \gg 1$ and $k_0 r \gg 1$, as required.

At a frequency of 200 kc, $51.8 \leq k_0 a^2 \leq 133$. Thus, even at the shortest range of 90 cm, $r/k_0 a^2$ is not very much less than unity and Mintzer's results may reasonably be expected to be valid. The single scattering approximation will be valid provided $k_0^2 a \alpha^2 r \ll 1$. For the extreme value of $\langle(\Delta T)^2\rangle^{\frac{1}{2}} = 0.1°\text{C}$, $a = 4.0$ cm, and $r = 240$ cm, $k_0^2 \alpha^2 a r = 2.5 \times 10^{-3}$. Consequently, we may expect only single scattering to be of importance.

According to Mintzer's theory the coefficient of variation V is a function of the square root of range, Eq. (7'); therefore, if α, a, and k_0 are held constant during range measurements, we expect that

$$V = Cr^{\frac{1}{2}}, \tag{11}$$

where $C = k_0 \alpha a^{\frac{1}{2}}$ (assuming an exponential correlation function). If the experimental points agree with Mintzer's theory they will follow a straight line (on a log-log graph) of slope $\frac{1}{2}$. Furthermore the intercept on the V axis will give the constant C.

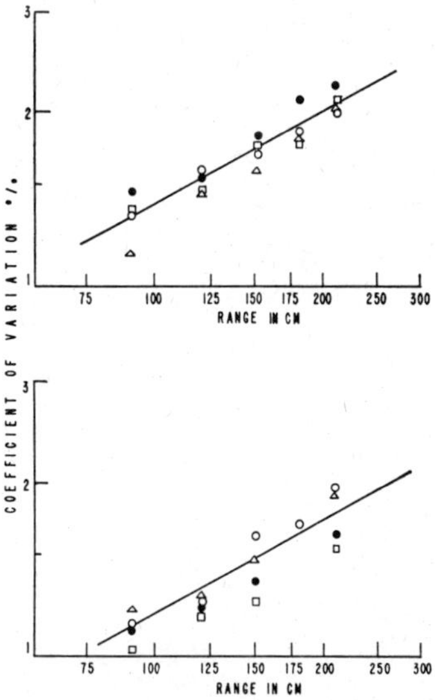

FIG. 2. Coefficient of variation as a function of range from source to receiver. The acoustic frequency is 208 kc and the power to the heater is 15 kw. Various symbols refer to individual experimental runs.

The determinations of the coefficient of variation with range made with a series of five hydrophones are shown in Figs. 2 and 3. In these cases a hydrophone was fixed at each of five ranges from the transducer as outlined above. Data taken in this manner are found to be quite consistent. Measurements were taken one foot above the heater at an acoustic frequency of 208 kc, pulse length of 0.15 msec, and repetition rate of one pulse every two sec. The data shown in Fig. 2 were taken at a heater input power of 15 kw and those in Fig. 3 at 18 kw input. Figure 2 is composed of eight separate sets of range-dependent data. Different symbols are used to identify the individual sets which were taken approximately 10 min apart; the data are presented on two graphs to show better the individual points. In a subsequent paper dealing with the dependence of the acoustic fluctuations on the temperature microstructure it will be shown that α, and consequently C, decreased approximately linearly with increasing ambient temperature $\bar{T}$ (over the temperature ranges investigated). By the nature of this experiment, $\bar{T}$ is increasing linearly with time, so that $C(\alpha)$ will decrease slowly with time. This dependence can be seen in Fig. 2 where the data in the upper half of the figure were taken during the first hour and those in the lower part of the figure were taken during the second hour of a data run. By extrapolating a line through the experimental points back to

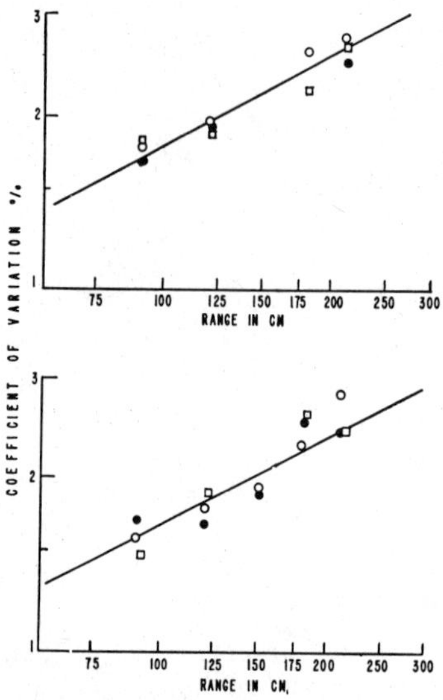

FIG. 3. Coefficient of variation as a function of range from source to receiver. The acoustic frequency is 208 kc and the power to the heater is 18 kw. Various symbols refer to individual experimental runs.

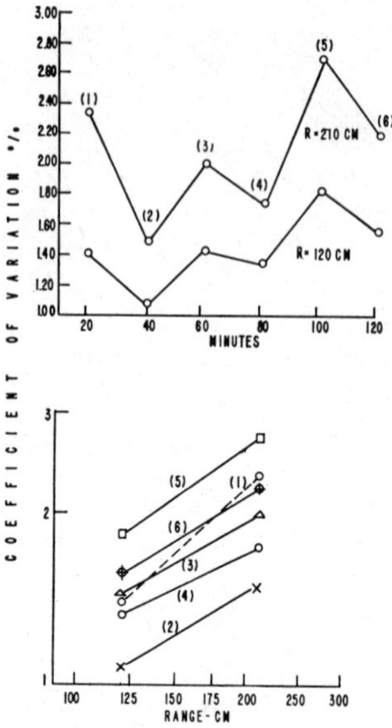

FIG. 4. Range dependence of the coefficient of variation during a period in which the temperature microstructure is changing rapidly. The upper figure shows V as a function of time for two ranges. The lower figure shows V as a function of range for the individually numbered sets.

$r=1$ cm, the intercept C is found to be smaller for the points shown in the bottom part of Fig. 2, than for those in the top part.

Similarly Fig. 3 is composed of separate sets of data, but for a power input of 18 kw. As in the case of the data presented in Fig. 2, there is again the same dependence of C on time. At this higher power input, however, the intercept and C are larger than in Fig. 2 since the larger power input increases both α and a.

The solid lines in these figures have been drawn with a slope of $\frac{1}{2}$ and positioned to pass through the experimental points. There is then excellent agreement between the experimental and theoretical range dependence over the ranges investigated.

By extrapolating the straight line shown in Figs. 2 and 3 back to one cm, the intercept on the log V axis will give the value of $k_0\alpha a^{\frac{1}{2}}$ since

$$\log V=\tfrac{1}{2}\log r+\log(\alpha k_0 a^{\frac{1}{2}}). \tag{12}$$

The intercept from Fig. 2 is found to lie between 0.10 and 0.12%. On the other hand, k_0 is known, $\langle(\Delta T)^2\rangle^{\frac{1}{2}}$, α, and a have been measured so that $\alpha k_0 a^{\frac{1}{2}}$ can be computed. For a frequency of 208 kc, k_0 is 8.6 cm^{-1}. In the present case the correlation distance a is approximately 3.5 cm, $\bar{T}$ is 22°C, and $\langle(\Delta T)^2\rangle^{\frac{1}{2}}$ is 0.08°C. The value of α is computed from[10]

$$\alpha=\frac{(421-7.4\bar{T})\langle(\Delta T)^2\rangle^{\frac{1}{2}}}{(141\,000+421\bar{T}-3.7\bar{T}^2)} \tag{13}$$

and gives 1.4×10^{-4}. On the other hand the value of α computed from the intercept[11] is 0.65×10^{-4}. This is in reasonable agreement considering the difficulty in specifying a definite value for α and a which is characteristic of the entire transmitting path and is constant over the time of the experiment. Furthermore, there is definite reason to believe that the value of $\langle(\Delta T)^2\rangle^{\frac{1}{2}}$ used above is too large. This is related to the difficulty in measuring the true rms temperature deviation from the mean temperature. A subsequent paper dealing with the dependence of V on the parameters of the temperature microstructure will discuss this point in detail.

Occasionally very rapid variations in microstructure are observed over a period of time; this is shown in Fig. 4. The coefficients of variation for ranges of 4 and 7 ft are plotted at successive time intervals, while the microstructure is changing. In Fig. 4 each set of points (1), (2), . . . , (6) has been plotted on log-log paper, and a line drawn between the experimental points at 4 and 7 ft for each set. We see that these experimental lines have slopes of approximately $\frac{1}{2}$ corresponding to the theoretical range dependence. This is a significant point since it shows that even during rapid fluctuations in microstructure, the square root of range dependence

[10] L. E. Kinsler and A. R. Frey, *Fundamentals of Acoustics* (John Wiley & Sons, Inc., New York, 1950), p. 127.

[11] In a subsequent paper showing the dependence of V on acoustic frequency, a value of $\alpha=0.62\times10^{-4}$ is found from the slope of the V vs frequency curve.

is obeyed. One may infer that during such rapid changes there may be small departures from statistical homogeneity, and, therefore, the range dependence may be of a more general applicability than is obvious from the theory. (The theory assumes homogeneity of the scattering medium.)

An experimental investigation of the range dependence at higher frequencies is now in progress. It is hoped that the range dependence can be examined in the transition and ray regions. Measurements of the coefficient of variation on a function of frequency seem to indicate that the transition region (at 120 cm) occurs for the frequency region above 250 kc, so that the experiment will be extended to frequencies above a Mc.

ACKNOWLEDGMENTS

The authors wish to express their thanks to Professor M. L. Wiedmann and Mr. M. J. Rosenblum for their part in the design of the experimental apparatus. The valuable contributions of Mr. Martin Sachs in carrying out experimental work and to Mrs. Sandra Cornell in data analysis are gratefully acknowledged. This work was supported by the Office of Naval Research.

The following paper compares the propagation losses in an underwater acoustic surface duct as calculated by ray theory and mode theory. The mode theory agrees better with experimental data than does the ray theory in regions of shadow zones and caustics but, in the image-interference region, both theories give almost identical results. The accuracy of the mode theory is limited at very close ranges by the number of modes treated.

Dr. Pedersen is a physicist. He has been engaged in theoretical and experimental research in the propagation of underwater sound since his employment by the Navy Electronics Laboratory in 1952.

Mr. Gordon is a mathematician. His principal interests are acoustic propagation theory and computer applications of the theory. He joined the Navy Electronics Laboratory in 1959.

This paper is reprinted with the permission of the authors and the Acoustical Society of America.

Reprinted from THE JOURNAL OF ACOUSTICAL SOCIETY OF AMERICA, Vol. 37, No. 1, 105-118, January 1965

Normal-Mode Theory Applied to Short-Range Propagation in an Underwater Acoustic Surface Duct

4

MELVIN A. PEDERSEN AND DAVID F. GORDON

U. S. Navy Electronics Laboratory, San Diego, California 92152
(Received 25 September 1964)

This paper presents detailed numerical comparisons between propagation losses as calculated by ray theory and by mode theory. Both theories are based on the bilinear surface-duct model. In the image-interference region, the theories give almost identical results, the accuracy of the mode solution being limited at very close ranges by the number of modes treated. As many as 40 modes are considered. Comparison with experimental data at 530 and 1030 cps indicates that the mode theory is definitely superior to ray theory in regions where ray theory predicts shadow zones and caustics. The nature of the mode solution is examined in detail. The manner in which the mode theory produces the image-interference pattern is controlled by the detailed amplitude and phase characteristics of a large number of modes and is not readily interpreted as the interference between a direct and a surface-reflected acoustic path.

INTRODUCTION

THIS paper presents the results of normal-mode theory as applied to short-range propagation of underwater sound in a surface duct (ranges less than 10 kyd). The normal-mode approach to surface ducts was applied to electromagnetic propagation by Furry[1] and later applied to underwater sound by Marsh.[2] Their interest was limited to a few modes at long range where ray theory breaks down. Our initial interest was to determine whether this mode theory could be extended to short ranges and overlap the region where ray theory applies, or whether there was a gap at intermediate ranges where neither theory was accurate.

With suitable improvements and corrections to the original approach, we have been able to apply normal-mode theory at short ranges in the image-interference region where ray theory is known to be valid. As is shown, there are range intervals over which the propagation losses of mode theory and ray theory are for all practical purposes identical. Thus, there is no range gap between valid solutions and, as a result, the whole normal-mode approach, which is exceedingly complicated and detailed as compared to ray theory, is substantiated. Furthermore, a comparison with experimental data at frequencies of 530 and 1030 cps indicates that the normal-mode-theory approach is definitely superior to ray theory at the somewhat longer ranges where ray theory predicts shadow zones and caustics.

Section I of this paper presents the normal-mode and ray-theory solutions. Section II compares these solutions with experimental propagation-loss data. Section III presents intermediate results of the mode theory, chosen to illustrate the nature of the solution in more detail.

All of the theoretical calculations appearing in this paper were made on a CDC-1604 computer.

I. THEORETICAL SOLUTIONS

In this section, some of the basic expressions necessary for the calculation of propagation loss are presented. We shall not treat derivations or the intricate mathematical problems that arise in the numerical evaluation of the functions that appear in the normal-mode solution.

A. Velocity-Depth-Profile Model

Both theories are based on the bilinear-gradient model of a surface channel, as shown in Fig. 1. The nomenclature for the model stems from the linear nature

[1] W. H. Furry, "Theory of Characteristic Functions in Problems of Anomalous Propagation," Mass. Inst. Technol. Lab. Rept. 680 (1945); "Methods of Calculating Characteristic Values for Bilinear *M* Curves," *ibid.* Rept. 795 (1946).

[2] H. W. Marsh, "Theory of the Anomalous Propagation of Acoustic Waves in the Ocean," U. S. N. Underwater Sound Lab. Rept. 111 (1950).

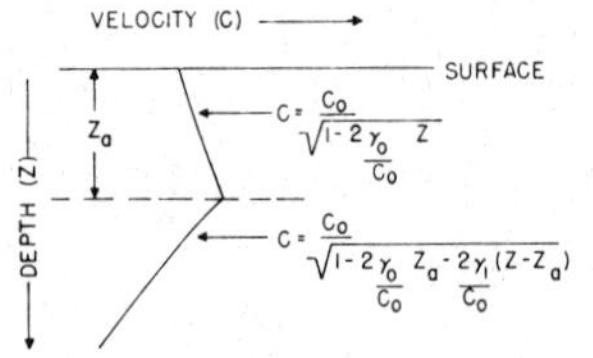

FIG. 1. Bilinear-gradient model of a surface channel.

of the index of refraction. The velocity profile consists of two curvilinear functions having the form as shown. In practice, these functions are very close to straight lines. This particular mathematical form is selected because the depth-dependent portion of the wave equation reduces exactly to Stokes' equation upon separation of the range and depth variables. Z_a is the depth of the channel, C_0 is the surface velocity, and γ_0 and γ_1, respectively, are measures of the gradient in and below the channel. In the general mode theory, γ_1 is negative, whereas γ_0 may be either positive or negative. This paper considers only the case of γ_0 positive—the condition for a surface sound channel.

There are three additional basic parameters that are determined by the experimental setup. These include the frequency f and the source and receiver depths, designated by Z_0 and Z, respectively.

B. Normal-Mode Solution

The general treatment follows Marsh,[2] with certain corrections, modifications, and extensions. For the present, those interested in derivations should consult Marsh[2] or Furry.[1] It should be noted, however, that Marsh[2] contains a considerable number of typographical errors that make comparison difficult with the solution presented here.

Before presenting the solution, let us point out some of the procedures necessary in the extension of Marsh's treatment (valid for long ranges) to short ranges. At long range, only a few low-order modes need be treated, whereas at short range many modes must be included. The eigenvalues for low-order modes were obtained by Marsh and Furry using certain asymptotic approximations. Since these approximations are not valid for higher-order modes, our approach must find the roots of the eigenvalue equation itself. Marsh and Furry also used asymptotic expansions that were valid only at long ranges. In contrast, our approach avoids mathematical approximations and all expressions are computed with as much accuracy as is feasible, using single-precision arithmetic.

Certain useful combinations of the basic parameters, given in the previous section, are

$$t = Z/Z_a, \quad t_0 = Z_0/Z_a, \quad \rho = (\gamma_0/\gamma_1)^{\frac{1}{3}}, \quad k = 2\pi f/C_0, \qquad (1)$$

and $M = (2k^2\gamma_0/C_0)^{\frac{1}{3}} Z_a$. Here, the real cube roots are taken. M is a dimensionless parameter that is a measure of the strength of the channel.

In this theory, the source and receiver depths are interchangeable; i.e., reciprocity holds. A set of major items necessary in the computation are the depth functions $U_n(t)$, where the subscript n indicates the mode number. These functions indicate the degree to which the mode excites (or is excited by) a particular receiver (source) at the particular depth. As we see later, these depth functions may be regarded as mode amplitudes before any dependence on range is considered. These depth functions are complex and, hence, contain phase information as well as amplitudes in the usual sense. When Marsh's[2] equations are evaluated in detail, the following form results for the product of the depth functions:

$$U_n(t)U_n(t_0) = (M/Z_a)F_n(t)F_n(t_0)/D_n. \qquad (2)$$

D_n is presented later. $F_n(t)$ takes one of two forms as follows:

Case I: When $t \leq 1$, corresponding to the receiver (source) in the channel,

$$F_n = \mathfrak{A}(t). \qquad (3a)$$

Case II: When $t \geq 1$, corresponding to the receiver (source) below the channel,

$$F_n = \mathfrak{A}(1)h_2(B_t)/h_2(B_1). \qquad (3b)$$

The function $\mathfrak{A}(t)$ and the argument B_t in (3a) and (3b) are given by

$$\mathfrak{A}(t) = h_1(Mx_n)h_2(Mx_n - Mt) - h_2(Mx_n)h_1(Mx_n - Mt) \qquad (4)$$

and

$$B_t = \rho^2(Mx_n - M) + (M - Mt)/\rho. \qquad (5)$$

$\mathfrak{A}(1)$ and B_1 are obtained by setting $t=1$ in (4) and (5). Expressions for $F_n(t_0)$ are obtained by replacing t by t_0 in Eqs. (3)–(5).

The h_1 and h_2 functions in Eqs. (3) and (4) are complex solutions of Stokes' equation and are referred to as modified Hankel functions of order one-third. These functions are computed by methods discussed by the Harvard Computational Laboratory.[3] First derivatives of these functions are designated as h_1' and h_2'.

The quantity Mx_n in Eqs. (4) and (5) is a complex eigenvalue and is a root of the characteristic equation

$$G = \rho h_2(B_1)\beta - h_2'(B_1)\mathfrak{A}(1) = 0, \qquad (6)$$

where

$$\beta = h_1(Mx_n)h_2'(Mx_n - M) - h_2(Mx_n)h_1'(Mx_n - M). \qquad (7)$$

Marsh treats the quantity x_n as a separate entity. This separation seems unnecessary as x_n always appears with M as a multiplier.

Equation (6) has an infinite number of roots, each

[3] *Tables of the Modified Hankel Functions of Order One-Third and of Their Derivatives* (Harvard University Press, Cambridge, Mass., 1945).

root corresponding to a normal mode. In the case treated here, the first mode corresponds to the root with the smallest real and imaginary parts, the second mode to the next larger real and imaginary parts, etc. Hence, the modes may be ordered by either $\mathrm{Re}Mx_n$ or $\mathrm{Im}Mx_n$. We have investigated other cases where the numbering of modes is ambiguous. In the case of γ_0 negative, a different ordering is obtained from a consideration of $\mathrm{Re}Mx_n$ than from consideration of $\mathrm{Im}Mx_n$. Another complication in ordering, which arises for small γ_1 is the appearance of an additional set of modes, entirely distinct from the set treated in this paper.

Equation (6) is solved by Newton's method in the complex plane. In brief, for the lowest-order modes, initial values for Newton's method are obtained from certain asymptotic approximations. For higher-order modes, initial values are obtained by extrapolation based on the three next-lower-order eigenvalues. The function D_n in (2) is given by

$$D_n=(\rho^3-1)[\beta^2+(Mx_n-M)\mathfrak{A}^2(1)]-2.124292605, \quad (8)$$

where the negative constant is the square of the Wronskian $W(h_1,h_2)$.

We now have presented all the expressions necessary to calculate the depth functions of (2) and are ready to introduce the range dependence. The propagation loss H is given as a function of range r by

$$H=-10\log\left|\sum_1^N H_0^2(\lambda_n r)U_n(t)U_n(t_0)\right|^2 -20\log\pi+\alpha_A r. \quad (9)$$

H_0^2 is the second Hankel function of order zero. The term λ_n is complex and is given by

$$\lambda_n=[k^2-Mx_n(M/Z_a)^2]^{\frac{1}{2}}. \quad (10)$$

The complex square root is taken so that λ_n lies in the fourth quadrant. N is the number of modes included in the computation. It is not always evident what this number should be but, as is shown later in an example, the number of modes required in the computation increases with decreasing range.

The result of the summation in (9) is complex. Hence, this complex number is changed to an absolute value before computing the logarithm. The item $\alpha_A r$ in (9) has nothing to do with normal-mode theory but represents the effect of physical absorption and scattering. If units of yards are used for all length dimensions throughout the computation, then H represents the decibel loss relative to 1 yd, with no additional scaling being necessary.

There are three important aspects of this normal-mode approach that contrast to many other normal-mode solutions to the wave equation: (1) The solution is valid even for short ranges, because the branch-line integral, often present in a complete solution, is zero for this model. (2) λ_n is a complex wavenumber and therefore the modes are damped. (3) There are no cutoffs in the frequency domain. Hence, in theory, an infinite number of modes must be treated, but in practice the higher-order modes are so highly damped that they contribute only at extremely short ranges.

Before continuing, we should note that the problem of the branch-line integral is an exceedingly complicated one and has not been investigated in detail by the authors. In a private communication, Marsh has indicated that the integral is zero for this model. This is demonstrated in some detail by Towne and Wilson[4] for the single layer case ($Z_a=0$). Furthermore, according to Stone[5] the appearance of a branch-line integral is only a consequence of the fact that the medium is unbounded in the Z direction and is not associated with boundary conditions at layer interfaces. Thus, we may conclude that the nature of the velocity profile in the infinite layer determines the presence of a branch-line integral. This view is substantiated by Towne and Wilson,[4] who state that the existence of branch points is associated with the fact that $C(Z)$ does not vanish at infinity. For the model of Fig. 1, $C(Z)$ does indeed vanish at infinity.

C. Simplified Form of the Mode-Theory Solution

The preceding section gives the rigorous and accurate mathematical expressions that are used in the actual computational work. However, these exact expressions are not readily interpreted. In contrast, this section discusses certain simplifying approximations that present an excellent qualitative description of the theoretical result and that can be readily understood.

It is convenient to decompose the λ_n of (10) into the form

$$\lambda_n=k-\sigma_n-i\tau_n. \quad (11)$$

In general, $k\gg|\sigma_n|$ or τ_n.

We now investigate the significance of σ_n and τ_n in the propagation-loss equation. Using only the first term of the asymptotic expansion for $H_0^2(\lambda_n r)$, we find that H may be written as

$$H\sim-10\log[(\sum_1^N A_n\cos\theta_n)^2+(\sum_1^N A_n\sin\theta_n)^2] -10\log(C_0/f)+10\log r+\alpha_A r, \quad (12)$$

where

$$A_n=|U_n(t)U_n(t_0)|\exp(-\tau_n r), \quad (13)$$

$$\theta_n=\Psi_n+\sigma_n r, \quad (14)$$

---

[4] D. H. Towne and K. G. Wilson, "Refraction and Diffraction of Explosive Pressure Pulses by Gradients in the Propagation Velocity," Pt. II, Woods Hole Oceanog. Inst. Ref. No. 57-45 (1957).

[5] J. L. Stone, "A Theoretical Analysis of Acoustic Wave Modes in Layered Liquids," Princeton Univ. Rept. No. 9, contract N6ONR-270 (1953).

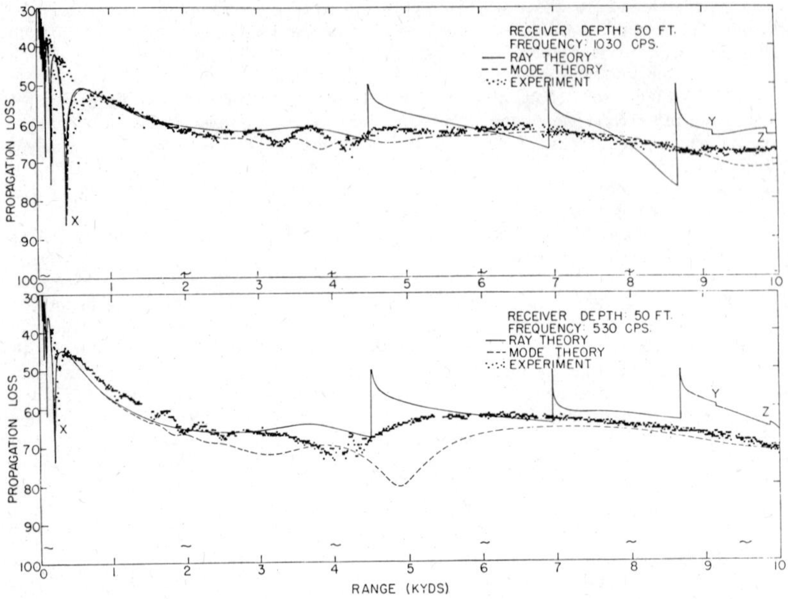

FIG. 2. Propagation losses for 50-ft receiver, 55-ft source.

and

$$\Psi_n = \arg\{U_n(t)U_n(t_0)\}. \tag{15}$$

The 10 log$r$ term of (12) corresponds to cylindrical spreading and represents the general trend with range. Superimposed on this general trend is the interaction between the modes. These modes may be thought of as damped sinusoidal waves whose damped amplitude and phase angles at range r are given by A_n and θ_n, respectively.

The value of a given A_n relative to the values of the other A_n's determines the extent to which a particular mode contributes to the final result. From (13), we see that there are two fundamentally different factors that determine how much a particular mode contributes to the result. The first, $|U_n(t)U_n(t_0)|$, depends on the source and receiver depth but not on range. The opposite is true of the second factor, $\exp(-\tau_n r)$. In describing the effect of range, it is necessary to note that τ_n always increases with increasing n. Thus, as the range increases, the relative contribution of a mode decreases with increasing mode number.

Now, it happens that k^2 is the dominant term in (10). A good approximation to λ_n is given by

$$\lambda_n \sim k - Mx_n(M/Z_a)^2/2k. \tag{16}$$

Thus,

$$\sigma_n \sim \Gamma \mathrm{Re} Mx_n \quad \text{and} \quad \tau_n \sim \Gamma \mathrm{Im} Mx_n \tag{17}$$

with $\Gamma = (\pi f \gamma_0^2)^{\frac{1}{3}}/C_0$. From (17), we see that the mode damping is determined primarily by the imaginary part of the eigenvalue, whereas the rate of change of phase with range is determined by the real part.

D. Ray-Theory Solution

In order to make a valid comparison between the mode theory and ray theory, the ray-theory approach must be based on the identical profile model of Fig. 1. In this section, the computing forms for a ray theory, based on the profile model

$$1/C^2 = a + bZ, \tag{18}$$

are presented.

The ray to be calculated is designated by the parameter C_m, the velocity at which the ray becomes horizontal. In this case, Snell's law becomes

$$\cos\theta_i = C_i/C_m, \tag{19}$$

where θ_i is the angle formed by the ray with the horizontal at interface i and C_i is the velocity at interface i. Then

$$\tan\theta_i = (C_m^2 - C_i^2)^{\frac{1}{2}}/C_i \tag{20}$$

and $\cot\theta_i$ is the reciprocal. (In this treatment, θ_i, itself, is never used so could be completely dispensed with. However, the notations $\tan\theta_i$ and $\cot\theta_i$ are physically meaningful and eliminate in the theoretical treatment the repeated use of the cumbersome forms by which they are actually computed.) Consider the

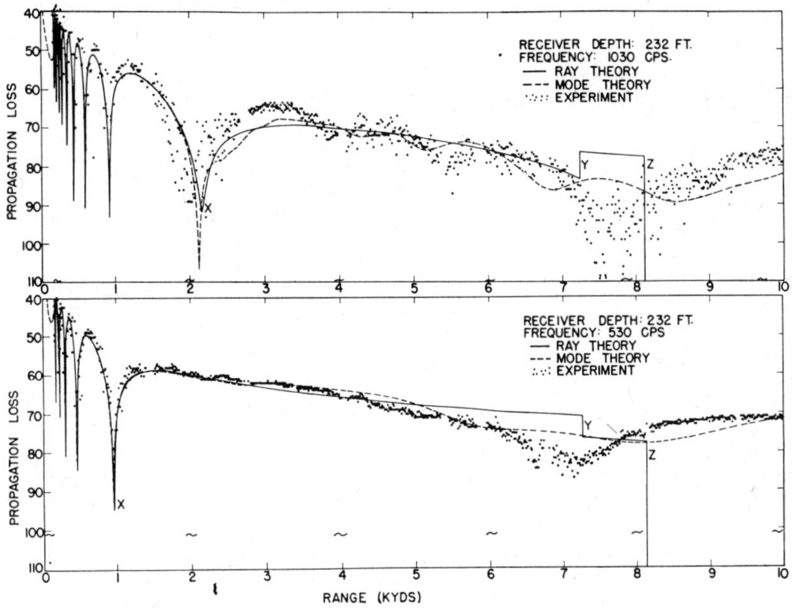

FIG. 3. Propagation losses for 232-ft receiver, 55-ft source.

profile layers ordered with increasing Z and that the velocity at the upper and lower interfaces of layer i are C_i and C_{i+1}, respectively. The horizontal range in layer i in case the ray passes completely through the layer is

$$R_i = 2(\tan\theta_{i+1} - \tan\theta_i)/bC_m^2. \tag{21}$$

The corresponding travel time is

$$T_i = 2(\tan^3\theta_{i+1} - \tan^3\theta_i)/3bC_m^3 + R_i/C_m. \tag{22}$$

The corresponding derivative term, necessary in the intensity formulation, is

$$B_i \equiv C_m dR_i/dC_m = 2(\cot\theta_{i+1} - \cot\theta_i)/bC_m^2 - R_i. \tag{23}$$

If the ray forms a nadir or an apex in the layer, then the first or second item in the parentheses of Eqs. (21)–(23) is omitted.

The complete range (R), travel time (T), and derivative (B) are obtained by summing expressions (21)–(23) over all layers as traversed along the ray path from source to receiver.

The ray intensity for a unit source is computed by the expression[6]

$$I/F = C_s|\cot\theta_s|\,|\cot\theta_h|/C_h RB, \tag{24}$$

where s and h refer to the source and receiver interfaces, respectively, and F indicates a measure of the source strength.

In the case of the ray that is horizontal at source or receiver, Eq. (24) is an indeterminate form. In this case, the general expressions given at the bottom of page 467 and top of 468 of Ref. 6 were used. Direct evaluation of the limit using (23) leads to the same result.

The propagation loss for the combination of n arrivals was calculated from

$$H = -10\log\left\{\left[\sum_1^n \left(\frac{I_j}{F}\right)^{\frac{1}{2}} \cos\theta_j\right]^2 + \left[\sum_1^n \left(\frac{I_j}{F}\right)^{\frac{1}{2}} \sin\theta_j\right]^2\right\} + \alpha_A r. \tag{25}$$

The phase angle is computed from

$$\theta_j = 2\pi f T_j + \pi P_j, \tag{26}$$

where T_j is the travel time of arrival j and P_j is the number of times the arrival has reflected from the surface. It should be noted that (I_j/F) and T_j must be evaluated at the same range for all j, whereas for the various arrivals the values of T and I/F as initially computed fall at random ranges, depending on the selection of the ray parameters. Values of I_j/F and T_j for the desired ranges are determined by curvilinear

[6] M. A. Pedersen, "Acoustic Intensity Anomalies Introduced by Constant Velocity Gradients," J. Acoust. Soc. Am. 33, 465–474 (1961).

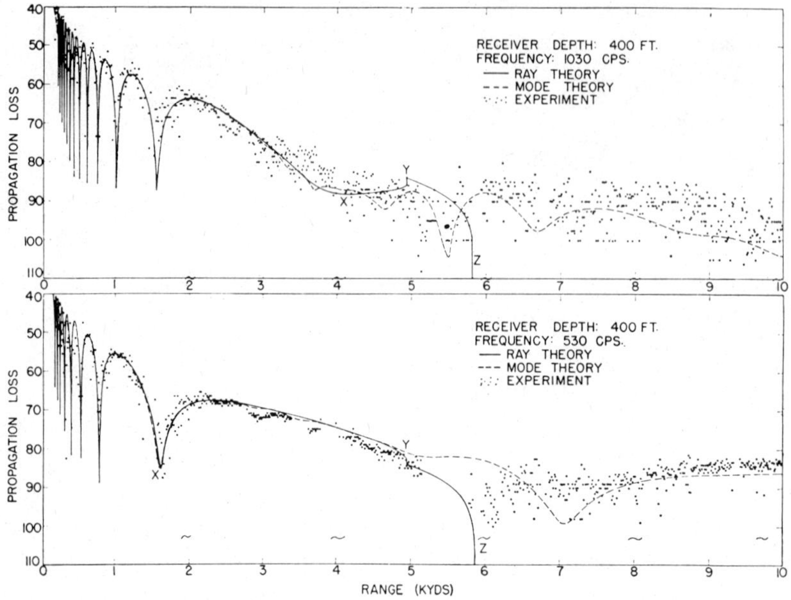

FIG. 4. Propagation losses for 400-ft receiver, 55-ft source.

interpolation based on the ray-theory computations for three successive values of the ray parameter. This complete process is done by computer, with the final output being propagation loss as a function of range.

II. COMPARISONS, THEORETICAL AND EXPERIMENTAL

In this section, we compare the ray- and mode-theory solutions with experimental results, the ray- and mode-theory solutions with each other, and the mode-theory solutions for two velocity profiles.

A. Comparison of Theories with Experiment

The experimental data are the same as discussed in a previous paper.[7] In brief, two sources (530- and 1030-cps frequency) were towed at depths of 55 ft. Signals were received on hydrophones at depths of 50, 232, and 400 ft. Values of the mean profile parameters as determined from bathythermograph data taken at the two ships are given in Table I, which uses the notation of Fig. 1. Values of the gradient in the layer are significantly less than that of an isothermal layer, which is 0.0182 sec^{-1}.

TABLE I. Parameters for profiles taken at the source (S.S.) and receiver (R.S.) ships.

	C_0(ft/sec)	Z_a(ft)	γ_0(sec^{-1})	γ_1(sec^{-1})
S.S	4940.94	326.67	0.0144	−0.438
R.S.	4940.79	313.33	0.0114	−0.423

Figures 2–4 present propagation loss as a function of range out to 10 kyd. Each dot in the experimental data represents measurements of an individual ½-sec pulse transmission. The tilde marks near the bottom of Figs. 2–4 indicate the propagation loss corresponding to the average noise level. Both the ray-theory and mode-theory calculations are based on the source-ship profile of Table I. The ray-intensity computations were based on 78 ray parameters, including all critical rays such as the ray grazing the bottom of the surface layer. At 1030-cps frequency, 40 modes were included in the normal-mode calculations while at 530 cps only 29 modes were included. (The 40-mode figure arises from storage limitation in the computer; the 29-mode figure arises from exponent overflow in the computer when attempting to evaluate modes of higher order.) In both theoretical approaches, α_A was taken to be 0.0127 and 0.345 dB/kyd at 530 and 1030 cps, respectively.

The letter Z indicates the dropout of the surface reflected ray that grazes the bottom of the surface channel. Note that the intensity change at range Z is much more abrupt in Fig. 3 than in Fig. 4. This results

[7] M. A. Pedersen, "Comparison of Experimental and Theoretical Interference in Deep-Water Acoustics," J. Acoust. Soc, Am. 34, 1197–1203 (1962).

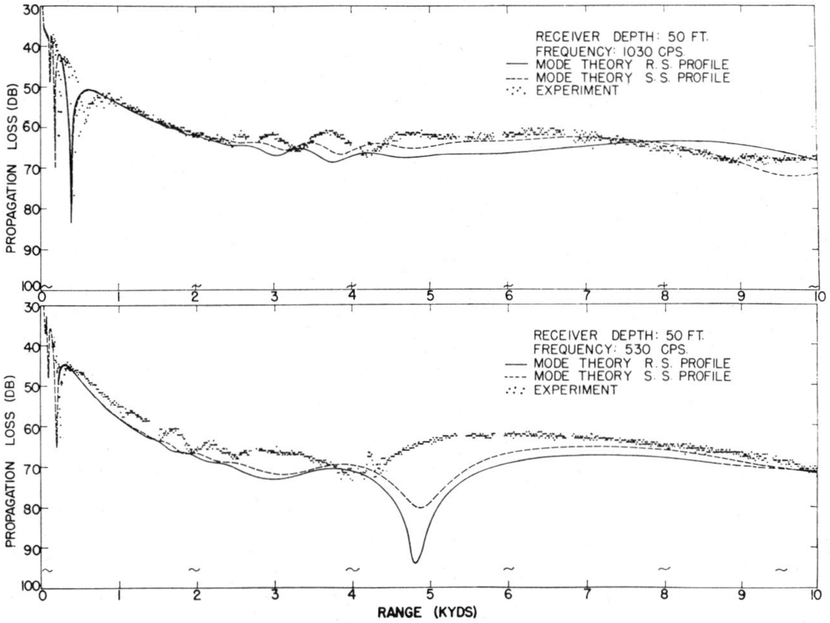

FIG. 5. Propagation losses for 50-ft receiver, 55-ft source.

from the fact that the slope discontinuity at the bottom of the channel does not affect the intensity of the limiting ray for a receiver in the channel. However, for a receiver below the layer, the discontinuity causes the intensity to go to zero in a continuous manner as the ray parameter approaches that of the limiting ray. The letter Y indicates the dropout of the direct ray, which grazes the bottom of the channel. The letter X indicates the range of the last null in the image interference pattern for ray theory. The discontinuities in the ray-theory calculation in Fig. 2, occurring at about 4.5, 6.9, and 8.7 kyd, are caustics, indicating the onset of additional arrivals. The minimum loss associated with these caustics has been arbitrarily terminated at 50 dB loss, although elementary ray theory yields infinite intensity at the caustic.

Note that, in all six examples given in Figs. 2–4, the mode-theory result is in better agreement with experiment than is the ray-theory result. In Fig. 2, there is no evidence of caustics or even convergence regions in the experimental data. For the 232-ft receiver of Fig. 3, ray theory predicts a shadow zone from 8.1 to 11.4 kyd. This contrasts to the experimental data in which the propagation losses are actually seen to be decreasing over the interval from 8 to 10 kyd. The mode theory shows such a decrease, but the last fade in the mode pattern is shifted out in range and is not as deep as the corresponding experimental fades occurring between 7 and 8 kyd. However, this last experimental fade cannot be explained at all by simple image interference, since the last null of ray theory occurs at range X. In Fig. 4, a ray-theory shadow zone exists at ranges beyond 5.8 kyd, although there is no such indication in the experimental data. Agreement between mode theory and experiment beyond this range is fairly good for the 530-cps frequency. It is only fair at 1030-cps frequency. This feature is discussed later in a further comparison between mode theory and experiment when the results of the mode theory for a different profile are presented.

B. Comparison of Ray- and Mode-Theory Solutions

The ray and mode theories give almost identical results over certain range intervals, which vary somewhat with depth and frequency. At extremely short ranges (as, for example, $R<500$ yd at 1030 cps in Fig. 4), there is considerable departure. As is shown later, this departure can be attributed to the omission of higher-order modes in the mode computation.

At short ranges then, agreement between mode and ray theory seems limited only by the number of modes that must be included. At somewhat longer ranges, where the rays are not so steep and where refraction

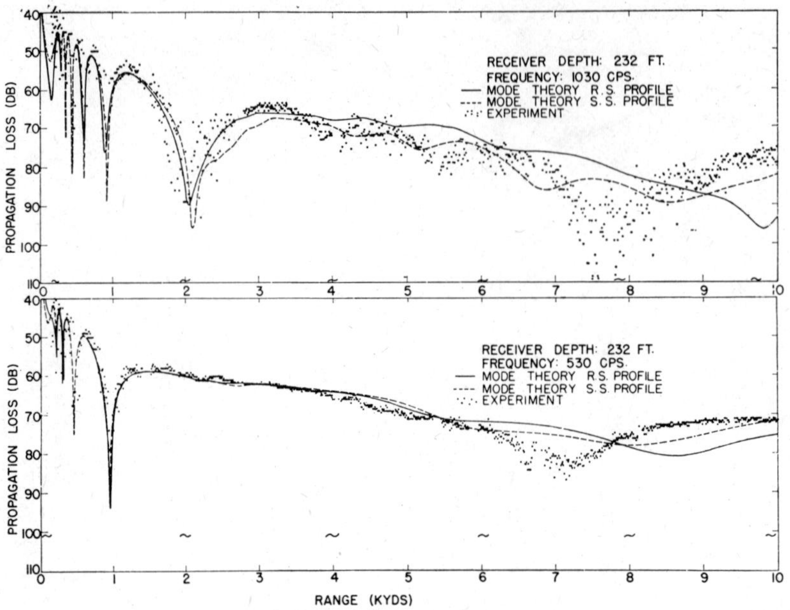

FIG. 6. Propagation losses for 232-ft receiver, 55-ft source.

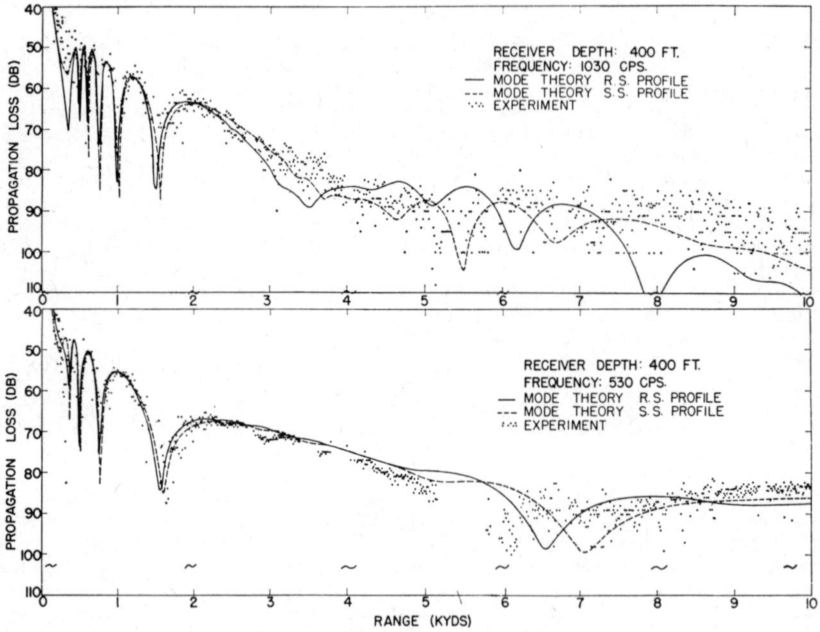

FIG. 7. Propagation losses for 400-ft receiver, 55-ft source.

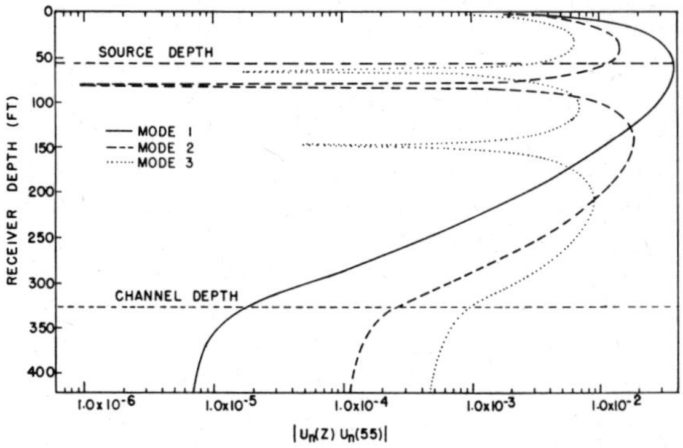

FIG. 8. Depth-function amplitude for modes 1–3 (1030 cps).

effects become more important, the solutions begin to depart. Note, however, in Figs. 3 and 4 that the solutions agree fairly well (the mode solution oscillating about the ray solution) clear out to the shadow zone where simple ray theory finally breaks down completely.

There are no shadow zones in Fig. 2. In this case, ray and mode theories are in substantial agreement until the ray-theory caustics appear. Beyond this range, the ray and mode theories show marked departures. Although the general level of propagation loss is comparable at 1030 cps, the mode-theory losses for 530 cps are systematically larger than the ray-theory losses. We believe that we could get better agreement between the ray and mode theories in the caustic region if we made the comparison for even higher frequencies as the ray-theory solution becomes a better approximation as the frequency is increased.

C. Comparison of Mode Theories for Two Profiles

Thus far, we have presented the mode-theory results for the source-ship profile. These results, as well as the experimental data, are duplicated in Figs. 5–7. However, the solid line in this case is the normal-mode solution for the receiver-ship profile of Table I. For this profile, 40 modes were included at 1030 cps and 25 modes at 530 cps.

Note that the greatest difference between the mode solutions for the two profiles occurs in Fig. 7 for 1030 cps at ranges beyond 5 kyd. (Note also that this is a region of great fluctuation in the experimental data, indicating a critical dependence on environmental conditions.) With this exception, the solutions are quite comparable in that the chief difference in many cases is a slight range shift in the patterns. This and the results of other investigations, not reported here, indicate that the mode phase is more sensitive to slight profile changes than is the mode amplitude. Thus, the detailed interference patterns of mode theory and experiment are often quite similar but shifted somewhat in range.

In general, agreement between theory and experiment is comparable to the agreement between the

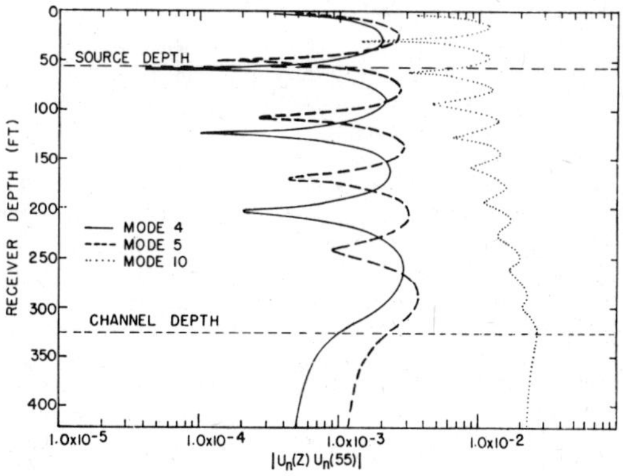

FIG. 9. Depth-function amplitude for modes 4, 5, and 10 (1030 cps).

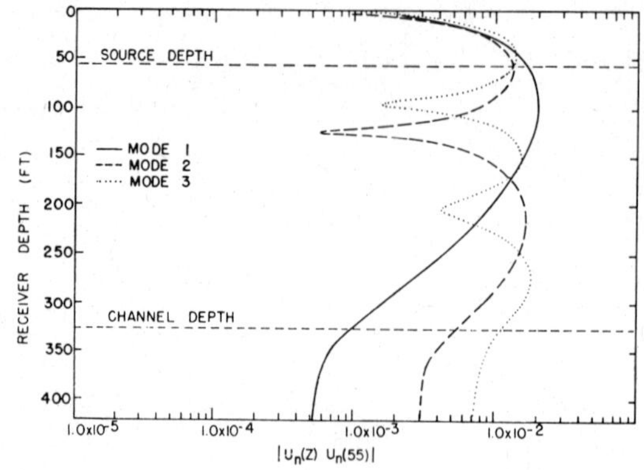

FIG. 10. Depth-function amplitude for modes 1–3 (530 cps).

mode-theory solutions for the two profiles. There are, however, exceptions, the most noteworthy being the pronounced null that appears in Fig. 5 at 530 cps in both theoretical solutions at 4.8 kyd, with little evidence of such in the experimental data.

The degree of over-all agreement between experiment and theory is probably more than we have any right to expect. In the first place, the theory is based on the simple two-layer model of Fig. 1 whereas the *BT* data of experiment showed a definite three-layer condition, the extra layer having a negative gradient and extending from the surface to a depth of 30 ft (Ref. 7). Secondly, the *BT* data showed considerable temporal and spatial variability during the experiment whereas the theoretical calculation is based on a fixed "average" profile.

III. INTERMEDIATE MODE-THEORY RESULTS

As is well known, the image-interference pattern is subject to a simple physical interpretation. It is formed by the interaction between a surface-reflected path and a direct path. Indeed, methods based on this physical concept with the assumption of isovelocity water provide an approximation that contains the essential features.[7] The detailed ray-theory approach follows this same basic concept, but with embellishments to account for refraction effects.

One might suppose then that the image-interference pattern of mode theory is produced by a simple interaction between modes and is subject to a simple physical interpretation. The purpose of this section is to indicate that this is not the case and that, although the mode theory indeed produces the pattern, it does so in an exceedingly complicated manner. We present examples that indicate not only the contribution of indivudal modes but also the dependence of this contribution on depth and range. This provides considerable insight into the general nature of the mode solution.

A. Initial Mode Amplitude and Phase

Consider first the dependence of the mode solution on the depth variable for profile S.S. of Table I. Figures 8–10 present the absolute value of the product of the depth function for modes 1–5 as well as 10 at 1030 cps and modes 1–3 at 530 cps. This product is the initial mode amplitude—i.e., the mode amplitude before any range dependence is introduced. In the theoretical section, this product is indicated as a function of the dimensionless parameters t and t_0. Here, we present the product as a function of Z (the receiver depth) for a fixed source depth ($Z_0=55$ ft). (Since reciprocity holds, we may also regard Figs. 8–10 as giving the initial mode amplitudes as a function of source depth for a fixed receiver depth of 55 ft.) Each decade change in amplitude corresponds to a 20-dB change in intensity level.

Although Figs. 8–10 are for a fixed source depth, the mode curve for any other source depth (Z_0) can be obtained by a simple graphical translation of the dependent variable. This is possible because the dependent variable is plotted on a logarithmic scale. The procedure is based on the identity,

$$\log|U_n(Z)U_n(Z_0)| = \log|U_n(Z)U_n(55)| + \log|U_n(Z_0)U_n(55)| - \log|U_n(55)U_n(55)|. \quad (27)$$

For a particular mode, one then measures the difference between the value for the receiver depth equal to the desired source depth and the value for the receiver depth equal to 55 ft. The entire mode curve is then translated by this difference.

In Figs. 8–10, the mode amplitudes are arbitrarily terminated near the surface. They do, in fact, go to zero at the surface as this is a boundary condition on the solution. Although the theoretical solutions above and below the channel depth have a different form, we see that the depth functions as well as their slope are continuous at the channel depth. These are also boundary conditions on the solution. In this treatment, which follows Marsh, the condition for continuity of slope at the layer interface is equivalent to the eigenvalue equation. The other boundary conditions are satisfied by the general form of the solution, as can readily be seen by examination of Eqs. (2)–(4). However, the slope condition will be satisfied only if the eigenvalue equation is solved exactly. In early investigation of this mode theory, we were made aware of the inadequacy of approximate solutions to the eigenvalue equation by the fact that slope discontinuities appeared at the channel interface. (Furry's[1] treatment differs in that the eigenvalue equation is equivalent to the condition that the solution be zero at the surface.)

These mode patterns bear a striking resemblance to those associated with elementary-wave-motion problems. Note in all cases that mode n has n nulls (nodes) and n peaks (antinodes) above the channel depth. We have found this to be true in all of our investigations thus far, although it may not apply for all conditions. This feature has been of help in identifying mode numbers in complicated cases where approximate roots to the eigenvalue equation are unavailable. We allow the computer to find roots more or less by chance; the depth functions associated with these roots are computed and plotted; and the mode number is determined from the number of nulls or peaks appearing in the plot.

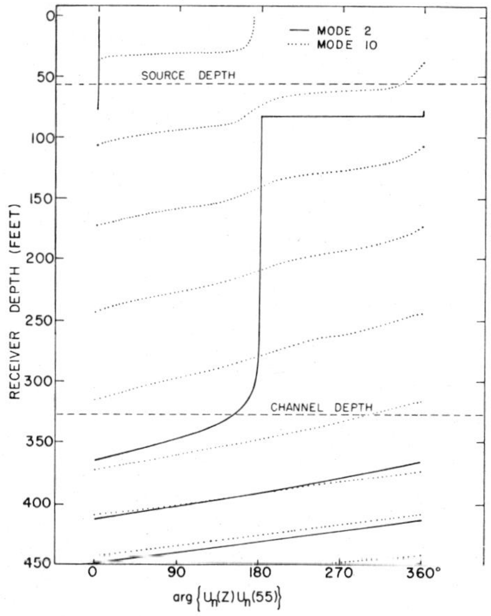

FIG. 11. Depth-function phase for modes 2 and 10 (1030 cps).

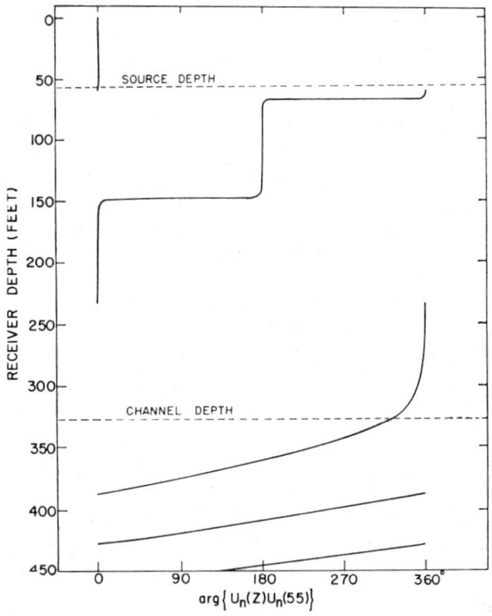

FIG. 12. Depth-function phase for mode 3 (1030 cps).

Thus, we can tell if any modes have been skipped and remain to be determined. Since the mode numbers are ordered by the magnitude of the eigenvalues, we then know in general where to look for the undertermined eigenvalues. For example, if mode 2 has been initially determined, then we need look for only one root that has a smaller real and a smaller imaginary part, i.e., the root for mode 1. Note that the amplitude at the successive peaks or nulls increases with increasing depth. We have found this to be true in other cases, although we have no proof that it will be true in general.

The mathematical explanation for the sharp nulls in Figs. 8–10 can best be treated in conjunction with a discussion of the initial mode phase. Values of this phase at 1030-cps frequency are shown for modes 2 and 10 in Fig. 11 and for mode 3 in Fig. 12. Values for mode 1, not shown here, lie within 0.1° of zero phase for receiver depths from the surface to a depth of about 265 ft. For low-order modes in the channel, the phase remains nearly constant at either 0° or 180° and changes rapidly but without discontinuity from one state to the other. The depths at which these changes occur correspond to the depths at which the nulls occur in Figs. 8–10. For example, in the case of mode 2 at 1030 cps, the phase changes by −178.8° and the mode amplitude varies through some 42 dB as the receiver depth is changed from 81 to 81.36 ft.

Let us now examine in more detail the mathematical nature of the solution for it indicates that the amplitude and phase patterns shown in Figs. 8–12 are characteristics of modes with small $\text{Im}Mx_n$ (small damping).

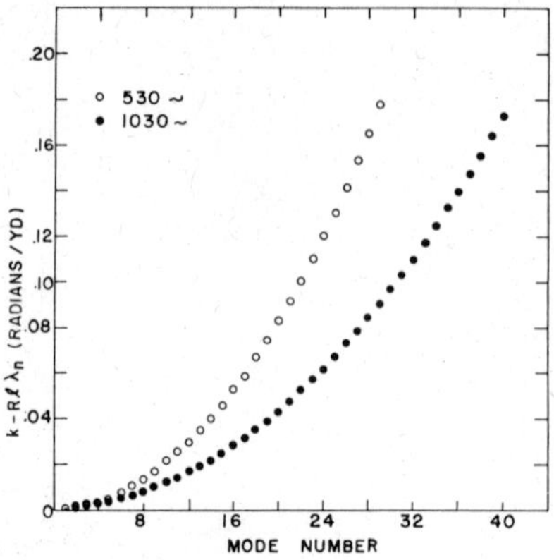

FIG. 13. Real part of the complex wavenumber vs mode number.

For receivers in the channel, there are three factors that determine the phase: $A_n(t)$, $U_n(t_0)$, and D_n. The latter two factors determine a fixed phase shift, i.e., independent of receiver depth. Consider first the effect of source depth. The pattern for source depths other than 55 ft can be obtained by a simple translation based on the identity,

$$\begin{aligned}\arg\{U_n(Z)U_n(Z_0)\} &= \arg\{U_n(Z)U_n(55)\}\\ &\quad+\arg\{U_n(Z_0)U_n(55)\}\\ &\quad-\arg\{U_n(55)U_n(55)\}. \quad (28)\end{aligned}$$

Note that, for most source depths in the channel, the phase diagram for the low-order modes will either be the same as those given in Figs. 11 and 12 or will differ by 180°. The phase will, in general, again be either 0° or 180°, with exceptions occurring if the source is placed at critical depth where the phase is changing rapidly from one state to the other.

Although it is not demonstrated here, D_n is almost a pure negative real number for modes with small $\mathrm{Im}Mx_n$.

The dependence on receiver depth is contained in the function $\mathcal{A}_n(t)$. We first note that $d\mathcal{A}_n(t)/dt$, evaluated at $t=0$, is equal to $-MW$. Where W is the Wronskian and is a negative pure imaginary constant. Since $\mathcal{A}_n(t)=0$, it follows then that $\arg\{\mathcal{A}_n(0)\}=90°$. This applies to all modes. In the case of modes for which $\mathrm{Im}Mx_n$ is small, we can expand the h functions of $\mathcal{A}_n(t)$ in a Taylor series about the real part of the arguments and obtain the following general form:

$$\mathcal{A}_n(t)=if_n(t)+\mathrm{Im}Mx_n g_n(t)+0(\mathrm{Im}Mx_n)^2. \quad (29)$$

Here, $f_n(t)$ and $g_n(t)$ are real functions. When $\mathrm{Im}Mx_n$ is small, then $f_n(t)$ will generally be the dominant term and $\arg\{\mathcal{A}_n(t)\}$ is either 90° or 270°, depending on the sign of $f_n(t)$. As t is increased from zero, $f_n(t)$ starts out positive and changes sign $n-1$ times. The peaks and nulls in the interference pattern for practical purposes correspond to the maxima and zeroes of $f_n(t)$. Note that the amplitude of $\mathcal{A}_n(t)$ at the nulls is proportional to $\mathrm{Im}Mx_n$. This explains why the nulls become less prominent as the mode number is increased since $\mathrm{Im}Mx_n$ increases with increasing mode number.

A. Complex Wavenumbers

Let us next examine in more detail the dependence of the mode solutions on the range variable for profile S.S. of Table I. In the computer program, the expression giving the range dependence $H_0{}^2(\lambda_n r)$ is calculated by an asymptotic expansion of many terms. However, as indicated in Eqs. (11)–(15), the essential behavior with range can be obtained by consideration of the complex wavenumber λ_n.

Since there is such a small difference between $\mathrm{Re}\lambda_n$ and k, it is convenient to consider $k-\mathrm{Re}\lambda_n$, which is the σ_n of Eqs. (11) and (14). Figure 13 presents σ_n as a function of mode number for frequencies of 530 and 1030 cps. $\mathrm{Re}\lambda_n$ is slightly less than k for mode 1 and decreases as the mode number increases. Since σ_n increases with increasing mode number, from (14) we see that the higher the mode number, the more rapidly the phase angle of that mode increases with range.

Figure 14 presents the mode attenuation as a function of n. This mode attentuation is given by $-8686\ \mathrm{Im}\lambda_n$ dB/kyd. Values for the first two modes at 530 cps and the first four modes at 1030 cps, too small to indicate in Fig. 14, are, respectively, 7.44×10^{-3}, 3.17×10^{-1}, 6.61×10^{-7}, 5.75×10^{-4}, 3.71×10^{-2}, and 3.74×10^{-1} dB/kyd. These are the modes that propagate to relatively long ranges and that are referred to in the literature as trapped modes. A generally accepted condi-

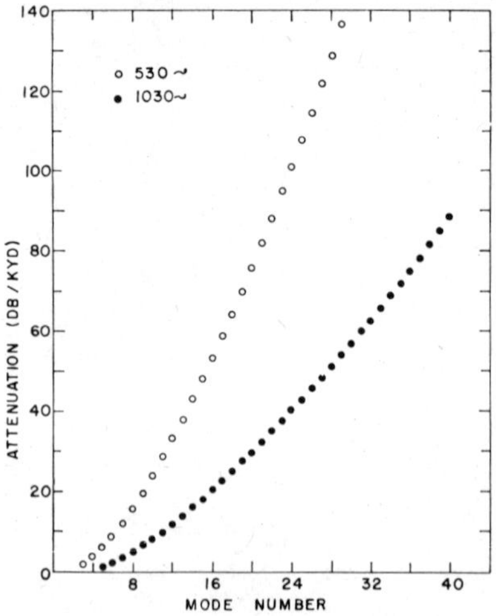

FIG. 14. Mode attenuation vs mode number.

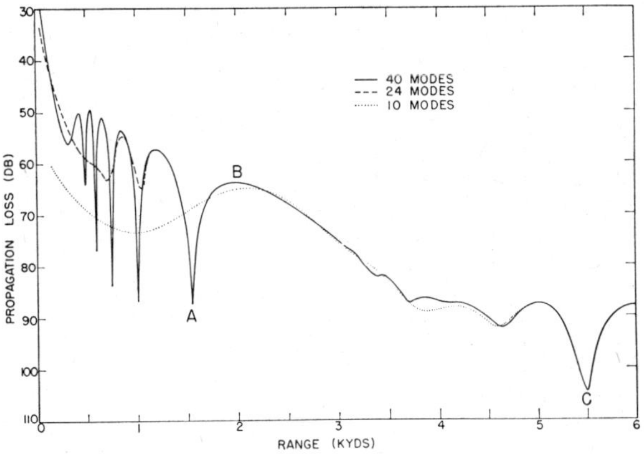

FIG. 15. Comparison of mode solutions obtained by including different numbers of modes (400-ft receiver, 1030 cps).

tion for trapping is that $Mx_n < M$. This condition, though convenient, is somewhat arbitrary since the onset of trapping for a particular mode is a continuous rather than a discrete process. In any case, the formerly held distinction between trapped and leaky modes is of small concern here since the leaky as well as the so-called trapped modes are included in the computations. For the condition of Fig. 14, the mode attenuation increases monotonically with mode number to as large as 136.5 dB/kyd for mode 29 at 530 cps and 88.2 dB/kyd for mode 40 at 1030 cps.

B. Vector Representation of Mode Contributions

Before examining in detail the nature of the mode combination that forms the image-interference pattern, consider how the number of modes included in the mode summation affects the resultant propagation loss. Figure 15 denotes the manner in which the inclusion of additional modes improves the mode solution. The three different lines represent the solution obtained by using different numbers of modes. At all ranges beyond 4.9 kyd, the three solutions agree to tenths of a decibel. The 10-mode solution is reasonably accurate beyond 2.3 kyd but departs markedly at shorter ranges. The 24-mode solution is accurate to 1.2 kyd. The 40-mode solution, which is identical to the dashed line in the upper part of Fig. 4, extends the familiar image-interference pattern in to somewhat less than 0.5 kyd. Comparison with ray theory indicates that the 40-mode solution is accurate in to a range of 550 yd.

In the mode theory, the basic dependence on range is cylindrical spreading with an additional attenuation term. The question then arises as to how this mode theory can be accurate in the image-interference region where the basic dependence on range is spherical spreading. The answer lies in the fact that the high attenuation of the high-order modes reduces their contribution very rapidly with range. Thus, at close range, the resulting propagation loss increases at a considerably greater rate than cylindrical spreading because of what is essentially a successive drop out of these high-order modes. At longer ranges, the contributions of these high-order modes is nil and the spreading is cylindrical with a relatively small attenuation term.

Figure 16 presents individual mode amplitudes in vector form with their resultants marked R. To be more specific, these vectors represent individual terms in the summation of Eq. (9) while the resultant represents the sum. The letters A to C refer to the range for which the terms were calculated. These ranges correspond to those similarly labeled in Fig. 15. A magnitude scale is indicated for each set with set C expanded to twice the scale of A and B. Modes not shown here make no essential contribution to the resultant. As can be seen in the mode-theory approach, the peaks and nulls of the image-interference pattern resul' from a much more complicated behavior than the relatively simple interaction between a source and its surface image as depicted by ray theory. For example, at range A, the amplitudes of modes 6–15 are all larger than that of the resultant. Investigation of Fig. 16 and other cases not shown here indicates that the peaks and nulls of the interference pattern are produced by a seemingly haphazard arrangement of vectors that amazingly reinforce or cancel each other at the proper ranges. It seems quite remarkable that the initial amplitudes and phase together with the attenuation and rate of change of phase operate for each mode in such a manner as to produce the image-interference pattern.

In discussing electromagnetic propagation, Furry[8] notes, ". . . completely trapped modes may exist separately, but leaky modes may not. The expansions of

[8] C. R. Burrows and S. S. Attwood, *Radio Wave Propagation* (Academic Press Inc., New York, 1949), p. 171.

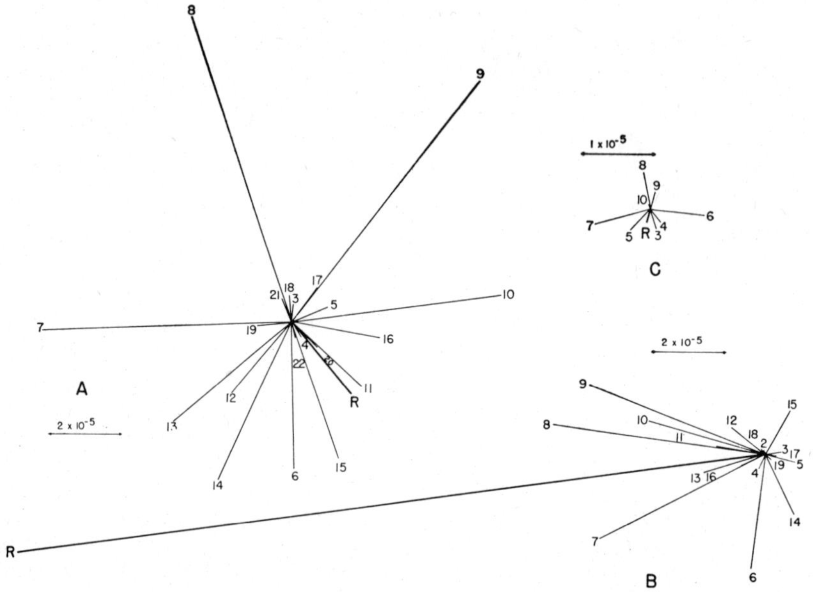

FIG. 16. Vector representation of the mode contributions at the three ranges designated by letters in Fig. 15.

fields in terms of leaky modes are thus essentially mathematical." Perhaps, then, we should not expect a complete interpretation of individual modes from a physical standpoint. If the problem has been stated and solved properly, then the mathematical solution consisting of the mode summation must of necessity produce the interference pattern despite the fact that the interaction of individual modes appears physically inexplicable.

IV. SUMMARY

Comparison with ray theory and experimental results indicates that normal-mode theory for a bilinear surface duct can be successfully applied in the image-interference region. The propagation losses of both theoretical approaches are practically identical with the accuracy of the mode solution at very close ranges, being limited only by the number of modes included in the computation. Comparison with experimental data indicates that the mode theory is superior to ray theory in the region where ray theory predicts shadow zones or caustics. Although the normal-mode solution faithfully reproduces the image-interference pattern, the process appears to be essentially mathematical and not amenable to a simple physical interpretation such as that inherent in the ray-theory approach.

ACKNOWLEDGMENTS

This investigation would not have been possible without the contribution of the late Alice Joy Keith in checking the theory, developing numerical procedures, and programming the computer. Grace Wofford drafted the final Figures.

A. O. Williams and W. Horne have made a theoretical study, based on normal-mode methods, of the focusing of the sound when a CW source and the receiver are on the sound channel axis. Due to the coherence of the modes, the focusing amounts to an enhancement of about 3 dB at a range R, which is about 26 km in the North Atlantic. Although enhancement is to be expected at $R/2$, none was found in this analysis, but the enhancement found at $2R$ is about 2 dB.

Dr. Williams is Professor of Physics at Brown University. He is a Fellow of the American Physical Society and the Acoustical Society of America. He has been consultant to the National Research Council since 1947, and to the Office of Naval Research since 1950.

This study was begun in an Honors Thesis submitted by William Horne in partial fulfillment of the degree of Sc.B. in Physics at Brown University in June 1964.

This paper is reprinted with the permission of the authors and the Acoustical Society of America.

Reprinted from THE JOURNAL OF THE ACOUSTICAL SOCIETY OF AMERICA, Vol. 41, No. 1, 189–198, January 1967

Printed in U. S. A.

5

Axial Focusing of Sound in the SOFAR Channel*

A. O. WILLIAMS, JR., AND WILLIAM HORNE

Department of Physics, Brown University, Providence, Rhode Island 02912

Approximating by a polynomial a standard "velocity profile" of sound speed near the axis of the SOFAR channel, we have used normal-mode methods to explore recurrent focusing when a receiver and a CW point source are exactly on the axis, for acoustic frequencies of the order of 100 cps. A certain range R exists—about 26 km in the North Atlantic—at which the few lowest and strongest modes of the sound field add nearly coherently. Focusing is to be expected near R and, decreasingly, near integral multiples of R. A velocity profile nearly symmetric about the axis would also produce focusing near $R/2$, $3R/2$, etc. Numerical calculations at 100 cps show no such effect at $R/2$, but a modest focusing enhancement of about 3 dB near R and 2 dB near $2R$. Our results disagree markedly with recent analysis by Hirsch, partly because he dealt with a pulsed source but more because he assumed a symmetric profile of sound speed. Axial focusing is evidently very sensitive to the profile. Experimental detection appears almost hopeless because the depths of axis, source, and receiver must be known with great precision.

INTRODUCTION

IN much of the deep ocean, there is a permanent submerged sound channel, often called the SOFAR channel. At the "axis" of this channel—some 4000 ft deep, in the North Atlantic—the speed of sound reaches a minimum value, under the joint influence of changing temperature and hydrostatic pressure. Sound emitted from a source at or near this depth is always refracted back toward the axis. Consequently, a good deal of this sound is confined and guided, with cylindrical spreading in range and without access to the dissipative influences of sea surface and bottom. If the acoustical frequency is low enough to make absorption of minor account, propagation to very great distances is achievable.

Many experiments, a great deal of ray analysis and a considerable amount of wave theory have been devoted to the study of SOFAR transmission, but much remains unknown, or at least unsettled. For example, to what extent is wave rather than ray analysis required at low frequencies? What aspects of SOFAR transmission depend intimately on details of the velocity profile—i.e., the plot of sound speed versus depth near the channel axis[1]? To what extent is sound from a source on the axis recurrently focused at points along the axis?

Hirsch[2,3] has recently discussed the question of focusing. He used wave theory, assuming acoustic pulses of finite sinusoidal trains and representing the velocity profile by a parabolic form:

$$c_0^2/c^2(z)=1-\alpha^2z^2, \tag{1}$$

where $c(z)$ is the speed of sound at a distance z measured vertically from the channel axis and c_0 and α are constants. The outcome was a prediction that, for frequencies around 200 cps (not limited closely to that figure), there would be strong focusing at intervals of about 6 NM (nautical miles) or 12 km, persisting to large multiples of this interval.

The investigation reported here was begun independently,[4] at about the same time. Wave theory is used, although in a different form. The acoustic source is assumed to be CW—a much simpler problem than the one solved by Hirsch, but in compensation a much better approximation to the velocity profile can be used. There are marked differences between our respective predictions, and the reasons can be traced.

I. NORMAL MODES FOR THE ACOUSTIC FIELD

We assume that the ocean is of constant depth H and constant density ρ_w. The speed of sound $c(z)$ varies only

* This study was begun in an Honors thesis submitted by W. Horne, in partial fulfillment of the requirements for the degree of Sc.B. in Physics, Brown University, June 1964.

[1] P. Hirsch and A. H. Carter, J. Acoust. Soc. Am. **37**, 90–94 (1965).

[2] P. Hirsch, J. Acoust. Soc. Am. **36**, 1998 (A) (1964).

[3] P. Hirsch, J. Acoust. Soc. Am. **38**, 1018–1030 (1965).

[4] A. O. Williams, Jr., and W. Horne, J. Acoust. Soc. Am. **36**, 1997 (A) (1964).

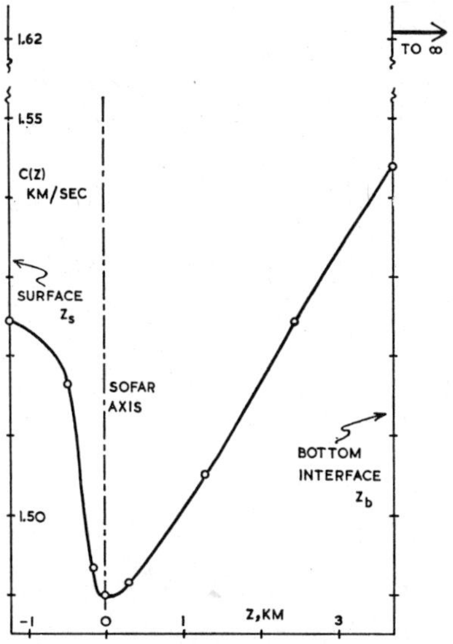

FIG. 1. The speed of sound $c(z)$ vs depth z in the ocean. ○: Standard values for the North Atlantic, from Ref. 5; c in the bottom is assumed to be 5% greater than in the adjacent water (note discontinuity in vertical scale). $z_s=-1.23$ km. $z_b=+3.71$ km.

with vertical position z, in a way shown by Fig. 1, which uses data of Ewing and Worzel.[5] The underlying bottom material is treated as a fluid with density $\rho_b>\rho_w$, and with a constant speed of sound 5% greater than that in the water just at the interface (the assumptions "fluid" and "5%" affect our results only slightly). This discontinuity in sound speed is shown in Fig. 1.

The source is taken as a point, radiating continuous waves with angular frequency ω, and lying on the channel axis where the sound speed is a minimum. Cylindrical coordinates are employed, measured from the source, r horizontally outward and z increasing downward. As dependent variable, we use the velocity potential Φ:

$$\Phi(r,z,t)=\varphi(r,z)e^{-i\omega t}. \qquad (2)$$

The negative gradient of Φ is particle velocity, and Φ itself is proportional to acoustic pressure.

The wave equation for this case is

$$\nabla^2\Phi-[c(z)]^{-2}\partial^2\Phi/\partial t^2=-(2\pi r)^{-1}\delta(r)\delta(z)e^{-i\omega t}, \qquad (3)$$

where $\delta(\ )$ is the Dirac delta function; the right hand side is the source term. Equation 3 is separable in the form

$$\Phi(r,z,t)=R(r)u(z)e^{-i\omega t}. \qquad (4)$$

[5] M. Ewing and J. D. Worzel, "Long Range Sound Transmission," Geol. Soc. Am. Mem. 27, 1–35 (1948).

The z equation is

$$d^2u/dz^2+[\omega^2/c^2(z)-k^2]u=0, \qquad (5)$$

with $-k^2$, the separation constant, to be determined. There are four boundary conditions—

1. As $r\rightarrow\infty$, only outgoing waves are allowed.
2. At the sea surface ($z=z_s$, $z_s<0$), the acoustic pressure and hence also Φ must vanish for all r and t; therefore $u(z_s)=0$.
3. At the bottom interface ($z=z_b$), acoustic pressure must be continuous, and so must the z component of particle velocity, which is proportional to $\partial u/\partial z$.
4. In the bottom, the acoustic field must vanish as $z\rightarrow\infty$, and therefore $u(z)\rightarrow 0$.

Conditions 2–4 together with Eq. 5 force the existence of eigenvalues $(k^2)_n$ and corresponding eigenfunctions $u_n(z)$ in Eq. 3. The set of functions $u_n(z)$ is orthonormal:

$$\int_{z_s}^{\infty} u_n(z)u_m(z)dz=\delta_{mn} \qquad (6)$$

with δ_{mn} the Kronecker delta. The set is also complete, so that other functions can be expanded in terms of it—notably, for our purposes, $\varphi(r,z)$:

$$\varphi(r,z)=\sum_n R_n(r)u_n(z). \qquad (7)$$

Each combination $[(k^2)_n,u_n(z)]$ is a *normal mode* of the acoustic problem.

Two complications have been passed over, because they affect our particular study only negligibly. Both difficulties have been discussed elsewhere.[6] First, if the u_n's are to be orthonormal, they must contain a factor $(\rho_{w,b})^{\frac{1}{2}}$, but we are concerned only with the acoustic field in the water, where $\rho_w^{\frac{1}{2}}$ is a constant factor that can be taken out of Eqs. 3, 7, etc., and ignored. Second, only a part of all the eigenvalues k^2 is a discrete set implied by the form of Eqs. 6 and 7. In the deep ocean, the number of discrete values $(k^2)_n$ is roughly $2.4f$, where f is the acoustic frequency cycles per second.[7] Each of these eigenvalues $(k^2)_n$ is real–positive; we hereafter write it as k_n^2 with $k_n>0$. In addition, there is a continuous spectrum of k^2 accompanied by functions $u(z)$ that vary infinitesimally from one to the next. This part of the normal modes is essential in fitting $\varphi(r,z)$ (Eq. 7) to the point source at $r=0=z$, but the corresponding portion of the acoustic field falls off rapidly with increasing range, and we shall ignore it. As a practical comment, at $f=100$ cps, Eq. 7 contains about 240 discrete terms, which together should be ample to represent all but the most misbehaving functions φ.

[6] E.g., A. O. Williams, Jr., J. Acoust. Soc. Am. 32, 363–371 (1960).
[7] (a) A. H. Carter, Doctoral dissertation, Brown Univ. (1963). (b) A. H. Carter and A. O. Williams, Jr., J. Acoust. Soc. Am. 34, 1985 (A) (1962).

As a result of Eqs. 2–7, $R_n(r)$ satisfies[8]

$$r^{-1}d/dr(rdR_n/dr)+k^2{}_nR_n=-(2\pi r)^{-1}\delta(r)u_n(0), \quad (8)$$

which has a solution

$$R_n(r)=(i/4)u_n(0)H_0{}^{(1)}(k_nr). \quad (9)$$

$H_0{}^{(1)}$ is the Hankel function of the first kind. For $k_nr\gg 1$ (which means that r must exceed a few acoustic wavelengths λ, since $k_n\approx 2\pi/\lambda$), the asymptotic form of the Hankel function can safely be used. Equation 7 becomes

$$\varphi(r,z)=(i/4)\exp(-i\pi/4)(2\pi)^{\frac{1}{2}}r^{-\frac{1}{2}} \times\sum_n\left[k_n^{-\frac{1}{2}}u_n(0)u_n(z)\exp(ik_nr)\right]. \quad (10)$$

If the suppressed factor $\exp(-i\omega t)$ is restored, all terms of the sum are seen to represent outgoing waves in the radial direction, satisfying Boundary Condition **1**.

Evidently, the main task is to find the normal modes—i.e., all combinations $[k_n{}^2,u_n(z)]$—from Eq. 5 with $c(z)$ specified, and from Boundary Conditions **2–4**. Thereafter, φ is known from Eq. 10 at all (r,z), except very close to the point source. In general, the task would be most formidable even with simplifying assumptions concerning $c(z)$, but it turns out that we need good values for only the first few terms of the sum in Eq. 10.

II. SOLUTION OF EIGENVALUE PROBLEM

Figure 1 shows that $c(z)$ varies by only a few percent of its mean value in the ocean and the bottom. Therefore, ω^2/c^2 in Eq. 5 will also change but little. It is convenient to add and subtract

$$\kappa^2=\omega^2/c_0{}^2 \quad (11)$$

in [ ] of Eq. 5; c_0 can be any representative value of the sound speed, and we choose for it the minimum value, at the channel axis where $z=0$. Equation 5 takes the form

$$d^2u/dz^2+[(\kappa^2-k_n{}^2)-(\kappa^2-\omega^2/c^2(z))]u(z)=0. \quad (12)$$

The form, except for a factor $2m/\hbar^2$, is that of Schrödinger's equation in one dimension. We adopt abbreviations based upon this resemblance:

$$E_n\equiv\kappa^2-k_n{}^2,\quad V(z)\equiv\kappa^2-\omega^2/c^2(z), \quad (13)$$

with E_n, $V(z)$ playing rôles analogous, respectively, to total energy and potential energy. It should be noted that E_n/κ^2 and V/κ^2 are independent of the acoustic frequency.

Figure 2 shows a "potential well" of $V(z)/\kappa^2$ versus z, with values of c taken from Fig. 1. Letting V/κ^2 go to $+\infty$ at $z\leq z_s$ ensures that $u(z)$ vanishes there, because, in Eq. 12, $u(z)$ and its second derivative, being proportional to acoustic pressure and gradient of particle velocity, must be finite everywhere. Two representative values of E_n/κ^2 are sketched in Fig. 2. Each crosses the curve of $V(z)/\kappa^2$ at two *turning points* (TP). Between the TP's, E_n exceeds $V(z)$ for any ω, and the corresponding solution u_n of Eq. 12 must in general be oscillatory. Outside the TP's, E_n is less than $V(z)$, and $u_n(z)$ must fall off in the outward direction, exponentially or faster, to satisfy Eq. 12 and Boundary Conditions **2–4**. We are interested only in values of E_n lower than $E_{n'}$ in Fig. 2. The corresponding functions $u_n(z)$ fall off so rapidly outside the TP's that z_s, z_b might as well be at $-\infty$, $+\infty$, respectively, and consequently the exact boundary conditions at z_s, z_b do not really matter (they would, however, for values of E_n/κ^2 lying near or above V_s/κ^2).

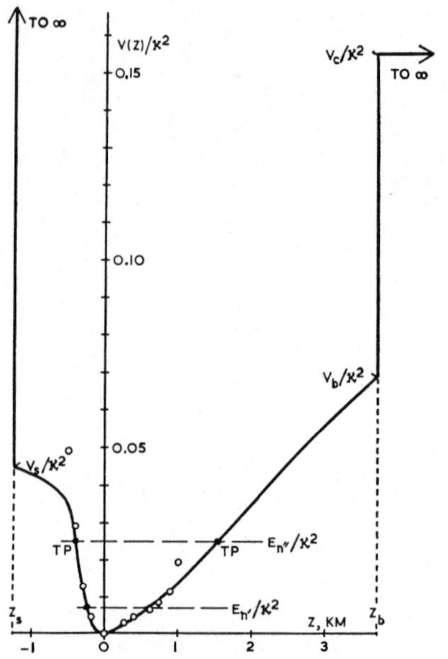

FIG. 2. The dimensionless function $V(z)/\kappa^2$ defined by Eqs. 11 and 13 vs depth z. ——: Data from Fig. 1. ○: Points computed from the polynomial approximation in Eqs. 15 and 16.

The general theory of such eigenvalue problems[9] shows that, for the lowest eigenvalue E_0, the function $u_0(z)$ has no node between TP's (it is roughly a Gaussian curve); $u_1(z)$ has one node; u_2 has two, and so on. With this guidance, the combinations $[E_n,u_n]$ can be found numerically, if numerical values for $c(z)$, H, and ω are given. We could, e.g., integrate $u(z)$ toward the right, from a value of 0 at z_s, and to the left from a simple decaying exponential beyond z_b, taking care to satisfy the acoustic boundary conditions when crossing $z=z_b$.

[8] P. M. Morse and H. Feshbach, *Methods of Mathematical Physics* (McGraw-Hill Book Co., Inc., New York, 1953), Pt. I, Chap. 7.

[9] Ref. 8, Pt. I, Sec. 6.3.

By trial and error, a value of E can be found to make these two functions $u(z)$ join smoothly somewhere between the TP's. The number of nodes then reveals n, and a new $[E,u]$ can be sought. There exist, of course, more sophisticated procedures and various approximate methods to speed the work.

Equation 12 can be solved exactly for a few special functions $V(z)$, most of which can be found in a text on quantum mechanics. The parabolic form (Eq. 1) assumed by Hirsch is one of these, but it is not a good fit to Fig. 2, except in the limit as z goes to zero.

III. DETERMINATION OF E_n/κ^2

We calculate the several lowest values of E_n/κ^2 by use of standard perturbation theory. $V(z)/\kappa^2$ in Fig. 2 can be fitted fairly well in its lowest part by the formula:

$$V(z)=\kappa^2[Az+Bz^2+Cz^3+Dz^4], \qquad (14)$$

where κ^2Bz^2 corresponds to Hirsch's assumption, and the terms in A, C, D are corrections. Determining a proper fit is tedious, because only after it has been tried and some values of E_n have been found can we decide how high up in Fig. 2 it is necessary to establish a good fit. After several trials, the following relation was chosen, with z and ξ in kilometers and κ in reciprocal kilometers:

$$V=\kappa^2[-4.2\times10^{-5}+0.0751\xi^2 -0.1576\xi^3+0.1043\xi^4]; \qquad (15)$$

$$\xi=z-0.0232. \qquad (16)$$

The "offset" from z to ξ automatically and exactly takes into account the correction Az (by completing the square of Bz^2+Az). The constant term in [ ] is very small, but in any event it merely causes an equal change in E_n. The terms in ξ^3 and ξ^4 remain as perturbation corrections to an "exact" solution of Eq. 12 for $V=a+b\xi^2$ (exact to the extent that we ignore the far-distant boundary conditions at z_s, z_b and merely ensure that $u_n(z)$ falls rapidly outside TP's).

The calculation of E_n from Eq. 12 (note that $d^2u/d\xi^2=d^2u/dz^2$) with $V(\xi)$ as given in Eq. 15 is a solved problem, of interest in the theory of molecular vibrations.[10] Some changes of symbols, and introduction of numbers from Eq. 15, yield

$$E_n=\kappa^2-k_n^2=(\kappa+k_n)(\kappa-k_n) \qquad (17)$$
$$=0.274\kappa-1.94+(0.548\kappa-6.18)n+6.18n^2.$$

For frequencies within an octave, at least, either way from 100 cps, $\kappa-k_n$ for n up to 10 and more is very small as compared with κ or k_n. Hence, in Eq. 17, we substitute 2κ for $(\kappa+k_n)$ and solve for k_n:

$$k_n=(\kappa-0.137+0.97/\kappa)-(0.274-3.09/\kappa)n +(3.09/\kappa)n^2. \qquad (18)$$

[10] L. I. Schiff, *Quantum Mechanics* (McGraw-Hill Book Co., Inc., New York, 1955), 2nd ed., pp. 305, 306.

Numerical values of ω and κ need not yet be specified, but at 100 cps, as an example, κ is 422 km^{-1}. Equation 18 shows that, for κ of this general size, changing n from 0 to 20 affects k_n by scarcely 1%.

We now return to Eq. 10, using k_n of Eq. 18, and make three changes. *First,* an average value of $k_n^{-\frac{1}{2}}$ can safely be factored out of the sum. *Second,* $\exp[i(\kappa-0.137+0.550/\kappa)r]$ can be factored. *Third,* we can lump together the existing multipliers before the sum in Eq. 10, and the two just mentioned, replacing them by φ_0, a complex function of r and ω with a magnitude that falls off as $r^{-\frac{1}{2}}$, and a phase that depends on r. Equation 10 becomes

$$\varphi(r,z)=\varphi_0\sum_n\{u_n(0)u_n(z)\exp[-i(0.274-3.09/\kappa)nr] \times\exp[i(3.09/\kappa)n^2r]\}. \qquad (19)$$

The acoustic field on the SOFAR axis is specified by setting $z=0$ in $u_n(z)$.

IV. CONDITIONS FOR FOCUSING

For $n>0$ and for r more than a very few kilometers, the first exp[ ] in Eq. 19 has an argument comparable with or greater than π. Hence, at fixed r, a change of unity in n will make the phase of the $(n+1)$st term differ by a large angle from that of the nth; over a number of terms, there will be almost a random scatter of phases. Then the acoustic amplitude specified by the sum in Eq. 19 will add almost incoherently—i.e., as intensities rather than as amplitudes. There is, however, an important exception. If the following condition is met—

$$|k_{n+1}-k_n|\approx\text{CONST}=M \qquad (20)$$

—for an appreciable subset of n, with k_n as in Eq. 18, there will exist a range R such that

$$|k_{n+1}R-k_nR|\approx MR\approx2\pi. \qquad (21)$$

Within this subset of n, successive terms in Eq. 19 have equivalent phases, differing by 2π, and they will add coherently. We observe that Eq. 20 would be satisfied exactly if the term in n^2 were missing from Eq. 18, or negligibly small.

Equation 20 is not sufficient to produce strongest focusing. The functions $u_n(0)$, $u_n(z)$ in Eq. 19 are oscillatory between TP's, and as n increases the amount of oscillation increases. Hence, in general, the product $u_n(0)u_n(z)$ may almost randomly be positive or negative and large or small. Focusing will be strongest when Eqs. 20 and 21 are satisfied and when the values of z in both functions u_n are the same; for then the product becomes $u_n^2(\cdots)$, which is nonnegative and, for most values of n, not zero. We now put $z=0$ in Eq. 19, satisfying this latter condition and specifying the acoustic field on the channel axis.

Equations 18 and 20 yield

$$|k_{n+1}-k_n|=(0.274-6.18/\kappa)-(6.18/\kappa)n\approx M. \qquad (22)$$

When n is small, Eq. 22 does approximate a constant value of M for κ larger, say, than 200 ($f>50$ cps). Equation 21 then gives a value of R—depending slightly upon frequency, via κ—of about 24 km or 12 NM, near which some focusing can be expected. Still, $|k_{n+1}-k_n|$ depends somewhat upon n. The portion of Eq. 19 that prevents perfectly coherent addition at R is $\exp[i(3.09/\kappa)n^2R]$. As n ranges from 0 to some maximum n_{max}, the argument of this exponential ranges from $i0$ to $i(3.09/\kappa)n_{max}^2R$. By readjustment of M and R in Eqs. 20 and 21, the phase deviations can be split on each side of some intermediate value of n, but the extreme swing cannot be allowed to exceed, roughly, $-i\pi/2$ to $+i\pi/2$, if any reasonable approximation of coherent addition is to remain. We thus arrive at an estimate of n_{max}:

$$(3.09/\kappa)n_{max}^2R\approx\pi. \qquad (23)$$

To go farther, we must specify a frequency, thereby fixing κ and n_{max}. Next, we must find values of $u_n^2(0)$, $0\leq n\leq n_{max}$, and then estimate the background effect of all modes with $n>n_{max}$, modes that should add practically incoherently in Eq. 19. First, however, there are four general points to make.—

1. If we use a parabolic fit for $V(z)$, i.e., κ^2Bz^2, Eq. 12 becomes the linear-oscillator problem,[11] and values of k_n^2 (not k_n) would be equally spaced in n. Solving for k_n, by an approximation one stage better than was necessary to get Eq. 18, shows that k_n is nearly linear in n but has a term in n^2 almost 100 times smaller than the corresponding one in Eq. 18. Hence, with this idealized parabolic representation for $V(z)$, the value of n_{max} rises severalfold over that given by Eq. 23, and much stronger focusing should result. In passing, we observe that $V(z)$ can be chosen in a form that, for small z, is not much different from a parabola, such that k_n is exactly linear in n. This case has been discussed in the literature, but the actual highly skewed shapes of Figs. 1 and 2 near $z=0$ make such choices of V rather artificial for the SOFAR channel.

2. If R satisfies Eq. 21, any integral multiple νR also satisfies it approximately. We see from Eq. 23 that replacing R by νR, with $\nu>1$, reduces the size of n_{max}. Therefore, any focusing achieved at R will be progressively degraded at $2R$, $3R$, and so on.

3. There is a possibility of focusing at $R/2$. If V were κ^2Bz^2 exactly, the eigenfunctions u_n would have definite parity, alternately even and odd.[12] At $z=0$, $|u_n|$ would have a maximum for all even n, but would vanish for all odd n. In this case, Eq. 20 is replaced by

$$|k_{n+2}-k_n|\approx 2M, \quad n \text{ even}. \qquad (24)$$

With $2M$ instead of M, Eq. 21 is satisfied by $R/2$ and by integral multiples of it. The skewed shape of the real $V(z)$ tends to upset this possibility, by destroying definite parity for the functions u_n, but detailed calculations would be necessary to find exactly what does happen near $R/2$. Hirsch[3] found focusing at intervals of some 6.3 NM, approximately our $R/2$, because he assumed $V\propto z^2$ alone—i.e., a symmetric V.

4. If parity of u_n exists, as in Point 3 just above, a source placed at $z'\neq 0$ will yield focusing near the point $(r=R/2,\ z=-z')$, again at $(R,\ +z')$, etc. This prediction results from considering the phase changes introduced both by Eq. 21 at the range $R/2$ and by the signs of $u_n(z')$, $u_n(-z')$.

[11] Ref. 10, pp. 60–66.
[12] Ref. 10, pp. 38, 39.

V. ENHANCEMENT FACTOR AT FOCI

We define an *enhancement factor* ϵ to measure the strength of focusing:

$$\epsilon^2=[I_c(0\leq n\leq n_{max})+I_i(n_{max}+1\leq n\leq N)]/I_i(0\leq n\leq N), \qquad (25)$$

where N is the number of the highest discrete mode (about 2.4 f, with f in cycles per second, as we have said). The symbol I means "intensity," in the sense of squared amplitude; subscripts c and i refer, respectively, to coherent addition in Eq. 19 by amplitudes, before squaring, and to incoherent addition of squared terms:

$$I_c=|\sum_0^{n_{max}} u_n^2(0)\exp\{-i[\ ]r\}|^2, \qquad (26)$$

$$I_i=\sum^N u_n^4(0), \qquad (27)$$

lower limit $n_{max}+1$, or 0. The denominator of Eq. 25 represents an average of intensities to be expected far from any focus. Evidently, $|\varphi_0|^2$ of Eq. 19—including its factor r^{-1}—will cancel throughout Eq. 25. The quantity ϵ^2 is therefore defined relative to an intensity that falls, on the average, cylindrically with range r.

There is a similar yet distinct focusing caused by some modes of much higher n. For our present parameters, with source and receiver at the same depth z, it occurs at repeated range intervals of about 35 NM (Ref. 7). For very small values of z, it is the familiar surface-zone convergence. We ignore this effect in the present study.

With the help of Eq. 25, we can examine one other aspect of Eq. 23. It may appear that increased frequency and therefore increased κ would enhance focusing by making n larger, with more modes adding coherently. But n_{max} rises only as $\omega^{\frac{1}{2}}$, so that I_c in Eq. 25 would increase roughly as ω. Simultaneously, the total number of modes N increases in the same proportion and therefore so do the two terms I_i in Eq. 25. An increase in the total number of modes requires a correspondingly diminished intensity in each mode separately, and so to a first approximation the value of ϵ^2 is not much affected by moderate changes in frequency, except possibly through changes in the values of $u_n^2(0)$.

VI. CALCULATION OF $u_n^2(0)$

Numerical evaluation of u_n^2 requires specification of acoustic frequency. We choose 100 cps ($\omega=628$ sec^{-1}, $\kappa=422$ km^{-1}). Equation 23, for $R\approx 24$ km, points to $n_{\max}\approx 4$ or 5. Numerical values of $u_n(z=0)$, $0\leq n\leq 5$, were found by two methods. The first was standard perturbation theory,[13] starting from the basis $V=\kappa^2(0.0751)\xi^2$ in Eq. 15, for which the functions u_n are those of the linear oscillator.[11] Perturbation corrections due to ξ^3 were computed through second order, but for ξ^4 through first order only, because they were small. The functions u_n were evaluated at $\xi=-0.0232$, which (Eq. 16) corresponds to $z=0$. For $n>2$, details of the calculation showed that the process converged poorly and the corresponding perturbed functions u_n were unreliable. This shortcoming does not necessarily mean that perturbation theory is inadequate for computing values of E_n and k_n (Eqs. 17 and 18), because the latter calculation involves only averages of the u's, not their values point by point.

More reliable values of $u_n^2(0)$ were sought by use of the Wentzel–Kramers–Brillouin (WKB) approximation,[14] after replacing the lower part of $V(z)$, Fig. 2, with a bilinear profile. Since the same procedure was employed in estimating some of the higher modes, to get I_i in Eqs. 25 and 27, we outline the calculation. Figure 3 defines the problem. The dotted curve is the bottom part of Fig. 2, and E_n is the location of some eigenvalue corresponding to Eqs. 17 and 18. The heavy line shows a bilinear approximation to the dotted curve, with a tip on the axis at $z=0$. Hereafter, $V(z)$ is defined by such a bilinear plot, and is taken to be zero at $z=0$; a constant change in $V(z)$ of Eqs. 12 and 13 is always allowed, provided that E be changed by the same amount.

Assuming that we know where the tip of the new $V(z)$ is, relative to the curved $V(z)$, and for brevity dropping all subscripts n, we can write the WKB solution[14] for u, between the TP's:

$$u(z)\approx\gamma(E-V)^{-\frac{1}{4}}\cos\left[\int_{z=-h_1}^{z}(E-V)^{\frac{1}{2}}dz-\tfrac{1}{4}\pi\right]. \quad (28)$$

Here, γ is a normalization constant, chosen to satisfy Eq. 6, V is the bilinear profile, and E is the eigenvalue corresponding to *this* V and *this* $u(z)$. To get an approximate value of γ we use a standard device, squaring $u(z)$ in Eq. 28, assuming that $\cos^2[\;]\approx\frac{1}{2}$, and neglecting entirely the exponential tails of $u(z)$ outside the TP's. Equation 6 becomes

$$\int_{-h_1}^{0}(\tfrac{1}{2}\gamma^2)(E-V)^{-\frac{1}{2}}dz+\int_{0}^{h_2}(\tfrac{1}{2}\gamma^2)(E-V)^{-\frac{1}{2}}dz\approx 1. \quad (29)$$

The integrals can be evaluated exactly by a change of variable:

$$w=[E-V(z)]^{\frac{1}{2}},\quad dz=-2(dV/dz)^{-1}wdw. \quad (30)$$

We see from Fig. 3 that

$$(dV/dz)^{-1}=-h_1/E,\quad z<0;$$
$$(dV/dz)^{-1}=h_2/E,\quad z>0. \quad (31)$$

Use of Eqs. 30 and 31 lets us find the constant γ^2 from Eq. 29:

$$\gamma^2\approx E^{\frac{1}{2}}/h. \quad (32)$$

Next, E_n is found by the WKB method[14]:

$$\int_{-h_1}^{h_2}(E_n-V)^{\frac{1}{2}}dz=(n+\tfrac{1}{2})\pi. \quad (33)$$

Splitting the integral at $z=0$ and again using Eqs. 30 and 31 lead to the result

$$E_n=[3\pi(n+\tfrac{1}{2})/(2h)]^{\frac{2}{3}}, \quad (34)$$

in which, we remember, h depends on E_n for a given $V(z)$. The same methods are used to evaluate the integral for the argument of cos[ ] of Eq. 28, between limits $-h_1$ and 0. As to the terms multiplying cos[ ], γ is known from Eq. 32 and $V=0$ at $z=0$. The result for $u_n^2(0)$ is

$$u_n^2(0)\approx h^{-1}\cos^2[(n+\tfrac{1}{2})(h_1/h)\pi-\tfrac{1}{4}\pi]. \quad (35)$$

To compute any one $u_n^2(0)$, $0\leq n\leq 5$, we choose n; E_n is calculated for the "curved" $V(z)$ in Fig. 2 from Eqs. 17 and 18; h_1, h_2, and h are found by marking this E_n on an enlarged plot of Fig. 2; and Eq. 35 is used with the value of n already chosen. In effect, we are carrying out a quantitative replacement of the true $V(z)$ by a bilinear V as shown in Fig. 3, and a different replacement for each value of n. For example, E_n of Eq. 34,

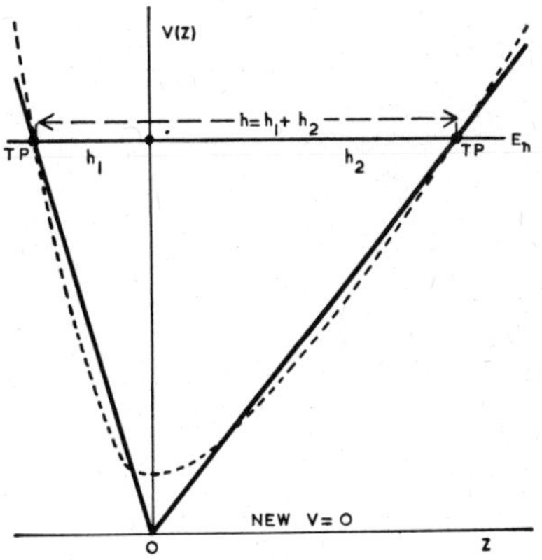

FIG. 3. Bilinear $V(z)$ (——) substituted for the lowest part of the curve in Fig. 2 (- - - -), and preserving the "skewness."

[13] Ref. 10, pp. 151–154.
[14] Ref. 10, pp. 184–192.

which pertains to the bilinear well, will not equal that found from Eqs. 17 and 18, yet we have chosen in place of the exact eigenfunction u_n another u, fitting a reasonably similar plot of $V(z)$, with the same spacing of turning points and the same n.

The WKB method provides at best an approximation to E_n, u_n, and one that is usually poorer for smaller values of n. Moreover, we have made two additional approximations, substituting for the curved $V(z)$ of Fig. 2 a bilinear profile (Fig. 3) and computing γ_n^2 rather roughly by Eqs. 29 and 32. Estimates of the resulting errors, and empirical corrections, can be found by applying the same approximations to an exactly known case, the linear oscillator, with $V=\kappa^2Bz^2$. For this symmetric V, $u_n(0)$ vanishes for odd n; for even n, the exact value of $u_n^2(0)$ exceeds that found from Eq. 35 by 13% when $n=0$ and by 27% when $n\geq 2$. We therefore increase the values of all $u_n^2(0)$ from Eq. 35 by corresponding fractions. The values thus corrected come much nearer to those perturbation results, for $n=0$, 1, 2, that appeared to be fairly good.[15]

Table I shows the values of $u_n^2(0)$ finally adopted (corrected WKB); perturbation values are listed in parentheses, for comparison.

VII. ESTIMATING HIGHER NORMAL MODES

For use in Eq. 25, we estimate the values of $u_n^4(0)$ from $n=n_{\max}+1$ to N in two steps. First, the lower and intermediate parts of $V(z)/\kappa^2$ in Fig. 2 are replaced by a bilinear profile that reaches V_b/κ^2 at both z_s and z_b and has its lower tip on the axis $z=0$. By calculations like those in Sec. VI, the tip is set at a point that makes E_5 for this new profile coincide in location on Fig. 3 with E_5 from Eqs. 17 and 18. Values of h in the new profile are proportional to E_n (the lower tip being defined anew as $V=0$):

$$h=3.85\times 10^{-4}E_n=0.205(n+\tfrac{1}{2})^{\frac{2}{3}}, \tag{36}$$

the last step from Eq. 34. Taking h as 4.94 km, the width of the well at V_b, yields the maximum n_b for this part of the profile:

$$n_b\approx 118. \tag{37}$$

We use Eq. 36 to calculate I_{i1}, a part of I_i in the numerator of Eq. 25, by getting $u_n^4(0)$ from Eq. 35 corrected as for Table I:

$$I_{i1}\approx 38\sum_{n=6}^{118}\{(n+\tfrac{1}{2})^{-\frac{1}{3}}\cos^4[\ ]\}. \tag{38}$$

Over small groups of neighboring values of n, the term

TABLE I. Values of $u_n^2(0)$ computed by use of the corrected WKB approximation (first line), and less reliably by standard perturbation theory (second line, entries in parentheses). All values are normalized. The assumed acoustic frequency is 100 cps.

n	0	1	2	3	4	5
$u_n^2(0)$, WKB	5.06	1.80	0.006	1.37	1.89	0.025
$u_n^2(0)$, pert.	(5.76)	(2.02)	(0.080)	(2.31)	(6.01)	(0.035)

$\cos^4[\ ]$ should approximate its normal average of $\frac{3}{8}$. Factoring this average and transforming the remaining sum to an integral over n lead to the result

$$I_{i1}\approx 14. \tag{39}$$

To estimate I_{i2} for the remaining modes with values of E_n between V_b and V_c in Fig. 2, we replace the whole plot of V by a "square well" of width $H=4.94$ km, with its bottom lowered a depth ΔV below V_b. Once more, we denote V as zero at the bottom of this new well. The eigenfunctions and eigenvalues are known[6] for a square well; between TP's,

$$u_n(z)=\Gamma_n\sin[E_n^{\frac{1}{2}}(z-z_s)], \tag{40}$$

where Γ_n is a normalization constant. Comparison with Eqs. 4, 5, 23, and 24 of Ref. 6 shows that

$$E_n^{\frac{1}{2}}=(x_n/H)\approx n\pi/H \tag{41}$$

for large values of n, which are all that concern us here. The extra depth ΔV of this square well, below V_b in Fig. 2, is fixed by requiring that the new E_n at the location V_b in Fig. 2 shall correspond to $n=118$ in Eq. 41, thus agreeing with Eq. 37 for the bilinear well. From Eq. 41, we find the greatest value N of n, for which E_N corresponds to the location of V_c in Fig. 2:

$$N\approx 229. \tag{42}$$

This result is in reasonable agreement with more precise calculations,[7] which gave $N\approx 238$.

Γ_n of Eq. 40 is determined from Eq. 6 by squaring u_n, equating $\sin^2[\ ]$ to $\frac{1}{2}$, and integrating u_n^2 from z_s to z_b (once more neglecting the exponential tail of u_n^2 to the right of z_b):

$$\Gamma_n^2\approx 2/H, \text{ regardless of } n. \tag{43}$$

It follows from Eqs. 27, 40, 42, and 43 that

$$I_{i2}=\left(\frac{4}{H^2}\right)\sum_{119}^{229}\sin^4[\ ]. \tag{44}$$

Replacing $\sin^4[\ ]$ by its average, $\frac{3}{8}$, we obtain

$$I_{i2}\approx 7. \tag{45}$$

Our assumption that the speed of sound in the bottom material exceeds by 5% the speed in the adjacent water (Fig. 1) affects only the value of I_{i2}, via the upper limit of the sum in Eq. 44, and no other part of the calculations. Use of Eqs. 41 and 44 shows that I_{i2} increases

[15] The discrepancies between WKB and perturbation values of u_0^2 and u_1^2 are in the right direction, and of reasonable magnitudes, to be blamed mainly on the imperfect normalization achieved in the perturbation method; this failing becomes more serious when, as here, the perturbation sum does not converge rapidly. As to u_2, it is evidently near a node, and small differences in approximations will markedly affect u_2^2; the important thing is that both calculations display the neighborhood of a node.

somewhat less than proportionally to this assumed percentage.

To complete the evaluation of I_i in Eqs. 25 and 27, we use Table I to get

$$\sum_{0}^{5} [u_n^4(0)] = 34.3. \tag{46}$$

Therefore, the denominator of Eq. 25 and the term I_i in the numerator are given by

$$I_i(0 \leq n \leq N) \approx 55; \quad I_i(n_{max}+1 \leq n \leq N) \approx 21. \tag{47}$$

Numerous as the higher modes are, their incoherent contributions are smaller than those of modes 0–5. Formally, the reason lies in factors h^{-1} and H^{-1} in Eqs. 35 and 43. More fundamentally, the main parts of the eigenfunctions u_n are confined to ever narrower layers of the ocean as n decreases from its maximum value. All functions u_n are normalized (Eq. 6) mainly over these same layers, and so the average amplitude of u_n near $z=0$ must increase as n decreases.

VIII. CALCULATION OF I_c IN EQ. 25

I_c was computed for a number of values of r bracketing the nominal focal distances $R/2$, R, and $2R$, at a frequency of 100 cps ($\kappa=422$). Equation 26 was used; the term exp{ } was the product of the two exponentials in Eq. 19. Values of $u_n^2(0)$ were from Table I and of k_n from Eq. 18. Summation over n was nominally from 0 to 5, but the terms for $n=2$ and 5 were omitted as being negligible. For comparison with estimates of I_i,above, the extremes of I_c encountered were 3 and 95. The latter is close to the result of perfectly coherent addition, which is the square of the sum of $u_n^2(0)$ for $0 \leq n \leq 5$, or about 103. The other value, 3, represents a close approach to complete destructive interference. The larger extreme of I_c, taken together with values of I_i from Eq. 47, shows that ϵ^2 is rather insensitive to the assumed contrast of sound speeds across the bottom interface, except at ranges r where nearly destructive interference occurred among the lowest modes.

Figure 4 shows ϵ, the enhancement factor for amplitudes, plotted versus range r in three regions, near the nominal values $R/2$, R, $2R$ estimated from Eqs. 20, 21, ff. Eleven equally spaced values of r were used for each of the three portions. The line of abscissas corresponds to $\epsilon=1$, for which the total amplitude equals that found with entirely incoherent addition of all modes. All three curves are highest some distance beyond the nominal R, etc. These displacements result from our having neglected the phase term in n^2 when estimating R; Eqs. 21 and 22 show that a smaller value of M and a larger value of R would have been obtained by using some average n between 0 and n_{max}.

Near $R/2$, about 6 NM from the source, Fig. 4 displays no focusing, but instead a preponderance of destructive interference among modes 0–5. We saw earlier that focusing would be expected at $R/2$ under conditions that make $u_n(0)$ vanish for odd n. Table I indicates how badly this condition is violated, because of the skewed shape of $V(z)$ in Fig. 2.

Near R, there is some focusing; ϵ is maximum—about 3 dB up from incoherent addition—at $r \approx 26$ km or 13 NM. The region of focus spreads over 2 or 3 km, and there is structure on either side of the broad maximum.

Near $2R$, there is some focusing also, but the greatest ϵ is only about 2 dB above incoherent addition; the focus is more structured and more spread out than it is near R.

Although the fact is not obvious from Fig. 4, the trend of numerical computations showed that the three distinctive patterns near R and $2R$ would quickly die out beyond each end of the plotted portions, to be replaced by irregular traces with small mean swing. Including more terms, with $n>5$, in I_c would not much affect Fig. 4, except to superimpose a nearly random fine structure.

Before the decision to abandon the perturbation values of $u_n^2(0)$—Table I—and turn to the WKB method, a graph equivalent to the present Fig. 4 had been computed. Near $R/2$, the two results differed very little. Near R and $2R$, the shapes were similar, as were the locations of $\epsilon=1$; but ϵ computed with the perturbation u's displayed greater swings about the axis, an effect caused almost entirely by the very large value of u_4^2. The increase in focusing, from R to $2R$, was also evident. This comparison tends to show that any reasonable set of approximations to the u_n's (the perturbation u_4 was not reasonable) would produce something very like the present Fig. 4.

IX. PULSED SIGNALS; GROUP VELOCITY

If the source had been pulsed, rather than CW, the signals carried in each mode would have traveled at

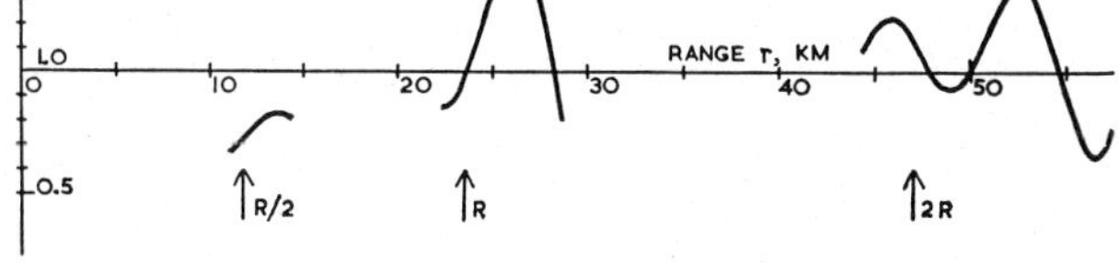

FIG. 4. Enhancement factor ϵ (Eq. 25) computed for a frequency of 100 cps, near the nominal focal distances $R/2$, R, $2R$ (Eq. 21), along the SOFAR axis. The corrected WKB values of u_n^2, in Table I, were used.

group velocities U_n given by[16]

$$1/U_n = dk_n/d\omega = c_0^{-1} dk_n/d\kappa. \qquad (48)$$

To the approximation of Eq. 18, which chiefly means that n^2 shall not be very large, we find

$$U_n \approx c_0[1+0.550/\kappa^2+(3.09/\kappa^2)(n^2+n)]. \qquad (49)$$

In our units, $c_0 \approx 1.5$ and $3.09/\kappa^2 \approx 2\times 10^{-5}$ at 100 cps. Evidently, U_n increases with n, although slowly when n is small. This result agrees with the usual belief about SOFAR transmission, that the "axial arrival" travels most slowly, for in the language of normal modes the axial arrival is predominantly the mode strongest near $z=0$, i.e., that for $n=0$.

Had we used for V a parabolic form, $\kappa^2 Bz^2$, in place of the skewed profile of Eqs. 14 and 15, the known[11] values of E_n would have yielded, as an exact result,

$$k_n = [\kappa^2 - 2\kappa B^{\frac{1}{2}}(n+\tfrac{1}{2})]^{\frac{1}{2}}. \qquad (50)$$

Obtaining U_n^{-1} exactly, from Eqs. 48 and 50, and then differentiating it with respect to n, shows that, for a parabolic profile, the group velocity U_n always *decreases* with increasing n—behaving oppositely to U_n of Eq. 49—as long as k_n is real-positive. Real-positive k_n is required to ensure outgoing acoustic waves; and k_n is real unless $c(z)$ rises to infinity, which would be physically meaningless. An alternative procedure is to expand the right-hand side of Eq. 50 in a binomial series through terms in n^2. This produces a formula for U_n much like that of Eq. 49, but the constant that multiplies (n^2+n) is now negative, still proportional to κ^{-2}, and nearly 100 times smaller than its counterpart in Eq. 49: the rate of change of U_n for small values of n is far less than that deduced from Eq. 49. (See Ref. 1 for related discussions.)

The change of U_n with n, whether an increase or a decrease, makes discussion of axial focusing rather different in the two cases of pulsed and CW sources. Given short enough pulses and great enough range, the parts of the whole acoustic signal that are carried by two different modes—different values of n, and of U_n—will reach the receiver at different times. This separation of arrivals is greatest for modes of highest number, to the extent that the approximate Eq. 49 is valid, because of the dependence of U_n on n^2. If separation occurs, the incoherent background of higher modes represented by I_i in our Eq. 25 would be much reduced. The signal that reaches range r at time t would be weaker, because only a few modes would "overlap" there, but it would display much greater swings between places of focusing and of defocusing.

Much of the discrepancy between Hirsch's strong focusing of pulsed signals and our weak focusing with a CW source probably arises in this way. Nevertheless, this difference is somewhat artificial, since it is related to the choice of profile for $V(z)$. From Eq. 49, we can estimate the difference in time of arrival Δt at range r for signals carried by the zeroth and nth modes. Measured in acoustic cycles, the time difference is $f\Delta t$; we find that

$$f\Delta t \approx n^2 r/(10f), \qquad (51)$$

with f in cycles per second and r in kilometers. Hirsch's numerical examples[3] assumed a pulse made up of two sine cycles at 233 cps. So short a pulse must, in our present language, be represented as a fairly wide spectrum, and our rough estimates about group velocities cannot replace his thorough analysis. Still, it appears from Eq. 51 that, for $r \gtrsim 2R$, the components of such a pulse that are carried by the tenth and higher modes will be fairly well separated in time from the components carried by the zeroth mode. This result is based on our assumed profile of sound speeds, not on Hirsch's; a much greater range and/or a much larger value of n must be used to separate mode components under his assumption. At much higher values of n, our Eq. 49 is no longer valid (the deviations can be estimated from Eqs. 13 and 36), and we have seen that, at larger values of r, focusing becomes degraded even among modes that still overlap.

[16] C. L. Pekeris, "Theory of Propagation of Explosive Sound in Shallow Water," Geol. Soc. Am. Mem. **27**, 1–117 (1948).

X. CONCLUSIONS

We have fitted a skewed curve to experimental sound speeds near the SOFAR axis and, assuming a CW point source on that axis, have used normal-mode methods to calculate the amount of focusing encountered as a receiver moves along the axis. It has been shown that some focusing is to be expected near integral multiples of a certain range R, because the few lowest normal modes, which contribute the greater part of the sound field near the axis, are nearly in phase agreement at R, and therefore add coherently. For the parameters that we have used, which correspond to the North Atlantic, R is about 26 km or 13 NM. Focusing is enhanced when, as we have assumed here, both source and receiver are at the same depth (for the real ocean, a better statement is "both are on the SOFAR axis," which will vary in depth, geographically). In principle, focusing might also occur near half-integral values of R, but more elaborate numerical calculations are needed to settle the point. The extent of any focusing will decrease as the range r increases. These results have been deduced for low frequencies—nominally, some 50–200 cps, although the limits are conservative.

An "enhancement factor" ϵ has been defined to show by how much the acoustic amplitude at or near a focal point differs from the average amplitude at other points where there is no phase coherence for the lowest modes.

Numerical results for ϵ are given at a frequency of 100 cps, near the ranges $R/2$, R, and $2R$. No focusing is found at $R/2$; this is an outcome of the strongly skewed profile of sound speed near the SOFAR axis. Near R, enhancement of about 3 dB is found, and near $2R$, about 2 dB. The expected degradation of focusing as range increases is thus corroborated.

These amounts of enhancement do not agree very well with our first report,[4] which predicted only a fraction of a decibel. There are three reasons for the discrepancy. (1) In the present calculation, a better fit to the profile of sound speed—a more skewed one—has been found and used; it increased the contributions to the sound field made by odd-numbered modes, without much changing those of even-numbered ones. (2) The amplitudes of all the lowest modes have now been more accurately calculated and, as it happens, the effect has been much as in (1), above. (3) Our earlier estimate of the acoustic background contributed by the incoherent higher modes was rather crude, and the background was considerably exaggerated, with consequent lowering of the enhancement factor.

Although these revisions change our previous results somewhat toward agreement with those of Hirsch, there remain strong discrepancies, both qualitative and quantitative. We find focusing only at twice his range intervals—in our notation, near integral multiples of R, not of $R/2$. In such focal regions, we find much less strong focusing, and much more degradation of focusing as range increases. As Hirsch first pointed out, one reason for disagreement lies in the different postulates of pulsed and CW sources. The existence of different group velocities for the various normal modes ensures that, with pulsed transmission at sufficiently large range, the incoherent background of higher modes is diffused in time and space, and so its blurring of focusing is much reduced. This distinction is a significant one, corresponding to different experimental procedures. We believe, however, that the selection of different profiles to represent sound speed is the more important cause of disagreement. Our own choice is not necessarily very good, yet it corresponds much better to the average experimental findings. Comparison of Hirsch's results with our present ones is therefore illuminating because it shows how sensitive the phenomenon of axial SOFAR focusing is to the initial choice of a velocity profile.

For this and other reasons, our numerical results in Fig. 4 must not be regarded as quantitative measures of the axial sound field. Our analytic approximation to the velocity profile, embodied in Eq. 15, is quite good enough to substantiate the general deductions in the first paragraph of this section, but other approximations to $V(z)$ would undoubtedly make some changes in our Fig. 4. So also would improvements on our subsequent calculations; e.g., $V(z)$ could be replaced by a numerical Table, and numerical solution could be used to find better values of $u_n(z)$ and hence $u_n^2(0)$, together with more accurate values for the first few k_n's (replacing Eq. 18, and possibly yielding a somewhat different result for n_{max} in Eq. 23). It would not be hard, either, to refine our estimates of the terms I_i in Eq. 25, by use of a computer, although it seems unlikely that important changes would result. We believe that Fig. 4 gives a good rough estimate of the location, strength, qualitative structuring, and degradation with increasing range, of focusing along the SOFAR axis with a CW source at 100 cps on the axis.

A comment about feasibility of related experiments can be offered. It is inherent in the present calculations that the vertical locations of source and receiver, relative to the SOFAR axis, must be quite well known. In Table I, the term $u_0^2(0)$ is evidently much the largest contribution to the axial sound field. The half-breadth of $u_0(z)$ extends roughly 100 m above and below the exact axis. Hence, an error of this size in the vertical source or receiver would lower u_0^2 in Table I to half its listed value, probably reducing focusing to a very small effect. As a more extreme measure of required accuracy, we consider $u_4(z)$, which also makes an important contribution. From Eq. 35, or from the exact eigenfunctions of a parabolic $V(z)$, we find that adjacent zeros of u_4 lie about 50–60 m above and below $z=0$. The other important functions, u_1 and u_3, are of intermediate breadth near $z=0$. Evidently the results in Fig. 4, or of any comparable calculation, will be markedly affected if either source or receiver is misplaced vertically by 50 m. There remains the problem of knowing where the SOFAR axis really is. Either ours or Hirsch's assumed profile of sound speeds shows that, for z equal to ± 50 m from the exact axis, the fractional change of sound speed from the exact minimum c_0 is about 10^{-4}. This figure corresponds to measuring a change in the speed of sound of about 0.5 ft/sec. Modern velocimeters are capable of this precision, but they have revealed fluctuations in the speed of sound, near the SOFAR axis, of at least this magnitude. It would appear hopeless to attempt experimental measurements of axial focusing in this frequency range around 100 cps.

ACKNOWLEDGMENTS

This work was supported in part by the U. S. Office of Naval Research.

Some of the methods of calculation employed in this article were developed earlier[7] by A. H. Carter in collaboration with one of us (A.O.W., Jr.).

The senior author is indebted to the Acoustics Group, Department of Physics, Imperial College of Science and Technology, London, for hospitality and for technical assistance in preparing this manuscript.

Errata: The 15th line from the bottom of column 2, p. 48, should start, "For sources and receiver very near the surface, it is the . . . " and the next to last sentence in Section VIII should start, "The decrease in focusing . . . ".

In the early measurements of fluctuation of sound propagation in the ocean, the instrumentation available was not adequate to determine the true magnitude of the fluctuation or the factors which were responsible for it.

The experiments reported in the following paper utilize instrumentation to measure accurately the fluctuation as a function of depth and simultaneously to measure temperature and pressure so that the sound velocity as a function of depth can be accurately determined. The results show definitely that propagation loss varies greatly with small changes in depth and that this variation in propagation loss can be explained in terms of the vertical temperature profile in the medium.

D. C. Whitmarsh and W. J. Leiss are both physicists and were engaged in underwater weapons research at the Harvard Underwater Sound Laboratory. Since 1945, they have been engaged in similar research at the Ordnance Research Laboratory, The Pennsylvania State University. Much of their research has been concerned with problems related to sound propagation in the ocean.

This paper is reprinted with the permission of the authors and the Acoustical Society of America.

Reprinted from The Journal of the Acoustical Society of America, Vol. 43, No. 5, 1036–1040, May 1968

Fluctuation in Horizontal Acoustic Propagation over Short Depth Increments

6

D. C. Whitmarsh and W. J. Leiss

Ordnance Research Laboratory, The Pennsylvania State University, University Park, Pennsylvania 16802

Underwater acoustic-propagation measurements have been made off Key West, Fla., for the past 3 yr. Two or three deep-submergence vehicles were allowed to sink freely and simultaneously through the water to depths as great as 5000 ft while transponding acoustically with each other. Pulses at frequencies of 10, 20, and 40 kHz and horizontal ranges of 1700 to 8000 yd were used in the experiment. Because direct acoustic signals showed amplitude fluctuations of approximately 15 dB for depth changes of 15–20 ft, a method of rapid pulsing was installed. This allowed the emission of discrete signals 0.25 sec apart over a period of 5 sec periodically during an operation. Amplitude fluctuations for depths changes of only a few inches could be measured. Pulse-to-pulse amplitude differences of up to 15 dB were found for depth changes of as little as 10 in. These fluctuations are discussed in terms of transmission loss and are compared to the losses suffered for the larger depth changes and for those predicted by ray theory on the basis of the measured sound-velocity structure of the water. It is concluded that refraction conditions can account for the fluctuation.

INTRODUCTION

The Acoustics of the Medium project of the Ordnance Research Laboratory has been making acoustic-propagation measurements in the waters off Key West, Fla., periodically over the past 3 yr. The instrumentation system used in making these measurements is known as DIVEAR (Diving Instrumented Vehicles for Environmental and Acoustic Research). It consists of three self-contained submerging vehicles that sink freely through the water to depths as great as 5000 ft while transponding acoustically with each other.

Figure 1 is a schematic illustration of the operating system at sea. The two or three vehicles are dropped simultaneously from separate boats on the surface. One vehicle acts as a master and emits an acoustic pulse, which, when it is received by the other two vehicles, triggers their acoustic transmitters so that, after a short delay, they each emit a pulse in return. One of these return pulses, determined on the basis of pulse width, is then answered by the master. This sequence continues as the vehicles sink to a preset depth where the electronic package in the vehicle is shut down, a ballast weight is dropped, and the vehicle floats to the surface. Each vehicle contains a digital magnetic-tape recorder that records the level of the incoming acoustic signal at 400-μsec intervals. At 0.1-sec intervals, it records another parameter such as water pressure, water temperature, and some internal quantities that indicate the vehicle orientation in space. The sinking rates are usually about 3 ft/sec. Acoustic frequencies of 10, 20, and 40 kHz have been used over ranges of 1700–8000 yd, but, for the results reported here, only 10- and 20-kHz signals were employed.

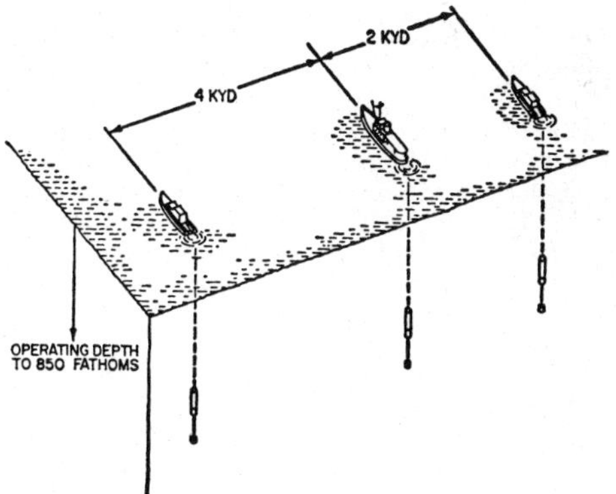

Fig. 1. Schematic illustration of the DIVEAR system in operation at sea.

I. FIRST EXPERIMENTS

Typical results using the DIVEAR system are shown in Fig. 2. On the left, in this Figure, is the sound velocity-depth profile computed from the temperature-depth profile measured by the vehicle during its descent. On the right, plotted as a function of depth, are the amplitudes of the direct (nonreflected) 20-kHz signals as received at one vehicle from another, at a range of 2600 yd. The amplitude of the received signal is plotted as a transmission or propagation loss. Because of the travel time of the signal during a transponding cycle, one data point is recorded about once in each 4-sec period. During this time interval between pulses, the vehicles will have sunk 12–15 ft. From one pulse to the next, it is possible to have as much as 10–15 dB difference in signal level. Most of the large fluctuations in signal level occur in the depth region where the sound-velocity profile is changing the most rapidly and is the most complex.

In an attempt to understand these results, ray theory was used to calculate the losses that might be predicted under the conditions of the experiment. A computer program was developed that would search for rays leaving the source and arriving at the receiver location. To make the program search reasonable from both the practical and mathematical point of view, the receiver was specified as a cube in space with dimensions of 0.1 yd on each side. The experimentally determined sound-velocity-depth profile was used to determine the ray paths; and it was assumed that the thermal structure was constant between locations. The acoustic loss along each ray was then computed. Such a ray-path search-and-loss calculation was made for each pulse emitted and received during an operation. Figure 3 shows the result of the calculation for the 0–3000-ft portion of one run. The temperature and velocity profiles are shown on the left of the figure. The central section is a plot of the measured losses, and the right section is a plot of the calculated losses. It can be seen that there is a considerable correlation between the two loss plots and that the theory predicts the same sort of loss fluctuations as those measured. This would indicate that the fluctuation is due primarily to the refraction conditions in the water. The agreement between the measured and calculated losses might have been better if two assumptions that are open to question had not been included in the calculation: (1) the sound-velocity depth profile was the same throughout the ocean in the region of interest and (2) the measured depths were accurately known—the resolution of the depth-measuring system was about 7 ft.

In the course of running the computer program for acoustic path search, it was found that quite often in some operations there were several paths connecting the source and receiver, as is illustrated by Fig. 4, which was taken from a different operation. This condition was not found to be a factor in the operation discussed above. In the particular case illustrated in Fig. 4, five direct paths were found. Since this multiple-path condition is due to the refraction conditions inside the medium, it would be expected that the different paths would encounter different acoustic losses. In this particular case, the computed losses for these paths, in order downward in depth, are: 92.4, 107.3, 94.6, 86.2, and 105.1 dB. There is a total spread of 21 dB here and also a difference of about 6 dB between the path with the least loss and the one with the next least. The arrival times along these paths are the same within 1 msec so that, at the receiver, all arrivals are mixed together at unknown phases. This would tend to cause some fluctuation in the received pulse if at least two of the arrivals were of similar amplitude. In this case, the two of highest amplitude are 6 dB apart, which means that mixing would change the amplitude by 6 dB at most. In a case where two paths had almost equal

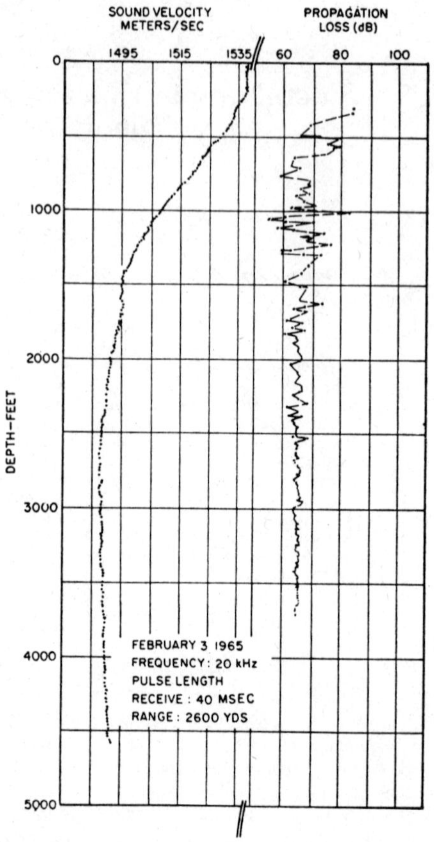

FIG. 2. Typical transmission-loss measurements as a function of depth. Measured sound-velocity profile is at the left.

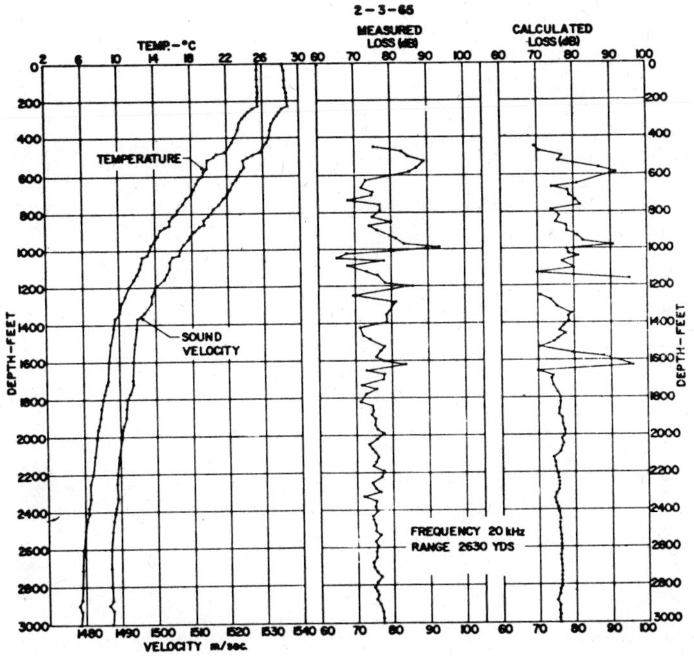

FIG. 3. Comparison of measured and calculated transmission losses as a function of depth. The measured temperature and sound-velocity profile are at the left.

losses, the resulting mixed signal could have quite drastic differences in amplitude.

II. SHORT-DEPTH-INCREMENT EXPERIMENT

Since large fluctuations in amplitude were measured over depth intervals of 12 to 15 ft, it was felt that perhaps such fluctuations were occurring over even smaller intervals and that this possibility should be investigated. On later field trips, a rapid pulser was incorporated into the acoustic control system of the master vehicle. This pulser interrupted the transponding sequence about once each minute and programmed the master unit to transmit a series of pulses at $\frac{1}{4}$-sec intervals for 5 or 6 sec. This meant a pulse for every 8–12-in. change in depth for these few seconds. Figure 5 shows a plot of relative amplitudes for four series of such pulses as received from the master by a second vehicle at a range of 1950 yd at four different depths. The measured sound-velocity-versus-depth profile is shown on the left. The large fluctuations in these examples occur at the 750- and 1000-ft depth series. By the time the vehicles had sunk to 1700 ft, the fluctuation was quite small, and by 2300 ft almost nonexistent. In the 500–1500-ft region, the sound-velocity profile structure is the most complex, while below 1500 ft, the profile is quite simple. The fluctuations on a pulse-to-pulse basis are about as large as they were for comparable depths in Fig. 3.

In an attempt to find an explanation for these short-term fluctuations, a set of ray-path search-and-loss calculations was made for some groups of these rapid pulses. Figure 6 illustrates one of these calculations. On the left, the measured losses are plotted for a series of 19 pulses transmitted over a time interval of about 5 sec and a depth interval of 12 ft from 764 to 776 ft. This is part of the same operation as that illustrated in Fig. 5. To the right are shown the calculated losses for the same set of pulses. It can be seen that there are four or five paths involved on each direct pulse. The losses along different paths are different, but the loss from pulse to pulse along the same path is quite consistent. Since the arrival times of the various paths are the same to within a millisecond, it would appear that the fluctuation in this case is due to the mixing of several signals at differing phases.

The groups of pulses at the deeper depths show little or no fluctuation, owing, undoubtedly, to the fact that the thermal conditions are steady and there is only one path involved in the direct transmission. Any variation then will be due to the slight changes in the medium along the path in the time between pulses.

During this same operation, a surface vessel was monitoring the signals from the sinking vehicles. The signals were received by a hydrophone suspended about 100 ft below the surface and at a horizontal range of 3800 yd from the launch point of the vehicle. Figure 7

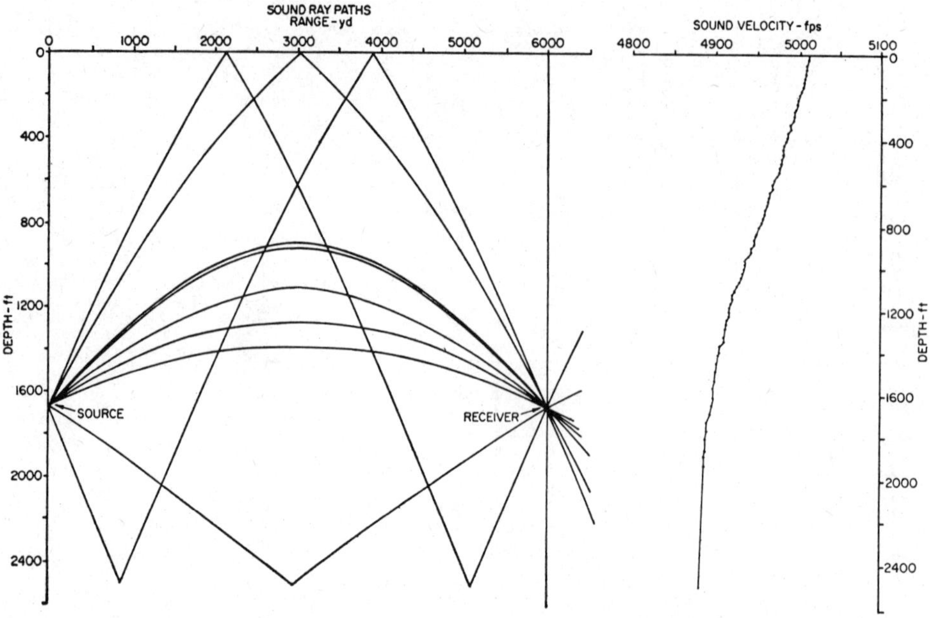

FIG. 4. Example of several possible paths from source to receiver. Sound-velocity profile is at the right.

is a plot of the relative amplitudes of the same four series of pulses as was illustrated in Fig. 5 but as received at the surface vessel. This plot was made from a playback from an analog magnetic-tape recording of the hydrophone output. It can be noticed that there is a tendency for the fluctuation to increase with depth. To try to explain this fluctuation, the computer was used to determine a set of paths from source to receiver and to calculate the loss along each path. This was done for the same series of pulses as was used in Fig. 6. The search indicated that there were only two acoustic paths between the source and receiver: one direct and one surface reflected. The calculated variation in loss along both paths from one pulse to the next was a fraction

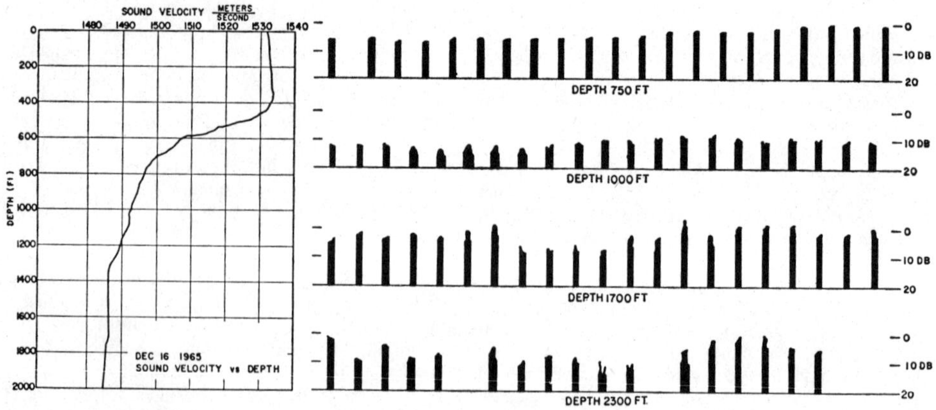

FIG. 5. Plots of measured relative losses as received by DIVEAR vehicle for series of pulses transmitted over 5-sec intervals at four depths. The sound-velocity profile is at the left.

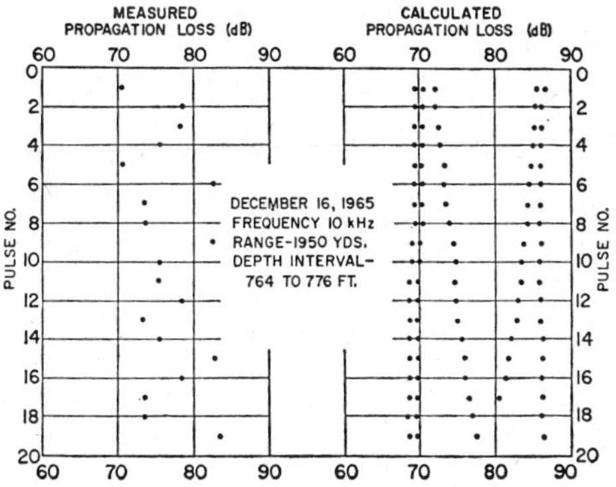

FIG. 6. Measured and calculated propagation losses for one series of pulses.

sea state 1, with some swell and a wind of about 5 kt. The surface roughness and the gradual change in depth of the vehicle during the transmission of this set of pulses would account for path-length differences changing approximately one wavelength and thereby account for the fluctuation.

III. SUMMARY

To summarize these results, it can be said that the acoustic amplitude fluctuations that were measured between vehicles at similar ocean depths over depth intervals of 15 to 20 ft can be explained by the refractive properties of the medium. The variations of sound velocity with depth are sufficient to cause this fluctuation. The large amplitude fluctuation measured over the very short depth intervals appears to be due to multipath mixing of more than one signal at unknown phase. These conclusions are not meant to rule out the fluctuations caused by changes due to such phenomena

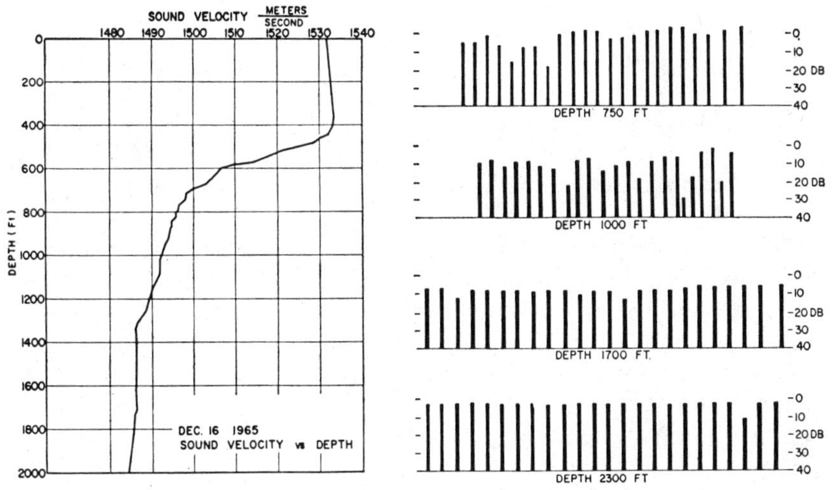

FIG. 7. Plot of relative amplitudes of series of pulses of Fig. 5 but as received by a hydrophone 100 ft under a surface vessel at a range of 3800 yd.

of a decibel if no surface loss is assumed when reflection occurs. Under these conditions, the difference in arrival times between the two paths was about 1 msec, which is about 10 wavelengths at 10 kHz, the frequency used. It would appear that the fluctuation recorded during this operation was due to the mixing of two signals at various phases. This is borne out somewhat by the rather sinusoidal amplitude variation evidenced in the figure. The ocean condition during this run was about

as internal waves and water currents that occur in the medium during a set of measurements. These results indicate that fluctuations are present and can be accounted for when the medium is stable.

ACKNOWLEDGMENT

This work was supported by the U. S. Naval Ordnance Systems Command.

A method is developed, in the following paper, for determining the true distance and direction to a sound source in the ocean where there is a constant sound speed gradient. The determination is made in terms of parameters which can be measured at the location of the observer.

Mr. Wood is a mathematician at the Naval Underwater Systems Center, New London, Connecticut. He has been at the Naval Underwater Systems Center since 1963, where he has been concerned with research on high-frequency solutions to wave propagation.

This paper is reprinted with the permission of the author and the Acoustical Society of America.

Reprinted from THE JOURNAL OF THE ACOUSTICAL SOCIETY OF AMERICA, Vol. 47, No. 5, (Part 2), 1448–1452, May 1970

13.3; 11.2

Received 4 August 1969

7

Refraction Correction in Constant-Gradient Media

D. H. WOOD

Navy Underwater Sound Laboratory, New London, Connecticut 06320

Let g be the magnitude of the gradient and c_0 be the sound speed at the observer. The apparent distance (c_0t) and direction (θ_0) of a sound source determine the true distance (R) and direction (θ), which are given by

$$R=(2c_0/g)[\cos^2\theta_0+(\sin\theta_0+\coth\tfrac{1}{2}gt)^2]^{-\frac{1}{2}} \quad \text{and} \quad \theta=\theta_0+\text{Tan}^{-1}[\cos\theta_0/(\sin\theta_0+\coth\tfrac{1}{2}gt)].$$

A compact graph for generating angular correction is based on the fact that $\theta-\theta_0$ is the argument of the complex number $1+ie^{-i\theta_0}\tanh\frac{1}{2}gt$. Taking c_0/g to be unit distance and g^{-1} to be unit time, only one ray-wavefiont diagram is needed to represent all unbounded constant-gradient media. Scaled distance (Rg/c_0) and true direction (θ) can be read directly from this diagram. Intensity (I) in an unbounded constant-gradient medium, measured in decibels, differs from 20 log$R$ by a term that depends only on travel time:

$$-10\log I=20\log R+20\log\cosh\tfrac{1}{2}g^{t}.$$

INTRODUCTION

In a medium with variable sound speed, refraction makes it difficult to determine the location of an object by echo ranging. Correcting for refraction may be done by ray tracing, but this paper gives simpler methods, which are valid for constant-gradient media.

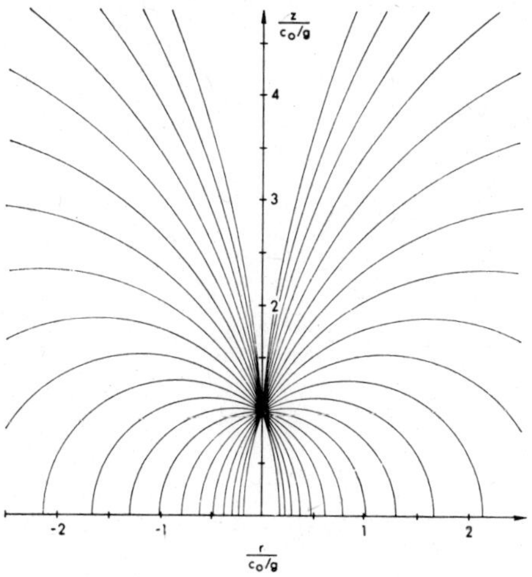

FIG. 1. Nondimensional ray diagram.

Our results are expressed in terms of these parameters: the sound speed c_0 at the position of the observer, the one-way travel time t, the apparent direction θ_0 of the echo, and the gradient g (assumed positive). These

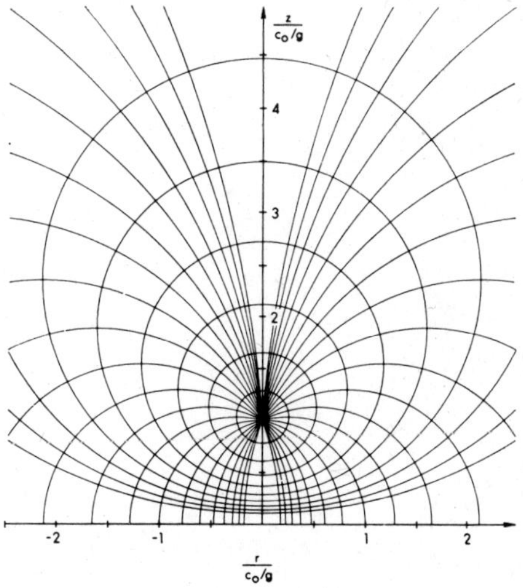

FIG. 2. Nondimensional ray-wavefront diagram. The wavefronts correspond to $gt=0.25$, 0.50, 0.75, $\cdots$, 2.25.

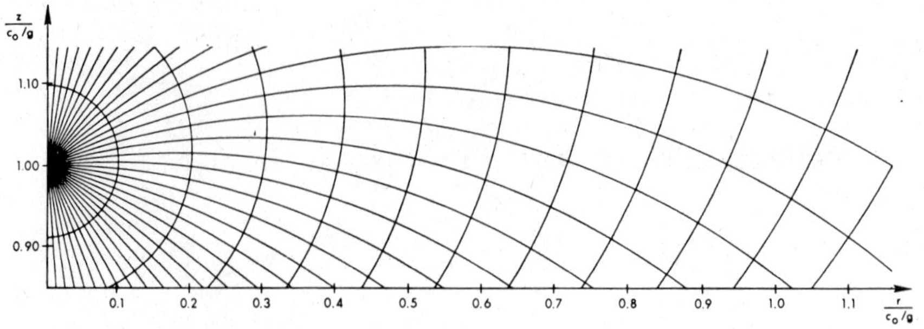

FIG. 3. Nondimensional ray diagram for underwater sound. The rays are every 5° and the wavefronts correspond to $gt=0.1$, 0.2, 0.3, ···, 1.1.

are the natural parameters for echo ranging because they may all be measured at the location of the observer.

I. RAYS AND WAVEFRONTS

The rays from a given point in a constant-gradient medium are known to be a family of arcs of circles with centers on a common line a distance c_0/g from the source.[1] Two families of rays can differ only by having different c_0/g. If we take c_0/g to be unit distance, only one ray diagram is needed to represent the rays in all possible constant-gradient media. Figure 1 shows this ray diagram[2] and introduces a coordinate system that we use. For upward refraction (equivalent to negative g), turn Fig. 1 upside down.

The contours of equal travel time (we call these contours "wavefronts") must be the family of curves orthogonal to the rays. The members of this family are circles that have centers on the z axis, known as the circles of Apollonius.[3] We may draw these circles, if we find where they meet the z axis. For sound that goes straight up, solving the differential equation

$$dz/dt=c=gz, \tag{1}$$

with the initial condition $z=c_0/g$ when $t=0$, gives

$$z=(c_0/g)e^{gt}. \tag{2}$$

Similarly, sound that goes straight down reaches $z=(c_0/g)e^{-gt}$ in time t. The center of the wavefront circle is at the average of these two values of z, $(c_0/g)\cosh gt$, and the radius is half the difference, $(c_0/g)\sinh gt$.

An alternate derivation uses the fact that the circles of Apollonius, with limit points $z=c_0/g$ and $z=-c_0/g$, have the equations[3]

$$[r^2+(z-c_0/g)^2]^{\frac{1}{2}}[r^2+(z+c_0/g)^2]^{-\frac{1}{2}}=\text{CONSTANT}. \tag{3}$$

The constant in Eq. 3 is evaluated by substituting the point $r=0$, $z=(c_0/g)e^{gt}$, which is known to be on the

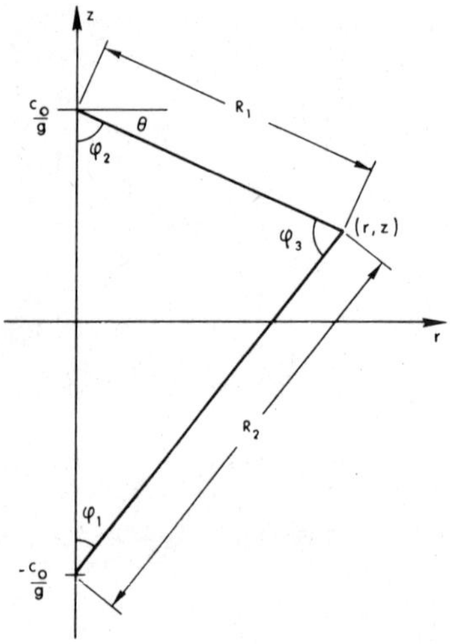

FIG. 4. The Pekeris triangle. (After C. L. Pekeris, Ref. 5.)

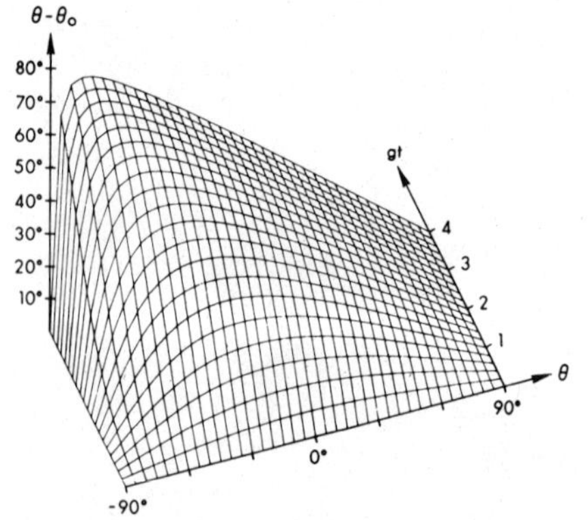

FIG. 5. Angular correction as a function of θ_0 and gt.

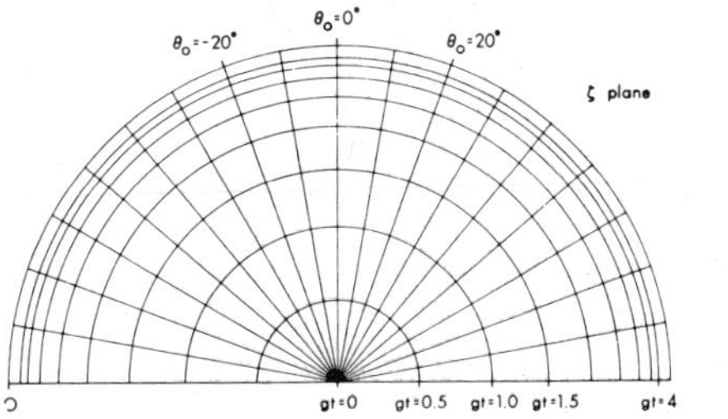

FIG. 6. Angular correction generator. The real axis and a line connecting the intersection of a θ_0 line and a gt circle to the origin O form the acute angle $\theta-\theta_0$. Use the negative of this angle for correcting upward refraction.

wavefront:

$$\text{CONSTANT}=(1-e^{gt})/(1+e^{gt})=\tanh\tfrac{1}{2}gt. \quad (4)$$

Solving Eqs. 3 and 4 for t gives

$$t(r,z)=2g^{-1}\tanh^{-1}\{[r^2+(z-c_0/g)^2]^{\frac{1}{2}} \times[r^2+(z+c_0/g)^2]^{-\frac{1}{2}}\}, \quad (5)$$

which is the known travel-time function for an unbounded constant-gradient medium.[4]

The rays and wavefronts are shown in Fig. 2, with the space dimensions scaled by a factor of c_0/g.[2] The wavefronts now depend only on scaled travel time, gt. With these scalings, Fig. 2 is the only possible ray-wavefront diagram for an unbounded constant-gradient medium. Figure 2 can be used to solve the refraction correction problem; however, only a small portion is needed for that purpose. Our sound-speed function, $c=gz$, can be written as

$$(c_0-c)/c_0=1-zg/c_0, \quad (6)$$

which says that zg/c_0 can be no further from 1 than the maximum fractional change in sound speed. An allowance of 0.15 for the latter quantity should be ample in the case of underwater sound. The corresponding maximum scaled range and travel time are $rg/c_0=1.16$ and $gt=1.1$, respectively. The portion of Fig. 2 that satisfies the above constraints is shown in Fig. 3.[2]

II. THE PEKERIS TRIANGLE FORMULAS

Another approach to the refraction-correction problem is based on an observation of Pekeris. Figure 4 shows the Pekeris triangle and introduces the notation that we use. Pekeris observed[5] that, for all points (r,z) on a given ray, the angle φ_3 is constant and that, for all points (r,z) on a given wavefront, the ratio R_1/R_2 is constant.

Let the depression angle to the point (r,z) be given by $\theta=\pi/2-\varphi_2$. Since φ_3 is constant on a given ray, letting φ_1 go to zero shows that the initial depression angle θ_0 of the ray is given by

$$\theta_0=\varphi_3-\pi/2. \quad (7)$$

Since the sum of the three angles of a triangle is π,

$$\theta-\theta_0=\varphi_1. \quad (8)$$

From the law of sines,

$$R_2\sin\varphi_1-R_1\sin\varphi_2=0, \quad (9)$$

and

$$\sin\varphi_3=\left(\frac{2c_0/g}{R_1}\right)\sin\varphi_1. \quad (10)$$

Adding up the projections on the z axis gives

$$R_2\cos\varphi_1+R_1\cos\varphi_2=2c_0/g. \quad (11)$$

Multiplying Eq. 9 by $(1/R_1)\sin\varphi_1$ and Eq. 11 by $(1/R_1)\cos\varphi_1$ and adding the two equations, we obtain

$$R_2/R_1+\cos(\varphi_1+\varphi_2)=\left(\frac{2c_0/g}{R_1}\right)\cos\varphi_1. \quad (12)$$

Dividing Eq. 10 by Eq. 12 yields

$$\tan\varphi_1=\frac{\sin\varphi_3}{\cos(\varphi_1+\varphi_2)+R_2/R_1}, \quad (13)$$

which in terms of θ and θ_0 is

$$\tan\varphi_1=\tan(\theta-\theta_0)=\frac{\cos\theta_0}{(\sin\theta_0+R_2/R_1)}. \quad (14)$$

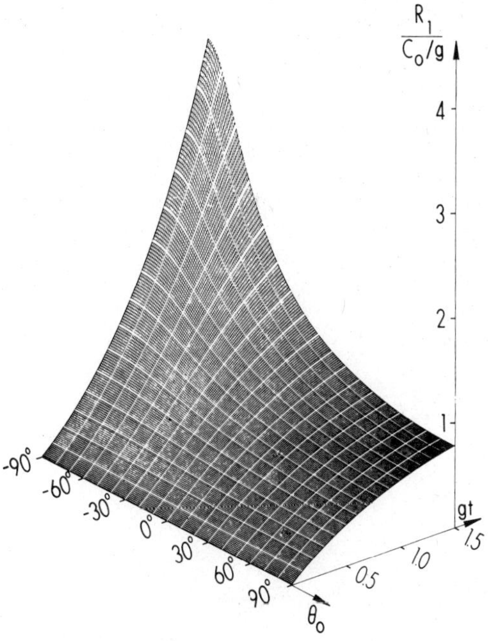

FIG. 7. Scaled range Rg/c_0 as a function of θ_0 and gt.

We know that R_2/R_1 is constant on any given wavefront; in fact, Eqs. 3 and 4 show that

$$R_2/R_1 = \coth\tfrac{1}{2}gt. \qquad (15)$$

Using this result, we obtain the angular correction

$$\theta - \theta_0 = \mathrm{Tan}^{-1}[\cos\theta_0/(\sin\theta_0 + \coth\tfrac{1}{2}gt)]. \qquad (16)$$

This correction function is graphed in Fig. 5.[6]

Equation 16 shows that the complex number

$$\zeta = 1 + ie^{-i\theta_0}\tanh\tfrac{1}{2}gt \qquad (17)$$

has argument (phase) $\theta - \theta_0$. The curves in the complex ζ plane where θ_0 and gt are constant are shown in Fig. 6.

We solve Eq. 10 for R_1 and use Eq. 7 to obtain

$$R_1 = \frac{(2c_0/g)\sin\varphi_1}{\cos\theta_0}. \qquad (18)$$

We can use Eq. 14 to obtain

$$R_1 = (2c_0/g)[\cos^2\theta_0 + (\sin\theta_0 + \coth\tfrac{1}{2}gt)^2]^{-\frac{1}{2}}. \qquad (19)$$

Figure 7 is a graph of R_1g/c_0.[6]

An estimate for R_1 is c_0t. The scaled correction to this estimate is graphed in Fig. 8.

III. INTENSITY

Pekeris derived the response to a point harmonic source of unit amplitude in an unbounded constant-gradient medium. The square of the modulus of that response is[5]

$$I = 4(c_0/g)zR_1^{-2}R_2^{-2}, \qquad (20)$$

provided that the frequency of the source is greater than $g/2$. I is what we call "intensity." It is easy to verify that

$$I = R_1^{-2} - R_2^{-2}, \qquad (21)$$

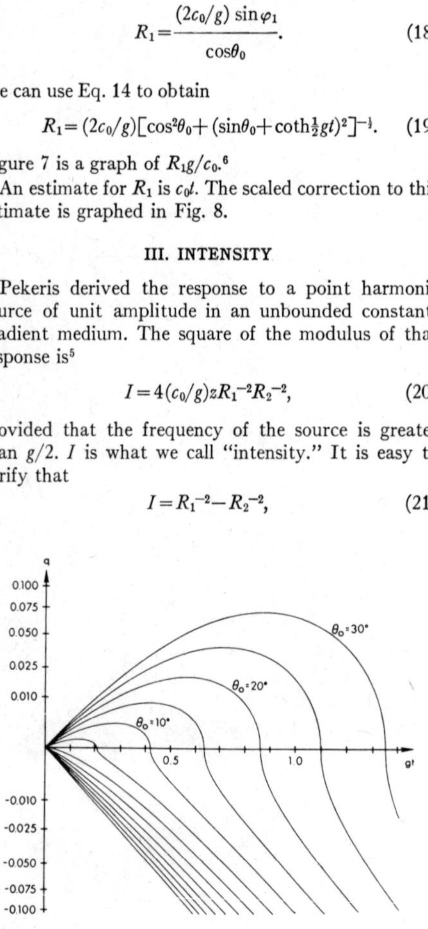

FIG. 8. Scaled correction to the estimate $R_1 \approx c_0t$. True range is given by $R_1 = c_0t + (c_0/g)q$, where q is given by the above graph.

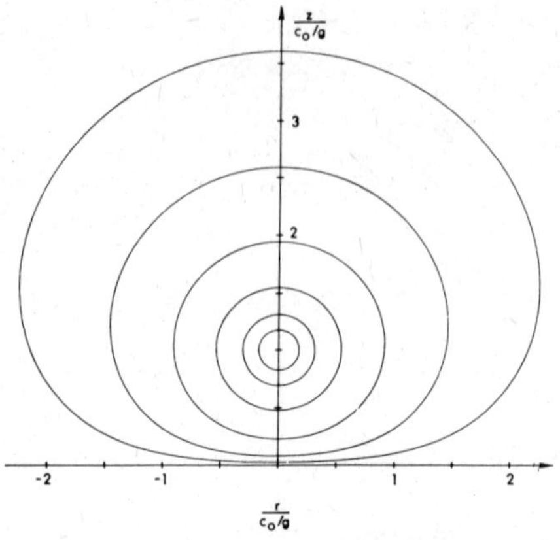

FIG. 9. Curves of constant intensity. The curves are given by $-10\log I - 20\log(c_0/g) = -15, -10, -5, 0, 5, 10$ dB.

and, therefore,

$$I = (R_1\cosh\tfrac{1}{2}gt)^{-2} \qquad (22)$$

by Eq. 15. Note that $I < R_1^{-2}$, where R_1^{-2} would be the intensity in a medium with uniform sound speed. Curves of constant intensity are shown in Fig. 9 (Ref. 2) and are clearly *not* wavefronts as they are in a medium with uniform sound speed. However, I is the product of R_1^{-2} with $\cosh^{-2}\tfrac{1}{2}gt$, which is constant on wavefronts; hence,

$$-10\log I = 20\log R_1 + 20\log\cosh\tfrac{1}{2}gt. \qquad (23)$$

Using common logarithms, the second term of Eq. 23 has the values 0.011, 0.043, 0.097, 0.173, 0.269, 0.385, 0.521, 0.677, 0.851, 1.043, and 1.252 on the wavefronts shown in Fig. 3 and 0.269, 1.043, 2.243, 3.768, 5.522, 7.430, 9.438, 11.509 on the gt contours in Figs. 6 and 10, the smaller values corresponding to the smaller values of gt.

Since Eq. 21 gives $I = R_1^{-2} - R_2^{-2}$, it is worth noting that $(R_2g/c_0)^{-1}$ can be obtained from Fig. 6. From Eq. 17, the modulus of ζ is seen to be

$$|\zeta| = (\tanh\tfrac{1}{2}gt)[(\sin\theta_0 + \coth\tfrac{1}{2}gt)^2 + \cos^2\theta_0]^{\frac{1}{2}}, \qquad (24)$$

which is seen to be

$$|\zeta| = 2(R_2g/c_0)^{-1} \qquad (25)$$

by Eqs. 15 and 19. To obtain a graph that will produce $(R_2g/c_0)^{-2}$, we map the graph in the ζ plane onto the χ plane with the map $\chi = \zeta^2/4$, giving the result shown in Fig. 10.

IV. INTRODUCING BOUNDARIES

In Fig. 2, we may introduce horizontal boundaries. We then follow a given ray until we have crossed the given number of wavefronts or until the ray hits a

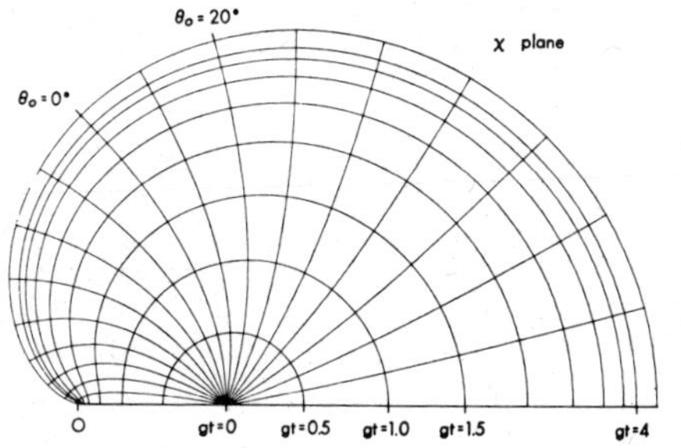

FIG. 10. Anomaly generator. The complex number at the intersection of given θ_0 and gt contours has modulus $(R_2g/c_0)^{-2}$ and phase $2(\theta-\theta_0)$.

boundary. In this case, we reverse direction along the ray and count wavefronts accumulatively until gt is all used up or until we again hit a boundary, etc. When we have crossed the given total number of wavefronts, gz/c_0 is read directly from the vertical axis, and gr/c_0 is determined by the total horizontal distance traversed in the above process.

Introducing vertical boundaries in Fig. 2 corresponds to a medium with a constant range gradient in the sense of Warfield and Jacobson.[7] If Fig. 2 is drawn on a transparent sheet and folded at these boundaries, the result is the ray-wavefront diagram for a constant range gradient medium with these boundaries. Naturally, the result is more suggestive if the range (z) axis is rotated to a horizontal position.

V. DISCUSSION

The refraction-correction formulas, Eqs. 16 and 19, can be derived in many ways: Fig. 2 presents half of a bipolar coordinate system with well-known formulas for inversion[8]; Fig. 2 presents half of a Steiner net that is the image of the uniform sound-speed ray-wavefront diagram under a certain complex bilinear mapping[3]; parameters can be eliminated from the ray equations[1]; etc. All methods must, of course, lead to the same formulas. The refraction-correction formulas are easily programmed on a desk computer. Equation 19 is slightly preferable to Eq. 18 as it is correct in all cases, whereas Eq. 18 fails when $\theta_0=\pm\pi/2$.

Angular correction can be obtained from Fig. 6 or from Fig. 10, where $\theta-\theta_0$ is doubled and small corrections can be read more accurately. Since intensity anomaly in decibels is a function of gt only, it is easy to obtain by using Fig. 3, 6, or 10 and the values given following Eq. 23.

Since much echo ranging is done in constant-gradient water, the above methods should be useful to those not equipped for ray tracing.

ACKNOWLEDGMENT

The author is indebted to Andrew Lesick, of the Navy Underwater Sound Laboratory, who devoted special effort to generating Figs. 5 and 7.

[1] C. B. Officer, *Introduction to the Theory of Sound Transmission* (McGraw-Hill, New York, 1958), p. 60.

[2] This figure was generated by a modified version of a computer subroutine given by H. R. Lewis in "Subroutines for Computing and Graphing Level Curves of Functions of Two Variables," Clearinghouse for Federal Sci. Tech. Inform., Nat. Bur. Std. (US) Rep. No. LA3877.

[3] L. V. Ahlfors, *Complex Analysis* (McGraw-Hill, New York, 1953), Chap. 1, Sec. 3.5.

[4] D. H. Wood, "Green's Functions for Unbounded Constant Gradient Media," J. Acoust. Soc. Amer. **46**, 1333–1339 (1969), Eq. 8.

[5] C. L. Pekeris, "Theory of Propagation of Sound in a Half Space of Variable Sound Velocity under Conditions of Formation of a Shadow Zone," J. Acoust. Soc. Amer. **18**, 295–315 (1946), Sec. 2.

[6] This figure was generated by a version of a computer subroutine given by A. A. Lesick and J. P. Shores in "Three Dimensional Isometric Plots," USL Tech. Mem. No. 2070-254-69.

[7] J. T. Warfield and M. J. Jacobson, "Acoustic Propagation in a Channel with Range-Dependent Sound Speed," J. Acoust. Soc. Amer. **45**, 1145–1155 (1969).

[8] P. Moon and D. E. Spencer, "Field Theory Handbook" (Springer, Berlin, 1961), p. 110.

By measuring the attenuation of sound in solutions of magnesium sulfate, it was demonstrated that the anomalous attenuation of sound in the sea is the result of the relaxation frequency of the magnesium sulfate. An abstract of the paper follows; a more detailed paper was published later.[1]

O. B. Wilson is a physicist who has worked in the fields of physical acoustics and underwater acoustics. He is Chairman of the Physics Department of the Naval Post Graduate School, Monterey, California.

Robert W. Leonard (deceased) was a physicist at the University of California at Los Angeles. He worked on propagation of sound in polyatomic gases, liquids, and solutions. He also studied the acoustic properties of porous materials. He was President of the Acoustical Society of America in 1962–1963.

This abstract is reprinted with the permission of the senior author and the Acoustical Society of America.

---

[1]O. B. Wilson, Jr., and R. W. Leonard, Measurements of sound absorption in aqueous salt solutions by a resonator method, *J. Acoust. Soc. Amer.* **26**, 223–226 (1954).

Journal of the Acoustical Society of America, vol. 23, 1951, p. 624A

8

Abstract from

Sound Absorption in Aqueous Solutions of Magnesium Sulfate and in Sea Water*

O. B. WILSON, JR.
and R. W. LEONARD

A reverberation method has been used to measure the sound absorption in $MgSO_4$ solutions, relative to that in water, under a wider range of conditions of temperature and concentration than for results previously reported. The relaxation frequency of the absorption process, approximately 150 kc at 25° C, showed no marked change when the concentration ranged from 0.003 to 0.02 molal. Dependence of the absorption concentration, m, is variable, seeming to be very approximately m^2 at $m = 0.003$ and approximately $m^{1.3}$ at $m = 0.02$. For $m = 0.01$, change in relaxation frequency when the temperature was varied from 4.5–43° C indicates an apparent activation energy for the reaction of approximately 7.5 kcal per mole. The large relative absorption present for temperatures near 4° C would indicate that the absorbing mechanism is not caused by a relaxation effect in the specific heat. Similar measurements on natural and synthetic sea waters agree well with each other, but give values somewhat smaller than do results of direct measurements in the sea.

*Research supported by Contract N6onr-27507.

Velocity of Sound

II

The equation for the speed of sound in the sea was developed from data on the effect of temperature, pressure, and salinity on the speed of sound in water. This equation makes possible the accurate computation of the speed of sound under any condition of temperature pressure and salinity.

Mr. Wilson has worked at the Naval Ordnance Laboratory since 1955. He has done research on the measurement of the speed of sound in water and he has spent the last four years designing, constructing, and testing small sonars.

He holds four patents on crystal growth and two patents on sonar devices.

This paper is reprinted with the permission of the author and the Acoustical Society of America.

9

Reprinted from THE JOURNAL OF THE ACOUSTICAL SOCIETY OF AMERICA, Vol. 32, No. 10, 1357, October, 1960

Equation for the Speed of Sound in Sea Water

WAYNE D. WILSON
U. S. Naval Ordnance Laboratory, White Oak, Silver Spring, Maryland
(Received July 18, 1960)

An equation is given for the speed of sound in sea water. The experimental data used to determine this equation were obtained in the temperature, pressure, and salinity ranges $-4°C < T < 30°C$, $1\ kg/cm^2 < P < 1000\ kg/cm^2$, and $0°/_{00} < S < 37°/_{00}$. The standard deviation of the experimental data was found to be 0.30 m/sec.

THE equation for the speed of sound in sea water presented earlier[1] was based on measurements of sound speed obtained in the temperature range $-4°C < T < 30°C$, the pressure range $1\ kg/cm^2 < P < 1000\ kg/cm^2$, and the salinity range $33°/_{00} < S < 37°/_{00}$. The standard deviation from the mean of the experimental data in this earlier work was 0.22 m/sec. It has been found that this equation will not predict sound speeds accurately for salinities outside the salinity range considered.

Additional measurements of sound speed have been obtained at salinities of $10°/_{00}$, $20°/_{00}$, and $30°/_{00}$ and for the same range of temperature and pressure considered in the earlier work. These data, the data used to obtain the earlier equation, and the data obtained in distilled water[2] were then used to obtain a new equation for the speed of sound in sea water over a wider salinity range.

The new equation is:

$$V = 1449.14 + V_T + V_P + V_S + V_{STP}$$

$$V_T = 4.5721T - 4.4532\times10^{-2}T^2 - 2.6045\times10^{-4}T^3 + 7.9851\times10^{-6}T^4$$

$$V_P = 1.60272\times10^{-1}P + 1.0268\times10^{-5}P^2 + 3.5216\times10^{-9}P^3 - 3.3603\times10^{-12}P^4$$

$$V_S = 1.39799(S-35) + 1.69202\times10^{-3}(S-35)^2$$

$$\begin{aligned} V_{STP} = {} & (S-35)(-1.1244\times10^{-2}T + 7.7711\times10^{-7}T^2 \\ & + 7.7016\times10^{-5}P - 1.2943\times10^{-7}P^2 + 3.1580\times10^{-8}PT \\ & + 1.5790\times10^{-9}PT^2) + P(-1.8607\times10^{-4}T \\ & + 7.4812\times10^{-6}T^2 + 4.5283\times10^{-8}T^3) \\ & + P^2(-2.5294\times10^{-7}T + 1.8563\times10^{-9}T^2) \\ & + P^3(-1.9646\times10^{-10}T). \end{aligned}$$

In this equation the units of temperature, pressure, salinity, and sound speed are °C, kg/cm^2, parts per thousand, and meters/sec, respectively. The standard deviation from the mean is 0.30 m/sec for all data obtained in the ranges $-4°C < T < 30°C$, $1\ kg/cm^2 < P < 1000\ kg/cm^2$, and $0°/_{00} < S < 37°/_{00}$. The new equation may consequently be used to predict sound speeds to within 0.30 m/sec over the indicated temperature, pressure, and salinity ranges.

[1] W. D. Wilson, J. Acoust. Soc. Am. **32**, 641 (1960).
[2] W. D. Wilson, J. Acoust. Soc. Am. **31**, 1067 (1959).

Dr. Del Grosso has done much research in measurement of the speed of sound in water. This paper provides tables of precise determinations of the speed of sound in pure water and sea water made by an interferometric method.

Dr. Del Grosso is a physicist with the Acoustics Division of the Naval Research Laboratory. He recently spent a year studying the propagation of sound in liquid helium with the Acoustics Group at the University of California at Los Angeles.

He developed the acoustic interferometer, with concomitant theory, into an extremely precise instrument. Equations fitted to his data obtained with this equipment are believed to be the most accurate extant and are valid for all ocean variables.

This paper is reprinted with the permission of the author and the Acoustical Society of America.

Reprinted from THE JOURNAL OF THE ACOUSTICAL SOCIETY OF AMERICA, Vol. 47, No. 3, (Part 2), 947–949, March 1970

Sound Speed in Pure Water and Sea Water

VINCENT A. DEL GROSSO
Naval Research Laboratory, Washington, D. C. 20390

New interferometric measurements of the speed of sound in pure water and standard sea water are presented. The pure-water results are reasonably consistent with other recent measurements while the sea-water results cast additional suspicion on the concept of salinity as a characterization of sea water.

AN EARLIER NOTE[1] CAUTIONED AGAINST THE TEMPTATION TO APPLY the demonstrated errors in the NOL[2] pure-water sound-speed data to either[3,4] of the NOL sea-water sound-speed equations. Since then new pure-water values have been reported,[5] which unfortunately, like earlier BTL[6] and University of Glasgow[7] data, are limited to temperatures above 25°C.

In the same note,[1] the results of our latest measurements were promised and are now presented. The measurements were made with the NRL ultrasonic interferometer at a frequency of 5 MHz, utilizing a 1-MHz x-cut quartz crystal (faces plane and parallel to better than 1 arc sec and crystallographic axis deviation less than 10 arc sec) of $ka=125\pi$ and set parallel to a plane reflector to better than 1 arc sec. This parallelism is maintained over a 0–4-cm source-reflector separation, but all measurements were made from 2 to 4 cm separation, where the guided-mode dispersion calculations[8] (contrasted to free-field diffraction calculations) indicate that resultant errors may be limited to 1 cm/sec for measurements in a cylindrical container of twice the source diameter.

The impedance of the quartz crystal is measured simultaneously with an impedance plotter and a vector voltmeter. Reflector displacements are measured with a laser interferometer operating with a retroreflector mounted within the hollow and evacuated reflector shaft (this replaces the fringecount micrometer and quartz gauge rod previously employed). Temperatures are measured within the interferometer cell by means of a platinum thermometer and Mueller bridge (the quartz thermometers previously utilized are now regulated to rapid, approximate temperature indication).

The raw measurements of sound speed in pure water are given in Table I, while Tables II–IV present the measured sound speeds in standard sea water (P_{46} 4/12 1966, 19.377 0/00 Cl) of 30.127 0/00 S, 35.005 0/00 S, and 41.298 0/00 S. The low and high salinities were prepared from the standard sea water by dilution with pure water and by evaporation (not under vacuum). Equations have not yet been fit to this data. A feeling for the sensitivity of the measurement may be obtained by perusual of the data. The

TABLE 1. Sound-speed measurements (raw) in pure water.

		PURE H_2O			
T (°C)	C_{NRL} (cm/sec)	C_{NBS} (cm/sec)	Δ_{NBS} (cm/sec)	C_{NOL} (cm/sec)	Δ_{NOL} (cm/sec)
0.056	140267.3	140301.7	34.4	140329.4	62.1
0.061	140269.5	140304.2	34.7	140331.9	62.4
0.064	140270.5	140305.7	35.2	140333.4	62.9
0.068	140272.6	140307.7	35.1	140335.4	62.8
0.072	140274.7	140309.7	35.0	140337.4	62.7
4.991	142611.5	142645.5	34.0	142671.0	59.5
4.994	142612.6	142646.8	34.2	142672.3	59.7
4.995	142612.9	142647.3	34.4	142672.8	59.9
4.996	142613.1	142647.7	34.6	142673.2	60.1
9.996	144723.4	144757.8	34.4	144783.3	59.9
10.000	144724.9	144759.4	34.5	144784.9	60.0
10.007	144727.6	144762.2	34.6	144787.7	60.1
10.016	144730.7	144765.7	35.0	144791.3	60.6
19.927	148209.1	148243.3	34.2	148269.4	60.3
19.928	148209.6	148243.6	34.0	148269.7	60.1
19.929	148210.2	148243.9	33.7	148270.0	59.8
24.990	149663.6	149697.3	33.7	149721.9	58.3
24.994	149664.6	149698.4	33.8	149723.0	58.4
29.991	150908.1	150941.8	33.7	150963.3	55.2
29.993	150908.9	150942.3	33.4	150963.8	54.9
34.981	151975.2	152008.3	33.1	152025.6	50.4
34.991	151976.8	152010.2	33.4	152027.5	50.7
34.997	151978.1	152011.4	33.3	152028.7	50.6
39.983	152882.0	152914.8	32.8	152926.8	44.8
39.985	152882.3	152915.1	32.8	152927.1	44.8
39.988	152882.7	152915.6	32.9	152927.6	44.9
39.995	152883.7	152916.8	33.1	152928.8	45.1
50.006	154254.5	154287.1	32.6	154288.9	34.4
50.023	154256.3	154289.0	32.7	154290.7	34.4
50.047	154259.1	154291.6	32.5	154293.3	34.2
50.057	154260.2	154292.7	32.5	154294.3	34.1
60.029	155099.9	155132.0	32.1	155127.3	27.4
60.031	155099.9	155132.2	32.3	155127.4	27.5
74.003	155514.4	155546.7	32.3	155540.7	26.3
74.019	155514.4	155546.7	32.3	155540.7	26.3
74.029	155514.5	155546.7	32.2	155540.7	26.2

TABLE 2. Sound-speed measurements in standard sea water of 30.127 0/00 S.

T (°C)	C_{NRL} (cm/sec)	C_{1952} (cm/sec)	Δ_{1952} (cm/sec)	C_{NOL1} (cm/sec)	Δ_{NOL1} (cm/sec)	C_{NOL2} (cm/sec)	Δ_{NOL2} (cm/sec)
0.126	144306.0	144306.0	0.0	144132.9	−173.1	144311.0	5.0
0.132	144308.1	144308.8	0.7	144135.7	−172.4	144313.8	5.7
0.138	144311.2	144311.6	0.4	144138.4	−172.8	144316.5	5.3
0.146	144314.4	144315.3	0.9	144142.1	−172.3	144320.2	5.8
9.998	148381.3	148418.7	37.4	148237.2	−144.1	148415.9	34.6
10.001	148383.4	148419.9	36.5	148238.2	−145.2	148417.0	33.6
10.006	148384.6	148421.7	37.1	148240.1	−144.5	148418.8	34.2
10.013	148387.7	148424.2	36.5	148242.6	−145.1	148421.3	33.6
19.965	151585.5	151671.3	85.8	151461.9	−123.6	151644.7	59.2
19.975	151588.6	151674.2	85.6	151464.8	−123.8	151647.7	59.1
19.988	151591.8	151677.9	86.1	151468.5	−123.3	151651.5	59.7
20.008	151597.0	151683.6	86.6	151474.1	−122.9	151657.1	60.1
29.994	154041.8	154165.3	123.5	153926.5	−115.3	154068.3	26.5
30.004	154043.9	154167.4	123.5	153928.6	−115.3	154070.4	26.5
30.015	154046.0	154169.8	123.8	153930.9	−115.1	154072.7	26.7
30.026	154048.1	154172.1	124.0	153933.2	−114.9	154075.0	26.9
40.048	155837.9	156012.7	174.8	155751.7	−86.2	156022.2	184.3
40.049	155839.0	156012.8	173.8	155751.8	−87.2	156022.4	183.4

TABLE 3. Sound-speed measurements in adjusted standard sea water of 35.005 0/00 S.

T (°C)	C_{NRL} (cm/sec)	C_{1952} (cm/sec)	Δ_{1952} (cm/sec)	C_{NOL1} (cm/sec)	Δ_{NOL1} (cm/sec)	C_{NOL2} (cm/sec)	Δ_{NOL2} (cm/sec)
0.141	144961.1	144925.0	−36.1	145003.7	42.6	144995.1	34.0
0.149	144964.1	144928.6	−36.5	145007.4	43.3	144998.7	34.6
0.154	144967.0	144930.9	−36.1	145009.7	42.7	145001.0	34.0
0.167	144972.7	144936.8	−35.9	145015.6	42.9	145006.9	34.2
10.000	148974.4	148978.5	4.1	149044.2	69.8	149039.8	65.4
10.002	148975.0	148979.2	4.2	149044.9	69.9	149040.5	65.5
10.007	148976.5	148981.0	4.5	149046.7	70.2	149042.3	65.8
10.012	148978.1	148982.8	4.7	149048.5	70.4	149044.1	66.0
19.936	152112.8	152170.3	57.5	152202.0	89.2	152195.6	82.8
19.938	152112.8	152170.9	58.1	152202.5	89.7	152196.2	83.4
29.970	154541.8	154622.0	80.2	154611.1	69.3	154577.0	35.2
29.975	154542.8	154623.1	80.3	154612.2	69.4	154578.1	35.3
29.981	154543.9	154624.3	80.4	154613.4	69.5	154579.3	35.4
29.994	154546.0	154627.1	81.1	154616.1	70.1	154582.0	36.0
40.027	156311.5	156440.5	129.0	156379.4	67.9	156477.3	165.8
40.031	156311.5	156441.1	129.6	156380.0	68.5	156478.0	166.5

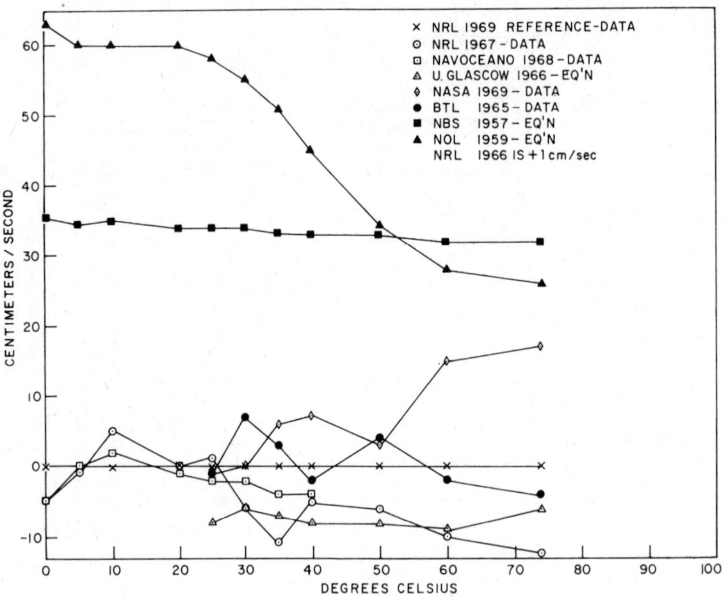

FIG. 1. Sound-speed excesses of recent data above these present measurements for pure water.

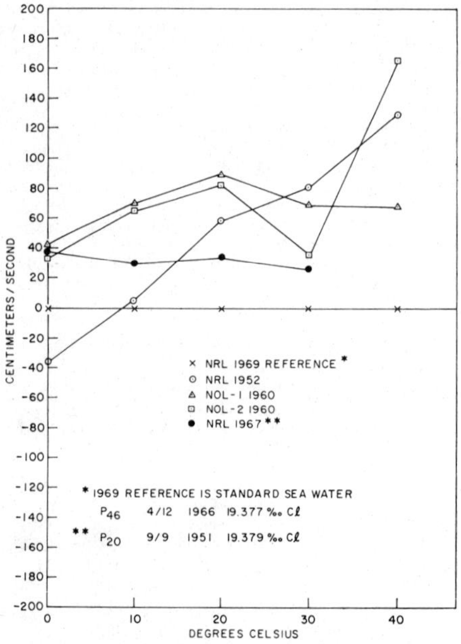

FIG. 2. Sound-speed excesses of recent data above these present measurements for adjusted standard sea water of 35.005 0/00 *S*.

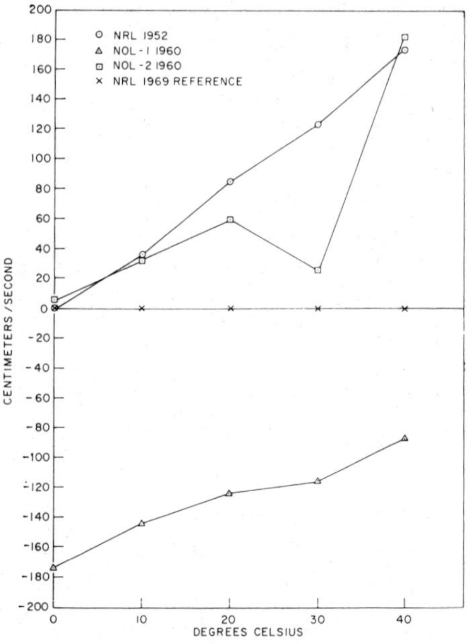

FIG. 3. Sound-speed excesses of recent data above these present measurements for standard sea water of 30.127 0/00 *S*.

sensitivity was degraded by an unexpected vibration, which rendered fringe interpolation difficult, but it is believed that the precision is somewhat better than 1 cm/sec.

Figure 1 is a plot of sound-speed excesses in centimeters/second of other data above these present measurements. We are currently reinvestigating the region between 0°–5°C, as we still have indications of an anomaly there. The present data is believed superior to the earlier because of the improvements mentioned above.

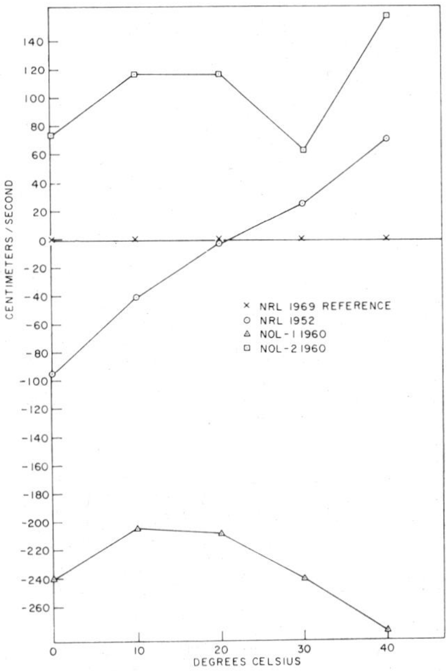

FIG. 4. Sound-speed excesses of recent data above these present measurements for adjusted standard sea water of 41.298 0/00 *S*.

TABLE 4. Sound-speed measurements in adjusted standard sea water of 41.298 0/00 *S*

T (°C)	C_{NRL} (cm/sec)	C_{1952} (cm/sec)	Δ_{1952} (cm/sec)	C_{NOL1} (cm/sec)	Δ_{NOL1} (cm/sec)	C_{NOL2} (cm/sec)	Δ_{NOL2} (cm/sec)
0.136	145804.6	145708.4	−96.2	145564.9	−239.7	145878.4	73.8
0.143	145807.7	145711.6	−96.1	145568.0	−239.7	145881.5	73.8
0.152	145811.3	145715.6	−95.7	145572.0	−239.3	145885.5	74.2
0.165	145816.6	145721.5	−95.1	145577.9	−238.7	145891.3	74.7
9.971	149727.2	149685.5	−41.7	149523.0	−204.2	149845.2	118.0
9.976	149728.8	149687.3	−41.5	149524.8	−204.0	149847.0	118.2
9.978	149729.3	149688.0	−41.3	149525.5	−203.8	149847.9	118.6
19.915	152818.4	152815.4	−3.0	152610.5	−207.9	152935.6	117.2
19.925	152821.4	152818.2	−3.2	152613.3	−208.1	152938.3	116.9
19.932	152822.5	152820.2	−2.3	152615.1	−207.4	152940.2	117.7
29.995	155195.1	155219.8	24.7	154954.9	−240.2	155257.1	62.0
29.996	155195.1	155220.0	24.9	154955.1	−240.0	155257.3	62.2
29.998	155195.2	155220.4	25.2	154955.5	−239.7	155257.7	62.5
40.002	156919.0	156989.7	70.7	156639.0	−280.0	157077.0	158.0
40.006	156920.1	156990.3	70.2	156639.5	−280.6	157077.7	157.6
40.010	156920.1	156990.9	70.8	156640.1	−280.0	157078.4	158.3

Figure 2 is a plot of sound-speed excesses above the present data for the two NOL equations, the old NRL equation, and the 1967 NRL raw data. The baseline shift between the 1967 and 1969 data is puzzling. It perhaps indicates that standard sea water, while offering a standard solution permitting relatively easy checking among various laboratories, may not in fact maintain a standard salinity to sound-speed relationship. This point is also being investigated.

Figures 3 and 4 are the same as Fig. 2 except for salinities of 30.127 0/00 *S* and 41.298 0/00 *S*, respectively. No discussion of these plots seems necessary. We are also measuring intermediate salinities so that a new equation may be developed. The pressure-coefficient data will follow shortly in another note.

[1] V. A. Del Grosso, J. Acoust. Soc. Amer. **45**, 1287–1288 (1969).
[2] W. D. Wilson, J. Acoust. Soc. Amer. **31**, 1067–1072 (1960).
[3] W. D. Wilson, J. Acoust. Soc. Amer. **32**, 641–644 (1960).
[4] W. D. Wilson, J. Acoust. Soc. Amer. **32**, 1357 (1960).
[5] R. C. Williamson, J. Acoust. Soc. Amer. **45**, 1251–1257 (1969).
[6] H. J. McSkimin, J. Acoust. Soc. Amer. **37**, 325–328 (1965).
[7] A. J. Barlow and E. Yazgan, Brit. J. Appl. Phys. **17**, 807–819 (1966).
[8] V. A. Del Grosso, Naval Res. Lab. Rep. No. 6852 (1968).

This paper describes the sing-around ultrasonic velocimeter developed by Greenspan and Tschiegg which has formed the basis of the velocimeters manufactured commercially for measuring the velocity of sound in the ocean. These instruments can be lowered in the ocean so that the velocity of sound as a function of depth can be measured directly with a high degree of precision.

Mr. Greenspan, the senior author, is a physicist at the National Bureau of Standards. He has been concerned with research in ultrasonics, sound, mechanics, and electronics. He received the Meritorious Service Award from the Department of Commerce in 1949 and 1961. He was President of the Acoustical Society of America in 1966–1967.

C. E. Tschiegg is a physicist in the Sound Section of the National Bureau of Standards. He received the Meritorious Service Award from the Department of Commerce in 1961. He has contributed in the fields of ultrasonics and underwater sound.

This paper is reprinted with the permission of the senior author and the Director of Publications of the American Institute of Physics.

Reprinted from THE REVIEW OF SCIENTIFIC INSTRUMENTS, Vol. 28, No. 11, 897–901, November, 1957
Printed in U. S. A.

Sing-Around Ultrasonic Velocimeter for Liquids*

11

MARTIN GREENSPAN AND CARROLL E. TSCHIEGG
National Bureau of Standards, Washington 25, D. C.
(Received April 11, 1957; and in final form, August 16, 1957)

A "sing-around" velocimeter of high stability and precision has been developed and tested. The instrument, which is of the fixed-path type, automatically measures and/or records the speed of sound in nondispersive liquids. Stabilities of about 50 ppm over a 24-hr period can be achieved.

I. INTRODUCTION

IN this paper we describe an automatic recording instrument for the precise measurement of the speed of sound in liquids. This "sing-around velocimeter" is restricted to use with liquids which show no appreciable frequency dispersion except possibly at very high frequencies. It must be adjusted and calibrated for a particular class of liquids within which the total variation of the speed of sound is not more than perhaps 15%, and within which the variation of attenuation is not too large. A liquid of which the temperature, pressure, or composition is changing is such a class. Sea water *in situ* is an important example, in fact the most common practical application of the instrument at present is the measurement of the speed of sound and its gradients in the sea. The velocimeters to be described have high stability and are thus especially adapted to differential measurements. We have used one to measure the temperature coefficient of the speed of sound in water near the turning point[1] and to determine the effect of dissolved air on the speed of sound in water.[2]

The sing-around velocimeter combines the accuracy of the instrument against which it is calibrated with the advantages of automatic, almost instantaneous operation and recording. Its limitations are that the operating range is relatively small and that it requires calibration for the particular class of liquids with which it is to be used.

II. GENERAL CONSIDERATIONS

A sing-around velocimeter is outwardly similar to the ultrasonic delay line employed in digital computers for information storage. It may be thought of as a cylindrical tank of which the ends are electroacoustic transducers, the whole filled with the liquid under test. A voltage pulse is applied to the "sender;" a corresponding pulse of sound travels through the sample liquid and is received and converted to an electrical pulse by the receiver. In order to define uniquely the time interval between the pulses, some characteristic, still recognizable after the pulse has been distorted by the transmission through the liquid, and by the band width limitations of the transducers, must be selected to specify the location in time of the pulse. In the present instrument the pulse position is specified by the instant at which it begins to rise from the noise. This choice has several important consequences. To begin with, a pulse-modulated carrier has now no advantage over the much simpler video pulse even though the distortion of the former would be much less.[3] Further, it becomes essential that the pulses rise rapidly. This is no problem so far as the input pulse is concerned, but the output pulse rises relatively very slowly. The fast rise is restored by amplification; nevertheless there is introduced an unknown delay equal to the time which the output pulse spends below the noise. This delay depends on the attenuation characteristics of the liquid. It is for this reason, mostly, that the velocimeter must be calibrated and used on a class of liquids within which the attenuation characteristics are not too variable.

The timing is automatic. The received pulse after suitable amplification and reshaping is again applied to the sender; thus the device regenerates and the pulse repetition frequency (prf) depends on the speed of sound in the liquid and to some extent on the electrical and other delays. The principle is not new. The earliest description we have found is in a patent[4] filed in 1937 by Shepard. Similar systems are described in later patents by Kock[5] and by Larsen.[6] The designation "sing-around" appears to have been coined by Hanson.[7] Hanson, Barrett and Suomi,[8] and several others (for references see a recent paper by Ficken and Hiedemann[9]) have constructed apparatus similar to that described here. These instruments were not of high precision.

A major source of difficulty is the existence of multiple echoes between the transducers. The various sets of

* This work was supported in part by the Office of Naval Research under contract NA-onr-70-48. The preliminary work was supported by the Office of Basic Instrumentation, National Bureau of Standards.

1 Greenspan, Tschiegg, and Breckenridge, J. Acoust. Soc. Am. **28**, 500 (1956).

2 M. Greenspan and C. E. Tschiegg, J. Acoust. Soc. Am. **28**, 501 (1956).

3 However, the choice of a video pulse restricts operation to nondispersive liquids. An instrument based on a pulse-modulated carrier could in principle be used on dispersive liquids although at a single frequency only.

4 F. H. Shepard, Jr., U. S. Patent 2,333,688, November 9, 1943.

5 W. E. Kock, U. S. Patent 2,400,309, May 14, 1946.

6 M. J. Larsen, U. S. Patent 2,580,560, January 1, 1952.

7 R. L. Hanson, J. Acoust. Soc. Am. **21**, 60 (1949).

8 E. W. Barrett and V. E. Suomi, J. Meteorol. **6**, 273 (1949).

9 G. W. Ficken, Jr. and E. A. Hiedemann, J. Acoust. Soc. Am. **28**, 921 (1956).

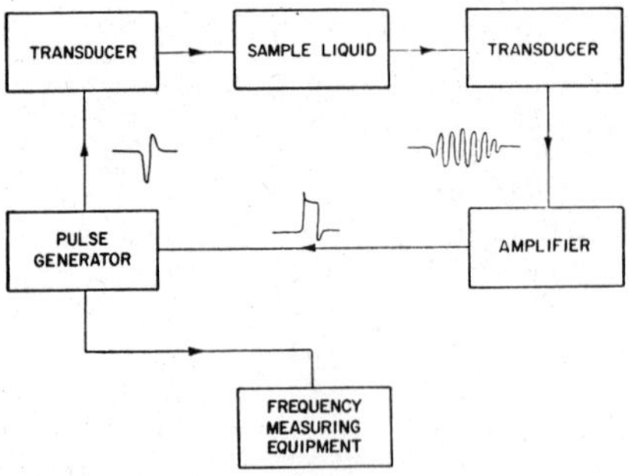

FIG. 1. Block diagram showing "sing-around" principle.

echoes, each set arising from a different primary pulse, are not synchronous because of the electrical time delay. Various means of eliminating the reflections have been used. In the case of a straight path the transducers are tilted slightly out of parallel so that all received pulses but the first are lost in the noise. In the case of a bent path, where the sound is reflected back nearly on itself, the transducers and reflector occupy their geometrically correct positions, but the reflection coefficient is rather small. The first received pulse is attenuated by reflection once, and the next three times; the result is that all received pulses after the first are negligible. Hard rubber, Teflon, and perforated metal are suitable materials for the reflector.

A bent path minimizes errors which arise from mass motion of the liquid and is preferred for field models and in the laboratory in cases where a large volume of sample, which necessitates vigorous stirring for maintenance of thermal equilibrium, is used. In cases where the liquid is contained in a small tank which is immersed in a temperature-controlled bath the straight path is satisfactory.

We recapitulate briefly the principle of operation. A block diagram is shown in Fig. 1. The input transducer is energized by a trigger-type pulse forming circuit which produces short fast pulses. This circuit is adjusted to run free at a prf somewhat less than the minimum operating prf to be expected. The pulses of pressure produced by the input transducer travel down the sample liquid in a time l/c, where c is the speed of sound and l is the path length. The received pulses are amplified and shaped and are used to synchronize the original pulse-forming circuit. If t_e is the sum of the electrical delays and the time lost in the noise, the total time delay is

$$1/f = t_e + l/c. \tag{1}$$

The prf, f, is measured and perhaps recorded. Both t_e and l are obtained by direct calibration on a liquid in which the speed of sound is known. If the velocimeter is for use in the sea, for example, a suitable calibration liquid is distilled water. Readings of f [Eq. (1)] on distilled water at various temperatures between 0 and 60°C cover the range which would be obtained in the sea where the extremes of temperature are 0 and 40°C and the salinity reaches perhaps 4%. Corresponding to each temperature of the distilled water is a known speed c and an observed prf, f. These determine the unknowns t_e and l in Eq. (1). It is also possible to determine t_e by measuring f for two different known values of l. This method is both more cumbersome and less accurate; the length l in Eq. (1) is only an effective length and is difficult to define in an absolute senes especially in the case where the receiving transducer is not accurately parallel to a wave front.

III. APPARATUS

A. The Transducers and the Path

Good results have been obtained with both x-cut quartz and with ceramic transducers. The transducers are made as thin as is consistent with ease of handling; thicknesses corresponding to resonant frequencies of from 1.5 to 5 Mc are satisfactory.

Figure 2 shows one type of transducer mount. The transducer shown is a $Ba_{0.80}Pb_{0.08}Ca_{0.12}TiO_3$ disk of diameter about 12 mm and thickness 0.66 mm. Flatness and parallelism are held to 25 μ. The electrodes are painted on using a solution of gold chloride in alcohol thickened with oil of lavender or collodion (equivalent paints are available commercially and are as good or better) and then fired at 500°C for 1 hr. Electrodes of this type are much more durable than evaporated electrodes, especially under water. The outer (ground) electrode covers one face completely; the inner (hot) electrode is about 6-mm in diameter. The disks are then polarized at about 25 kv/cm in mineral oil, cooling from 135°C to room temperature in about 1 hr.

The inner surface of the transducer bears solidly against a flat surface machined on the mount. This is a composite surface; it shows the hot electrode, about 6-mm in diameter, concentric with the housing which serves as ground, and separated from it by an insulating

FIG. 2. Transducer mounts.

FIG. 3. Two transducer mounts and a hard-rubber reflector mounted on an Invar plate.

cylinder, limestone in this case. Before assembly a very small amount of air-drying conductive paint is placed on the center electrode of the mount to provide good electrical contact to the corresponding electrode on the transducer. The Neoprene O-ring and the cap complete the seal. The small space between the cap and the ground electrode on the transducer is filled with the same conductive paint; the resistance across the gap should not exceed 0.2 ohm. (The low-resistance contacts are unnecessary in the case of a quartz transducer.) The mounts shown in Fig. 2 can be subjected to a pressure greater than 25 000 psi without damage.

A complete bent-path assembly including a hard-rubber reflector is shown in Fig. 3. The mounting plate is of Invar to minimize thermal expansion of the sound path. This particular assembly is part of an underwater velocimeter built by the Naval Ordnance Laboratory. The underwater part of the instrument ("fish") is shown in Fig. 4. Electronic circuits occupy the bulk of the fish. The instrument is for use at moderate depths (up to 250 ft) and therefore does not have the high-pressure transducer mounts shown in Fig. 2, although the design is similar. It can be adapted to laboratory use. For this purpose it is mounted with the axis vertical and the transducers on top. A metal can is attached to hold the liquid, and heaters, a stirrer, and a thermostat are provided. Greater precision is obtained if the hard rubber reflector is replaced by a metal reflector of adjustable effective area or by a perforated metal reflector, as described in Sec. II, if the instrument is to used at temperatures exceeding about 60°C.

B. The Pulse Generator

The input pulse is generated by a blocking oscillator wired as shown in Fig. 5 (*V*6). The pulse transformer *T*1 is chosen so that under load conditions (that of the sending transducer and connecting cable) the pulse amplitude is about -100 v, and the rise time less than 0.05 μsec. The optimum pulse width is 0.1 to 0.2 μsec, although a somewhat wider pulse is tolerable so long as the fast rise is maintained. Pulse widths in excess of about 0.5 μsec tend to excite transverse modes of the transducer; these send sound pulses through the solid parts of the apparatus. Suitable transformers are commercially available.

The free running rate of the blocking oscillator is set to the proper value (somewhat below the lowest operating rate) by adjustment of grid-circuit time-constant components R1 and C1 and of the cathode bias potentiometer R2.

C. The Amplifier

An amplifier is required to build up the rise rate of the received pulse. The first two stages (*V*1 and *V*3) shown in Fig. 5 act to do this in the same way that an ordinary RC-coupled amplifier is used to produce a square wave from a sine wave. However the low-frequency response is made purposely poor, mostly through the use of small coupling condensers, to shorten the pulse duration and suppress possible oscillation. The diode *V*2 clips undershoots and removes, by rectification, any oscillations produced by ringing of the crystals.

The stage *V*4 has a very fast response to fast negative-going signals but almost no response to low-frequency signals or to positive-going signals. For the component values shown in Fig. 5 the grid is at about $+0.8$ v and the plate is bottomed in the quiescent state. The input impedance at the grid is only about 3.5K on account of the considerable grid current; the input time constant determined by this and the coupling condenser *C*2 (100 $\mu\mu$f) is thus about 0.35 μsec. As a result low-frequency signals are greatly attenuated and have little effect on the output. As for fast signals, those which are positive going are blocked because the plate is bottomed; negative-going signals on the other hand are greatly amplified, as the transconductance of the tube in the grid conduction region is very high. Further, as the signal becomes more negative the grid current decreases and the input resistance therefore increases

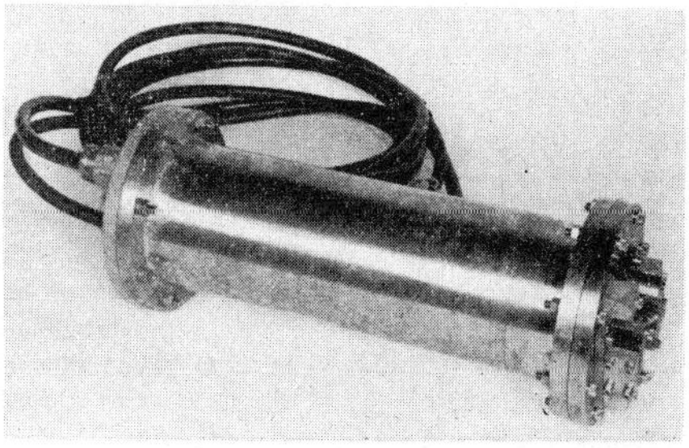

FIG. 4. An underwater velocimeter built by N.O.L.

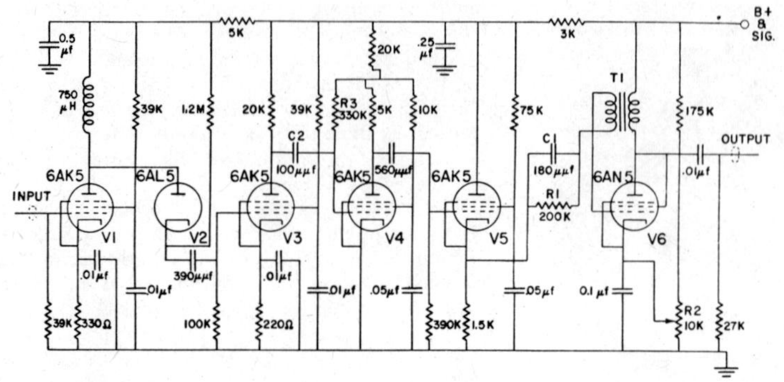

FIG. 5. Circuit suitable for use with ceramic transducers. The component values shown are approximate; the optimum values for a particular application are gotten by trial.

until finally, as the grid potential reaches cutoff the input resistance is about equal to that of *R*3 (330K). The time constant is now ten times as great and almost all of the signal present at the plate of the preceding stage is available at the grid of *V*4. When the grid potential reaches about −2 v, the tube is cut off and the output pulse reaches its saturation value. This positive pulse synchronizes the blocking oscillator *V*6 through the cathode follower *V*5.

The circuit of Fig. 5 is suitable for use with ceramic transducers. Quartz transducers require two additional stages like *V*3, together with two suitably oriented diodes like *V*2, all ahead of the stage *V*4. In either case *V*4 responds to the first negative pulse which the receiving transducer delivers to stage *V*1. As the input pulse from the blocking oscillator to the sending transducer is also negative, the crystal faces in contact with the liquid should ideally be of same polarity. However, ringing of the transducers causes the output of the receiver to build up slowly; the second half-cycle is some ten times bigger than the first. The instrument operates much more reliably if the circuit recognizes the second half-cycle and therefore it is our practice to reverse the polarity of one transducer. In this arrangement the ceramic transducers should be of a composition which has a low-temperature coefficient of frequency, otherwise the additional time delay which amounts to one-half period of the transducer, will vary too much with temperature.

D. Frequency Measurement

The prf of the system may be measured by any of the standard techniques depending on the particular requirements and the available equipment. In the laboratory an electronic counter is very convenient although a stable radio receiver tuned to a high-order harmonic (say 200) of the prf gives equally good results.

In many field applications automatic operation and/or recording is required. In such cases it is convenient to convert to an audio-frequency which may be fed to an ordinary frequency meter having a dc output suitable for operation of a recorder. The local oscillator is preferably crystal controlled.

IV. PERFORMANCE

A. Stability

1. *Supply-Voltage Fluctuations and Tube Changes*

The degree of stability to be expected against changes of supply voltages is here illustrated by citing test results for 5 instruments of the type shown in Fig. 4. A change of ±10% in the *B*+ supply changed the output frequency of each instrument by less than 1 part in 15 000. A 1-v (12%) change in the bias had a somewhat larger effect, about 1 part in 13 000. Replacement of any tube except the 6AN5 in the blocking oscillator (*V*6 in Fig. 5) produced a frequency change of less than 1 part in 10 000. Replacement of *V*6 sometimes produced changes of 1 in 5000, but spares could easily be preselected to hold the change to less than one part in 10 000.

2. *Frequency Stability*

The term "frequency stability" here denotes the degree to which the prf of the velocimeter remains constant if the speed of sound in the liquid is constant. Stability tests were made on the five instruments just discussed. The liquid was water held at 30°C by a conventional mercury type thermoregulator and electric heater arrangement. The velocimeter was then operated over-night while the prf was measured on the 180th harmonic using a communications receiver and recording arrangement. The results are shown in Table I.

TABLE I. Stability of five field velocimeters.

Serial No.	Duration of test hr	Peak-to-peak variation ppm
1	16	29
2	18	40
3	24	59
4	24	33
5	24	33

The calibration oscillator of the receiver, which was used as a local oscillator, is subject to drifts of the order of 10 ppm in 24 hr, and a drift of 0.01°C in the water temperature changes the speed of sound in water at 30°C by about 17 ppm. The stability test just described is therefore marginal even for field instruments.

The speed of sound in water is a maximum at about 74°C.[1] At this temperature a variation of 1°C changes the speed of sound by only 13 ppm, and a variation of 0.1°C by only 0.1 ppm. A stability test on a laboratory model velocimeter was made in water at 74±0.1°C; the conditions of test were otherwise the same as before. However, the prf (about 15 kc) was measured by means of an electronic counter. The duration of each measurement was 10 sec (about 150 000 counts); the measurements were repeated about 3 times per minute, and the results recorded in sequence on the tape of an automatic printer. The test was run for 3 hr one day and for 7.5 hr the next day. In all, 1864 measurements were recorded. Table II gives the distribution of counts obtained.

From Table II it is seen that the peak-to-peak variation was 4 counts or 26 ppm and the standard deviation was about 0.7 count or less than 5 ppm. It should be borne in mind that these figures include the inherent ±1 count error of the counter and the errors due to accidental counts.

B. Calibration

As an example of the procedure, we describe a set of calibrations made on a laboratory instrument fitted with ceramic transducers and a perforated metal reflector. The standard liquid was outgassed distilled water. The water, rapidly stirred, was held briefly at each of several temperatures between 0 and 80°C while the prf, f, of the instrument was measured by counting for 10 sec. At the same time the temperature of the water was measured to the nearest 0.01°C with a platinum resistance thermomenter.

TABLE II. Stability of a laboratory velocimeter.

Count[a]	Number of counts
153 384	16
153 385	257
153 386	969
153 387	587
153 388	35

[a] Average count—153 386.197. Standard deviation—0.723 count.

TABLE III. Calibration of a laboratory velocimeter.

Run No.	1	2	3	Combin
No. of points	18	21	18	57
Temperature extremes, °C	1.0–80.1	0.8–74.1	1.8–83.2	0.8–83.2
Effective length, l				
Value, cm	14.6948	14.6959	14.6923	14.6941
Std. dev., cm	0.0013	0.0009	0.0020	0.0010
Time delay, t_e				
Value, μsec	0.7158	0.7059	0.7301	0.7188
Std. dev., μsec	0.0005	0.0003	0.0008	0.0004
Std. dev. of $1/f$ data				
μsec	0.0012	0.0010	0.0021	0.0018
ppm	12	10	21	18

Accurate values of c are known from recent work[10] in this laboratory. The corresponding values of c and f were fitted by least squares to Eq. (1). Table III shows the values of l and t_e obtained in three different runs made several days apart, and also the values (denoted "combined") obtained from the combined data.

It is easily calculated that, for any value of f within the range, the four values of c predicted by the four linear equations described in Table III will agree within 35 ppm at worst. It should be borne in mind, in this connection, that the uncertainty in f due to the inherent counter error is ±10 ppm and that, at the lowest temperatures, an uncertainty of 0.01°C corresponds to an error in f of 30 ppm.

Although, taking the worst case, the effective length l for run 2 is 0.0036 cm greater than for run 3, the time delay t_e is 0.0242 μsec shorter and the discrepancies compensate. The compensation is effective over the entire range of c or f because the range is only about 10% of the mid-range value; this short range, conversely, makes the least-square values of l and t_e very sensitive to small variations in the data.

V. ACKNOWLEDGMENTS

The authors are indebted to the staff of the Chesapeake Bay Institute of Johns Hopkins University who made their facilities and personnel available for field tests, and to the U. S. Naval Underwater Sound Laboratory, New London, Connecticut, which carried out extensive field tests in the laboratory and in the ocean. The services of our section shop support group headed by Henry A. Schmidt, Jr., were invaluable.

[10] M. Greenspan and C. E. Tschiegg, J. Research Natl. Bur. Standards 59, 249 (1957).

The following paper is of great value to anyone concerned with the use of velocimeters in the ocean. Procedures for the use and calibration of velocimeters to achieve maximum possible precision and methods of correcting for the effects of pressure on the measurements of sound velocity are given.

Mr. Mackenzie has conducted undersea warfare research for 26 years and has published numerous scientific papers on propagation, reverberation, sound speed, and instrumentation. He is a supervisory physicist with the Acoustic Propagation Division, Ocean Sciences Department, at the Naval Undersea Research and Development Center (formerly Navy Electronics Laboratory), San Diego, California.

In 1963, he made the first seven of ten dives aboard the nation's only bathyscaph, *Trieste*, locating and photographing debris of the submarine *Thresher* at a depth of 8400 feet.

He was awarded the Certificate of Merit from the Office of Scientific Research, the Certificate for Exceptional Service from the Bureau of Ordnance, and the Navy Unit Commendation to the *Trieste* group in 1963.

This paper is reprinted with the permission of the author and the Acoustical Society of America.

Reprinted from: The Journal of the Acoustical Society of America

Received 24 March 1971

12

A Decade of Experience with Velocimeters

Kenneth V. Mackenzie

Naval Undersea Research and Development Center, San Diego, California 92132

Over a decade of experience with velocimeters aboard deep submergence vehicles has demonstrated that accuracies of 0.1 m/sec or better can be obtained with proper procedures. Calibrations and distilled water equations are discussed, along with the new International Temperature Scale adopted in 1968. Calibration drift is described for three types of velocimeters to underscore that accuracy requires frequent calibrations as well as intercomparisons between several instruments during deep dives. Pressure corrections are computed. Errors due to thermal lags of earlier velocimeters were observed while trans. :itting thermoclines. Laboratory measurements with a velocimeter in 50 gal of natural seawater of 27.405‰ salinity gave good results, closer to those of V. A. Del Grosso [J. Acoust. Soc. Amer. 47, 947 (1970)], than to W. D. Wilson [J. Acoust. Soc. Amer. 32, 1357(L) (1960)].

INTRODUCTION

The purpose of this paper is to relate more than 10 years experience with the performance of velocimeters aboard deep submersibles, during which time considerable sound speed data have been amassed to derive the best equations for living seawater. All equipment manifested calibration drifts, thermal lags, and often pressure effects. Supporting pressure and temperature experiences have been previously published.[1,2] Attributes and limitations of velocimeters are reviewed because thorough analyses are necessary to complete the building blocks for a future study in which accumulated measurements and seawater sound speed equations will be discussed.

In-situ sound speed studies have been conducted aboard deep vehicles, including the bathyscaph *TRIESTE I* near San Diego in 1959 and off Guam to 5700 m in 1960[3,4]; the French bathyscaph *ARCHIMEDE* in Japanese waters to 7000 m in 1962[5] and 1967[6,7]; and *DEEPSTAR-4000* on many series of dives off San Diego and Central America from 1966 through 1968.[8]

Objectives of the experiments have been targeted toward acquiring precision deep-sea measurements and verifying laboratory samples for sound speed, temperature, salinity, and especially *in-situ* depth dependence. Accuracy requires (1) meticulous calibrations before, between, and after dives; (2) delicate handling of all equipment; (3) intercomparison of several instruments; (4) consistently careful measurements; and (5) correct assessment of thermal lags and pressure effects.

Manned submersibles offer the capability of slow, constant descent rates seldom achieved by surface ship lowerings of sensors. *DEEPSTAR-4000*, with propulsion secured, descended vertically as slowly as possible in a horizontal position, which permitted virgin water flow normal to the instruments.

Figure 1 illustrates instruments utilized aboard *DEEPSTAR*, including a salinometer, Fjarlie bottles with reversing thermometers, two temperature sensors, and three NUS (formerly ACF) velocimeters mounted on a special outrigger brow designed to minimize hydrodynamic effects of the vehicle. Center-to-center separation of the TR-4 velocimeters was 4.06 m, and 4.17 m between the Dymec quartz temperature sensors. Each pair of these instruments was mounted in close proximity, to eliminate effects of horizontal and vertical gradients.

Some early velocimeters exhibited a frequency jump equivalent to a change in path length as the pressure increased.[4] However, improvements have been incorporated to overcome this problem during the past several years and newer devices are now on the market.

The TR types are herein emphasized because no other velocimeters in the past offered acceptable performance. An accuracy of 0.1 m/sec is possible when calibration drift is considered and two or more instruments are intercompared. When calibration drift is negligible, accuracies of 0.05 m/sec can be realized. Although many NUS TR-2's and TR-3's are still at sea, TR-4 and TR-5 types are the most numerous. With proper care, even good laboratory measurements can be accomplished with some of these instruments; with special efforts, performance can surpass manufacturers' specifications.

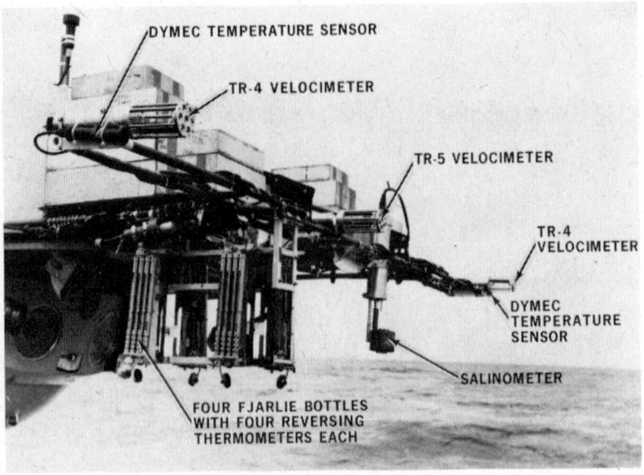

FIG. 1. Equipment mounted on brow of *DEEPSTAR-4000*.

I. GENERAL DESCRIPTION

Velocimeters, precision instruments for directly measuring speed of sound in water, evolved from the first successful sing-around prototype developed in 1957 by Greenspan and Tschiegg[9] of the National Bureau of Standards (NBS) and used by the author for early *TRIESTE* dives.[3]

Two electroacoustic transducers and a reflector are mounted to produce a sound path of fixed length in water, which along with necessary amplifiers and a blocking oscillator form the essential operating elements of a sing-around circuit. All-silicon semiconductor components are now employed to permit stable operation over a wide temperature range.[10] The electronic package is contained within a stainless-steel pressure-proof housing that can withstand maximum ocean depths.

A 3.6-MHz pulse of acoustic energy is transmitted, reflected, received, and amplified to generate another pulse. Figure 2 depicts sound paths for the NUS TR-3, TR-4, and TR-5. The interval between pulses is the reciprocal of the sing-around frequency (f in hertz) and is the sum of travel time in water plus an effective electronic time delay (B in seconds).

For sound speeds (c in meters/sec),

$$\frac{1}{f}=\frac{A(1+\alpha T+\beta T^2)}{c}+B, \tag{1}$$

where A is the effective path length in meters at 0°C, α and β are the thermal expansion coefficients, and T is temperature in degrees centigrade.

Generally, β is taken as zero and Eq. 1 yields

$$c=\frac{A(1+\alpha T)}{(1/f)-B}. \tag{2}$$

The sound path of the TR-3 is similar to the original NBS prototype, and thermal expansion α of the 316 stainless-steel end plate is 1.46×10^{-5}/°C. Values of A and B are ascertained by calibration in distilled water.

Velocimeters were enhanced by a new design which included three invar rods in the standoff posts for the reflector and almost eliminated temperature effects in the TR-4 and TR-5. Each 0.80-cm-diameter invar rod measures 11.191 cm in length and is joined at both ends to short mounting rods of stainless steel 0.912 cm long. A neoprene jacket, approximately 1.10 cm o.d. and 12 cm long, overlaps the stainless steel at both ends to seal the invar and prevent corrosion. The median length of the transducer mount is 2.753 cm.

If thermal expansion of the invar were 1.21×10^{-6}/°C, temperature compensation would be attained and α would be zero. However, heat treatment or cold work changes α for invar[11] considerably, and values commonly vary from 0.64×10^{-6}/°C to 2.01×10^{-6}/°C.

For $\alpha=0$, Eq. 2 becomes

$$c=\frac{A}{(1/f)-B}, \tag{3}$$

which is in general use. However, if a temperature dependence exists, Eq. 2 must be applied.

The electronic time delay B is about 2×10^{-7} sec, and an approximation of Eq. 2 is

$$c=kf(1+\alpha T). \tag{4}$$

The manufacturer supplied values of A, B, and k determined by calibrations in distilled water at 0°, 12°, 25°, 37°, and 50°C as well as values for α applicable to the TR-2 and TR-3.

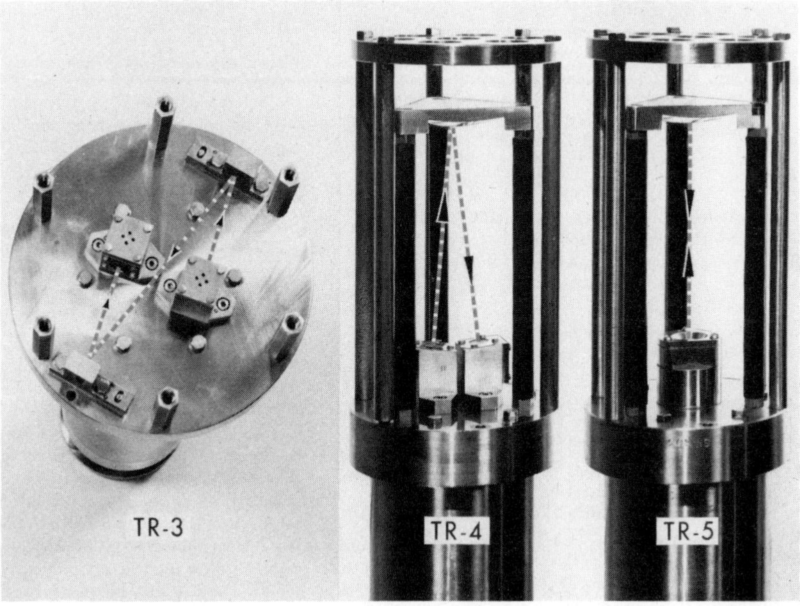

FIG. 2. Sound paths for NUS TR-3, TR-4, and TR-5 velocimeters.

II. PRESSURE CORRECTIONS

Precision deep-ocean measurements with velocimeters require small corrections for path length changes. Pressure causes volume compression of components and bowing of the endplate which serves as a cap for the steel cylinder housing the electronics. Crystal assemblies are designed to be insensitive to pressure.

Ideally, velocimeters should be calibrated in distilled water under pressure to determine the total effective pressure correction, but suitable facilities are not yet available nor are laboratory measurements in good agreement. Therefore, these corrections are computed, with the assumption that crystal assemblies are pressure independent, until experimental methods are further established.

Effects of compression and bowing are expressed as $\Delta A/A$ and Eq. 2 becomes

$$c=\frac{A(1+\Delta A/A)(1+\alpha T)}{(1/f)-B}. \tag{5}$$

Pressure corrections discussed by Willig[12] in an unpublished engineering memo agree with the following volume-compression effects, but assess corrections for bowing differently.

A. Volume Compression

Since each component exposed to pressure compresses, unit change in length L is expressed as

$$\Delta L/L=-[(1-2\sigma)/E]P. \tag{6}$$

Young's modulus E is 1.93×10^{11} in pascals (10^5 dyn/m^2) for 316 stainless steel, and 1.41×10^{11} Pa for invar; Poisson's ratio σ is 0.3 for both; P is pressure in pascals.

Equation 6 applies directly to the TR-3 and for compression

$$\Delta L/L=-2.07\times10^{-12}P.$$

For TR-4 and TR-5 velocimeters, Eq. 6 must be applied separately to the invar, stainless-steel rod extensions, and transducer mounts. The invar and rod extensions compress to decrease the path length which is increased by compression of the transducer mountings.

Due to the folded acoustic path, the effective change ΔL is twice the algebraic sum of these compressions. For an effective path length A of 20.6 cm for the TR-4 and TR-5

$$\Delta L/L=-2.90\times10^{-12}P.$$

B. Bowing

A sufficient approximation for path length changes caused by bowing of the mounting plate under pressure can be achieved by applying an equation[13] for a thick circular plate simply supported about a radius and uniformly loaded. Maximum deflection w at the center is

$$w=\frac{3a^4}{16Eh^3}\left[(5+\sigma)(1-\sigma)+0.4(8+\sigma+\sigma^2)\left(\frac{h}{a}\right)^2\right]P, \quad (7)$$

where a is the radius of the support and h is the plate thickness. For velocimeters included in this paper, a, the radius of the supporting outside edge of the **O**-ring groove, is 3.332 cm, and h is 1.588 cm for the TR-3 and 2.540 cm for the TR-4 and TR-5.

Sound paths of the TR-2 and TR-3 are shortened as the reflectors move when the plate bows. To a first approximation,

$$\Delta L/L=-\mu(2d+h)ew/a^2A, \quad (8)$$

where d is the distance between the axis of the sound beam and the surface of the plate, e is the distance between reflectors, and μ is a factor to allow for stiffening caused by bolting and the protective endplate (removed in Fig. 2). For the TR-3, $d=1.60$ cm, $e=11.33$ cm, and Eq. 8 gives $\Delta L/L=-31.72\times10^{-12}\mu P$. The factor μ was determined empirically to be 0.477 by intercomparing the TR-3 with the TR-4 near the surface and at the bottom during *DEEPSTAR* dives to 1200 m, and $\Delta L/L$ is then $-15.13\times10^{-12}P$.

Since the TR-4 path length is increased by bowing, for source and receiver both 1.09 cm off-center, the effect is

$$\Delta L/L=2(a-1.09)w/aA=+2.70\times10^{-12}P. \quad (9)$$

For the TR-5, increase in path length is essentially $+2w$ and

$$\Delta L/L=2w/A=+4.02\times10^{-12}P. \quad (10)$$

Volume compression and bowing are additive for the TR-3 to give

$$\Delta A/A=(-2.07\times10^{-12}-15.13\times10^{-12})P$$
$$=-17.20\times10^{-12}P.$$

For the TR-4, where the two effects have opposite signs,

$$\Delta A/A=(-2.90\times10^{-12}+2.70\times10^{-12})P$$
$$=-0.20\times10^{-12}P,$$

and for the TR-5

$$\Delta A/A=(-2.90\times10^{-12}+4.02\times10^{-12})P$$
$$=+1.12\times10^{-12}P.$$

When Eq. 2 is used rather than Eq. 5, a higher indicated sound speed is obtained when pressure shortens the path length.

Corrections for sound speed in meters/sec per kilometer of depth are approximately

$$\text{TR-3}=-2.54\times10^{-1},$$
$$\text{TR-4}=-2.95\times10^{-3},$$
$$\text{TR-5}=+1.65\times10^{-2},$$

which is another reason, besides temperature dependence, why the author stopped using the TR-3 when TR-4's became available in the fall of 1966.

III. TEMPERATURE SCALES

Sound speed measurements are reported for temperatures to the closest 0.001°C by a number of researchers. Published data to the present time have been based on a now-obsolete scale which agrees with the triple point of distilled water at 0.01°C and at 100°C, but is 0.010 degree greater than the new scale at 35°C. Old data should be converted to the new scale if accuracies of 0.001°C are required for future intercomparisons.

The International Practical Temperature Scale of 1968 (IPTS-68) was adopted by the Comité International des Poids et Mesures at the October 1968 meeting.[14] The new scale replaced IPTS-48, which had been promulgated by NBS as the official measure. NBS calibrations for platinum resistance thermometers, liquid-in-glass thermometers, and thermocouples have generally been based on the new scale since 1969.

Conversion of T_{48} to T_{68} is discussed by Douglas,[15] whose Eq. 80 for 0° to 630.74°C, combined with his Eq. 77, yields

$$T_{68}-T_{48}=\frac{4.904\times10^{-7}\times T_{68}(T_{68}-100)}{1-2.939\times10^{-4}T_{68}}$$
$$+0.045\left(\frac{T_{68}}{100}\right)\left(\frac{T_{68}}{100}-1\right)\left(\frac{T_{68}}{419.58}-1\right)$$
$$\times\left(\frac{T_{68}}{630.74}-1\right), \quad (11)$$

where 419.58° and 630.74°C are the freezing points of zinc and antimony, respectively.

Equation 11 is cumbersome, but can be approximated for the range 0° to 100°C by

$$T_{68}=(1-4.9454\times10^{-4})T_{48}+6.5416\times10^{-6}T_{48}^2$$
$$-1.5962\times10^{-8}T_{48}^3, \quad (12)$$

with differences within $\pm3\times10^{-5}$°C, and over the range for seawater 0° to 35°C by

$$T_{68}=(1-4.88\times10^{-4})T_{48}+5.80\times10^{-6}T_{48}^2, \quad (13)$$

with differences within $\pm4\times10^{-5}$°C.

IV. DISTILLED WATER SOUND SPEEDS

Velocimeters, when accurately calibrated and properly handled, are capable of accuracies of 0.05 m/sec or

TABLE I. T_{48} sound speed differences in meters/sec from least-squares fit to Del Grosso (March 1970).

	Temperature (°C)							
	0	10	20	30	40	50	60	70
Greenspan and Tschiegg	0.35	0.34	0.34	0.34	0.33	0.33	0.33	0.33
Wilson's equation $P=14.7$ psi	0.63	0.60	0.60	0.55	0.45	0.34	0.28	0.26
Least-squares fit to Wilson's data	0.56	0.29	0.33	0.44	0.49	0.45	0.35	0.24

better Since instruments are usually calibrated in distilled water, sound speed as a function of temperature must be known reliably for satisfactory results.

The most recent published tables of measured sound speeds are from Del Grosso,[16] whose Fig. 2 intercompares laboratory data of several scientists on T_{48}. Wilson[17,18] is approximately 0.6 m/sec too high for temperatures below 20°C, and a noticeably different dependence is exhibited for higher temperatures.

Since intercomparisons between equations are more exact than graphical displays, least-squares fits (LSF) were applied to Del Grosso's results as well as to Wilson's data for atmospheric pressure of 14.7 psi.

Differences from Del Grosso[16] are presented in Table I on the common basis of T_{48} for investigators whose equations are used for velocimeter calibrations. Greenspan and Tschiegg[19,20] have practically the same temperature trend, but are about 0.34 m/sec too high. Brooks[21] obtained values lower than Greenspan and Tschiegg; perhaps this can be explained by a diffraction effect. Ilgunas *et al.*[22] studied the problem and found sound speeds 0.4 m/sec lower than Greenspan and Tschiegg. Wilson's equation deviates from a least-squares fit to his own atmospheric pressure data, which appear to be somewhat better than his equation.

Data of Wilson,[18] Del Grosso,[16,23] and Carnvale[24] were corrected to T_{68} by Eq. 12, and least-squares fits were made to polynomials

$$c_c=a_0+a_1T+a_2T^2+a_3T^3+\cdots+a_nT^n.$$

When a least-squares fit is made to a polynomial, the standard error of estimate, s, is

$$s=\left[\frac{(c_c-c_0)^2}{p-q}\right]^{\frac{1}{2}}, \tag{14}$$

where c_0 is distilled water sound speed, p is the number of data points, and q is the number of independent coefficients equal to $n+1$.

The following equations resulted.

(1) Wilson (1960),[18] for 10 values 0.906° to 91.266°C (excluding the questionable point at 10.203°C):

$$c=1402.9676+4.979711T-5.439236\times10^{-2}T^2+2.464158\times10^{-4}T^3-6.041346\times10^{-7}T^4, \tag{15}$$

with $s=0.064$ m/sec. A fifth-degree polynomial showed no improvement.

(2) Del Grosso (1969),[23] for 19 values 0.100° to 89.997°C, with $s=0.021$ m/sec:

$$c=1402.3683+5.050052T-5.915762\times10^{-2}T^2+3.634154\times10^{-4}T^3-1.817832\times10^{-6}T^4+4.565699\times10^{-9}T^5. \tag{16}$$

(3) Carnvale (1968),[24] for 9 values 0.688° to 40.061°C:

$$c=1402.3760+5.041375T-5.832370\times10^{-2}T^2+3.317363\times10^{-4}T^3-1.217159\times10^{-6}T^4, \tag{17}$$

with $s=0.0023$ m/sec. A fifth-degree polynomial was not feasible.

Del Grosso (1970),[16] for 36 values 0.056° to 74.022°C, combined with 112 of his unpublished recent measurements from 0.001° to 95.128°C, which he kindly furnished, gives

$$c=1402.3876+5.037111T-5.808522\times10^{-2}T^2+3.341988\times10^{-4}T^3-1.478004\times10^{-6}T^4+3.14643\times10^{-9}T^5, \tag{18}$$

with $s=0.0029$ m/sec. The maximum of Eq. 18 occurs at 74.172°C with a value of 1555.147 m/sec.

For ease of intercomparison, T_{68} sound speed differences from Eq. 18 are presented in chronological order in Table II. Equations 15–18 were computed with the temperatures indicated. Least-squares fits to unconverted data of Brooks[21] and Neubauer and Dragonette[25] were determined with quadratics; to that of McSkimin[26] and Williamson,[27] with quartics. These equations, as well as those of Greenspan and Tschiegg, Wilson, and Barlow and Yazgan[28] were computed for exact T_{48} corresponding to integer T_{68}. Results since 1964 agree except for Williamson at higher temperatures, where emergent stem corrections were as much as 0.2°C for his mercury-in-glass thermometers.

Table II demonstrates that recent data substantiate the advisability of employing Eq. 18 for calibrations and computations when precision is required.

Obviously, accurate calibrations are the cornerstone of good measurements. Early velocimeter calibrations are described in Ref. 3. The majority of the author's recent calibrations have been contracted to Ramsay Engineering Company, Anaheim, California. All instruments are periodically calibrated in distilled water every 1°C as temperature is decreased from 32° to 1°C. The triple point of water is employed as a reference. Temperature measurements are recorded to 0.001°C with a Rosemount Engineering Company platinum resistance thermometer and a Dauphine direct-reading temperature bridge manufactured by Guildline Company.

TABLE II. T_{68} sound speed differences in meters/sec from Eq. 18 in chronological order.

	Temperature (°C)												
	0	10	20	25	30	40	50	60	70	75	80	90	95
Greenspan and Tschiegg equation	0.35	0.34	0.34	0.34	0.33	0.33	0.33	0.33	0.33	0.33	0.34	0.35	0.35
Least-squares fit to Brooks	...	...	...	−0.13	...	...	...	...	...	...	...	...	...
Wilson's equation 14.7 psi	0.63	0.60	0.60	0.58	0.55	0.45	0.34	0.28	0.26	0.26	0.24	0.06	−0.18
Least-squares fit to Wilson data: Eq. 15	0.58	0.30	0.34	0.39	0.44	0.49	0.45	0.35	0.24	0.20	0.16	0.09	...
Least-squares fit to Neubauer and Dragonette	...	...	−0.03	...	...	...	...	...	...	...	...	...	...
Least-squares fit to McSkimin	...	...	...	−0.05	0.00	0.02	−0.01	−0.03	−0.04	−0.05	...	...	...
Barlow and Yazgan equation	...	...	...	−0.09	−0.06	−0.08	−0.09	−0.09	−0.08	−0.08	−0.09	...	...
Least-squares fit to Carnvale: Eq. 17	−0.01	0.01	−0.01	−0.02	−0.03	−0.03	...	...	...	...	...	...	...
Least-squares fit to Del Grosso, June 1969: Eq. 16	−0.02	0.03	−0.01	−0.03	−0.05	−0.07	−0.08	−0.09	−0.12	−0.14	−0.16	−0.16	...
Least-squares fit to Williamson	...	...	...	−0.03	−0.01	0.03	0.06	0.10	0.14	0.14	...	...	...

The Greenspan and Tschiegg unmodified equation for distilled water was employed by the author until June 1966, and since then his published results[8] have been reduced by 0.33 m/sec, in concurrence with Del Grosso's[29] laboratory measurements.

If T_{68} had been used directly with Greenspan and Tschiegg, differences (Table II) at 0°, 10°, 20°, 30°, 40°, and 50°C would have been 0.35, 0.33, 0.32, 0.31, 0.31, and 0.32 m/sec, with an average difference of 0.32 m/sec. Consequently, adequate calibrations for use at sea can be obtained with the data of Greenspan and Tschiegg minus 0.34 m/sec for T_{48} and −0.32 m/sec for T_{68}, since their temperature trend agrees sufficiently with Eq. 18.

Some investigators use Greenspan and Tschiegg unmodified; others[30,31] advocate subtracting 0.35 to 0.40 m/sec. Many, including the National Oceanographic Instrumentation Center, continue to use Wilson's equation. However, NOIC provides temperature and frequency data for the user to enable independent computations. This fundamental information, in addition to computed coefficients, should be supplied with every calibration.

Cross checks were carried out to ascertain that no temperature measurement differences existed. Ramsay temperatures were checked against the Naval Undersea Research and Development Center (NUC) calibration tank by a quartz thermometer with excellent agreement—only one point differed by as much as 0.001°C. A Dymec temperature sensor was calibrated by Ramsay and recalibrated by NOIC with very good accordance.

Although calibrations of the author's velocimeters to date have been based on T_{48}, all 32-point calibrations recently have been converted to T_{68} with Eq. 12. Fits were applied to Eq. 1 assuming $\beta=0$ and α was as specified by the manufacturer. Double precision is required for dependable least-squares results. Values of A and B are substituted into Eq. 2 for computed sound speeds, c_c. Though fits were made to Eq. 1, the standard error of estimate, s, is assumed to be given by Eq. 14 with $q=2$.

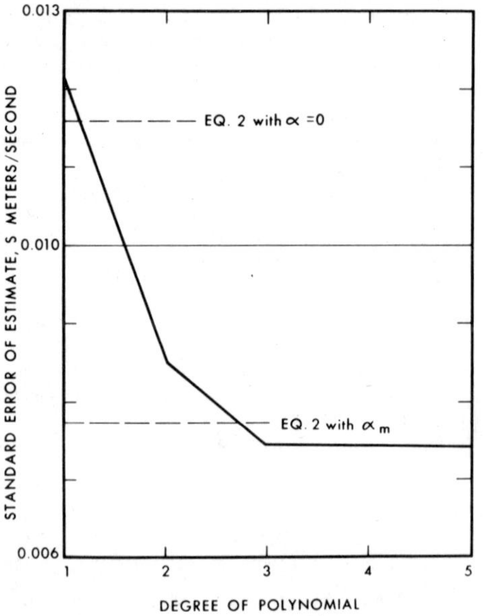

FIG. 3. Standard error of estimate versus degree of polynomial for least-squares fits to Eq. 19 for TR-4 No. 17 calibration of 14 August 1968.

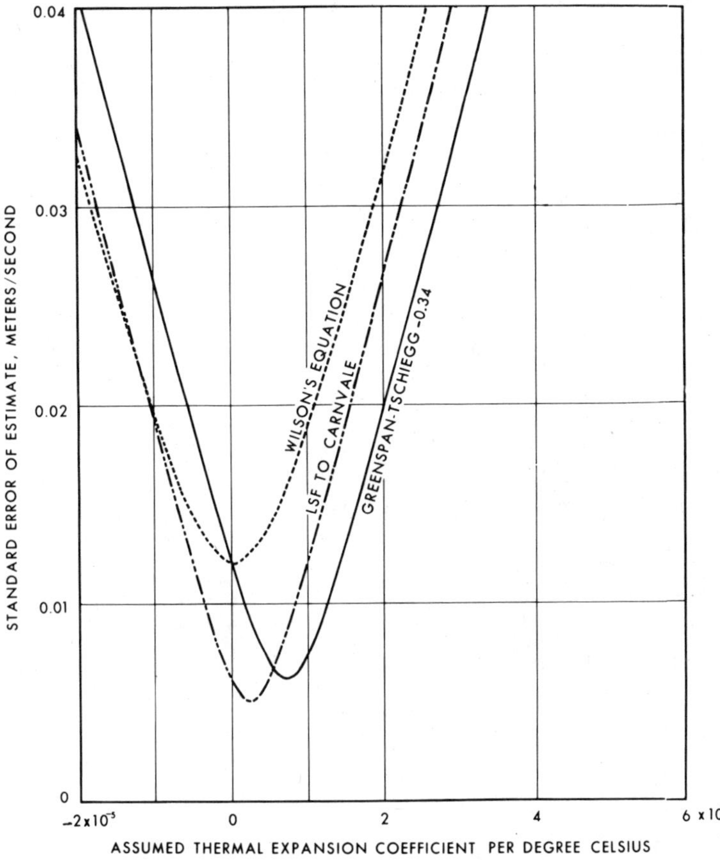

FIG. 4. Assumed thermal expansion coefficient versus standard error of estimate for least-squares fits to Eq. 2 for TR-4 No. 17 calibration of 12 July 1966.

Equation 2, while adequate for general use, did not fit as well as a third-degree polynomial. Fits were made to a general polynomial

$$c_c = a_0 + a_1 f + a_2 f^2 + \cdots + a_n f^n. \qquad (19)$$

Results of a typical 32-point calibration are displayed in Fig. 3. Equation 4, employed by a number of investigators, gave a poor fit, with a standard error of estimate of 0.047 m/sec.

The TR-4 was designed for $\alpha=0$, which yields a standard error of about the same magnitude as for a first-degree polynomial. Equation 2 was programmed for automatic at-sea data reduction for the TR-3 and a fourth-degree polynomial was exploited to reduce TR-4 and TR-5 data. Equation 2 must be used for instruments with $\alpha \neq 0$.

V. VELOCIMETER EQUATIONS

The reason why the proper Eq. 2 failed to fit any of the calibrations as well as the polynomial Eq. 19 was examined. Because the thermal expansion coefficient might not be specified correctly—or some temperature effect in the crystal assembly could exist—fits were run to determine the standard error of estimate s as a function of α. Figure 4, based on T_{48}, shows typical s-vs-α curves for several distilled-water sound speed relationships. Minimum values α_m with Eq. 2 give standard error of estimates as low as with a cubic polynomial. The fairly reasonable α_m was about $7\times10^{-6}/°\text{C}$ larger than the nominal value. However, the electronic time delay B was negative in all cases for α_m and thus failed to represent a physically real condition, although mathematically a good fit resulted.

The next try, consisting of two calibrations each for TR-3 No. 1 and TR-4 No. 17, let $B=b_0+b_1 f$, since a

TABLE III. Intercomparison of TR-4 No. 17 least-squares-fit sound speeds relative to a fourth-degree polynomial with Eq. 18 in meters/sec.

	Temperature (°C)	Eq. 4 $c=kf$	Eq. 19 $n=1$	Eq. 19 $n=4$	Eq. 2 $\alpha=0$	Eq. 2 α_m
Del Grosso, Eq. 18	1	0.74	−0.02	Ref.	−0.02	0.01
	10	0.74	0.01	Ref.	0.01	0.00
	20	0.71	0.01	Ref.	0.01	0.00
	30	0.60	−0.02	Ref.	−0.02	0.02
Carnvale, Eq. 17	1	0.74	−0.01	−0.01	−0.01	0.00
	10	0.74	0.01	0.01	0.01	0.01
	20	0.70	0.00	−0.01	0.00	0.00
	30	0.65	−0.04	−0.03	0.04	−0.03
Greenspan and Tschiegg −0.32 m/sec	1	0.75	0.00	0.03	0.00	0.03
	10	0.75	0.02	0.01	0.02	0.01
	20	0.71	0.01	0.00	0.01	0.00
	30	0.66	−0.02	0.00	−0.02	0.00
Wilson's Eq. 1	1	1.29	0.59	0.62	0.59	0.60
	10	1.31	0.60	0.58	0.60	0.59
	20	1.29	0.58	0.58	0.58	0.57
	30	1.25	0.53	0.53	0.53	0.54

frequency dependence was logical. Again, A, b_0, b_1, and α_m were determined. Even with nominal α, B was minus in three out of four calibrations. The minima were very broad and α_m varied from -27 to $+15\times10^{-5}/°C$. B was negative once and too large the other times; consequently, these fits for B were ruled out.

Actually, thermal expansion is nonlinear, and β is small but not zero as has been assumed for velocimeters. Equation 1 is awkward for least-squares fits for A, B, α, and β, but because β is a second-order quantity

$$\frac{1}{f}=\frac{A(1+\alpha T)}{c}+\frac{\delta T^2}{c}+B \tag{20}$$

was employed where A, B, and δ were determined for given values of α. The value β was then computed from

$$\beta=\delta/A.$$

Minima were even broader, and the nominal values of $\alpha_n=1.46\times10^{-5}/°C$ for TR-2 and TR-3 and zero for TR-4 and TR-5 gave good fits with B positive, and plausible values for β of $+9\times10^{-8}$ per degree squared for the TR-3 and $+6\times10^{-8}$ for the TR-4. Least-squares fits with Eq. 20 for α_m were not notably better and had unreasonable values for α_m, with B again negative for three out of four calibrations; values for β were also improbable. Nothing further could be extracted from the data.

The conclusion is that Eq. 2 should be modified to include a βT^2 term with nominal values of α and β for the materials employed. Consideration of β as well as α could lead toward an optimum design of velocimeters to achieve true temperature independence.

The practical decision of which velocimeter equation and which distilled-water sound speed versus temperature relationship should be employed was examined with two 32-point T_{68} calibrations for the TR-4 No. 17. Individual least-squares fits were applied to Eqs. 2, 4, and 19 for various distilled-water sound speeds.

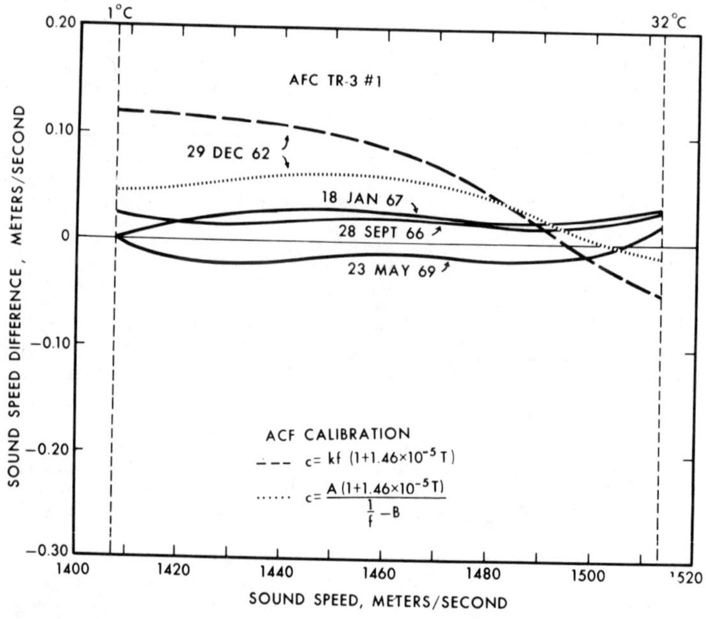

FIG. 5. Calibration drift of TR-3 No. 1 velocimeter relative to 14 July 1966.

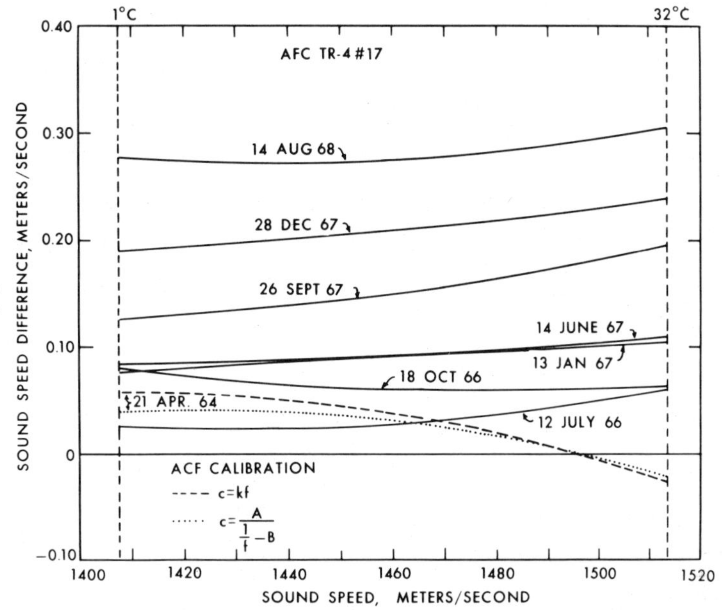

FIG. 6. Calibration drift of TR-4 No. 17 velocimeter relative to 21 April 1966.

The fourth-degree polynomial with Eq. 18 was selected for comparison. Average differences between regression values in Table III are tabulated at 1°, 10°, 20°, and 30° because temperature trends exist and differences are not the same as in Table II. Previous fits based on T_{48} gave almost the same conclusions. Even for operations when accuracies limited to 0.1 m/sec will suffice, $C=kf$ is inadequate and a first-degree polynomial is barely acceptable. The fourth-degree polynomial gives essentially the same results as Eq. 2 with α nominal or α_m. Table III confirms that Eqs. 17, 18, or the Greenspan–Tshiegg data minus 0.32 agree; but Wilson's Eq. 1 gives substantial errors.

VI. CALIBRATION DRIFT

Calibration drift has been observed from earliest times for all instruments operated at sea. NBS prototypes in 1959 exhibited path length changes caused by jarring of bolted reflectors.[4] Although for the NUS TR-2 and TR-3 both bolted and pinned reflectors were specified, as well as a protecting post to minimize effects of possible impacts, TR-3 No. 1 showed a drift of calibration with time. Drifts were determined by comparing fifth-degree polynomials for each 32-point calibration with one calibration chosen as the reference. Computations for frequencies corresponded to every 1°C, and the Greenspan–Tschiegg equation was applied.

Intercomparisons of TR-3 No. 1, TR-4 No. 17, and TR-5 No. 603 are plotted in Figs. 5–7. The dashed and dotted curves were computed from calibration constants supplied by the manufacturer, who also used the Greenspan–Tschiegg equation.

Three options appeared to result from fits to Eq. 19: (1) the effective path length A changed with time and the effective time delay B remained fairly constant; (2) A remained reasonably fixed and B drifted as electronic components aged; or (3) A and B both shifted in a systematic manner. In reality, plots of A and B (computed with Eq. 1, $\beta=0$) versus time revealed a random behavior.

Attention is called to the agreement of the two calibrations of the TR-4 No. 17 on 13 January 1967 and 14 June 1967. No dives transpired during this period; however, the instrument had been airshipped round trip to Denmark between those dates.

Other drifts might have been caused by pressure cycling once each dive to 1200 m, or by shake-table effects on instruments aboard *DEEPSTAR* when she was cradled on the fantail of the support ship.

Instruments were mounted prior to departure and velocimeters, temperature and pressure sensors, as well as salinometers remained in place until the conclusion of a series of two or three dives. Each velocimeter was deactivated between dives by removing a jumper wire connected to the internal battery, which had a life of about 100 operating hours; new batteries were used for each series of dives.

Severe vibration occurred while transiting in and out of port to dive areas and between sites. Reversing

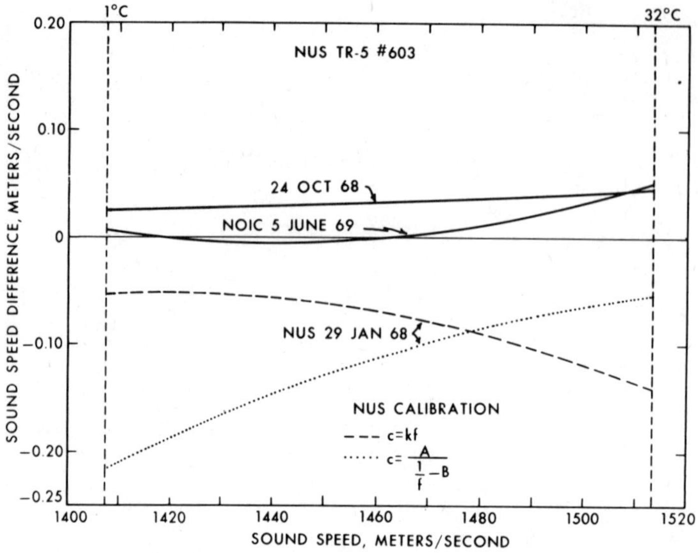

FIG. 7. Calibration drift of TR-5 No. 603 velocimeter relative to 12 February 1968.

thermometers had to be removed immediately after a dive before getting underway or vibration would have shaken the mercury down within a few minutes—which actually happened twice before the procedure of removal was instituted.

At least two velocimeters were employed, and sound speeds were computed with before and after dive series calibrations. Intercomparisons unquestionably improved precision and often pinpointed a velocimeter that was experiencing difficulties and yet appeared to be behaving normally. One, for example, agreed within 0.2 m/sec near the surface but wandered off—rather than jumped—to a difference of 2.5 m/sec. When recalibration was attempted, the instrument failed to operate in distilled water but would function in tap water. Investigation revealed that the outer electrode plated on the receiver crystal was absent—no doubt owing to a grounding problem. Seawater probably acted as an outer electrode to allow partial operation during the dives.

Another time, the TR-4 No. 12 was calibrated by NOIC and recalibrated by Ramsay with apparently good results. Frequencies appeared normal during a subsequent dive, but intercomparison of reduced data exposed an error of 0.6 m/sec, which was traced to a leak behind the crystal transducer. Only careful intercomparisons could detect such a discrepancy.

VII. THERMAL LAGS

Equation 2 applies to equilibrium conditions. When a velocimeter is transported through a gradient such as the thermocline, a thermal lag occurs. Descent rates during *DEEPSTAR-4000* dives were executed at minimum rates—usually less than 12 m/min. Even at this reduced speed, significant thermal-lag errors were observed between descending and ascending data for temperature probes[2] and the TR-3 velocimeter.[32] Equation 2 was programmed to compute sound speed measured by the TR-3, whose output frequency together with an indicated temperature were inputted.

No substantial differences between ascending and descending records were detected through strong thermoclines for the TR-4. Although a certain amount of bootstrapping was involved, the dives were apparently slow enough for the TR-4 to read correctly despite metal masses and neoprene insulators protecting the invar. Figure 8 is a sound speed profile with elapsed time of 7 h between descent and ascent in slightly different locations.

Indicated temperature errors due to thermal response time—troublesome while transiting the thermocline—were determined by assuming as correct both the sound speed measured with the TR-4 and the output of the salinometer. No thermal lag was found for the salinometer, and salinity effects are small anyway.

Sound speed profiles were computed with Wilson's October equation[33] for indicated temperatures and salinities and TR-4 outputs were shifted to agree, on the average, with the computed profile near a depth of 300 m, where conditions are more stable. Indicated temperatures were then adjusted as necessary to obtain agreement between the TR-4 and computed profiles at shallow depths. Differences between indicated and

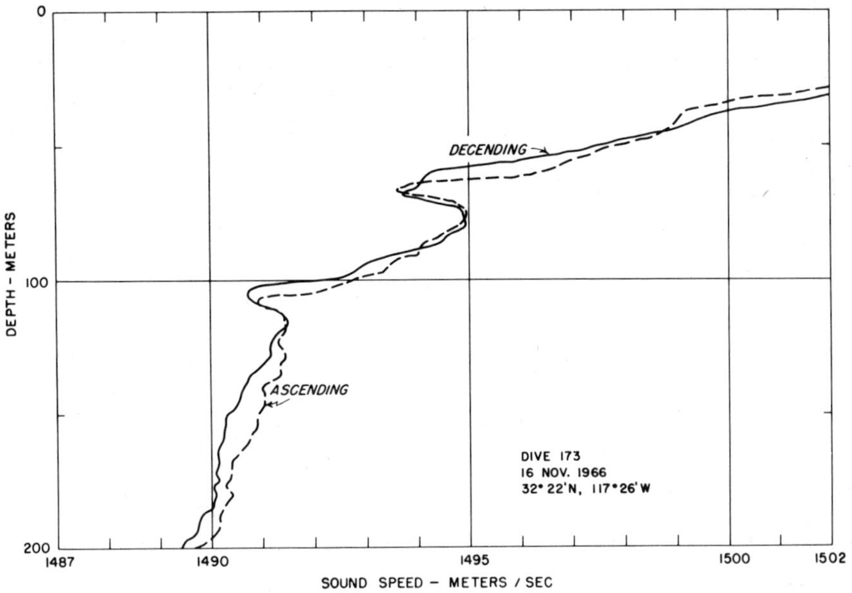

FIG. 8. Sound speed versus depth for *DEEPSTAR-4000* dive 16 November 1966.

adjusted temperatures formed the basis for calculating temperature errors.[2]

Advantageously, since computed pressures change only a fractional amount with a small change in adjusted temperature, nearly correct results were attained with the first recalculated sound speed profile, and accurate temperature corrections were achieved with the second iteration.

An opportunity arose to compare TR-3 and TR-4 velocimeters descending and ascending through the thermocline for two successive short dives, each aborted at 360 m. The thermal-lag error in meters/sec was found to be approximately

$$(\text{TR-3})-(\text{TR-4})=-6.2\Delta T/\text{sec}, \quad (21)$$

where ΔT/sec was the change in temperature as the instrument moved through the thermocline.

Errors due to thermal lag could have been greatly reduced, of course, if the vehicle had moved more slowly—2 m/min for depths less than 200 m where strong thermoclines are found, and 5 m/min for greater depths. Such buoyancy control, with propulsion off, was beyond the capability of *DEEPSTAR-4000* at the time of the dives.

Great quantities of data at sea are gathered from surface vessels at considerably faster lowering rates—commonly 60 or more meters per minute—and thermal-lag errors are seldom checked by the method of recording data, for later comparison, while raising instruments back to the surface. Consequently, precision is limited by undetermined lag errors and smearing of profile details.

VIII. LABORATORY SEAWATER MEASUREMENTS

Laboratory measurements have been conducted with high precision on filtered seawater, generally diluted with distilled water to yield lower salinity samples, or concentrated by evaporation to furnish higher salinities.[16,33]

Searching for the most optimum relationship between sound speed, temperature, and salinity for natural seawater required collecting and measuring large samples while the water was still fresh, to verify laboratory results. A well-calibrated velocimeter offered a convenient means for accomplishing this.

The Danish Defence Research Board Laboratory cooperated with the author for measurements at Copenhagen. Fifty gallons of clean water were transported in large plastic bottles from the Kattegat area and placed in the laboratory calibration tank. Sound speed data were acquired with the TR-4 No. 17 from 4.237° to 25.164°C; temperatures were measured with quartz thermometers. A least-squares fit[32] gave, in meters/sec,

$$c=1438.815+4.69236T_{48}-5.4843\times10^{-2}T_{48}^{2}+2.77\times10^{-4}T_{48}^{3}, \quad (22)$$

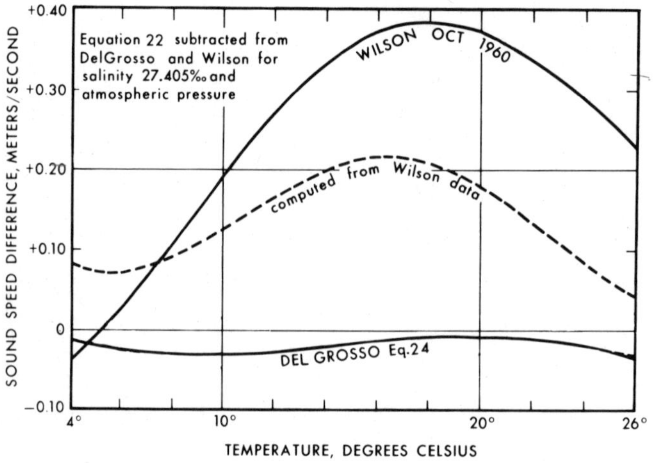

FIG. 9. Sound-speed differences from Eq. 22 versus temperature.

or to update to the new temperature scale,

$$c = 1438.813 + 4.69477T_{68} - 5.4926\times10^{-2}T_{68}^2 + 2.78\times10^{-4}T_{68}^3, \quad (23)$$

with an estimated accuracy of ±0.05 m/sec. Salinity measured 27.405‰.

An equation furnished by Del Grosso[34] covering his most recent measurements for salinities 31‰–39‰ from 0° to 30°C yields

$$c = 1449.0634 + 4.57462T_{48} - 5.27147\times10^{-2}T_{48}^2 + 2.46419\times10^{-4}T_{48}^3 + (S-35)(1.34455 - 1.32888\times10^{-2}T_{48} + 1.0444\times10^{-4}T_{48}^2), \quad (24)$$

with rms deviation of 0.013 m/sec.

For intercomparison on a basis of T_{48}, Eq. 22 was selected as a standard and subtracted to prepare Fig. 9. Differences between Eq. 23 with T_{68} and Eq. 22 are small. Agreement with Del Grosso in Eq. 24 is good, despite some extrapolation, and is certainly well within the experimental accuracy of Eq. 22.

Wilson's October 1960 equation[33] displays a definite temperature trend, with a maximum difference of 0.39 m/sec. Raw Wilson data[18] at atmospheric pressure, processed for an interpolated least-squares fit, lie closer to Eq. 22 and Del Grosso's Equation 24—affirming a similar but less pronounced temperature trend with a maximum difference of 0.21 m/sec—still significantly different from Del Grosso.

Of historical interest, formulas developed by the author[35] to fit Kuwahara's tables differ in trend from Eq. 22 by only +0.025 m/sec per °C. These older formulas were utilized by computers at the U. S. Navy Electronics Laboratory from 1952 until mid-1960.

IX. CONCLUSIONS

TR-4 and TR-5 velocimeters are capable of precision measurements if operated in accordance with proper laboratory procedures. Frequent calibration is essential, and Eq. 18 is recommended as the standard.

Thermal-lag errors must be accurately assessed for all sensors for precision measurements, and rates of lowering and raising controlled to contain errors within acceptable limits.

Future studies should be reported on the new International Temperature Scale T_{68} and earlier work updated from T_{48} as required.

[1] K. V. Mackenzie, "Accurate Depth Determinations During *DEEPSTAR-4000* Dives," J. Ocean Technol. **2**, 61–67 (1968).
[2] K. V. Mackenzie, "Accurate Temperature Determinations During *DEEPSTAR-4000* Dives," Marine Technol. Soc. Trans., Marine Temperature Measurements Symp. (1969), pp. 47–69.
[3] K. V. Mackenzie, "Sound Speed Measurements Utilizing the Bathyscaph *TRIESTE*," J. Acoust. Soc. Amer. **33**, 1113–1119 (1961).
[4] K. V. Mackenzie, "Further Remarks on Sound Speed Measurements Aboard *TRIESTE*," J. Acoust. Soc. Amer. **34**, 1148–1149 (1962).
[5] K. V. Mackenzie, "In-Situ Sound Speed Measurements Aboard the French Bathyscaph *ARCHIMEDE* at Japan," J. Acoust. Soc. Amer. **34**, 1974(A) (1962).
[6] K. V. Mackenzie, "Sound Speed Measurements Aboard French Bathyscaph *ARCHIMEDE* in the Japan Trench," J. Acoust. Soc. Amer. **42**, 1210(A) (1967).
[7] K. V. Mackenzie, "Sound Speed Measurements Aboard Deep Submersibles," Rep. Office Naval Res. Code 468 (Apr. 1966) (unpublished).
[8] K. V. Mackenzie, "Precision In-Situ Sound Speed Measurements," NEL Deep Submergence Log No. 3, 79–99 (1967).
[9] M. Greenspan and C. E. Tschiegg, "Sing-Around Ultrasonic Velocimeter for Liquids," Rev. Sci. Instrum. **28**, 897–901 (1957).
[10] E. M. Zacharias, "Instruments for the Direct Measurement of Sound Speed in Sea Water," Rep. Office Naval Res. Code 468 (Oct. 1966) (unpublished).
[11] M. A. Hunter, "Low-Expansion Alloys," in *Metals Handbook* (Amer. Soc. for Metals, Metals Park, Ohio, 1961), 8th ed., pp. 817–819.

[12] D. E. Willig, "Analytical Expressions for the Effect of Hydrostatic Pressure on the Acoustic Path Length of Sing-Around Velocimeters," ACF Industries Tech. Mem. No. 8279 (July 1963).
[13] A. E. H. Love, *A Treatise on the Mathematical Theory of Elasticity* (Dover, New York, 1944).
[14] "The International Practical Temperature Scale of 1968," Metrologia **5**, 35–52 (1969).
[15] T. B. Douglas, "Conversion of Existing Calorimetrically Determined Thermodynamic Properties to the Basis of the International Practical Temperature Scale of 1968," J. Res. Nat. Bur. Stand. **73A**, 451–470 (1969).
[16] V. A. Del Grosso, "Sound Speed in Pure Water and Sea Water," J. Acoust. Soc. Amer. **47**, 947–949 (1970).
[17] W. D. Wilson, "Speed of Sound in Distilled Water as a Function of Temperature and Pressure," J. Acoust. Soc. Amer. **31**, 1067–1072 (1959).
[18] W. D. Wilson, "Ultrasonic Measurement of the Velocity of Sound in Distilled and Sea Water," U. S. NOL Rep. 6746 (Jan. 1960).
[19] M. Greenspan and C. E. Tschiegg, "Speed of Sound in Water by a Direct Method," J. Res. Nat. Bur. Stand. RP 2795, **59**, 249–254 (1957).
[20] M. Greenspan and C. E. Tschiegg, "Tables of the Speed of Sound in Water," J. Acoust. Soc. Amer. **31**, 75 (1959).
[21] R. Brooks, "Determination of the Velocity of Sound in Distilled Water," J. Acoust. Soc. Amer. **32**, 1422–1425 (1960).
[22] V. Ilgunas, O. Kubilyunene, and A. Yapertas, "A Precision Interferometer for Measuring the Speed of Ultrasound in Liquids in the Range 1–12 Mc," Sov. Phys. Acoust. **10**, 44–48 (1964).
[23] V. A. Del Grosso, "Remarks on Absolute Sound Speed Measurements in Water," J. Acoust. Soc. Amer. **45**, 1287(L) (1969).
[24] A. Carnvale, P. Bowen, M. Basileo, and J. Sprenke, "Absolute Sound Velocity Measurement in Distilled Water," J. Acoust. Soc. Amer. **44**, 1098–1102 (1968).
[25] W. G. Neubauer and L. R. Dragonette, "Experimental Determination of the Freefield Sound Speed in Water," J. Acoust. Soc. Amer. **36**, 1685–1690 (1964).
[26] H. J. McSkimin, "Velocity of Sound in Distilled Water for the Temperature Range 20° to 75°C," J. Acoust. Soc. Amer. **37**, 325–328 (1965).
[27] R. C. Williamson, "Echo Phase-Comparison Technique and Measurement of Sound Velocity in Water," J. Acoust. Soc. Amer. **45**, 1251–1257 (1969).
[28] A. J. Barlow and E. Yazgan, "Phase Change Method for Measurement of Ultrasonic Wave Velocity and a Determination of the Speed of Sound in Water," Brit. J. Appl. Phys. **17**, 807–819 (1966).
[29] V. A. Del Grosso, "Problems in the Absolute Measurement of Sound Speed," Rep. Office Naval Res. Code 468 (Oct. 1966) (unpublished).
[30] J. R. Lovett, "Performance of SVTP Instruments at NOTS," Rep. Office Naval Res. Code 468 (Oct. 1966) (unpublished).
[31] J. R. Lovett, "Comments Concerning the Determination of Absolute Sound Speeds in Distilled and Seawater and Pacific Sofar Speeds," J. Acoust. Soc. Amer. **45**, 1051–1053(L) (1969).
[32] K. V. Mackenzie, "Oceanographic Data Acquisition System for Undersea Vehicles," Trans. Marine Technol. Soc. (June 1970), pp. 1079–1098.
[33] W. D. Wilson, "Equation for the Speed of Sound in Sea Water," J. Acoust. Soc. Amer. **32**, 1357(L) (1960).
[34] V. A. Del Grosso, personal communication (July 1970).
[35] K. V. Mackenzie, "Formulas for the Computation of Sound Speed in Sea Water," J. Acoust. Soc. Amer. **32**, 100–104 (1960).

Ambient Noise in the Sea

III

This paper presents a summary of a more comprehensive report prepared by the authors under the direction of Section 6.1 of the National Defense Research Committee. Although this work was published in 1948, the "Knudsen curves" are still the standard reference for ambient noise level as a function of the parameters, sea state and frequency.

Dr. V. O. Knudsen, the senior author, is a physicist with important contributions in physiological acoustics, hearing of speech in auditoriums, architectural acoustics, audiometry, and noise control. He is a Fellow of the American Physical Society and of the Acoustical Society of America and a past President of the Acoustical Society of America.

This paper is reprinted with the permission of the senior author and the Editors of the *Journal of Marine Research.*

Journal of Marine Research, 1948, V. 3, p. 410–429.

13

UNDERWATER AMBIENT NOISE[1]

By

VERN O. KNUDSEN

University of California, Los Angeles, California

AND

R. S. ALFORD AND J. W. EMLING

Bell Telephone Laboratories, New York

INTRODUCTION

Underwater sound continues to be a problem of high priority in naval warfare. Since atomic bombing may render obsolete most types of surface ships, the relative if not absolute importance of the submarine certainly has increased. Thus, the U. S. Navy logistics chief recently told a congressional committee, "By far the most important and difficult problem which confronts the navy in so far as ship characteristics and fleet operating technique are concerned is the problem of undersea warfare."

Quite apart from its military applications, the purely scientific investigation of sound in the sea presents a fascinating "front" between the known and the unknown which invites further exploration. The scientific advances along this front that were incidental to the researches in subsurface warfare of World War II are of considerable significance, and it is almost entirely from these researches that we have gained our present knowledge of underwater acoustics.

The present report deals with a small but significant part of underwater acoustics; it surveys and compiles the principal available data (to March 15, 1944) on underwater ambient noise. It is condensed from a much longer survey report[2] which contains 76 pages of text and 146 charts and figures, prepared by the authors as a comprehensive

[1] Based on a survey prepared by the authors under the direction of Section 6.1 of the National Defense Research Committee, Survey of Underwater Sound: Report No. 3—Ambient Noise, OSRD Report No. 4333, Sec. No. 6.1-NDRC-1848, September 26, 1944. The complete survey report has been declassified; photostatic or microfilm copies are procurable at the Publication Board, Scientific and Industrial Reports, Office of Technical Services, U. S. Department of Commerce, Washington, D. C. (The Pub. Board number of this report is 31021).

[2] See footnote 1.

Reprint from SEARS FOUND. JOURN. MAR. RES.

reference work for naval and civilian groups concerned with underwater sound problems during World War II. The major part of the material of the original survey report was based on reports prepared by the following organizations: U. S. Naval Ordnance Laboratory; Massachusetts Institute of Technology—Bureau of Ships, U. S. Navy; Columbia University, Division of War Research; and University of California, Division of War Research. A complete list of reports consulted, together with a short summary of technical information applying to each, is given in the original survey.

TABLE I. General Locations at Which Ambient Noise Was Measured

Boston Harbor and off Massachusetts
Long Island Sound
Off Block Island
New York Harbor and approaches
Lower Chesapeake Bay and off Cape Henry
Off Beaufort, North Carolina
Off Florida and the Bahamas
Puget Sound
San Francisco Harbor and approaches
Off San Diego
Off Oahu and Midway Islands
Off Portsmouth, England
Loch Goil, Scotland
Off Hebrides Islands

Table I lists the areas in which the measurements of underwater noise were made. The data reported are confined largely to the frequency range of 100 c.p.s. to 25 kc. They are expressed in terms of pressure levels in db relative to 0.0002 dynes/cm^2. Often the results are given in terms of the "pressure level spectra," *i. e.*, the pressure level in a band one cycle wide, as a function of frequency. These values are derived, since the measurements are usually made with filters having a band width of about one octave, although in some measurements the band width is either less or greater than one octave.

QUALITATIVE NATURE OF AMBIENT NOISE

The term *underwater noise* is used to describe unwanted underwater sounds which tend to impair the operation of acoustically operated devices. *Ambient noise*, sometimes referred to as background noise, is the sound normally prevailing in water, usually from a multiplicity of sources such as water motion, marine life, and unwanted ship sounds. Usually it is not possible to specify quantitatively the contribution of sound from each of the sources present, but frequently a particular type of source is known to be preponderant. In such cases it is convenient to refer to the ambient noise more specifically as water noise, noise from marine life, ship noise, etc.

There are three main sources of ambient noise:

(1) *Water Motion*—Usually the prevailing source of noise in open and deep water. The magnitude is largely determined by the motion of the sea surface, particularly the number and prominence of breaking waves and whitecaps. Therefore, noise from this source is related to weather conditions.

(2) *Marine Life*—Produced by a large number of vertebrates and crustacea. Usually found in shallow tropical or semitropical waters. Diurnal and seasonal variations are common.

(3) *Ship and Man-made Sources*—Found in busy harbors and connecting waters. The magnitude depends on local conditions and may vary greatly from time to time.

Water noise is caused by a large number of widely distributed sources at the water surface; for this type of noise the field is believed to be essentially isotropic. An isotropic distribution is also to be expected in the center of large areas containing marine life. At a distance from such areas, or near localized concentrations of marine life, noise levels will vary with orientation and with the distance from the localized source. A similar situation will exist where the noise is produced primarily by a few near-by ships.

Variability is an outstanding characteristic of ambient noise. The magnitude of the noise from an individual source usually varies from moment to moment, and many sources exhibit large diurnal and seasonal variations. Some sources, such as ships, change their geographical position with time. Ambient noise is usually produced by many sources, some near the point of measurement and others located at a distance. Consequently, variations in the transmission loss in the medium are also a cause of variability. For these reasons ambient noise cannot be specified as a constant quantity but must be described in statistical terms. That is, an estimate can be made of the most probable (or average) amount of noise that is to be expected under given circumstances. In addition, it is frequently possible to estimate the degree of variability in terms of a frequency distribution *vs.* time, or in terms of the standard deviation when the distribution is known to follow the normal law. A knowledge of the average noise and the magnitude of the variation from the average is essential for the correct application of quantitative data on ambient noise.

NOISE FROM WATER MOTION

The term *water noise* is used to designate the underwater noise (at a sufficient distance from the shore to avoid the sound of waves breaking near the shore) produced by the agitation of the sea surface. This agitation is usually produced by wind. Water noise is the principal type of noise encountered in open and deep sea water. In practically all parts of the ocean, water noise is an important component of the noise, at least in some portion of the spectrum.

Noise is also produced by breaking surf, by the impact of rain and hail on the water surface, and occasionally by a movement of the bottom material, such as rock, gravel, shells, etc. Underwater springs, volcanoes, and gas vents also have been suggested as possible sources of noise. Surface ice and icebergs are subjected to stresses which often give rise to audible sounds. No satisfactory measurements of such noises have been made.

It is generally assumed that water noise is produced principally by breaking wave crests at the sea surface. It is believed that the unbroken undulations of the surface water do not contribute appreciably to underwater noise. The magnitude of water noise probably is related in a complex manner to such variables as height of waves, steepness of waves, and the number and magnitude of whitecaps present. Ideally, the estimation of water noise levels should be based on the combined effects of all these variables. Practically, this is not possible

TABLE II. METEOROLOGICAL SCALES

State of Sea			Beaufort Wind Force			
Scale No.	*Description*	*Height of Waves, Crest to Trough (Ft)*	*Scale No.*	*Description*	*Velocity m.p.h.*	*Velocity Knots*
0	Calm	0	0	Calm	<1	<1
1	Smooth	<1	1	Light Air	1–3	1–3
2	Slight	1–3	2	Light Breeze	4–7	4–6
3	Moderate	3–5	3	Gentle Breeze	8–12	7–10
4	Rough	5–8	4	Moderate Breeze	13–18	11–16
5	Very Rough	8–12	5	Fresh Breeze	19–24	17–21
6	High	12–20	6	Strong Breeze	25–31	22–27
7	Very High	20–40	7	Strong Wind	32–38	28–33
8	Precipitous	>40	8	Fresh Gale	39–46	34–40
			9	Strong Gale	47–54	41–47
			10	Whole Gale	55–63	48–55
			11	Storm	64–75	56–65
			12	Hurricane	>75	>65

with the data now available. Fortunately, however, it has been found that a reasonably good estimate of noise level due to water motion can be based on either wave height or wind velocity. A summary chart for making these estimates (see Fig. 4) will be presented after considering some selected but typical results of the noise from water motion.

Table II presents some meteorological scales which are in current use, and which are used in this report, for describing the state of the sea (in terms of wave height) and the Beaufort wind force (in terms of wind velocity in knots or m.p.h.).[3]

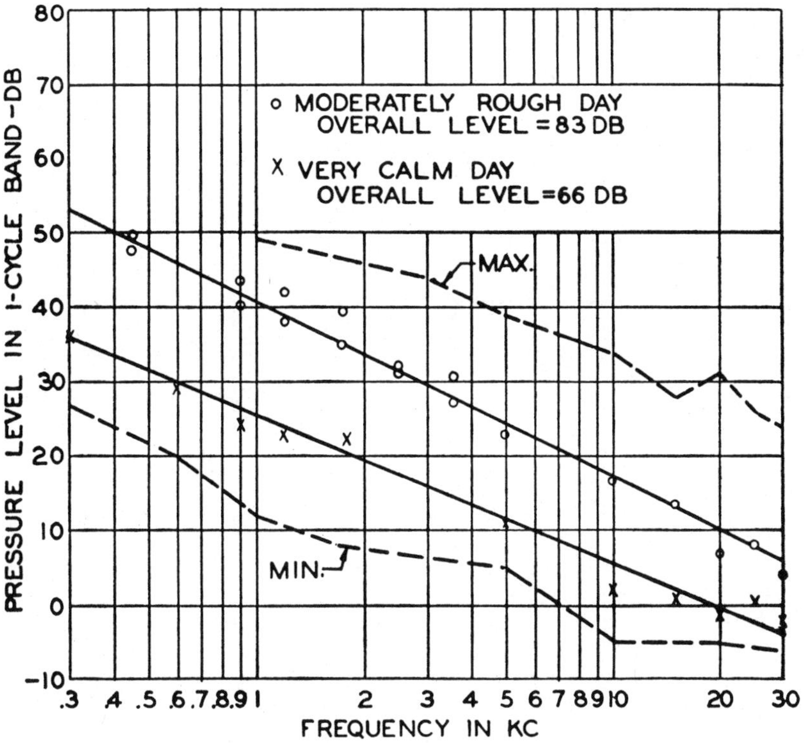

Figure 1. Pressure level spectra of ambient noise in Long Island Sound.

[3] In the original survey report, world charts prepared by the U. S. Weather Bureau are presented, one for each of the four seasons of the year, showing the distribution of average wind velocity, from which the average water noise can be estimated (with the use of Fig. 4). Thus, in one region of the North Atlantic, during December, January, and February, the average wind velocity is 24 knots, and, according to Fig. 4, the expected over-all (0.1-kc) noise level would be about 81 db.

Figs. 1–3 are illustrative of many results from measurements of noise, due mostly to water motion. Thus, Fig. 1 gives pressure level spectra measured in Long Island Sound on a calm day and on a moderately rough day. The over-all pressure levels (0.1–10 kc) corresponding to these spectra are 66 and 83 db, respectively, assuming a straight line extension of the curves to 100 c.p.s. On the same figure are also shown for comparison the maximum and minimum noise pressure levels measured from July to November 1942 in the waters adjacent to New York Harbor.

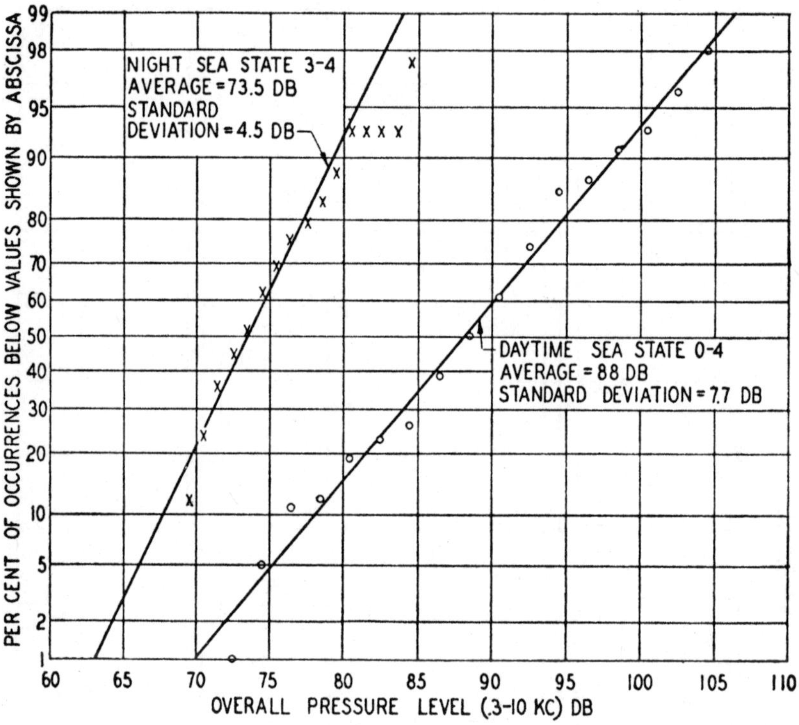

Figure 2. Ambient noise in approaches to New York Harbor; distributions of over-all pressure levels.

Fig. 2 is based on a noise survey of the area between Rockaway Beach and Sandy Hook conducted in March–April 1943. Pressure levels were measured throughout the frequency range of 150 c.p.s. to 25 kc, and over-all pressure levels were measured in a 0.3–10 kc band. Only the cumulative distribution of occurrence of over-all pressure levels for a series of daytime measurements (when ship sounds were predominant) and for a series of nighttime measurements (when ship sounds were almost completely absent) is shown in Fig. 2, plotted in the usual manner for determining the average value and the standard deviation. The curve for the nighttime measurements, when the sea state was estimated by the observers to be between 3 and 4, is believed to be free from man-made sounds, and probably therefore is representative of water noise. If the over-all pressure level measurements had been made with a band width of 0.1–10 kc instead of 0.3–10 kc, the pressure levels would have been about 2 db higher than those shown in Fig. 2. A comparison of the two curves of Fig. 2 reveals the marked difference between the daytime and nighttime underwater noise levels in the approaches to a busy harbor.

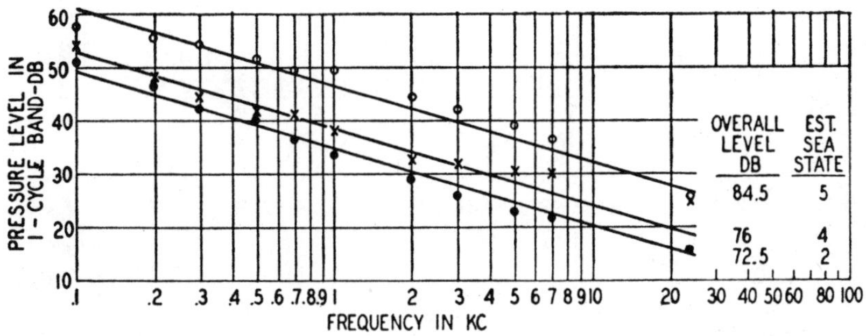

Figure 3. Pressure level spectra of ambient noise in open water about 700 feet deep.

Fig. 3 gives pressure level spectra of water noise in open and deep water five miles off Fort Lauderdale, Florida, for sea states 2, 4, and 5. The measurements were made by using a sensitive hydrophone having a low system noise. The hydrophone was carefully suspended from a cable by means of damped springs so that the movement of the hydrophone was negligible even in a very rough sea. The depth of submergence of the hydrophone was varied from 25 to 300 ft. No systematic variation of the over-all pressure level with hydrophone depth was observed; the small variations appeared to be random and probably resulted from the changing roughness of the sea and possibly from occasional sounds of distant ships or from marine life. Each point shown in Fig. 3 is the average of the data taken at all depths. There is a slight indication of a possible change in slope of the spectra for different states of sea, but the change is not systematic, and a reasonably good fit to the data is obtained with curves having the same slope as that of the spectrum determined by averaging all the data.

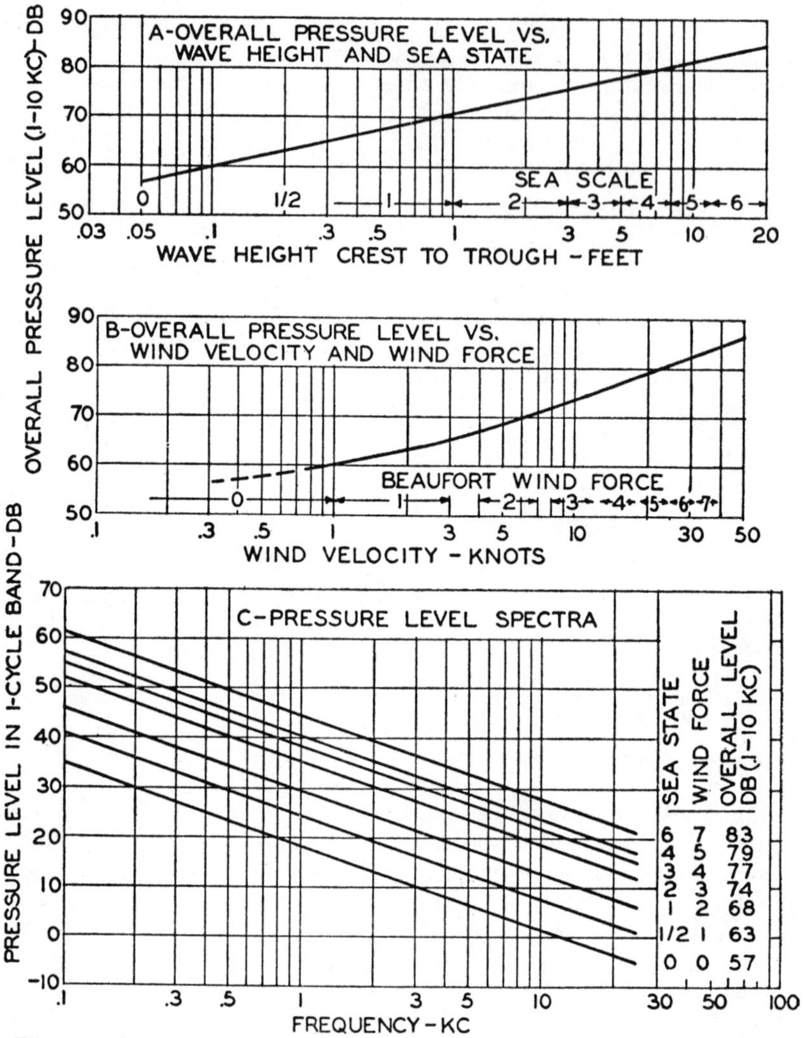

Figure 4. Ambient noise from water motion. Over-all pressure levels and pressure level spectra as a function of sea and wind conditions.

Fig. 4 presents composite summaries of ambient noise from water motion. The relation between over-all noise level and wave height (crest to trough) is shown in Part A. The "state of sea" scale corresponding to various wave heights is also shown along the abscissa. This sea scale provides a useful and an approximate method for describing the height of the waves.

The relation between over-all noise level and wind velocity is shown in Part B of Fig. 4. The Beaufort scale of wind force, which is frequently used for specifying approximate wind velocities, is also shown along the abscissa. The curves in Parts A and B are for average values. For a given wave height or wind velocity, considerable variation from the average noise level is to be expected. For example, noise levels higher than average are probable when whitecaps and spray are unusually prominent. On the other hand, noise levels lower than average are probable when the sea consists of long smooth crested waves. The standard deviation of observed noise levels with respect to the average values specified by the curves is of the order of 4 to 5 db. The variation is less at high wave heights and wind velocities than at low.

Frequently, when both wave height and wind velocity are known for a given location, the noise estimate based on wind and waves will differ. In such a case, an average of the two predicted noise levels usually will be more reliable than either one alone. An exception would occur near the lee shore of a large land mass. In this case, the estimate based on wave height will be the more accurate.

The average pressure level spectra to be expected for various wind and sea conditions are shown in Part C of Fig. 4. The slope of the spectrum appears to be independent of wind and sea and averages −5 db/octave. Experimental evidence indicates that random departures from this slope may occur, but usually the slope will not be more than −6 db or less than −4 db/octave. Neither noise level nor spectrum varies greatly with water depth so long as the water is sufficiently deep to prevent breaking of the waves. In deep water the average noise level and spectrum are essentially independent of the depth (20 to 300 ft.) at which the noise is measured. There is, however, a difference in the character of the noise. Near the surface, the noise from individual waves and whitecaps can be discerned, and the momentary variations in noise level are greater than at a greater depth.

Noise from Surf, Rain, Hail, Tide Rips, and Movement of Bottom Material

The roar of breaking surf, which often sounds very loud to a submerged swimmer, is clearly indicated or registered on underwater listening or sound recording equipment. However, there are almost no quantitative data on the noise produced by surf. Observers at Cape Henry reported that during "rough weather" the over-all pressure level (0.1–10 kc) measured near the bottom, 300 yards offshore, was 78 db (corresponds to sea state 3–4).

The impact of rain and hail is believed to produce a noticeable amount of noise in the absence of other prominent noise sources. There are no data available on the effects of hail and only a small amount of data on the effects of rain. One set of measurements indicates that with a sea of state 1, or greater, the impact of rain has a negligible effect at frequencies below 1 kc but, owing to the presence of noise from marine life, the effect above 1 kc could not be determined quantitatively. Measurements in the Thames River at New London, Connecticut, showed an over-all pressure level (0.1–20 kc) of 75 to 76 db in the presence of a "steady but not torrential" rain as compared to 57 db prior to the rain. This agrees qualitatively with observations made by swimmers in lakes and rivers while they were swimming under water during summer showers of rain and hail.

One set of measurements made near a tide rip (current 3.5 knots) in a calm sea showed a rise in the pressure level spectrum beginning at about 3 kc. Levels of about 38 db (in a one cycle band) were obtained at frequencies of 5–10 kc (corresponds to sea state 8–9).

Sounds attributed to the movement of gravel on the bottom have been reported, but there appear to be no quantitative data on the noise from this source.

NOISE FROM MARINE LIFE[4]

Many forms of marine life are capable of producing underwater noise. Table III lists some pertinent information regarding the noise-producing characteristics of vertebrate specimens investigated at Shedd Aquarium, and of other soniferous specimens of marine life. Thus, the croaker, by means of drumming muscles on its air bladder, gives bursts of sound in rapid succession, each burst having a short duration and a principal frequency of about 250 c.p.s.; the porpoise "barks" and follows this with a "gobble, gobble" sound, similar to that of a turkey cock; the toadfish emits intermittent "boops"; the spot produces a series of raucous "honks"; snapping shrimp, by snapping closed an enlarged pincerlike claw, emit a "crackling" sound, not unlike that of burning dry twigs, or, as heard from a distance, like the sizzle of frying fat. Among fresh water soniferous fish, the giant boom, or singing catfish, when played on an angler's taut line, is re-

[4] Several others have reported results of measurements on noise from marine life. See D. P. Loye and D. A. Proudfoot, Underwater noise due to marine life, J. Acous. Soc. Amer., *18:* 446–449 (1946); F. A. Everest, R. W. Young, and M. W. Johnson, Acoustical characteristics of noise produced by snapping shrimp, J. Acous. Soc. Amer., *20:* 137–142 (1948); and M. W. Johnson, F. A. Everest, and R. W. Young, The role of snapping shrimp (Crangon and Synalpheus) in the production of underwater noise in the sea, Biol. Bull. Woods Hole, *93* (2): 122–138 (1947).

TABLE III. Noise-Producing Marine Life

Vertebrate Noise-Producing Specimens at Shedd Aquarium

Name	Description of Sound	Sound Producing Mechanism	Principal Freq. CPS* 1	2	3	Length of Pulse Sec.*
Croaker	Bursts of drumming in rapid succession	Drumming muscles on air bladder	250	—	—	0.022
Spot-fin Croaker	Individual drum beats	Drumming muscles on air bladder				
Black Drum	Isolated groans	Drumming muscles on air bladder	50	80	150	0.270
Red Drum	Isolated groans	Drumming muscles on air bladder				
Garibaldi	Clicking, rasping	Pharyngeal teeth	7,400	1,000	150	0.011
Sea-anemone	Drumming, tapping	Pharyngeal teeth				
Single-striped Damozel	Drumming, tapping	Pharyngeal teeth	700	—	—	0.023
Coral-reef Damozel	Drumming, tapping	Pharyngeal teeth				
Common Triggerfish	Rasping, hissing, spitting	Pharyngeal teeth	700			
Surgeonfish	Sucking, rasping	Pharyngeal teeth				
Long-finned Pompano	Clicking, grinding	Hitting together dorsal fins				
Sheepshead	Crunching, grinding	Extrinsic food crushing	5,800	1,900	—	0.026
Razorfish	Low crunching	Extrinsic food crushing				
Hagfish	Loud crunching, grinding	Extrinsic food crushing				
Muttonfish	Crunching, sucking	Extrinsic food crushing				
Sting Ray	Crunching, grinding	Extrinsic food crushing				
Red Grouper	Clicking, grinding	Snapping teeth while eating				
Nassau Grouper	Clicking, grinding	Snapping teeth while eating				
Spadefish	Thump from rapid motion	Mechanical disturbance of water caused by quick rotation of body	220	150	—	0.150
French Angelfish	Thump from rapid motion					
Black Angelfish	Thump from rapid motion					
Diamond Flounder	Grating, scraping	Disturbance of gravel				
Sea Catfish	Popping, drumming	?				

Other Reported Noise Producers

Vertebrates: Name	Vertebrates: Remarks	Crustacea: Name	Crustacea: Remarks
Porpoise	Bark and gobble (observed off Cape Henry)	Snapping Shrimp	Crackling (principally *Crangon* [*Alpheus* and *Synalpheus*)
Toadfish	Intermittent "Boops" (Beaufort, N. C.)		
Foolfish			
Grunt		Crabs	Especially *Cancer* and *Portunus*. Noise only while feeding
Sand Perch			
Squeteague	Only males have drumming muscles		
Sargo		Barnacles	Occasional clicks of low intensity
Midshipman		Mantis Shrimp	Sharp click
Gray Trout	Probably like croaker	Spiny Lobster	Grating
Spot	Series of raucous honks (Beaufort, N. C.)		
Sea Robin	Squawk, squeal or cackle. Similar to croaker but more rasping (Beaufort, N. C.)		

* From photomicro analysis of phonograph recording grooves.
Note:—The list given above may not be complete, not all noise producers being of equal importance; several other crustacea produce noise.

puted to sing "deep purring music that wanders up and down four full tones." Two of the forms of marine life, the snapping shrimps (*Crangon* and *Synalpheus*) and the croaker, are of great practical importance because they produce sustained noises of high level.

Snapping Shrimp. Snapping shrimp can be expected throughout the oceans at locations where environmental conditions are favorable. These conditions are:

Temperature—Limited by the 52° F winter surface isotherm. Some period of the year with about 60° F also required.

Depth—Generally less than 180 feet. The highest noise levels appear to occur in water between 30 and 150 feet deep.

Bottom—Rock, coral, shell, weed, or other material providing ready concealment. Relatively uncommon on mud or sand bottoms which are free from sheltering material.

The region in which shrimp are to be expected when depth and bottom conditions are favorable is roughly defined by a belt around the earth extending from latitudes 40° N to 40° S. Along the west coast of Europe extending north to Lands End, England, there is an additional area where the temperature is favorable. The inhabitable range along the west coast of South America south of Latitude 10° S and along the east coast south of Latitude 30° S is not known.

Since snapping shrimp are nonmigratory animals they can be considered as a constant characteristic of any region in which they have once been found. The noise produced by shrimp is continuous. There appears to be no pronounced seasonal variation and but slight diurnal variation. During the night hours, the noise level is usually a few db higher than during the daytime, and there is a peak in the noise level (3 to 4 db above daytime level) just before sunrise and another after sunset.

The sounds from snapping shrimp have been reported by others (ftn. 4, p. 419) and therefore they will not be considered further here, except for reference to Fig. 5 which gives summary curves typical of the ambient noise spectra to be expected in waters where snapping shrimp are prevalent. The data for this figure indicate that for frequencies up to about 1 or 2 kc the noise spectrum is largely determined by water noise. Above 2 kc the crackling of shrimp is the major source. Part A of Fig. 5 shows noise spectra with average shrimp noise and various sea conditions. Average shrimp noise is representative of the 24-hour average noise level found at water depths of 30 to 150 feet in the following regions: Off Florida, below Latitude 27° N, off Bahama and Cay Sal Banks, off San Diego, and off Oahu and Midway Islands.

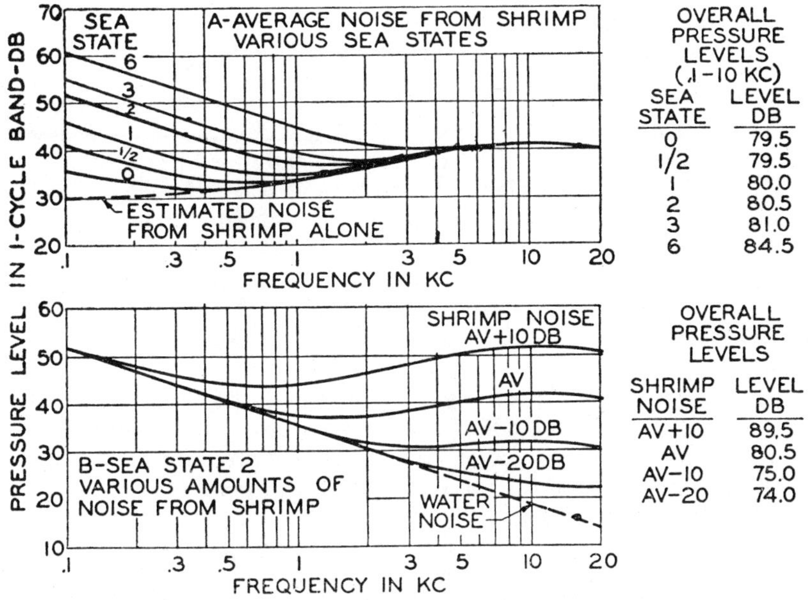

Figure 5. Pressure level spectra of ambient noise in the presence of snapping shrimp.

The noise levels at depths between 30 and 150 feet in these regions had a standard deviation of 5 db.

At depths greater than about 150 feet, shrimp crackle depends largely on the transmitted noise from near-by shallow water areas. At a distance of a nautical mile from such a shallow water area, the noise from shrimp is about 20 db lower than in the shallow water area. The colonies of shrimp appear to be smaller and more scattered at the colder margins of the areas which they inhabit. For example, along the eastern coast of the United States, shrimp have been found and their noise observed as far north as Latitude 35° N, but at offshore locations above Latitude 27° N the noise levels seem to be 15 to 20 db lower than the average level shown in Fig. 5. However, large colonies apparently occur in the harbors at Beaufort and Morehead City, N. C. Other isolated large colonies are probably to be found above Latitude 27° N.

The highest observed level of shrimp noise was measured near a pier at Kaneohe, Island of Oahu. The level at this location, at frequencies above 3 kc, was approximately 20 db above the average spectrum shown on Fig. 5. This is believed to be an exceptional condition caused by the presence of a large colony of shrimp living in the fouling material on the piling. Levels of this magnitude have not been observed in open water.

Spectra representative of a sea of state 2 and various amounts of noise from shrimp are shown in Part B of Fig. 5.

Croakers. Croakers (*Micropogon undulatus*), a variety of drumfish, are found during the late spring and early summer months in great numbers in Chesapeake Bay (estimated population 300 to 400 million). They also occur in considerable numbers at other east coast locations below Chesapeake Bay. The larger fish migrate to sea during the winter months. Different species occur on the west coast of the United States below Point Conception but apparently not in as great concentrations as along the eastern coast. Most of the available noise data have been obtained at east coast locations, and the discussion will be confined to the species found there.

Croakers produce noise by the contraction of drumming muscles attached to the air bladder. The sounds produced by an individual croaker consist of a series of "taps" that continue for about one and one-half seconds at a rate of about seven taps per second. The series is repeated at intervals of 3 to 7 seconds. The sounds resemble the tapping of a woodpecker on a dry pole. In an area with a high concentration of croakers, the sound during a period of great activity is a continuous roar, and the sounds of individuals may be heard only infrequently over the chorus of the whole population.

Croaker noise occurs principally during the feeding period, which starts in the evening as the bottom begins to darken. This, together with the migratory habits of the fish, accounts for the very pronounced diurnal and seasonal variation in croaker noise discussed in the following paragraph. Croakers are believed to feed in the shallow water on the slopes of banks.

Fig. 6 presents typical data on the seasonal and diurnal variations of over-all noise pressure levels from croakers at Cape Henry, Virginia. Note that the average over-all level in early June is 110 db, although levels as high as 119 db have been observed.

Typical spectra of ambient noise for various sea states in the presence of croakers (lower Chesapeake Bay) during hours of maximum activity are shown in Fig. 7 for late May to early June and for early July.

Other Marine Life. Other forms of marine life which may be of importance are toadfish, sea robins, some species of drumfish other than croakers, porpoises,[5] and a number of unidentified sources; one

[5] Some marine zoologists believe that the sounds here attributed to porpoises were produced by dolphins.

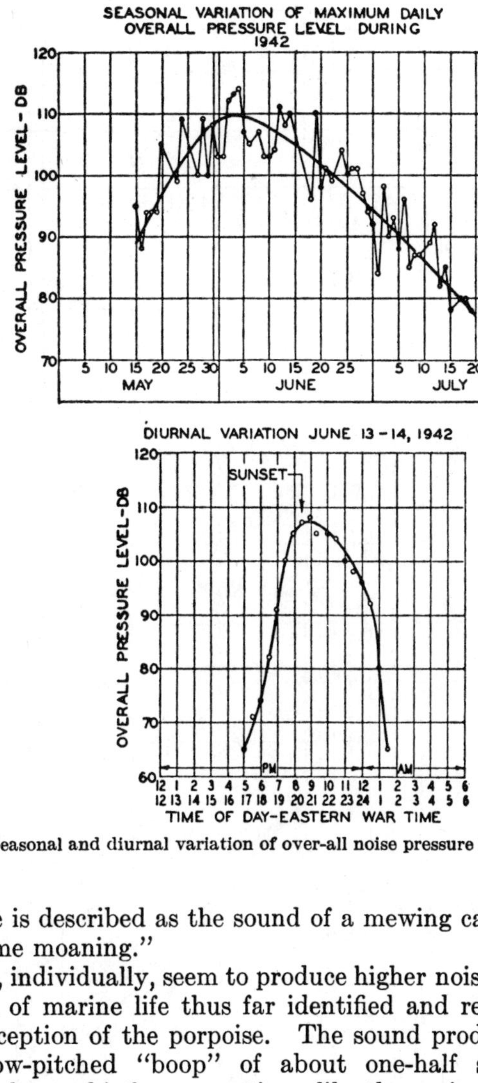

Figure 6. Seasonal and diurnal variation of over-all noise pressure levels from croakers.

such source is described as the sound of a mewing cat and another as an "awesome moaning."

Toadfish, individually, seem to produce higher noise levels than any other form of marine life thus far identified and reported, with the possible exception of the porpoise. The sound produced is an intermittent, low-pitched "boop" of about one-half second duration, similar to a boat whistle or sometimes like the cooing of a dove. The spectrum shown in Fig. 8 was measured very close to a single specimen. It will be appreciated that the levels decrease rapidly as the distance from the fish increases. These fish are shallow water bottom dwellers which nest under rocks, tin cans, and similar debris. Toadfish are not gregarious and apparently do not occur in sufficient concentration

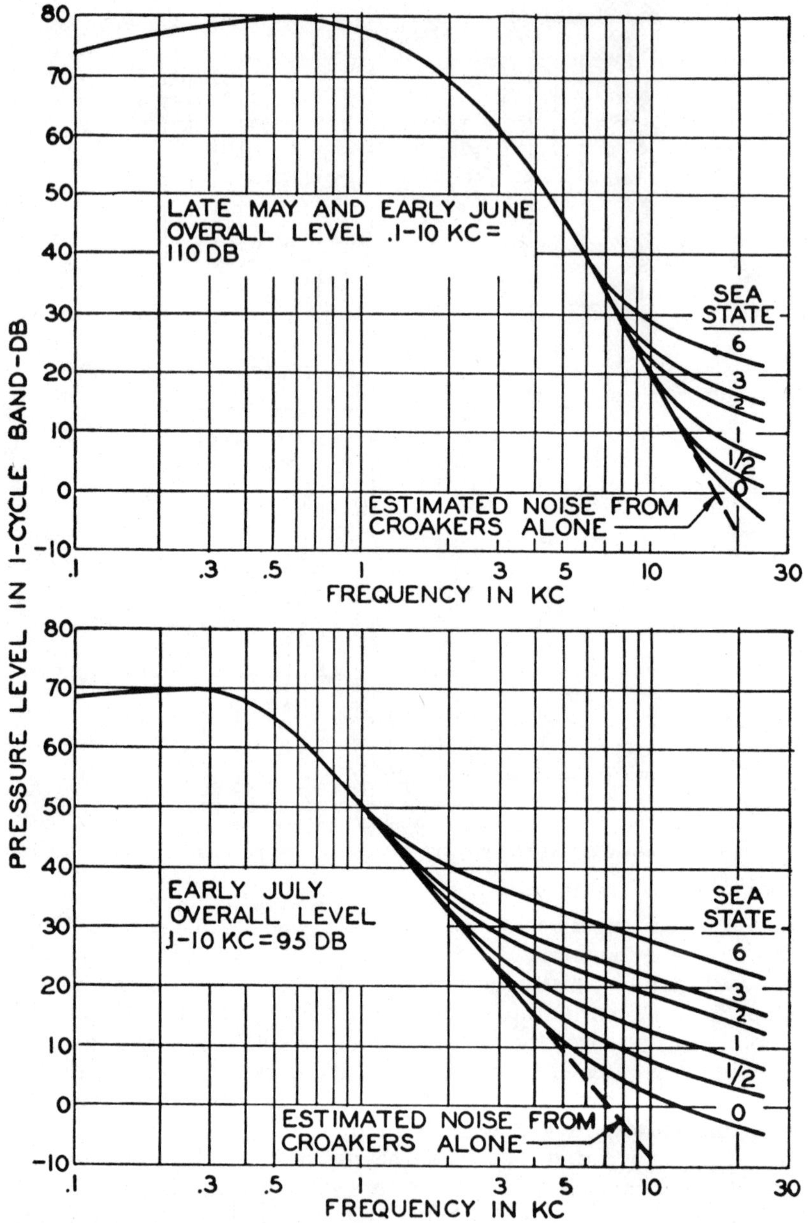

Figure 7. Typical prsssure level spectra of ambient noise in the presence of croakers.

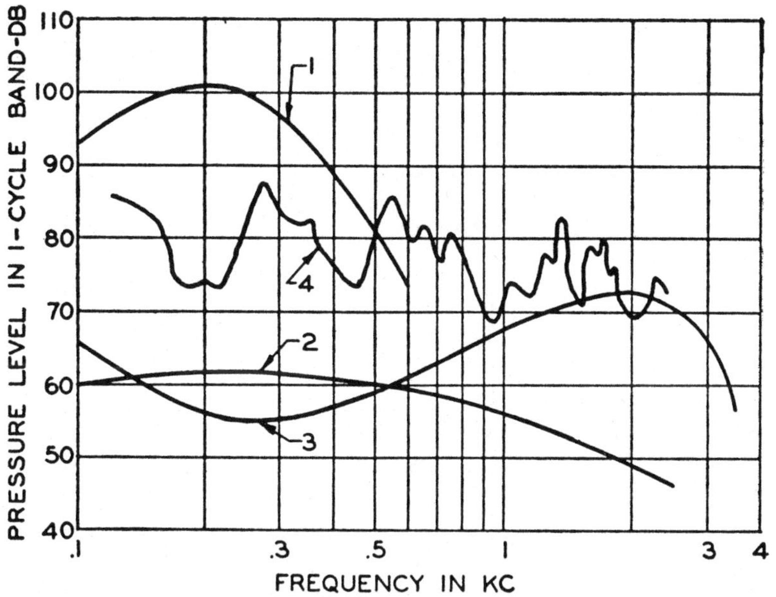

Figure 8. Illustrations of ambient noise spectra obtained in the presence of miscellaneous marine life. Curves: 1. Individual toadfish very close to hydrophone. 2. Principally from sea robins. 3. Unidentified "high pitched drumfish," possibly bastard trout. 4. Porpoises.

to produce the continuous roar generated by croakers during peak activity. Toadfish noise is subject to little diurnal variation. Other spectra shown in Fig. 8 are for (2) sea robins, (3) bastard trout, and (4) porpoises.

SHIP AND MAN-MADE SOURCES OF NOISE

In and near busy harbors and industrial centers, ships and manufacturing activities are the principal sources of ambient noise.

A wide variety of noise may be produced by industrial activities, and it is usually impossible to predict the noise from such sources in specific locations.

The noise produced by ships is more uniform in character than industrial noise, but it varies in magnitude over a wide range. It obviously depends on the number of ships present, their speed, their distance from the point of measurement, and the sound transmission characteristics of the water. Measurements in a number of east coast harbors showed an average over-all (0.1–10 kc) noise level of about 80 db. Location averages varied greatly, ranging from 65 db to 103 db. The standard deviation of the location averages was 8 db.

The average level of 80 db appears to be representative of locations with a moderate amount of ship traffic, such as the area near ship lanes at the upper end of Long Island Sound. An over-all pressure level (0.1–10 kc) of 90 db is more nearly representative of locations with a large amount of ship traffic, as, for example, in the approaches to New York Harbor during the daytime.

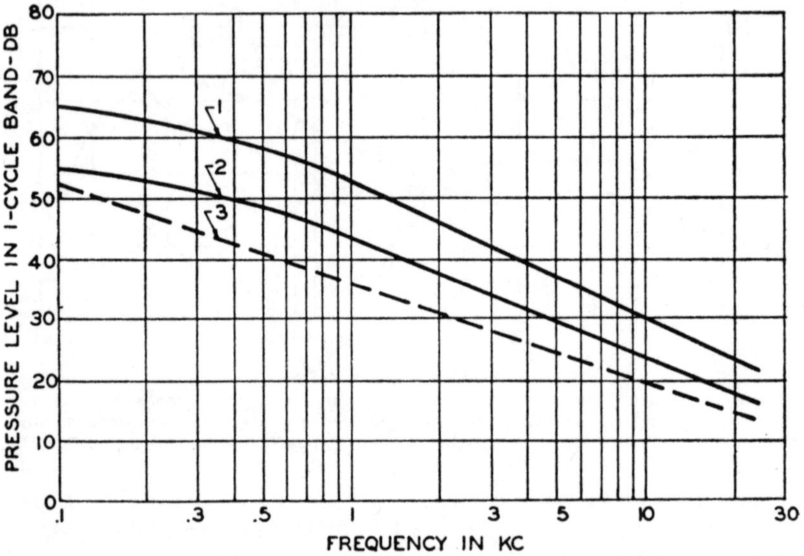

Figure 9. Pressure level spectra of ambient noise in the presence of ship sounds. Curves: 1. High noise locations; example, entrance to New York Harbor in daytime—over-all level 90 db. 2. Average locations; example, upper Long Island Sound near ship lanes—over-all level 80 db. 3. Water noise, sea state 2—over-all level 74 db.

Pressure level spectra representative of moderate and busy locations are shown in Fig. 9. A spectrum of water noise with a sea of state 2 is shown for comparison. In locations with light ship traffic, the average noise level is not likely to be lower than that produced by a sea of state 2.

PREDICTION OF NOISE CONDITIONS

The data compiled in the present report can be used for predicting the noise conditions to be expected in oceans, under various situations and in certain locations. Obviously, if survey data are available for a specific location they are preferable to estimates based on the general characteristics of the location. The principles which should be followed in predicting noise are outlined briefly as follows:

Open and Deep Water. Water noise is the major component. When sea state (wave height) and/or wind force are known, over-all pressure levels and spectra can be estimated from the data in Fig. 4. If, for a given wave height, the sea is unusually choppy with exceptionally prominent whitecaps and spray, the noise levels probably will be somewhat higher (4 to 8 db) than the average values shown in Fig. 4. If, on the other hand, the sea consists largely of long swells with exceptionally few whitecaps and little spray, the noise levels probably will be lower than average. When the prediction is to be made for a short time in the future, wind and sea conditions can be estimated from a series of weather maps. For prediction far in the future, the probable wind conditions can be estimated from U. S. Hydrographic Office Pilot Charts.

Shallow Coastal Waters. Noise from three sources is to be expected —water motion, marine life, and ships. Occasionally, surf and the movement of bottom material will contribute to the noise, but data for estimating the noise levels from these sources are not available. Water noise is estimated as outlined above. Noise from marine life is probable in tropical or semitropical waters. Snapping shrimp are to be expected at locations where the winter surface temperature is higher than about 52° F, where the water depth is less than 180 feet and the bottom consists of rock, coral, or similar material. Charts showing the nature of the bottom and water depth will be useful in estimating the prevalence of shrimp. Noise levels and spectra to be expected in the presence of shrimp are shown in Fig. 5. Croakers are a likely source of noise during the evening hours between late May and August in Atlantic coastal regions between Chesapeake Bay and Cape Canaveral. No doubt there are other regions where croaker noise is likely to be of local importance. The spectra shown in Fig. 7 are probably the best *a priori* estimates of noise conditions. Croaker noise is likely to be variable, and local surveys are advisable at locations where croakers are to be expected. The likelihood of ship sounds can be estimated from a knowledge of ship traffic. The data in Fig. 9 are probably representative of the underwater noise near active coastal ship lanes. During the absence of ships, water noise and marine life will, of course, determine the ambient noise level.

Bays, Rivers, and Harbors. The three main noise sources are possible contributors. The procedure for estimating noise conditions is the same as outlined above for coastal waters with these exceptions: In warm waters near piers with old piling, noise from shrimp may be 10 to 20 db above average. Recent dredging operations may reduce the shrimp population. In busy harbors ship noise probably will be the predominant source during the hours of shipping activity. Average noise conditions in harbors comparable with the approaches to New York City can be represented by the "High Noise" spectrum of Fig. 9. Local conditions such as industrial development at the shore or the presence of unusual man-made noise sources may necessitate a local noise survey.

The spectra of the ambient noises illustrated in Figs. 1 to 9 are typical of some of the noise levels that have been measured. They indicate the wide range and diversity of underwater ambient noise.

SUMMARY

The forgoing report presents the results of a survey of available data (to March 15, 1944) on underwater ambient noise at frequencies from about 100 c.p.s. to 25 kc. The data are presented in the form of pressure levels (for a wide band of frequencies or for a band one cycle wide) in db relative to 0.0002 dynes/cm^2. The principal sources of ambient noise are water motion, marine life, and ship sounds. In the absence of sounds from ships and marine life, the noise level in deep water is largely determined by the state of the sea; the over-all (0.1 to 10 kc) pressure level varies from about 57 db in a "calm sea" to about 83 db in a sea under a strong wind (about 35 m.p.h.). Similarly, the underwater noise from croakers may reach over-all levels as high as 119 db, and the sounds of ships in busy harbors produce underwater noises that often exceed 90 db. Data on the average pressure levels of the principal sources of ambient noise are summarized. The actual noise encountered may differ appreciably from the average values; standard deviations vary from about 4 to 8 db. Principles are deduced for predicting the noise levels that may be expected under usual conditions in specific locations.

The following paper reports on measurements made in the spring, summer, and winter on the ambient noise under the arctic-sea ice. The noise is generally impulsive and is generally related to the mechanical activity of the ice cover. Such measurements are difficult to make, since the observers must transport their equipment to ice stations in the far North.

The authors are members of the staff of the Pacific Naval Laboratory, at Esquimalt, B.C., of the Defence Research Board of Canada. They have made extensive measurements of the ambient noise and sound propagation under the arctic-sea ice.

This paper is reprinted with the permission of the authors and the Acoustical Society of America.

Reprinted from THE JOURNAL OF THE ACOUSTICAL SOCIETY OF AMERICA, Vol. 36, No. 5, 855–863, May 1964

14

Ambient Noise under Arctic-Sea Ice

A. R. MILNE AND J. H. GANTON

Pacific Naval Laboratory, Defense Research Board of Canada, Victoria, British Columbia, Canada
(Received 22 January 1964)

Underwater ambient-noise spectra and amplitude distributions are described for data acquired during field experiments made in the spring, late summer, and winter within the Canadian Arctic Archipelago. Measurements in each case were made using a bottom-mounted hydrophone. The ice cover was fast to the shore during the spring and winter experiments. In the summer experiment, the ice cover, although 10/10, was free to move. The ambient noise was generally impulsive and at times highly non-Gaussian. The significant noise was the result of mechanical activity associated with the ice cover. For shore-fast spring and winter ice, surface cracks as a result of thermal stresses were important; for late summer ice, relative motion of the floes was important.

INTRODUCTION

SOME characteristics of underwater noise beneath sea ice in the Canadian Archipelago are described. Spectrum levels versus frequency and amplitude distributions in selected bandwidths are shown for measurements obtained in April and September of 1961 and January of 1963. The appropriate field trips were termed ICE PACK 1, 2, and 3, respectively. Figure 1 indicates the geographical locations of the recording sites of these field trips within the Archipelago. Figures 2 and 3 are more-detailed maps of the recording sites. During the winter and spring, the sea ice becomes fast to the shore of the islands within the Archipelago; therefore, the recordings of underwater noise made during April and February were not affected by the gross motion of the ice. Measurements made during early September, on the other hand, were from close-packed floes of summer ice. The most striking aspect of the underwater noise recorded in winter and early spring was its apparent dependence on the mechanical properties of the ice cover. High rms noise levels were related to the frequency of occurrence of tension cracks at the surface of the sea ice. These, in turn, were related to variations in the air temperature. Under these conditions, the noise was generally impulsive. In winter, the individual tension cracks appeared to have more energy than in the warmer springtime, with the result that the winter noise was more highly impulsive.

Descriptions, results, and discussions of experiments ICE PACK 1, 2, and 3 are presented in turn. These experiments are also termed "Springtime," "Summer," and "Winter" experiments.

I. ICE PACK 1—April 1961

Measurements were made, at position 78°29′N, 105°15′W, from a camp on shore-fast ice 5 miles west of Ellef Ringnes Island (see Fig. 2). Figure 4 is a photograph of the ice cover on Prince Gustaf Adolph Sea near the recording site. The ice field consisted of old, polar ice covering 95% of the area, amalgamated by leads of one-year ice up to several miles in extent. Surrounding these flat frozen leads were extensive pressure ridges separating the one-year ice from the polar ice. The surface of the polar ice undulated as hills and valleys, with many isolated hummocks jutting upward 30 to 40 ft., and an occasional fairly broad expanse that was relatively flat. These flats were in depressions, the top layer being of clear blue ice 2–24 in. thick. These pools were of virtually fresh-water ice resulting from summer runoffs. The old ice beneath had a minimum thickness of 12 ft. The one-year ice was uniformly 8 ft thick. The leads of one-year ice had very little snow cover, with large patches completely bare. The snow tended to accumulate around the pressure ridges and among the surface irregularities in the pack ice.

Underwater ambient noise between 0.2 cps and 20 kc/sec was recorded during a 30-h period beginning at 1600 h, local time, on 27 April. Since the ice was shorefast, a bottom-mounted hydrophone was used. The water depth was 1580 ft. The hydrophone was a barium

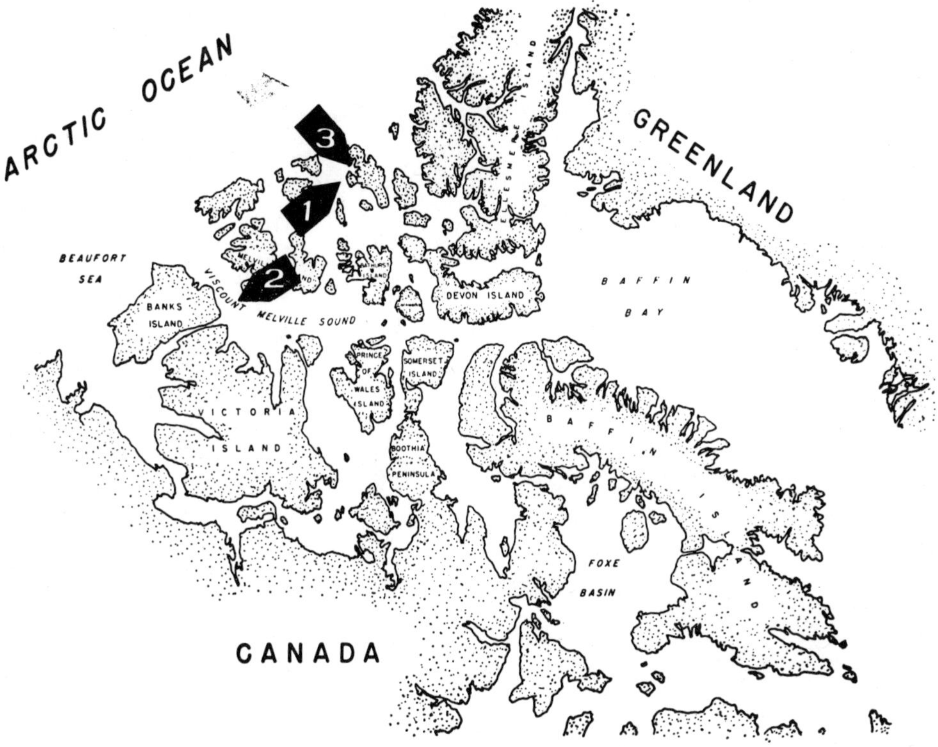

FIG. 1. Positions of experiments ICE PACK 1, 2, and 3.

titanate cylinder (Channel Industries Ltd., Type B 17 CR.), 8 in. in diameter, with a sensitivity of −65 dB *re* 1V/μbar. Recordings were made for 5 min every hour, using battery power.

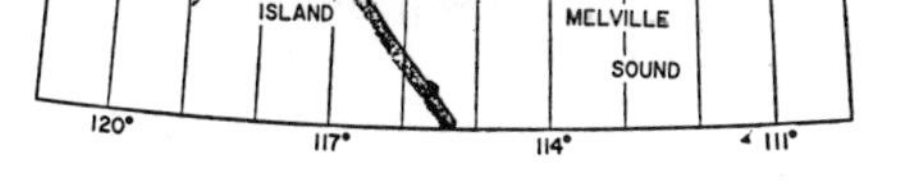

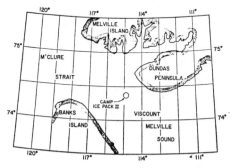

FIG. 2. Position of experiment ICE PACK 2.

A. Field Observations

The recorded noise, for frequencies above 10 cps, appeared to come from the ice cover. Aurally, the noise varied in character between booms and crashes, with their reverberations, to infrequent crackling. With declining air temperatures, the crackling approached almost a hiss, accompanied by "richochet" noises. Air temperatures during the recording periods varied between 0°F at midday and −18°F at midnight. However, during the previous week, air temperatures had been much lower, ranging between −20° and −40°F. Qualitative observations made during prior acoustic work indicated that cooling at lower mean temperatures resulted in higher noise levels. Therefore, the recorded "noisiest" levels of ambient noise almost certainly did not represent the maximum obtainable.

B. Results

The results show the change in noise level versus time, typical noise spectra, and typical amplitude distributions in selected bandwidths.

FIG. 3. Positions of experiments ICE PACK 1 and 3.

Figure 5 shows the change of noise-spectrum level in dB *re* 1 μbar in a 1-cps band versus local time for selected octave bandwidths between 200 cps to 20 kc/sec. Shown in Fig. 6 are spectrum levels versus frequency at the two times indicated by the vertical dashed lines, marked "quiet" and "noisy," in Fig. 5. Figure 6 also shows a zero sea-state spectrum[1] plotted for comparison purposes. Amplitude distributions for selected octaves of the "noisy" spectrum are shown in Fig. 7. The distributions are plotted on probability graph paper, which is designed so that a Gaussian distribution will appear as a straight line. The ordinate is the percentage of time that the instantaneous noise voltage exceeds a threshold voltage V_t. The abscissa is V_t normalized with respect to the rms-noise voltage. Measurements were made in samples of 4 min duration for the 12.5- to 25- and 200- to 400-cps bands, and of 1 min duration for the 3.2- to 6.4-kc/sec band. Results for the two lower frequency bands are significantly more impulsive than Gaussian. The ambient noise from the "quiet" spectrum was too contaminated by instrument noise for significant amplitude distributions to be measured.

[1] V. O. Knudsen, R. S. Alford, and J. W. Emiling, J. Marine Res. **7**, 410 (1948).

C. Discussion

Although the ambient-noise measurements were made within a 30-h time interval, prior underwater-sound-transmission and seismic-refraction measurements indicated that the ambient-noise levels varied with the time of day. During the latter measurements, it was possible to use smaller explosive charges during the quieter, warming period of the day than during cooling. As described earlier, the leads and tips of hummocks were free of snow and were, therefore, subjected to direct radiation from the sky. The angle of the sun's direct radiation varied roughly between 25° at noon to 0° at midnight, which resulted in changes in the heat loss and gain at the immediate surface of the ice. The resultant increase of tensile stress on cooling apparently caused cracking and abrupt increases in noise toward evening (see Fig. 5). Cracking noises were audible to the unaided ear. Furthermore, it was noticed that propagating cracks in the leads penetrated only to a depth of a few inches—

FIG. 4. Spring Pack Ice; ICE PACK 1.

sufficient to relieve the tensile stress within the ice in proximity to the air.

The snow during the measurements was hard-packed and drifted little with the wind. There appeared to be no discernable relationship between wind speed and noise levels.

In Fig. 6, the noise level corresponding to the spectrum marked "quiet" is seen to be well below that for zero-state seas. In the 20- to 300-cps band, the noise level was almost identical to the lowest reported by McPherson.[2]

The spectrum marked "noisy" exhibited an essentially constant spectrum level from 20 cps to 8 kc/sec. For other "noisy" periods, a similar characteristic "flat" was observed. This constant spectrum level would be a consequence of frequency-independent sound transmission in this band from impulsive sources of sound. For higher frequencies, the ambient-noise spectrum under rough ice cover is largely the result of ice noises propagating to the hydrophone by direct ray paths. Contributions from a distance will be at lower frequencies, with wavelengths in excess of the scale size of the under-

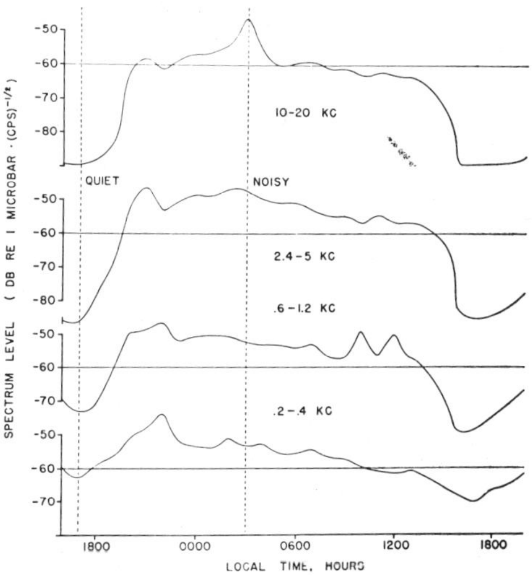

FIG. 5. Spectrum levels of ambient noise versus time of day: ICE PACK 1.

[2] J. D. Macpherson, J. Acoust. Soc. Am. 34, 1149 (1962).

ice projections.[3,4] The "noisy" spectrum of Fig. 6 shows that these frequencies are below 10 cps. Figure 8 is a ray diagram showing direct sound paths from surface sources to the hydrophone on the bottom. The bottom-limiting ray grazed the ice-water surface at an angle of 9.5° at a range of $4\frac{1}{3}$ miles, to enclose a circular area of 59 square miles. Other ray paths between the surface and the hydrophone within and beyond a $4\frac{1}{3}$-mile range reflect at least once at the underside of the rough ice, and, consequently, are severely scattered.

Analysis of an explosive-sound-transmission test made at the same location a week earlier indicated the severity of the reflection loss at the rough under-ice surface, and also its frequency independence from 160 cps to 16 kc/sec. Frequency independence is demonstrated in Fig. 9, which shows on a relative decibel scale, the energy detected by a hydrophone 300 ft deep, for 1-lb shots fired at depths of 300 and 800 ft at increasing horizontal ranges. Corrections have been made to take account of the effects of cylindrical spreading and absorption in the water layer by the addition of 10 log R $+0.011f^2R$ dB, where R is the range in nautical miles and f is the geometric mean frequency in kc/sec. The mean slopes of the broken lines are identical for each bandwidth from 160 cps to 16 kc/sec, to a range of 15 miles. From 15 to 18 miles, the slope is essentially zero for shots fired under a smooth frozen lead.

An estimate of the loss per reflection at the underside of the ice was obtained by consideration of the allowed ray paths between the shots and the hydrophone. It was evident that the only significant energy arrived via ray paths undergoing the fewest under-ice reflections.

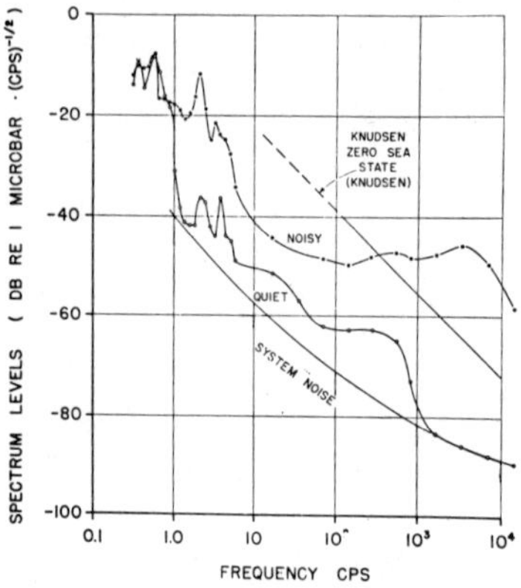

FIG. 6. Spectra of ambient noise corresponding to the local times marked "quiet" and "noisy" in Fig. 5: ICE PACK 1.

[3] H. Kutschale, J. Geophys. Res. **66**, 2189 (1961).
[4] K. Hunkins and H. Kutschale, J. Acoust. Soc. Am. **35**, 542 (1963).

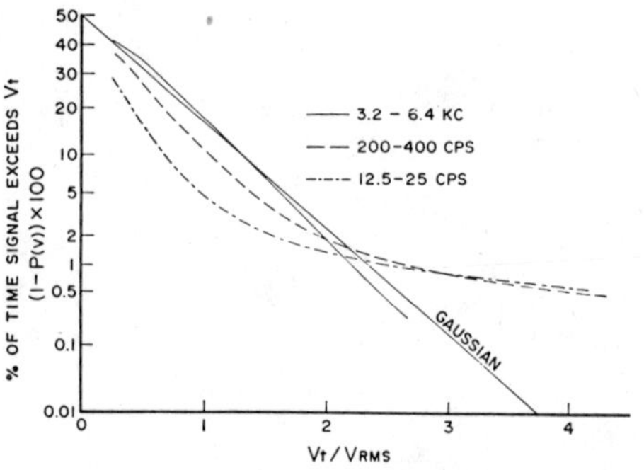

FIG. 7. Amplitude distributions in bands 12.5–25 cps, 200–400 cps, and 3.2–6.4 kc/sec for the time marked "noisy" in Fig. 5: ICE PACK 1.

On a statistical basis, the loss per under-ice reflection was 15 dB for grazing angles of between ±7.5° to ±8.5°. In this estimate, contributions from bottom-reflected sounds from distant shots were ignored, since these sounds made many more under-ice reflections than sounds confined to refracted paths within the water layer.

Figure 7 shows amplitude-distribution functions for the "noisy" spectrum of Fig. 6. The superposition of sounds from the impulsive events that resulted in these distributions indicates that the number of events per unit time increased with frequency but that the average amplitude of each event decreased with frequency. This resulted in essentially Gaussian noise above 1 kc/sec. A hypothesis that explains this phenomenon is that the high-frequency sounds were produced by a profusion of small-scale cracks on the hummocks and angular projections of the pack ice, while the low-frequency sounds were generated by longer cracks on the more uniform surfaces of the frozen leads. The only significant sources of ambient noise appeared to be caused by ice cracks resulting from thermal stresses.

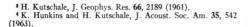

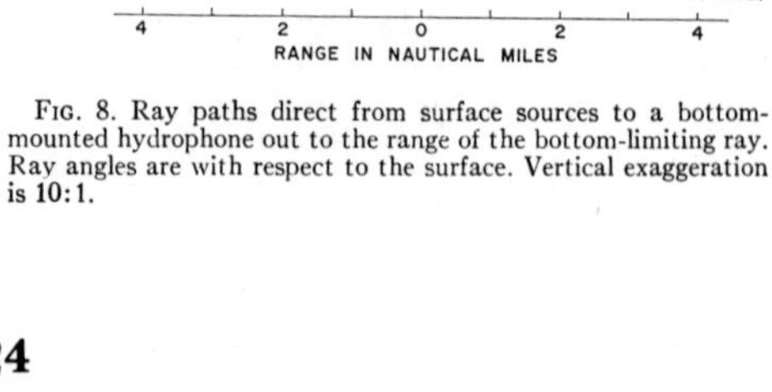

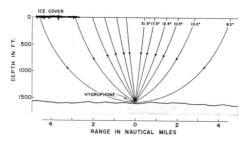

FIG. 8. Ray paths direct from surface sources to a bottom-mounted hydrophone out to the range of the bottom-limiting ray. Ray angles are with respect to the surface. Vertical exaggeration is 10:1.

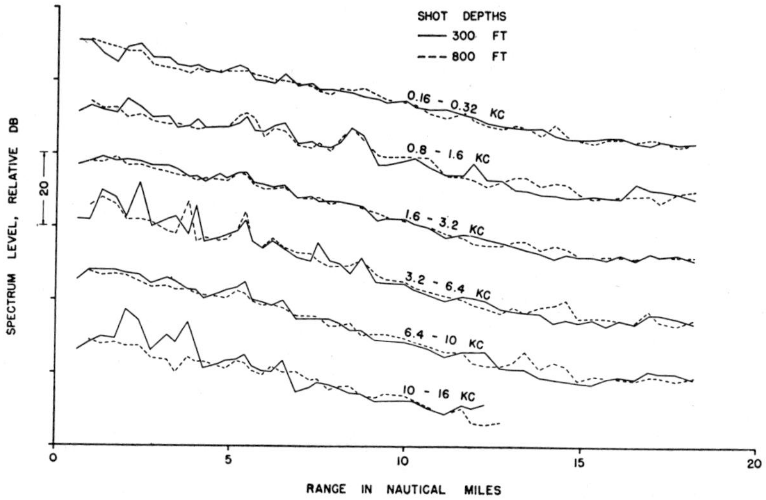

FIG. 9. Transmission loss versus range in frequency bands between 160 cps and 16 kc/sec. Absolute levels have been omitted to permit separation of the curves. Cylindrical spreading and absorption corrections have been applied: ICE PACK 1.

II. ICE PACK 2—September 1961

Measurements were made near position 74°28′N, 115°06′W, from a camp established on a polar floe in M'Clure Strait by the icebreaker *CGS LABRADOR*. Figure 10 shows the general nature of the nearly 10/10 ice cover, which was formed of about 70% one-year ice, the remainder being polar floes. The surfaces of the polar floes were well-weathered, with hummocks extending to 6 ft above sea level and pools of melt water between. The characteristic light-blue color of these pools showed the polar ice beneath. The floes displaced between 10 and 12 ft of water and were porous to the extent that all the pools were at sea level. In contrast, the one-year ice was between 3 and 5 ft thick, and was so rotten that it could hardly be considered to be continuous ice.

Recordings of ambient noise were made for 5 min every hour for a 25-h period beginning at 1500 h local time on 2 September. To obtain useable measurements down to 0.5 cps, the hydrophone was lowered onto the bottom at the commencement of each recording period, after which an extra 400 ft of wire was paid off the winch to allow for movement of the floe. The water depth varied not more than 20 ft from a constant 1590 ft. The hydrophone was identical to the 8-in.-diam barium titanate cylinder used during ICE PACK 1.

A. Field Observations

For the hours during which the underwater noises were being recorded, wind velocities seldom exceeded 15 Kt (knots) and air temperatures declined slowly from −3° to −6°C. During this time, new ice was slowly forming and was nearly 2 in. thick by 3 September.

Only one recording, at 1100 h on 3 September, exhibited significantly more noise than all the others, amounting to at most a 10-dB increase over all. At this time, parts of the ice field were undergoing slow relative motions that caused rafting of the ice rind. Aurally, these sounds were akin to those produced by intermittently escaping steam. Bumping and grinding noises were conspicuously absent.

B. Results

Ambient-noise spectra for four samples are shown in Fig. 11. Samples 2, 3, and 4 were typical of most recordings, Sample 1 being the noisiest. Figure 12 shows amplitude distributions for a typical noise sample in the bands 12.5–25 cps, 200–400 cps, and 3.2–6.4 kc/sec. Distributions were computed for a 4-min sample in the two lower frequency bands, and for a 1-min sample in the 3.2 to 6.4-kc/sec band.

C. Discussion

No physical explanation is put forward to explain the "flat" in the ambient-noise spectra below 100 cps (Fig. 11). Except near the surface, the sound-velocity versus depth curve for ICE PACK 2 was identical to the spring curve for ICE PACK 1; therefore, since the depth was nearly identical, the ray diagram of Fig. 8 applies with respect to the direct arrival of surface noises at the hydrophone. Owing to the rough ice cover, the primary sources of the noise at frequencies subject to severe surface-scattering would, therefore, be confined within a $4\frac{1}{3}$-mile radius.

The amplitude distributions of the summer ambient noise (Fig. 12) are an interesting contrast to those of the "noisy" springtime (Fig. 7). The 12.5- to 25-cps-band

Fig. 10. Summer ice cover: ICE PACK 2.

distributions were similar and non-Gaussian, both resulting from highly impulsive noise. Differences occurred in the 200- to 400-cps and 3.2- to 6.4-kc/sec bands. In the summer results, the 200- to 400-cps band exhibited a nearly Gaussian distribution, in contrast to an impulsive springtime distribution. Also, in the summer, the noise in the 3.2- to 6.4-kc/sec band was non-Gaussian and impulsive, in contrast to a Gaussian distribution in the springtime. These differences in amplitude distributions between the springtime and summer suggest fundamental differences in the mechanisms of noise generation in the ice cover. This difference was emphasized by the completely different aural character described previously under "Field Observations."

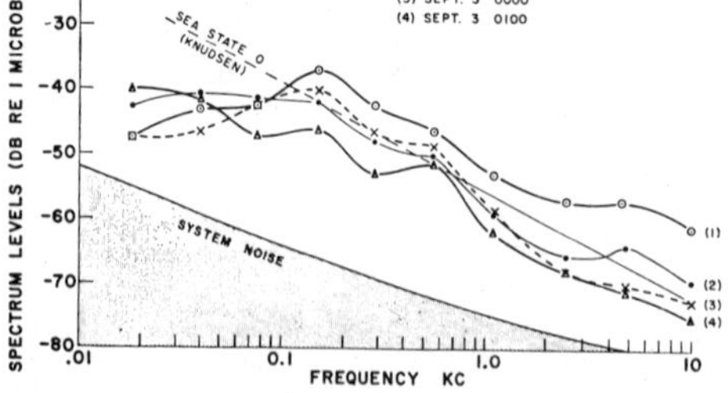

Fig. 11. Spectra of ambient noise in late summer: ICE PACK 2.

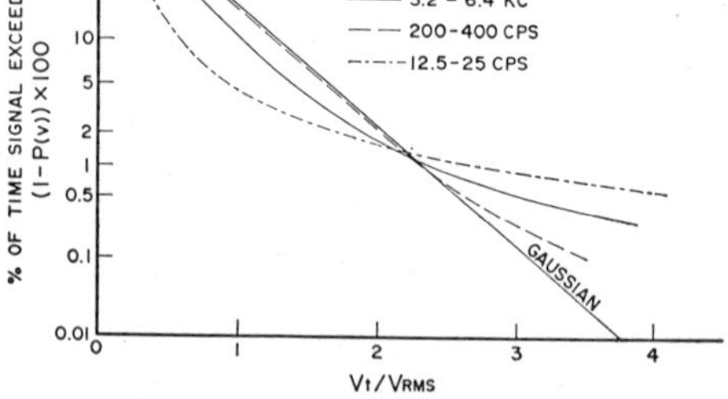

Fig. 12. Amplitude distributions in bands 12.5–25 cps, 200–400 cps, and 3.2–6.4 kc/sec for a typical noise sample: ICE PACK 2.

III. ICE PACK 3—February 1963

Measurements of underwater noise from pack ice in winter were made not far from the springtime location of ICE PACK 1 (see Fig. 3). A description of the ice cover would parallel that of ICE PACK 1 and was similar to that shown in Fig. 4. Again, the ice was shore-fast. The intention had been to proceed into deeper water in Prince Gustaf Adolph Sea to the westward of Ellef Ringnes Island. However, due to tractor unserviceability, the measurements were made, instead, near the mouth of Deer Bay over water 325 ft deep. Four hydrophones in a 400-ft-square array were laid on the bottom so that the horizontal directional properties of the ambient noise could be investigated.

Between 10 and 17 February, recordings were made for 15-min periods at intervals mainly determined by changes in noise levels observed on an oscilloscope monitor. Since solar radiation was absent, no diurnal change in noise levels was expected. The hydrophones were similar to those used on the two experiments previously described, but equipped with lower-noise preamplifiers.

A. Field Observations

Air temperatures varied between −18° and −55°F and winds between 0–35 mph with the passage of weather systems over the area. Temperature rises were usually accompanied by increasing winds and, conversely, decreasing temperatures usually accompanied decreasing winds. The granular snow tended to accumulate in low drifts that moved slowly under wind action to cover and uncover the surface of the ice on the frozen leads and to redistribute itself in the pack ice. During cooling, shallow surface cracks could be observed on bare lead ice, which would fill with snow. During temperature rises, this snow would be extruded from the cracks to be worn away by the wind. The cracks would appear to heal and to occur elsewhere during a subsequent reduction of air temperature.

The ambient noise appeared to be composed of two distinct components. There existed a "background" level, well above instrument noise, that was free of cracking noises and appeared to be caused by the motion of drifting granular snow over the sea ice. The second consisted of high-energy impulsive events that, during periods of decreasing air temperatures, had peak pressures of 40 dB or more above the background level.

B. Results

Field observations of the directional properties of the underwater noise in the horizontal plane in the 10- to 100-cps band were made by using a clipper correlator.[5] With an averaging time of 8 sec, no significant correlations were obtained between hydrophone pairs. Inconclusive results have also been obtained by considering the distribution of individual noise impulses in azimuth. For the noise field observed, it was tentatively concluded that there was no azimuthal preference.

[5] J. J. Faran, Jr., and R. Hills, Jr., "Correlator for Signal Reception," Harvard Univ. Div. Appl. Sci. Acoust. Res. Lab. Tech. Mem. No. 27 (15 Sept. 1952).

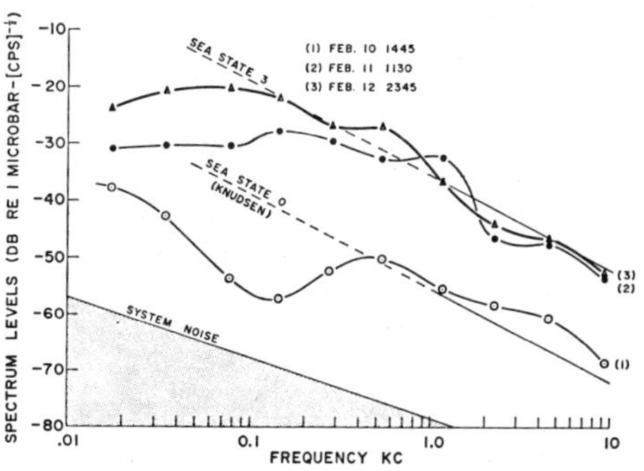

FIG. 13. Spectra of ambient noise in winter: ICE PACK 3.

Figure 13 shows spectra of underwater-noise samples recorded on 10, 11, and 12 February, respectively. Spectra numbered 2 and 3 in this Figure could be considered typical spectra of noise recorded during the more-"noisy" periods, whereas spectrum number 1 is for the quietest period observed. Figure 14 shows the change of spectrum level versus time for the 200- to 400-cps band. Circles in Fig. 13 mark the times for which the extended spectra are displayed.

Figure 15 shows amplitude distributions for selected octaves of the "noisy" spectrum labeled number 2 in Fig. 13. Amplitude distributions for the "quiet" spectrum, labeled number 1, were Gaussian to within the accuracy of the measuring apparatus and, therefore, are not shown.

C. Discussion

The mechanism of ice-cracking caused by thermal stresses was obviously operating during the winter ex-

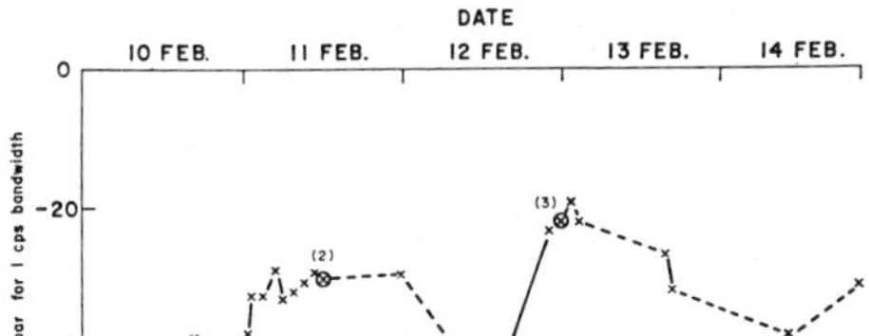

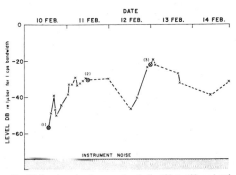

FIG. 14. Spectrum levels versus time in the 200- to 400-cps band. Complete spectra corresponding to times designated numbers 1, 2, and 3 are shown in Fig. 13: ICE PACK 3.

periment. A short record of air temperature versus spectrum level in the 200- to 400-cps band is shown in Fig. 16. Although the air temperature was not an independent variable, it appeared to be the dominant one during the February experiment.

A significant difference in the "noisy" winter and "noisy" spring recordings of underwater noise existed in the low-frequency (12.5- to 25-cps) band. In this band, the winter noise was close to Gaussian (Fig. 15), in contrast to the highly impulsive non-Gaussian noise of the spring (Fig. 7). The reasons for this are not clear, but may be associated with the shallower-water recording site in the winter, with a consequent loss of higher propagating modes of sound in this bandwidth (ICE PACK 3, Fig. 3).

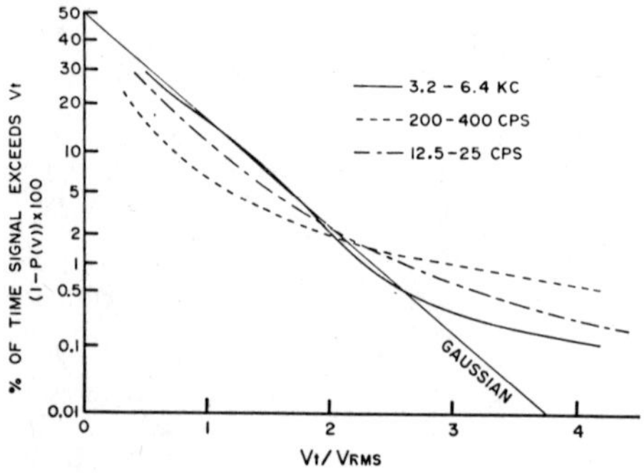

FIG. 15. Amplitude distributions in bands 12.5–25 cps, 200–400 cps, and 3.2–6.4 kc/sec for noise sample designated number 2 in Fig. 14: ICE PACK 3.

In the winter experiment, there existed a second noise-producing mechanism. This was apparent during air-temperature rises, when the impulsive ice-fracturing noises disappeared. Pending further investigation, this mechanism has been attributed to the motion of wind-driven granular snow moving over and impinging on the rough sea ice. The mechanism resulted in Gaussian noise, well above instrument noise (spectrum number 1, Fig. 13).

IV. CONCLUSIONS

A prediction of noise spectra and amplitude distributions under sea ice cannot be made on the basis of available information. The measurements obtained during three field trips made in the spring, summer, and winter were insufficient to characterize underwater noise at other times and other locations. These serve instead to emphasize its variability. Several outstanding characteristics do emerge: (a) the noise measured was generally impulsive and at times highly non-Gaussian in contrast to oceanic noise of icefree seas[6]; (b) the significant underwater noise was the result of mechanical activity associated with the ice cover. For shore-fast winter and spring pack ice, surface cracks as a result of thermal stresses were important. For late-summer ice, relative motion of floes was important; (c) most spectra exhibited noise levels that were not exceptionally low, particularly in winter when spectra equivalent to a state-three sea spectrum[1] were commonly measured.

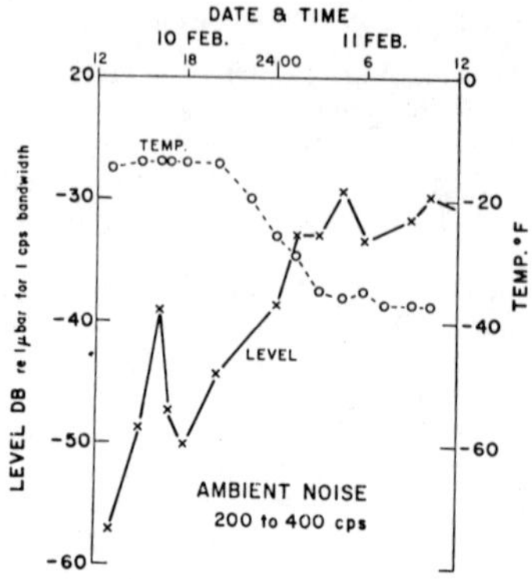

FIG. 16. Spectrum levels of ambient noise in the 200- to 400-cps band versus air temperature: ICE PACK 3.

ACKNOWLEDGMENTS

Participants in various phases of the field work included W. H. M. Burroughs, J. R. Brown, T. Hughes, R. G. Robson, J. A. O'Malia, and C. E. Kelly, of the Pacific Naval Laboratory, and R. H. Herlinveaux of the Pacific Oceanographic Group, Nanaimo, B. C. The able assistance and cooperation of these personnel in the field and later in processing the data is very much appreciated. Agencies participating also included RCAF Transport Command, Department of Transport Marine Service, and the Department of Mines and Technical Survey's Polar Continental Shelf Project.

[6] A. H. Green, "A Study of the Bivariate Distribution of Ocean Noise," Bell Telephone Labs. Tech. Rept. No. 10 (13 Apr. 1962).

The noise spectra reported by Knudsen *et al.* are the results of measurement and have been verified many times. In this paper, Dr. Marsh develops a theory, assuming that the ambient noise is due to surface waves; this explains the spectral distribution of the ambient noise energy as observed experimentally in the open ocean.

Dr. Marsh organized the staff and directed the research activity of the AVCO Corporation, Marine Electronics Department, at New London, Connecticut. He is now at the Naval Underwater Systems Center at New London, Connecticut. He has contributed a number of papers in the field of underwater sound and oceanography. He received the Ph.D. degree in physics at Brown University in 1950.

This paper is reprinted with the permission of the author and the Acoustical Society of America.

Reprinted from THE JOURNAL OF THE ACOUSTICAL SOCIETY OF AMERICA, Vol. 35, No. 3, 409–410, March, 1963

15

Origin of the Knudsen Spectra

H. W. MARSH
AVCO Corporation, New London, Connecticut
(Received 10 December 1962)

A theory is presented, ascribing the ambient noise spectra of Knudsen to the generation of sound by surface waves. A quantitative formula, based on Longuet-Higgins theory of surface waves in heavy, compressible fluids and the sea-surface spectra of Neumann-Pierson, Burling, and Marsh, is given, which accounts rather well for the numerical values in the Knudsen curves.

THE title of this letter refers to the ambient noise spectra in the ocean according to Knudsen[1] *et al.* The original measurements have been repeatedly verified, and the spectra have occupied a place equivalent to that reserved for universal constants for two decades. The mechanisms at work in producing this noise, however, have remained a mystery, although the noise is generally regarded as associated with the sea surface.

During the past few years, I have had occasion to consider interactions between the sea surface and the sound field. It has been possible to develop a rather accurate theory of surface reflection and scattering, using the recent descriptions of the sea surface, according to Neumann-Pierson[2] and Burling.[3] This theory[4] accounts for observed values of surface-propagation loss and of surface reverberation.

It is now possible to give an account of surface-generated acoustic noise, using the same sea-surface descriptions. The theory required is that of the pressure waves at depth in a compressible fluid, caused by oscillations of the fluid surface (waves). Although many have written on the subject, the only directly applicable work is the very excellent development of Longuet-Higgins,[5] who obtained the following expression for the sound pressure:

$$p=2\rho a_{+}a_{-}f^2\cos 2\pi ft. \qquad (1)$$

In this equation the fluid pressure p is independent of depth, ρ is the static fluid density, f the frequency of the (sinusoidal) surface waves, t the time, and a_{+}, a_{-} the amplitude of waves proceeding

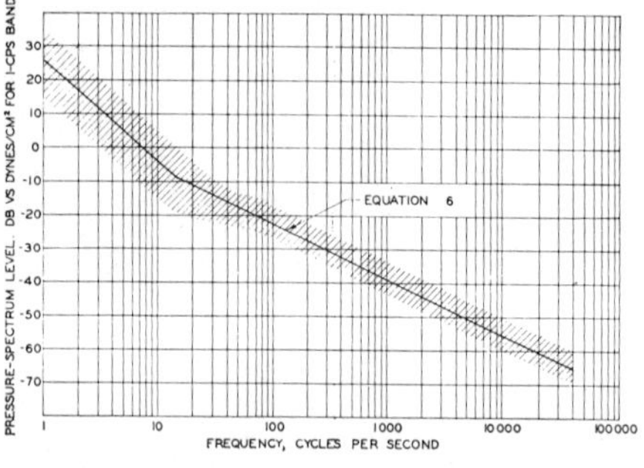

FIG. 1. Knudsen's noise spectrum vs frequency for mean-wave height of 2.5 ft. The shaded area represents the average range of measurements reported by Knudsen, but is extrapolated below 100 cps.

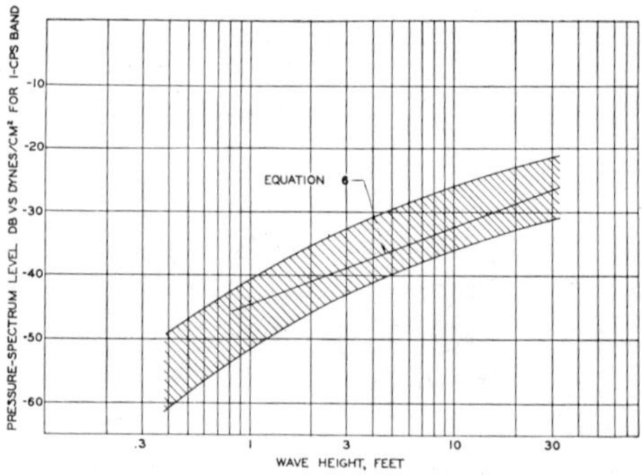

FIG. 2. Knudsen's noise spectrum vs wave height for acoustic frequency of 1000 cps. The shaded area represents the average range of measurements reported by Knudsen.

in opposite directions. We can generalize this result to apply to surface waves in two dimensions having a continuous frequency distribution and obtain for the pressure spectrum

$$p_f^2=\frac{\rho^2}{(2\pi)^3(2\alpha-\pi)}\int_0^\infty\int_{\frac{1}{2}\pi}^{\alpha}\omega_1^2A^2(\omega_1)(\omega-\omega_1)^2A^2(\omega-\omega_1)\times\cos(\pi-\theta)d\omega_1 d\theta. \qquad (2)$$

$A^2(f)$ is the one-dimensional power spectrum and α the "beam width,"[6] which we may take to be nearly π. Using the Neumann-Pierson spectra, we have finally, for frequencies above a few cycles per second,

$$p_f^2=\frac{2\rho^2}{3\pi^4}\left(\frac{9\pi c_0^2}{8192}\right)^{1/5}h^{6/5}\omega^2A^2(\omega). \qquad (3)$$

In Eq. (3), p_f^2 is the mean-square pressure (dyn/cm²) per unit-frequency interval; c_0 is the Pierson constant, 4.8×10^4 cm² sec⁻⁵; ρ is the water density, g cm⁻³; h is the rms wave amplitude, cm; f is the acoustic frequency, cps.

Pierson's spectrum is necessary only to evaluate approximately the integral in Eq. (2) since it accounts for nearly all of the surface-wave energy. To evaluate Eq. (3) we have, for the different frequency ranges,

$$A^2(\omega)=\beta g^2\omega^{-5}, \qquad \omega\leqslant\omega \qquad (4)$$

$$=\mu g\gamma^{1/3}\omega^{-11/3}, \qquad \omega\geqslant\omega. \qquad (5)$$

Equation (4), according to Burling, is applicable somewhat above the modal frequency of the surface-wave spectrum, while Eq. (5), according to Marsh,[7] is applicable above the frequency of minimum surface-wave velocity. In these equations, $\beta=7.4\times10^{-3}$, $\mu=\beta/4^{1/3}$, g is the acceleration of gravity, and γ the ratio of surface tension to density (74 cm^3 sec^{-2} for clean water).

In practical units, we have finally

$$\begin{aligned} p_f^2 &= 94H^{6/5}f^{-3}, \quad 1\leqslant f\leqslant 13.5 \\ &= 2.9H^{6/5}f^{-5/3}, \quad 13.5\leqslant f \end{aligned} \tag{6}$$

for f in cycles per second, and H the mean crest to trough wave height in feet.

The form of Knudsen's spectra is well known, and for verification of Eq. (6) we need only show the wave height and frequency variation, as in the figures. Here we plot p_f^2 vs frequency for $h=2.5$ ft (sea state 2) and vs wave height for a frequency of 1000 cps.

We show also the experimental results of Knudsen, with associated variability, and indicate their extension to lower frequencies. This extension must be considered semiquantitative at this time, since there is an extreme scarcity of published experimental results below 100 cps. However, the extension represents reasonably those values which I have been able to locate in unpublished form, and may be compared with the reader's own experience.

It must be emphasized that the account above is only approximate, and that there is a number of useful features yet to be developed. The directional properties of the noise field may be easily developed, as well as the depth dependence, although in the latter case the sea bottom will play an important role. Studies of these features are in progress and will be reported in detail.

In summary, it would appear that the theory sketched here provides an acceptable account of the cause of noise characterized by Knudsen's spectra, and that further properties of that noise field can be calculated from the sea-surface spectra.

This work was supported by the Acoustics Programs of the Office of Naval Research.

[1] V. O. Knudsen, R. S. Alford, and J. W. Emling, J. Marine Res. **7**, 410 (1948).
[2] W. J. Pierson, Jr., G. Neumann, and R. W. James, *Practical Methods for Observing and Forecasting Waves by Means of Wave Spectra and Statistics* (U. S. Navy Dept., Hydrographic Office, Publication No. 603, Washington, D. C., 1955).
[3] R. W. Burling, "Wind Generation of Waves on Water," PhD dissertation, Imperial College, University of London (1955).
[4] H. W. Marsh, "Sound Reflection and Scattering from the Sea Surface," J. Acoust. Soc. Am. **35**, 240 (1963).
[5] M. S. Longuet-Higgins, "A Theory of the Origin of Microseisms," Trans. Roy. Soc. (London) **A243**, 1 (1950).
[6] C. S. Cox and W. Munk, "Slopes of the Sea Surface Deduced from Photographs of Sun Glitter," Bulletin, Scripps Inst. of Oceanography **6**, 401 (1956).
[7] H. W. Marsh, "Sea Surface Statistics Deduced from Underwater Sound Measurements," manuscript submitted to New York Academy of Sciences (1962).

The following is a review paper which brings together the results of research on ambient noise in the ocean prior to 1962. The author then uses these data to arrive at an explanation of the ambient noise spectra in various regions of the ocean, based on the sources available in each region.

Mr. Wenz became a member of the U.S. Navy Electronics Laboratory staff in 1945, where he started the study of the noise output of ships and submarines. He had a major part in the design, development, and installation of the Carr Inlet Acoustic Range at the Puget Sound Naval Shipyard.

Since 1955, he has been primarily concerned with the investigation of acoustic ambient noise in the ocean, with special interest in low-frequency, long-term, and large-sample measurements. He transferred to the U.S. Naval Undersea Research and Development Center, San Diego, California, when it was established in 1967, and retired in July 1970. He now resides in San Diego, California.

This paper is reprinted with the permission of the author and the Acoustical Society of America.

Reprinted from THE JOURNAL OF THE ACOUSTICAL SOCIETY OF AMERICA, Vol. 34, No. 12, 1936–1956, December, 1962

Acoustic Ambient Noise in the Ocean: Spectra and Sources

16

GORDON M. WENZ
U. S. Navy Electronics Laboratory, San Diego 52, California
(Received May 25, 1962)

The results of recent ambient-noise investigations, after appropriate processing, are compared on the basis of pressure spectra in the frequency band 1 cps to 20 kc. Several possible sources are discussed to determine the most probable origin of the observed noise. It is concluded that, in general, the ambient noise is a composite of at least three overlapping components: *turbulent-pressure fluctuations* effective in the band 1 cps to 100 cps; wind-dependent *noise from bubbles and spray* resulting, primarily, from surface agitation, 50 cps to 20 kc; and, in many areas, *oceanic traffic*, 10 cps to 1000 cps. Spectrum characteristics of each component and of the composite are shown. Additional sources, including those of intermittent and local effects, are also discussed. Guidelines for the estimation of noise levels are given.

INTRODUCTION

THE summary work of Knudsen, Alford, and Emling[1,2] discussed the nature of underwater acoustic ambient noise in the frequency range from 100 cps to 25 kc. While a few results from remote open-ocean areas were available, a large part of the source data for their study was taken in off-shore areas, and in the vicinity of ports and harbors. Three main sources of underwater ambient noise were identified: water motion, including also the effects of surf, rain, hail, and tides; manmade sources, including ships; and marine life. The "Knudsen" curves showing the dependence of the noise from water motion on wind force and sea state are well known. Increased levels due to nearby shipping and industrial activity have been observed. For several marine-life sources,[3–6] e.g., snapping shrimp and croakers, the noise characteristics and the times and places of occurrence have been indicated.

A number of ambient-noise studies has been made since 1945, including some investigation of the frequency range below 100 cps and some additional measurements in deep-water open-ocean areas. A great deal of the underwater ambient-noise information is the result of investigations conducted by U. S. Navy Laboratories, and by university and commercial laboratories operating under contract with government agencies, usually with the Office of Naval Research.

While most of the later results have been in general agreement with the Knudsen *et al.* data, there appear to be some significant differences from the earlier summary data and among the recent data. For example, as will be shown, in several studies it was observed that, at frequencies below 500 cps, the dependence of the underwater ambient-noise levels on wind speed and sea state decreased as the frequency decreased, and at 100 cps and below, little or no dependence was seen; while other observers have reported a substantial wind-speed dependence extending to frequencies as low as 50 cps.

As might be expected, differing procedures have been used in obtaining and processing data, and in defining results. Data from the various sources cannot always be compared directly but must be given additional treatment in many cases.

This review is designed to bring together for comparison, after being appropriately processed, the results of recent investigations; to show that many of the observed differences as well as similarities can be explained by certain plausible assumptions as to source and source characteristics; and to indicate how to apply this information in estimating the ambient-noise levels for a given situation.

1. AMBIENT-NOISE SPECTRA

The main purpose of this paper is to discuss the more widespread and prevailing characteristics of ambient noise in the ocean. Obvious noise from marine life, nearby ships, and other sources of intermittent and local noise is not included in the data considered in this section.

As has been shown,[1,2] in the absence of sounds from ships and marine life, underwater ambient-noise levels are dependent on wind force and sea state, at least at frequencies between 100 cps and 25 kc. Therefore, wind dependence was made the starting point for the analysis of the results of recent investigations.

The processing of data reported in differing terms included the following: conversion of levels to dB *re* 0.0002 dyn/cm^2 and a 1-cps bandwidth; estimation of wind force from stated sea states (see Table I); the derivation of spectra corresponding to the means of the Beaufort-scale wind-speed ranges, from given equations or graphs relating level to wind speed at various frequencies; and the computation of average spectrum levels corresponding to Beaufort-scale groupings. When the sampling was small, graphical smoothing and interpolation were often employed.

Each datum point used in determining the ambient-

[1] V. O. Knudsen, R. S. Alford, and J. W. Emling, "Survey of Underwater Sound, Report No. 3, Ambient Noise," 6.1–NDRC–1848 (September 26, 1944) (P B 31021).
[2] V. O. Knudsen, R. S. Alford, and J. W. Emling, J. Marine Research **7**, 410 (1948).
[3] E. O. Hulburt, J. Acoust. Soc. Am. **14**, 173 (1943).
[4] D. P. Loye and D. A. Proudfoot, J. Acoust. Soc. Am. **18**, 446 (1946).
[5] M. W. Johnson, F. A. Everest, and R. W. Young, Biol. Bull. **93**, 122 (1947).
[6] F. A. Everest, R. W. Young, and M. W. Johnson, J. Acoust. Soc. Am. **20**, 137 (1948).

TABLE I. Approximate relation between scales of wind speed, wave height, and sea state.

Sea criteria	Beaufort scale	Wind speed: Range knots (m/sec)	Wind speed: Mean knots (m/sec)	12-h wind: Wave height[a,b] ft (m)	Fully arisen sea: Wave height[a,b] ft (m)	Fully arisen sea: Duration[b,c] h	Fully arisen sea: Fetch[b,c] naut. miles (km)	Sea-state scale
Mirror-like	0	<1 (<0.5)						0
Ripples	1	1–3 (0.5–1.7)	2 (1.1)					½
Small wavelets	2	4–6 (1.8–3.3)	5 (2.5)	<1 (<0.30)	<1 (<0.30)			1
Large wavelets, scattered whitecaps	3	7–10 (3.4–5.4)	8½ (4.4)	1–2 (0.30–0.61)	1–2 (0.30–0.61)	<2.5	<10 (<19)	2
Small waves, frequent whitecaps	4	11–16 (5.5–8.4)	13½ (6.9)	2–5 (0.61–1.5)	2–6 (0.61–1.8)	2.5–6.5	10–40 (19–74)	3
Moderate waves, many whitecaps	5	17–21 (8.5–11.1)	19 (9.8)	5–8 (1.5–2.4)	6–10 (1.8–3.0)	6.5–11	40–100 (74–185)	4
Large waves, whitecaps everywhere, spray	6	22–27 (11.2–14.1)	24½ (12.6)	8–12 (2.4–3.7)	10–17 (3.0–5.2)	11–18	100–200 (185–370)	5
Heaped-up sea, blown spray, streaks	7	28–33 (14.2–17.2)	30½ (15.7)	12–17 (3.7–5.2)	17–26 (5.2–7.9)	18–29	200–400 (370–740)	6
Moderately high, long waves, spindrift	8	34–40 (17.3–20.8)	37 (19.0)	17–24 (5.2–7.3)	26–39 (7.9–11.9)	29–42	400–700 (740–1300)	7

[a] The average height of the highest one-third of the waves (significant wave height).
[b] Estimated from data given in U. S. Navy Hydrographic Office (Washington, D. C.) publications HO 604 (1951) and HO 603 (1955).
[c] The minimum fetch and duration of the wind needed to generate a fully arisen sea.

noise spectra is an average of several samples from a single locality. In many cases, shipborne systems were used and usually only a very few samples were obtained at each of several stations in the same general area. The mean spectra derived from such measurements comprise average values of data from all stations in the same general area.

[1.1] 20 cps to 10^4 cps

The spectra[7] resulting from measurements made in five different shallow-water areas are shown in Fig. 1. [Shallow water is defined as water less than 100 fathoms (183 m) in depth.] Deep-water ambient-noise spectra from five different areas are presented in Fig. 2. It is evident that spectra corresponding to the same scale of wind speed or sea state can exhibit considerable differences in spectrum shape and level. However, as is indicated by the arrangement of the figures, groupings can be made within which the spectra are roughly the same, and between which distinct differences are evident.

The wind-dependent spectra at the left in Fig. 1 [parts (a), (c), and (e)] are characterized by broad maxima, the highest value occurring at a frequency between 400 and 800 cps. None of the other spectra demonstrates this spectrum shape clearly, although there are suggestions of it in some, being indicated by a flattening between 200 and 1000 cps. The maxima appearing in Figs. 1(b) and 1(d) and Figs. 2(b) and 2(d) occur at frequencies 100 cps and below, and the levels in the neighborhood of the spectrum maximum are not wind-dependent.

The wind-dependent aspects of the ambient noise appear to be greatly influenced by non-wind-dependent components. In the spectra at the right in each figure, little wind dependence is evident below about 200 cps and the levels of the non-wind-dependent noise are as high as or higher than the levels shown for the highest wind speeds in the left-hand graphs in each figure. The levels of this non-wind-dependent noise decrease rapidly, 8 to 10 dB per octave, at frequencies above 100 cps.

In the areas associated with Figs. 1(a) and 1(c), a relatively high residual noise limits wind dependence to wind speeds more than 5 to 10 knots (Beaufort 2 to 3), depending on frequency.

At frequencies above 500 cps, where the noise levels show wind dependence in nearly every case, the spectra have the same general shape and approach a spectrum slope of about −6 dB per octave above 1000 cps.

[7] Unless otherwise indicated, all spectra are given in terms of pressure–spectrum level in dB *re* 0.0002 dyn/cm², the reference bandwidth being 1 cps as included in the definition of the term "spectrum level" in Sec. 2.8, American Standard Acoustical Terminology, ASA-S1.1-1960 (American Standards Association, May 25, 1960).

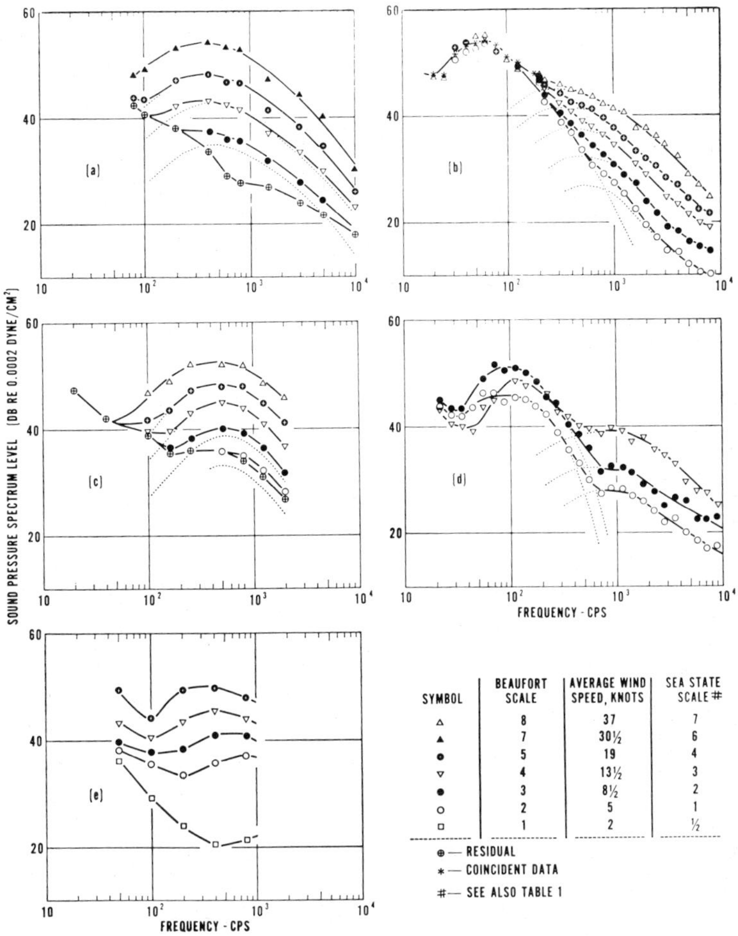

FIG. 1. Shallow-water ambient-noise spectra, showing average spectrum levels for each of several Beaufort-scale wind-speed groupings, as measured in five different areas. The dotted curves define component spectra according to an analytical interpretation of the observed spectra.

SYMBOL	BEAUFORT SCALE	AVERAGE WIND SPEED, KNOTS	SEA STATE SCALE #
△	8	37	7
▲	7	30½	6
●	5	19	4
▽	4	13½	3
◉	3	8½	2
○	2	5	1
□	1	2	½

⊕ — RESIDUAL
✳ — COINCIDENT DATA
— SEE ALSO TABLE 1

However, the shallow-water levels (Fig. 1) are in general about 5 dB higher than the corresponding deep-water levels (Fig. 2) at the same frequency and wind speed. The figures show average values. The variability is such that the higher deep-water levels for a given condition are about the same as the lower shallow-water levels; that is, the distributions overlap.

For frequencies between 20 cps and 10 kc, the range of the data in Figs. 1 and 2, the diverse ambient-noise spectra can be explained by assuming various combinations of a wind-dependent component and a non-wind-dependent component. The application of this interpretation is demonstrated in Figs. 1 and 2 by dotted curves which indicate the probable spectra of the components which have combined in each case to produce the observed spectrum. The wind-dependent component is assumed to have a spectrum with a broad maximum between 100 cps and 1000 cps, like those in Figs. 1(a), 1(c), and 1(e). In general, the spectrum of the non-wind-dependent component is assumed to peak at 100 cps or lower, and to fall off steeply above 100 cps, as seen in Figs. 1(b), 1(d), 2(b), 2(d),[8] and 2(e). As the figures illustrate, quite reasonable combinations of such spectra result in spectra like the observed spectra.

In Figs. 1(a) and 1(c), the spectrum of the residual noise does not decrease rapidly with frequency, and the wind dependence is altered at the higher frequencies as well as at low frequencies. In this case the observed

[8] The peak near 60 cps in Fig. 2(d) is not caused by a self-noise "hum" component from system–power sources. The maximum appears to be real but may be accentuated by variations in the response of the measurement system not revealed by the calibration data.

FIG. 2. Deep-water ambient-noise spectra, showing average spectrum levels for each of several Beaufort-scale wind-speed groupings, as measured in five different areas. The dotted curves define component spectra according to an analytical interpretation of the observed spectra.

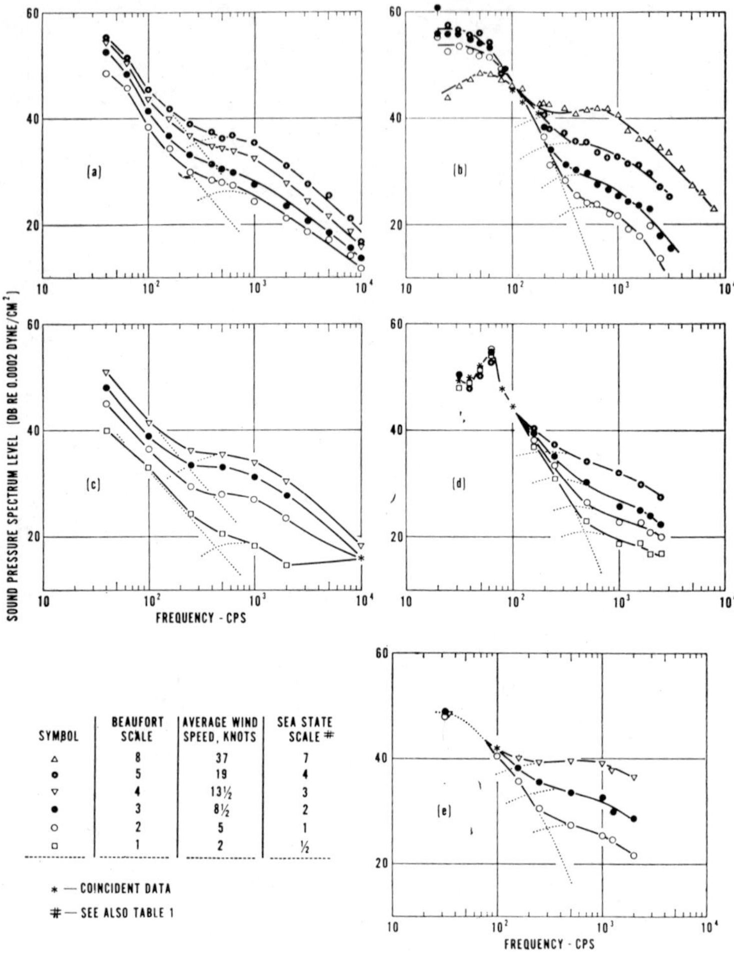

SYMBOL	BEAUFORT SCALE	AVERAGE WIND SPEED, KNOTS	SEA STATE SCALE #
△	8	37	7
⊙	5	19	4
▽	4	13½	3
●	3	8½	2
○	2	5	1
□	1	2	½

* — COINCIDENT DATA

— SEE ALSO TABLE 1

spectra are a combination of the wind-dependent spectra with the spectrum of a residual-noise component which prevails at the lower wind speeds.

In Figs. 1(e), 2(a), and 2(c), wind dependence appears to be universal. Also, minima or inflection points appear in the spectra between 100 and 500 cps. This spectrum shape suggests the possibility, at least, of two different wind-dependent sources or mechanisms. In the system used for the measurements from which the data in Figs. 2(a) and 2(c) were obtained, the hydrophones were not well isolated from the effects of surface fluctuations, and there is a strong possibility that the low-frequency wind-dependent noise is a form of system self-noise. This does not apply to the data in Fig. 1(e), however.

When measured in the same area, or even in the same place with the same system and at the same wind speed, there is often considerable variation in the observed levels of the wind-dependent noise as measured at different times. Such differences are illustrated in Fig. 3, which shows spectra comprising levels averaged over two different time periods at each of two different locations. While the spectra are of the same general shape, for the same wind speeds, the wind-dependent September levels in Fig. 3(a) run about 8 dB below those for January in Fig. 3(b), and the wind-dependent June–July levels in Fig. 3(c) are about 5 dB below those for September–October in Fig. 3(d). If it is assumed that the source of the wind-dependent noise is in the surface agitation resulting from the effects of the wind,

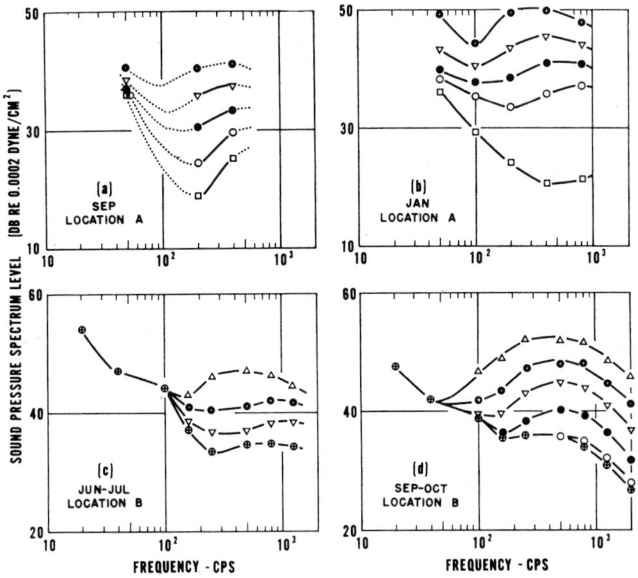

SYMBOL	BEAUFORT SCALE	AVERAGE WIND SPEED, KNOTS	SEA STATE SCALE#
△	8	37	7
◉	5	19	4
▽	4	13½	3
●	3	8½	2
○	2	5	1
□	1	2	½

⊕ — RESIDUAL

— SEE ALSO TABLE 1

FIG. 3. Ambient-noise spectra, illustrating differences in the averages of levels measured in the same area and at the same wind speeds, but during different periods of time.

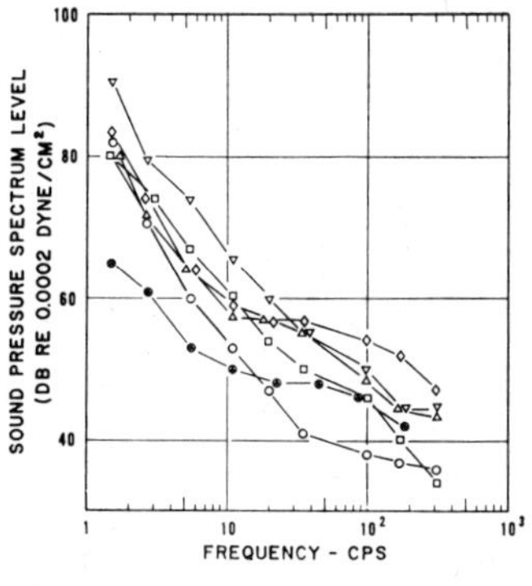

FIG. 4. Low-frequency ambient-noise spectra, comparing the averages of a number of measurements made in each of five different areas. The open triangles and the inverted triangles represent data taken in the same general area but at different depths.

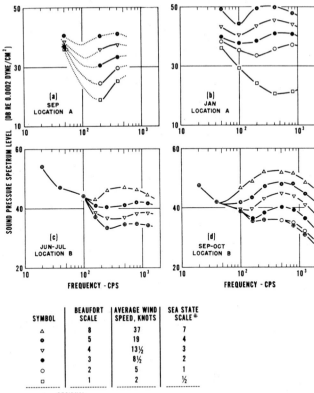

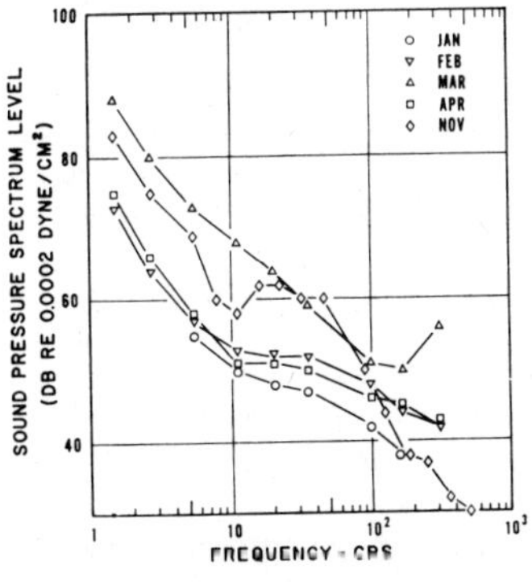

FIG. 5. Low-frequency ambient-noise spectra, comparing the averages of levels measured during different months of the year at the same location.

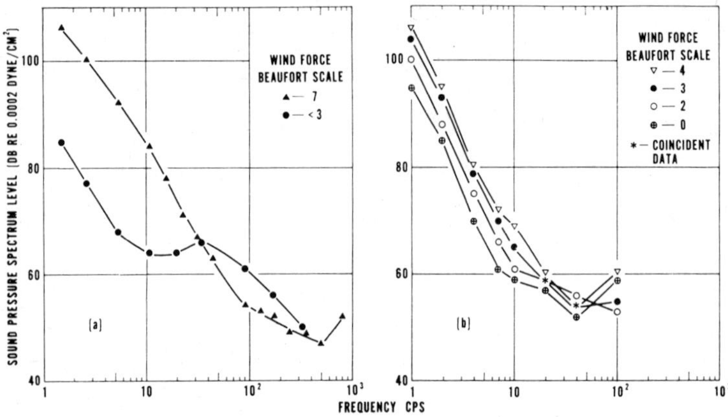

FIG. 6. Low-frequency ambient-noise spectra showing wind-dependence in very shallow water (less than 25 fathoms or 46 m) at two widely separated locations.

variations of this nature are not entirely unexpected. Wind speed alone is only a crude and incomplete measure of the surface agitation which depends also on such factors as the duration, fetch, and constancy of the wind, and its direction in relation to local conditions of swell, current, and, in near-shore areas, topography. Subjective estimates of sea state are not necessarily an improvement over wind speed as a measure of the pertinent surface agitation.

[1.2] 1 cps to 100 cps

The amount of data available from measurements at very low frequencies is relatively small. No consistent wind dependence has been reported, except for very shallow water. Some obvious effects from nearby shipping have been observed. Exclusive of these obvious effects, rather wide variations in the very low-frequency noise levels have been experienced.

In Fig. 4, the individual curves compare the averages of a number of measurements made in each of five different areas. All of the data in Fig. 5 were obtained at the same location, the different curves showing averages of data taken during different months of the year. Wind dependence in very shallow water (less than 25 fathoms or 46 m) at two widely separated locations is demonstrated in Fig. 6.

Some generalizations can be made about the results shown in Figs. 4–6. The very low-frequency noise may differ in level by 20 to 25 dB from one place to another, and from one time to another. The spectrum shape below 10 cps is nearly always the same, and has a slope of −8 to −10 dB per octave. Between 10 cps and 100 cps, the spectrum often flattens and may even show a broad maximum, but in some instances the spectrum slope shows little or no change from the slope below 10 cps.

1.3 Minimum Levels

The lowest levels encountered in the data available to the author are shown in Fig. 7. The solid symbols represent measurements made in an inland lake. The remainder of the data pertains to measurements in the ocean. Data from the same set of measurements are connected by dashed lines. Symbols shown with downward-pointing arrows designate equivalent system-noise levels which mark an upper limit to the ambient-noise levels existing at the time of the measurement. The solid curve defines levels which will almost always be exceeded by observed levels.

As indicated by the individual sets of measurements, it is unlikely that the very low levels will be encountered in every part of the spectrum at the same time. This may

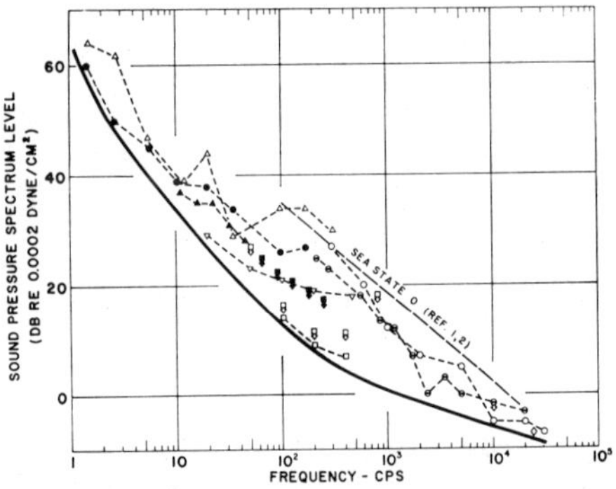

FIG. 7. The empirical lower limit of ambient-noise spectra (solid curve), as determined by the lowest of observed levels. The solid symbols refer to measurements made in an inland lake, the open symbols to those in the ocean. Symbols with downward-pointing arrows designate equivalent system-noise levels which mark an upper limit to the ambient-noise level existing at the time of the measurement. The sea state 0 curve from references 1 and 2 is shown for comparison.

be interpreted as an indication that the different regions of the spectrum are dominated by different components which combine to produce the observed spectra, but which are not all at a minimum at the same time.

The solid curve is an estimate of the minimum levels existing in the ocean. This estimate is probably high since the curve is to some extent an indication of the state of the measurement art. In the results of one investigation, it was stated that during a period of measurement covering 44 h of data, 40% of the time the noise levels at 200 cps were too low to measure. The limiting equivalent system-noise spectrum level at 200 cps was 10 dB *re* 0.0002 dyn/cm². The data represented by the symbols with downward-pointing arrows are also an indication of the need for improved measurement techniques.

1.4 Interpretation

In the absence of noise from marine life and nearby ships, the underwater ambient-noise spectrum between 1 cps and 10 kc may be resolved into several overlapping subspectra: A *low-frequency spectrum* with a −8 dB to −10 dB per octave spectrum-level slope, in the range 1 to 100 cps; a *"non-wind-dependent" spectrum* in the range 10 cps to 1000 cps with a maximum between 20 and 100 cps and falling off rapidly above 100 cps (sometimes not observed); and a *wind-dependent spectrum* in the range 50 cps to 10 kc with a broad maximum between 100 cps and 1000 cps and a −5-dB- or −6-dB-per-octave slope above 1000 cps.

At low frequencies, the ambient noise is dominated by the component, or components, characterized by a −8-dB- to −10-dB-per-octave slope, which extends in some cases to frequencies as high as 100 cps. Above 500 cps, wind-dependent noise nearly always prevails. The frequency band between 10 cps and 1000 cps, being the region of overlap, is a highly variable one in which each observed spectrum depends on a combination of the three overlapping component spectra, each of which may vary independently with time and place.

Minimum levels are determined in some cases, mostly in shallow water, by local residual-noise components, such as those illustrated in Figs. 1(a) and 1(c), whose levels exceed the levels of the more general limiting noise indicated in Fig. 7.

2. SOURCES OF NOISE IN THE OCEAN

2.1 Thermal Agitation

The effects from the thermal agitation of a medium determine a minimum noise level for that medium. For the ocean, the equivalent thermal-noise sound-pressure level is given, for ordinary temperatures between 0° and 30°C, by the relation[9]:

$$L_t \approx -101 + 20 \log f, \tag{1}$$

where L_t is the thermal-noise level in dB *re* 0.0002 dyn/cm² for a 1-cps bandwidth, and f is the frequency in cps. According to Eq. (1), the thermal-noise spectrum has a slope of +6 dB/octave and a level of −10 dB at 35 kc. At the upper frequency limit of the ambient-noise data shown in Fig. 7, around 20 or 30 kc, the minimum ambient-noise levels are about the same as the thermal-noise levels. It is obvious from Fig. 7, however, that, at frequencies below 10 kc, even the lowest of the observed ambient-noise levels are well above the thermal-noise limit, and other noise sources must be found to explain the observed spectra.

[9] R. H. Mellen, J. Acoust. Soc. Am. 24, 478 (1952).

2.2 Hydrodynamic Sources

A wide variety of hydrodynamic processes is continually taking place in the ocean, even at zero sea state. It is known that the radiation of sound often results from these processes.

2.2.1 Bubbles

An oscillating bubble is an effective sound source. Both free and forced oscillations of bubbles occur in the ocean, particularly in the surface agitation resulting from the effects of wind.

Pertinent information concerning the radiation of sound by air bubbles in water has been given by Strasberg.[10,11] The sound pressures associated with the higher modes of oscillation of the bubbles are negligible so that only simple volume pulsations (zeroth mode) need be considered. In the case of forced oscillations, the sound energy tends to be concentrated at the natural frequency of oscillation of the zeroth mode also, but this tendency may be altered if the frequencies associated with the environmental pressure fluctuations are much below the natural frequency of oscillation of the bubbles.

The natural frequency of oscillation for the zeroth mode is

$$f_0 = (3\gamma p_s \rho^{-1})^{\frac{1}{2}} (2\pi R_0)^{-1}, \tag{2}$$

where γ is the ratio of specific heats for the gas in the bubble, p_s is the static pressure, ρ the density of the liquid, and R_0 is the mean radius of the bubble. The amplitude of the radiated sound pressure at a distance d from the center of the bubble is

$$p_0 = 3\gamma p_s r_0 d^{-1}, \tag{3}$$

r_0 being the amplitude of the zeroth mode of oscillation. It is assumed that the amplitude of the bubble oscillation is relatively small so that the various modes are independent of each other.

The natural frequency is inversely proportional to the bubble size, and the radiated sound-pressure amplitude is directly proportional to the bubble-oscillation amplitude. There is a practical limit to bubble size, and it is quite probable, also, that in many cases there would be a predominance of bubbles of nearly one size only. It is

[10] M. Strasberg, J. Acoust. Soc. Am. 28, 20 (1956).
[11] H. M. Fitzpatrick and M. Strasberg, David Taylor Model Basin Rept. 1269 (January 1959).

to be expected, therefore, that in general the spectrum has a maximum at some frequency associated with either a predominant bubble size or a maximum bubble size, the exact shape depending on the distribution of bubble sizes and amplitudes of oscillation.

Franz[12] has measured the sound energy radiated by air bubbles formed when air is entrained in the water following the impact of water droplets on the surface of the water. His results are given in the form of one-half-octave-band sound-energy spectra which exhibit maxima. The decline toward lower frequencies is sharp (8 to 12 dB per octave, in terms of energy-spectrum level) and is attributed to an almost complete absence of bubbles larger than a certain size. A more gradual decline toward higher frequencies (−6 dB to −8 dB per octave) was found and was interpreted as being the result of a decrease in the radiated sound energy per bubble rather than a decrease in the prevalence of bubbles.

Data on bubble size and environmental conditions are not available in sufficient detail for making exact predictions concerning the bubble noise in the ocean. However, a rough appraisal can be made. According to Eqs. (2) and (3), a spherical air bubble of mean radius 0.33 cm, in water at atmospheric pressure, oscillating with an amplitude one-tenth the mean radius ($r_0=0.1R_0$), has a simple source-pressure level[13] referred to 1 m, of about 133 dB above 0.0002 dyn/cm^2 at a frequency of approximately 1000 cps. For a frequency of 500 cps, the mean-bubble radius is about 0.66 cm, and, for the same amplitude-to-size ratio, the source level is 6 dB higher.

These source levels are some 75 to 100 dB above the observed ambient-noise spectrum levels at these frequencies. The noise from such bubble sources could be observed at a considerable distance. The maxima in the observed wind-dependent ambient-noise spectra (see preceding Sec. 1.1 and Figs. 1 and 2) occur at frequencies between 300 cps and 1000 cps, which correspond to bubble sizes of 1.1 cm to 0.33 cm in mean radius, a reasonable order of magnitude.

The characteristic broadness of the maxima in the wind-dependent ambient-noise spectra can be explained by the reasonable assumption that in the surface agitation the bubble size and energy distributions are not sharply concentrated around the averages. The ambient-noise high-frequency spectrum slope above the maximum, approximately −6 dB/octave, agrees with that of the bubble noise.

The nature of cavitation noise has been described by Fitzpatrick and Strasberg.[11] According to the acoustic theory, the sound-pressure spectra have maxima at frequencies corresponding approximately to the reciprocal of the time required for growth and collapse of the vapor cavities. At low frequencies the predicted spectrum slope is 12 dB per octave, but at high frequencies the spectrum is determined by details of very rapid changes in sound pressure which are not given correctly by the acoustic theory.

The noise produced by a stirring rod 2 in. long and $\frac{1}{16}$ in. in diameter rotating at 4300 rpm in the Thames River (New London, Connecticut) was measured by Mellen.[14] His results are given in the form of a sound-pressure spectrum which shows a maximum near 1000 cps and slope of approximately −6 dB per octave at higher frequencies. The spectra of noise from cavitating submerged water jets as reported by Jorgensen[15,16] show a slope of approximately 12 dB per octave at low frequencies, in agreement with the acoustic theory, and a slope of about −6 dB per octave at high frequencies, in agreement with Mellen's data. Observed spectra of noise radiated by submarines exhibit characteristics which are in general agreement with these data and which have been attributed to cavitation effects.[17]

The spectrum shape of cavitation noise is similar to that of the air-bubble noise, which, as has been pointed out, resembles the spectrum shape of the wind-dependent ambient noise (see Figs. 1 and 2). For cavities of comparable size one would expect higher noise levels from cavitation than from the simple volume pulsations of gas bubbles since the amplitude of oscillation is usually greater.

From the foregoing, it is concluded that air bubbles and cavitation produced at or near the surface, as a result of the action of the wind, could very well be a source of the wind-dependent ambient noise at frequencies between 50 cps and 10 kc.

Bubbles are present in the sea (or lakes) even when the wind speeds are below that at which whitecaps are produced. Bubbles are created, not only by breaking waves, but also by decaying matter, fish belchings, and gas seepage from the sea floor. Furthermore, there is evidence of the existence of invisible microbubbles in the sea, and of the occurrence of gas supersaturation of varying degree near the surface. These conditions provide a favorable environment for the growth of microbubble nuclei into bubbles as a result of temperature increases, pressure decreases, and turbulence associated with currents and internal waves, as well as with surface waves.[18–20] As the bubbles rise to the surface, growing in size (because of the decreasing hydrostatic pressure), they are subjected to transient pressures which induce the oscillations which generate the noise. Even on quiet

[12] G. J. Franz, J. Acoust. Soc. Am. **31**, 1080 (1959).

[13] Source-pressure level is defined as the sound-pressure level at a specified reference distance in a specified direction from the effective acoustic center of the source.

[14] R. H. Mellen, J. Acoust. Soc. Am. **26**, 356 (1954).

[15] D. W. Jorgensen, David Taylor Model Basin Rept. 1126 (November 1958).

[16] D. W. Jorgensen, J. Acoust. Soc. Am. **33**, 1334 (1961).

[17] NDRC Summary Tech. Repts. Div. 6, Vol. 7, Principles of Underwater Sound, Sec. 12.4.5. (Distributed by Research Analysis Group, Committee on Undersea Warfare, National Research Council.)

[18] E. C. LaFond and P. V. Bhavanarayana, J. Marine Biol. Assoc. India **1**, 228 (1959).

[19] W. L. Ramsey, Limnology and Oceanography **7**, 1 (1962).

[20] E. C. LaFond and R. F. Dill, NEL TM-259 (1957) (unpublished technical memorandum).

days and in the absence of wind, bubbles have been seen to emerge from the water, sometimes persisting for a time as foam, and then to burst. These oceanographic data support the hypothesis that bubble noise may still be an important component of underwater ambient noise, even when there is little or no surface agitation from the wind.

Thus, there is evidence of the presence of bubbles in the ocean both when winds are high and when winds are low, or even during a calm. Oscillating and collapsing bubbles are efficient and relatively high-level noise sources. Both the level and shape of the observed wind-dependent ambient noise can be explained by the characteristics of bubble noise and cavitation noise.

2.2.2 *Water Droplets*

The underwater noise radiated by a spray of water droplets at the surface of the water has been investigated by Franz.[12] The noise from such splashes appears to be made up of noise from the impact and passage of the droplet through the free surface. In many cases, air bubbles are entrained so that the total noise includes contribution from the bubble oscillations as well. The sound-energy spectrum has a broad maximum near a frequency equal to twice the ratio of the impact velocity to the radius of the droplets. Towards lower frequencies, the spectrum density decreases gradually at a rate of 1 or 2 dB per octave. At frequencies above the maximum, the slope approaches -5 or -6 dB per octave. The impact part of the radiated sound energy increases with increase in droplet size and impact velocity. The relation is modified somewhat by the bubble noise, particularly at intermediate velocities.

Franz estimated the sound-pressure spectrum levels to be expected from the impact of rain upon the surface of the water. He concluded that rain exceeding a rate of 0.1 in./h would be expected to raise ambient-noise levels and flatten the spectrum at frequencies above 1000 cps under sea-state-1 conditions. The measurements of ambient sea noise made by Heindsmann *et al.*[21] during periods of rainfall are in fair agreement with the estimates made by Franz.

The noise from splashes of rigid bodies, such as from hail or sleet, is in general similar to that from water droplets, but is modified by the effects of resonant vibrations of the bodies.

In addition to the effects of precipitation, there is the possibility that noticeable contribution to the ambient sea noise may come from spray and spindrift, especially at the higher wind speeds.

2.2.3 *Surface Waves*

The fluctuations in the elevation of the surface of a body of water cause subsurface pressure fluctuations which, whether controlled by compressibility or not, affect the transducer of an underwater system.

[21] T. E. Heindsmann, R. H. Smith, and A. D. Arneson, J. Acoust. Soc. Am. **27**, 378 (1955).

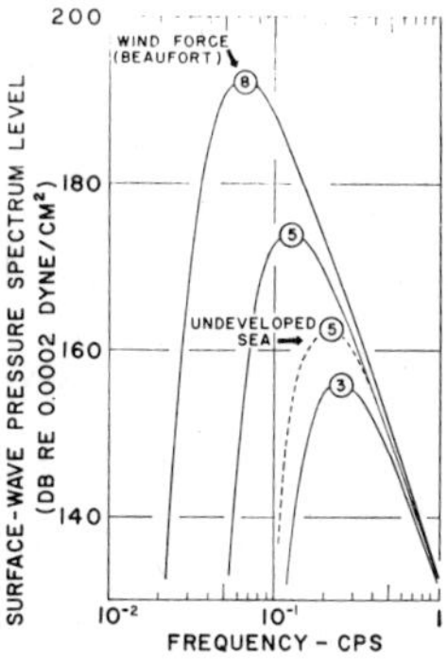

FIG. 8. Surface-wave pressure-level spectra, derived from Neumann-Pierson surface-wave elevation spectra (see references 22 and 23).

The spectra of surface waves, that is, the spectrum densities of the time variation of the surface elevation at a fixed point, according to Neumann[22] and Pierson,[23] are represented by the relation

$$\bar{h}^2(\omega)=C\omega^{-6}\exp(-2g^2\omega^{-2}v^{-2}), \qquad (4)$$

where $\bar{h}^2(\omega)$ is the mean-square elevation of the surface per unit bandwidth at the angular frequency ω, g is the acceleration of gravity, and v is the wind speed. In cgs units the constant C, determined from empirical data, is equal to 4.8×10^4 cm² sec⁻⁵.

Surface-wave elevation spectra for winds of force 3, 5, and 8 (about 5, 10, and 20 m/sec) were computed using Eq. (4). The spectra shown in Fig. 8 are in terms of the pressure-spectrum levels corresponding to the mean square of the variation in the surface elevation in reference to a 1-cps bandwidth. The maximum of spectrum energy occurs at frequencies below 0.5 cps, and the band of maximum energy moves to lower frequencies as wind speed increases.

Equation (4) applies to a fully developed sea. When the sea is not fully developed, the high-frequency part of the spectrum is unchanged, but the larger low-frequency waves have not yet been produced, and the spectrum is cut off at the lower end as roughly exemplified by the dashed curve in Fig. 8. The cutoff frequency depends on the duration and fetch of the wind.

For frequencies above 1 cps, the value of the exponential function in Eq. (4) is very nearly unity, and the spectrum densities decrease as f^{-6} (-18 dB per octave). However, the relation was derived from measurements of the larger waves of frequencies below 0.5 cps, and extrapolation to frequencies above 0.5 cps is not certain.

The higher frequency surface fluctuations are in the form of small gravity waves and capillaries. Phillips[24] discusses an equilibrium region for the small gravity waves for which the spectrum is given by the relation

$$\bar{h}^2(\omega)\sim7.4\times10^{-3}g^2\omega^{-5}. \qquad (5)$$

[22] G. Neumann, Department of the Army, Corps of Engineers, Beach Erosion Board Tech. Mem. No. 43 (December 1953).
[23] Willard J. Pierson, Jr., Advances in Geophysics **2**, 93 (1955).
[24] O. M. Phillips, J. Marine Research **16**, 231 (1957–1958).

According to Eq. (5), the pressure level corresponding to $\tilde{h}^2(2\pi)$ (frequency 1 cps) is 132.6 dB, essentially the same as the corresponding values derived from Eq. (4) (see Fig. 8). A relation is given for capillaries[25] in which the surface-displacement spectrum is proportional to $\omega^{-7/3}$. Details of the capillary-gravity wave system are not known, although Cox[26] has presented evidence from wave-slope observations which suggests that capillary waves become important only for wind speeds above 6 m/ sec (11 to 12 knots).

Kinsman[27] has summarized a number of surface-wave spectrum measurements and his results indicate that the slope is between −13.5 and −16.5 dB per octave in the frequency band from 0.7 to 2.1 cps.

The first-order pressure fluctuations induced by the surface waves are attenuated with depth, the attenuation being frequency-dependent.[28–30] The characteristics of the "depth filter" are shown in Fig. 9. The depth filter is a low-pass filter with a sharp cutoff and generally limits significant first-order pressure effects from surface waves to frequencies below 0.2 to 0.3 cps, and to depths less than a few hundred feet.

In Fig. 10 low-frequency pressure spectra obtained from "ocean-wave" measurements are compared with spectra resulting from shallow-water "ambient-noise" measurements. The ocean-wave spectra were derived from data reported by Munk *et al.*[31] The ambient-noise spectra were selected from the sources for Fig. 6 (of this paper). The effect of the depth filter is indicated by the ocean-wave spectra. Considering the difference in measurement methods and the variety of environmental conditions involved, the two sets of data merge remarkably well.

Because of the steep negative slope of the surface-wave spectra, and because of the steep cutoff of the depth filter, it is doubtful that the ambient noise at frequencies above 1 cps includes any significant contribution from the first-order pressure fluctuations induced by surface waves. At frequencies below 0.3 cps, approximately, these pressure fluctuations will very likely comprise a large part of the "ambient noise" observed *at very shallow depths* (<300 ft or 100 m) with pressure transducers.

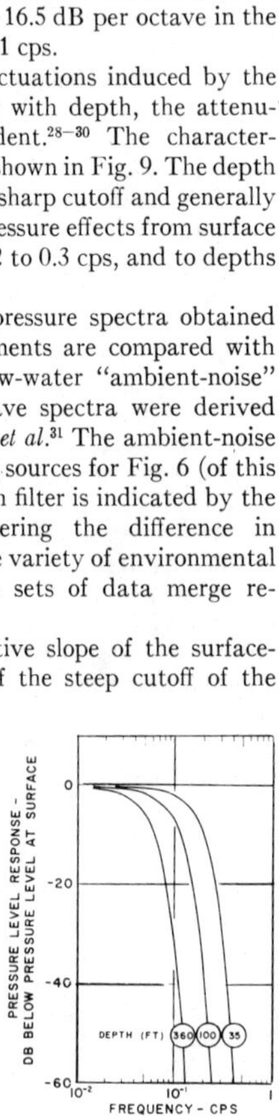

FIG. 9. Depth-filter characteristics, showing the attenuation of first-order pressure fluctuations as a function of frequency at three selected depths.

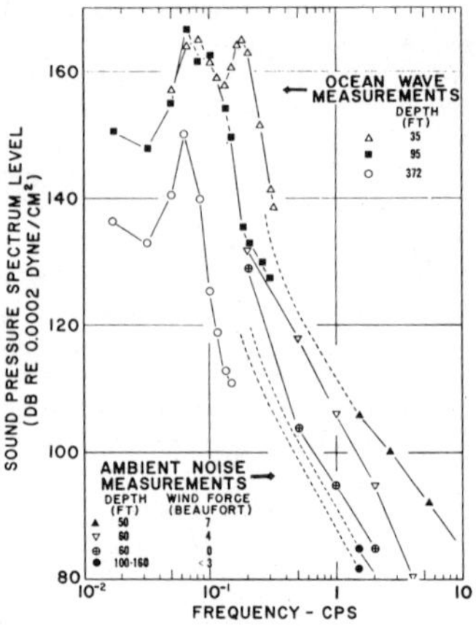

FIG. 10. Pressure-level spectra comparing results of ocean-wave measurements (derived from reference 31) and ambient-noise measurements. The dashed curves are extrapolations.

Longuet–Higgins[32] has called attention to a second-order pressure variation which is not attenuated with depth. These second-order pressure variations occur when the wave trains of the same wavelength travel in opposite directions. The resulting pressure variation is of twice the frequency of the two waves with an amplitude proportional to the product of the amplitudes of the two waves. When the depth is of the same order as, or greater than, the length of the compression wave, compression waves are generated.

The conditions for the Longuet–Higgins effect are met in the open ocean where the winds associated with a cyclonic depression produce waves travelling in opposite directions. Opposing waves occur when waves are reflected from the shore. The vagaries of local winds may also produce the required patterns in the high-frequency capillary-gravity wave system, which, though short-lived, may be numerous and frequent.

One may postulate that surface waves are a source of low-frequency ambient noise by way of these second-order pressure variations. A comparison of the observed ambient-noise levels in Fig. 10 with the estimated surface-wave spectrum levels in Fig. 8 indicates that the pressure variations are of sufficient magnitude. The observed ambient-noise low-frequency spectrum slope

[25] O. M. Phillips, J. Marine Research **16**, 229 (1957–1958).
[26] Charles S. Cox, J. Marine Research **16**, 244 (1957–1958).
[27] Blair Kinsman, J. Geophys. Research **66**, 2411 (1961).
[28] S. Rauch, University of Calif. Department of Engineering, Fluid Mechanics Lab. Tech. Rept. HE–116–191 (November 29, 1945).
[29] R. L. Wiegel, University of Calif. Department of Engineering, Fluid Mechanics Lab. Mem. HE–116–108 (September 8, 1948).
[30] W. H. Munk, F. E. Snodgrass, and M. J. Tucker, Bull. Scripps Inst. of Oceanog. Univ. Calif. **7**, 283 (1959), Fig. 5.
[31] Reference 30, charts 2.1, 4.1, and 5.1.
[32] M. S. Longuet–Higgins, Phil. Trans. Roy. Soc. (London) **A243**, 1 (1950).

of −8 to −10 dB per octave could be accounted for by the assumption of a suitable combination of the capillary and gravity waves in the wave system for the frequency range 0.3 to 10 cps.

It is concluded that the *second-order pressure variations* resulting from surface waves may sometimes be a significant part of the ambient noise at frequencies below 10 cps (see Sec. 1.2 and Figs. 4–6).

2.2.4 *Turbulence*

The state of turbulence is mainly one of unsteady flow with respect to *both* time and space coordinates. When the fluid motion is "turbulent," irregularities exist relative to a point moving with the fluid, as well as relative to a fixed point outside the flow.

Turbulence may occur in a fluid as a result of current flow along a solid boundary and also when layers of the fluid with different velocities flow past or over one another. Turbulence may be expected in the ocean at the water ocean-floor boundary, particularly in coastal areas, straits, and harbors; at the sea surface because of the movement and agitation of the surface; and within the medium as a result of the horizontal and vertical water movements, such as advection, convection, and density currents.

Noise resulting from turbulence created by relative motion between the water and the transducer is considered to be self-noise of the system rather than ambient noise in the medium.

Much of the energy input into the ocean occurs at frequencies too low to be of direct consequence to the ambient noise at frequencies above 1 cps. In the processes of turbulence,[33] the largest-scale eddies and the lowest wavenumbers correspond to the region of energy input. The largest eddies break up into smaller and smaller eddies, some of the energy being transferred to higher and higher frequencies. This is precisely the kind of mechanism which could transfer low-frequency energy in the ocean to higher frequencies.

However, the generation and radiation of noise from turbulence are a very inefficient process. Radiated noise levels derived from the relations formulated by Lighthill,[34] using values of turbulent-velocity fluctuations and dissipation rates estimated by Pochapsky,[35] for the ocean are many orders of magnitude below the observed ambient-noise levels. The levels corresponding to experimental values found for regions of strong currents in an inland passage by Grant *et al.*[36] are also low by many orders of magnitude. Even when there is comparatively violent turbulence, such as the case of a turbulent jet,[37] the level of the radiated pressure fluctuations is low compared to ambient-noise levels. It is concluded that the *radiated noise* from turbulence does not contribute to the observed ambient noise, except possibly under specific local conditions.

The pressure fluctuations of the turbulence itself are of much greater magnitude than those of the radiated noise.[38] A pressure-sensitive hydrophone[39] in the turbulent region responds to these pressure fluctuations as it does to any pressure fluctuations, whether they are those of propagated sound energy or not.

According to experimental results and the generally accepted theory of turbulence,[33] the following relations may be used for rough estimates of the turbulent velocity and pressure fluctuations:

$$\tilde{u}\approx 0.05\,\bar{U}, \qquad (6)$$

$$\tilde{u}^2(k)\propto k^4 \qquad (k\ll 0.01\ \text{cm}^{-1}), \qquad (7a)$$

$$\tilde{u}^2(k)\propto k \qquad (k<0.01), \qquad (7b)$$

$$\tilde{u}^2(k)\approx \text{maximum} \qquad (0.01<k<0.1), \qquad (7c)$$

$$\tilde{u}^2(k)\propto k^{-5/3} \qquad (k>0.1), \qquad (7d)$$

$$\tilde{u}^2(k)\propto k^{7} \qquad (k\gg 0.1), \qquad (7e)$$

$$f=k\bar{U}(2\pi)^{-1}, \qquad (8)$$

$$\tilde{p}\approx\rho\tilde{u}^2. \qquad (9)$$

The first of these relations states that, if and when turbulence exists, the rms turbulent velocity $\tilde{u}$ is on the average about five percent of the mean-flow velocity $\bar{U}$. The general features of the turbulent-velocity spectra are indicated by the set (7a)–(7e), showing the approximate dependence of the turbulent-velocity spectrum density function $\tilde{u}^2(k)$ on the wavenumber k in different parts of the wavenumber range. Equation (8) for the frequency f is a reminder that it is the *mean-flow velocity* that relates wavenumber and frequency in this case. An estimate of the rms turbulent-pressure fluctuation may be derived from the mean-square turbulent velocity according to expression (9) in which ρ is the density of the fluid.

Pochapsky[35] has estimated that the magnitude of $\tilde{u}$ is no more than 2 cm/sec for the horizontal velocity components of the "ambient" oceanic turbulence. By Eq. (6), the corresponding current speed is 40 cm/sec (0.8 knot). These estimates are indicative of an upper limit to the background turbulence. Flow rates of as much as 200 cm/sec (3.9 knot) are observed in the swifter ocean currents, and of 600 cm/sec (11.7 knot) or more in straits and passages where very strong tidal

[33] Several of the classical papers may be found in the book, *Turbulence*, edited by S. K. Friedlander and L. Topper (Interscience Publishers, Inc., New York, 1961). More recent experimental and theoretical results are included in the book, J. O. Hinze, in *Turbulence* (McGraw–Hill Book Company, Inc., New York, 1959).

[34] M. J. Lighthill, Proc. Roy. Soc. (London) **A222**, 1 (1954).

[35] T. E. Pochapsky, Columbia University Hudson Laboratories Tech. Rept. 67 (March 1, 1959).

[36] H. L. Grant, R. W. Stewart, and A. Moilliet, Pacific Naval Laboratory, Esquimalt, B. C., Canada Rept. 60–8 (1960).

[37] Reference 11, p. 268.

[38] The possible significance of the pressure fluctuations near and within the turbulent regions was suggested to the author by Paul O. Laitinen.

[39] The so-called "velocity" transducer is not excepted since such devices are sensitive to pressure gradients and will respond to the pressure gradients of the turbulence.

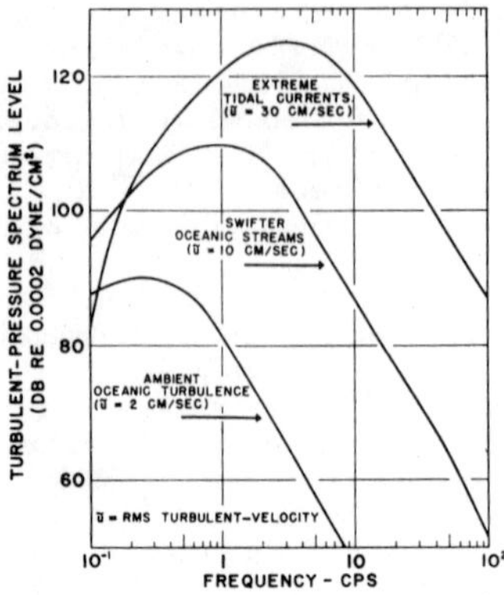

FIG. 11. Turbulent-pressure-level spectra, derived from theoretical and experimental relations [see Eqs. (6) through (9)].

currents are experienced. Corresponding rms turbulent velocities are 1υ cm/sec and 30 cm/sec.

Equations (6) through (9) were used to derive turbulent-pressure spectra for each of the three flow conditions mentioned in the preceding paragraph. These spectra are shown in Fig. 11 in terms of turbulent-pressure spectrum levels. The estimates are rough and variations of at least one order of magnitude are probable. A comparison with the spectra in Figs. 4–6 reveals that the estimated oceanic turbulent-pressure spectra agree quite well in both slope and level with the ambient-noise spectra below 10 cps, and between 10 and 100 cps in some instances.

For a hydrophone to respond to the turbulent-pressure fluctuations it must be in the region of turbulence. There is evidence that the movement of the water masses in the ocean is mostly chaotic and turbulent.[40] The dimensions of flow are nearly always large so that turbulent conditions may prevail, even though the flow velocity is often small. Turbulence is spread out and maintained in the ocean volume by a continual source of turbulent energy at the boundaries, particularly the sea-surface boundary. There is reason to believe that there exists in the ocean an essentially universal "ambient" turbulence which varies widely in intensity with both time and place.

According to the "elementar" current theory,[41] which assumes a homogenous ocean, the variation of velocity with depth depends on the mutual effect of wind-induced "drift currents" and "gradient currents" which result from the pressure differences produced by sea-surface slopes. The magnitude of the pure gradient current is constant with depth, except near the bottom boundary where frictional forces decrease the magnitude logarithmically to zero. The speed of the drift current decreases with depth and becomes negligible at a depth, dependent on the surface velocity, of the order of 45 to 200 m (25 to 110 fathoms). The speed of the elementar current, therefore, does not change greatly with depth except near the surface and the bottom, and in shallow water, where the bottom is near the surface. The characteristics of the elementar current are modified by the effects of density currents resulting from internal forces.[42] Vertical circulatory patterns may occur which produce velocity maxima and minima between the surface and the bottom.[43]

The theoretical and conjectural aspects of the preceding discussion have experimental support. For example, measurements[44] made in water 45 m (25 fathoms) in depth showed a logarithmic decrease of velocity with depth from about 40 cm/sec to 10 cm/sec between 160 cm and 20 cm above the bottom, and conditions of turbulence were found to exist. Recent deep-water measurements[45–47] made with the Swallow neutrally buoyant float indicate that deep currents are faster and more variable than was anticipated, with no evidence for a decrease in speed with depth. Average speeds of 6 cm/sec were observed at 2000-m depth (approximately 1100 fathoms), while at 4000-m depth (2200 fathoms) the average was 12 cm/sec, with as much as 42 cm/sec being observed in two cases.[47] The existence of eddies with a typical diameter of as much as 185 km (100 nautical miles) was implied by the observed fluctuations of the deep currents.

From Eqs. (6) and (9), one would ordinarily expect the turbulent-pressure levels to increase and decrease as the flow velocities increase and decrease. In the limited amount of low-frequency ambient-noise data available, there is no evidence of consistent depth dependence. This could be explained in part by the probable variety of patterns in the vertical velocity structure of the currents. Since the speed of the drift current decreases with depth, a decrease in turbulent-pressure level with depth should be observed in shallow water and at shallow depths in deep water, except when strong density or local currents are present.

The characteristics of the "drift" current, which is wind-dependent, could quite reasonably explain the wind dependence of the ambient noise in very shallow water, which is illustrated in Fig. 6, and also, possibly, the wind dependence at frequencies below 100 cps in Fig. 1(e).

Whenever local boundaries are involved, particularly if there are sharp edges and rough surfaces, the local scales of motion are generally small, but the velocities of

[40] Albert Defant, *Physical Oceanography* (Pergamon Press, New York, 1961), in particular Vol. I, part II.
[41] Reference 40, Vol. I, p. 413.
[42] Reference 40, Vol. I, Chap. XV.
[43] Reference 40, Vol. I, Chaps. XVI and XXI.
[44] R. M. Lesser, Trans. Am. Geophys. Union **32**, 207 (1951).
[45] J. C. Swallow, Deep Sea Research **4**, 93 (1957).
[46] J. C. Swallow and L. V. Worthington, Nature (London) **179**, 1183 (1957).
[47] M. Swallow, Oceanus (Woods Hole Oceanographic Institution) **VII** (3), 2 (1961).

the local turbulence may be large, and the local effects may be intense.

A great deal of the low-frequency ambient-noise data has been acquired using systems employing fixed bottom-mounted hydrophones. With such systems it is sometimes difficult to separate the self-noise produced by turbulence resulting from water motion past the stationary transducer from noise which is characteristic of the medium. Similar considerations apply to ship-borne systems when differential drift between the ship and the hydrophone causes the hydrophone to be towed by the ship, or when any forces, such as buoyancy and gravity, produce relative motion between the hydrophone and the water.

On the basis of spectrum shape and level, there is good support for the hypothesis that one component of low-frequency ambient noise is turbulent-pressure fluctuations. There is good reason to believe that turbulence of varying degree is a general circumstance throughout the ocean. By the mechanisms of turbulence, a portion of the energy which is introduced into the ocean at very low frequencies (large eddies, low wavenumbers) is transferred to range of higher frequencies, which, according to the curves of Fig. 11, extends above 1 cps. Some aspects of depth and wind dependence (and the lack of it) are explained by the hypothesis, but the data are inconclusive.

The conclusion is that, while noise *radiated* by turbulence does not greatly influence the ambient noise, the turbulent-pressure fluctuations are probably an important component of the noise below 10 cps, and sometimes in the range from 10 to 100 cps.

2.3 Oceanic Traffic

The ambient noise may include significant contributions from two types of noise[48] from ships. *Ship noise* is discussed briefly in Sec. 2.6. *Traffic noise* is the subject of this section.

The degree to which traffic noise influences the ambient noise depends on the particular combination of *transmission loss*, *number of ships*, and the *distribution of ships* pertaining to a given situation. For instance, a significant contribution could result from widely scattered ships if the average transmission loss per unit distance were relatively small such as might be the case at a deep-water location in an open-ocean area crossed by transoceanic shipping lanes. A significant contribution could also result even when transmission losses per unit distance are high if there were a comparatively large concentration of ships at relatively close range, such as might be the case in shallow water near a harbor and coastal shipping lanes. The traffic-noise characteristics are determined by the mutual effect of the three factors.

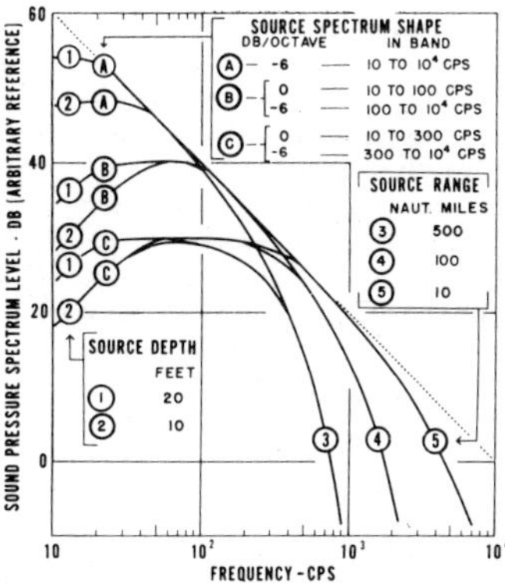

FIG. 12. Traffic-noise spectra deduced from ship-noise source characteristics and attenuation effects. Several variations are shown. For example, the curve 1B4 defines the expected spectrum shape at 100 nautical miles (185 km) from a source whose noise spectrum is flat up to 100 cps and decreases −6 dB per octave above 100 cps, the effective source depth being 20 ft (6m).

Traffic-noise characteristics also depend on the kinds of ships involved, that is, upon the nature of the source. In general the base is broad so that individual differences blend into an average source characteristic.

A study of the noise from surface ships[49] indicates that on the average, when measured at distances of about 20 yards, the sound-pressure-level spectra have a slope of about −6 dB per octave. The spectrum is highly variable at frequencies below 1000 cps, and, under some circumstances, the slope tends to flatten in the neighborhood of 100 cps. This source-spectrum shape is altered in transmission by the frequency-dependent attenuation part of the transmission loss. According to Sheehy and Halley,[50] the attenuation is 0.033 $f^{\frac{3}{2}}$ dB per kiloyard or 0.066 $f^{\frac{3}{2}}$ dB per nautical mile, where f is the frequency in kc. At long ranges, the attenuation increases rapidly with frequency above 500 cps.

For most surface ships, the effective source of the radiated noise is between ten and thirty feet below the surface. Up to frequencies of about 50 cps, the source and its image from surface reflection operate as an acoustic doublet radiating noise with a spectrum slope of +6 dB per octave relative to the spectrum of the simple source.

To obtain some notion of the probable shape of traffic-noise spectra, the foregoing information was used in deriving the curves shown in Fig. 12. Variations in

[48] *Ship noise* is the noise from one or more ships at close range. It may be identified by short-term variations in the ambient-noise characteristics, such as the temporary appearance of narrow-band components and a comparatively rapid rise and fall in noise level. Ship noise is usually obvious and therefore generally can be and is deleted from ambient-noise data.

Traffic noise is noise resulting from the combined effect of all ship traffic, excepting the immediate effects of ship noise as defined in the preceding paragraph. Traffic noise is usually not obvious as such.

[49] M. T. Dow, J. W. Emling, and V. O. Knudsen, "Survey of Underwater Sound, Report No. 4, Sounds from Surface Ships," 6.1–NDRC–2124 (1945).

[50] M. J. Sheehy and R. Halley, J. Acoust. Soc. Am. 29, 464 (1957).

the spectra caused by differences in source depth, differences in the shape of the source-noise spectrum, and differences in the attenuation at different ranges are indicated by the composite set of curves. The effect of the source depth at low frequencies is shown by the curves numbered (1) for a depth of 20 ft, and (2) for 10 ft. The choices of source-noise spectrum shape, based on the data reported by Dow,[49] are described in terms of the slopes of the sound-pressure-level spectra, and the resulting curves are identified as follows: (A) -6 dB per octave; (B) 0 dB per octave up to 100 cps and -6 dB per octave above 100 cps; and (C) 0 dB per octave up to 300 cps and -6 dB per octave above 300 cps. The change in spectrum shape as the range varies, a consequence of attenuation, is shown by curve (3) representing a range of 500 miles (926 km), curve (4) a range of 100 miles (185 km), and curve (5) a range of 10 miles (185 km). The spectrum corresponding to a particular set of conditions may be found by following the curves identified by the relevant numbers and letter. For example, the curve 1B4 is the spectrum form which would be observed at 100 miles from a source located at a depth of 20 ft, and whose noise spectrum is flat up to 100 cps and decreases at 6 dB per octave above 100 cps.

There is a remarkable similarity between the synthetic traffic-noise spectra of Fig. 12 and the spectra of the non-wind-dependent component of the observed ambient noise, which are discussed in Sec. 1.1. In each case, the maximum is in the vicinity of 100 cps, and the spectrum falls off steeply above 100 cps.

The high-frequency "cutoff" occurs at lower frequencies in the deep-water spectra of Figs. 2(b), 2(d), and 2(e) than in the shallow-water spectra of Figs. 1(b) and 1(d). This effect can be explained, or even anticipated, by assuming that the average range of the effective traffic-noise sources is generally less for shallow- than for deep-water locations.

Measurements of the noise radiated by surface ships have been reported by Dow *et al.*[49] Corresponding to these data, for surface ships the equivalent simple source-pressure levels in a 1-cps band at 100 cps at a distance of 1 yard are between 125 dB and 145 dB (*re* 0.0002 dyn/cm^2) in most cases. Hale[51] has shown that experimental results from long-range transmission in deep water do not fit the free-field, spherical divergence law very well, and that better agreement with experiment does result if boundaries and sound–velocity structure are taken into account. According to this theory and experiment, 105 dB is a reasonable estimate of the average transmission loss at 100 cps for a range of 500 miles. Accordingly, the spectrum level (*re* 0.0002 dyn/cm^2) at 100 cps from one "average" ship source at 500 miles is 20 to 40 dB; from 10 "average" ships all at 500 miles, 30 to 50 dB (assuming power addition); and from 100 ships, 40 to 60 dB. At a range of 1000 miles the levels would be only 3 to 6 dB lower. The spectrum levels at 100 cps of the aforementioned non-wind-dependent component of the ambient noise are between 40 and 55 dB, according to Figs. 1(b), 1(d), 2(b), 2(d), and 2(e).

It is apparent from this evaluation that the effective distance for traffic-noise sources in the deep-open ocean can be as much as 1000 miles or more.

The spectra in Figs. 1 and 2 were arranged originally on the basis of spectrum shape. A similar arrangement would result from traffic-noise considerations. At the locations for Figs. 1(a) and 1(e), some ship noise is encountered (deleted from the data when detected), but usual transmission ranges are too short for the composite effect of traffic noise. Figure 1(c) pertains to an isolated area where ship noise is infrequent. The data in Figs. 1(b) and 1(d) were obtained in the midst of both coastal and transoceanic shipping lanes, and illustrate the case of a comparatively large concentration of sources at relatively close range. While Figs. 2(a) and 2(c) represent measurements in deep water, i.e., greater than 100 fathoms in depth, the locations were not in the open ocean in the sense of being open to long-range transmission. The spectra shown in Figs. 2(b), 2(d), and 2(e) were derived from measurements made in the deep ocean open to long-range transmission and crossed by transoceanic shipping lanes.

The evidence is strong that the non-wind-dependent component of the ambient noise at frequencies between 10 cps and 1000 cps is traffic noise. It is concluded that, while there are many places which are isolated from this noise, in a large proportion of the ocean, traffic noise is a significant element of the observed ambient noise and often dominates the spectra between 20 and 500 cps.

2.4 Seismic Sources

As a result of volcanic and tectonic action, waves are set up in the earth. Even when the point of origin is distant from the ocean boundary, appreciable amounts of the energy may find their way into the ocean and be propagated as compressional waves in the water (T phase).[52–55] (Similar effects often result from artificial causes such as manmade explosions.)

When observed at close range, waterborne noise of seismic (volcanic) origin has been reported[56] as including observable energy at frequencies up to at least 500 cps. The spectrum characteristics depend on the magnitude of the seismic activity, the range, and details of the propagation path, including any land or sea-floor segments. Experimental data[57,58] indicate that in

[51] F. E. Hale, J. Acoust. Soc. Am. **33**, 456 (1961).

[52] I. Tolstoy, M. Ewing, and F. Press, Columbia University Geophysical Lab. Tech. Rept. 1 (1949), or Bull. Seismol. Soc. Am. **40**, 25 (1950).

[53] R. S. Dietz and M. J. Sheehy, Bull. Geol. Soc. Am. **65**, 1941 (1954).

[54] D. H. Shurbet, Bull. Seismol. Soc. Am. **45**, 23 (1955).

[55] D. H. Shurbet and Maurice Ewing, Bull. Seismol. Soc. Am. **47**, 251 (1957).

[56] J. M. Snodgrass and A. F. Richards, Trans. Am. Geophys. Union **37**, 97 (1956).

[57] Allen R. Milne, Bull. Seismol. Soc. Am. **49**, 317 (1959).

[58] J. Northrop, M. Blaik, and I. Tolstoy, J. Geophys. Research **65**, 4223 (1960).

general the spectrum has a maximum between 2 and 20 cps and that noticeable waterborne noise from earthquakes may be expected at frequencies from 1 to 100 cps. The noise is manifest as a single transient, or a series of transients, of relatively short duration and infrequent occurrence. However, in some areas and during some periods of time the frequency of occurrence may be as often as several times an hour.

A seismic background of continuous disturbance of varying strength is also observed, being attributed to the aftereffects of the more transient events, and to the effects of storms, and of waves and swell at the coastal boundary, with local contributions from winds, waterfalls, traffic, and machinery. The spectrum of the vertical ground-particle displacements of the background noise as observed on land (and excluding noise from obvious local and transient sources) has a maximum between 0.1 and 0.2 cps, with amplitudes from 2×10^{-2} to 20 μ.[59,60] The amplitudes decrease approximately in inverse proportion to frequency between 1 and 100 cps. At 1 cps, the amplitudes range from 10^{-3} to 10^{-1} μ, and at 100 cps from 10^{-6} to 10^{-3} μ. Vertical and horizontal velocity spectra show maxima at the same frequencies as the displacement spectra.[61] In the neighborhood of 0.5 cps, the upper frequency limit of the data, the velocity spectra begin to flatten, suggesting a flat velocity spectrum above 1 cps, as might be expected from the inverse frequency dependence of the displacement spectra.

A crude estimate of noise in the ocean associated with the continuous seismic disturbances may be obtained by assuming, in the absence of specific data, that the seismic spectrum characteristics on the sea floor are about the same as those on land and that the vertical components of the particle displacements and velocities of the water at the boundary are the same as those of the sea floor. For such conditions, the pressure-level spectrum is essentially flat between 1 and 100 cps, varying in level between 45 and 95 dB *re* 0.0002 dyn/cm^2. The peak levels between 0.1 and 0.2 cps are from 65 to 120 dB. The spectrum shape does not agree with the observed ambient-noise spectra (see Sec. 1), but the levels are of sufficient magnitude to suggest the possibility that some of the variability in ambient-noise spectra may be a consequence of the seismic background activity.

Measurements at frequencies between 4 and 400 cps directly comparing background seismic velocity components and waterborne sound pressures as measured at the sea bottom in shallow water[62,63] show order-of-magnitude agreement when continuity across the interface is assumed. The velocities which were measured fall within the range of the results of seismic measurements made on land,[59–61] which were mentioned in the preceding discussion. However, the sea-floor velocity spectra showed a slope of -5 or -6 dB per octave.

The experimental data indicate a close relation between the seismic background and the nearby pressure fluctuations in the water. A pertinent question is: Which is the cause and which the effect? Possibly some equilibrium process exists, the direction of net energy transfer depending on the relative energy levels existing in the two media in a particular situation.

From this brief survey, it is concluded that noise from earthquakes does dominate the ambient noise at frequencies between 1 and 100 cps, but such effects are transient and highly dependent on time and location. Significant noise from lesser, but more or less continuous, seismic disturbances is possible particularly when current velocities and turbulence are at a minimum, but additional data are needed for a more definite evaluation.

The seismic disturbances may have a direct effect on bottom-mounted transducers. For this reason, the vibration sensitivities of the transducer should be known and taken into consideration in the design of such a system, and in the interpretation of results.

2.5 Biological Sources

Many species of marine life have been identified as noise producers. Noise of biological origin has been observed at all frequencies within the limits of the systems used, which, in aggregate, have covered from 10 cps to above 100 kc.[1–6,64,65] The individual sounds are usually of short duration, but often frequently repeated, and include a wide variety of distinctive types such as cries, barks, grunts, "awesome moans," mewings, chirps, whistles, taps, cracklings, clicks, etc. Pulse-type sounds which change in repetition rate, sometimes very quickly, have been identified with echo location by porpoise.[66–68] Repetitive pulse sounds have also been attributed to whales. Continuous (in time) biological noise is frequently encountered in some areas when the sounds of many individuals blend into a potpourri, such as the crackling of shrimp and the croaker chorus.[1–6]

The contribution of biological noise to the ambient noise in the ocean varies with frequency, with time, and with location, so that it is difficult to generalize. In some cases diurnal, seasonal, and geographical patterns may be predicted[1,2,5] from experimental data, or from the habits and habitats, if known, of known noisemakers.

[59] J. N. Brune and J. Oliver, Bull. Seismol. Soc. Am. **49**, 349 (1959).
[60] G. E. Frantti, D. E. Willis, and J. T. Wilson, Bull. Seismol. Soc. Am. **52**, 113 (1962).
[61] R. A. Haubrich and H. M. Iyer, Bull. Seismol. Soc. Am. **52**, 87 (1962).
[62] E. G. McLeroy (unpublished data).
[63] R. D. Worley and R. A. Walker (unpublished data).
[64] W. N. Kellog, R. Kohler, and H. N. Morris, Science **117**, 239 (1953).
[65] M. P. Fish, University of Rhode Island, Narragansett Marine Lab., Kingston, Rhode Island Reference 58 8 (1958).
[66] W. N. Kellog, Science **128**, 982 (1958).
[67] W. N. Kellog, J. Acoust. Soc. Am. **31**, 1 (1959).
[68] K. S. Norris, J. H. Prescott, P. V. Asa-dorian, and Paul Perkins, Biol. Bull. **120**, 163 (1961).

Noises having the distinctive nature of biological sounds are readily detected in the ambient noise, but the biological source is not always certain.

The data presented in Sec. 1 exclude noise of known or suspected biological origin.

2.6 Additional Sources

Various other sources of intermittent and local effects include ships, industrial activity, explosions, precipitation, and sea ice.

Ship noise is the noise from one or more ships at close range, and, as used here, is differentiated from traffic noise.[48] Ship noise causes short-term variations in the ambient noise characterized by the temporary appearance of narrow-band components at frequencies below 1000 cps, and broad-band cavitation noise extending well into the kilocycle region, often with low-frequency modulation patterns.

Industrial activity on shore, such as pile driving, hammering, riveting, and mechanical activity of many kinds, can generate waterborne noise. The characteristics of the noise depend on the particular situation. Noise from industrial activity may predominate at times in particular near-shore areas.

Noise from *explosions* is very much like that from earthquakes (see Sec. 2.4). At close range, the effects cover a wide range of frequencies, but at longer range the spectrum has been modified by propagation, and the larger part of the energy is usually at frequencies below 100 cps.

Precipitation noise is basically noise from a spray of water droplets (rain) and rigid bodies (hail) (see Sec. 2.2.2). The effects of precipitation are most noticeable at frequencies above 500 cps, but may extend to as low as 100 cps if heavy precipitation occurs when wind speeds are low.

The various kinds of *sea–ice* movements are a source of noise which at times covers a wide range of frequencies at high level. The noise originates in the straining and cracking of the ice from thermal effects, and in the grinding, sliding, crunching, and bumping of floes and bergs.

It is difficult to generalize on the characteristics of the various kinds of intermittent and local noise, since such noise is dependent to a great degree on the particular time and place of concern.

3. COLLIGATION

The experimental and theoretical data which have been presented in the foregoing lead to the conclusion that the general spectrum characteristics of the prevailing ambient noise in the ocean are determined by the combined effect of several components, which, though of widespread and continual occurrence, vary each in its own way with time and location. A composite picture is given in Fig. 13, which summarizes the characteristics and probable causes in the various parts of the spectrum between 1 cps and 100 kc.

On the basis of spectrum characteristics, the principal components of the prevailing ambient noise in the ocean are:

(a) a low-frequency component characterized by a spectrum-level slope of approximately −10 dB per octave and wind dependence in very shallow water, the most probable source being ambient turbulence (turbulent-pressure fluctuations);

(b) a high-frequency component characterized by wind dependence, a broad maximum between 100 and 1000 cps and a slope of approximately −6 dB per octave at frequencies above the maximum, with levels, on the average, some 5 dB less in deep water than in shallow (most probable source—bubbles and spray from surface agitation);

(c) a medium-frequency component characterized by a broad maximum between 10 and 200 cps, and a steep negative slope at frequencies above the maximum band (most probable source–oceanic traffic); and

(d) a thermal noise component characterized by a +6 dB per octave slope.

Component (a) nearly always predominates at frequencies below 10 cps, and its influence may be evident up to frequencies as high as 100 cps in the absence of component (c). Component (b) nearly always predominates at frequencies above 500 cps. Component (c) frequently predominates in the band from 20 to 200 cps, but is not observed in isolated[69] areas. Component (d) is a factor at frequencies above 20 kc.

The general similarity between the observed levels and the spectrum shape of the ambient noise and of the spectra estimated for the turbulent-pressure fluctuations leads to the conclusion that the low-frequency ambient noise, component (a), consists primarily of the turbulent-pressure fluctuations (which are much greater in magnitude than the radiated sound pressures generated by turbulence). It may be said further that turbulence is a process by which some of the energy introduced into, or originating in, the ocean at very low frequencies is transformed into energy of consequence to the ambient noise at frequencies above 1 cps. The not unreasonable assumption of a widespread ambient turbulence in the ocean is required. (See Sec. 2.2.4.)

A small amount of low-frequency data taken in very shallow water indicates a dependence of level, but not of spectrum slope, on wind speed (see Fig. 6). This wind dependence is explained by the influence of wind-caused drift currents on the turbulence. The data are not of sufficient quantity to establish even a tentative quantitative relationship between level and windspeed; so none is shown in Fig. 13.

Components of the seismic background noise in the

[69] "Isolated" either because of geographic location or because of propagation conditions, or both.

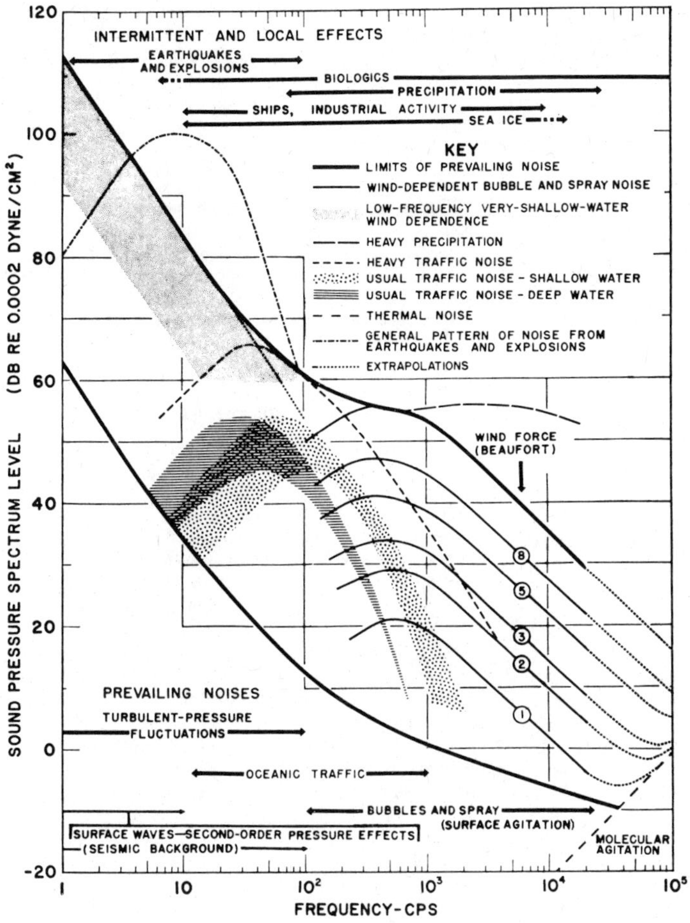

FIG. 13. A composite of ambient-noise spectra, summarizing results and conclusions concerning spectrum shape and level and probable sources and mechanisms of the ambient noise in various parts of the spectrum between 1 cps and 100 kc. The key identifies component spectra. Horizontal arrows show the approximate frequency band of influence of the various sources. An estimate of the ambient noise to be expected in a particular situation can be made by selecting and combining the pertinent component spectra.

earth, including the sea floor, are often attributed to the effects of pressure fluctuations in the ocean rather than vice versa. There is probably some sort of a give-and-take process by which in some areas the seismic waves are produced by waterborne pressure fluctuations, such as the Longuet–Higgins second-order pressure effects from surface-wave trains, while in other areas the seismic activity causes waterborne pressure fluctuations. Such a process would allow hydrodynamic pressure effects to be transmitted from one place to another via seismic processes. The estimated noise from seismic background activity differs in spectrum shape from that of the low-frequency component (a), but seismic noise could account for some of the variability observed at low frequencies and could become significant where current velocities and turbulence are at a minimum. (See Sec. 2.4.)

Because of the steep negative slope of surface-wave spectra, and because of the rapid attenuation with depth, it is doubtful that the first-order pressure effects from surface waves are of much significance to the ambient noise at frequencies above 1 cps. However, it is probable that the Longuet–Higgins second-order effects contribute to the noise at frequencies up to 10 cps, particularly in or near storm areas and certain coastal areas. (See Sec. 2.2.3.)

The wind dependence and the "on-the-average" -5 to -6 dB per octave slope of the high-frequency component (b) are well established. The high-frequency wind-dependent curves shown in Fig. 13 represent over-all averages, including both deep- and shallow-water data. (The comparable shallow-water averages are 2 to 3 dB higher than the over-all averages, the deep-water averages 2 to 3 dB lower.) The hypothesis that gas bubbles, cavitation, and spray in the surface agitation are the sources of the high-frequency wind-dependent component (b) is well supported by the observed spectrum levels, spectrum shape, and band of maximum levels. (See Secs. 2.2.1 and 2.2.2.)

The assumption that the medium-frequency component (c) is the result of the combined effect of many ships at relatively long range is shown to be a reasonable one by the general agreement between the observed spectra and the spectra estimated for traffic noise (see Sec. 2.3). The high-frequency "cutoff" begins at lower frequencies in deep water, as compared to shallow according to the observed data (Sec. 1). This effect is to be expected from the curves in Fig. 12, if one assumes that, while in both cases the ranges are comparatively long, the average range of the effective noise sources is greater for deep water than for shallow.

The shape of the lower-limit curve in Figs. 7 and 13 is not greatly different from that which would result from a combination of a turbulent-pressure spectrum (Sec. 2.2.4), a low-level bubble-noise spectrum due to a remanent distribution of bubbles (which, as was shown in Sec. 2.2.1, may exist even at zero sea state), and the thermal noise spectrum. It is quite likely that the empirical lower-limit curve is, to some degree, an indication of the state of the art of measuring very low-pressure levels in the ocean. With a sensitive system, one might expect to encounter even lower levels, particularly at low frequencies in a region of minimum current and turbulence, where the probable source of residual noise is seismic background activity.

The relatively high residual or threshold noise observed in some shallow-water locations [Figs. 1(a) and 1(b)] apparently arises from local conditions. The shape of the residual spectra suggests that the noise is a combination of the average effects from such sources as local turbulence and cavitation (particularly if rough boundaries and high currents are involved), industrial activity, marginal traffic noise, and the more subtle types of self-noise, such as that which may result from relative motion between the hydrophone and the water.

The prevailing features of the ambient noise are altered by intermittent and local effects. The noise from earthquakes and explosions perturbs the spectrum mostly at frequencies around 10 cps, the extent of the band of influence being determined, to a large degree, by the range. Disturbance from biological sources may show up at almost any frequency, often conforming to diurnal, seasonal, and geographic patterns, but with distinctive characteristics. The effects of precipitation are most apparent at frequencies above 500 cps, but may extend to lower frequencies when heavy precipitation occurs at low wind speeds. Major ship-noise effects occur at frequencies above 10 cps and often include narrow-band components below 1000 cps. Noise from industrial activity occurs in particular near-shore locations. Noise resulting from sea–ice movement is distinctive in character, covering a wide range in frequency, but is restricted to particular geographic areas.

4. APPLICATION NOTES

The provisional picture of ambient noise in the ocean which has been developed shows that the ambient noise depends on several variables, some of which, unfortunately, are as difficult to estimate or measure as the ambient noise itself. However, a few general guidelines can be given for estimating and predicting the ambient noise which prevails in the absence of intermittent and local effects.

[4.1] 500 cps to 20 kc

Estimates of the prevailing ambient-noise levels in the ocean at frequencies between 500 cps and 20 kc can be made from a knowledge of the wind speed. Estimates of wind speed can be obtained from the *Atlas of Climatic Charts of the Oceans* issued by the U. S. Weather Bureau (1938), if more specific information is lacking. Table I, or similar tables, may be used to estimate wind speed from wave-height or sea-state information. Knowing the wind speed, one may determine "on-the-average" levels by reference to the high-frequency wind-dependent curves in Fig. 13 (which, including the influence of more recent data, are 2 to 3 dB lower in level than the Knudsen[1,2] averages).

For example, a wind speed of 7 knots (3.6 m/sec) falls at the lower limit of the range for Beaufort wind force 3. According to the wind-force-3 curve in Fig. 13, the estimated spectrum level at 1 kc (to the nearest dB) is 32 dB (*re* 0.0002 dyn/cm^2). The curves represent average levels and may be considered as corresponding to the mean of the Beaufort wind-speed grouping. Interpolations may be made for estimates of levels at other wind speeds. For the example of 7 knots, an interpolated level estimate at 1 kc is 30 dB. Similar interpolations may be made for wind speeds falling in the wind-force-4, -6, and -7 groups, for which curves are not shown in Fig. 13.

The curves in Fig. 13 are averages over both deep- and shallow-water data. For a closer deep-water estimate, the over-all average level should be lowered by 2 or 3 dB. Thus, for deep water and a wind speed of 7 knots, the estimated level at 1 kc is 27 or 28 dB. For shallow water, the over-all levels are raised by 2 or 3 dB. (The shallow-water estimates are about the same as those of the Knudsen curves, while the deep-water estimates are about 5 dB lower in level.)

As it happens, the wind-dependent spectra follow ***approximately*** an empirical "rule of fives," which is stated as follows:

> In the frequency band between 500 cps and 5 kc the ambient sea-noise spectrum levels decrease 5 dB per octave with increasing frequency, and increase 5 dB with each doubling of wind speed from 2.5 to 40 knots; the spectrum level at 1 kc in deep water is equal to 25 dB (5×5) *re* 0.0002 dyn/cm^2 when the wind speed is 5 knots, and is 5 dB higher in shallow water.

The "rule of fives" is fairly accurate up to a frequency of 20 kc.

To illustrate the use of the rule of fives: The estimated spectrum level at 4 kc in deep water when the wind speed is 20 knots (10.3 m/sec) is equal to 25 dB (1-kc level for 5 knots) plus 10 dB (wind speed doubled twice from 5 knots) minus 10 dB (2 octaves above 1 kc), which adds up to 25 dB. The corresponding shallow-water level is 30 dB, and the mean between the deep and shallow values is $27\frac{1}{2}$ dB. This value is about the same as the value indicated at 4 kc by the wind-force-5 curve in Fig. 13, showing that the rule-of-fives result agrees reasonably well with the average data.

It is believed that the procedures which have been given will lead to useful estimates of the ambient sea noise in the frequency band from 500 cps to 20 kc. But it should be remembered that considerable departure may sometimes be expected, since wind speed is not a precise measure of the actual surface agitation (see Sec. 1.1 and Fig. 3), nor are estimates of sea state. If information is known about duration, fetch, and topography, a judicious use of the kind of information given in Table I will improve the reliability of the estimate. Ideally, information is needed as to the details of the surface agitation, such as the relation between meteorological and oceanographic conditions and the distributions of the sizes of bubbles and droplets, and the degree of cavitation. At low wind speeds, the effects of other influences may be expected to increase the variability.

[4.2] 1 cps to 10 cps

An "on-the-average" spectrum-level slope of −8 to −10 dB per octave in the frequency range from 1 to 10 cps may be predicted with some assurance, since the observed data are quite consistent in this respect. This slope persists down to 0.05 cps, and perhaps to lower frequencies, except when the first-order effects of surface waves are encountered at shallow depths.

It is to be generally anticipated that the levels will fall between the lower-limit curve and the bottom of the shaded area (upper left) in Fig. 13. According to the conclusions reached in Sec. 2.2.4, the spread in the observed low-frequency ambient-noise levels is principally the result of the variation in the oceanic turbulent-pressure fluctuations, and the higher levels are to be expected in or near the major oceanic currents, or anywhere else where relatively high water motion is known to exist. One might also expect high levels to accompany high winds and storms because of a general increase in the turbulent energy, and because of the Longuet-Higgins second-order pressure effects.

When observations are made at very shallow depths, less than 25 fathoms (roughly 50 m), one should be prepared for high and wind-dependent levels, as indicated by the shaded (upper left) area in Fig. 13. The data are insufficient to establish a quantitative relation between level and wind speed which can be applied generally.

Although these qualitative approximations do not permit very definite predictions of the ambient noise at the low frequencies, they should be useful in defining specific areas for investigation.

[4.3] 10 to 500 cps

In the frequency band between 10 and 500 cps, the noise is influenced by at least three components. To predict the noise, one must know or assume the spectrum for a low-frequency component, a traffic-noise component, and a high-frequency wind-dependent component, and then combine the component spectra. The high-frequency wind-dependent component and the low-frequency component were discussed in Secs. 4.1 and 4.2, respectively.

The experimental data indicate that the traffic-noise levels usually fall within the medium-frequency shaded areas shown in Fig. 13. The difference between shallow-water levels (area shaded by large dots) and deep-water levels (area shaded by lines) is to be noted. The more extreme higher and lower traffic-noise levels correspond respectively to relatively nearby concentrations of shipping and to the more remote[69] areas. The classification as a "remote" area may be time-dependent, the result of daily, seasonal, or other variations in propagation or traffic patterns. As was brought out in Sec. 2.3, the effective distance for traffic-noise sources in the deep ocean can be as much as 1000 miles or more, and a rather widely observed traffic-noise component is to be expected. A comparison of the traffic-noise spectra with the other spectra in Fig. 13 indicates that the traffic-noise component generally predominates at frequencies between 20 and 200 cps.

In shallow-water areas isolated from general shipping activity in very remote deep-water areas, and in some deep-water areas which are isolated by underwater topography or oceanographic conditions, the effect of the traffic-noise component is very small, and levels should be estimated by combining only the appropriate low-frequency component and the high-frequency wind-dependent component.

4.4 Example

The guidelines suggested by the foregoing remarks were used to estimate three spectra, which are shown in Fig. 14, corresponding to three frequently encountered situations. The three spectra are: (1) that expected in

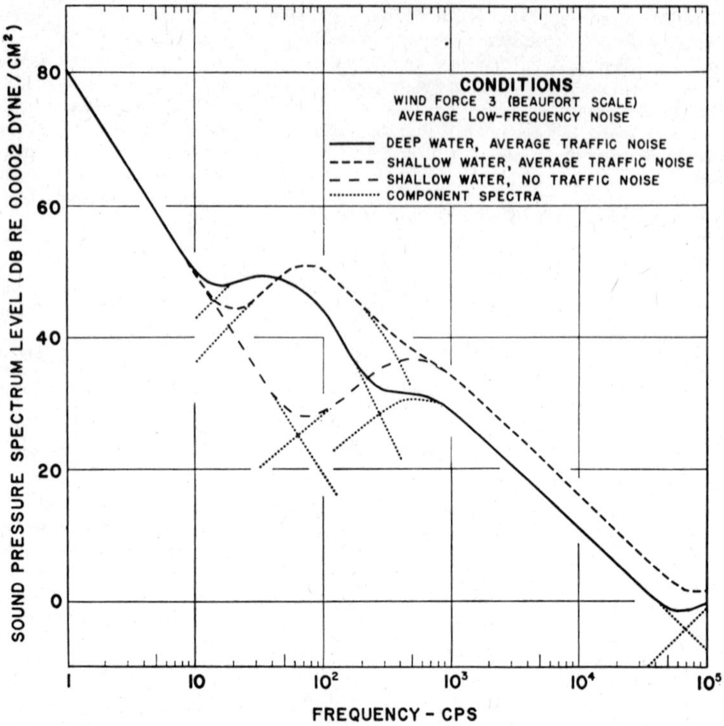

FIG. 14. Ambient-noise spectra estimated for three frequently encountered situations. The dotted-line extensions indicate, in the regions of overlap, the individual component spectra from which the estimated spectra were derived.

deep water with average traffic noise, (2) that expected in shallow water with average traffic noise, and (3) that expected in isolated shallow water (no traffic noise); when in each case average low-frequency noise conditions (average oceanic ambient turbulence) and force-3 winds prevail.

The shapes of the individual components, in the region of overlap before combination, are indicated by the dotted extensions. The low-frequency component,[70] in Fig. 14, is at a level approximately halfway between the lower-limit curve and the bottom of the low-frequency shaded area (upper left) in Fig. 13. The medium-frequency curves in Fig. 14 are the approximate medians of the deep- and shallow-water shaded traffic-noise areas in Fig. 13. The shallow-water high-frequency component is about $2\frac{1}{2}$ dB above the wind-force-3 curve shown in Fig. 13, the deep-water component $2\frac{1}{2}$ dB below.

[70] It is of interest to note that this approximation of the average low-frequency noise-level spectrum derived from observed ambient-noise data is almost identical to the turbulent-pressure-level spectrum estimated for ambient oceanic turbulence as shown in Fig. 11. This nearly exact agreement is fortuitous, but is a good indication of the similarity between observed ambient-noise spectra at the low frequencies and the probable oceanic turbulent-pressure-level spectra.

These components were then combined by power addition. The resulting deep-water spectrum is shown by the solid curve, the two shallow-water spectra by the dashed lines.

5. CONCLUDING REMARKS

The scope of this review has been limited, primarily, to comparisons based on averaged spectrum characteristics. Other features, for which existing data are very meager, and which should be investigated further, include: temporal characteristics such as hourly, daily, seasonal, and like time patterns; spatial characteristics, such as the directional properties of the ambient-noise sound field; other statistical properties such as amplitude distributions; and the factors influencing such characteristics.

The accumulation of ambient-noise data is not yet large in amount nor comprehensive in scope when considered in relation to the number and range of the variables. The generalizations which have been made are subject to further test. The discussion of sources and mechanisms has some verisimilitude, and patterns illustrated in Sec. 1 are frequently repeated. But more direct experimental data and more precise quantitative relationships are needed. The results and conclusions should be useful for immediate needs and for guidance in future investigations.

ACKNOWLEDGMENTS

The writer thanks Elaine A. Kyle, Mel A. Calderon, and H. M. Linnette for assistance in processing the data and in preparing the figures, and Dr. Robert W. Young for comments and suggestions regarding the manuscript.

An essential part of the source material was found in unpublished data resulting from the work of the following: Arthur D. Arneson, Alan Berman, D. Gradner, R. Halley, W. G. Houser, Henry R. Johnson, R. H. Nichols, Jr., F. R. Nitchie, Elliot Rhian, Joseph D. Richard, Jr., Donald Ross, A. J. Saur, R. A. Walker, and R. D. Worley.

ADDITIONAL REFERENCES

In addition to the references cited in the footnotes, the following are also pertinent:

Berman, Alan and Saur, A. J., "Ambient Noise as a Function of Depth," J. Acoust. Soc. Am. **32**, 915(A) (1960) (Hudson Laboratories Contribution No. 77).

Dietz, F. T., Kahn, J. S., and Birch, W. B., "Effect of Wind on Shallow Water Ambient Noise," J. Acoust. Soc. Am. **32**, 915(A) (1960).

Dietz, F. T., Kahn, J. S., and Birch, W. B., "Nonrandom Associations between Shallow Water Ambient Noise and Tidal Phase," J. Acoust. Soc. Am. **32**, 915(A) (1960).

Frosch, R. A., "How to Make an Ambient Noise in the Ocean," J. Acoust. Soc. Am. **32**, 915(A) (1960) (Hudson Laboratories Contribution No. 81).

Lomask, Morton R. and Saenger, R. Alfred, "Ambient Noise in a Deep Inland Lake," J. Acoust. Soc. Am. **32**, 878 (1960).

Lomask, Morton and Frassetto, Roberto, "Acoustic Measurements in Deep Water Using the Bathyscaph," J. Acoust. Soc. Am. **32**, 1028 (1960).

Milne, Allen R., "Shallow Water Under-Ice Acoustics in Barrow Strait," J. Acoust. Soc. Am. **32**, 1007 (1960).

Walkinshaw, H. M., "Low-Frequency Spectrum of Deep Ocean Ambient Noise," J. Acoust. Soc. Am. **32**, 1497(A) (1960).

Milne, A. R., "Ambient Noise Under Old Shore-Fast Sea Ice," J. Acoust. Soc. Am. **33**, 1675(A) (1961).

Saenger, R. Alfred, "An Estimate of the Offshore Ambient Noise Spectrum Produced by Pounding Surf," J. Acoust. Soc. Am. **33**, 1674(A) (1961).

Willis, Jack and Dietz, Frank T., "Preliminary Study of Low-Frequency Shallow Water Ambient Noise in Narragansett Bay, R. I.," Narragansett Marine Laboratory, University of Rhode Island Reference 61–4 (November 1961).

Flow Noise

IV

The following two papers by Dr. Lighthill, although concerned with the aerodynamic problem, provided the theoretical basis for all of the later theoretical work on underwater flow noise.

Dr. Lighthill is a physicist. He was Director of the Royal Aircraft Establishment, Farnborough, from 1959 to 1964. He was concerned with research in aerodynamics of aircraft, theory of shock waves, and theory of jet noise. He is now Professor of Mathematics at Cambridge University.

This paper is reprinted with the permission of the author and the Executive Secretary of the Royal Society, London.

Proc. Roy. Soc., 1952, V. A211, p. 564–587.

On sound generated aerodynamically

17

I. General theory

BY M. J. LIGHTHILL

Department of Mathematics, The University, Manchester

(*Communicated by M. H. A. Newman, F.R.S.—Received* 13 *November* 1951)

A theory is initiated, based on the equations of motion of a gas, for the purpose of estimating the sound radiated from a fluid flow, with rigid boundaries, which as a result of instability contains regular fluctuations or turbulence. The sound field is that which would be produced by a static distribution of acoustic quadrupoles whose instantaneous strength per unit volume is $\rho v_i v_j + p_{ij} - a_0^2 \rho \delta_{ij}$, where ρ is the density, v_i the velocity vector, p_{ij} the compressive stress tensor, and a_0 the velocity of sound outside the flow. This quadrupole strength density may be approximated in many cases as $\rho_0 v_i v_j$. The radiation field is deduced by means of retarded potential solutions. In it, the intensity depends crucially on the frequency as well as on the strength of the quadrupoles, and as a result increases in proportion to a high power, near the eighth, of a typical velocity U in the flow. Physically, the mechanism of conversion of energy from kinetic to acoustic is based on fluctuations in the flow of momentum across fixed surfaces, and it is explained in §2 how this accounts both for the relative inefficiency of the process and for the increase of efficiency with U. It is shown in §7 how the efficiency is also increased, particularly for the sound emitted forwards, in the case of fluctuations convected at a not negligible Mach number.

1. INTRODUCTION

The subject of this paper is sound generated aerodynamically, that is, as a by-product of an airflow, as distinct from sound produced by the vibration of solids. The airflow may contain fluctuations as a result of instability, giving at low Reynolds numbers a regular eddy pattern which is responsible for the sound produced by musical wind instruments, and at high Reynolds numbers an irregular turbulent motion which is responsible for the roar of the wind and of jet aeroplanes; or they may be inherent in the mechanism for producing flow, as in the siren, or in machinery containing rotating blades. Since the pressure fluctuations within the airflow are in the main balanced by fluctuations of fluid acceleration, it is not clear even to within a very large factor what proportion of their energy is radiated as sound. It is true that the whole pressure fluctuation may play its part in generating sound if it is provided with a solid sounding-board;* but all such effects requiring the vibration of solid boundaries are here excluded, and not classed as aerodynamic.

Earlier studies of sound generated aerodynamically are almost all concerned with frequency. Experiments have been directed towards showing that the frequencies in the flow are identical with those of the sound produced, and towards relating them with other constants of the flow; theory has been concerned with explaining the production of such frequencies in the flow by instability. The theory of the instability was further illuminated by the experiments showing how oscillations

* For example, a loose panel in the wall of a wind-tunnel may produce very intense noise in sympathy with pressure fluctuations in the boundary layer, greater by a large factor than the noise emitted when the wall is rigid.

introduced acoustically at the orifice of a jet or flame may be rapidly amplified under certain conditions; in this case the fluid flow shows a very sensitive response to sound waves from an external source.

But the evolution of a general procedure for estimating the *intensity* of the sound produced in terms of the details of the fluid flow, which this paper attempts, is one fundamental question, perhaps the only one, which escaped the attention of the great physicists who in the last century created the science of sound. It is true that, with many instruments which generate sound aerodynamically, the acoustic power output is known as a function of the various conditions of operation; but, nevertheless, a study of its relation to the actual flow has not been attempted, largely because in most cases the latter has not been seriously investigated. The problem is a fundamental one because it is concerned with uncovering the mechanism of conversion of energy between two of its forms, namely, the kinetic energy of fluctuating shearing motions and the acoustic energy of fluctuating longitudinal motions.

This paper is concerned with the general problem: given a fluctuating fluid flow, to estimate the sound radiated from it. Part II will take up the question of turbulent flows proper, with special reference to the sound field of a turbulent jet, for which a comparison with experiment is possible.

The problem's utility may be questioned on the grounds that we never know a fluctuating fluid flow very accurately, and that therefore the sound produced in a given process could only be estimated very roughly. Indeed, one could hardly expect, even with the great advances in knowledge of turbulent flow which have lately been made, that such a theory could be used with confidence to predict acoustic intensities within a factor of much less than 10. But, on the other hand, one could certainly make no confident estimate even to within a factor of 1000 on existing knowledge, and further, the range of intensities in which one is interested is at least 10^{14}. Also, nothing is known at present concerning how different sorts of changes in a flow pattern may be expected to alter the sound produced, and this is a serious impediment to those experimenting in novel fields of aerodynamic sound production. Clearly such knowledge can arise only from a process in two parts, the first considering what sort of fluctuating flow will be generated, and the second what sound the flow will produce.

The proposed method of attack, in which first the details of the flow are to be estimated, from aerodynamic principles not concerned with the acoustic propagation of fluctuations in the flow, and secondly the sound field is to be deduced, precludes the discussion of phenomena where there is a significant back-reaction of the sound produced on the flow field itself. But such back-reaction is only to be expected when (as in wind instruments) there is a resonator close to the flow field. All the evidence of experiment, and of the theory to be developed below, is that the sound produced is so weak relative to the motions producing it that no significant back-reaction can be expected unless there is such a resonator present to amplify the sound.

Actually it seems likely from the theory in its present form that quantitative estimates of the sound field will be obtainable (as will be seen in part II) only for the sound radiated into free space; and thus they will neglect not only neighbouring resonators but also all effects of reflexion, diffraction, absorption or scattering by solid boundaries. But the general result of these effects could often be sketched in subsequently, in the light of existing knowledge. Again, the estimates refer only to the energy which actually *escapes* from the flow as sound, and to its directional distribution; thus the departures from inverse square-law radiation which are to be expected within a very few wave-lengths of the flow, due to a standing wave pattern, will *not* appear in the estimates, although they are implicit in the general theory. As a final restriction the theory is effectively confined in its application to completely subsonic flows, and could hardly be used to analyze the change in character of the sound produced which is often observed on transition to supersonic flow, at least if, as Cave-Browne-Cave suggests, it is due to high-frequency emission of shock waves.

Because the material of parts I and II is likely to be of interest to workers in acoustics, turbulence theory, gas dynamics and aeronautical engineering, and because it draws ideas from a number of other parts of the physical sciences, the author has thought it desirable to develop the theory less briefly than is customary, so as not to assume an intimate knowledge on the part of the reader of the less salient points of any of these subjects. Since also the method of part I is intended to supply a fundamental basis for further work on the subject (of which part II will be only a first *ad hoc* attempt), as well as to answer the question concerning the mechanism of energy conversion, it has been thought desirable to devote a further introductory section towards justifying the choice of method and giving a preliminary explanation of it.

The author gratefully acknowledges the assistance in writing parts I and II which he obtained from discussions with G. K. Batchelor, H. Cohen, J. H. Gerrard, W. A. Mair, I. Proudman, M. Schwarzschild, H. B. Squire, Sir Geoffrey Taylor, E. Wild and A. D. Young.

2. Description and justification of the approach used

Considering a fluctuating fluid flow occupying a limited part of a very large volume of fluid of which the remainder is at rest. Then the equations governing the fluctuations of *density* in the real fluid will be compared with those which would be appropriate to a *uniform acoustic medium at rest*, which coincided with the real fluid outside the region of flow. The difference between the two sets of equations will be considered as if it were the effect of a fluctuating external force field, known if the flow is known, acting on the said uniform acoustic medium at rest, and hence radiating sound in it according to the ordinary laws of acoustics.

This scheme has two advantages. First, since we are not concerned (see § 1) with the back-reaction of the sound on the flow, it is appropriate to consider the sound as produced by the fluctuating flow after the manner of a forced oscillation. Secondly, it is best to take the free system, on which the forcing is considered to occur, as a uniform acoustic medium at rest, because otherwise, after the sound produced has been estimated, it would be necessary to consider the modifications due to its convection with the turbulent flow and propagation at a variable speed within it, which would be difficult to handle. But by the method just described all these effects are replaced by equivalent forcing terms and incorporated in the hypothetical external force field.

The comparison is made most easily if we write the equation expressing conservation of momentum in the form first used extensively by Reynolds. This is equivalent simply to considering the momentum contained within a fixed region of space as changing at a rate equal to the combined effect of (i) the stresses acting at the boundary and (ii) the flow across the boundary of momentum-bearing fluid. It is easy to see that the latter part is equivalent to an additional stress system. This may be represented symbolically by $\rho v_i v_j$, and referred to either as the 'momentum flux tensor' (i.e. the rate at which momentum in the x_i direction crosses unit surface area in the x_j direction; it is worth noting that similarly the real stresses* are simply a mean momentum flux tensor for the peculiar motions of the molecules) or as the 'fluctuating Reynolds stresses' (to distinguish them from their mean value, minus the product of means $\overline{\rho v_i v_j}$, which is called 'Reynolds stress' in the theory of turbulence).

Thus the Reynolds momentum equation already expresses that the momentum in any fixed region of space changes at a rate exactly the same as if the gas were *at rest* under the combined action of the real stresses,* say p_{ij}, and the fluctuating Reynolds stresses $\rho v_i v_j$. On the other hand, a *uniform acoustic medium* at rest would experience stresses only in the form of a simple hydrostatic pressure field, whose variations would be proportional to the variations in density, the constant of proportionality being the square a_0^2 of the speed of sound. Hence the density fluctuations in the real flow must be exactly those which would occur in a uniform acoustic medium subject to an external stress system given by the *difference*

$$T_{ij} = \rho v_i v_j + p_{ij} - a_0^2 \rho \delta_{ij} \qquad (1)$$

between the effective stresses in the real flow and the stresses in the uniform acoustic medium at rest.

It is important to notice (as the equations will show in § 3) that this analogy approach to the problem of aerodynamic sound production is an exactly valid one. It merely assumes that the mass in a given region changes at a rate equal to the total inward normal momentum, and that the momentum changes at a rate equal to the total inward normal component of the complete stress system of Reynolds. It makes no simplifying assumption concerning the real stresses and their relation to rates of strain. The external stress system T_{ij} incorporates not only the generation of sound, but its convection with the flow (in part of the term $\rho v_i v_j$), its propagation with variable speed and gradual dissipation by conduction (each in part of the difference between pressure variations and a_0^2 times the density variations), and its gradual dissipation by viscosity (in the viscous contribution to the stress system p_{ij}).

In practice the dissipation of acoustic energy into heat, by viscosity and heat conduction, is a slow process; in the atmosphere only half the energy is lost in the first mile of propagation even at the frequency (4 kc/s) of the top note of the piano-

* These are made up of hydrostatic pressure $p\delta_{ij}$ and viscous stresses.

forte, while for other frequencies the required distance varies as their inverse square. The contribution of the viscous stresses to T_{ij} is therefore probably unimportant, at least for phenomena on a terrestrial scale of distance. For flows in which the temperature departs little from uniformity the differences between the exact pressure field $p\delta_{ij}$ and the approximate one $a_0^2\rho\delta_{ij}$ are similarly unimportant, and then the principal generators of sound are the fluctuating Reynolds stresses, corresponding to variable rates of momentum flux across surfaces fixed in the fluctuating fluid flow.

Now at this stage the question concerning the mechanism of conversion of energy from the kinetic energy of fluctuating shearing motions into the acoustic energy of fluctuating longitudinal motions is already effectively answered. The answer is most illuminating if one lists three different ways in which one can cause kinetic energy to be converted into acoustic energy, as follows:

(i) By forcing the mass in a fixed region of space to fluctuate, as with a loudspeaker diaphragm embedded in a very large baffle.

(ii) By forcing the momentum in a fixed region of space to fluctuate, or, which is the same thing, forcing the rates of mass flux across fixed surfaces to vary; both these occur when a solid object vibrates after being struck.

(iii) By forcing the rates of momentum flux across fixed surfaces to vary, as when sound is generated aerodynamically with no motion of solid boundaries.

This is a linear sequence of methods of energy conversion, concerning which two facts are important. First, each is less efficient than the preceding one. Secondly, this statement becomes increasingly true as the frequency is lowered, or more precisely as the wave-length of the sound produced is increased.

These two facts, as far as the relation between methods (i) and (ii) is concerned, are now a commonplace of acoustics, having been first fully realized and understood by Stokes, whose explanation of the reduced acoustic power of a bell when hydrogen is added to the air in which it is sounded (so that the wave-length is augmented) is quoted at length in Rayleigh's *Theory of sound*. In brief the physical explanation is that any forcing motion on a scale comparable with the wave-length is balanced partly by a local reciprocating motion, or standing wave, and partly by compressions and rarefactions of the air whose effect is propagated outwards. The larger the wave-length in comparison with the scale of the forcing motion, the more completely can the motion be fully reciprocated by the local standing wave. Mathematically, the difference is between the fields of an acoustic source and an acoustic dipole, which represent fluctuating rates of production at a point of mass and momentum respectively, so that motions of types (i) and (ii) can be represented as due to distributions of sources and dipoles respectively.

For method (iii), the Stokes effect is even more marked. Indeed, Stokes himself showed that if the surface of a sphere vibrates in such a way that its volume and the position of its centroid remain fixed, and the associated wave-length is twice the circumference, the acoustic power is about $\frac{1}{1000}$ of what it would be if motion near the sphere were forced to be purely radial (so that the lateral reciprocating motion could not be set up and the forcing were like that due to a number of independent point sources). The corresponding factor for a *rigid* vibration is $\frac{1}{13}$. A similar effect occurs in the more complicated conditions of aerodynamic sound production; it is the radiation due to the minute fraction of the fluctuation in momentum flux which is *not* balanced by a local reciprocating motion that we must seek to determine. Mathematically, we have to find the radiation field of a distribution of acoustic quadrupoles.

Now the approach described above, in which the sound is shown to be that which would be produced by the externally applied stresses (1), gives at once the local quadrupole strength per unit volume. For since a dipole is equivalent to a concentrated fluctuating force, it follows that a stress field, which produces equal and opposite forces on both sides of a small element of fluid, is equivalent to a distribution of quadrupoles whose instantaneous strength per unit volume is proportional to the local stress. The radiation field of this quadrupole distribution can at once be written down. These deductions will be given at greater length in § 4.

Thus in this approach the influence of the minute fraction of the fluctuations which is not balanced is at once isolated. The author believes that any approach which differed from his up to this stage, by approximating too early, might well throw away the one small part of the sound field which it is desirable to keep, or swamp it with much larger terms, as, for example, if eddies were supposed to emit sound like vibrating rigid spheres.

One important conclusion from the above qualitative discussion concerns those flows, referred to in § 1, the fluctuations in which are inherent in the mechanisms for producing them, these mechanisms being fluctuating sources of matter for the siren, or of momentum for machinery containing rotating blades. The conclusion in question is that the *strictly* aerodynamic contribution to the sound produced, namely, the quadrupole field arising from the oscillations in momentum flux across fixed surfaces in the flow, will be less important than the direct sound due to the source or dipole distributions corresponding to the puffs of air or the motion of the blades respectively. This conclusion is borne out by the success of theories of propeller noise (at least as regards the more intense, low frequency, part) which simply replace the propeller by a rotating line of dipoles. The conclusion could perhaps bear further investigation, but nevertheless, on the basis of it, the remainder of part I and, later, part II will treat only of flows with fixed boundary conditions, i.e. flows in which the fluctuations are solely the result of instability.

In part I the formula for sound radiation is not applied to a particular case, but only used to make a general dimensional analysis of the intensity field in terms of typical velocities and lengths in the flow. The result of this analysis is strongly affected by the considerations just given concerning quadrupole fields and the Stokes effect. The amplitude of the quadrupole strength per unit volume T_{ij} is evidently proportional to the square of a typical velocity U in the flow; but the amplitude of the radiation field due to a quadrupole is proportional to its strength multiplied by the square of its frequency (the Stokes effect). Since in many cases a typical frequency will be roughly proportional to the velocity U, it follows that the amplitude of the sound generated by a given fluctuating flow will increase roughly like the *fourth* power of a typical velocity U in the flow. Hence the intensity will increase roughly like U^8. This is the cardinal result of § 6, which also considers the effect of Reynolds number and Mach number upon it, and discusses the acoustic efficiency.

Lastly, in §7, the formulae for the radiation field are rewritten for the case in which it is convenient to describe the fluctuating flow in terms of a frame of reference which is not stationary with respect to the undisturbed atmosphere; this work is somewhat analogous to the Lienard-Wiechert theory of the field of a moving electron. It is found that for fluctuations in momentum flux across a surface moving through the undisturbed atmosphere the Stokes effect may be greatly mitigated in respect of radiation emitted forwards. This theory will be made use of in part II to estimate the influence of Mach number in promoting departures from the power laws indicated by dimensional analysis.

3. The equivalent external stress field

The propagation of sound in a uniform medium, without sources of matter or external forces, is governed by the equations

$$\frac{\partial \rho}{\partial t}+\frac{\partial}{\partial x_i}(\rho v_i) = 0, \quad \frac{\partial}{\partial t}(\rho v_i)+a_0^2\frac{\partial \rho}{\partial x_i} = 0, \quad \frac{\partial^2 \rho}{\partial t^2}-a_0^2\nabla^2\rho = 0, \tag{2}$$

of which the first is the exact equation of continuity, the second is an *approximate* equation of momentum, and the third follows by eliminating the momentum density ρv_i from the other two. Here ρ is the density, v_i the velocity in the x_i direction, a_0 the speed of sound in the uniform medium, and any suffix repeated in a single term is to be summed from 1 to 3.

On the other hand, the *exact* equation of momentum in an arbitrary continuous medium under no external forces is

$$\frac{\partial}{\partial t}(\rho v_i)+\frac{\partial}{\partial x_j}(\rho v_i v_j+p_{ij}) = 0, \tag{3}$$

in Reynolds's form. Here p_{ij} is the compressive stress tensor, representing the force in the x_i direction acting on a portion of fluid, per unit surface area with inward normal in the x_j direction. Equation (3) is most simply derived from the physical argument given in §2. Alternatively, it can be obtained from the momentum equation in the more familiar Eulerian form by adding a multiple of the equation of continuity.

Hence the equations of an arbitrary fluid motion can be rewritten, as suggested in §2, as the equations of the propagation of sound in a uniform medium at rest due to externally applied fluctuating stresses, namely, as

$$\frac{\partial \rho}{\partial t}+\frac{\partial}{\partial x_i}(\rho v_i) = 0, \quad \frac{\partial}{\partial t}(\rho v_i)+a_0^2\frac{\partial \rho}{\partial x_i} = -\frac{\partial T_{ij}}{\partial x_j}, \quad \frac{\partial^2 \rho}{\partial t^2}-a_0^2\nabla^2\rho = \frac{\partial^2 T_{ij}}{\partial x_i \partial x_j}, \tag{4}$$

where the instantaneous applied stress at any point is

$$T_{ij} = \rho v_i v_j+p_{ij}-a_0^2\rho\delta_{ij}. \tag{5}$$

Equations (4), of which the middle one is simply (3) rewritten, are chosen as the basic equations of the theory of aerodynamic sound production for reasons fully discussed in §2.

For a Stokesian gas the stress tensor p_{ij} is given in terms of the velocity field by the equations

$$p_{ij} = p\delta_{ij} + \mu\left\{-\frac{\partial v_i}{\partial x_j} - \frac{\partial v_j}{\partial x_i} + \frac{2}{3}\left(\frac{\partial v_k}{\partial x_k}\right)\delta_{ij}\right\}, \tag{6}$$

where μ is the coefficient of viscosity, and the pressure p is related to the other thermodynamic variables as for the gas at rest. Atmospheric air may be taken as a Stokesian gas for practical purposes.

It is emphasized again that all effects such as the convection of sound by the turbulent flow, or the variations in the speed of sound within it, are taken into account, by incorporation as equivalent applied stresses, in equations (4); this fact is evident, since the equations are exactly true for any arbitrary fluid motion. However, for an airflow embedded in a uniform atmosphere at rest (a case important in practice) the stress system (5) can be neglected outside the flow itself. For there the velocity v_i consists only of the small motions characteristic of sound, and it appears quadratically in (5). Also the viscous stresses in p_{ij}, and the conduction of heat (which causes departures of $p-p_0$ from $a_0^2(\rho-\rho_0)$, where the suffix zero signifies atmospheric values) are both very small effects. (Actually the solution of the equations of sound, taking into account these effects of viscosity and heat conduction, was effected by Kirchhoff. His analysis, given in Rayleigh's *Theory of sound*, shows that the stresses equivalent to the effects of viscosity and heat conduction simply cause a damping of the sound due to the conversion of acoustic energy into heat by these processes, which as indicated in § 2 is negligible except for very large-scale phenomena.) Thus outside the airflow the density satisfies the ordinary equations of sound (2), and the fluctuations in density, caused by the effective applied stresses within the airflow, are propagated acoustically.

If it is assumed that the viscous stresses in T_{ij} (see (6)) can also be ignored in the flow (this point will be returned to), then it should be noted that at low Mach number, provided that any difference in temperature between the flow and the outside air is due simply to kinetic heating or cooling (that is, heating by fluid friction or cooling by rapid acceleration), T_{ij} is approximately $\rho_0 v_i v_j$, with a proportional error of the order of the square of the Mach number M. This results from the fact that relative changes in density under these conditions are known to be of order M^2, while the ratio of fluctuations in pressure to fluctuations in density departs from a_0^2 by a proportional error of order M^2. The resulting approximate form

$$T_{ij} \simeq \rho_0 v_i v_j \tag{7}$$

of the equivalent applied stress field might have been found from an approach which made approximations in the equations of motion right from the start, but there would then have been no guarantee that, in obtaining this main contribution to the quadrupole field, source and dipole fields of small strength might not have been neglected, whose contribution to the sound radiated might (§ 2) be relatively large.

4. Source, dipole and quadrupole radiation fields

The theory of the quadrupole radiation field due to the equivalent applied stresses will be given at some length. This is because quadrupole radiation has hitherto played a relatively small part in physics, and therefore the subject is not covered adequately in books, and is probably unfamiliar to most readers. The confusion concerning the subject is demonstrated particularly by the fact that the term 'quadrupole radiation' is used in at least two senses besides its proper one: namely, as the part of the intensity field which falls off as the inverse fourth (or, sometimes, sixth) power of the distance from the source, and as the part of the amplitude field expressible in terms of spherical harmonics of the second order. The proper meaning is of course a certain limiting case of the field of four sources; at large distances this obeys the inverse *square* law of radiation; and its amplitude field includes spherical harmonics of zero order as well as of the second order.

But to analyze the subject clearly and relate it to the generation of sound by externally applied stresses it is convenient to recall parts of the theory of the generation of sound by simpler mechanisms. First, if fluctuating sources of additional matter are continuously distributed throughout part of the medium, so that a mass $Q(\mathbf{x},t)$ per unit volume per unit time is introduced at $\mathbf{x}$ at time t, then equations (2) are modified by an additional term Q on the right of the first equation, and hence an additional term $\partial Q/\partial t$ on the right of the third equation. If the sources of matter are concentrated into a point, where the total rate of introduction of mass is $q(t)$, and if the medium is unbounded, then the density field is given by the equation

$$\rho-\rho_0=\frac{1}{4\pi a_0^2}\frac{q'(t-r/a_0)}{r}, \tag{8}$$

where r is the distance from the source. When the sources of matter are not so concentrated, the density field is given by a volume integral of terms such as (8), namely,

$$\rho-\rho_0=\frac{1}{4\pi a_0^2}\int\frac{\partial}{\partial t}Q\left(\mathbf{y},t-\frac{|\mathbf{x}-\mathbf{y}|}{a_0}\right)\frac{d\mathbf{y}}{|\mathbf{x}-\mathbf{y}|}, \tag{9}$$

where the integral is taken over all space, and is of the kind referred to in electromagnetic theory as a retarded potential. But if solid boundaries are present, reflected and diffracted waves must be added to (8) and (9).

Since in (8) it is only the rate of change with time of the rate of introduction of mass $q(t)$ which affects the sound produced, it is this derivative $q'(t)$ which will here be called the instantaneous *strength* of the concentrated source. Similarly, $\partial Q/\partial t$ is called the source strength per unit volume; note that this source strength density is precisely the term appearing on the right of the *third* of equations (2) when continuously distributed fluctuating sources of matter are present. The nomenclature just introduced is not by any means standard, but (at least in this paper) is very convenient, especially since it means that the instantaneous strength of a dipole is equal to the equivalent applied force, as will shortly be seen.

If, in fact, sources of matter are absent, and the sound is generated instead by a fluctuating external force field F_i per unit volume in part of the medium, then equations (2) are modified by an additional term F_i on the right of the second equation, and so by an additional term $-\partial F_i/\partial x_i$ on the right of the third equation. Thus such a fluctuating force field is equivalent (in the density fluctuations it produces) to a source distribution whose strength per unit volume is equal to the *flux of force inwards*, $-\partial F_i/\partial x_i$.

But it would be most misleading to base any estimate of the acoustic power output of the fluctuating force field on the order of magnitude of this equivalent source strength per unit volume. This is because, at any one instant, the total source strength is zero, as being the integrated flux of a quantity which vanishes outside a limited region. Hence the sound given by (9) at large distances (where $|\mathbf{x}-\mathbf{y}|^{-1}$ can effectively be replaced by x^{-1}) is simply due to the fact that the values of Q therein do not quite cancel out because they are not *simultaneous* values throughout the fluctuating force field. Evidently if F_i varies with time only slowly, that is, if the frequencies are low, this may mean relatively little energy radiation (compare § 2).

Actually of course the sound field generated is a dipole field, and the arguments of the last paragraph merely show that it is essential to take this into account before any approximate estimation of acoustic power output be made. For example, the term $-\partial F_1/\partial x_1$ in the source distribution is equivalent (in the limit as $\epsilon \to 0$) to a distribution $\epsilon^{-1}F_1(x_1, x_2, x_3)$ and a distribution $-\epsilon^{-1}F_1(x_1+\epsilon, x_2, x_3)$; so that any one *value*, say $\epsilon^{-1}F_1(x_1, x_2, x_3)$, occurs with positive sign at (x_1, x_2, x_3) and negative sign at $(x_1-\epsilon, x_2, x_3)$. These two together constitute in the limit a dipole of strength F_1 with axis in the positive x_1 direction.

It follows that the force field is equivalent to a field of dipoles with axis in the x_1 direction and strength F_1 per unit volume, together with two similar fields with axes in the x_2 and x_3 directions. But the whole may of course be regarded, from a more fundamental point of view, as a single dipole field whose strength per unit volume is a *vector* F_i (whose direction as well as magnitude may fluctuate). One may see at once the significance of this vector strength density by choosing, at any instant, the x_1-axis in the direction of F_i, in which case the argument of the last paragraph shows that the dipole is of strength equal to the magnitude of F_i and has axis in the direction of F_i. A force field F_i per unit volume emits sound like a volume distribution of dipoles, whose strength vector per unit volume is F_i.

If the force is concentrated at a point, with value $f_i(t)$, and if the medium is unbounded, the density field is given by

$$\rho-\rho_0 = -\frac{1}{4\pi a_0^2}\frac{\partial}{\partial x_i}\left(\frac{f_i(t-r/a_0)}{r}\right), \tag{10}$$

as follows immediately from (8) when values of q' equal to $\pm\epsilon^{-1}f_1$ are placed at $(0, 0, 0)$ and $(-\epsilon, 0, 0)$ (as in the argument just given), and similarly for f_2 and f_3, and ϵ is allowed to tend to zero. It follows from (10) that for the general volume distribution of dipoles

$$\rho-\rho_0 = -\frac{1}{4\pi a_0^2}\frac{\partial}{\partial x_i}\int F_i\left(\mathbf{y}, t-\frac{|\mathbf{x}-\mathbf{y}|}{a_0}\right)\frac{d\mathbf{y}}{|\mathbf{x}-\mathbf{y}|}. \tag{11}$$

(As a mathematical point it may be noticed that the identity of (11) with (9), if in the latter $\partial Q/\partial t$ at $\mathbf{y}$ be replaced by the flux $-\partial F_i/\partial y_i$, can also be deduced from the divergence theorem as applied to the vector integrand of (11).)

Now in carrying out the differentation with respect to x_i in (11), one sees that the part due to differentiating the $|\mathbf{x}-\mathbf{y}|$ in the denominator falls off like the inverse square of the distance from the field of fluctuating forces. But the part due to differentiating F_i itself falls off only like the inverse first power of this distance. Hence at large distances from the field of flow the density fluctuations are dominated by this latter part, namely,

$$\rho - \rho_0 \sim \frac{1}{4\pi a_0^2}\int \frac{x_i - y_i}{|\mathbf{x}-\mathbf{y}|^2}\frac{1}{a_0}\frac{\partial}{\partial t}F_i\left(\mathbf{y}, t - \frac{|\mathbf{x}-\mathbf{y}|}{a_0}\right)d\mathbf{y}. \tag{12}$$

Since fluctuations of $\partial F_i/\partial t$ differ from fluctuations of F_i by a factor of the order of magnitude 2π times a typical frequency, it is easily seen that the term neglected in (12) is truly negligible if x is at a distance from the field of flow large compared with $(2\pi)^{-1}$ times a typical wave-length. It is then in the radiation field of each dipole separately. One may note that if, in addition, this distance were large compared with the dimenions of the field of flow itself, then by choosing an origin within the field of flow one could approximate to the fraction $(x_i - y_i)/|\mathbf{x}-\mathbf{y}|^2$ in (12) by $x_i/|\mathbf{x}|^2$ and take it outside the integral. This approximation would be perfectly sufficient for obtaining the radiation field of the flow as a whole, and hence the acoustic power output, since only terms of order x^{-2} in the amplitude are neglected.

Equation (12) also shows explicitly how the sound radiated to large distances depends on the fact that rays of sound reaching a distant point simultaneously were *not emitted simultaneously*. For it depends only on the rate of change of dipole strength with time.

After these preliminaries the properties of the generation of sound by applied fluctuating stresses T_{ij} are understood more easily. As equations (4) show, the stresses produce a force per unit volume equal to their flux inwards $-\partial T_{ij}/\partial x_j$. Hence they generate sound like a dipole field of strength $-\partial T_{ij}/\partial x_j$ per unit volume.

But again it is essential that the sound radiated be not estimated from the order of magnitude of the dipole strength per unit volume, since at any instant the total dipole strength is zero. Actually the sound produced is quadrupole field. For example, the term $-\partial T_{i1}/\partial x_1$ is equivalent (in the limit as $\epsilon \to 0$) to a dipole field $\epsilon^{-1}T_{i1}(x_1, x_2, x_3)$ and a second dipole field $\epsilon^{-1}T_{i1}(x_1+\epsilon, x_2, x_3)$, so that any one *value*, say $\epsilon^{-1}T_{i1}(x_1, x_2, x_3)$, occurs with positive sign at (x_1, x_2, x_3) and negative sign at $(x_1-\epsilon, x_2, x_3)$. The two together, in the limit, may be said to constitute a quadrupole whose strength is the magnitude of the vector T_{i1}, and whose axes are in the direction of T_{i1} and in the x_1 direction. (Generally, if a quadrupole is formed by equal and opposite dipoles with axes in one direction, whose relative position is in another direction, these two directions are called the axes of the quadrupole. When the axes coincide the quadrupole may be called longitudinal, and when they are perpendicular it may be called lateral.)

It has been shown that the stress field emits sound like three quadrupole fields, namely, the one just described and two others similarly associated with T_{i2} and T_{i3}. But the whole may be regarded, from a more fundamental point of view, as a single quadrupole field *whose strength per unit volume is the stress tensor* T_{ij}. The division into three quadrupoles with specified axes which has just been mentioned is merely one of many possible analyses of the quadrupole into simpler elements. First, each of the nine elements of the tensor T_{ij} is a quadrupole whose strength is the scalar quantity T_{ij} and whose axes are in the x_i and x_j directions. This gives an analysis into three longitudinal quadrupoles, e.g. T_{11} with both axes in the x_1 direction, and three lateral quadrupoles, e.g. $2T_{23}$ with axes in the x_2 and x_3 directions. Secondly, if the principal axes of stress are used locally and instantaneously, then the three lateral quadrupoles disappear and only three mutually orthogonal longitudinal quadrupoles remain (whose orientations as well as strengths will in general be fluctuating).

Thirdly (and perhaps most significantly for the present problem), an analysis of the stress T_{ij} into a pressure and a single pure shearing stress may be made, which leads to an analysis of the quadrupole into three *equal* mutually orthogonal longitudinal quadrupoles, each of strength $T = \frac{1}{3}T_{ii}$, and one lateral quadrupole. The three longitudinal quadrupoles add up to form a simple source of strength $a_0^2 \partial^2 T/\partial t^2$, at least as far as the effect they produce *outside* the stress field is concerned. For if a *source* distribution of strength T per unit volume produces a field $\rho = R$, then a longitudinal quadrupole distribution of strength T per unit volume and axes in the x_1 direction produces a field $\rho = \partial^2 R/\partial x_1^2$. Hence the three longitudinal distributions together produce a field $\rho = \nabla^2 R$. But outside the stress field

$$\nabla^2 R = a_0^{-2} \partial^2 R/\partial t^2,$$

which is the density field due to a source distribution $a_0^{-2} \partial^2 T/\partial t^2$ per unit volume.

Thus the sound field, due to the applied fluctuating stresses T_{ij}, may be split up into a source field of strength

$$\frac{1}{3}\left(\frac{1}{a_0^2}\frac{\partial^2 T_{ii}}{\partial t^2}\right), \tag{13}$$

due to the equivalent applied fluctuating pressures, which for the stress field (5) are

$$\tfrac{1}{3}T_{ii} = \tfrac{1}{3}\rho v_i^2 + p - a_0^2 \rho, \tag{14}$$

and a field of lateral quadrupoles due to the applied fluctuating shearing stresses. The latter are due to lateral momentum flux (fluctuating Reynolds shearing stresses) and to viscous stresses.

Note that the fact that quadrupoles can combine to form a source does not contradict the important principle that quadrupole radiation is less effective, especially for large wave-lengths, than source radiation. For as (13) shows the equivalent source strength is inversely proportional to the square of the wave-length.

From the point of view of spherical harmonic analysis the result is due to the fact that the field of a single longitudinal quadrupole like T_{11} can be analyzed into a term proportional to a spherical harmonic of the second order, and a term proportional to one of zero order, corresponding to the field of a source of strength $\frac{1}{3}a_0^{-2}\partial^2 T_{11}/\partial t^2$. On the other hand, a lateral quadrupole field is simply proportional to a spherical harmonic of the second order.

Now when the medium is unbounded, an expression analogous to (10) can at once be written down for the field of a concentrated quadrupole. The expression analogous to (11) for the field of a continuous distribution with tensor strength density T_{ij} can then be deduced in the form

$$\rho - \rho_0 = \frac{1}{4\pi a_0^2}\frac{\partial^2}{\partial x_i \partial x_j}\int T_{ij}\left(\mathbf{y}, t - \frac{|\mathbf{x}-\mathbf{y}|}{a_0}\right)\frac{d\mathbf{y}}{|\mathbf{x}-\mathbf{y}|}. \tag{15}$$

Note that mathematically (15) could have been deduced from the retarded potential solution (namely (9) with $\partial^2 T_{ij}(\mathbf{y},t)/\partial y_i \partial y_j$ for $\partial Q(\mathbf{y},t)/\partial t$) to the third of equations (4), by the relatively short process of applying the divergence theorem twice. The author prefers the arguments of the present section as being more illuminating as well as more general (showing the field to be a quadrupole field even in the presence of solid boundaries).

At points far enough from the flow to be in the radiation field of each quadrupole, that is, at a distance large compared with $(2\pi)^{-1}$ times a typical wave-length, the differentiation in (15) may be applied (compare (12)) to T_{ij} only, giving

$$\rho-\rho_0 \sim \frac{1}{4\pi a_0^2}\int \frac{(x_i-y_i)(x_j-y_j)}{|\mathbf{x}-\mathbf{y}|^3}\frac{1}{a_0^2}\frac{\partial^2}{\partial t^2}T_{ij}\left(\mathbf{y},t-\frac{|\mathbf{x}-\mathbf{y}|}{a_0}\right)d\mathbf{y}. \tag{16}$$

This formula for the sound radiation field must give an exact result for the total energy radiated and its directional distribution, since only terms falling away more rapidly than the inverse first power of the distance are excluded. The formula, depending principally on the second time derivative of the equivalent applied stress T_{ij}, is the basic result of this paper.

It is worth emphasizing again that because in this problem the total effective dipole strength (i.e. the integral of $-\partial T_{ij}/\partial x_j$ over all space) is zero, the radiation field in the form (12) is only non-zero because the values of $\partial F_i/\partial t$ in the integrand are not *simultaneous* values. In fact, rays of sound from different dipoles reaching a distant point simultaneously were not in general emitted simultaneously. For this reason, as (16) shows, the true form of the radiation field involves a further differentiation with respect to time.

Provided that the quadrupole distribution T_{ij} is not itself approximately a space derivative and therefore approximately replaceable by an octupole field, it will be possible to estimate the radiation field (16) if the order of magnitude of the fluctuations of $\partial^2 T_{ij}/\partial t^2$ in the flow are known. (Certainly there is no general theoretical reason why at any instant the total quadrupole strength should be practically zero, and in all the problems so far considered by the author and co-workers (as will be seen in part II) there is substantial evidence that it is not so. But the possibility should always be borne in mind.)

Actually the total quadrupole strength arising from the contribution of the *viscous* stresses to T_{ij} (see (5) and (6)) is certainly very small, for clearly their integral over the whole flow is of the order μ times a typical velocity *outside* the flow. This gives support to the suggestion made in § 3 that the viscous stresses are just as unimportant inside the flow as they are known to be outside it—independently of the fact that in most flows the macroscopic momentum flux will greatly exceed the viscous stress.

At distances large compared with the dimensions of the flow one may approximate x_i-y_i by x_i in (16), provided that the origin is taken within the flow, without neglecting any terms of order x^{-1} (where x is the magnitude of $\mathbf{x}$). This gives the simpler form

$$\rho-\rho_0 \sim \frac{1}{4\pi a_0^2}\frac{x_i x_j}{x^3}\int \frac{1}{a_0^2}\frac{\partial^2}{\partial t^2}T_{ij}\left(\mathbf{y},t-\frac{|\mathbf{x}-\mathbf{y}|}{a_0}\right)d\mathbf{y} \tag{17}$$

for the radiation field of the complete flow. But for extensive flows the replacement of $x_i - y_i$ by x_i may be inadequate at distances from the flow at which one can conveniently make measurements.

5. Intensity and frequency analysis of the sound field

The quantities which can be estimated by the human ear, or measured by other phase-insensitive instruments, are the intensity at any point, and its frequency spectrum. The intensity of sound at a point where the density is ρ is a_0^3/ρ_0 times the mean-square fluctuation of ρ (i.e. with bars signifying average values at a point, times $\overline{(\rho-\bar{\rho})^2}$, which is also called the variance of ρ and written $\sigma^2\{\rho\}$, where $\sigma\{\rho\}$ is the standard deviation). In symbols, then, the intensity is

$$I(\mathbf{x}) = \frac{a_0^3}{\rho_0}\sigma^2\{\rho(\mathbf{x}, t)\}. \tag{18}$$

In the radiation field the intensity signifies the rate at which energy is crossing unit surface area at the point. To obtain the total acoustic power output of the field of flow one must integrate the intensity over a sphere of radius large compared with the dimensions of the flow, that is, one so large that the formula (17) may be used for the variations in density on it.

The natural units for intensity are watts/sq.cm, but a rather more convenient scale for most purposes is the decibel scale. The intensity level on the decibel scale is now usually defined as $10\log_{10}[I/(10^{-16}\,\text{watts/sq.cm})]$..

Now in fields with a discrete frequency spectrum the variance $\sigma^2\{\rho\}$ would be simply half the sum of the squares of the amplitudes in the various frequencies. For such problems the theoretical worker would be well advised to concentrate on determining these amplitudes, and to leave the determination of intensity and its distribution over the various frequencies until the final question arises of displaying his results and comparing them with experiment.

But at higher Reynolds numbers, when the flow field is fully turbulent, so that presumably the sound field partakes of the same chaotic quality, there is no meaning to be attached to the concept of 'amplitude' for any frequency, or even for any band of frequencies however small. This is because the phase is completely random and discontinuous, so that the fluctuating physical quantities themselves (as distinct from the intensity) possess no spectral density, because the limiting process which is used to define this concept does not converge. Hence intensity and its spectrum are all that can be given a meaning in such cases, not simply all that we need to know.

Now since the value of $\rho-\rho_0$ in the radiation field, by (16), is a time-derivative, its mean with respect to time at any point is 0 and the mean density $\bar{\rho}$ is simply ρ_0. Hence the variance $\sigma^2\{\rho\}$ is simply the mean value at a point of the square of (16). To write down the square of (16) one may simply write it down twice, but it is necessary to use different symbols in the two cases for the variables of summation and integration i, j and $\mathbf{y}$ if confusion is to be avoided. The two integrals can then be combined into a double integral. Then the intensity field derived from (16) and (18) is

$$I(\mathbf{x}) \sim \frac{1}{16\pi^2 \rho_0 a_0^5} \iint \frac{(x_i - y_i)(x_j - y_j)(x_k - z_k)(x_l - z_l)}{|\mathbf{x}-\mathbf{y}|^3 |\mathbf{x}-\mathbf{z}|^3} \times \overline{\frac{\partial^2}{\partial t^2} T_{ij}\left(\mathbf{y}, t - \frac{|\mathbf{x}-\mathbf{y}|}{a_0}\right) \frac{\partial^2}{\partial t^2} T_{kl}\left(\mathbf{z}, t - \frac{|\mathbf{x}-\mathbf{z}|}{a_0}\right)} \, d\mathbf{y} \, d\mathbf{z}. \tag{19}$$

The mean product, or 'covariance',* in (19) is the type of quantity which can be measured by hot-wire techniques in a turbulent flow, though it is a lot more complicated than any such quantity which has yet been studied. The physical interpretation of (19), and the methods for approximating to it by simpler expressions, especially by making use of the fact that the covariance is negligibly small except when the points **y** and **z** in the turbulent flow are rather close together, are all postponed to part II.

However, a general expression for the total acoustic power output of the field will here be obtained. For this one need only use the simplified form (17) for the ultimate radiation field, and integrate its mean square (which is (19), with the fraction inside the integral replaced by $x_i x_j x_k x_l / x^6$) over the surface of a large sphere Σ. Now the fraction $x_i x_j x_k x_l / x^6$ may easily be integrated over Σ; the answer is evidently zero unless i, j, k, l are equal in pairs; it is $4\pi/15$ if the pairs are unequal, and $4\pi/5$ if the pairs are equal. In fact

$$\int_\Sigma \frac{x_i x_j x_k x_l}{x^6} \, dS = \frac{4\pi}{15} (\delta_{ij}\delta_{kl} + \delta_{ik}\delta_{jl} + \delta_{il}\delta_{jk}). \tag{20}$$

Hence the total acoustic power output P is given as

$$P = \frac{1}{60\pi\rho_0 a_0^5} \iint \left[\overline{\frac{\partial^2}{\partial t^2} T_{ii}\left(\mathbf{y}, t - \frac{|\mathbf{x}-\mathbf{y}|}{a_0}\right) \frac{\partial^2}{\partial t^2} T_{kk}\left(\mathbf{z}, t - \frac{|\mathbf{x}-\mathbf{z}|}{a_0}\right)} + 2 \overline{\frac{\partial^2}{\partial t^2} T_{ij}\left(\mathbf{y}, t - \frac{|\mathbf{x}-\mathbf{y}|}{a_0}\right) \frac{\partial^2}{\partial t^2} T_{ij}\left(\mathbf{z}, t - \frac{|\mathbf{x}-\mathbf{z}|}{a_0}\right)} \right] d\mathbf{y} \, d\mathbf{z}$$

$$= \frac{1}{60\pi\rho_0 a_0^5} \left[\sigma^2 \left\{ \int \frac{\partial^2}{\partial t^2} T_{ii}\left(\mathbf{y}, t - \frac{|\mathbf{x}-\mathbf{y}|}{a_0}\right) d\mathbf{y} \right\} + 2 \sum_{i=1}^{3} \sum_{j=1}^{3} \sigma^2 \left\{ \int \frac{\partial^2}{\partial t^2} T_{ij}\left(\mathbf{y}, t - \frac{|\mathbf{x}-\mathbf{y}|}{a_0}\right) d\mathbf{y} \right\} \right]. \tag{21}$$

It should be noticed that in the second term in (21) each of the nine quadrupole fields T_{ij} ($i = 1$ to 3, $j = 1$ to 3) makes its contribution to the power output independently. The first term is connected with the equivalent source strength (13) when the field is split up into a source field and a field of lateral quadrupoles; it involves cross-terms between the different longitudinal quadrupole fields T_{11}, T_{22}, T_{33}. It may also be noticed that the time derivatives may be taken outside the

* Of which a 'correlation coefficient' is a non-dimensional form.

integrals in (21). This gives a rough method of simplification for the purposes of estimation, by making use of the general result that

$$\sigma^2\{f'(t)\} = 4\pi^2 N^2 \sigma^2\{f(t)\}, \tag{22}$$

where N is a frequency somewhere within the band of frequencies appearing in the oscillations of $f(t)$.

Now, most of the work of part I and of part II will be concerned with the intensity field and total power output, but it is also desirable to be able to predict their frequency spectra. To modify the theory for this purpose the following principle may be used: If, from the quadrupole strength density T_{ij}, all fluctuations with frequency outside a certain band were removed,* then the intensity field of the modified quadrupole distribution would be nothing else but the integral of the intensity spectrum over the said band in the original field. This tells us a definite procedure to be adopted if the latter quantity is to be estimated.

Many will find the above principle intuitively obvious, but for the benefit of others it may be noted that a proof in terms of the retarded potentials solutions is possible, using the statistical theorems that the part of a fluctuating function $f(t)$ in the frequency band from 0 to $a/2\pi$ is $\int_{-\infty}^{\infty} f(t+s)\sin(as)\frac{ds}{\pi s}$, and that the portion of its mean square relating to frequencies in the same band is $\int_{-\infty}^{\infty} \overline{f(t)f(t+s)}\sin(as)\frac{ds}{\pi s}$.

One use of the principle might be that in experimental comparison of the properties of T_{ij} (or its approximation (7)) in turbulent flow with those of the sound field generated, it would be permissible to use instruments which excluded fluctuations of T_{ij} outside a certain band of frequencies, provided that they were compared with the sound intensities in this band.

Notice that the frequency spectrum of T_{ij} in turbulent motion, which the above analysis shows to be dominant in determining that of the sound field, is different from that of the velocity field since it depends on it quadratically. Thus it will include summation and difference tones of the velocity frequencies, and generally may be expected to be a considerably flatter spectrum.

If one is interested in a relatively small band of frequencies, the use of the principle (22) for estimating expressions like (21) becomes much more accurate.

6. Dimensional analysis of aerodynamic sound production

Certainly the simplest, and perhaps at first the most practically useful deduction which can be made from the theory of §§ 3 to 5 is an analysis of the dependence of the sound field, for geometrically similar mechanisms of flow production, on a typical velocity U in the flow and a typical linear dimension l, and also on constants of the gas such as ρ_0, a_0 and the kinematic viscosity ν_0. Only in the light of some such analysis can experiments be co-ordinated.

* And it is known that *this* kind of spectral analysis *is* (theoretically) possible for a turbulent flow, even though the fluctuating flow quantities do not, perhaps, have a spectral *density* in the strict sense.

Now by (5) the amplitude of the fluctuations in T_{ij} will, at corresponding points in similar flows, be in the main proportional to $\rho_0 U^2$, although there will be some additional dependence on Reynolds number $R = Ul/\nu_0$, Mach number U/a_0, and on the ratio of a typical temperature in the flow to its atmospheric value; actually changes with Reynolds number are usually very gradual, and changes with Mach number are small unless it approaches 1. But to infer from these facts anything concerning the dependence of the density variations (16) on U, l, ρ_0, a_0 and ν_0 it is necessary to know how typical frequencies of the flow depend on these quantities, so as to be able to relate the fluctuations in $\partial^2 T_{ij}/\partial t^2$ to those in T_{ij}.

Since in § 2 it was decided that the discussion could be confined to flows whose fluctuations are generated by instability rather than by any direct external cause, the dominant frequency or band of frequencies must of course vary in an ascertainable manner with the other constants of the flow. Now in all cases, if n is a typical frequency, the non-dimensional product nl/U (sometimes known as the Strouhal number) has been found to vary far less with changing conditions than n itself. For example, at low Reynolds numbers, when a regular eddy pattern appears, the product nl/U rises only slowly with Reynolds number; in particular for the eddies shed by a wire of diameter l in a stream of speed U, nl/U is about $0{\cdot}2 - 4R^{-1}$ for $40 < R < 40000$. At the upper end of this range of R the frequency spectrum spreads out more and more, while the most prominent frequencies have a slightly higher value of nl/U. The appearance of frequencies very much less than $0{\cdot}2U/l$ is impeded by the scale of the system, but the appearance of higher frequencies is limited only by viscous damping, and so the range of values of nl/U continues to grow at its upper end as R increases. However, the turbulent energy continues to be borne principally by frequencies with nl/U less than 1, although the fluctuations of velocity *gradient* are greatest for rather higher frequencies. For example, the mean square vorticity is carried principally by motions with nl/U proportional to $R^{\frac{1}{2}}$, but these motions have relatively little energy and a significant rate of energy loss by viscous dissipation.

These considerations indicate that to obtain a preliminary rough idea of how the sound produced varies with the constants of the flow, one may take frequencies as proportional to U/l on the whole, and so take the fluctuations in $\partial^2 T_{ij}/\partial t^2$ as roughly proportional to $(U/l)^2 \rho_0 U^2$. One may then conclude that at a distance x from the centre of the flow, in a given direction, the density variations (16) are roughly proportional to the product

$$\frac{1}{a_0^2}\frac{1}{x}\frac{1}{a_0^2}\left(\frac{U}{l}\right)^2 \rho_0 U^2 l^3 = \rho_0 \left(\frac{U}{a_0}\right)^4 \frac{l}{x}. \tag{23}$$

The most striking fact about this formula is the dependence of the density changes in the sound radiation field on the *fourth* power of the Mach number $M = U/a_0$. By contrast, density changes in the flow itself (where l/x is of order 1) are known to be of order $\rho_0 M^2$. The additional factor M^2 at distances large compared with $(2\pi)^{-1}$ wave-lengths, showing that sound radiation is a 'Mach number effect', is due entirely to the quadrupole nature of the field (see § 2).

From (18) it follows that the intensity is roughly proportional to a_0^3/ρ_0 times the square of (23), i.e. to

$$\rho_0 U^8 a_0^{-5} \left(\frac{l}{x}\right)^2, \tag{24}$$

and hence that the total acoustic power output is roughly proportional to

$$\rho_0 U^8 a_0^{-5} l^2. \tag{25}$$

The prediction that sound intensities increase like some high power, near the eighth, of a typical velocity U in the flow, is borne out by experiment (as will be seen in part II).

On the other hand, in a careful experimental study, one would expect to be able to detect departures from laws such as (25), because of the approximate character of the arguments used to support them. Hence, to get the maximum benefit from such experiments, by laying bare the precise nature of such departures, the data should be expressed in terms of an 'acoustic power coefficient'

$$K = \frac{\text{acoustic power}}{\rho_0 U^8 a_0^{-5} l^2}. \tag{26}$$

The dependence of this quantity K on Reynolds number and Mach number could be studied by varying U and l independently, while retaining geometrically similar flow-producing mechanisms. The dependence of K on the ratio of a typical temperature in the flow to the atmospheric value should also be investigated by preheating or pre-cooling the flow. Finally, analyses of K may be made both with respect to direction of propagation and with respect to frequency. (Some of these procedures will be used in part II in analyzing certain data obtained by Gerrard.) Here again a frequency analysis of K would have most value if made in terms of the non-dimensional frequency parameter nl/U, especially if such analyses were performed at different Reynolds numbers, when changes of the shape of the spectrum might give important information as to which aspects of the turbulent flow contribute most to the sound produced.

At this stage it is impossible to make predictions concerning the variations discussed above. It might be thought that K will increase with Reynolds number because as explained above the frequencies bearing the major fluctuations of derivatives like $\partial^2 T_{ij}/\partial t^2$ tend to grow gradually (relative to U/l) with Reynolds number. But this is (at least partly) counteracted by the fact that the eddy-sizes corresponding to these frequencies, and hence the range of values of $|\mathbf{y}-\mathbf{z}|$ for which the covariance in (19) is not negligible, are correspondingly smaller.

To conclude this section it may be noted that in a steadily maintained flow the energy per unit volume will be roughly proportional to $\rho_0 U^2$, and the total rate of supply of energy to $(\rho_0 U^2)(Ul^2)$. Hence the ratio of the acoustic power output to the supply of power, which ratio can be described as the *efficiency* of aerodynamic sound production, will satisfy (to the same sort of accuracy as (25), that is only very roughly)

$$\eta \propto M^5. \tag{27}$$

Of course acoustic efficiencies are always very low indeed, and doubtless that of aerodynamic sound production, even at Mach numbers near the top of the range in which (27) is expected to have some validity (i.e. with M approaching 1) is no exception. (The experiments of Gerrard indicate an order of magnitude 10^{-4} for the coefficient η/M^5.) But (27) makes it clear that turbulence at *low* Mach numbers is a quite exceptionally inefficient producer of sound.

7. Modifications required when flow is analyzed with respect to a moving frame

One can imagine various applications of the theory of §§3 to 5 in which the necessary time-derivatives, and space-integrals, could be estimated more easily, and accurately, if they referred to co-ordinate axes in uniform motion relative to the undisturbed atmosphere. As an example one may quote flow fields carried along with a moving aircraft; and a more complicated application, involving the use of different frames of reference for different parts of a turbulent flow, will be given in part II. In any case the necessary modifications to the theory are easily made, and are therefore worth giving here.

The symbol T_{ij} will continue to have its old significance (5), with the velocities v_i measured relative to the undisturbed atmosphere. However, it will here be considered as given in terms of a co-ordinate system (of constant orientation) whose origin moves with uniform velocity $a_0\mathbf{M}$, where $M < 1$. Thus T_{ij} refers to momentum flux across surfaces moving uniformly through the fluid. It will be particularly interesting to compare the acoustic effect of fluctuations in *this* with that of identical fluctuations of momentum flux across surfaces fixed in the fluid.

Now, if the moving axes are chosen to coincide with the fixed axes at time t, then the retarded value of T_{ij} appearing in the fundamental expression (15) must be rewritten as

$$T_{ij}\left(\boldsymbol{\eta}, t-\frac{|\mathbf{x}-\mathbf{y}|}{a_0}\right), \quad \text{where} \quad \boldsymbol{\eta} = \mathbf{y}+\mathbf{M}\,|\mathbf{x}-\mathbf{y}|. \tag{28}$$

This is because the axes have moved on a distance $\mathbf{M}\,|\mathbf{x}-\mathbf{y}|$ during the time taken for a ray of sound to go from $\mathbf{y}$ to $\mathbf{x}$.

When the integral (15) is transformed into the $\boldsymbol{\eta}$ space the element of volume is altered by a factor equal to the Jacobian of the transformation (28). One easily calculates that

$$d\boldsymbol{\eta} = d\mathbf{y}\left(1-\frac{\mathbf{M}.(\mathbf{x}-\mathbf{y})}{|\mathbf{x}-\mathbf{y}|}\right), \tag{29}$$

and so by (15)

$$\rho-\rho_0 = \frac{1}{4\pi a_0^2}\frac{\partial^2}{\partial x_i \partial x_j}\int T_{ij}\left(\boldsymbol{\eta}, t-\frac{|\mathbf{x}-\mathbf{y}|}{a_0}\right)\frac{d\boldsymbol{\eta}}{|\mathbf{x}-\mathbf{y}|-\mathbf{M}.(\mathbf{x}-\mathbf{y})}. \tag{30}$$

Now at points far enough from the flow to be in the radiation field of each quadrupole, the differentiation in (30) may be applied to T_{ij} only. This requires a knowledge of the derivative of $|\mathbf{x}-\mathbf{y}|$ with respect to x_i, keeping $\boldsymbol{\eta}$ constant, where x and y are related as in (28). This is easily calculated as

$$\frac{\partial}{\partial x_i}|\mathbf{x}-\mathbf{y}| = \frac{x_i-y_i}{|\mathbf{x}-\mathbf{y}|-\mathbf{M}.(\mathbf{x}-\mathbf{y})}. \tag{31}$$

Hence (30) becomes

$$\rho-\rho_0 \sim \frac{1}{4\pi a_0^2}\int \frac{(x_i-y_i)(x_j-y_j)}{\{|\mathbf{x}-\mathbf{y}|-\mathbf{M}.(\mathbf{x}-\mathbf{y})\}^3}\frac{1}{a_0^2}\frac{\partial^2}{\partial t^2}T_{ij}\left(\boldsymbol{\eta}, t-\frac{|\mathbf{x}-\mathbf{y}|}{a_0}\right)\mathrm{d}\boldsymbol{\eta}. \tag{32}$$

Comparing (32) with the sound field (16) due to fluctuations in momentum flux across *fixed* surfaces, we see that given fluctuations, when they take place across moving surfaces, will send an increased quantity of sound in directions making an acute angle with the direction of motion, and a decreased quantity in those making an obtuse angle with it. If the said angle is θ then the factor multiplying the amplitude, due to motion of the axes, is $(1-M\cos\theta)^{-3}$. (Note that the added sound emitted forwards is more than the reduction in sound emitted backwards.)

There are two physical factors contributing to these differences, corresponding to the two mathematical factors emerging from the above (the Jacobian of the transformation and the effect of double differentiation). These factors are the position and time of origin (respectively) of those rays of sound from a single eddy* which arrive at a distant point simultaneously. First, the positions of origin of such rays fill a greater *volume* when the rays are emitted forward, since the foremost parts of the eddy have moved on before they need emit. Secondly, the cancelling out of rays from different parts of a quadrupole is less effective for rays emitted forward, owing to the *time* interval between emission from fore and aft parts being increased. Speaking more crudely, waves emitted forward by an object in motion pile up (the Doppler effect) and this makes cancelling of successive waves less effective.

The *intensity* field derived from (32) takes the form (19), with T_{ij} a function of $\boldsymbol{\eta}$ rather than $\mathbf{y}$, and similarly T_{kl} of $\boldsymbol{\zeta}$, and the integrations taken over $\boldsymbol{\eta}$ and $\boldsymbol{\zeta}$ space. Also the denominator $|\mathbf{x}-\mathbf{y}|^3$ is replaced by $\{|\mathbf{x}-\mathbf{y}|-\mathbf{M}.(\mathbf{x}-\mathbf{y})\}^3$, and similarly with $|\mathbf{x}-\mathbf{z}|^3$. Hence to obtain the power output, as in § 5, one must evaluate not the integral (20) but rather the integral

$$\int_\Sigma \frac{x_i x_j x_k x_l}{(x-\mathbf{M}.\mathbf{x})^6}\,\mathrm{d}S. \tag{33}$$

It is convenient for this purpose to choose the x_1-axis in the direction of the vector $\mathbf{M}$. Then it is clear that the integral (33) vanishes, by symmetry, unless i, j, k, l are equal in pairs. The evaluation when they *are* is somewhat tedious, but straightforward if spherical polars are used. One finds that the corresponding values for the case $M = 0$ are multiplied by

$$(a)\ \frac{1}{(1-M^2)^3},\quad (b)\ \frac{1+5M^2}{(1-M^2)^4},\quad (c)\ \frac{1+10M^2+5M^4}{(1-M^2)^5}, \tag{34}$$

according as (*a*) neither, (*b*) one or (*c*) both of the pairs of suffixes is 1.

This means that the power output is given by an expression precisely of the form (21), but with each term multiplied by one of the factors (34), according as it results

* Here the reader may understand 'eddy' as meaning simply a small (imaginary) volume carried along with velocity $a_0 M$.

from quadrupoles with (a) neither, (b) one or (c) both of their axes in the direction of motion. The cross-terms between two longitudinal quadrupoles are multiplied by factor (b) when one is in the direction of motion and by factor (c) when neither is.

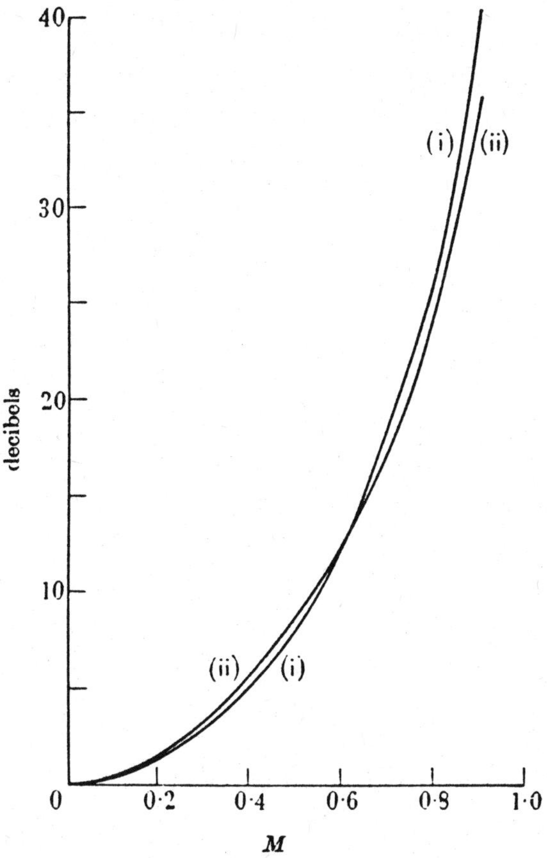

FIGURE 1. Change, as a result of translation at a Mach number M, in the total acoustic power output of (i) three equal mutually orthogonal longitudinal quadrupoles and (ii) a lateral quadrupole with one axis in the direction of translation.

To illustrate the above theory, the increase in power output, expressed in decibels, is shown as a function of Mach number in figure 1 for (i) three equal longitudinal quadrupoles (which for $M = 0$ would produce a simple source field, see § 4), (ii) a single lateral quadrupole of which one of the axes is in the direction of motion. The effect of translation of the fluctuating field in case (ii) is multiplication by the factor (b), so that it is 10 times the logarithm to base 10 of factor (b) which is plotted. In case (i), by combining the effects of the factors (34) according to the laws given above, one finds that the intensity is multiplied by

$$\frac{1+2M^2+\frac{1}{5}M^4}{(1-M^2)^5}. \tag{35}$$

The increase in intensity level in cases (i) and (ii) is further analyzed in figures 2 and 3 respectively, by a plot of variation in intensity level with direction θ (measured from the direction of motion*) and with Mach number M. The quantities plotted are ten times the logarithm to base 10 of

$$\text{(i)}\ \frac{1}{(1-M\cos\theta)^6},\quad \text{(ii)}\ \frac{4\sin^2\theta\cos^2\theta}{(1-M\cos\theta)^6} \tag{36}$$

in figures 2 and 3 respectively (so that in each case the maximum of the curve for $M = 0$ is arbitrarily chosen as the intensity level 0). These figures illustrate the

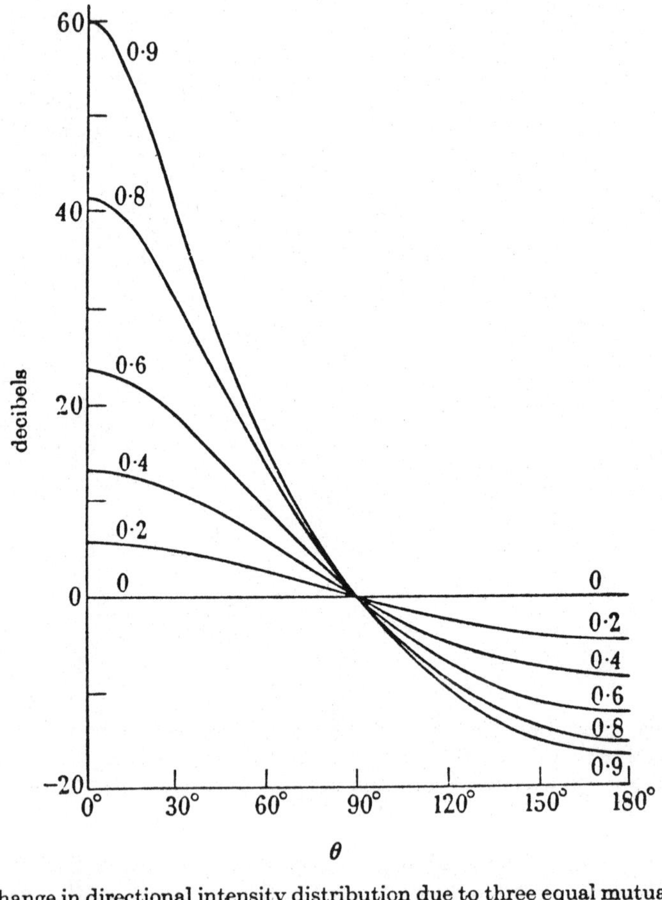

FIGURE 2. Change in directional intensity distribution due to three equal mutually orthogonal longitudinal quadrupoles as a result of translation at a Mach number, the value of which is indicated against each curve, the direction θ being measured from the direction of translation.

change with Mach number of the directional intensity field of a moving fluctuating eddy, at points far from it compared both with $(2\pi)^{-1}$ times a typical wave-length and also with the largest distance between points $\mathbf{y}$ and $\mathbf{z}$ in the flow such that a

* Note that the direction θ is measured relative to the position of the eddy when it emits, not its position when the sound arrives at the point $\mathbf{x}$.

covariance such as that in (19) is significant. For when this is so the difference between $x_i - y_i$ and $x_i - z_i$ in (19) can be neglected, and both replaced by

$$|\mathbf{x}-\mathbf{y}|\cos\theta \quad \text{or} \quad |\mathbf{x}-\mathbf{y}|\sin\theta$$

according as i is 1 or not.

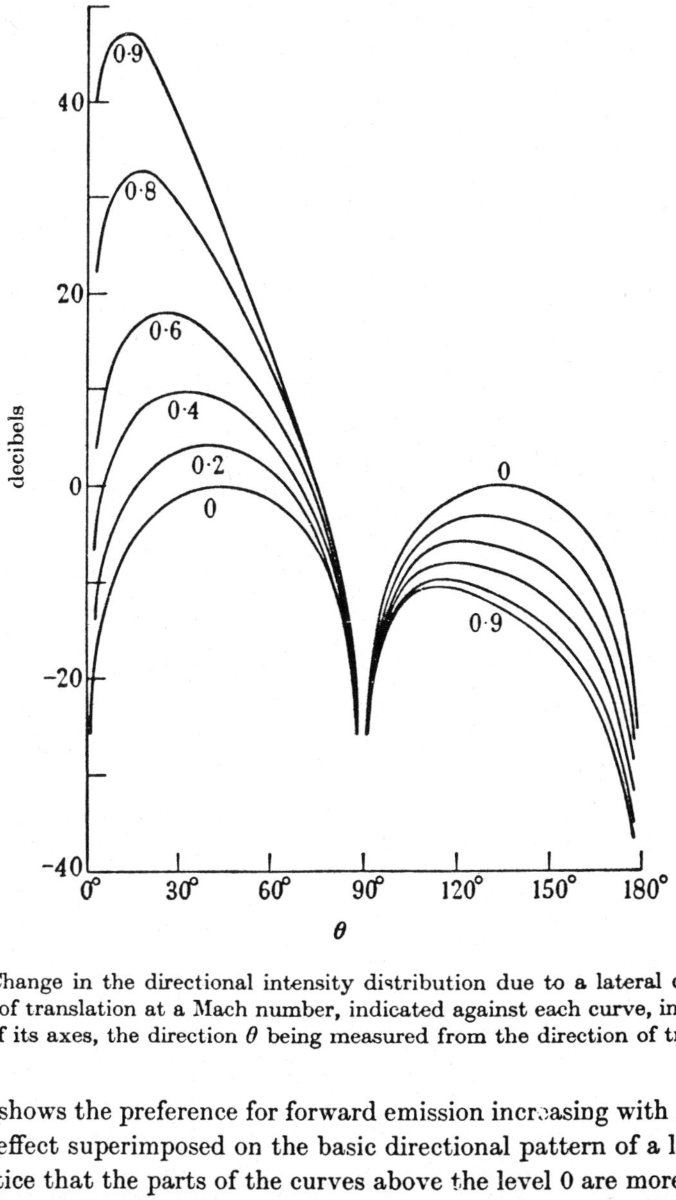

FIGURE 3. Change in the directional intensity distribution due to a lateral quadrupole as a result of translation at a Mach number, indicated against each curve, in the direction of one of its axes, the direction θ being measured from the direction of translation.

Figure 2 shows the preference for forward emission increasing with M. Figure 3 shows this effect superimposed on the basic directional pattern of a lateral quadrupole. Notice that the parts of the curves above the level 0 are more effective in increasing the power outputs (figure 1) than are the parts below in decreasing them, owing to the logarithmic scale used in plotting the curves.

Finally, the frequency analysis of the sound produced by fluctuations of T_{ij} at points in motion relative to the atmosphere will be governed by the principle stated at the end of § 5 only if it is modified to include the Doppler effect. Thus the band of frequencies retained in $T_{ij}(\eta, t)$ will be responsible for the sound radiation to a distant point x with frequencies in a band obtained by multiplying those in the original band by

$$\left(1-\frac{\mathbf{M}.(\mathbf{x}-\mathbf{y})}{|\mathbf{x}-\mathbf{y}|}\right)^{-1}. \tag{37}$$

The frequency is of course increased for sound emitted forwards, and decreased for sound emitted backwards.

The drawing of practical conclusions from the results of this section is postponed to part II.

Proc. Roy. Soc., 1954, V. 221, p. 1–32.

On sound generated aerodynamically
II. Turbulence as a source of sound

18

BY M. J. LIGHTHILL, F.R.S.
Department of Mathematics, The University of Manchester

(*Received* 21 *July* 1953)

The theory of sound generated aerodynamically is extended by taking into account the statistical properties of turbulent airflows, from which the sound radiated (without the help of solid boundaries) is called aerodynamic noise. The theory is developed with special reference to the noise of jets, for which a detailed comparison with experiment is made (§ 7 for subsonic jets, § 8 for supersonic ones).

The quadrupole distribution of part I (Lighthill 1952) is shown to behave (see § 3) as if it were concentrated into independent point quadrupoles, one in each 'average eddy volume'. The sound field of each of these is distorted, in favour of downstream emission, by the general downstream motion of the eddy, in accordance with the quadrupole convection theory of part I. This explains, for jet noise, the marked preference for downstream emission, and its increase with jet velocity. For jet velocities considerably greater than the atmospheric speed of sound, the 'Mach number of convection' M_c may exceed 1 in parts of the jet, and then the directional maximum for emission from these parts of the jet is at an angle of $\sec^{-1}(M_c)$ to the axis (§ 8).

Although turbulence without any mean flow has an acoustic power output, which was calculated to a rough approximation from the expressions of part I by Proudman (1952) (see also § 4 below), nevertheless, turbulence of given intensity can generate more sound in the presence of a large mean shear (§ 5). This sound has a directional maximum at 45° (or slightly less, due to the quadrupole convection effect) to the shear layer. These results follow from the fact that the most important term in the rate of change of momentum flux is the product of the pressure and the rate of strain (see figure 2). The higher frequency sound from the heavily sheared mixing region close to the orifice of a jet is found to be of this character. But the lower frequency sound from the fully turbulent core of the jet, farther downstream, can be estimated satisfactorily (§ 7) from Proudman's results, which are here reinterpreted (§ 5) in terms of sound generated from combined fluctuations of pressure and rate of shear in the turbulence. The acoustic efficiency of the jet is of the order of magnitude $10^{-4}M^5$, where M is the orifice Mach number.

However, the good agreement, as regards total acoustic power output, with the dimensional considerations of part I, is partly fortuitous. The quadrupole convection effect should produce an increase in the dependence of acoustic power on the jet velocity above the predicted U^8 law. The experiments show that (largely cancelling this) some other dependence on velocity is present, tending to reduce the intensity, at the stations where the convection effect would be absent, below the U^8 law. At these stations (at 90° to the jet) proportionality to about $U^{6.5}$ is more common. A suggested explanation of this, compatible with the existing evidence, is that at higher Mach numbers there may be *less turbulence* (especially for larger values of nd/U, where n is frequency and d diameter), because in the mixing region, where the turbulence builds up, it is losing energy by sound radiation. This would explain also the slow rate of spread of *supersonic* mixing regions, and, indeed, is not incompatible with existing rough explanations of that phenomenon.

A consideration (§ 6) of whether the terms other than momentum flux in the quadrupole strength density might become important in heated jets indicates that they should hardly ever be dominant. Accordingly, the physical explanation (part I) of aerodynamic sound generation still stands. It is re-emphasized, however, that whenever there is a fluctuating *force* between the fluid and a solid boundary, a dipole radiation will result which may be more efficient than the quadrupole radiation, at least at low Mach numbers.

1. INTRODUCTION

Part I (Lighthill 1952) was concerned with building up a physical theory of aerodynamic sound production in general. In this paper the subject is specialized to the study of *turbulence* as a source of sound, and a more quantitative approach, involving comparison with the existing experimental data, is attempted.

The sound produced by turbulence (in the absence of any fluctuating forces between the fluid and solid boundaries)* may be called aerodynamic *noise*, for certainly, like the turbulence itself, it has always a broad frequency spectrum. The most familiar aerodynamic noise is from natural winds, but industrial workers are aware of a wide range of noise arising from turbulent boundary layers and wakes, and from turbulent flow in ducts and free jets. Most *hydrodynamic* noise is of a different character, being associated with cavitation, but there is a residual noise in hydraulic machinery, when cavitation has been eliminated, which may be ascribable to turbulence; and the noise of turbulence in the blood stream is of importance in clinical medicine.

However, fundamental experiments on aerodynamic noise have begun only recently, largely as a result of the public odium with which the aircraft industry is threatened on account of the noise of jet aeroplanes. In a review of the various components of aircraft noise Fleming (1946) spoke of the importance of the aerodynamic noise, which he expected to become predominant at the higher Mach numbers, but said that nothing was known concerning it. In the years that followed, the firms making jet aircraft confined their attention at first to noise data for whole engines, and a valuable correlation of these data has recently been given by Mawardi & Dyer (1953), but it is never clear in these experiments how much of the noise produced was aerodynamic in the sense here used (rather than, say, having its origin in the combustion chamber),† nor can the effects of the velocity and temperature of the jet be separated from one another. Recently, however, five studies of the noise of the cold jet, which is obtained by simply evacuating a reservoir of compressed air through a suitable orifice, have been published (Fitzpatrick & Lee 1952; Gerrard 1953; Lassiter & Hubbard 1952; Powell 1952*b*; Westley & Lilley 1952); in all these an effort was made to eliminate noise other than the strictly aerodynamic noise of the jet itself.

The results of these experiments are particularly full for the subsonic cold jet. Many points concerning it are well corroborated by being found by more than one investigator; these are as follows:

(i) The acoustic power output varies as a high power, near the eighth, of the jet velocity; as a consequence it becomes too small at low velocities for any measurements to have been found possible at Mach numbers below 0·3. A rough order of magnitude for the acoustic efficiency of the subsonic cold jet is $10^{-4}M^5$, where M is the orifice Mach number.

* It was perhaps not sufficiently stressed in the mathematical theory of part I, although it is evident from the physical discussion given there, that such forces can produce important dipole sound fields, not only when the solid boundaries move (e.g. sounding-boards, rotating blades), but also when they remain rigid; thus sound can be generated in this way by the fluctuating lift on a rigid cylinder in a uniform stream. Mathematically, this is because where boundaries are present the retarded potential solution must be supplemented by a surface integral, whose physical interpretation is easily verified to be the dipole radiation associated with the force between the solid boundary and the fluid.

† They find acoustic efficiencies increasing rapidly with jet power above a value of about 2000 kW, and remaining roughly constant below this value. It seems reasonable (in the light of the later work) to conclude from this that aerodynamic noise was dominant at the higher jet speeds, and that a source whose output bears a ratio to energy expended independent of jet Mach number (presumably, combustion noise) was dominant at the lower speeds.

(ii) The spectrum is a very broad one indeed, of the order of seven octaves. The peak frequency is always in the neighbourhood of $U/2d$, where U is the jet velocity and d the orifice diameter, but the peak is not sufficiently clearly defined for consistent results on its variation with U and d to have been obtained; probably it increases less rapidly than U.

(iii) Almost all the sound is radiated in directions making an *acute* angle with the jet; in fact, measurements behind the orifice were subject to some danger of error, because the relatively small intensity there was liable to be smothered by reflexions of the sound which had been radiated forward.

(iv) The directional maximum for the higher frequency sound is at an angle of 45°, or slightly less, to the jet axis. This higher frequency sound appeared to be emitted principally from near the orifice.

(v) The lower frequency sound has a directional maximum at a much smaller angle to the jet axis; this angle decreases as the frequency is reduced, or the Mach number increased, and is often below the limit (about 18°) for which measurements can be made without gusts blowing on to the microphone. The sound intensity at these small angles to the jet axis increases like a higher power of the jet velocity than the average sound intensity. The lower frequency sound appears to be emitted principally from the part of the jet of the order of five to twenty diameters downstream of the orifice.

In view of this substantial body of information on which the different experimenters are in agreement, it has been natural to develop the theory of aerodynamic noise with an eye constantly on this application, and to regard the comparison with those experimental results as an essential test of the theory. In §7, general agreement with all the points noted above will be found, and detailed reference will also be made to particular results obtained in individual experiments.

The information on supersonic jets is much less complete, but Powell (1952*b*) and Westley & Lilley (1952) have made valuable studies of different aspects of the subject, and a section on it is included below (§8). There is little precise evidence of the effect of heating a jet, but some comments on the theoretical aspects of this are made in §§6, 7 and 8.

2. Physical introduction to the theory of aerodynamic noise

It was established in part I that any turbulent flow can be regarded as equivalent, acoustically, to a distribution of quadrupoles, whose strength per unit volume at any instant is

$$T_{ij} = \rho v_i v_j + (p_{ij} - a_0^2 \rho \delta_{ij}), \tag{1}$$

placed in a uniform medium at rest. Here ρ is the density, v_i the fluid velocity, p_{ij} the compressive stress tensor, and a_0 the velocity of sound in the atmosphere. Only the first term in (1), representing momentum flux, is expected to be important in a 'cold' flow, i.e. one where no marked temperature differences exist.

The principal developments of the theory of part I, which are made in this paper, are as follows:

(i) It is necessary to make use of the statistical character of turbulence, and in particular of the fact that the values of the momentum flux at points with no eddy in common are uncorrelated, while its values at points with many eddies in common are well correlated. This fact leads to an expression for the sound field which can be interpreted by saying that separate 'average eddy volumes' can be thought of as giving out their quadrupole radiation independently.

(ii) A further consequence of the statistical character of turbulence is important when the turbulence is superimposed on a mean flow whose Mach number is not negligible. As explained under heading (i) of §1, this includes all the experiments on jets (a Mach number of 0·3 not being negligible for these purposes). The fact that the rate of distortion of the local eddy pattern is least when viewed in a frame of reference moving with a certain 'local eddy convection velocity' is found to imply that the radiation field of those eddies will be modified, approximately, as was calculated in §7 of part I for a quadrupole distribution analyzed with respect to a frame moving through the atmosphere (with that local convection velocity). One result of this conclusion (explained physically on p. 583 of part I, and more fully in §3 below) is the observed fact that jets radiate far more sound forwards than backwards (heading (iii) of §1).

(iii) Although turbulence without a mean flow has an acoustic power output, which Proudman (1952) has estimated for isotropic turbulence from the expressions of part I (with results which are summarized in §4 below), nevertheless turbulence of given intensity can generate more sound in the presence of a large mean shear. Also the terms which then dominate the sound radiation field are considerably simpler than in the general case. These results might be expected from the fact that the fluctuations in momentum flux are increased when they consist principally of a shaking to and fro of the large *mean* momentum of the fluid by the smaller *fluctuations* in velocity; a process significant only when the mean momentum has a large gradient. Actually, the important term in the rate of change of momentum flux at a point can be shown to be the product of the pressure and the rate of strain. This is because high pressure at a point causes momentum to tend to flow in the direction of greatest expansion of a fluid element (see figure 2). Accordingly, the quadrupoles in a shear layer are mostly oriented along the principal axes of mean rate of strain, that is, at 45° to the motion. However, this lateral quadrupole field must be distorted (see (ii) above) as in figure 3 of part I when the eddy convection velocity is at non-negligible Mach number. In the case of a jet the heavy shear is near the orifice, and the turbulence there is probably responsible for the high-frequency noise, with its directional maximum at 45° or slightly less, noticed under heading (iv) of §1.

At the beginning of part I the presence of solid sounding-boards to amplify the sound resulting from turbulent pressure fluctuations was excluded, but it is now seen that such amplification is provided also by a heavy mean shear, which may therefore be described loosely as an 'aerodynamic sounding-board'.

To some extent a similar amplification may arise from large inhomogeneities of temperature or fluid composition in the flow, resulting in a speed of sound notably different in the turbulent flow from that in the atmosphere. The second term in (1) is then important, because the pressure fluctuations in the turbulence are only partly balanced by a_0^2 times the density fluctuations. The unbalanced pressure fluctuations produce, according to the results of part I, §4 (and, indeed, as any externally applied pressures must do), a source field whose instantaneous strength per unit volume is a_0^{-2} times their second time derivative. However, the amplification described here is studied in detail in §6, where it is shown that even when the whole pressure fluctuation is unbalanced the effect may still be a small one because in turbulence the momentum flux fluctuates more vigorously than the pressure. Accordingly, it may be necessary to resist the temptation to use the result to explain the commonly observed excess noise of hot jets over cold jets at the same velocity, which may be due rather to causes like combustion noise (see §1). If the effect is ever important, the discussion of §6 indicates that it will be in the case of high-frequency sound from a turbulent mixing region, especially where heavy gases are used.

In interpreting the results for the cold jet, it should be noted that the lower frequency sound (which carries somewhat more than half the acoustic energy) comes principally (see §1, heading (v)) from that part of the jet where the turbulence is known to be most intense (see, for example, Corrsin & Uberoi 1949). Here the shear is not great enough for the terms described in (iii) above to predominate, and probably the sound field is a mixture of lateral quadrupole radiation, due to the interaction of the turbulence and the mean shear, and less strongly directional radiation due to the turbulence alone (which can be estimated roughly from Proudman's results for isotropic turbulence). The combination of these, when modified to allow for convection (see figures 2 and 3 of part I), would account for the directional maximum being at an angle considerably less than 45° and decreasing to very low values as the Mach number increases or the frequency decreases.

Paradoxically enough, the feature of the experimental results which is the hardest of all to explain, in the light of the theories briefly sketched above, is the accuracy with which the acoustic power output varies as U^8 (where U is the mean exit velocity), although this law was suggested in §6 of part I. For the conclusion of (ii) above (which seems to be inescapable, both on theoretical grounds and because it is impossible to think of any other reason for the greatly reduced sound behind the jet orifice) implies a further increase of power output with Mach number, over and above the U^8 law, as figure 1 of part I makes clear. Now the Mach number at which the principal sound-producing eddies are convected is probably not more than half the Mach number at the jet orifice; it is a familiar idea that eddies in the thin mixing region near the orifice travel at about half the speed of the jet (which one can infer by thinking of them as rigid rollers: observed values of the ratio (Brown 1937) are slightly less); again, the mean velocity on the axis in the region of intensest turbulence ten diameters from the orifice is about two-thirds of the exit velocity, and drops to one-third at twenty diameters. But even if the correction due to convection at only half the exit velocity U is put in, the expected slope of power output against U on a logarithmic scale, based on values between $M = 0{\cdot}5$ and $0{\cdot}9$, is increased from 8 to 9·5.

On the other hand, the measured slope does not exceed 8. Only Fitzpatrick & Lee (1952) have measured acoustic power output directly (by a reverberation chamber method), and they get a slope of almost exactly 8·0, especially at the higher Mach numbers. This deficiency in slope is borne out by measurements of intensity by other investigators, which give slopes comparably below the value expected by the theory as modified by convection (see §7). The fact (§1, heading (v)) that the slope is greater at smaller angles makes it appear likely not that the convection effect (which would cause such a trend) is absent, but that the factors in the *turbulence* which cause sound production are reduced as the velocity increases.

This could be either a Reynolds number or a Mach number effect; the experimental evidence is inconclusive. From the theoretical point of view, however, one could hardly expect increased Reynolds number (which means less viscous damping) to *reduce* the acoustic output, and, indeed, there are arguments for expecting its effect to be small at the already relatively high Reynolds numbers (around 5×10^5) of the jets used. There are, however, two possible explanations of the discrepancy as an additional *Mach number* effect:

(*a*) The size of the effective quadrupoles in the jet (comparable by (i) above to an average eddy size in a certain sense, see §3 below) may become comparable with the acoustic wave-length at high subsonic Mach numbers; this would lead to a reduction in the Stokes effect (part I, §2) by which the efficiency of quadrupole radiation increases as the fourth power of the frequency (and hence as U^4). However, the expected values of the relevant eddy sizes at a given frequency do not quite confirm this view (see §3).

(*b*) Possibly the intensity of the turbulence itself decreases at high subsonic Mach numbers. There is no definite evidence on this point, although it is well known that *supersonic* jets mix much more slowly with the outer air, and so presumably have less intense turbulence in the mixing region. (Typical observations are that in subsonic flow the total included angle of the turbulent mixing region is 15°, and that in supersonic flow it is 5°.) If the damping of turbulence in the mixing region increases *continuously* with Mach number, and already militates significantly against the rate of gain of energy from the mean shearing flow at Mach numbers slightly above 1, then presumably the resulting intensity of turbulence in the fully turbulent part of the jet would be reduced already at high subsonic Mach numbers. Reduced sound radiation from this part would then result.

Now when one inquires why there is greater damping in the turbulent mixing region of a supersonic jet, one is told that it is because disturbances in the jet boundary set up pressure distributions, characteristic of supersonic flow, which tend to reduce the disturbances. But since the linearized theory of supersonic flow is equivalent to the theory of sound, and the phenomenon is unsteady in any case, we see that the suggested cause of damping is in reality the sound field. This leads to the suggestion that the damping is really radiation damping resulting from the energy lost as sound. The present paper shows that turbulence in the layer of heavy shear would radiate an exceptional quantity of sound, increasing rapidly with Mach number, and it is possible that even at high subsonic Mach numbers it may thus lose sufficient energy (in the crucial region where the turbulence is being created most-rapidly) to reduce the general level of the turbulence in the jet. (This would be in contradiction to the suggestion of part I, §1, that the sound radiated would have no significant back-reaction on the flow.) Certainly the observed acoustic power output of the mixing region, at high subsonic Mach numbers, is great enough (see §7) to render plausible that it might reduce the rate of growth of turbulence in this crucial region to some measurable extent. It may be noted, too, that most of the experimental work indicates a fall in the dimensionless frequency parameter nd/U of the peak in the sound spectrum as the Mach number increases (in other words the peak frequency increases less rapidly than U); and this might be because the radiation damping mechanism is more effective at the higher frequencies, as the theory of §5 below would lead us to expect.*

To sum up, it seems likely that at higher Mach numbers the level of turbulence is reduced, especially for higher values of nd/U, where n is frequency and d diameter, because in the mixing region, where the turbulence builds up, it is losing energy by sound radiation. This is invoked to explain both the overall reduction in the intensity field from what would be expected on the convection theory, and also the change in the spectrum, based on the dimensionless frequency nd/U, as M increases.

The remainder of the paper will exhibit, in greater detail than has been possible above, first the extensions to the theory and then the comparison with experiment.

3. Effects of the statistical character of turbulence on the sound it produces

At a given instant the velocity (and hence also the momentum flux, which is the dominant term in the quadrupole strength density T_{ij}) has a highly random variation with position throughout a turbulent flow; and although its values at two points which are fairly close together show some statistical correlation, which, of course, tends to unity as the points tend to coincide, it is well established that there is a distance, say D (crudely, the 'diameter of the largest eddies'), such that the values of the velocity (and hence, doubtless, of T_{ij}) at points separated by distances greater than D are effectively uncorrelated. The first necessity is to make use of this fact in the statistical analysis of the radiation field of the quadrupoles.

The theory takes its simplest form in the case of small Mach number, which will be considered first. But since (§1) the experiments on jets were found possible only at Mach numbers exceeding 0·3, which are not small for the present purposes, it will be found that an extension of the theory to include the principal effects of Mach number is needed to obtain agreement.

A rough argument, which suggests the character of the necessary statistics, follows from the fact that, if the strengths of two point quadrupoles are statistically uncorrelated (in Rayleigh's terminology, 'unrelated in phase'), then the ***intensity*** field due to both together is the sum of the intensity fields of each separately.

* Of course any mechanism explaining a general reduction in frequency (relative to U/d) with increasing Mach number would explain also a falling away from the U^8 rule. Dr Powell has suggested to the author that such a mechanism might be connected with some statistical dependence of eddy creation at the orifice on the sound field of the jet. However, the author would expect rather a general increase of turbulence level, as the sound effects increased with increasing Mach number, to result from such a mechanism, whereas a decrease at the higher frequencies is what is required.

To apply this result one starts by replacing the correlation curve for T_{ij} by a square form; thus the values of T_{ij} within a certain volume V around a point are taken as perfectly correlated with the value there, and the values outside as perfectly uncorrelated. Here V is a sort of 'average eddy volume', whose value might vary through the field of flow. Now imagine the field divided up into a set of non-overlapping regions which fill it completely, each with volume equal to the local average eddy volume V. All the quadrupoles in one such region fluctuate in phase with the one at the centre; so they can be replaced by a single point quadrupole at the centre, of strength V times the local strength density. But, by hypothesis, all the resulting point quadrupoles are completely uncorrelated, so the intensity field is the sum of the intensity fields due to each separately. Further, each intensity field is proportional to the square of the strength of the quadrupole, and so to V^2. Hence the intensity field resulting from *unit* volume of turbulence is proportional to V, and, indeed, is V times that of a point quadrupole of strength equal to the local strength density T_{ij}.

This crude argument indicates at once that the acoustic power output from a given volume element of turbulence is proportional to some local average eddy volume. This is confirmed by equation (19) of part I for the intensity $I(\mathbf{x})$, namely,

$$I(\mathbf{x}) \sim \frac{1}{16\pi^2\rho_0 a_0^5}\iint \frac{(x_i-y_i)(x_j-y_j)(x_k-z_k)(x_l-z_l)}{|\mathbf{x}-\mathbf{y}|^3\,|\mathbf{x}-\mathbf{z}|^3}$$

$$\times\overline{\frac{\partial^2}{\partial t^2}T_{ij}\left(\mathbf{y},t-\frac{|\mathbf{x}-\mathbf{y}|}{a_0}\right)\frac{\partial^2}{\partial t^2}T_{kl}\left(\mathbf{z},t-\frac{|\mathbf{x}-\mathbf{z}|}{a_0}\right)}\,d\mathbf{y}\,d\mathbf{z}, \quad (2)$$

which shows that the intensity field of the volume element $d\mathbf{y}$ is proportional to the volume integral of the covariance of the value of $\partial^2 T_{ij}/\partial t^2$ at the element with its value at all points $\mathbf{z}$ of the turbulent field. Since the covariance (of which a correlation coefficient is a non-dimensional form) may be expected to fall from a maximum when $\mathbf{z}=\mathbf{y}$ to zero outside a sphere of radius D, the first factor in the integral has negligible variation with $\mathbf{z}$ from its value for $\mathbf{z}=\mathbf{y}$ when $x \gg D$. Further, the integral of the second factor with respect to $\mathbf{z}$ may evidently be regarded as a product of a sort of average eddy volume with the value of the integrand when $\mathbf{z}=\mathbf{y}$ (which is the mean square of $\partial^2 T_{ij}/\partial t^2$). This suggests a way of estimating the intensity field.

The last conclusion would evidently be false if there were a substantial region where the correlation of T_{ij} with its value at the element $d\mathbf{y}$ were negative, and especially if the integrated correlation over this part came near to balancing the integrated correlation over the part where it is positive. Then, too, the previous 'rough' argument would be invalidated, since the correlation curve could not reasonably be replaced by a square form. Actually the sound field would then have to be regarded as essentially an octupole field, as arguments on p. 576 of part I indicate. (In particular, if T_{ij} were of the form $\partial W_{ijk}/\partial x_k$, whose integrated covariance is zero because it can be written as the covariance of the value at a point with the surface integral of the normal component of W_{ijk} over a surface far from the point, then the equivalent octupole strength density would be W_{ijk}.) But the evidence of what follows (§5) will be that the difficulty does not appear to arise in practice.

More seriously, the argument as it stands neglects the differences between the times of emission from different points **y**, **z** of waves which reach simultaneously a given distant point x—between the 'retarded times', in fact, in equation (2). This, however, is permissible at small Mach numbers. For the covariance in (2) is negligible except for points **y**, **z** separated by distances less than D, and for such points the difference between the retarded times is certainly less than D/a_0. This time interval must be negligible compared with times significant in the turbulent fluctuations* if the velocity of sound a_0 is sufficiently great compared with the flow velocities. The covariance can then be treated as a 'simultaneous' covariance (mean product of values at two points at the *same* instant), as has been tacitly assumed above.

But if the Mach number is not small the analysis as given above would run into two difficulties. First, the difference in retarded times would not be negligible, implying (as §4 of part I makes clear) the presence of important octupole fields. Secondly, if eddies were being convected at an appreciable mean Mach number, then the intensity field at a distant point in the downstream direction would for statistical reasons be greater than on the foregoing theory; for the correlation between the values at the two points must then exceed the simultaneous correlation, because the retarded time of the downstream point is later, and at that time the eddy has moved a considerable distance towards the point from its position at the retarded time of the upstream point.

To obtain a better approximation to what happens under these conditions, the considerations of part I, §7, are used. For one can minimize the importance of the difference between the retarded times by analyzing the turbulence in a given neighbourhood in the frame of reference for which the time scale of the fluctuations of $\partial^2 T_{ij}/\partial t^2$ is greatest. This frame will move at what may be called the local 'eddy convection velocity', and its use will also remove the statistical difficulty noted above.

In a region where the mean velocity varies little from some average value, a frame moving with this average velocity would be used; note that in this frame the time scale of the turbulent fluctuations would be of the order of an eddy size divided by a typical *departure* of the velocity from its mean, and this would be large compared with D/a_0, provided only that the Mach number of the velocity fluctuations is small, which will almost always be the case in practice. On the other hand, even in a heavily sheared turbulent layer, as in the mixing region of a jet, it is generally believed that an 'eddy convection velocity' V_c exists, not in the sense that eddies are convected downstream with this velocity unchanged, but that they alter slowest when viewed by an observer moving with this velocity. A typical value for this velocity observed by Brown (1937) at low Reynolds numbers was $0{\cdot}4U$, where U is the mean velocity at the orifice. The phenomenon may not be as different as one might expect at high Reynolds numbers, even with turbulent oncoming flow, because jets have frequently been observed to form their large-scale eddies in the mixing region, due to dynamic instability, in a manner relatively uninfluenced by the small eddies already present. If the time scale of fluctuations in this new frame of reference

* Put another way, this means that the eddy size is small compared with the acoustic wave-length (see heading (*a*) of §2).

is large compared with D/a_0, as is probably the case, the difference between retarded times can be safely ignored, and the difficulty raised under heading (*a*) of §2 does not arise.

The effect of the use of the moving frame, as obtained in part I, §7, is to multiply the intensity field of an element dy by a factor

$$\left[1-\frac{\mathbf{M}_c.(\mathbf{x}-\mathbf{y})}{|\mathbf{x}-\mathbf{y}|}\right]^{-6}, \tag{3}$$

where $\mathbf{M}_c$ (the 'Mach number of convection') signifies the eddy convection velocity divided by the *atmospheric* speed of sound a_0.

The question at once arises, however, whether one is justified in using the results of part I, §7, in view of the fact that the eddy convection $a_0\mathbf{M}_c$ is not in general uniform, but varies over the field of flow. One might reasonably claim to do so, provided that $\mathbf{M}_c$ varies only a little in a distance of order D, on the grounds that two points in the turbulent flow have been shown to radiate sound independently if the distance between them exceeds D. However, a further discussion is desirable, and it is shown in the appendix that, provided $\mathbf{M}_c$ is a solenoidal vector field, then by referring T_{ij} to the Lagrangian co-ordinates appropriate to the steady velocity field $a_0\mathbf{M}_c$ (this procedure is the equivalent, for a non-uniform field, of using moving axes), then the integral (2) transforms into an integral of the same form with an additional factor

$$\left[1-\frac{\mathbf{M}_c(\mathbf{y}).(\mathbf{x}-\mathbf{y})}{|\mathbf{x}-\mathbf{y}|}\right]^{-3}\left[1-\frac{\mathbf{M}_c(\mathbf{z}).(\mathbf{x}-\mathbf{z})}{|\mathbf{x}-\mathbf{z}|}\right]^{-3}. \tag{4}$$

Provided that $\mathbf{M}_c$ varies only a little in a distance D the two factors in (4) can be equated, giving the result expressed in (3).*

If the angle between the direction of convection and the direction of emission is θ, the factor (3) may be written

$$(1-M_c\cos\theta)^{-6}. \tag{5}$$

This factor can signify a great increase in intensity for small θ, which was explained physically in part I as due to two causes:

(i) The positions of origin of those rays of sound from a single eddy which arrive at a distant point simultaneously fill a greater *volume* when the rays are emitted forward, since the foremost parts of the eddy have moved on before they need emit. This accounts for one factor $(1-M_c\cos\theta)^{-1}$ in the amplitude, and hence for two in the intensity.

(ii) The time interval between emission from the fore and aft parts of an eddy is greater for rays emitted forward and arriving simultaneously at a distant point; hence the 'Stokes effect' (cancelling out of the separate source fields because their

* At first the author felt inclined to use the result (4) directly, with $a_0\mathbf{M}_c$ taken as the local mean velocity (which for a subsonic jet would be very closely solenoidal). There is no theoretical objection to this, but it is questionable whether the extra complication in the mixing region (due to the rapid variation of $\mathbf{M}_c$) corresponds to any improvement of accuracy, owing to the commonly observed fact that there exists some sort of eddy convection velocity for the mixing region as a whole.

total strength is zero) is greatly mitigated in respect of those rays. This accounts for two factors $(1 - M_c \cos\theta)^{-1}$ in the amplitude field, and hence for four in the intensity.

The arguments up to this point (including that by which the distribution of quadrupoles in an average eddy volume was combined into a single one) can be summarized graphically by regarding an eddy as four sources spread out over it to form a quadrupole, which is convected along at the local eddy-convection velocity.

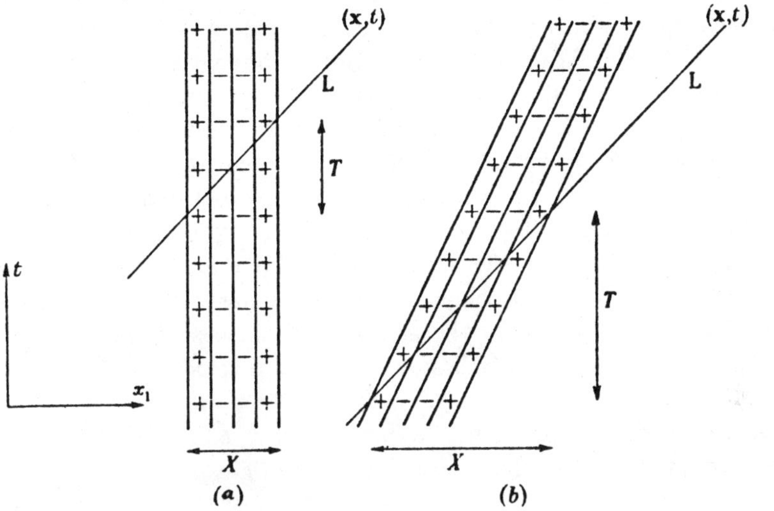

FIGURE 1. Effect of eddy convection in increasing the sound emitted forwards. L = locus of points emitting rays which reach **x** at time t. The quadrupole representing the eddy is stationary in (*a*), but in (*b*) it moves towards **x** at a Mach number $M_c = 0{\cdot}5$. This motion increases the effective volume (proportional to X) of the quadrupole, and so increases its strength (for given strength per unit volume T_{ij}). The motion also increases the time delay T between emissions from the four sources, which, by reducing cancellations, increases the intensity at **x** for given quadrupole strength.

The two points just made can then be represented in a simple diagram. Consider for simplicity the case of a longitudinal quadrupole with its axes in the direction of motion, and consider emission in that direction ($\theta = 0$). Figure 1 shows the motion of the quadrupole in the (x, t) plane, and shows also the locus of points which emit rays forward to reach a distant point simultaneously, in the two cases $M_c = 0$ and $M_c = 0{\cdot}5$. In the latter case the increased quadrupole strength due to increased volume and the reduced cancelling of the four sources due to increased time intervals between emission from each, are evident from the figure.

The reader should be warned against two possible misconceptions relating to this way of treating aerodynamic noise at non-negligible Mach number.

(*a*) It has been stressed that the general theory of part I, in representing the acoustic properties of turbulent flow by means of a distribution of quadrupoles, effectively 'reduces the flow to rest', and it may be wondered how the translation of quadrupoles by the mean motion of the fluid is compatible with this idea. The answer is clear when the situation is stated more exactly: the sound generated is that which the quadrupole distribution T_{ij} per unit volume would emit if placed in a uniform acoustic medium at rest—so that no effects of refraction by the varying velocity and temperature of the stream need be considered in calculating it*—but the variation of the quadrupole distribution with time may, as a result of convection, take the form of a combination of a translation of the whole with other changes which are more random but also more gradual, and when this is so the directional character of the radiated sound field will be affected. In the model, in fact, the quadrupoles can move, but not the fluid.

(*b*) It may be thought that the transformation will cause a great reduction in the value of the integral (which would be inadmissible, at least when M_c is small so that the factor $(1-M_c\cos\theta)^{-6}$ can make little corresponding increase) because the time variations described by $\partial^2 T_{ij}/\partial t^2$ will be much smaller in a frame of reference which moves along with the eddy convection velocity. In other words the false time variations, as measured by a stationary hot wire—which are really space variations—are then omitted. But these space variations would not in any case contribute significantly to the integral, for they constitute an octupole field whose radiation must be small compared with the quadrupole field due to the time variations, at least at low Mach number—in which limiting case the transformation therefore becomes an identity. Thus only the genuine frequencies of the turbulence, as measured in a frame moving with the eddy convection velocity, are relevant to the sound production.

The conclusions of this section can be summarized in a formula for the intensity field, per unit volume of turbulence situated at the origin. This is

$$i(\mathbf{x}) \sim \frac{x_i x_j x_k x_l}{16\pi^2 \rho_0 a_0^5 (x-\mathbf{M}_c.\mathbf{x})^6} \int \overline{\frac{\partial^2}{\partial t^2} T_{ij}(0,t) \frac{\partial^2}{\partial t^2} T_{kl}(\mathbf{z},t)}\, d\mathbf{z}, \tag{6}$$

where $i(\mathbf{x})$ has been written for the intensity at $\mathbf{x}$ per unit volume of turbulence at the origin. Equation (6) is obtained by putting $\mathbf{y} = 0$ in equation (2), dropping the integration with respect to $\mathbf{y}$, and neglecting the difference between $\mathbf{x}-\mathbf{y}$ and $\mathbf{x}-\mathbf{z}$ on the assumption that $\mathbf{x}$ is far from the origin compared with an average eddy size (as well as far compared with $(2\pi)^{-1}$ times an average acoustic wave-length as assumed in part I). The difference in times between the two values of $\partial^2 T_{ij}/\partial t^2$ is also neglected, for the reasons discussed above, and the additional factor $(1-\mathbf{M}_c.\mathbf{x}/x)^6$ is introduced into the denominator, changing x^6 into $(x-\mathbf{M}_c.\mathbf{x})^6$, on the understanding that T_{ij} is specified in a frame moving with the local eddy convection $a_0\mathbf{M}_c$ (so that it is what is called S_{ij} in the appendix).

4. Proudman's estimate of the intensity field of isotropic turbulence

Proudman (1952) has made an approximate calculation of the expression (6) in the case of isotropic turbulence without mean motion, taking the approximate form $\rho_0 v_i v_j$ for T_{ij} as in equation (7) of part I, and of course $M_c = 0$. Actually, he uses a slightly more accurate form than (6), because in writing down the covariance

* Actually, the extent to which this is true in practice was perhaps exaggerated in part I. If one neglects terms in T_{ij} resulting from the sound itself, and in particular the product of a mean-flow velocity at a point with the velocity associated with sound waves from more distant points in the flow, one may be neglecting phenomena which can be regarded essentially as refraction by the mean flow (cf. Lighthill 1953). But it appears that these refractions do not have as marked an effect on the directional distribution of (say) jet noise as does the quadrupole convection effect here discussed.

(which is the mean product of departures from the mean) he takes into account the fact that isotropic turbulence is necessarily decaying, so that time derivatives like $\partial^2 T_{ij}/\partial t^2$ may have a non-zero *mean*, due to the gradual reduction of the turbulent energy. However, this effect would be expected to be very small, and actually in his final estimate Proudman concludes that it is of the order of 1 % of the whole. So it is certainly permissible to neglect it, and to do the same by analogy in problems of noise due to shear flow turbulence, in which again the turbulence, viewed (as was found necessary in § 3) in a moving frame, may be decaying.

With the decay terms omitted, Proudman's transcription of equation (6) becomes

$$i(\mathbf{x}) \sim \frac{\rho_0}{16\pi^2 a_0^5 x^2} \int \overline{\frac{\partial^2}{\partial t^2}(v_x^2(0,t))\frac{\partial^2}{\partial t^2}(v_x^2(\mathbf{z},t))}\, d\mathbf{z}, \tag{7}$$

where $x_i v_i/x$ has been rewritten as v_x, the velocity component in the direction of emission. Clearly the integral in (7) is independent of the direction of $\mathbf{x}$ for isotropic turbulence, so that v_x can be replaced by an arbitrary component of $\mathbf{v}$, say v_1, and the directional distribution is that of a simple source field, whose power output (say p_i) per unit volume is

$$p_i = \frac{\rho_0}{4\pi a_0^5} \int \overline{\frac{\partial^2}{\partial t^2}(v_1^2(0,t))\frac{\partial^2}{\partial t^2}(v_1^2(\mathbf{z},t))}\, d\mathbf{z}. \tag{8}$$

The covariance in (8) is of the fourth degree in the velocities, but Proudman approximates to it, using an idea of Batchelor (1950), as that combination of covariances of the second degree in the velocities to which it would be equal if the values of v_1 and of its first two derivatives at 0 and $\mathbf{z}$ had a normal joint-probability distribution. There is good evidence that this should give a correct order of magnitude result. Further, with Heisenberg's form of the correlation function for v_1, which has reasonably good experimental support for large Reynolds numbers, and using the Batchelor type of approximation several more times in the analysis, the resulting integral is computed and expression (8) reduced to

$$p_i = \frac{38\rho_0(\overline{v_1^2})^{\frac{5}{2}}\epsilon}{a_0^5}, \tag{9}$$

where ϵ is the mean rate of dissipation of energy per unit mass.

Proudman concludes also, from the details of his calculations, that at large Reynolds numbers the principal contribution to this power output comes from the larger, non-dissipating eddies; this is presumably because although the mean square of expressions like $\partial^2(v_1^2)/\partial t^2$ would be expected to be greatest for energy-dissipating eddies, the greatly reduced average volume of those eddies renders their total contribution negligible, for reasons made clear in § 3. The question why the coefficient 38 in (9) should be large compared with unity is discussed in § 5.

In the case of isotropic turbulence which is convected through the atmosphere at a mean velocity $a_0 \mathbf{M}_c$, one must expect the intensity field to be modified by a factor $(1 - M_c \cos\theta)^{-6}$ to become

$$i(\mathbf{x}) \sim \frac{p_i x^4}{4\pi(x - \mathbf{M}_c . \mathbf{x})^6}, \tag{10}$$

with power output (cf. (35) of part I)

$$p_i \frac{1 + 2M_c^2 + \frac{1}{5}M_c^4}{(1 - M_c^2)^5}. \tag{11}$$

5. The amplifying effect of shear

It is fairly evident that turbulence of given intensity can generate more sound in the presence of a large mean shear than it would in the absence of a mean flow. For there can be far wider variations of momentum flux if there is a large mean momentum to be shaken about by the turbulent fluctuations in velocity, and conversely a large mean velocity to transport turbulent fluctuations in momentum; in symbols, $\rho v_i v_j$ can fluctuate most widely if the fluctuations of ρv_i are amplified by a heavy mean value $\bar{v}_j$.

At first sight this argument might be thought to apply even if the mean velocity were merely uniform, but this is not so. For the time derivative of $(\rho v_i)\,\bar{v}_j$ is $\bar{v}_j$ times the rate of change of momentum per unit volume, so its integral over all space would be zero if $\bar{v}_j$ were uniform (since total momentum is conserved). The field would therefore be reduced to an octupole field, by §4 of part I (or by §3 above). Even if the mean velocity were nearly uniform, not everywhere (which indeed is impossible, since it is assumed in the basic theory to become negligible outside a limited region), but only over an average eddy volume, then the effect would still be absent, because there is a negligible interaction, as far as the generation of sound is concerned, between the values of the quadrupole strength at points a distance greater than D apart (§3).

However, where there is a large mean velocity *gradient* (causing considerable changes of velocity across a single eddy), large non-cancelling values of the quadrupole strength density may occur, as we shall see. These are presumably connected with the fact that momentum flux is always most important when it takes place along a large gradient of mean velocity.

To make the arguments precise, the momentum flux $\rho v_i v_j$ is now thrown into a form which brings out directly the importance of velocity gradient. More precisely, the *time derivative* of $\rho v_i v_j$ is used, because this is what is needed in equation (6) and because it was this whose integrated value was shown above to be zero in the case of uniform v_j.

Now evidently
$$\frac{\partial}{\partial t}(\rho v_i v_j) = v_j \frac{\partial(\rho v_i)}{\partial t} + v_i \frac{\partial(\rho v_j)}{\partial t} - v_i v_j \frac{\partial \rho}{\partial t}, \tag{12}$$
and the derivatives on the right-hand side can be expressed as space derivatives by the equations of motion (see part I, equations (2) for the equation of continuity and (3) for Reynolds's form of the momentum equation). In fact
$$\begin{aligned}\frac{\partial}{\partial t}(\rho v_i v_j) &= -v_j \frac{\partial}{\partial x_k}(\rho v_i v_k + p_{ik}) - v_i \frac{\partial}{\partial x_k}(\rho v_j v_k + p_{jk}) + v_i v_j \frac{\partial}{\partial x_k}(\rho v_k) \\ &= p_{ik}\frac{\partial v_j}{\partial x_k} + p_{jk}\frac{\partial v_i}{\partial x_k} - \frac{\partial}{\partial x_k}(\rho v_i v_j v_k + p_{ik} v_j + p_{jk} v_i).\end{aligned} \tag{13}$$

The last of the three terms in (13) is a space derivative and so represents an octupole field (again, by §4 of part I or §3 above), and must be expected to radiate relatively little sound, especially at the lower Mach numbers.* The quadrupole

* If the convection effect is non-negligible, the left-hand side of (13) should be replaced by the rate of change in a frame of reference moving with velocity $a_0\mathbf{M}_c$. In the original co-ordinates, this would be $(\partial/\partial t + a_0\mathbf{M}_c.\boldsymbol{\nabla})\,(\rho v_i v_j)$. Since $\mathbf{M}_c$ has been assumed solenoidal, the extra term is also a space derivative (the divergence of $\rho v_i v_j a_0 \mathbf{M}_c$), and in fact helps to reduce the strength of the octupole field, which may be an important consideration at the higher Mach numbers.

terms that remain emphasize the influence of velocity *gradient*, and will clearly fluctuate widely when there is a large mean velocity gradient.

Physically, equation (13) is an expression of the fact (foreshadowed in part I, §§2 and 4) that although there is conservation of mass and momentum—and in particular $\partial\rho/\partial t$ and $\partial(\rho v_i)/\partial t$ are exact space derivatives, so that their integral over a sphere of radius $> D$ is uncorrelated with their value at the centre—there is no similar conservation of momentum *flux*, whose time derivative per unit volume (13) indeed contains residual terms which are not space derivatives. Accordingly, genuine fluctuations in the total momentum flux in such a volume are possible (i.e. integrals of the kind occurring in (6) are not identically zero), leading to a true quadrupole radiation field.

Not only the last term in (13), but also the *viscous* contribution to the stresses p_{ik} and p_{jk} in the first two terms, will here be ignored. For the *mean* viscous stress is always small compared with the turbulent shearing stress $\rho\overline{v_1' v_2'}$, to which, however, the pressure fluctuations (the other term in the stress tensor) are known to be comparable. Again, the *fluctuations* in viscous stress are of order $\mu(\overline{v_1'^2})^{\frac{1}{2}}/\lambda$, where λ is Taylor's dissipation length, and this must be small compared with pressure fluctuations of order $\rho\overline{v_1'^2}$ if the Reynolds number $R_\lambda = \rho(\overline{v_1'^2})^{\frac{1}{2}}\lambda/\mu$ is large compared with 1, which is known to be true in fully developed turbulence.

Retaining therefore only the parts due to pressure fluctuations in the first two terms of (13), they become

$$p\left(\frac{\partial v_i}{\partial x_j}+\frac{\partial v_j}{\partial x_i}\right) = pe_{ij}, \tag{14}$$

where e_{ij} is the rate-of-strain tensor. Thus as stated under heading (iii) of §2, *the important term in the rate of change of momentum flux at a point is the product of the pressure* and the rate of strain.*†

The reason for this becomes particularly clear if the rate-of-strain tensor is viewed in terms of its principal axes. (These are the axes along which a rod-shaped element of fluid is being simply extended or contracted parallel to itself.) Now, high pressure at a point will tend to increase the flow along an extending element of momentum parallel to itself, and to decrease the flow of such momentum along

* Throughout this paper and part I, the pressure p is understood to be measured from the atmospheric pressure as zero.

† One immediate consequence of this is that the 'source field' (see (13) of part I) resulting from the 'trace' ρv_i^2 of the momentum flux is normally negligible; for the trace of (14) is $2p\,\partial v_i/\partial x_i$, and the divergence of the velocity field is small compared with the individual velocity gradients in subsonic flows. Now (part I, p. 575, where, however, the word 'single' was a mistake) the part of T_{ij} other than the source term can be expressed as a combination of lateral quadrupoles. Thus Proudman's omnidirectional radiation from isotropic turbulence is not in any genuine sense a source field, but is rather a statistical assemblage of lateral quadrupole fields of all orientations, due to the combination of fluctuating pressures and rates of shear in the turbulent flow. (Note that, of course, no correlation between pressure and rate of shear at a point is necessary for the fluctuation of their product to be important.)

The full source field (including the part due to viscous terms in p_{ij}) arises essentially from local fluctuations in the rate of energy dissipation. Many different arguments indicate that these must be associated with sound generation, but it follows from the arguments presented above that this is unimportant compared with the sound associated with fluctuations in lateral momentum flux.

a contracting element. Hence there should be a term in the rate of change of momentum flux with the same principal axes as e_{ij}, and proportional both to it and to the pressure, as has been found above.

Before the result is applied to shear flows, it will first be used to throw some light on the significance of Proudman's results (§4) on isotropic turbulence without a mean flow. Now, replacement of $\partial T_{ij}/\partial t$ in Proudman's theory by pe_{ij} as in (14) indicates (after the usual simplifications resulting from isotropy) that the power output p_i per unit volume in isotropic turbulence is

$$p_i = \frac{1}{\pi \rho_0 a_0^5} \int \overline{\frac{\partial}{\partial t}\left(p\frac{\partial v_1}{\partial x_1}\right)_{\mathbf{x}=0} \frac{\partial}{\partial t}\left(p\frac{\partial v_1}{\partial x_1}\right)_{\mathbf{x}=\mathbf{z}}}\, d\mathbf{z}. \tag{15}$$

Now, in the course of the calculations described by Proudman (1952) he estimated an integral somewnat similar to that in (15), and obtained the result*

$$\int \overline{\left(\frac{\partial p}{\partial t}\right)_{\mathbf{x}=0} \left(\frac{\partial p}{\partial t}\right)_{\mathbf{x}=\mathbf{z}}}\, d\mathbf{z} \doteqdot 1{\cdot}6\rho_0^2(\overline{v_1^2})^3 L, \tag{16}$$

where L is a scale of turbulence related to the energy-dissipation rate ϵ of equation (9) by the equation $(\overline{v_1^2})^{\frac{3}{2}}/\epsilon = L$; also, to within the accuracy of Proudman's results, L is identical with the so-called 'integral scale' of the turbulence, $\int_0^\infty f(r)\, dr$ in the usual notation. Hence we can obtain the ratio of the two integrals in (15) and (16) (taking the value of the former from equation (9)) as

$$\frac{\int \overline{\frac{\partial}{\partial t}\left(p\frac{\partial v_1}{\partial x_1}\right)_{\mathbf{x}=0} \frac{\partial}{\partial t}\left(p\frac{\partial v_1}{\partial x_1}\right)_{\mathbf{x}=\mathbf{z}}}\, d\mathbf{z}}{\int \overline{\left(\frac{\partial p}{\partial t}\right)_{\mathbf{x}=0} \left(\frac{\partial p}{\partial t}\right)_{\mathbf{x}=\mathbf{z}}}\, d\mathbf{z}} \doteqdot \frac{38\pi}{1{\cdot}6} \frac{\overline{v_1^2}}{L^2}. \tag{17}$$

One may reasonably infer from this that the order of magnitude of the fluctuations in velocity gradient that contribute most to the effectiveness of expression (14) in generating sound is the square root of expression (17),† that is, about $8(\overline{v_1^2})^{\frac{1}{2}}/L$. The interest of this result is its indication that, although velocity gradients typical of those in the 'energy-dissipating eddies' are *not* effective in generating sound (see §4), nevertheless, the effective velocity gradients are greater by a factor‡ of order 8 than those typical of the largest 'energy-bearing eddies'. It is an intermediate size of eddy that plays the most important role.

Now, in isotropic turbulence, velocity gradients in all directions contribute to expression (14), and the resulting noise field is omnidirectional. But in turbulence

* The second figure in the coefficient is not claimed to be significant.

† It might be thought that this argument is faulty because the covariance in the numerator would have a zero integral if the p's were removed, and hence (even though they are not) would be expected to be positive in some regions and negative in others. But approximate estimates indicate that the pressure covariance is already small at distances where the velocity-derivative covariance, integrated round a sphere, is negative, and so the 'negative' regions are relatively unimportant.

‡ Strictly, part of this large factor inferred from (17) may be due to the relevant *frequencies* of $\partial v_1/\partial x_1$ exceeding those of p.

with a large *mean* shear (say, a gradient in the x_2 direction of the $\bar{v}_1$ component of mean velocity) a single term $p\bar{e}_{12}$ will dominate expression (14). In words, the mean shear tends to orient the bulk of the quadrupoles along the principal axes of rate of strain, which for a shearing motion are at 45° to the direction of motion. Further,

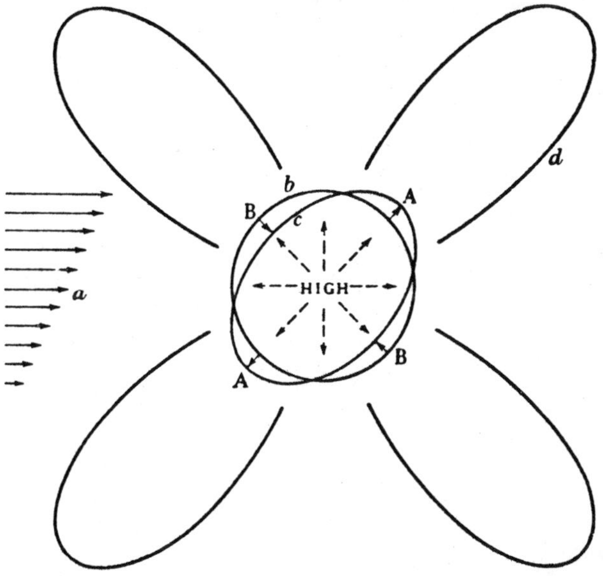

FIGURE 2. Basic lateral quadrupole resulting from combined excess pressure and shear at a point. Shear (a) in small time interval deforms spherical fluid element (b) into ellipsoid (c), with principal axes at 45° to the flow. Excess pressure (HIGH) causes momentum outwards from it to increase (broken-line arrows). At A momentum is being created in the direction of deformation, so there is *increasing* momentum flux outwards. At B momentum is being created in the direction opposite to that of deformation, so there is *decreasing* momentum flux outwards. Fluctuations in this pattern of changing momentum flux, due to fluctuations of either pressure or shear, produce sound radiation with the polar intensity diagram (d). (For the distortion of this directional pattern due to convection see part I, figure 3.) The work of §5 shows that even sound from isotropic turbulence can be thought of as formed from elements like the one illustrated; but a large mean shear makes them conspicuous by giving them a predominant orientation.

one would expect the sound resulting from given pressure fluctuations to be amplified in this case, at least if $\bar{e}_{12}$ is large enough.* This is the result referred to in §2 by describing shear as an 'aerodynamic sounding-board'.

The arguments which have shown that pressure fluctuations in shearing flow generate a lateral quadrupole field, with maximum rate of change of momentum

* To estimate how large, note that, according to the indications of the preceding discussion of isotropic turbulence, the *fluctuating* velocity gradients which are effective are of the order $8(\overline{v_1'^2})^{\frac{1}{2}}/L$, where L is the length scale of the energy-bearing eddies. However, there is also a factor $\sqrt{15}$ due to the fact that where only one (lateral) quadrupole orientation is present the sound energy radiated is $\frac{1}{15}$ as much as when all orientations are present in equal strength. (In symbols, $\int (x_1^2 x_2^2/x^6)\,\mathrm{d}S = \frac{1}{15}\int x^{-2}\mathrm{d}S$, where the integrals are taken over a large sphere.) Finally, then, amplification will result if the mean shear exceeds about $30(\overline{v_1'^2})^{\frac{1}{2}}/L$ (unless, perhaps, the inference from the discussion of isotropic turbulence is inapplicable to heavily sheared turbulence).

flux (and, in consequence, maximum radiated sound) at 45° to the direction of motion, are illustrated graphically in figure 2.

We now rewrite expression (6) for the intensity field, per unit volume of turbulence situated at the origin, on the assumption that the expression (14), which is used to replace $\partial T_{ij}/\partial t$, is dominated by the term

$$p\frac{\partial \bar{v}_1}{\partial x_2} = p\bar{e}_{12}. \tag{18}$$

It becomes

$$i(\mathbf{x}) \sim \frac{x_1 x_2 \bar{e}_{12}(0)}{8\pi^2 \rho_0 a_0^5 (x - M_c x_1)^6} \int x_k x_l \bar{e}_{kl}(\mathbf{z}) \overline{\frac{\partial p(0,t)}{\partial t}\frac{\partial p(\mathbf{z},t)}{\partial t}}\, d\mathbf{z} \tag{19}$$

(where the direction of eddy convection has been assumed parallel to the only significant mean velocity $\bar{v}_1$). It is clearly not too bad an approximation in (19) to neglect all components of $\bar{e}_{kl}(z)$ except $\bar{e}_{12}$ and $\bar{e}_{21}$, because the covariance in (19) is significant only when $\mathbf{z}$ is fairly close to the origin, so that in most cases the mean shears at $\mathbf{z}$ and the origin would be somewhat alike in orientation; and further, the approximation at most slightly distorts the *directional* dependence of the sound produced; it cannot alter the power output because the integral of

$$x_1 x_2 x_k x_l/(x - M_c x_1)^6$$

is zero unless k and l are 1 and 2 in some order.

Making this approximation, and writing $\tau(\mathbf{z})$ for $\bar{e}_{12}(\mathbf{z})$, the component of mean shear *in the direction of the shear at the origin*, (19) can be written as

$$i(\mathbf{x}) \sim \frac{x_1^2 x_2^2}{(x - M_c x_1)^6} \frac{\tau^2(0)\,\overline{(\partial p/\partial t)^2}\, V}{4\pi^2 \rho_0 a_0^5}, \tag{20}$$

where in accordance with the ideas of §3 an 'average eddy volume' has been introduced, whose precise definition is

$$V = \int \frac{\tau(\mathbf{z})}{\tau(0)} \frac{\overline{\partial p(0,t)/\partial t\, \partial p(\mathbf{z},t)/\partial t}}{\overline{(\partial p(0,t)/\partial t)^2}}\, d\mathbf{z}. \tag{21}$$

The power output p_s per unit volume, under these shear flow conditions, is obtained by integrating (20) over a large sphere centre $\mathbf{x} = 0$ (the integration was already carried out in part I, §7) as

$$p_s = \frac{1 + 5M_c^2}{(1 - M_c^2)^4} \frac{\tau^2(0)\,\overline{(\partial p/\partial t)^2}\, V}{15\pi \rho_0 a_0^5}. \tag{22}$$

The gain in power due to convection of the lateral quadrupole at the Mach number M_c was plotted in part I, figure 1, curve (ii); the directional distribution at different Mach numbers of convection M_c was shown in part I, figure 3.

The mode of expression of the results as in (20) and (22), which was convenient in the case of isotropic turbulence, has disadvantages in the case of shear-flow turbulence. To use it requires an approximate knowledge of $\overline{(\partial p/\partial t)^2}$ and V. The former might be estimated by measurements with a microphone at points successively closer to the mixing region, followed by an extrapolation to the mixing region itself (the microphone must not, of course, be placed in the flow).* It is difficult to do anything with V, however, except equate it to some other average eddy volume, measured more simply by integrating velocity correlations. The velocity correlation measurements could be carried out for convenience in a low Mach number flow (though such measurements have not yet been made, to the author's knowledge, in the mixing region of a round jet).

However, the procedure described suffers from a serious inaccuracy; it ignores the fact that the $\tau(\mathbf{z})/\tau(0)$ term in the integral (21) weights it in favour of places near the origin where the mean shear is greatest.† Also, it is unsuitable for giving the total power output of a shear layer because p_s will vary so much across the layer. A more satisfactory approximation for estimating the *total* power output of a thin shear layer, namely,

$$\frac{1+5M_c^2}{(1-M_c^2)^4}\frac{1}{15\pi\rho_0 a_0^5}\iint \tau(\mathbf{y})\,\tau(\mathbf{z})\,\overline{\frac{\partial p(\mathbf{y},t)}{\partial t}\frac{\partial p(\mathbf{z},t)}{\partial t}}\,d\mathbf{y}\,d\mathbf{z}, \tag{23}$$

is to assume that the covariance varies much more slowly, as $\mathbf{y}$ and $\mathbf{z}$ cross the layer, than $\tau(\mathbf{y})$ and $\tau(\mathbf{z})$ do. The integrations across the layer can then be carried out directly; if U is the change in velocity across the layer, (23) becomes

$$\frac{1+5M_c^2}{(1-M_c^2)^4}\frac{U^2}{15\pi\rho_0 a_0^5}\iint \overline{\frac{\partial p(\mathbf{y},t)}{\partial t}\frac{\partial p(\mathbf{z},t)}{\partial t}}\,dS_{\mathbf{y}}\,dS_{\mathbf{z}}, \tag{24}$$

where $dS_{\mathbf{y}}$ is an element of area of the shear layer in $\mathbf{y}$ space, and similarly with $dS_{\mathbf{z}}$. This gives a result

$$\Pi_s = \frac{1+5M_c^2}{(1-M_c^2)^4}\frac{U^2\overline{(\partial p/\partial t)^2}\,S}{15\pi\rho_0 a_0^5} \tag{25}$$

for the power output per unit *area* of turbulent shear layer, where

$$S = \int \frac{\overline{\partial p(\mathbf{y},t)/\partial t\,\partial p(\mathbf{z},t)/\partial t}}{\overline{(\partial p(y,t)/\partial t)^2}}\,dS_{\mathbf{z}} \tag{26}$$

is an 'average eddy area' at the point $\mathbf{y}$. There would be somewhat fewer sources of error in an estimate of the total acoustic power output of a shear layer by means of (25) than in one based on (22). Data not available in the existing literature, in particular $\overline{(\partial p/\partial t)^2}$ and S (the latter to be replaced, presumably, by a simpler definition of average eddy area), are needed, but neither presents insuperable difficulties of measurement.

* Note that, since the time derivative is one in a frame of reference moving with the eddy convection velocity, one cannot obtain $\overline{(\partial p/\partial t)^2}$ by attaching a differentiating circuit to the stationary microphone. Rather one would have to measure $\overline{p^2}$, and then (cf. part I, §5) multiply by $(2\pi N)^2$, where N is a frequency typical of the radiation from the shear layer.

† Another procedure, if this term is ignored, would be to take $\overline{(\partial p/\partial t)^2}\,V$ as given by Proudman's estimate (16) of the resulting integral in the case of isotropic turbulence. This, however, is likely to be a serious underestimate for another reason. In a mixing region, pressure fluctuations are almost certainly not of order $\rho\overline{v_1'^2}$ (as they are in isotropic turbulence, and as (16) implies) but rather are more closely comparable with $\rho\bar{v}_1 v_1'$, just as for small disturbances to laminar flow.

It is easily shown that the result (25) applies not only to a plane shear layer, but also (for example) to the annular mixing region of a round jet, provided that in (26) area is taken to mean projected area on to the tangent plane at **y**. (This is because $\tau(\mathbf{z})$ must be taken as the component of the shear in the direction of the shear at **y**.) Further, as a result of the combination of the lateral quadrupoles in the (x_1, x_2) and (x_1, x_3) planes, the directional distribution is proportional to

$$\frac{x_1^2(x_2^2+x_3^2)}{(x-M_c x_1)^6} = \frac{\sin^2\theta\cos^2\theta}{x^2(1-M_c\cos\theta)^6}, \tag{27}$$

as in (36) of part I, where θ is the angle between the direction of emission and the jet axis.

Now the indication of the experimental work is that the sound from the annular mixing region is dominated by the lateral quadrupole terms discussed above at all subsonic Mach numbers (see §7 for the detailed evidence). This sound is at high frequency (note that frequencies in shear-flow turbulence are expected to be of the order of magnitude $(2\pi)^{-1}$ times the mean shear, because of the term $v_2 \mathrm{d}\bar{v}_1/\mathrm{d}x_2$ in the acceleration $\mathrm{D}v_1/\mathrm{D}t$), and so is easily distinguished from the sound emanating from the 'core' of the jet. Without this experimental confirmation it would have been difficult to predict with confidence that the octupole term in (13) would not become important at the higher Mach numbers (but see the footnote following that equation for some considerations relevant to this issue).

One more point needs to be considered. Does the integrand in either the average eddy volume V (equation (21)) or the average eddy area S (equation (26)) remain, on the whole, positive, as was observed to be necessary in §3 if the methods of estimation of sound production are to be valid? Now, in these integrals, the covariance would be expected to remain positive in *incompressible* flow. For, first, the pressure covariance remains positive according to Batchelor's calculations (1950) for isotropic turbulence, and Proudman's (1952) indicated that this is probably also true of $\partial p/\partial t$. And, secondly, there is no known mechanism in incompressible flow tending to associate an increasing pressure at one point with a decreasing pressure at any neighbouring point. But for a *compressible* fluid the pressure trends are associated with density trends, and total conservation of mass implies that the integrated covariance of $\partial p/\partial t$ is zero. However, the distance between points such that increasing pressure at one is correlated with decreasing pressure at the other is of the order of half the acoustic wave-length. Hence, when this exceeds the size of the larger eddies, the central volume where the covariance has large positive values is surrounded by a widespread annular region where it is very small but negative, in such a way that the total volume integral is zero. But the integrand in (21) is weighted by the term $\tau(\mathbf{z})/\tau(0)$ and the integral (16) is only a surface integral. In both cases the values of the integrand which are relevant are confined to the thin shear layer. Hence the cancellation does not occur in either integral, because only a thin slice of the region of negative covariance, which is required to cancel the central core of positive covariance, is available. (Similar considerations may apply to equation (30) in §6 below.)

6. The amplifying effect of inhomogeneities of temperature or composition

This section treats of the sound associated with the terms in T_{ij} other than the momentum flux. These become significant only when the local mean velocity of sound varies widely in the fluid, as a result of inhomogeneities of temperature or of fluid composition. Thus, such inhomogeneities may be regarded as amplifying the sound due to turbulence, just as shear was shown to do in §5. But evidently this amplification would be important only if the additional sound were at least comparable with the sound resulting from momentum flux; the theory given below indicates that in many important cases this is not so. In particular, it will be concluded that any large apparent extra noise from hot jets is probably due to causes (such as combustion noise, or simply higher speed) other than the direct amplification due to heat discussed in this section; the amplification, if observed at all, should affect principally the high-frequency components of the jet noise.

The main term in T_{ij} (see (1)) other than momentum flux is $(p-a_0^2\rho)\delta_{ij}$. (The unimportance of the viscous contribution to p_{ij} was fully discussed in part I.) The sound field resulting from this term has a simple physical interpretation. It is the source field due to that part of the turbulent pressure fluctuations which is *not* balanced by the multiple $a_0^2\rho$ of the local density fluctuations accompanying it in the *free* acoustic vibrations of the atmosphere. Evidently the part not so balanced can produce forced oscillations.

At a point where the mean velocity of sound is $\bar{a}$, the pressure fluctuations, being nearly adiabatic, would be approximately $(\bar{a})^2$ times the fluctuations in density. Hence they lead to a term in T_{ij} which is approximately $(1-(a_0/\bar{a})^2)p\delta_{ij}$, corresponding by part I, p. 575, to a source strength per unit volume

$$\left(1-\left(\frac{a_0}{\bar{a}}\right)^2\right)\frac{1}{a_0^2}\frac{\partial^2 p}{\partial t^2}. \tag{28}$$

It is interesting to note that the simple-minded idea that the sound emitted from turbulent flow would include 'the source field corresponding directly to the turbulent pressure fluctuations'—whose strength per unit volume, in the sense used in part I, p. 572 (*rate of change* of the rate at which mass is introduced), would be $\partial^2/\partial t^2$ of the corresponding density fluctuations, or approximately

$$\frac{1}{a_0^2}\frac{\partial^2 p}{\partial t^2} \tag{29}$$

—is correct only when $(\bar{a})^2 \gg a_0^2$, as, for example, in the limiting case of flows much hotter than the atmosphere into which they radiate sound. But even in this case it will be seen below that the noise due to momentum flux may be more important.

A larger effect is obtainable if values of $\bar{a}$ considerably *less* than a_0 are present; these are achieved most conveniently by the use of a gas of high molecular weight. Thus for SF_6 (Freon) $a_0/\bar{a}$ is about 2·3, so that (28) would be $-4{\cdot}3$ times (29), and the associated intensity fields would be in the ratio 18:1. The importance of the pressure terms relative to the momentum flux terms would be enhanced by this factor. Probably this is the main application in which the pressure terms would be really important. (Note that also the higher density of SF_6 increases the sound output, but to the same extent for both sets of terms.)

The power output of the source field (28), per unit volume of turbulence at the origin, is easily approximated by the methods of §3 as

$$\frac{1}{4\pi\rho_0 a_0^5}\left[1-\left\{\frac{a_0}{\bar{a}(0)}\right\}^2\right]\int\left[1-\left\{\frac{a_0}{\bar{a}(\mathbf{z})}\right\}^2\right]\overline{\frac{\partial^2 p(0,t)}{\partial t^2}\frac{\partial^2 p(\mathbf{z},t)}{\partial t^2}}\,d\mathbf{z}, \tag{30}$$

where the factor due to quadrupole convection is omitted for simplicity. The basic argument, indicating that (30) is normally smaller than the power output due to momentum flux (at least in the cases where $\bar{a} > a_0/\sqrt{2}$, so that the square-bracketed factors have modulus less than 1), depends on the fact that the latter output takes a form in which, essentially, $\partial(pe_{ij})/\partial t$ replaces the $\partial^2 p/\partial t^2$ of (30). (See, for example, (15), in the case of isotropic turbulence.) Hence it will be larger if $(2\pi)^{-1} e_{ij}$ exceeds in magnitude the dominant frequencies for pressure fluctuations. If, in addition, the dominant frequencies for e_{ij} exceed those for the pressure, it will be larger still. Both suppositions are likely to be true, though more so in turbulence which is qualitatively similar to isotropic turbulence than in heavily sheared turbulence.

For isotropic turbulence one can be fairly precise. Replacing the square-bracketed factors in (30) by 1 (as in the case of very hot flows), the integral becomes expression (16), whose value is about $1{\cdot}6\rho_0^2(\overline{v_1^2})^3 L$, but with two time differentiations instead of one. Also one can estimate the integral without any time differentiations, from Batchelor's (1950) results, as $2\rho_0^2(\overline{v_1^2})^2 L^3$. These two results indicate that the dominant frequency for pressure fluctuations is of order $(\overline{v_1^2})^{\frac{1}{2}}/L$, and permits an estimate of the integral in (30) as $\rho_0^2(\overline{v_1^2})^4/L$ times a coefficient of order 1. Thus in this case (30) is less than the power output of isotropic turbulence (9) by a factor less than 10^{-2}. Even in the extreme case of Freon mentioned above (which introduces an extra factor of 18), the term (30) must still remain negligible.

These conclusions for isotropic turbulence can be assumed to hold qualitatively for the sound emitted from the 'core' of a jet, but not to that emanating from the heavily sheared mixing region. There the frequencies with which the pressure fluctuates may not fall so far below $(2\pi)^{-1}$ times the mean shear $\bar{e}_{12}$; further, the frequencies with which e_{ij} fluctuates are no longer relevant. In addition, the limitation to a lateral quadrupole field (as against a source field) gives the radiation due to $\bar{e}_{12}\partial p/\partial t$ a disadvantage by a factor $\frac{4}{15}$ (as a comparison of (22) and (30) shows). As an upper-bound estimate of the amount of amplification, due to high temperature, of the noise emanating from the mixing region, one may therefore take $10\log_{10}(1+\frac{15}{4}) = 6$ decibels, but one would regard a lower figure as more likely. The amplification will be greatest in the *highest* frequency bands of this noise (which is itself the high-frequency part of the *total* jet noise), because the relative importance of $\partial^2 p/\partial t^2$ to $\bar{e}_{12}\partial p/\partial t$ is greatest in those bands.

In the light of these estimates there is evidently some possibility of amplification due to the use of a Freon jet being really large (up to 19 db) at the higher frequencies, quite apart from the influence of the larger density. The evidence of Lassiter &

Hubbard's (1952) measurements with a Freon jet is somewhat negative on this point, but not conclusive since no frequency analysis was recorded (see §8 for a further discussion).

Generally, the arguments of this section indicate that pressure fluctuations are likely to be of small importance in generating aerodynamic noise through their direct source field. (It is fairly evident that this conclusion would remain true even in the case of pressure fluctuations which were not, for some reason, approximately adiabatic, since it has been shown that the source field (29) is relatively ineffectual even when it is not mitigated by corresponding density fluctuations.) Accordingly one may reaffirm the general conclusion of part I, that where no fluctuating external forces or fluctuating sources of matter are present the principal source of noise generation is the fluctuation in the momentum flux across fixed surfaces.

7. Discussion of the experimental work on the cold subsonic jet

The experimental work (Fitzpatrick & Lee 1952; Gerrard 1953; Lassiter & Hubbard 1952; Westley & Lilley 1952) on the cold subsonic jet, referred to in general terms in §§1 and 2, will here be discussed in detail. The different aspects referred to in §§1 and 2 will be considered in the order adopted there.

Direct measurements of acoustic power output have been made only by Fitzpatrick & Lee (1952), who obtained them from measurements of intensity in a reverberation chamber which had been previously calibrated, in certain of the octave bands used, by means of a source of similar directional characteristics. The result (their figure 5) is close to an eighth-power law variation with the jet exit velocity U. The acoustic power coefficient

$$K = \frac{\text{acoustic power}}{\rho_0 U^8 a_0^{-5} d^2} \tag{31}$$

(which the authors adopt from (26) of part I) takes values between $0{\cdot}6 \times 10^{-4}$ and $1{\cdot}2 \times 10^{-4}$ as U varies by a factor of $2\frac{1}{2}$ and the jet diameter d by a factor of 2. There is very little evidence that the scatter is systematic, although the authors suggest tentatively a Reynolds number effect. For the larger Mach numbers, K was consistently close to $0{\cdot}9 \times 10^{-4}$.

Other authors measure only intensity, at various positions round the jet. Thus Lassiter & Hubbard (1952) give intensity as a function of the azimuth angle θ (measured from the orifice as origin) at a jet exit velocity $U = 1000$ ft./s, for $d = \frac{3}{4}$ to 12 in. Performing a rough integration of their values, and making a guess at the values behind the orifice (which were not measured, but which other workers have found to be low), one obtains an average value of the acoustic power coefficient K, over the cases considered, as $0{\cdot}5 \times 10^{-4}$. For the smallest values of the orifice diameter d the value of K was sensibly lower, at $0{\cdot}3 \times 10^{-4}$.

But such a small variation of K over a large range of d indicates that acoustic power output is very closely proportional to d^2; in fact, Lassiter & Hubbard find intensities closely proportional to $(d/r)^2$ for large r/d, and other authors who have varied both d and r are in accord. Also, in two experiments, Lassiter & Hubbard varied the density of the jet from the atmospheric value ρ_0 to a value ρ_1 (that of He and SF_6). The intensities changed by a factor $(\rho_1/\rho_0)^2$. This is what might be expected from the theory (e.g. equation (2)), since T_{ij} contains a factor ρ_1. Thus, where the density of flow ρ_1 differs from that, ρ_0, of the atmosphere into which it radiates sound,* a modified acoustic power coefficient

$$K = \frac{\text{acoustic power}}{\rho_1^2 \rho_0^{-1} U^8 a_0^{-5} d^2} \tag{32}$$

is expected to remain approximately constant (at least if the amplification effects discussed in § 6 are unimportant).

Gerrard (1953) integrates his measurements of acoustic intensity round a large sphere, to obtain the acoustic power coefficient as a function of mean Mach number M at the jet orifice. He finds that $K = 0{\cdot}4 \times 10^{-4} M^{-0{\cdot}8}$ for a jet from a circular pipe 36 in. long, and $K = 0{\cdot}8 \times 10^{-4} M^{-0{\cdot}3}$ for a jet from a pipe 54 in. long (for both pipes the diameter $d = 1$ in.). Thus he finds that acoustic power increases as the (7·2)th and (7·7)th powers of the velocity in the two cases.

From these three bits of evidence (with varying degrees of turbulence in the oncoming stream, from that produced in the converging nozzle of Lassiter & Hubbard, through that of Gerrard's two straight pipes, up to that in Fitzpatrick & Lee's straight nozzle preceded by a sudden divergence) we have general agreement that in the range of Mach numbers covered ($0{\cdot}3 < M < 1$) the acoustic power coefficient K is between $0{\cdot}3 \times 10^{-4}$ and $1{\cdot}2 \times 10^{-4}$, with $0{\cdot}6 \times 10^{-4}$ as a typical value. The acoustic efficiency

$$\eta = \frac{\text{acoustic power}}{\frac{1}{2}\rho_0 U^3 \frac{1}{4}\pi d^2} = \frac{8}{\pi} K M^5 \tag{33}$$

is usually slightly greater than $10^{-4} M^5$, but this is certainly its order of magnitude, and the evidence is inadequate to justify stating more than this.

The spectrum of the sound is very broad; an estimate of this is given by the width of the frequency band in which the intensity per octave remains within 20 db of its peak value. From the curves of Fitzpatrick & Lee, performing a slight extrapolation, we may estimate this as seven octaves. Westley & Lilley (1952) obtain six octaves at small angles to the jet axis, and higher values at larger angles, for which, however, the extrapolation becomes impossible since measurements were not taken at high enough frequencies. Lassiter & Hubbard (1952) obtain between six and seven octaves.

The 'peak' of the spectrum is very flat. Gerrard (1953) obtains values of acoustic power within 2 db over a range of three octaves around the peak. Westley & Lilley (1952) obtain a similar result for their intensity measurements at 60 and 75° to the jet axis; they find more pronounced peaks at the higher frequencies in the range in question for larger angles, and at the lower frequencies for smaller angles. Fitzpatrick & Lee (1952) obtain sharper peaks, with a variation of 4 to 7 db over the top three-octave band, but one might in this matter prefer the verdict of the other authors, whose calibration procedure is to be preferred for such a purpose.

Evidently the peak frequency n_p is in all cases ill-defined by the measurements, but rough values can be obtained. It is natural to express the results in terms of a 'Strouhal number' $n_p d/U$, on the assumption that the frequency is related to

* A heated jet is an important case of this.

frequencies characteristic of the energy-bearing eddies in the turbulence, which would be closely proportional to U/d. Peak Strouhal numbers obtained by Fitzpatrick & Lee (1952) are between 0·3 and 0·6. Lassiter & Hubbard (1952) publish spectra with clear peaks, whose Strouhal number remains constant at 0·5 as U is varied; they remark, however, that these were found under conditions when the stream emerging from the orifice was more turbulent than in their main series of experiments, in which they state that less increase of peak frequency with velocity was observed. The peaks are especially ill-defined in the curves of Gerrard (1953) and Westley & Lilley (1952), which have a 'plateau' rather than a peak, but in each case the 'plateau' contains the frequency corresponding to Strouhal number 0·5.

As a summary of the situation we can say that the peak frequency is always *in the neighbourhood of* $U/2d$, but the evidence regarding its variation with U is inconclusive. The indications of the results of Gerrard and of Westley & Lilley are against proportionality to U. Fitzpatrick & Lee's spectra for $d = 1{\cdot}53$ in. are compatible with it to within the experimental scatter, but those for $d = 0{\cdot}765$ in. are interpretable in this way only if some systematic error be assumed; they are much more easily interpreted as indicating a peak frequency (4000 c/s) independent of jet velocity. Lassiter & Hubbard state that their spectra showing direct proportionality to velocity were not typical of experiments on jets issuing directly from a reservoir through a converging nozzle. Thus the balance of evidence is in favour of a slower increase of frequency with velocity than is given by direct proportionality, but the matter remains in doubt.* (The issue was discussed theoretically at the end of § 2.)

Passing to the directional distribution, the most striking fact, at once evident on walking round one of the jets, was that almost all the sound was radiated in directions making an acute angle with the jet. This is also well known for jet engines running on the ground (if the high-pitched 'compressor whine' is ignored).

Thus, for a jet at Mach number 1, Westley & Lilley (1952) obtain intensities about 15 db lower behind the jet (at angles $\theta > 90°$ to the jet axis) than at the directional maximum, and this is persistently true in each frequency band except the two highest used. Gerrard (1953) obtains similar results at high subsonic Mach numbers, with smaller differences at lower Mach numbers.

The detailed directional distribution in the forward region ($\theta \leqslant 90°$) is best understood for the sound of higher and lower frequencies separately. Lassiter & Hubbard (1952) show in their figure 10 that, when $M = 0{\cdot}8$, the band of Strouhal numbers $nd/U > 1{\cdot}5$ (where n = frequency) has a directional maximum at $\theta = 40°$, that the band from 0·4 to 0·8 has a maximum at $\theta = 30°$, and that the band from 0·2 downwards has intensities increasing monotonically as θ is reduced, at least as far as $\theta = 15°$. Westley & Lilley (1952), with $M = 1$, find a maximum at $\theta = 40°$ for Strouhal numbers $> 0{\cdot}7$, and a maximum at $\theta = 30°$ in the band 0·3 to 0·7, with monotonically increasing intensities in bands below this. Gerrard (1953), at the higher Mach numbers, finds a maximum at $\theta = 40°$ for a Strouhal number exceeding

* Even if the peak frequency is independent of velocity, as much of the evidence indicates, results must still, if they are to be useful, be referred to the only significant non-dimensional parameter nd/U. The result mentioned then signifies that the intensity increases more slowly with U (for the Mach number range of the experiments) at higher values of nd/U than at lower ones.

0·6, one at $\theta = 35°$ for a Strouhal number of 0·5, and a maximum at 20° or less from 0·3 downwards.

These observations indicate that a fairly definite division exists between 'high-frequency' sound (with frequencies sensibly above the peak frequency, say $0{\cdot}7U$ or more, and with a directional maximum slightly less than 45°, as expected theoretically from a heavily sheared flow), and 'low-frequency' sound (with frequencies sensibly below the peak frequency, say $0{\cdot}3U/d$ or less, and with a preference for forward emission). The phrases high and low frequency will be used in this sense henceforth.

In the light of the theory it would be natural to assume that the high-frequency sound, with its 'lateral quadrupole' type of directional distribution, emanates from the heavily sheared mixing region near the orifice; and that the low-frequency sound, with its distribution characteristic of omnidirectional radiation (as modified, of course, by the convection of the quadrupole sources), emanates from the region of more nearly isotropic turbulence farther downstream in the 'core' of the jet. Sound of intermediate frequency would be a combination of both types.

This interpretation is especially likely to be correct because it is known that the highest frequency turbulence occurs in the mixing region. A direct check is not however, easy. Actual plots of contours of equal intensity, for a given frequency band, as made by Gerrard (1953) and by Westley & Lilley (1952), are not very helpful for this purpose, because the measurements near the jet contain sound other than that of the radiation field, and because the far field results are hard to interpret for this purpose owing to the large total extent of the source. Perhaps the best approach is on the lines hinted at by Lassiter & Hubbard (1952), who state without details that they tackled the problem by using 'a directional type of microphone and various recording equipment' (perhaps correlation measurements at two distant points in line with a point of the jet). They found (though again they give no details) that 'the higher frequency components emanate from the region immediately outside the jet pipe'.

Gerrard (1953) determines a 'source' of the higher frequency sound by studying the contours of equal intensity at large distances and drawing a *common perpendicular* to them (at about 40°, the directional maximum) to cut the axis at the source position. This varies from one to nine diameters downstream of the orifice as M increases from 0·3 to 1·0 at a frequency of 8500 c/s. (It is worth remarking that the largest value of M corresponds to a Strouhal number of 0·65 and so is barely in the high-frequency range.) The same approach applied to Westley & Lilley's measurements at $M = 1$ in the frequency range 6400 to 12800 c/s gives a 'source' eight diameters downstream, which agrees with Gerrard. On the other hand, their intensity measurements at these frequencies in the near field are greatest at about five diameters from the orifice. Taken together the results quoted are sufficiently in accord with the general pattern suggested above.

Gerrard (1953) has also attempted to determine the 'eddy-convection Mach number' M_c (introduced in §3 above) for this high-frequency sound. To fit his observed radiation field he assumes a combination of a lateral quadrupole (as postulated in §5) with such a pair of longitudinal quadrupoles as will give an axi-symmetrical intensity field whose directional maximum is at $\theta = 90^\circ$. (Some term of this kind is necessary because without it there would be no sound at all at $\theta = 90^\circ$.) He modifies the sum of the two intensity fields by the factor $(1 - M_c \cos\theta)^{-6}$ and then gets best fit by taking $M_c = 0{\cdot}4M$, where M is the mean Mach number at the orifice. Such a result for the Mach number M_c of convection of the eddies is in good agreement with observation of the motion of eddies in a low-speed jet mixing region (Brown 1937).

The source of the low-frequency sound appears to be very extended; contour lines in the near field (Gerrard 1953; Westley & Lilley 1952) are indicative of a line source extending from say five to twenty diameters downstream of the orifice. Westley & Lilley conclude explicitly from their measurements that 'the jet contains a distribution of sound sources the extent of which increases in the downstream direction as the frequency is decreased'. Lassiter & Hubbard (1952) state that they located the source 'several' diameters downstream, in contrast to their findings on the high-frequency sound.

They also located the source of the *total* noise as being similarly placed, indicating that what we have called the low-frequency noise constituted the greater part of the total noise. The figures quoted above concerning the directional distribution in different frequency bands also indicated that rather (though not much) more than half the total energy is in the low-frequency noise.

It is appropriate therefore to seek to explain the observed order of magnitude $10^{-4}M^5$ for the acoustic efficiency in terms of the sound emitted from the core of the jet, where the turbulence is not strongly sheared, and so not qualitatively different from isotropic turbulence. Then it is reasonable to use Proudman's estimate $38(\overline{v_1'^2})/a_0^5$ (see (9)) for the ratio of the rate at which energy is radiated at a point to the rate at which it is dissipated. Now measurements by Corrsin & Uberoi (1949), of the mean velocity and turbulent energy distributions at a cross-section of a round jet twenty diameters downstream of the orifice, show that at that cross-section three-quarters of the initial energy flux has been dissipated. (Of the remaining quarter, about 20 % is in the form of turbulent energy.) Accordingly, one may insert for $\overline{v_1'^2}$ in Proudman's estimate a typical value in the range zero to twenty diameters downstream, to obtain the order of magnitude of the acoustic efficiency. Inserting a value $(\overline{v_1'^2})^{\frac{1}{2}} = 0{\cdot}08U$ (which Corrsin & Uberoi (1949) find is exceeded only for $6{\cdot}5 < x/d < 17$ on the axis, where it rises to a maximum of $0{\cdot}11U$, but with smaller values taken away from the axis, at least at $x/d = 20$, though possibly less so for smaller x/d), we obtain

$$38(0{\cdot}08M)^5 = 1{\cdot}2 \times 10^{-4}M^5 \tag{34}$$

for the acoustic efficiency, which is in agreement with the experiments.* Of course no more than order of magnitude agreement is significant here.

* Similarly, one would like to make use of (25) to explain that a fair proportion, perhaps one-third, or about $2 \times 10^{-5}\rho_0 U^8 a_0^{-5} d^2$, of the total acoustic power output is radiated from the shear layer. But, as noted in § 5, one is prevented at present by total ignorance of the value of $\overline{(\partial p/\partial t)^2}$ in the shear layer. Taking the average eddy area S as d^2, and the area of the shear layer as $\pi d \times 5d$, one would obtain the power output stated (neglecting the convection effect) if the root mean square of $\partial p/\partial t$ were $0{\cdot}008\rho_0 U^3/d$. Whether this is actually a correct order of magnitude result in the mixing region is, however, a matter of pure conjecture at present.

These considerations on acoustic efficiency do not take into account the effects predicted as due to quadrupole convection. As indicated in §2 these would be expected to produce a further addition to the power output (over and above the U^8 rule) at the higher subsonic Mach numbers, which is not observed in practice. Hence one must scrutinize the varying distribution of directional radiation to determine whether its trend is consistent with an explanation on which the translation effect is present at the higher Mach numbers but is balanced, as far as its effect on total power output is concerned, by some general reduction, either in turbulence level, or in those aspects of the turbulence which generate sound.

That this is so is at once indicated by results on the variation of intensity with velocity in the direction $\theta = 90°$, at which the convection effect must be absent since $(1-M_c\cos\theta)^{-6} = 1$. Measurements in this direction happen to be especially numerous, and yield intensities varying as powers of the velocity considerably lower than those typical of the total noise. Lassiter & Hubbard (1952) get an exponent 5·5 for their $\frac{3}{4}$ in. jet, 7·0 for their 3 in. jet, and 6·3 for their 12 in. jet (fair scatter would be expected in these as each value is based on only four points). Westley & Lilley get 6·3 (note that their $\mathfrak{P}$ varies like the (2·45)th power of the velocity at such a typical subsonic Mach number as 0·7). Gerrard gets from 6·3 to 6·8 under different conditions. These results indicate that the basic acoustic output is increasing less rapidly than U^8, and that it is only the quadrupole convection effect, causing more rapidly increasing intensities for smaller θ, which restores the measured power output to proportionality to U^8 or slightly less. Roughly, the intensity at $\theta = 90°$ (in the range of Mach numbers treated) is less by about a factor of $U^{1\cdot5}$ than what is expected, which corresponds as pointed out in §2 to the factor by which U^8 is less than the expected total acoustic power output, including the convection terms.

The few measurements of the simultaneous dependence of intensity on velocity U and angle θ for lower values of θ support this conjecture. Gerrard (1953) finds (as do Westley & Lilley) that proportionality to $U^{8\cdot2}$ is typical for $\theta \leqslant 30°$. It is significant, too, that while his measurements for $\theta = 30°$ show almost perfect proportionality to such a power of U, those for $\theta = 90°$ indicate that the intensity was increasing as a lower power of U for higher Mach numbers than for lower ones. Since the convection factor $(1-M_c\cos\theta)^{-6}$ which is present for $\theta < 90°$ increases as a higher power for the higher Mach numbers than for the lower ones, such a result is in accordance with the theory. Gerrard notes also that the behaviour which he observes for $\theta > 90°$ is qualitatively in accord with what would be expected from the presence of a factor $(1-M_c\cos\theta)^{-6}$ in the intensity.

To sum up, it seems likely that the interpretation of the directional distribution of intensity in terms of quadrupole convection is correct. Refraction by the mean flow in the jet may affect finer details, but it does not appear to be fundamental. However, the analysis leading to the proportionality of intensity to U^8 in the absence of the convection effect is at fault, at least at the higher subsonic Mach numbers. The possible sources of this error have been fully discussed in §2.

8. Some notes on the noise of supersonic jets

The information on the noise of supersonic jets is less systematic than in the subsonic case, and although some trends can be observed there is more divergence between results of different experiments.

Thus Powell (1952*b*) obtained a sound field dominated by a 'whistle or screech' with a more or less well-defined note. Westley & Lilley (1952) and Lassiter & Hubbard (1952) obtained no such effect. However, in the latter case the supersonic speed was obtained by the use of a helium jet, the jet speed being less than the speed of sound in helium, though considerably greater than that in air. Accordingly the shock wave structure, to which Powell (1952*b*) attributes the whistle, was absent. It is harder to see why Westley & Lilley (1952) obtained no whistle; the geometry in their experiments was similar to that in Powell's, except that a cylindrical pipe of length two or three diameters intervened in their case between the contraction cone and the orifice.

Lassiter & Hubbard measured intensity of noise from their helium jet principally at $\theta = 90°$, where no effect of quadrupole translation would be expected. Accordingly the sound intensity, after correction for the density of the jet (it was explained in §7 that it would vary as the square of this density), would be expected to lie on a continuous curve through the subsonic points. This it does, and the curve takes the form of proportionality to $U^{5\cdot2}$. But it is surprising that this exponent is so much below other measured values for $\theta = 90°$. The points on this curve obtained for Freon, air and helium could all be explained on the basis of a higher velocity exponent (say $U^{6\cdot3}$) if dependence on a higher power of the density (say $\rho^{2\cdot4}$) could be assumed, but this is very difficult to believe, especially for the low-density fluid. A tentative explanation could be that the velocity exponent really falls to 5·2 only at the *higher* speeds, owing to the radiation damping effect discussed in §2, and that it appears to be so low in the case of the low-speed Freon jet only because this experiences the amplification discussed at the end of §6.

Lassiter & Hubbard give also (for one jet speed, $U = 2620$ ft./s) a polar diagram for the intensity of noise from their helium jet. It is very strongly directional, with a maximum at $\theta - 42°$, where the intensity level is 15 db higher than at $\theta = 90°$.

Now the explanation of this maximum from the point of view of this paper bears no relation to the explanation of the directional maximum of the high-frequency sound from subsonic jets in terms of a lateral quadrupole field. The sound measured in the experiments was not of high frequency in this sense. The suggested explanation is rather that the phenomenon results from *supersonic convection of quadrupoles.* An eddy convected at Mach number M_c would produce, on the simple theory of §3, infinite sound at an angle $\theta = \sec^{-1} M_c$. The infinity is not, of course, really present, because the theory contains approximations (especially the neglect of certain differences in times in the integral for the intensity) which if omitted would leave a finite value; but presumably this finite value would be large. Now $\sec^{-1} M_c$ would be 42° if M_c were 1·35, which is $0\cdot58M$, where $M = 2\cdot33$ is the ratio of the jet speed to the atmospheric speed of sound. Such a value is of the right order of magnitude.

Physically, the explanation given means that the circular wave fronts emitted by an eddy as it moves downstream have a conical envelope whose semi-angle is the well-known Mach angle, and whose direction of propagation is the complement, $\sec^{-1} M_c$, of this Mach angle. The intensity is greatest at such an envelope of wave fronts because a large number of sound pulses are heard simultaneously,* and because the times of emission are spread over a large interval, so that cancelling of different pulses is minimized.

Mr Lilley has shown the author unpublished Schlieren photographs of supersonic jets and their instantaneous sound-wave patterns, which show a strong maximum at an angle of about 30° to the jet, which he interprets in terms of the 'ballistic shock waves' emitted by the eddies while they are travelling supersonic. This explanation is very closely similar to that given above.†

Passing to the features of supersonic jets which are perhaps more closely associated with the steady shock-wave pattern in the jet (and so would be absent in Lassiter & Hubbard's helium jets, or in hot jets at speeds between the speed of sound in the atmosphere and that in the jet), one may notice first an additional sound maximum at $\theta = 100°$ found in unpublished work by Lilley & Peduzzi. The author would tentatively ascribe this to sound radiated as a result of turbulence passing through the stationary shock-wave pattern in the jet. In a recent paper he gave a theory of this sound radiation (Lighthill 1952), according to which the energy radiated would be a certain fraction, depending on the shock strength, of the kinetic energy of the turbulence. For a shock of pressure ratio 1·3 the fraction would be 6 %, and it would increase with shock strength. but unfortunately the theory was not expected to be accurate for pressure ratios much above 1·3. It is fairly evident, however, that the process might lead to acoustic efficiencies, for a supersonic jet, comparable with those already discussed. For the flux of turbulent energy through shocks in the jet might well amount to about 1 % of the total energy flux at the orifice, and if 6 % of this quantity was radiated, the efficiency would be 6×10^{-4}, which already exceeds the efficiency of a jet at $M = 1$. (Note, too, that the pressure ratio across any *normal* shock exceeds 1·3 for $M > 1{\cdot}12$.)

Directionally, the theory (Lighthill 1953) indicates that, in a frame of reference in which the air ahead of the shock is at rest, the sound radiated has a strong preference for emission in directions nearly parallel to the direction of shock-wave propagation (for reasons closely connected with the quadrupole convection effect discussed in this paper). Indeed, it forms part of a greater volume of sound, of which the rest is directed so close to the direction of motion of the shock that it catches up with it, and is largely absorbed by it. The source of this sound is at a distance behind the shock comparable with the length scale L of the turbulence. Evidently the 'freely scattered' sound which just misses the shock must, in a frame of reference in which the shock is stationary, be radiated in directions nearly parallel with the shock or making small angles with it in the upstream direction. The part associated

* Compare the well-known explanation of the so-called 'sonic bangs'.

† Again, hot jets from turbojet engines often show such a maximum when the jet speed sufficiently exceeds the atmospheric speed of sound, and the explanation probably holds good in this case too.

with the normal or nearly normal shocks (which are the strongest shocks in the jet) would then be expected to have a directional maximum close to the 100° observed by Lilley & Peduzzi, although evidently some of the sound in question would be refracted more into the upstream direction as it passed through the edge of the jet. The wave-length of the sound would be expected to be comparable with the scale of the turbulence, and this too agrees with the visual observations of Lilley & Peduzzi. However, this suggested explanation must remain very tentative.

To conclude this section some reference must be made to the analysis by Powell (1952*b*) of the 'whistle or screech' referred to above, although Powell's theory is already so complete that no further contribution is required. Briefly, the theory is that under favourable conditions the shock pattern in the jet may introduce a resonant frequency. For any specially strong eddy, created at the orifice, may (after amplification as it sweeps along the shear layer) send out a sound wave as it passes through the first shock, which, when it reaches the orifice, may cause the appearance of another strong eddy. This mechanism yields the observed frequency, and also suggests conditions for the appearance of the whistle which have some experimental support. The mechanism bears an obvious analogy to the mechanism of edge tones, which has recently received some attention (Curle 1953; Powell 1952*a*). The shock wave replaces the edge. As with edge tones the presence of the phenomenon greatly amplifies the sound produced, so that the first step towards reducing supersonic jet noise is to devise methods for avoiding the whistle, which Powell has successfully done.

References

Batchelor, G. K. 1950 *Proc. Camb. Phil. Soc.* **47**, 359.
Brown, G. B. 1937 *Proc. Phys. Soc.* **49**, 493.
Corrsin, S. & Uberoi, M. 1949 *Tech. Notes Nat. Adv. Comm. Aero., Wash.*, no. 1865.
Curle, N. 1953 *Proc. Roy. Soc.* A, **216**, 412.
Fitzpatrick, H. M. & Lee, R. 1952 *Rep. David W. Taylor Model Basin, Wash.*, no. 835.
Fleming, N. 1946 *J. R. Aero. Soc.* **50**, 639.
Gerrard, J. H. 1953 Ph.D. thesis, Manchester.
Lassiter, L. W. & Hubbard, H. H. 1952 *Tech. Notes Nat. Adv. Comm. Aero., Wash.*, no. 2757.
Lighthill, M. J. 1952 *Proc. Roy. Soc.* A, **211**, 564 (part I).
Lighthill, M. J. 1953 *Proc. Camb. Phil. Soc.* **49**, 531.
Mawardi, O. K. & Dyer, I. 1953 *J. Acoust. Soc. Amer.* **25**, 389.
Powell, A. 1952*a* *Rep. Aero. Res. Comm., Lond.*, no. 15333 (limited circulation), to appear in *Acoustica*.
Powell, A. 1952*b* *Rep. Aero. Res. Comm., Lond.*, no. 15623 (limited circulation), to appear in *Proc. Phys. Soc.*
Proudman, I. 1952 *Proc. Roy. Soc.* A, **214**, 119.
Westley, R. & Lilley, G. M. 1952 *Rep. Coll. Aero. Cranfield*, no. 53.

Appendix. Effect of non-uniformity of the eddy convection velocity on the distortion of the directional pattern due to convection

As stated in §4, we refer T_{ij} to the Lagrangian variable associated with the steady velocity field $a_0 M_c(\mathbf{y})$. Thus we put

$$T_{ij}(\mathbf{y}, t) = S_{ij}(\mathbf{a}, t), \tag{A1}$$

where the relation between $\mathbf{y}$ and $\mathbf{a}$ (for given t) is that the solution of

$$\frac{d\boldsymbol{\xi}}{dt} = a_0 \mathbf{M}_c(\boldsymbol{\xi}), \quad \boldsymbol{\xi} = \mathbf{a} \text{ at } t = 0, \tag{A2}$$

is $\boldsymbol{\xi} = \mathbf{y}$ at time t. If $\mathbf{M}_c$ is a solenoidal vector field then $d\mathbf{y} = d\mathbf{a}$, and also

$$\left(\frac{\partial}{\partial t}\right)_{\mathbf{a}} = \left(\frac{\partial}{\partial t}\right)_{\mathbf{y}} + a_0 \mathbf{M}_c . \frac{\partial}{\partial \mathbf{y}}. \tag{A 3}$$

Now the integral in (2) is the mean square of the integral

$$\int \frac{(x_i - y_i)(x_j - y_j)}{|\mathbf{x}-\mathbf{y}|^3} \frac{\partial^2}{\partial t^2} T_{ij}\left(\mathbf{y}, t - \frac{|\mathbf{x}-\mathbf{y}|}{a_0}\right) d\mathbf{y}. \tag{A 4}$$

When T_{ij} is replaced by S_{ij}, the argument of S_{ij} will be the Lagrangian variable $\boldsymbol{\eta}$ associated with $\mathbf{y}$ at the time $t - |\mathbf{x}-\mathbf{y}|/a_0$. Now to obtain the Jacobian $d\boldsymbol{\eta}/d\mathbf{y}$, we can ignore the part *independent* of y in the time shift $t - |\mathbf{x}-\mathbf{y}|/a_0$ separating $\mathbf{y}$ from $\boldsymbol{\eta}$, because volumes are unchanged in the solenoidal velocity field $a_0 \mathbf{M}_c$, and concentrate only on the small variations in time shift in the small volume $d\mathbf{y}$. But for these the mapping is simply the change to moving axes studied in part I, §7, and so

$$\frac{d\boldsymbol{\eta}}{d\mathbf{y}} = 1 - \frac{\mathbf{M}_c.(\mathbf{x}-\mathbf{y})}{|\mathbf{x}-\mathbf{y}|}. \tag{A 5}$$

Hence (A 4) becomes

$$\int \frac{(x_i - y_i)(x_j - y_j)}{|\mathbf{x}-\mathbf{y}|^3} \left(1 - \frac{\mathbf{M}_c.(\mathbf{x}-\mathbf{y})}{|\mathbf{x}-\mathbf{y}|}\right)^{-1} \left(\frac{\partial}{\partial t} - a_0 \mathbf{M}_c . \frac{\partial}{\partial \boldsymbol{\eta}}\right)^2 S_{ij}\left(\boldsymbol{\eta}, t - \frac{|\mathbf{x}-\mathbf{y}|}{a_0}\right) d\boldsymbol{\eta}. \tag{A 6}$$

In (A 6) the $\partial/\partial\boldsymbol{\eta}$ does not apply to the term $t - |\mathbf{x}-\mathbf{y}|/a_0$, although that varies with $\boldsymbol{\eta}$ for fixed t. Rewriting (A 6) so that it does apply to that term too, the

$$\partial/\partial t - a_0 \mathbf{M}_c . \partial/\partial \boldsymbol{\eta}$$

becomes

$$\left(1 - \mathbf{M}_c . \frac{\partial}{\partial \boldsymbol{\eta}} |\mathbf{x}-\mathbf{y}|\right) \frac{\partial}{\partial t} - a_0 \mathbf{M}_c . \frac{\partial}{\partial \boldsymbol{\eta}}. \tag{A 7}$$

The second term in (A 7) can now be neglected, because $\partial \mathbf{M}_c/\partial\boldsymbol{\eta} = 0$ and using the divergence theorem. (Note that in this process no derivatives of terms on the first line of (A 6) will arise, because they vanish in the radiation field.)

But it may easily be calculated that

$$\mathbf{M}_c . \frac{\partial}{\partial \boldsymbol{\eta}} |\mathbf{x}-\mathbf{y}| = - \frac{\mathbf{M}_c.(\mathbf{x}-\mathbf{y})}{|\mathbf{x}-\mathbf{y}| - \mathbf{M}_c.(\mathbf{x}-\mathbf{y})}. \tag{A 8}$$

Substituting, we obtain finally as the equivalent (in the radiation field) of (A 4).

$$\int \frac{(x_i - y_i)(x_j - y_j)}{|\mathbf{x}-\mathbf{y}|^3} \left(1 - \frac{\mathbf{M}_c.(\mathbf{x}-\mathbf{y})}{|\mathbf{x}-\mathbf{y}|}\right)^{-3} \frac{\partial^2}{\partial t^2} S_{ij}\left(\boldsymbol{\eta}, t - \frac{|\mathbf{x}-\mathbf{y}|}{a_0}\right) d\boldsymbol{\eta}. \tag{A 9}$$

On taking the mean square, by writing down the same integral with a different letter $\mathbf{z}$ instead of $\mathbf{y}$ (and consequentially a different letter $\boldsymbol{\zeta}$ instead of $\boldsymbol{\eta}$), and averaging their product, we obtain the result stated in §3, equation (4).

The authors of the following paper consider theoretically and experimentally the power spectrum of the wall pressure that would be measured by transducers of vanishingly small size and the corrections which must be made for transducers of various finite sizes and shapes. This is an important problem in the understanding of the phenomenon of flow noise.

Dr. Willmarth is an aeronautical engineer. He has contributed in the fields of condensation of gases, transonic flow, turbulent boundary layer, unsteady aerodynamics, and aerodynamic sound. He is Professor of Aerospace Engineering at the University of Michigan.

This paper is reprinted with the permission of the senior author and the Editor of the *Journal of Fluid Mechanics.*

J. Fluid Mech. (1965), vol. 22, part 1, pp. 81–94
Printed in Great Britain

Resolution and structure of the wall pressure field beneath a turbulent boundary layer

By W. W. WILLMARTH AND F. W. ROOS
Department of Aeronautical and Astronautical Engineering,
The University of Michigan

(Received 2 July 1964)

The power spectrum of the wall pressure that would be measured by a transducer of vanishingly small size and the corrections to the power spectra measured by finite-size transducers are determined from the spectra measured by four transducers of different diameters. The root-mean-square wall pressure measured by a transducer of vanishingly small size is $\sqrt{\overline{p^2}}/\tau_w = 2{\cdot}66$, approximately 13 % higher than the root-mean-square pressure measured by the transducer used in the earlier investigations of Willmarth & Wooldridge (1962). Corrections to the power spectrum measured by a finite-size transducer are computed using the theory of Uberoi & Kovasznay (1952, 1953). The computations require information about the correlation of the wall pressure for very small spatial separation of the transducers. Unfortunately, these measurements have never been made. Corcos's (1964) similarity of the cross-spectral density is assumed to represent the missing information, but the computed corrections fail at high frequencies because the similarity expression is not valid when the spatial separation is small. The range of validity of the similarity is determined, and the average radial derivative of the cross-spectral density is inferred from the measured power spectra.

1. Introduction

Measurements of the statistical properties of a random field made with transducers of finite size will be affected by the shape and size of the transducer. It is the purpose of this paper to investigate the effect of a finite-size circular transducer on measurements of the wall pressure beneath a turbulent boundary layer. The investigation depends upon the fact that a complete and accurate set of wall-pressure measurements can be corrected for the effect of the transducer if the transducer responds linearly to the pressure.

In the body of the paper the theory (Uberoi & Kovasznay 1952) for correcting the measurements is outlined. Computations of the corrections to the pressure are attempted using Corcos's (1964) similarity model for the measurements of Willmarth & Wooldridge (1962). The computations are inaccurate at high frequencies and cannot be improved because the set of measured data upon which everything depends is not complete (the pressure correlation at small spatial separations, $|\zeta| < 0{\cdot}7\delta^*$, of the transducers is not known).

The power spectra measured with transducers of four different diameters (Willmarth & Wooldridge 1963) are used to determine the corrections to the measured spectra. Analysis of the experimentally measured corrections of the power spectra shows that Corcos's (1964) similarity model for the cross-spectral density is not valid for spatial separations less than $0{\cdot}7\delta^*$. The average radial derivative of the cross-spectral density is determined. The paper ends with some general remarks about the effect of a finite-size transducer on the measurement of wall pressure.

2. Relations between measured and true wall pressure

The mapping of a stationary stochastic function of several variables by an instrument with a linear response has been treated by Uberoi & Kovasznay (1952, 1953) and by Liepmann (1952). The result of their work is the conclusion that, if the properties of the instrument and the mapped stochastic function are completely known, the properties of the original stochastic function can be recovered.

We present these relations in a notation appropriate to the problem of the resolution of a pressure field by a flat transducer of finite size mounted flush with a wall. The mapped stochastic property that has been measured is the pressure correlation

$$R_{pm}(\boldsymbol{\zeta},\tau) = \langle p_m(\mathbf{x},t)\, p_m(\mathbf{x}+\boldsymbol{\zeta},t+\tau)\rangle, \tag{1}$$

where the brackets indicate an average. The vectors $\boldsymbol{\zeta}(\xi,\eta)$ and $\mathbf{x}(x,y)$ lie in the plane of the wall and the subscript m denotes a measured quantity. The pressure has zero mean and is homogeneous in the plane of the wall. The measured pressure p_m is related to the true pressure p by

$$p_m(\mathbf{x},t) = \int_\infty p(\mathbf{s},t)\, K(\mathbf{s}-\mathbf{x})\, dA(\mathbf{s}), \tag{2}$$

where $K(\mathbf{s}-\mathbf{x})$ is the response kernel characterizing the spatial response of the pressure transducer. (The transducer is assumed to respond instantly to the applied pressure, and K depends only on $\mathbf{s}-\mathbf{x}$.) For a statistically homogeneous pressure field, the measured pressure correlation may then be expressed as

$$R_{pm}(\boldsymbol{\zeta},\tau) = \int_\infty\int_\infty R_p(\boldsymbol{\zeta}+\boldsymbol{\epsilon},\tau)\, K(\mathbf{s})\, K(\mathbf{s}+\boldsymbol{\epsilon})\, dA(\mathbf{s})\, dA(\boldsymbol{\epsilon}). \tag{3}$$

The essence of the problem we shall consider is the solution of this integral equation (3). The original pressure correlation $R_p(\boldsymbol{\zeta},\tau)$ is to be determined from a knowledge of $K(\mathbf{s})$ and $R_{pm}(\boldsymbol{\zeta},\tau)$. We will not actually compute R_p but will be concerned with computing the temporal Fourier transform of $R_p(0,\tau)$ (the power spectrum of the pressure).

We define the spectrum of the pressure in space and time

$$E(\mathbf{k},\omega) = \frac{1}{8\pi^3}\int_\infty\int_\infty R_p(\boldsymbol{\zeta},\tau)\exp\left[-i(\omega t+\mathbf{k}.\boldsymbol{\zeta})\right] d\tau\, dA(\boldsymbol{\zeta}), \tag{4}$$

where ω is the circular frequency and $\mathbf{k}(k_1,k_2)$ is the wave-number vector in the plane of the wall. The corresponding inverse relation is

$$R_p(\boldsymbol{\zeta},\tau) = \int_\infty\int_\infty E(\mathbf{k},\omega)\exp\left[i(\omega\tau+\mathbf{k}.\boldsymbol{\zeta})\right] d\omega\, dA(\mathbf{k}). \tag{5}$$

We will also need the cross-spectral density

$$\Gamma(\boldsymbol{\zeta},\omega) = \frac{1}{2\pi}\int_\infty R_p(\boldsymbol{\zeta},\tau)\exp(-i\omega\tau)\,d\tau, \tag{6}$$

and the power spectrum

$$\Gamma(0,\omega) = \phi(\omega). \tag{7}$$

The corresponding measured quantities $E_m(\mathbf{k},\omega)$, $\Gamma_m(\boldsymbol{\zeta},\omega)$, and $\phi_m(\omega)$ are defined in the same way (with the measured correlation $R_{pm}(\boldsymbol{\zeta},\tau)$ substituted for $R_p(\boldsymbol{\zeta},\tau)$).

In order to solve for the original power spectrum of the pressure, we must invert equation (3). We define the 'correlation' of $K(\mathbf{s})$,

$$\theta(\boldsymbol{\epsilon}) = \int_\infty K(\mathbf{s})\,K(\mathbf{s}+\boldsymbol{\epsilon})\,dA(\mathbf{s}), \tag{8}$$

and insert it in equation (3)

$$R_{pm}(\boldsymbol{\zeta},\tau) = \int_\infty R_p(\boldsymbol{\zeta}+\boldsymbol{\epsilon},\tau)\,\theta(\boldsymbol{\epsilon})\,dA(\boldsymbol{\epsilon}). \tag{9}$$

We recognize that the Fourier transform of a function R_{pm} which is the convolution of two functions is the product of their respective transforms. Equation (9) becomes

$$E'_m(\mathbf{k},\tau) = 4\pi^2 E'(\mathbf{k},\tau)\,\psi(\mathbf{k}), \tag{10}$$

where

$$E'(\mathbf{k},\tau) = \frac{1}{4\pi^2}\int_\infty R_p(\boldsymbol{\zeta},\tau)\exp(-i\mathbf{k}.\boldsymbol{\zeta})\,dA(\boldsymbol{\zeta}), \tag{11}$$

and

$$\psi(\mathbf{k}) = \frac{1}{4\pi^2}\int_\infty \theta(\boldsymbol{\epsilon})\exp(-i\mathbf{k}.\boldsymbol{\epsilon})\,dA(\boldsymbol{\epsilon}). \tag{12}$$

We determine $E'(\mathbf{k},\tau)$ from equation (10) and use the inverse of equation (11) to determine

$$R_p(\boldsymbol{\zeta},\tau) = \frac{1}{4\pi^2}\int_\infty E'_m(\mathbf{k},\tau)\,[\psi(\mathbf{k})]^{-1}\exp(i\mathbf{k}.\boldsymbol{\zeta})\,dA(\mathbf{k}). \tag{13}$$

We shall be interested in the cross-spectral density obtained from the temporal Fourier transform of equation (13)

$$\Gamma(\boldsymbol{\zeta},\omega) = \frac{1}{4\pi^2}\int_\infty E_m(\mathbf{k},\omega)\,[\psi(\mathbf{k})]^{-1}\exp(i\mathbf{k}.\boldsymbol{\zeta})\,dA(\mathbf{k}), \tag{14}$$

and the power spectrum

$$\Gamma(0,\omega) = \phi(\omega) = \frac{1}{4\pi^2}\int_\infty E_m(\mathbf{k},\omega)\,[\psi(\mathbf{k})]^{-1}\,dA(\mathbf{k}). \tag{15}$$

$\psi(\mathbf{k})$ is the spatial 'spectrum' of the transducer and is easily computed from equation (8), where $\theta(\boldsymbol{\epsilon})$ is expressed as the convolution of $K(\mathbf{s})$. The Fourier transform of $\theta(\boldsymbol{\epsilon})$ is

$$\psi(\mathbf{k}) = \frac{1}{4\pi^2}\left[\int_\infty K(\mathbf{s})\exp(-i\mathbf{k}.\mathbf{s})\,dA(\mathbf{s})\right]^2. \tag{16}$$

3. Computation of the original power spectrum from the measured pressure field

Equation (15) of the previous section gives the relation with which the original power spectrum of the pressure may be computed from quantities measured by the finite-size transducer. Measurements of the boundary-layer wall-pressure field have been reported by Willmarth & Wooldridge (1962) and more recently by Bull (1963). Corcos has used the measurements of Willmarth & Wooldridge in his work and proposed (Corcos 1962, 1963, 1964) the use of the similarity variable $\omega\boldsymbol{\zeta}/U_c$ to simplify the expressions for the measured cross-spectral density. He assumes

$$\Gamma_m(\boldsymbol{\zeta}, \omega) = \phi_m(\omega)\, A(\omega\xi/U_c)\, B(\omega\eta/U_c) \exp(-i\omega\xi/U_c), \tag{17}$$

where $A(\omega\xi/U_c)$, $B(\omega\eta/U_c)$, and $U_c(\omega\delta^*/U_c)$ are reproduced from Corcos (1963) in figures 1–3. Corcos (1962 or 1964) also shows by numerical integration that

$$R_{pm}(\boldsymbol{\zeta}, \tau) = \int_\infty \phi_m(\omega)\, A(\omega\xi/U_c)\, B(\omega\eta/U_c) \exp[i(\omega\tau - \omega\xi/U_c)]\, d\omega \tag{18}$$

gives a good representation of the measured space-time correlation for the special cases $R_{pm}(\xi, 0, \tau)$ and $R_{pm}(0, \eta, 0)$.

In order to evaluate equation (15), we will use equation (17), and we need $\psi(\mathbf{k})$ for the particular transducer used in the measurements. The transducer was circular with a radius R equal to $0{\cdot}166\delta^*$ (see Willmarth & Wooldridge 1962). Therefore, if we assume that the response of the transducer is instantaneous and spatially uniform (see equation (2)),

$$\left.\begin{aligned} K(\mathbf{s}) &= 1/\pi R^2, \quad |\mathbf{s}| < R, \\ &= 0, \qquad\quad |\mathbf{s}| > R. \end{aligned}\right\} \tag{19}$$

From equation (16) we determine

$$\psi(\mathbf{k}) = (1/\pi^2)\,[J_1(kR)/kR]^2; \; k^2 = k_1^2 + k_2^2, \tag{20}$$

and from equation (17) we determine

$$E_m(\mathbf{k}, \omega) = \phi_m(\omega) \frac{1}{4\pi^2} \int_\infty A(\omega\xi/U_c)\, B(\omega\eta/U_c) \exp[-i(\omega\xi/U_c + \mathbf{k}.\boldsymbol{\zeta})]\, dA(\boldsymbol{\zeta}). \tag{21}$$

In the dimensionless form that we used for our computations, equation (15) can be written

$$\phi(\omega)/\phi_m(\omega) = \left(\frac{U_c}{2\pi\omega\delta^*}\right)^2 \int_{-\infty}^{\infty}\int_{-\infty}^{\infty} F_m\left(\frac{k_1^* U_c}{\omega\delta^*}, \frac{k_2^* U_c}{\omega\delta^*}\right) \left[\frac{k^* R^*}{2J_1(k^*R^*)}\right]^2 dk_1^*\, dk_2^*, \tag{22}$$

where $k^{*2} = k_1^{*2} + k_2^{*2}$, $k_{1,2}^* = \delta^* k_{1,2}$, $R^* = R/\delta^*$ and

$$F_m = \int_{-\infty}^{\infty}\int_{-\infty}^{\infty} A(\gamma)\, B(\beta) \exp\left[-i\left(\frac{k_1^* U_c}{\omega\delta^*} + 1\right)\gamma - i\frac{k_2^* U_c}{\omega\delta^*}\beta\right] d\gamma\, d\beta. \tag{23}$$

By a change of variables it is easy to show that ϕ/ϕ_m is a function of $\omega R/U_c$ only.

In our computations we were not able to evaluate the integral of equation (22) exactly because the zeros of $J_1(k^*R^*)$ cause it to diverge. F_m of equation (23) does not have any zeros; however, it should have zeros at the same points as the zeros of $J_1(k^*R^*)$ because equation (10) shows that, when $\psi(\mathbf{k})$ is zero, the measured spatial spectrum $E'_m(\mathbf{k}, \tau)$ should also be zero.† (If $E'(\mathbf{k}, \tau)$ is finite, as it must be since R_p is not periodic and since $R_p = 0$ when $\boldsymbol{\zeta} \to \infty$, the ratio E'_m/ψ must also be finite.) The source of the difficulty is the measurement of $R_{pm}(\boldsymbol{\zeta}, \tau)$. All the measurements of $R_{pm}(\boldsymbol{\zeta}, \tau)$ (Willmarth & Wooldridge 1962) were made with $|\boldsymbol{\zeta}| > 0{\cdot}66\delta^*$ or $|\boldsymbol{\zeta}| > 3{\cdot}96R$. On the other hand, the transducer 'correlation', equation (8), whose spatial Fourier transform $\psi(\mathbf{k})$ produces the zeros of the measured spatial spectrum $E_m(\mathbf{k}, \omega)$, is non-zero only when the spatial separation of the transducers is less than $2R$, $|\boldsymbol{\epsilon}| < 2R$. It is apparent that $\Gamma_m(\boldsymbol{\zeta}, \omega)$, given in equation (17), does not contain enough experimental information about $R_{pm}(\boldsymbol{\zeta}, \tau)$ to correct properly the measured power spectrum $\phi_m(\omega)$.

Putting aside these remarks for the moment, we know from Corcos (1964), for instance, that his expression for the measured cross-spectral density, equation (17), gives an acceptable representation of $R_{pm}(\xi, 0, \tau)$ for $|\boldsymbol{\zeta}| > 4R$. We expect that at low frequencies the major contribution to $E_m(\mathbf{k}, \omega)$ will occur at small wave numbers because the wall pressure fluctuations at low frequencies are produced by an essentially frozen large-scale eddy pattern carried past the transducer at the speed U_c. The high-wave number contribution of $E_m(\mathbf{k}, \omega)$ to the low-frequency power spectrum will be negligible, and the incorrect divergent portions of the integral of equation (22), giving the corrected contributions to $\phi(\omega)$ from high wave numbers, can safely be ignored by limiting the range of integration in equation (22).

We have computed the quantity F_m of equation (23) by fitting a sum of exponential functions to Corcos's A and B. The expressions used were

$$A(\gamma) = \exp(-0{\cdot}1145|\gamma|) + 0{\cdot}1145|\gamma| \exp(-2{\cdot}5|\gamma|), \tag{24}$$

$$B(\beta) = 0{\cdot}155 \exp(-0{\cdot}092|\beta|) + 0{\cdot}70 \exp(-0{\cdot}789|\beta|) + 0{\cdot}145 \exp(-2{\cdot}916|\beta|) + 0{\cdot}99|\beta| \exp(-4{\cdot}0|\beta|). \tag{25}$$

The fit obtained is shown in figures 1 and 2. The last terms on the right-hand side of equations (24) and (25) were chosen to make $A'(0)$ and $B'(0)$ zero. In figures 1 and 2, showing $A(\gamma)$ and $B(\beta)$, the slope at the origin is negative. Our addition of the extra term is conservative and reduces the magnitude of the spatial spectrum $E_m(\mathbf{k}, \omega)$ at high wave numbers, but not significantly. We investigated the effect of omitting these two terms in the computation but found no significant change in the results or conclusions.

We have computed the integral for $\phi(\omega)/\phi_m(\omega)$ using equation (22) at four different frequencies, with $R^* = 0$, out to wave numbers $k^* \leqslant 16$ and $k^* \leqslant 18$ with the result shown in table 1.‡ The results show that at low frequencies the

† In actual experiment, $E'_m(\mathbf{k}, \tau)$ may not be exactly zero because unavoidable noise will always be present in the measuring apparatus or experimental environment. The consideration of noise does not concern us here because the experimental information upon which $E'_m(\mathbf{k}, \tau)$ is based is so limited (see below) that E'_m does not show any tendency to oscillate or have zeros.

‡ The integration of equation (22) was performed by weighing contours of constant F_m or $F_m[\psi(k)]^{-1}$ cut from paper of uniform thickness. The errors in this process are responsible for $\phi(\omega)/\phi_m(\omega) > 1$ at $R^* = 0$, $\omega\delta^*/U_\infty = 0{\cdot}5$.

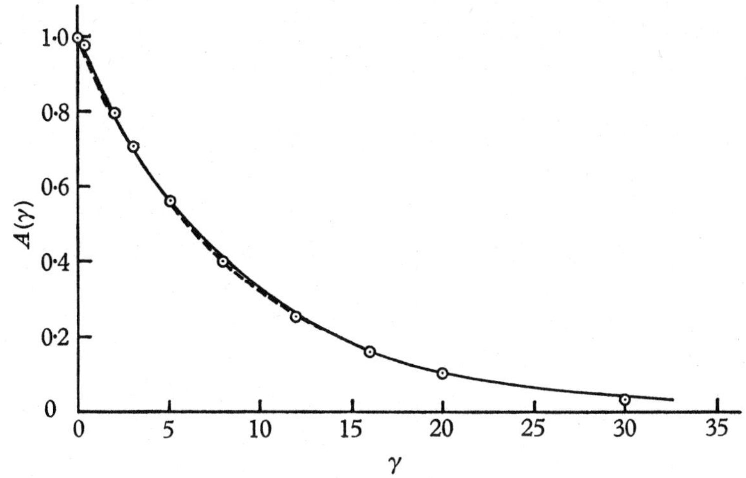

FIGURE 1. The function $A(\gamma)$, $\gamma = \omega\xi/U_c$. ———, $A(\gamma)$ determined by Corcos (1963); – – – –, exp $(-0{\cdot}1145\gamma)$; ⊙, equation (24).

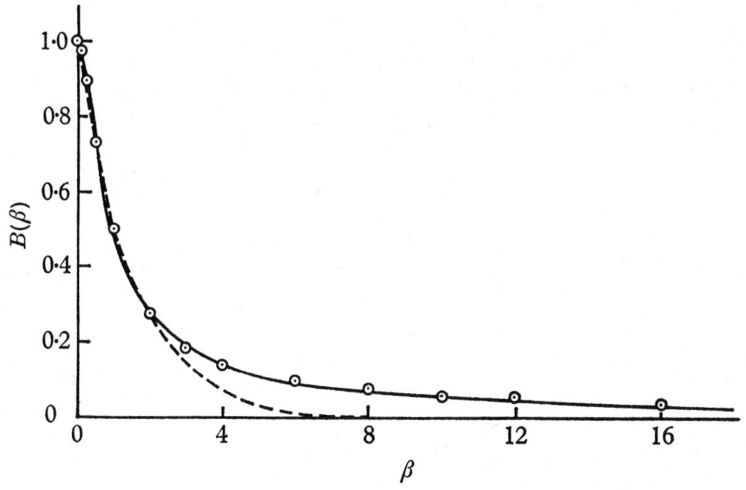

FIGURE 2. The function $B(\beta)$, $\beta = \omega\eta/U_c$. ———, $B(\beta)$ determined by Corcos (1963); – – – –, exp $(-0{\cdot}7\beta)$; ⊙, equation (25).

		$k^* \leqslant 16$		$k^* \leqslant 18$	
		$R^* = 0$	$R^* = 0{\cdot}166$	$R^* = 0$	$R^* = 0{\cdot}166$
$\omega\delta^*/U_\infty$	$\omega R/U_c$	ϕ/ϕ_m	ϕ_m/ϕ	ϕ/ϕ_m	ϕ_m/ϕ
0·5	0·109	1·023	0·957	1·024	0·952
2·0	0·468	0·970	0·758	0·980	0·700
4·0	1·006	0·824	0·627	0·870	0·488
6·3	1·660	0·626	0·535	0·694	0·389

TABLE 1. Results of computations of the ratio ϕ/ϕ_m using equation (22).

ratio $[\phi(\omega)/\phi_m(\omega)]_{R^*=0}$ is nearly one, but, as the frequency increases, more and more contributions of the spatial spectrum $E_m(\mathbf{k}, \omega)$ at high wave numbers are missing. We then computed the correction to $\phi_m(\omega)$ for $R^* = 0{\cdot}166$ and $k^* \leqslant 16$

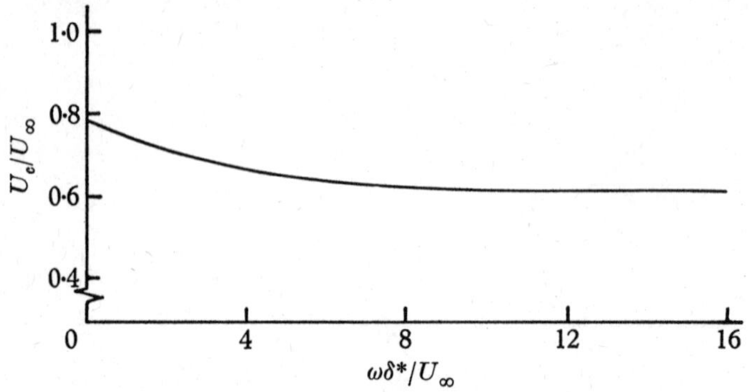

FIGURE 3. The dependence of convection velocity on frequency (Corcos 1963).

and $k^* \leqslant 18$ with the results also shown in table 1. We may place considerable confidence in the correction for $\omega\delta^*/U_\infty = 0{\cdot}5$ and possibly $2{\cdot}0$, because a correction has been applied to most of the measured spatial spectrum E_m which is included in the limited range of integration. Our confidence becomes uncertainty as $\omega\delta^*/U_\infty$ increases, and less of the measured spatial spectrum is included in the limited range of integration.

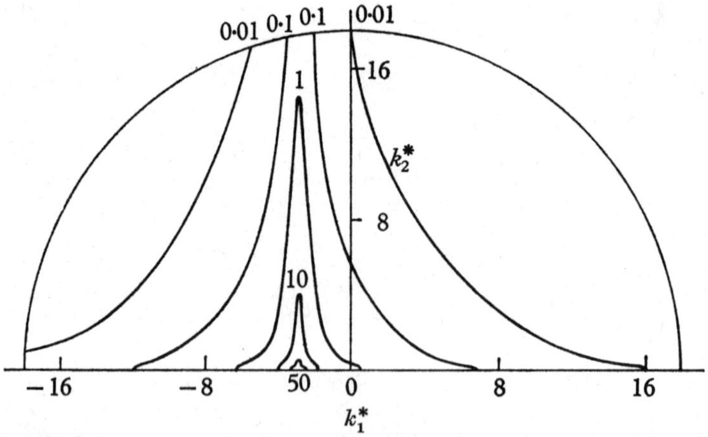

FIGURE 4. Representative contours of constant values of the integrand of equation (22) with $R^* = 0$.

An example of the character of the integrand of equation (22) is shown in figures 4 and 5 for just one frequency $\omega\delta^*/U_\infty = 2$. We see that there are ridges of large $F_m(k_1^*, k_2^*)$ extending in the k_1^* and k_2^* directions that receive a rather large correction for $k^* = 18$ (the first zero of $\psi(\mathbf{k})$ occurs at $k^* \doteqdot 23$). The error in the model for $\Gamma_m(\boldsymbol{\zeta}, \omega)$ has a very serious effect for $\omega\delta^*/U_\infty \geqslant 4$.

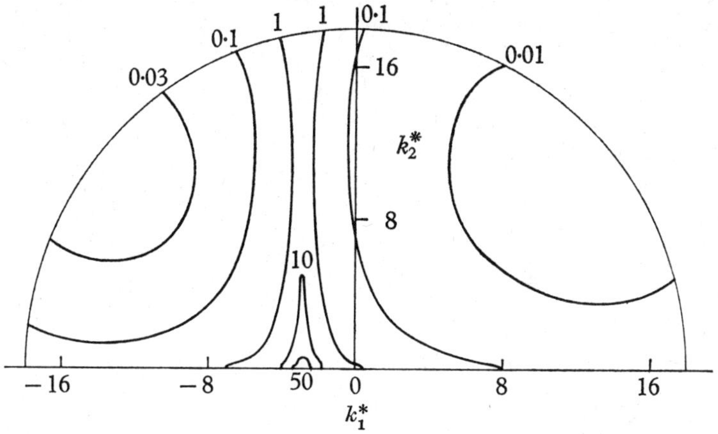

FIGURE 5. Representative contours of constant values of the integrand of equation (22) with $R^* = 0{\cdot}166$.

4. Experimental measurements with different size transducers

A suggestion by Corcos and our inability to correct properly earlier measurements of the power spectrum (Willmarth 1961) led us to measure the wall pressure spectrum (see appendix of Willmarth & Wooldridge 1963) with four circular transducers of different diameters. The results of the measurements are displayed in figures 6 and 7. The experimentally measured values of the various spectra are shown in figure 6. Smooth curves through these points were drawn, and at certain frequencies cross-plots of the spectral amplitude $\phi_m(\omega)$ as a function of transducer radius R were made from the faired curves. Extrapolated values† of $\phi(\omega)$ at $R = 0$ were obtained from the cross-plots and appear (the circled crosses) on figure 6. We have computed the area $\sqrt{\overline{p^2}}/\tau_w$ under each of the five faired spectra of figure 6 and displayed the values of $\sqrt{\overline{p^2}}/\tau_w$ on figure 7. Note that the value of $\sqrt{\overline{p^2}}/\tau_w = 2{\cdot}31$ at $R^* = 0{\cdot}166$ does not agree with our earlier value $\sqrt{\overline{p^2}}/\tau_w = 2{\cdot}19$ (Willmarth & Wooldridge 1962). We have re-examined our previous work and found errors, which we understand, in the methods of measuring $\sqrt{\overline{p^2}}/\tau_w$.‡ We attribute the scatter in the spectra at low frequencies, $\omega\delta^*/U_\infty < 1$ of figure 6, to unavoidable noise and disturbances in the wind tunnel, which had lost two turbulence screens and developed leaks and rattles that were not present previously (1962). The anomalous behaviour of the spectra above $\omega\delta^*/U_\infty = 20$ is caused by electrical noise in the amplifier of the transducer with $R/\delta^* = 0{\cdot}104$, which was made from barium titanate. The other transducers were made from lead zirconate (PZT-5) which is considerably more sensitive. The spectra of the three lead-zirconate transducers all come together above $\omega\delta^*/U_\infty = 20$ because the vibration of the tunnel wall or the pressure fluctuations from the background sound level in the wind tunnel begin to dominate those

† The variation of $\phi_m(\omega)$ with R was approximately linear as $R \to 0$. The extrapolation was not difficult nor did it appear uncertain.

‡ See the corrigendum Willmarth (1965). G. M. Corcos and M. K. Bull first pointed out this error to one of us (W. W. W.).

from the local turbulence in the boundary layer. The spatial extent of the background sound pressure or vibration field will be much greater than that of the turbulent pressure field in the boundary layer at high frequencies. Therefore, all transducers are able to resolve the sound of vibration field and the spectra should coalesce.

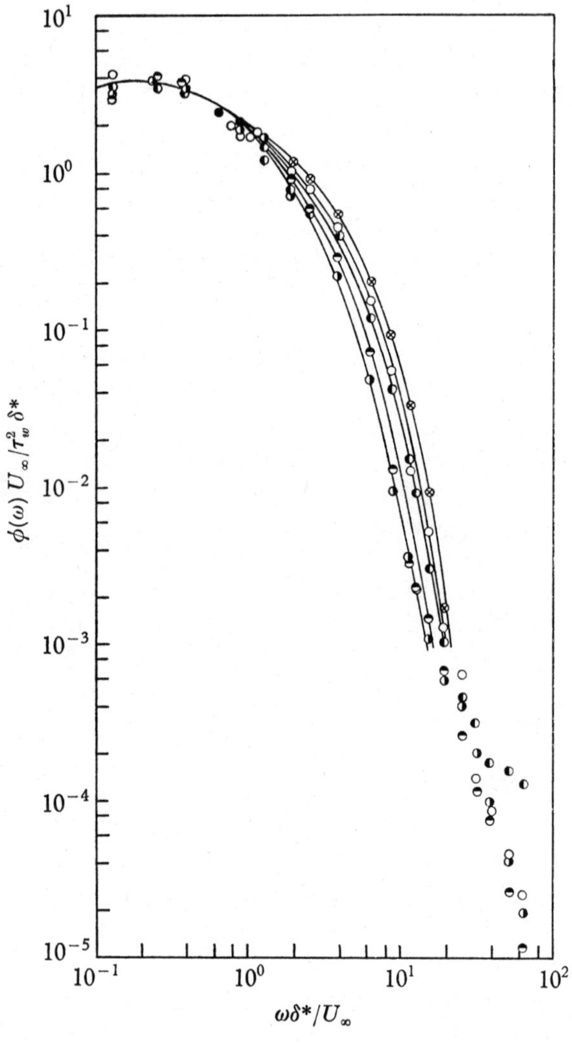

FIGURE 6. The measured and extrapolated dimensionless spectra of the wall pressure.

	R/δ^*	$\sqrt{\overline{p^2}}/\tau_w$
⊗	0·00	2·66
○	0·061	2·54
◐	0·104	2·46
◒	0·166	2·31
◑	0·221	2·20

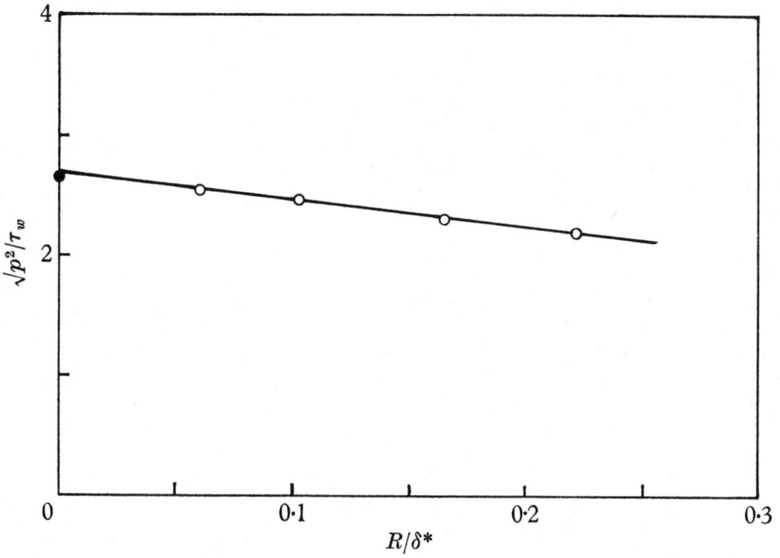

FIGURE 7. The root-mean-square wall pressure determined from the spectra of figure 6. ○, experiment; ●, from spectrum, $R/\delta^* = 0$.

5. Comparison of the experimental and computed corrections

In figure 8 we have collected the results from the computed corrections (§3, table 1) and the measured attenuation† (§4) of the wall-pressure spectra. The computed corrections for $R^* = 0{\cdot}166$ are uncertain at high frequencies, as

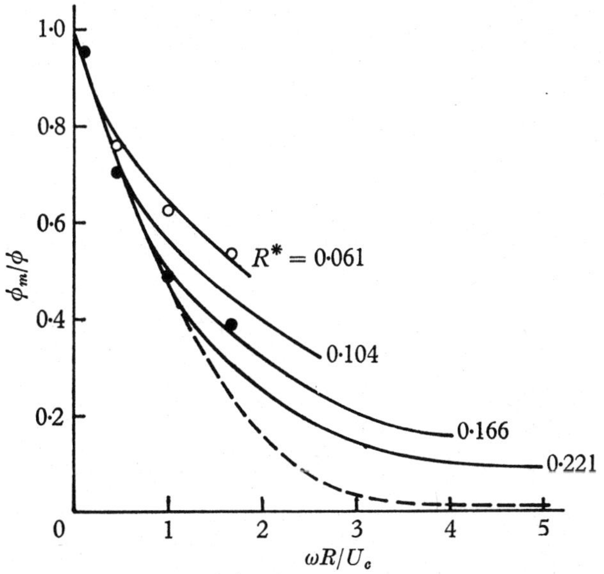

FIGURE 8. The ratio of the measured to actual power spectrum for various size transducers as a function of $\omega R/U_c$. ———, Experiment; – – – –, Corcos (1963); ○, equation (22), $R^* = 0{\cdot}166$, $k^* \leqslant 16$; ●, equation (22), $R^* = 0{\cdot}166$, $k^* \leqslant 18$.

† Data were taken from the faired curves of figure 6.

discussed in §3. The agreement between the computations with $k^* \leqslant 18$ and the experiments for $R^* = 0{\cdot}166$ at high frequencies is only a coincidence.

The experiments show the true attenuation of the wall-pressure spectra for various size transducers at the Reynolds number of the experiments, $R_\theta = 38{,}000$. The attenuation of the spectra ϕ_m/ϕ appears to be a function of $\omega R/U_c$ alone for small values of $\omega R/U_c$, but not for large values. In §3, below equation (22), we noted that ϕ_m/ϕ should be a function of $\omega R/U_c$ alone. Therefore, the assumption of a similarity expression, see equation (17), for the measured cross-spectral density that was used in equation (22) cannot be correct for all values of $\boldsymbol{\zeta}$ and ω. It is approximately correct to use the similarity for corrections to the spectra at low frequencies but not at high frequencies. More exact limitations on the similarity are discussed in §7. We note that the measured spatial spectrum $E_m(\mathbf{k}, \omega)$ should be zero for $k = \alpha_n/R^*$ at any frequency ω (see §3)†. It is clear that this behaviour cannot be obtained from the similarity expressed in equation (17).

6. Averaged properties of the derivative of the cross-spectral density, $\partial\Gamma(\boldsymbol{\zeta}, \omega)/\partial r$, for $\boldsymbol{\zeta} = 0$

Some information about the average radial derivative of the cross-spectral density can be obtained from the measurements displayed in figure 6. We consider the temporal Fourier transform of equation (9) with $\boldsymbol{\zeta} = 0$,

$$\Gamma_m(0, \omega) = \phi_m(\omega) = \int_\infty \Gamma(\boldsymbol{\epsilon}, \omega)\, \theta(\boldsymbol{\epsilon})\, dA(\boldsymbol{\epsilon}). \tag{26}$$

For a circular transducer $\theta(\boldsymbol{\epsilon})$ (the 'correlation' of $K(\mathbf{s})$ in equation (8)) is proportional to the overlapping areas of two circles whose centres are separated by a distance ϵ

$$\left.\begin{aligned} \theta(\boldsymbol{\epsilon}) &= \frac{2}{\pi^2 R^2}\left[\cos^{-1}\frac{\epsilon}{2R} - \frac{\epsilon}{2R}\sqrt{\left\{1-\left(\frac{\epsilon}{2R}\right)^2\right\}}\right], \quad |\boldsymbol{\epsilon}| = \epsilon \leqslant 2R, \\ &= 0 \qquad\qquad\qquad\qquad\qquad\qquad\qquad\qquad \epsilon > 2R. \end{aligned}\right\} \tag{27}$$

Equation (26) becomes

$$\phi_m(\omega, R) = \frac{8}{\pi^2}\int_0^{2\pi}\int_0^1 \Gamma(2R\alpha, \theta, \omega)\, [\cos^{-1}\alpha - \alpha\sqrt{(1-\alpha^2)}]\, \alpha\, d\alpha\, d\theta, \tag{28}$$

and the derivative with respect to R at $R = 0$

$$\left[\frac{\partial\phi_m(\omega, R)}{\partial R}\right]_{R=0} = \frac{32}{\pi}\left\{\int_0^1 [\cos^{-1}\alpha - \alpha\sqrt{(1-\alpha^2)}]\, \alpha^2\, d\alpha\right\}\left\{\frac{1}{2\pi}\int_0^{2\pi}\left\langle\frac{\partial\Gamma}{\partial r}\right\rangle_{r=0} d\theta\right\}, \tag{29}$$

where $r^2 = \xi^2 + \eta^2$. Evaluating the integral and normalizing, we obtain

$$\left[\frac{\partial\phi_m(\omega)/\phi(\omega)}{\partial R}\right]_{R=0} = \frac{384}{135\pi}\left\langle\frac{\partial \ln \Gamma}{\partial r}\right\rangle_{r=0} \doteqdot 0{\cdot}906\left\langle\frac{\partial \ln \Gamma}{\partial r}\right\rangle_{r=0}. \tag{30}$$

We show $\langle\partial \ln \Gamma/\partial(\omega r/U_c)\rangle_{r=0}$, determined from experimental values of $\partial\phi_m/\partial R$ using equation (30) and figure 6, in figure 9. In figure 9 we have also shown

† α_n $(n = 1, 2, \ldots)$ are the zeros of $J_1(\alpha)$.

$\langle \partial \ln \Gamma / \partial r / \delta^* \rangle_{r=0}$ in order to display this quantity as a function of a length characteristic of the boundary layer rather than the wavelength $2\pi U_c/\omega$, whose meaning and interpretation is doubtful when ζ is small (see §7).

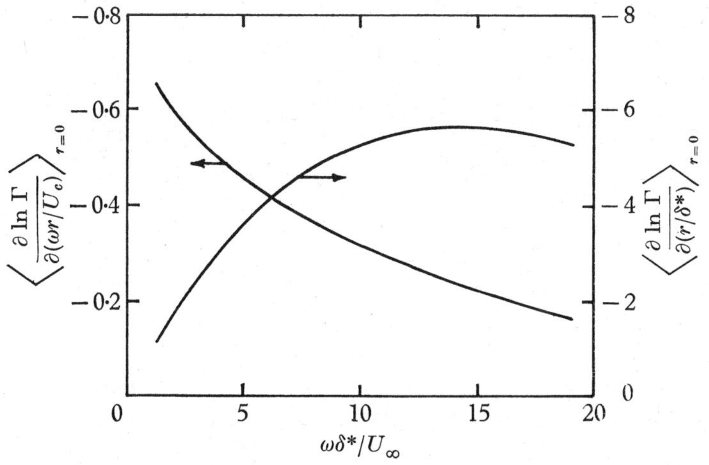

FIGURE 9. Experimentally determined average value of the derivative of the cross-spectral density at $\zeta = 0$ as a function of dimensionless frequency.

7. Range of validity of the Corcos similarity hypothesis

The ratio $\phi_m(\omega)/\phi(\omega)$ measured with different diameter transducers is not a function of $\omega R/U_c$ alone. If the measured cross-spectral density is similar, in the manner proposed by Corcos (1964) and expressed in equation (17), the attenuation of the power spectrum ϕ_m/ϕ would be a function of $\omega R/U_c$ alone (see § 5).

From the measurements shown in figure 8, the lack of similarity (dependence on $\omega R/U_c$ alone) occurs at larger values of $\omega R/U_c$ when the transducer diameter increases. $2\pi U_c/\omega$ is the characteristic wavelength λ of the component of the convected pressure field that produces a pressure fluctuation of frequency $\omega/2\pi$. The value of the dimensionless parameter $\omega R/U_c$ at which the curve $\phi_m(\omega)/\phi(\omega)$ of figure 8, for a given transducer, departs from the common curve describing the attenuation of the remaining transducers indicates the frequency above which the transducer attenuation no longer obeys the similarity hypothesis. The approximate wavelength of the convected pressure pattern below which similarity is no longer obtained is then $\lambda/\delta^* \doteqdot 2\pi R^*[\omega R/U_c]^{-1}$. From figure 8 this value of λ/δ^* is 1·4 for $R^* = 0{\cdot}166$, 1·3 for $R^* = 0{\cdot}104$, and 1·5 for $R^* = 0{\cdot}061$. From an average of these values of λ/δ^*, we conclude that for spatial separations of the order $\frac{1}{2}\lambda = 0{\cdot}7\delta^*$ or less, Corcos's similarity hypothesis is not valid when (as in the above discussion) the dimensionless frequency is greater than approximately three, $\omega\delta^*/U_\infty > 3$. We do not mean to imply that there is any basic contradiction to Corcos's work (Corcos 1964), since the experimental evidence for the similarity is obtained from measurements with $|\zeta| > 0{\cdot}66\delta^*$, and Corcos (1964) discusses the fact that for small $\omega\xi/U_c$ there is no strong support for the similarity.

Another way to demonstrate the lack of similarity for small $|\boldsymbol{\zeta}|$ is to note that, if the original cross-spectral density were similar in the sense of equation (17), the average normalized radial derivative $\langle \partial \ln \Gamma / \partial r \rangle_{r=0}$ can be computed. Using equation (17) and the last term on the right of equation (29), the average radial derivative becomes

$$\left\langle \frac{\partial \ln \Gamma}{\partial (\omega r/U_c)} \right\rangle_{r=0} = \frac{2}{\pi}[A'(0)+B'(0)]. \tag{31}$$

Equation (31) is a constant as a direct consequence of the similarity expressed in equation (17). From the results shown in figure 9 it is apparent that

$$\langle \partial \ln \Gamma / \partial(\omega r/U_c) \rangle_{r=0}$$

is not constant indicating that the similarity does not exist for vanishingly small $\boldsymbol{\zeta}$ and frequencies above $\omega\delta^*/U_\infty = 1{\cdot}0$.

To complete this discussion of the validity of the similarity, we note that Bull (1963) and Bull *et al.* (1963) have measured the wall pressure correlation in narrow frequency bands for a wide range of frequencies and for $|\boldsymbol{\zeta}| \geqslant 0{\cdot}82\delta^*$. They found (Bull 1963, figures 16 and 17, and Bull *et al.* 1963) that for low frequencies the similarity was not valid. According to the data there were departures from similarity for dimensionless frequencies $\omega\delta^*/U_\infty$ less than approximately one and spatial separations greater than δ^*.

Consider the quarter-infinite region $0 < \omega\delta^*/U_\infty < \infty$, $0 < |\boldsymbol{\zeta}|/\delta^* < \infty$. From our results and the measurements by Bull (1963) and Bull *et al.* (1963) the similarity expressed by equation (17) is not valid at the origin ($\omega = \boldsymbol{\zeta} = 0$) and in the regions $\omega\delta^*/U_\infty > 3$, $|\boldsymbol{\zeta}|/\delta^* < 0{\cdot}7$ and $\omega\delta^*/U_\infty < 1$, $|\boldsymbol{\zeta}|/\delta^* > 1$.

8. Additional remarks about the effect of finite size

It is natural to ask whether the similarity (expressed in equation (17)) of the original cross-spectral density $\Gamma(\boldsymbol{\zeta}, \omega)$ would be destroyed by measuring Γ with a finite-size transducer. We suppose that Γ possesses the similarity expressed in equation (17) and that A and B are exponential functions; $\gamma, \beta > 0$. Γ_m is determined from the temporal Fourier transform of equation (9). Then, because A and B are even functions of their arguments and because $\theta(\boldsymbol{\epsilon})$ is a function of ϵ/R (being zero for $\epsilon \geqslant 2R$, see equation (27)), the original similarity of Γ is destroyed for spatial separations $|\xi| < 2R$ or $|\eta| < 2R$ but is retained if $|\xi| > 2R$ and $|\eta| > 2R$. If A and B are not represented by exponential functions the original similarity is lost for all spatial separations. It is not possible to make a general statement about the amount of destruction of the original similarity. One must consider a specific expression for $\Gamma(\boldsymbol{\zeta}, \omega)$.

We can state that it will be difficult to learn much about the cross-spectral density when the separation is small, because the experimental measurements must be made with specially-shaped (intersecting circular or square) transducers and then corrected for the effects of the finite size of the transducer. It would be best to make the transducer size as small as possible for this type of investigation.

9. Conclusions

(1) The attenuation of the power spectrum of the wall pressure has been experimentally measured for circular transducers of four different diameters.

(2) The root-mean-square wall pressure that would be measured by a vanishingly small transducer is $\sqrt{\overline{p^2}} = 2{\cdot}66\tau_w$ when the Reynolds number based on the momentum thickness $R_\theta = 38{,}000$.

(3) Corcos's conclusion (Corcos 1963, 1964) that transducers used in contemporary measurements were unable to resolve a large fraction of the total pressure signal is incorrect. In one set of measurements (Willmarth & Wooldridge 1962), the unresolved root-mean-square pressure was 13 % of the total root-mean-square pressure signal.

(4) Existing measurements of the wall pressure correlation with finite-size transducers have not been made at small enough spatial separations between the transducers to allow accurate computation of the loss of resolution at high wave-numbers or frequencies.

(5) The average radial derivative of the normalized cross-spectral density has been determined from the measurements of the attenuation of the power spectrum and is not proportional to U_c/ω as required by the similarity proposed by Corcos (1964).

(6) The similarity proposed by Corcos (1964) is not valid for spatial separations less than approximately $0{\cdot}7\delta^*$, and dimensionless frequencies above three. From the work of Bull (1963, figures 16 and 17) and Bull *et al.* (1963) the similarity is not valid when the dimensionless frequency is less than approximately one, and the spatial separation greater than approximately δ^*.

The authors wish to thank Professor M. S. Uberoi for many valuable discussions and suggestions. This work was initiated at the University of Michigan and was supported by the Office of Naval Research under contract no. Nonr 1224(30). It was completed during a portion of the senior author's Visiting Fellowship at the Joint Institute for Laboratory Astrophysics.

REFERENCES

Bull, M. K. 1963 *University of Southampton A.A.S.U., Rep.* no. 234.

Bull, M. K., Wilby, J. F. & Blackman, D. R. 1963 *University of Southampton A.A.S.U., Rep.* no. 243.

Corcos, G. M. 1962 *University of California Inst. of Eng. Res. Rep., Series* 183, no. 2.

Corcos, G. M. 1963 *J. Acoust. Soc. Amer.* **35**.

Corcos, G. M. 1964 *J. Fluid Mech.* **18**, 353.

Liepmann, H. W. 1952 *Z. angew. Math. Phys.* **3**, 322.

Uberoi, M. S. & Kovasznay, L. S. G. 1952 *Johns Hopkins University Project Squid Tech. Rep.* no. 30.

Uberoi, M. S. & Kovasznay, L. S. G. 1953 *Quart. Appl. Math.* **10**, 375.

Willmarth, W. W. 1961 *WADC Tech. Rep.* no. 59–676 109.

Willmarth, W. W. 1965 *J. Fluid Mech.* **21**, 107.

Willmarth, W. W. & Wooldridge, C. E. 1962 *J. Fluid Mech.* **14**, 187.

Willmarth, W. W. & Wooldridge, C. E. 1963 *AGARD Rep.* no. 456.

Sound Scattering and Reverberation

V

In this paper, the author develops the theory of sound scattering by solid cylinders and spheres, and verifies the theory experimentally. This paper contains the essential results of a thesis submitted by the author for the Ph.D. degree at Harvard University.

Dr. Faran is a physicist. He has contributed in the fields of physical acoustics, vacuum tube circuit design, signal processing, semiconductor circuit design, and computer programming. He is Leader of an Engineering Programming Group at the General Radio Company, West Concord, Massachusetts.

This paper is reprinted with the permission of the author and the Acoustical Society of America.

Reprinted from THE JOURNAL OF THE ACOUSTICAL SOCIETY OF AMERICA, Vol. 23, No. 4, 405–418, July, 1951

20

Sound Scattering by Solid Cylinders and Spheres*

JAMES J. FARAN, JR.
Acoustics Research Laboratory, Harvard University, Cambridge, Massachusetts
(Received March 13, 1951)

The theory of the scattering of plane waves of sound by isotropic circular cylinders and spheres is extended to take into account the shear waves which can exist (in addition to compressional waves) in scatterers of solid material. The results can be expressed in terms of scattering functions already tabulated. Scattering patterns computed on the basis of the theory are shown to be in good agreement with experimental measurements of the distribution-in-angle of sound scattered in water by metal cylinders. Rapid changes with frequency in the distribution-in-angle of the scattered sound and in the total scattered energy are found to occur near frequencies of normal modes of free vibration of the scattering body.

I. INTRODUCTION

THE scattering of sound was first investigated mathematically by Lord Rayleigh.[1] However, because of the complexity of the mathematical solution, he only considered the limiting case where the scatterers are small compared with the wavelength. The solution for scattering by rigid, immovable circular cylinders and spheres, not necessarily small compared with the wavelength, was given in convenient form by Morse, who defined and tabulated values of phase-angles associated with the partial scattered waves, in order to simplify the complicated dependence on bessel functions.[2] Although most solid scatterers in air can be considered rigid and immovable, it is valid only in a few special cases to assume that a scatterer in a liquid medium is rigid and immovable. In general, the sound waves which penetrate the scatterer must be taken into account, as they can have a considerable effect on the distribution-in-angle of the scattered sound and on the total scattered energy. Morse, with Lowan, Feshbach, and Lax, later extended his solution to include the effects of compressional waves inside (fluid) cylindrical and spherical scatterers.[3] These results are also given in convenient form in terms of several additional phase-angles whose values are tabulated. The object of the research reported here has been to study sound scattering by cylinders and spheres of solid material (which will support shear waves in addition to compressional waves). The mathematical solution will be given first, after which experimental apparatus and results will be described.

II. THE MATHEMATICAL SOLUTION

List of Symbols

Most of the symbols used here are, in the appropriate sections of the analysis, the same as those used by Love and those used by Morse:

a = radius of cylinder or sphere;
a_n, b_n, c_n = expansion coefficients;

* This paper contains the essential results of a thesis submitted to the Faculty of Harvard University in partial fulfillment of the requirements for the degree of Doctor of Philosophy. This research has been aided by funds made available under a contract with the ONR.

[1] Lord Rayleigh, *The Theory of Sound* (Dover Publications, New York, 1945), first American edition.

[2] P. M. Morse, *Vibration and Sound* (McGraw-Hill Book Company, New York, 1936), first edition, and (1948), second edition.

[3] Mathematical Tables Project and M.I.T. Underwater Sound Laboratory, *Scattering and Radiation from Circular Cylinders and Spheres* (U. S. Navy Department, Office of Research and Inventions, Washington, D. C., 1946).

A = vector displacement potential;
A_z = z-component of vector potential;
A_ϕ = ϕ-component of vector potential;
c_1 = velocity of compressional waves in the scatterer;
c_2 = velocity of shear waves in the scatterer;
c_3 = velocity of sound in the fluid surrounding the scatterer;
E = Young's modulus;
j = $(-1)^{\frac{1}{2}}$;
$j_n(\ )$ = spherical bessel function of the first kind;
$J_n(\ )$ = bessel function of the first kind;
k_1 = ω/c_1;
k_2 = ω/c_2;
k_3 = ω/c_3;
n = order integer;
$n_n(\ )$ = spherical bessel function of the second kind;
$N_n(\ )$ = bessel function of the second kind;
p = pressure;
p_i = pressure in incident wave;
p_s = pressure in scattered wave;
$P_n(\cos\theta)$ = Legendre polynomial;
P_0 = amplitude of pressure in incident wave;
r, θ, z = cylindrical coordinates;
r, θ, ϕ = spherical coordinates;
$[rr], [r\theta], [rz]$ = stress components in cylindrical coordinates;
$[rr], [r\theta], [r\phi]$ = stress components in spherical coordinates;
t = time;
u = displacement;
u_r, u_θ = components of displacement in the solid;
$u_{i,r}$ = radial component of displacement in incident wave;
$u_{s,r}$ = radial component of displacement in scattered wave;
x, y, z = rectangular coordinates;
x_1 = k_1a;
x_2 = k_2a;
x_3 = k_3a;
$\alpha_n, \beta_n, \delta_n, \delta_n', \zeta_n, \eta_n$ = scattering phase-angles;
Δ = dilatation;
ϵ_n = Neumann factor; $\epsilon_0=1$; $\epsilon_n=2$, $n>0$;
λ, μ = Lamé elastic constants;
$2\bar{\omega}$ = rotation;
ρ_1 = density of the scatterer;
ρ_3 = density of the fluid surrounding the scatterer;
σ = Poisson's ratio;
Φ_n = boundary impedance scattering phase-angle;
Ψ = scalar displacement potential;
ω = angular frequency $(2\pi f)$.

Scattering by Solid Circular Cylinders

Plane waves of sound of frequency $\omega/2\pi$ in a fluid medium are incident upon an infinitely long circular cylinder of some isotropic solid material. Let the axis of the cylinder coincide with the z-axis of a rectangular coordinate system, and let the plane wave approach the cylinder along the negative x-axis, as shown in Fig. 1. As in the solutions given previously for rigid and fluid scatterers,[1–3] the wave motion external to the scatterer is assumed to consist of the incident plane wave and an outgoing scattered wave. It is desired to find the amplitude of the scattered wave as measured at large distances from the cylinder. The mathematical expressions for displacement and dilatation inside and for pressure and displacement outside the cylinder will be found in general form first, after which the application of the proper boundary conditions at the surface of the cylinder will lead directly to the solution.

The waves inside the cylinder will be represented by suitable solutions of the equation of motion of a solid elastic medium, which may be written[4]

$$(\lambda+2\mu)\nabla\Delta-\mu\nabla\times(2\bar{\omega})=\rho_1\partial^2\mathbf{u}/\partial t^2, \tag{1}$$

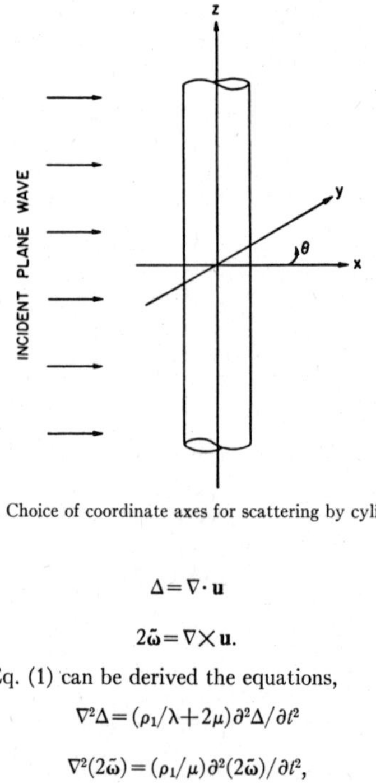

FIG. 1. Choice of coordinate axes for scattering by cylinders.

where

$$\Delta=\nabla\cdot\mathbf{u} \tag{2}$$

and

$$2\bar{\omega}=\nabla\times\mathbf{u}.$$

From Eq. (1) can be derived the equations,

$$\nabla^2\Delta=(\rho_1/\lambda+2\mu)\partial^2\Delta/\partial t^2 \tag{3}$$

and

$$\nabla^2(2\bar{\omega})=(\rho_1/\mu)\partial^2(2\bar{\omega})/\partial t^2, \tag{4}$$

which define the wave velocities

$$c_1=[(\lambda+2\mu)/\rho_1]^{\frac{1}{2}}=[E(1-\sigma)/\rho_1(1+\sigma)(1-2\sigma)]^{\frac{1}{2}} \tag{5}$$

and

$$c_2=(\mu/\rho_1)^{\frac{1}{2}}=[E/2\rho_1(1+\sigma)]^{\frac{1}{2}}. \tag{6}$$

Solutions of Eq. (1) can be found by assuming that the

[4] A. E. H. Love, *A Treatise on the Mathematical Theory of Elasticity* (Dover Publications, New York, 1944), fourth edition, p. 141.

displacement can be derived from a scalar and a vector potential:

$$\mathbf{u} = -\nabla\Psi + \nabla\times\mathbf{A}. \tag{7}$$

The displacement thus can be thought of as the sum of two displacements, one associated with compressional waves and the other with shear waves. If we assume that the potentials satisfy the equations,

$$\nabla^2\Psi = (1/c_1^2)\partial^2\Psi/\partial t^2 \tag{8}$$

and

$$\nabla^2\mathbf{A} = (1/c_2^2)\partial^2\mathbf{A}/\partial t^2, \tag{9}$$

we can show that $\Delta(=\nabla\cdot(-\nabla\Psi))$ satisfies Eq. (3), and that $2\tilde{\omega}(=\nabla\times\nabla\times\mathbf{A})$ satisfies Eq. (4). That these assumptions do lead to a valid solution of Eq. (1) may be seen by noting that the solutions we shall obtain satisfy Eq. (1) by direct substitution. If we now change to a cylindrical coordinate system defined by

$$x = r\cos\theta,\quad y = r\sin\theta,\quad z = z,$$

it can be seen that pressure and displacement must be symmetrical about $\theta=0$ (the direction of the positive x-axis). Moreover, because the cylinder is of infinite length, and the incident plane wave of infinite extent, there can be no dependence on z, and it is logical to assume that there is no displacement in the z-direction. Subject to these conditions, the solution of Eq. (8) can be written

$$\Psi = \sum_{n=0}^{\infty} a_n J_n(k_1 r)\cos n\theta. \tag{10}$$

(The time dependence factor $\exp(j\omega t)$ will be understood in all the expressions representing waves.) Examination of Eq. (9) shows that, subject to the conditions discussed above, the vector potential can have no component in the r- or the θ-direction. The vector Eq. (9) then reduces to a scalar equation in A_z, and its solution can be written

$$A_z = \sum_{n=0}^{\infty} b_n J_n(k_2 r)\sin n\theta. \tag{11}$$

Only sine terms appear here, because the vector potential must be anti-symmetrical about $\theta=0$ in order that the displacement derived from it shall be symmetrical about $\theta=0$. Now, by Eqs. (7) and (2),

$$u_r = \sum_{n=0}^{\infty}\left[\frac{nb_n}{r}J_n(k_2 r) - a_n\frac{d}{dr}J_n(k_1 r)\right]\cos n\theta, \tag{12}$$

$$u_\theta = \sum_{n=0}^{\infty}\left[\frac{na_n}{r}J_n(k_1 r) - b_n\frac{d}{dr}J_n(k_2 r)\right]\sin n\theta, \tag{13}$$

and

$$\Delta = k_1^2\sum_{n=0}^{\infty} a_n J_n(k_1 r)\cos n\theta. \tag{14}$$

The waves in the fluid surrounding the cylinder will be represented by suitable solutions of the wave equation for a (nonviscous) fluid medium, which can be written

$$\nabla^2 p = (1/c_3^2)\partial^2 p/\partial t^2.$$

The incident plane wave is represented by[5]

$$p_i = P_0\exp(-jk_3 x) = P_0\exp(-jk_3 r\cos\theta) = P_0\sum_{n=0}^{\infty}\epsilon_n(-j)^n J_n(k_3 r)\cos n\theta. \tag{15}$$

The radial component of displacement associated with this wave is

$$u_{i,r} = (1/\rho_3\omega^2)\partial p_i/\partial r = \frac{P_0}{\rho_3\omega^2}\sum_{n=0}^{\infty}\epsilon_n(-j)^n\frac{d}{dr}J_n(k_3 r)\cos n\theta. \tag{16}$$

The outgoing scattered wave must be symmetrical about $\theta=0$ and therefore of the form

$$p_s = \sum_{n=0}^{\infty} c_n[J_n(k_3 r) - jN_n(k_3 r)]\cos n\theta. \tag{17}$$

The radial component of displacement associated with this wave is

$$u_{s,r} = \frac{1}{\rho_3\omega^2}\sum_{n=0}^{\infty} c_n\frac{d}{dr}[J_n(k_3 r) - jN_n(k_3 r)]\cos n\theta. \tag{18}$$

The factors c_n are the unknown coefficients which must be evaluated.

The following boundary conditions are applied at the surface of the cylinder: (I) The pressure in the fluid must be equal to the normal component of stress in the solid at the interface; (II) the normal (radial) component of displacement of the fluid must be equal to the normal component of displacement of the solid at the interface; and (III) the tangential components of shearing stress must vanish at the surface of the solid. That is,

$$p_i + p_s = -[rr] \quad\text{at}\quad r=a, \tag{19}$$

$$u_{i,r} + u_{s,r} = u_r \quad\text{at}\quad r=a, \tag{20}$$

and

$$[r\theta] = [rz] = 0 \quad\text{at}\quad r=a. \tag{21}$$

In cylindrical coordinates,[6]

$$[rr] = \lambda\Delta + 2\mu\partial u_r/\partial r = 2\rho_1 c_2^2[(\sigma/1-2\sigma)\Delta + \partial u_r/\partial r],$$
$$[r\theta] = \mu[(1/r)(\partial u_r/\partial\theta) + (r\partial/\partial r)(u_\theta/r)],$$

and

$$[rz] = \mu[\partial u_r/\partial z + \partial u_z/\partial r].$$

By the conditions of symmetry, $[rz]=0$ everywhere. Upon substitution from Eqs. (15), (17), (14), (12), (16), (18), and (13), the boundary condition Eqs. (19), (20),

[5] See reference 2, second edition, p. 347.
[6] See reference 4, p. 288.

and (21) become, for the nth mode,

$$x_1 J_n'(x_1)a_n - nJ_n(x_2)b_n + (x_3/\omega^2\rho_3)[J_n'(x_3) - jN_n'(x_3)]c_n = -(P_0 x_3/\omega^2\rho_3)\epsilon_n(-j)^n J_n'(x_3), \quad (19a)$$

$$2\rho_1 c_2^2 x_1^2[(\sigma/1-2\sigma)J_n(x_1) - J_n''(x_1)]a_n + 2\rho_1 c_2^2 n[x_2 J_n'(x_2) - J_n(x_2)]b_n + a^2[J_n(x_3) - jN_n(x_3)]c_n = -P_0\epsilon_n(-j)^n a^2 J_n(x_3), \quad (20a)$$

and

$$2n[x_1 J_n'(x_1) - J_n(x_1)]a_n = [n^2 J_n(x_2) - x_2 J_n'(x_2) + x_2^2 J_n''(x_2)]b_n. \quad (21a)$$

Solving these equations simultaneously for c_n is laborious but straightforward. The result is

$$c_n = -P_0\epsilon_n(-j)^{n+1}\sin\eta_n \exp(j\eta_n), \quad (22)$$

where η_n, the phase-shift angle of the nth scattered wave, is defined by

$$\tan\eta_n = \tan\delta_n(x_3) \times[\tan\Phi_n + \tan\alpha_n(x_3)]/[\tan\Phi_n + \tan\beta_n(x_3)].$$

The intermediate scattering phase-angles

$$\delta_n(x) = \tan^{-1}[-J_n(x)/N_n(x)],$$
$$\alpha_n(x) = \tan^{-1}[-xJ_n'(x)/J_n(x)],$$

and

$$\beta_n(x) = \tan^{-1}[-xN_n'(x)/N_n(x)],$$

have been defined and their values tabulated previously.[3] The angle Φ_n, which is a measure of the boundary impedance at the surface of the scatterer, is given, for a solid scatterer, by

$$\tan\Phi_n = (-\rho_3/\rho_1)\tan\zeta_n(x_1, \sigma), \quad (23)$$

where the new scattering phase-angle $\zeta_n(x_1, \sigma)$ is given by

$$\zeta_n(x_1, \sigma) = \tan^{-1}\left[-\frac{x_2^2}{2}\frac{\dfrac{x_1 J_n'(x_1)}{x_1 J_n'(x_1) - J_n(x_1)} - \dfrac{2n^2 J_n(x_2)}{n^2 J_n(x_2) - x_2 J_n'(x_2) + x_2^2 J_n''(x_2)}}{\dfrac{(\sigma/1-2\sigma)x_1^2[J_n(x_1) - J_n''(x_1)]}{x_1 J_n'(x_1) - J_n(x_1)} + \dfrac{2n^2[x_2 J_n'(x_2) - J_n(x_2)]}{n^2 J_n(x_2) - x_2 J_n'(x_2) + x_2^2 J_n''(x_2)}}\right]. \quad (24)$$

For convenience in computing values of this function, it can be written in terms of the angle $\alpha_n(x)$:

$$\zeta_n(x_1, \sigma) = \tan^{-1}\left[-\frac{x_2^2}{2}\frac{\dfrac{\tan\alpha_n(x_1)}{\tan\alpha_n(x_1)+1} - \dfrac{n^2}{\tan\alpha_n(x_2) + n^2 - \frac{1}{2}x_2^2}}{\dfrac{\tan\alpha_n(x_1) + n^2 - \frac{1}{2}x_2^2}{\tan\alpha_n(x_1)+1} - \dfrac{n^2[\tan\alpha_n(x_2)+1]}{\tan\alpha_n(x_2) + n^2 - \frac{1}{2}x_2^2}}\right]. \quad (25)$$

Although ζ_n as written above is explicitly a function of x_1 and x_2, it can be considered a function of x_1 and σ, since the ratio of x_1 and x_2 is a function of σ only. Values of $\zeta_n(x_1, \sigma)$ computed from Eq. (25) for $\sigma = \frac{1}{3}$ are given in Table I. For convenience in finding the tangent, the value of the angle lying between $\pm 90°$ is given in the table. The dotted lines indicate that $\zeta_n(x_1, \sigma)$ passes through $\pm 90°$ between the adjacent entries, and thus serve to point out the infinities of $\tan\zeta_n(x_1, \sigma)$. It will be seen below that the infinities of $\tan\zeta_n(x_1, \sigma)$ occur at precisely the frequencies of those normal modes of free vibration of the scatterer which satisfy the conditions of symmetry of the scattering problem. The dotted lines in Table I thus mark the locations of the normal modes of vibration of the scatterer. For other values of Poisson's ratio the functions will be similar, the only difference being shifts in the locations of the normal modes.

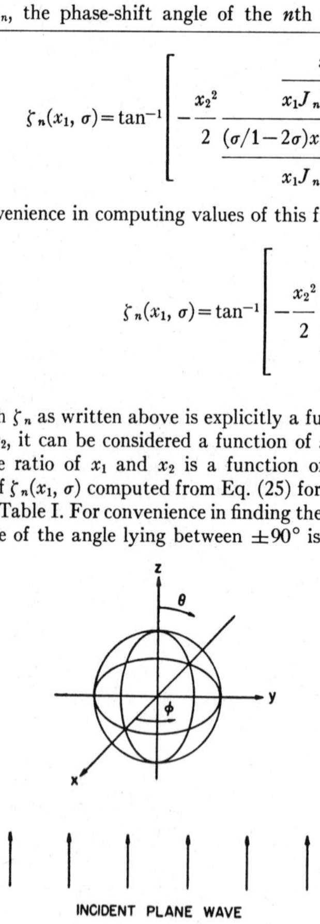

FIG. 2. Choice of coordinate axes for scattering by spheres.

The scattering pattern, or distribution-in-angle of pressure in the scattered wave at large distances from the cylinder, can be found from Eqs. (17) and (22), using the asymptotic expressions for the bessel functions for large arguments:

$$|p_s| \underset{r\to\infty}{\to} P_0\left(\frac{2}{\pi k_3 r}\right)^{\frac{1}{2}} \left|\sum_{n=0}^{\infty}\epsilon_n \sin\eta_n \exp(j\eta_n)\cos n\theta\right|. \quad (26)$$

Scattering by Solid Spheres

Let us assume that plane waves of sound in a fluid medium are incident upon a sphere of some isotropic

TABLE I. Values of $\zeta_n(x_1, \sigma)$ for the cylindrical case for $\sigma=\frac{1}{3}$.

x_1	$n=0$	$n=1$	$n=2$	$n=3$	$n=4$	$n=5$	$n=6$	$n=7$	$n=8$	$n=9$
0.0	0.00°	−45.00°	0.00°	0.00°	0.00°	0.00°	0.00°	0.00°	0.00°	0.00°
0.2	1.37	−44.13	3.59	2.06	1.52	1.20	1.01	0.86	0.76	0.67
0.4	6.27	−43.33	15.47	8.73	6.33	4.97	4.02	3.46	3.03	2.68
0.6	14.35	−41.18	36.24	20.10	14.07	11.15	9.12	7.80	6.80	6.04
0.8	25.60	−37.48	60.28	35.26	25.40	19.90	16.24	14.00	12.12	10.24
1.0	38.90	−31.24	79.03	51.42	37.95	30.52	25.18	21.23	18.52	16.84
1.2	52.22	−18.06	−88.71	65.40	50.79	41.33	34.96	30.31	26.58	23.56
1.4	63.81	+62.49	−79.71	76.03	61.87	52.07	45.11	38.96	34.61	31.22
1.6	73.08	−31.82	−74.29	83.85	70.71	61.11	53.83	47.93	42.73	39.11
1.8	80.33	−7.53	−68.94	89.73	77.52	68.61	61.51	55.74	50.71	46.45
2.0	86.05	+15.39	−63.54	−85.59	82.84	74.55	67.96	62.24	57.38	53.06
2.2	−89.25	36.63	−56.21	−81.57	87.06	79.25	73.05	67.88	63.03	59.43
2.4	−85.16	53.17	−49.71	−77.67	−89.43	83.05	77.21	72.25	67.97	64.10
2.6	−81.30	64.93	−37.81	−73.08	−86.35	86.22	80.71	76.15	72.17	68.51
2.8	−77.32	73.30	−19.43	−64.48	−83.42	88.92	83.58	79.25	75.52	72.15
3.0	−72.71	79.69	+6.92	+68.10	−80.28	−88.67	86.02	81.83	78.34	75.20
3.2	−66.65	85.83	34.29	−74.64	−76.17	−86.39	88.15	84.09	80.71	77.75
3.4	−57.35	74.49	54.21	−65.75	−68.06	−84.05	−89.92	86.04	82.78	79.94
3.6	−40.29	87.38	67.25	−55.95	−11.94	−81.35	−88.10	87.77	84.58	81.86
3.8	−6.92	−89.02	76.96	−40.24	+85.24	−77.55	−86.28	89.35	86.18	83.57
4.0	+34.15	−86.10	−87.62	−11.45	−83.85	−69.80	−84.30	−89.16	87.59	85.03
4.2	58.25	−83.24	+70.33	+28.36	−78.33	−27.54	−81.87	−87.69	88.91	86.34
4.4	70.32	−80.06	80.59	57.79	−73.27	+75.02	−78.22	−86.15	−89.85	87.54
4.6	77.19	−76.03	84.69	78.81	−66.94	88.55	−70.46	−84.38	−88.63	88.64
4.8	81.68	−70.05	87.52	14.69	−56.75	−86.36	−29.41	−82.10	−87.38	89.69
5.0	84.94	−60.60	89.87	68.66	−34.99	−82.97	+70.97	−78.49	−86.02	−89.28

solid material. Let the center of the sphere coincide with the origin of a rectangular coordinate system, and let the plane waves approach the sphere along the negative z axis, as shown in Fig. 2. The analysis is very similar to that for the cylindrical case. We transfer to spherical coordinates defined by

$$x=r\sin\theta\cos\phi,\quad y=r\sin\theta\sin\phi,\quad z=r\cos\theta.$$

Because the incident wave approaches along the axis of ϕ, there is no dependence on ϕ. It is logical to assume that there is no component of displacement in the ϕ-direction, and it follows that the only non-zero component of the vector potential in this case is A_ϕ. The potentials are then found to be of the forms,

$$\Psi=\sum_{n=0}^{\infty} a_n j_n(k_1 r)P_n(\cos\theta)$$

and

$$A_\phi=\sum_{n=0}^{\infty} b_n j_n(k_2 r)\frac{d}{d\theta}P_n(\cos\theta).$$

Pressure in the incident wave is represented by[7]

$$p_i=P_0\exp(-jk_3 z)=P_0\exp(-jk_3 r\cos\theta)$$
$$=P_0\sum_{n=0}^{\infty}(2n+1)(-j)^n j_n(k_3 r)P_n(\cos\theta).$$

The outgoing scattered wave will be of the form,

$$p_s=\sum_{n=0}^{\infty} c_n[j_n(k_3 r)-jn_n(k_3 r)]P_n(\cos\theta). \tag{27}$$

The same boundary conditions at the surface of the scatterer are applied to the expressions for displacement, pressure, and dilatation, which are either given above or derivable from the above. In spherical coordinates the stress components are

$$[rr]=\lambda\Delta+2\mu\partial u_r/\partial r=2\rho_1 c_2^2[(\sigma/1-2\sigma)\Delta+\partial u_r/\partial r],$$

$$[r\theta]=\mu\left[\frac{\partial u_\theta}{\partial r}-\frac{u_\theta}{r}+\frac{1}{r}\frac{\partial u_r}{\partial\theta}\right],$$

[7] See reference 2, second edition, p. 354.

and

$$[r\phi]=\mu\left[\frac{1}{r\sin\theta}\frac{\partial u_r}{\partial\theta}+\frac{\partial u_\phi}{\partial r}-\frac{u_\phi}{r}\right].$$

By carrying the analysis through as in the cylindrical case, we find that

$$c_n=-P_0(2n+1)(-j)^{n+1}\sin\eta_n\exp(j\eta_n), \quad (28)$$

where the phase-shift η_n of the nth scattered wave is defined by

$$\tan\eta_n=\tan\delta_n(x_3)[\tan\Phi_n+\tan\alpha_n(x_3)]/\tan\Phi_n+\tan\beta_n(x_3).$$

The intermediate angles,

$$\delta_n(x)=\tan^{-1}[-j_n(x)/n_n(x)],$$
$$\alpha_n(x)=\tan^{-1}[-xj_n'(x)/j_n(x)],$$
$$\beta_n(x)=\tan^{-1}[-xn_n'(x)/n_n(x)],$$

have been defined and their values tabulated previously.[8] The boundary impedance phase-angle Φ_n is defined by

$$\tan\Phi_n=-(\rho_3/\rho_1)\tan\zeta_n(x_1,\sigma), \quad (29)$$

where the new scattering phase angle $\zeta_n(x_1,\sigma)$ is given by

$$\zeta_n(x_1,\sigma)=\tan^{-1}\left[-\frac{x_2^2}{2}\frac{\dfrac{x_1j_n'(x_1)}{x_1j_n'(x_1)-j_n(x_1)}-\dfrac{2(n^2+n)j_n(x_2)}{(n^2+n-2)j_n(x_2)+x_2^2j_n''(x_2)}}{\dfrac{(\sigma/1-2\sigma)x_1^2[j_n(x_1)-j_n''(x_1)]}{x_1j_n'(x_1)-j_n(x_1)}-\dfrac{2(n^2+n)[j_n(x_2)-x_2j_n'(x_2)]}{(n^2+n-2)j_n(x_2)+x_2^2j_n''(x_2)}}\right].$$

This function can be expressed in terms of the angle $\alpha_n(x)$:

$$\zeta_n(x_1,\sigma)=\tan^{-1}\left[-\frac{x_2^2}{2}\frac{\dfrac{\tan\alpha_n(x_1)}{\tan\alpha_n(x_1)+1}-\dfrac{n^2+n}{n^2+n-1-\frac{1}{2}x_2^2+\tan\alpha_n(x_2)}}{\dfrac{n^2+n-\frac{1}{2}x_2^2+2\tan\alpha_n(x_1)}{\tan\alpha_n(x_1)+1}-\dfrac{(n^2+n)[\tan\alpha_n(x_2)+1]}{n^2+n-1-\frac{1}{2}x_2^2+\tan\alpha_n(x_2)}}\right]. \quad (30)$$

Values of this function computed from Eq. (30) for $\sigma=\frac{1}{3}$ are given in Table II. The dotted lines again indicate the infinities of $\tan\zeta_n(x_1,\sigma)$, that is, the normal modes of free vibration of the scatterer.

The distribution in angle of pressure in the scattered wave at large distances from the sphere is found from Eqs. (27) and (28) by means of the asymptotic expressions for the spherical bessel functions for large arguments:

$$|p_s|\underset{r\to\infty}{\longrightarrow}\frac{P_0}{k_3r}\left|\sum_{n=0}^{\infty}(2n+1)\sin\eta_n\exp(j\eta_n)P_n(\cos\theta)\right|. \quad (31)$$

III. EXPERIMENTAL APPARATUS

Measurements of the distribution-in-angle of sound scattered in water by metal cylinders were made for the purpose of checking the theory. These measurements were made in a large steel tank at or near a frequency of one megacycle per second. A sound projector in one end of the tank irradiated the scatterer with sound. A receiving hydrophone was mounted in such a way that it could easily be moved to any position lying on a circle concentric with the scatterer, and served to measure the distribution in angle of the pressure in the scattered wave. Short wave trains or "pulses" of sound were used in order that the measurement of each pulse could be effectively completed before sound reflected from the walls of the tank could reach the receiving hydrophone. A novel feature, frequency modulation of the pulse repetition rate, served to identify interfering pulses which, still reverberating in the tank from the previous transmitted pulse, happened to arrive at the receiver at the same time as the pulse to be measured. A small adjustment of the average pulse repetition rate was effective in controlling interference of this type. Both transducers employed x-cut quartz crystals operated at resonance. Serious distortion of the short (64 μsec) pulses by the transducers was prevented by lowering the Q of the quartz crystals by increasing the radiation loading. This was accomplished by inserting between the crystals and the water an acoustic quarter-wave transformer in the form of a thin disk of Plexiglas. The amplitude of the scattered sound pulses was measured by a modified substitution method, an oscilloscope being used as an indicator. The pulses were brought to a standard deflection on the oscilloscope, changes in the pulse amplitude being compensated by changes in the attenuation in the receiving system.

IV. COMPARISON OF THEORY AND EXPERIMENT

The experimental data were normalized so that they could be compared with scattering patterns computed from the theory. In order to do this, the amplitude of the pressure in the incident wave (P_0) was measured by moving the receiving transducer to the position of the

[8] See reference 3. Care must be taken to distinguish between the cylindrical and spherical cases, since the same symbols are used for the scattering phase-angles in both cases.

TABLE II. Values of $\zeta_n(x_1, \sigma)$ for the spherical case for $\sigma = \frac{1}{3}$.

x_1	$n=0$	$n=1$	$n=2$	$n=3$	$n=4$	$n=5$	$n=6$	$n=7$	$n=8$	$n=9$
0.0	0.00°	−45.00°	0.00°	0.00°	0.00°	0.00°	0.00°	0.00°	0.00°	0.00°
0.2	1.16	−44.72	3.05	1.94	1.41	1.13	0.96	0.81	0.73	0.64
0.4	4.66	−43.57	12.77	7.77	5.91	4.66	3.91	3.33	2.93	2.62
0.6	10.58	−41.62	29.97	17.88	13.38	10.38	8.73	7.50	6.58	5.88
0.8	18.89	−38.54	51.40	31.55	23.47	18.83	15.93	13.58	12.19	11.01
1.0	29.12	−33.76	70.41	46.79	35.44	28.67	24.25	20.74	18.32	16.09
1.2	40.28	−26.18	83.89	60.52	47.88	39.29	33.22	29.67	25.84	23.32
1.4	51.05	−12.80	−86.94	71.44	58.67	50.00	43.23	38.34	34.49	31.51
1.6	60.55	+16.87	−80.31	79.64	67.68	59.02	52.20	46.57	42.41	38.72
1.8	68.41	89.15	−75.00	85.78	74.68	66.42	59.69	54.53	49.80	45.77
2.0	74.78	−25.58	−70.20	−89.46	80.14	72.50	66.17	60.82	56.28	52.29
2.2	79.92	+9.84	−65.27	−85.58	84.46	77.29	71.50	66.61	62.27	58.30
2.4	84.14	30.55	−59.50	−82.18	87.97	81.17	75.75	71.13	67.13	63.16
2.6	87.71	48.97	−52.57	−78.94	−89.07	84.39	79.27	74.89	71.11	67.61
2.8	−89.15	61.45	−41.61	−75.41	−86.44	87.06	82.18	78.10	74.62	71.07
3.0	−86.24	70.09	−25.03	−70.56	−83.95	89.38	84.63	80.80	77.41	74.43
3.2	−83.36	76.48	−0.51	−55.48	−81.37	−88.53	86.75	83.03	79.87	77.00
3.4	−80.31	81.81	+27.37	−81.31	−78.27	−86.56	88.61	84.98	81.93	79.23
3.6	−76.72	88.64	49.05	−68.79	−73.44	−84.57	−89.70	86.69	83.75	81.23
3.8	−72.00	75.11	63.24	−59.85	−59.37	−82.35	−88.10	88.21	85.33	82.88
4.0	−64.84	85.64	72.99	−47.06	+65.07	−79.48	−86.50	89.61	86.71	84.30
4.2	−51.66	89.00	81.77	−24.13	−87.41	−74.70	−84.81	−89.07	87.99	85.62
4.4	−22.39	−88.45	−65.12	+13.38	−80.54	−60.64	−82.81	−87.76	89.17	86.81
4.6	+26.20	−86.12	+73.10	47.64	−75.60	+48.49	−80.07	−86.41	−89.71	87.88
4.8	56.58	−83.70	80.80	68.81	−70.16	84.65	−75.21	−84.90	−88.62	88.89
5.0	70.03	−80.89	84.41	−88.10	−62.36	−88.16	−60.25	−83.03	−87.50	89.85

scatterer. After normalization, it was still necessary to add a factor amounting to 1.9 db to the amplitude of the scattered sound in order to bring the experimental data into good agreement with the theory. This correction factor has been explained, and its value computed with good accuracy, by taking into account the fact that the illumination of the scatterer varies in phase and amplitude along its length.[9]

The part of Eq. (26) which was evaluated in computing the patterns was

$$\tfrac{1}{2}\left|\sum_{n=0}^{\infty} \epsilon_n \sin\eta_n \exp(j\eta_n) \cos n\theta\right|,$$

and the corresponding numerical scale is shown on all the patterns used as illustrations. The values of Poisson's ratio for the various scatterers were assumed, because of the difficulty of measuring this constant directly; but the values of Young's modulus were measured (to within ±5 percent) by finding the frequency of the first mode of flexual vibration of the cylindrical specimen mounted so that it could vibrate as a fixed-free bar. The value of x_1 was then determined by means of Eq. (5). In some cases where the pattern was very sensitive to frequency, it was necessary to choose a value of x_1 slightly different from that based on the Young's modulus measurement in order to bring the measured and computed patterns into agreement. Comparison of the value of Young's modulus corresponding to the assumed value of x_1 with the measured value serves in these cases to indicate the degree of agreement between experiment and theory.

Figures 3 through 13 are measured and computed scattering patterns for cylinders of various sizes. The pressure in the scattered wave is plotted linearly against scattering angle. In each case the arrow indicates the direction of the incident sound. The angle θ is measured from the top center of the graph, the incident sound coming from the direction $\theta = 180°$. For each size of

[9] J. J. Faran, Jr., *Sound Scattering by Solid Cylinders and Spheres*, Technical Memorandum No. 22 (March 15, 1951), Acoustics Research Laboratory, Harvard University, Cambridge, Massachusetts.

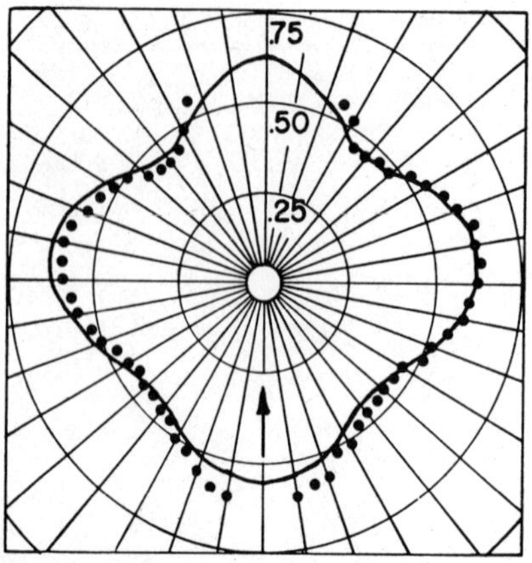

FIG. 3. Scattering pattern for brass cylinder 0.0322 in. in diameter at 1.00 mc/sec. *Points:* Measured amplitude of pressure in the scattered wave. The measured Young's modulus was 10.1 $\times 10^{11}$ dynes/cm². *Curve:* Computed pattern for $x_3=1.7$, $x_1=0.6$, $\sigma=\frac{1}{3}$, $\rho_1=8.5$ g/cm³.

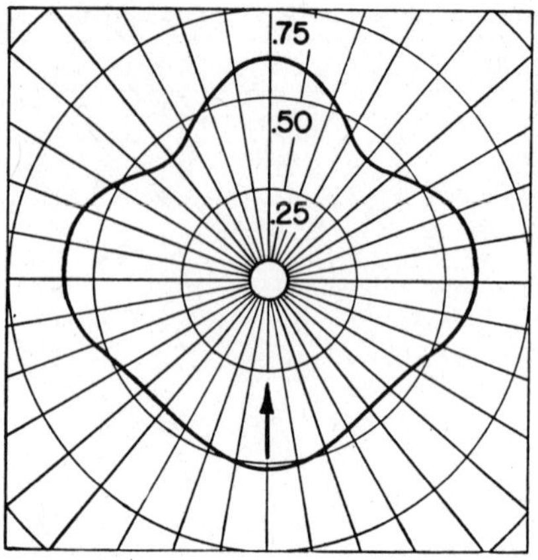

FIG. 5. Computed amplitude of pressure in wave scattered by a rigid, immovable cylinder for $x_3=1.7$.

scatterer, the pattern computed on the basis that the scatterer is rigid and immovable is included for comparison.

Figures 3 and 4 show scattering patterns for brass and steel (drill rod) cylinders of the same size, for each of which $x_3=1.7$. These patterns are both very similar to that for a rigid, immovable scatterer of the same size (Fig. 5).

Figures 6–8 show scattering patterns for cylinders of various materials twice as large in diameter, that is, $x_3=3.4$. The pattern for a brass cylinder of this size (Fig. 6) is somewhat unusual; the amplitude of sound scattered back in the direction of the source is nearly zero. This near-null in the back-scattered sound is fully explained by the mathematical solution in which the

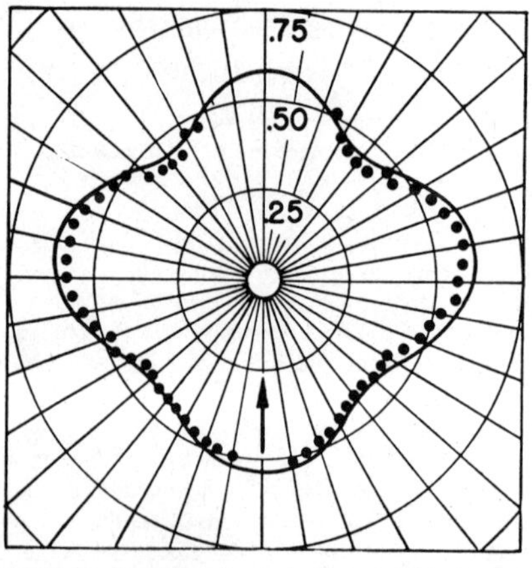

FIG. 4. Scattering pattern for steel cylinder 0.032 in. in diameter at 1.00 mc/sec. *Points:* Measured amplitude of pressure in the scattered wave. The measured Young's modulus was 20.0×10^{11} dynes/cm². *Curve:* Computed pattern for $x_3=1.7$, $x_1=0.45$, $\sigma=0.28$, $\rho_1=7.7$ g/cm³.

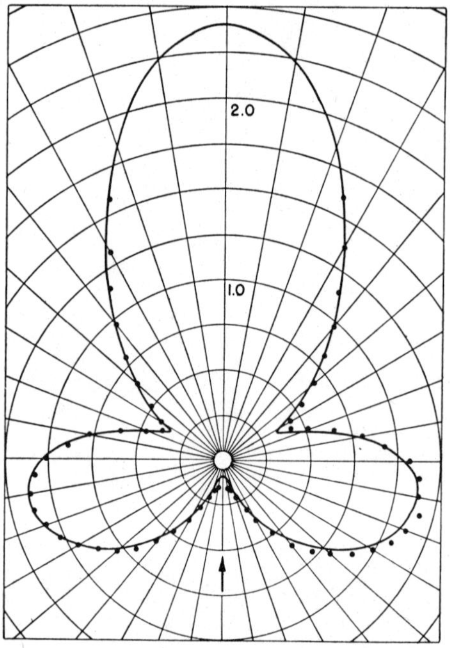

FIG. 6. Scattering pattern for brass cylinder 0.0625 in. in diameter at 1.02 mc/sec. *Points:* Measured amplitude of pressure in the scattered wave. The measured Young's modulus was 10.4×10^{11} dynes/cm². *Curve:* Computed pattern for $x_3=3.4$, $x_1=1.185$, $\sigma=\frac{1}{3}$, $\rho_1=8.5$ g/cm³ (corresponding to $E=10.5\times 10^{11}$ dynes/cm²).

term for $n=2$ in the series for the scattering pattern suddenly becomes very large in amplitude and of the proper phase to cancel the sum of all the other terms at $\theta=180°$. This, in turn, is brought about by the presence of an infinity in the $\tan\zeta_n(x_1, \sigma)$ function for $\sigma=\frac{1}{3}$ at $x_1=1.18\cdots$ (corresponding to a normal mode or resonance of the scatterer), in the neighborhood of which this function goes rapidly through a wide range of values causing the variations in the coefficient of the $n=2$ term. The value of x_1 for the computed pattern of Fig. 6 was chosen to give a deep notch at $\theta=180°$, and the frequency at which the experimental pattern was

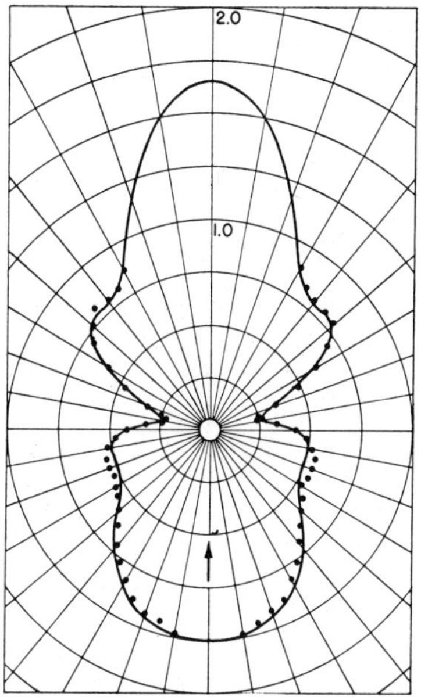

FIG. 7. Scattering pattern for copper cylinder 0.0625 in. in diameter at 1.00 mc/sec. *Points:* Measured amplitude of pressure in the scattered wave. The measured Young's modulus was 11.9×10^{11} dynes/cm². *Curve:* Computed pattern for $x_3=3.4$, $x_1=1.08$, $\sigma=\frac{1}{3}$, $\rho_1=8.9$ g/cm³ (corresponding to $E=12.7\times10^{11}$ dynes/cm²).

measured was chosen the same way. Figure 7 is the scattering pattern for a copper cylinder of the same size. The value of x_1 for the copper cylinder is near enough to 1.18 that the coefficient of the $n=2$ term is still large, but in this case it is of the opposite phase and causes the sound scattered in the direction $\theta=180°$ to be somewhat larger in amplitude than that scattered by a rigid, immovable cylinder of this size (Fig. 9). The velocity of sound in steel is so much higher than that in brass or copper that this scatterer behaves nearly as though it were rigid and immovable, and its scattering pattern (Fig. 8) is little different from that for the rigid, immovable case.

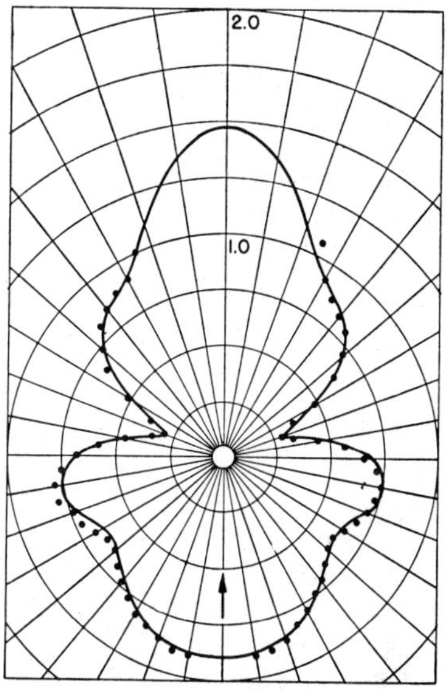

FIG. 8. Scattering pattern for steel cylinder 0.0625 in. in diameter at 1.00 mc/sec. *Points:* Measured amplitude of pressure in the scattered wave. The measured Young's modulus was 19.5×10^{11} dynes/cm². *Curve:* Computed pattern for $x_3=3.4$, $x_1=0.9$, $\sigma=0.28$, $\rho_1=7.7$ g/cm³.

Figures 10, 11, and 12 are scattering patterns for brass, steel, and aluminum cylinders for which $x_3=5.0$

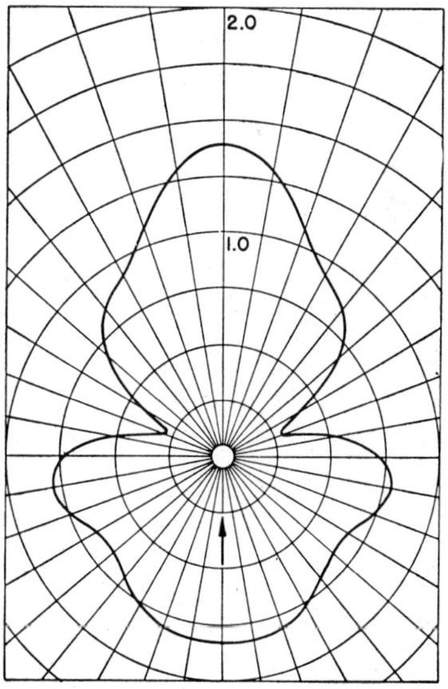

FIG. 9. Computed amplitude of pressure in wave scattered by a rigid, immovable cylinder for $x_3=3.4$.

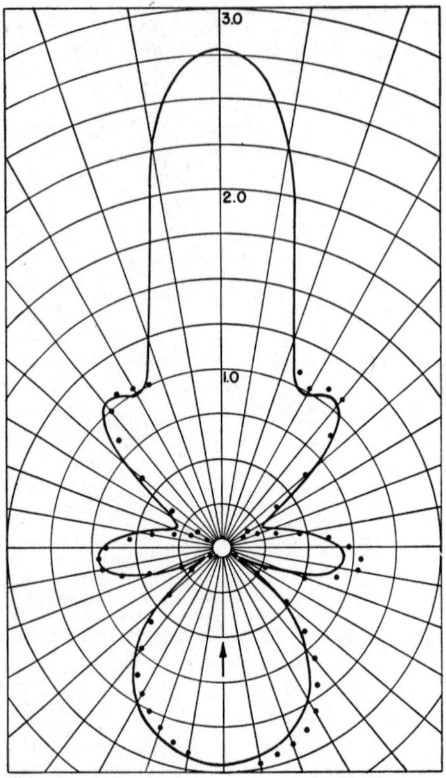

Fig. 10. Scattering pattern for brass cylinder 0.093 in. in diameter at 1.015 mc/sec. *Points:* Measured amplitude of pressure in the scattered wave. The measured Young's modulus was 10.0×10^{11} dynes/cm². *Curve:* Computed pattern for $x_3=5.0$, $x_1=1.78$, $\sigma=\frac{1}{3}$, $\rho_1=8.5$ g/cm³ (corresponding to $E=10.2\times10^{11}$ dynes/cm²).

The frequency of measurement of the pattern of the brass scatterer was chosen to give the deepest notch at 120°, and the value of x_1 was chosen to make the patterns agree. The choice of the value of x_1 is well substantiated by the measurement of the Young's modulus of this scatterer, since the value of E corresponding to the chosen value of x_1 is within 2 percent of the measured value. Figure 11 shows that, just as in the case of brass (Fig. 4), there is a near-null in the sound back-scattered from a steel cylinder at a frequency near that of the lowest-frequency normal mode which, for $\sigma=0.28$, occurs at $x_1=1.30\cdots$. Figure 12 shows that the same is true of an aluminum scatterer of the same size. Although the velocity of compressional waves in steel is not the same as that in aluminum, the values of Poisson's ratio differ sufficiently that this normal mode occurs in these two materials for the same physical size of the scatterers. These two patterns are so similar that they are seen to depend much more critically upon the value of x_1 than upon the density of the scatterer. The pattern for a rigid, immovable cylinder of the same size is shown in Fig. 13, and it is apparent that all these patterns for metal cylinders of this size bear little resemblance to this limiting case.

The theory thus verifies the existence of nulls in the back-scattered sound for cylinders of various metals, and at the proper frequencies; but a further test is to see whether it predicts properly the manner in which the amplitude of the back-scattered sound (and the shape of the entire pattern) changes with frequency. In order to test this, patterns were measured for the brass cylinder of Fig. 6 and the steel cylinder of Fig. 11 at two other frequencies, 3 percent below and above that at which the reference patterns were measured. The corresponding patterns predicted by the theory were computed by making a corresponding change in the values of the x parameters. In Fig. 14, the pattern of Fig. 6 is reproduced in the center, and those for 3 percent changes in frequency are shown at either side. In Fig. 15, the pattern of Fig. 11 is reproduced in the center, and the patterns for 3 percent changes in frequency are shown on either side. The theory is seen to predict the changes in the measured patterns with gratifying precision. These groups of patterns also emphasize the fact that the null

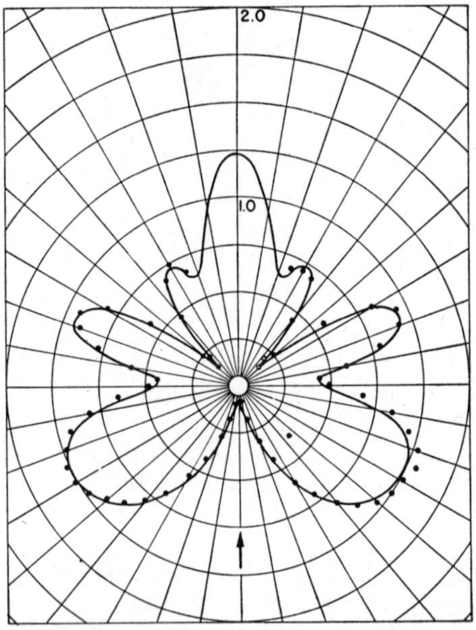

Fig. 11. Scattering pattern for steel cylinder 0.09375 in. in diameter at 0.99 mc/sec. *Points:* Measured amplitude of pressure in the scattered wave. The measured Young's modulus was 19.3×10^{11} dynes/cm². *Curve:* Computed pattern for $x_3=5.0$, $x_1=1.293$, $\sigma=0.28$, $\rho_1=7.7$ g/cm³ (corresponding to $E=19.7\times10^{11}$ dynes/cm²).

in the back-scattered sound is very sensitive to frequency.

Measurements of scattering by a few spheres were made with this apparatus. However, because the sound scattered by a sphere diverges in three dimensions (instead of two, as in the case of a long cylinder), the measurement was found to be very difficult, because of the reduced margin of signal to noise. The measurements (and also computations) indicate that, although rapid changes in the pattern do occur, there is no null in the sound back-scattered in water by a brass sphere, near its lowest-frequency normal mode of vibration.

V. REMARKS ON THE BEHAVIOR OF SOLID SCATTERERS

It is interesting to examine the behavior of certain of the functions which appear in the mathematical solution, especially the $\tan\zeta_n(x_1, \sigma)$ functions. As noted above, it can be shown that the infinities of the $\tan\zeta_n(x_1, \sigma)$ functions occur at precisely the frequencies of those normal modes of free vibration of the scattering body which satisfy the conditions of symmetry of the scattering problem. This can be done by applying boundary conditions to expressions for displacement and dilatation written in general form in terms of an unknown frequency. The boundary conditions, for free vibrations, are simply that the normal component of stress and the tangential components of shearing stress

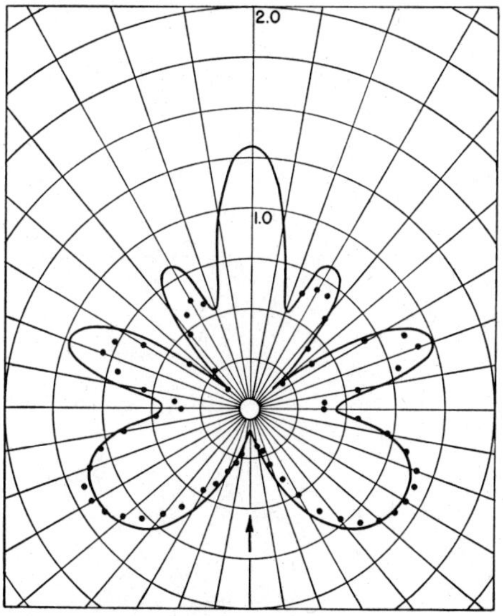

FIG. 12. Scattering pattern for aluminum cylinder 0.0925 in. in diameter at 1.00 mc/sec. *Points:* Measured amplitude of pressure in the scattered wave. The measured Young's modulus was 7.0×10^{11} dynes/cm^2. *Curve:* Computed pattern for $x_3=5.0$, $x_1=1.17$, $\sigma=\frac{1}{3}$, $\rho_1=2.7$ g/cm^3 (corresponding to $E=7.2\times10^{11}$ dynes/cm^2).

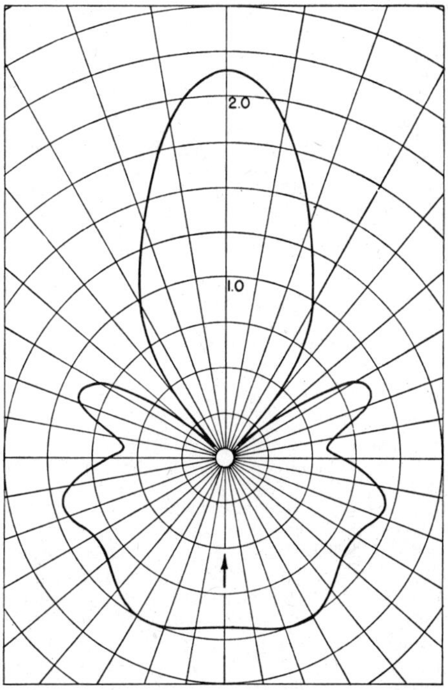

FIG. 13. Computed amplitude of pressure in wave scattered by a rigid, immovable cylinder for $x_3=5.0$.

at the surface of the body must both vanish. Solving the resultant equation for frequency (in terms of unknown x_1 and x_2 parameters) gives a condition which, in the cylindrical case, is identical to requiring the denominator of Eq. (24) to vanish.[10] For $\sigma=\frac{1}{3}$, the first few of these normal modes occur at the following values of the frequency parameter:

for $n=0$, $x_1=2.17\cdots$, $5.43\cdots$, $8.60\cdots$;
for $n=1$, $x_1=1.43\cdots$, $3.27\cdots$, $3.74\cdots$;
for $n=2$, $x_1=1.18\cdots$, $2.25\cdots$, $3.98\cdots$;
for $n=3$, $x_1=1.81\cdots$, $3.01\cdots$, $4.65\cdots$;
for $n=4$, $x_1=2.36\cdots$, etc.

The first normal modes for $n=1$, 2, and 3 occur for lower values of x_1 (lower frequencies) than that for $n=0$, contrary to what we might expect. The reason for this is that there are no shear waves associated with the $n=0$ normal modes. The complicated wave structure which comprises a normal mode can be realized at a much lower frequency with shear waves than without, because the velocity of shear waves is so much lower than that of compressional waves.

For fluid scatterers, the functions $\tan\zeta_n(x_1, \sigma)$ in Eqs. (23) and (29) are replaced[3] by the functions

[10] For details of this demonstration, see reference 9.

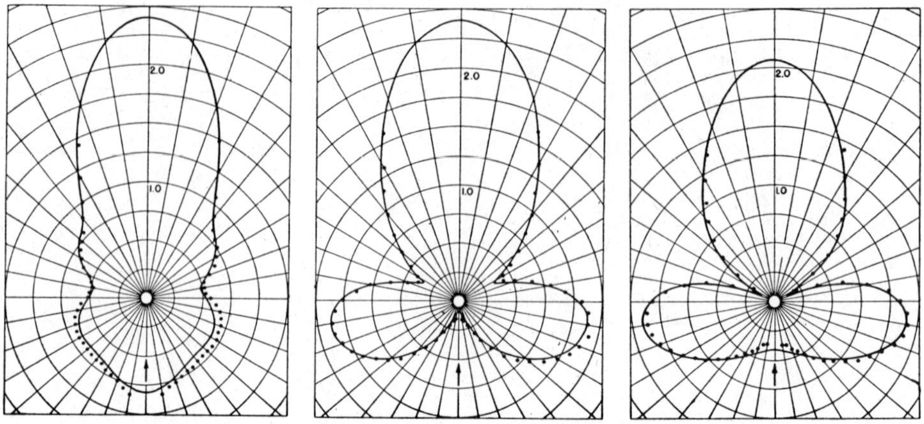

FIG. 14. The scattering pattern of Fig. 6 repeated for comparison with measured and computed patterns for frequencies 3 percent higher (right) and 3 percent lower (left).

$\tan\alpha_n(x_1)$. It is interesting to note that the infinities of these functions also correspond to frequencies of normal modes of free vibration of the (fluid) scatterer, since the infinities of $\tan\alpha_n(x_1)$ occur at the zeros of $J_n(x_1)$ or $j_n(x_1)$, in the cylindrical and spherical cases, respectively.

The coefficient c_n in the series for the scattering pattern does not attain its maximum value at exactly the frequencies of the normal modes of free vibration of the scatterer. Since the amplitude of c_n is proportional to $\sin\eta_n$, c_n reaches its maximum value when $\tan\eta_n$ becomes infinite. This represents a shift in the resonant frequency of the normal mode, and this shift is attributed to the reactive component of the acoustic impedance presented to the scatterer by the surrounding fluid, i.e., the reactive component of the radiation loading. In the case of solids having densities greater than that of the surrounding fluid, however, this frequency shift is usually small.

While measurements were being made with the experimental apparatus at frequencies near that of a normal mode, it was in some cases possible to observe "ringing" of that normal mode following the end of the pulse; that is, long transients could be observed at the end (and at the beginning) of the scattered pulse. By adjusting the frequency to give the maximum amplitude of the transient at the end of the pulse, it was thus possible to measure the frequencies of various normal modes. It was also possible to identify the order n of the excited mode, because the amplitude of the transient following the pulse was proportional to $\cos n\theta$. These transients were not noticeable in the case of the first normal mode for $n=2$. Apparently the damping by

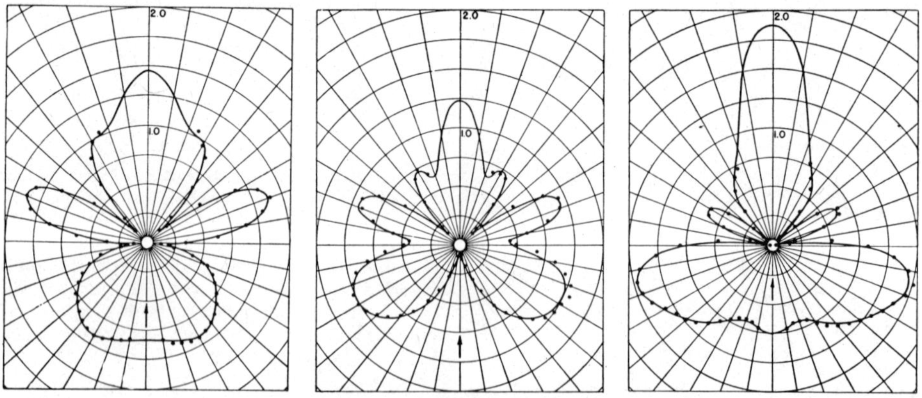

FIG. 15. The scattering pattern of Fig. 11 repeated for comparison with measured and computed patterns for frequencies 3 percent higher (right) and 3 percent lower (left).

radiation into the water was great enough to cause any ringing to die out quickly. However, the first normal modes for $n=0$, 1, and 3 were observed and identified for brass and steel cylindrical scatterers of appropriate sizes and showed good agreement with the frequencies predicted by the theory.

That there are sizeable shifts in the frequencies of the normal modes with changes in Poisson's ratio suggests that finding the frequencies of one or more of these normal modes of vibration might provide a method of measuring Poisson's ratio for cylindrical or spherical specimens. The variation of the frequencies of these normal modes with Poisson's ratio is illustrated in Figs. 16 and 17, where the values of x_1 at which the first normal modes for $n=0$, 1, 2, 3, and 4 occur are plotted as functions of Poisson's ratio. The variation of the second normal mode for $n=2$ is also shown in the graph for the spherical case. In this connection, as well as in the scattering problem itself, the potential utility of having the $\zeta_n(x_1, \sigma)$ functions computed for a wide range of values of Poisson's ratio will be evident. A computation program to yield these results appears to be justified. The frequencies of the normal modes cannot be computed explicitly, but can be found easily from the locations of the infinities of the $\tan\zeta_n(x_1, \sigma)$ functions.

It is interesting to compare the behavior of the $\tan\Phi_n$ functions for solid and fluid scatterers as x_1, the frequency parameter for the scatterer, approaches zero. For solid scatterers, either cylindrical or spherical, as $x_1 \to 0$,

$$\tan\Phi_n \to 0, \quad n \neq 1; \quad \tan\Phi_1 \to \rho_3/\rho_1;$$

while for fluid scatterers, where

$$\tan\Phi_n = (-\rho_3/\rho_1) \tan\alpha_n(x_1),$$

as $x_1 \to 0$,

$$\tan\Phi_n \to (\rho_3/\rho_1)n.$$

In neither case, by letting $x_1 \to 0$, do we realize the case of the rigid, immovable scatterer where $\tan\Phi_n = 0$ for all n. In order that $x_1 = \omega a/c_1 \to 0$ at finite frequencies in the solid case, the velocities of both the compressional and shear waves must become infinite, and the scatterer does indeed become rigid. The only term where $\tan\Phi_n$ does not vanish is that for $n=1$. This deviation from the rigid, immovable case is simply due to oscillation of the scatterer as a whole in synchronism with the incident sound field. Thus, by setting $x_1 = 0$ in the solution given here for solid scatterers, we can calculate the scattering from a rigid, *movable* cylinder or sphere of density ρ_1. To pass to the case of the rigid, immovable scatterer, we must also require that the density of the scatterer become infinite. In the case of a fluid scatterer, as $x_1 \to 0$, only $\tan\Phi_0$ approaches the value for the limiting case of a rigid, immovable scatterer. For $n=1$, $\tan\Phi_n$ behaves in the same way as in the case of the solid scatterer, and represents oscillation of the scatterer in synchronism with the incident sound. Now, for fluid scatterers, in order that $x_1 \to 0$ at finite frequencies, it is necessary that the fluid become incompressible; but as this happens, the scatterer does not necessarily become rigid to shear distortions. It must then be that, for $n=2$ and higher, shape distortions of the incompressible fluid scatterer make the components of the scattered wave different from what they would be if the scatterer were rigid. Because the fluid scatterer never becomes rigid as $x_1 \to 0$, one can only pass from this solution to the case of the rigid, immovable scatterer by letting the density become infinite.

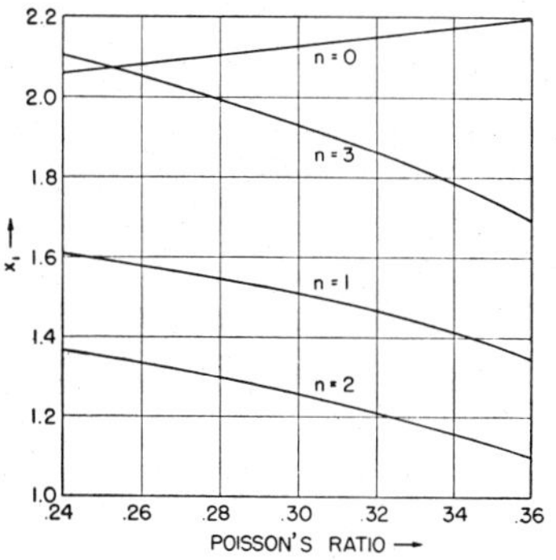

FIG. 16. The values of x_1 for the first few symmetrical normal modes of free vibration of a solid cylinder plotted as functions of Poisson's ratio.

Two summary comments can be added regarding the general features of scattering by solid cylinders and spheres. If the frequency of the incident sound is lower than that of the first symmetrical normal mode of free vibration of the solid scatterer, and if the density of the scatterer is greater than that of the liquid, there is little difference between the scattering pattern for the solid scatterer and that for a rigid, immovable scatterer. But, rapid changes in the shape of the scattering pattern and in the total scattered power (or scattering cross section)

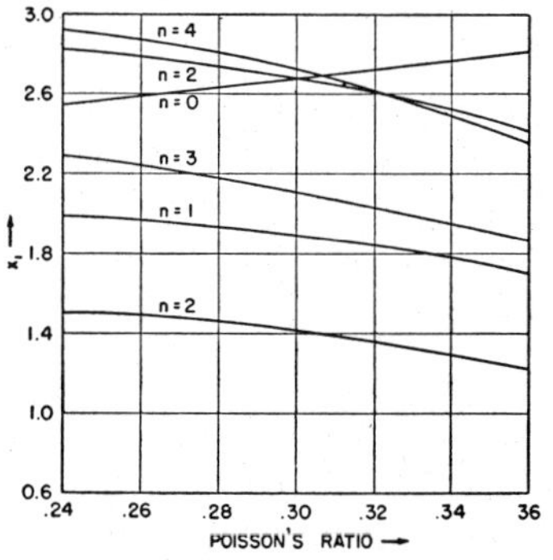

FIG. 17. The values of x_1 for the first few symmetrical normal modes of free vibration of a solid sphere plotted as functions of Poisson's ratio.

can occur with small changes in frequency in the vicinity of certain of the normal modes of free vibration of the solid scatterer. These changes include the appearance of deep minima in the scattering pattern at certain angles and may include, for cylinders, a near-null in the sound scattered back toward the source.

VI. ACKNOWLEDGMENTS

The author is indebted to Professor F. V. Hunt for his guidance and encouragement throughout this investigation. The assistance of Dorothea Greene, who performed the laborious computations for Figs. 16 and 17, is gratefully acknowledged.

In the first of the following papers, the author derives theoretically the mechanism of echo formation from a rigid body immersed in an ideal fluid medium using a method of analysis analogous to that of physical optics.

The second paper applies the results of the first paper to calculate the echo structure for several rigid bodies of simple shape under CW and pulse conditions.

Dr. Freedman is a research physicist. He received the Ph.D. degree from London University in 1962. He joined the British Navy Department in 1939. He has contributed in the fields of transducer and dome calibration, shock protection of sonar equipment, underwater echoes, acoustic echo formation mechanisms, acoustic radiator fields and radiation mechanisms, and acoustic array interactions. In 1964–1965, he spent a year as an exchange scientist at the U.S. Navy Electronics Laboratory at San Diego, California. He is a Fellow of the Institute of Physics, and is on the Editorial Board of the *Journal of Sound and Vibration*. He is a member of the Scientific Staff of the Admiralty Underwater Weapons Establishment, Portland, Dorset, England.

These papers are reprinted with the permission of the author and the Editor in Chief of *Acustica*.

Acustica, 1962, V. 12, p. 10–21.

21

A Mechanism of Acoustic Echo Formation

A. FREEDMAN

A MECHANISM OF ACOUSTIC ECHO FORMATION

by A. Freedman

Admiralty Underwater Weapons Establishment, Portland, England

Summary

Using a method of analysis analogous to that of physical optics, the direct backscattering of small amplitude acoustic waves from a rigid body, immersed in an ideal fluid medium is re-examined. The incident radiation is a pulse of general type, and at long ranges, no restrictions are imposed on the directivity patterns of the transmitting and receiving transducers. Clarification of the echo-formation mechanism applicable at small wavelengths is obtained.

The echo is shown to be composed of a number of discrete pulses, each a replica of the transmission pulse, and hence termed an 'image pulse'. An image pulse is generated whenever there is a discontinuity with respect to range, r, in $\mathrm{d}^n W(r)/\mathrm{d}r^n$, where $W(r)$ is the solid angle subtended at the transducers by that part of the scattering body within range r. n may be zero or any positive integer. It is shown that four types of echo envelope arise from varying degrees of overlap of these image pulses.

The combination of 'image pulse' and 'creeping wave' mechanisms is believed to account for the main scattering phenomena from rigid convex bodies, the former mechanism being paramount outside the shadow region at small wavelengths, the latter mechanism predominating at large wavelengths.

Sommaire

En utilisant une méthode d'analyse analogue à celles qu'on emploie en optique physique, on a examiné la dispersion en retour produite sur des ondes acoustiques de faible amplitude par un corps rigide immergé dans un milieu contenant un fluide parfait. Le rayonnement incident est une impulsion de type très général, à grands intervalles; on n'impose aucune réserve sur les diagrammes directifs des transducteurs émetteurs ou bien récepteurs. On obtient ainsi une explication du mécanisme suivant lequel se forme l'écho, explication qui s'applique dans le cas des ondes courtes.

On constate que l'écho est formé d'une série d'impulsions discrètes dont chacune est la répétition de l'impulsion émise et peut ainsi être désignée sous le nom d'image de l'impulsion. Une telle image de l'impulsion se produit en tout point où la quantité $\mathrm{d}^n W(r)\,\mathrm{d}r^n$ subit une discontinuité en fonction de la distance r. $W(r)$ désigne l'angle solide sous lequel on voit le transducteur à partir du corps qui produit la dispersion et se trouve à la distance r. Le nombre n peut être ou bien zéro ou bien un entier positif. On constate qu'il y a quatre types d'enveloppes d'échos qui diffèrent par la proportion dans laquelle se recouvrent les images de l'impulsion.

L'utilisation des mécanismes qui expliquent l'un la formation des images d'impulsion et l'autre la production d'ondes rampantes permet d'expliquer la plupart des phénomènes de diffraction produits par des corps solides rigides et de formes convexes. Le premier mécanisme se produit surtout pour les ondes courtes en dehors de la région d'ombre; le second mécanisme au contraire est important dans le cas des ondes longues.

Zusammenfassung

Die Rückstreuung von Schallwellen geringer Intensität von einem festen Körper, der sich in einer idealen Flüssigkeit befindet, wird mit analytischen Methoden in Analogie zur physikalischen Optik untersucht. Der einfallende Schall hat beliebige Pulsform. Bei größeren Entfernungen unterliegen die Richtcharakteristiken von Sender und Empfänger keinerlei Beschränkungen. Der Mechanismus der Entstehung von Echos bei kleinen Wellenlängen wird aufgeklärt.

Das Echo besteht aus einer Reihe von Einzelimpulsen. Diese Impulse sind Nachbildungen des Sendeimpulses; sie werden daher hier Nachbildungsimpulse genannt. Nachbildungsimpulse entstehen immer dann, wenn $\mathrm{d}^n W(r)/\mathrm{d}r^n$ sich mit der Entfernung r unstetig ändert. Dabei ist $W(r)$ der Raumwinkel, unter dem der innerhalb der Entfernung r gelegene Teil des Streukörpers vom Sender aus gesehen erscheint. $n = 0, 1, 2 \ldots$ Es wird gezeigt, daß durch die verschiedenartige Überlagerung der Nachbildungsimpulse vier Arten von Echos entstehen können.

Die wesentlichen Erscheinungsformen der Streuung an festen, konvexen Körpern lassen sich vermutlich durch das Zusammenwirken von „Nachbildungsimpulsen" und „Kriechwellen" erklären, wobei außerhalb der Schattenzone bei kleinen Wellenlängen die Nachbildungsimpulse, bei großen Wellenlängen die Kriechwellen vorherrschen.

1. Introduction

When acoustic or electro-magnetic waves are incident upon a body of given shape and acoustic or electro-magnetic properties, complicated spatial distributions of scattered radiation result. Such distributions have been investigated by many workers for various body shapes. Much of the work has been confined to determining these distributions in the continuous wave (C. W.) case, although in recent years growing attention has been given to the mechanisms underlying the scattering patterns. The discovery of creeping waves by Fock [1], the clarification by Franz and Deppermann [2], Deppermann and Franz [3] and Van de Hulst [4] of the contribution of these waves to the scattering from cylinders and spheres, and the extension of the concept of creeping waves to all wavelengths by Wu [5] are notable examples of this trend. The creeping wave mechanism is, however, not of importance to scattering outside the umbra when the wavelength is small in terms of the dimensions of the scattering body; for such 'high frequency' irradiation, clarification of the scattering mechanism as a time-resolved process is still called for. The present paper attempts to fill some of this gap in our knowledge, and, as a first step, the treatment will be restricted to the case of backscattering of small-amplitude, acoustic radiation from a simple model.

2. The model

The scattering body will be assumed to be rigid and immersed in an unbounded, non-dissipative, isotropic, fluid medium. Transducers, scattering body and medium will be deemed fixed in space.

As the scattering body is rigid, no internal penetration by radiation is involved and the means by which the scattered signal is built up can be divided, arbitrarily, into three main processes (Fig. 1).

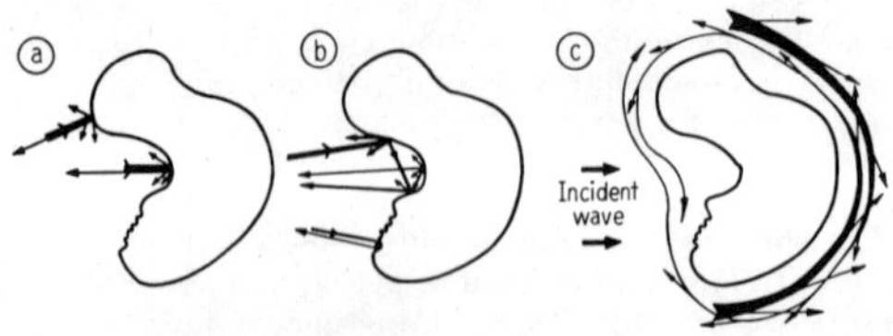

Fig. 1. Scattering processes which may contribute to the echo from a rigid body.
(a) direct scattering,
(b) multiple scattering,
(c) scattering due to creeping waves.

These are direct and multiple scattering, which are important at wavelengths small compared with the dimensions of the scattering body, and creeping waves (and, possibly, other forms of surface waves) which become important as the wavelength increases to a size comparable with the dimensions of the scattering body. In this paper, we shall concern ourselves only with the mechanism of direct scattering, so that the treatment will be principally of relevance to smooth bodies of a generally convex shape and will not give a full picture in the case of concave surfaces, for which multiple scattering may be a factor of prime importance. The treatment is, however, applicable to concave surfaces in the sense that it relates to the formation of the first of a chain of components which go to make up the total backscattering.

Deppermann and Franz [3] computed a curve showing the variation, with sphere radius, of the relative magnitudes of the creeping wave and direct scattering contributions to the echo from a rigid sphere. Their curve shows that the direct scattering component predominates for sphere radii down to the order of a wavelength. Although the relative magnitudes may differ somewhat for shapes other than the sphere, it seems likely that, for convex bodies, direct backscattering is the predominant mechanism for dimensions down to the order of several wavelengths.

In the present treatment, the dimensions and radii of curvature of the scattering body will, formally, be assumed large in terms of wavelengths, so that the Kirchhoff approximation for scattering may be used. For such a physical optics analysis [1], only the shape of that part of the scattering body within the directly irradiated region is of relevance, and the shape within the geometrical shadow region is deemed not to matter. However, a point yet to be resolved is whether the radii of curvature of the body at and just beyond the geometrical shadow boundary need to be large in terms of wavelengths, as the local curvature may conceivably affect the formation of creeping waves which, while making no significant direct contribution to the backscattering, may significantly modify the local field.

Although it has been shown by Meecham [7] that conditions for good accuracy when using the Kirchhoff approximation are a) large radii of curvature, and b) small maximum slope of the scattering surface relative to the incident wavefront, it has nevertheless been found that the physical optics solution frequently gives useful results in situations where the above assumptions are seriously violated.

In our analysis, the transducers will be treated, virtually, as nominally coincident point source and point receiver with superposed directivity patterns. Although the source is treated as having directivity, the approximation will be made of assuming that

[1] For a definition, see Van de Hulst [6].

the solutions of the wave equation for an isotropic spherical wave in a non-dissipative medium can be applied. At sufficiently large distances from the source this should involve little error, for the directivity term will only involve very small changes in pressure amplitude per wavelength travel over the wavefront.

The transducers may be either voltage sensitive or current sensitive devices, but for convenience of description they will be treated as if they were the former, although no loss of generality is implied.

3. The pressure incident on the scattering body

We first consider the case of a continuous, single-frequency transmission.

Let the electro-acoustic transmitter and receiver be coincident at the origin of a (r, θ, φ) polar co-ordinate system (Fig. 2). Let the transmitter sensitivity in the direction (θ, φ) be given by $P F_1(\theta, \varphi)$, where P is the pressure produced per applied volt at unit distance in the direction $\theta = \varphi = 0$, and let the corresponding receiver sensitivity be given by $H F_2(\theta, \varphi)$, where H is the sensitivity in the direction $\theta = \varphi = 0$.

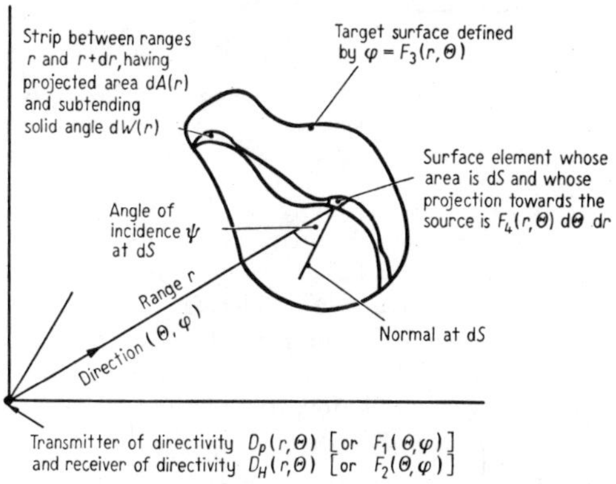

Fig. 2. Definition of the geometrical parameters.

Let the surface of the scattering body be defined by $\varphi = F_3(r, \theta)$, where φ will, in general, be double – (or many –) valued, corresponding to two (or more) points on the surface. Thus at this surface, $F_1(\theta, \varphi) = F_1(\theta, F_3(r, \theta)) = D_P(r, \theta)$, say, and, similarly, $F_2(\theta, \varphi) = D_H(r, \theta)$, say.

Measuring time from some arbitrary instant, let the voltage applied to the transmitter be $V \exp(\mathrm{i}\,\omega T)$, where V is real, ω is the angular frequency and T denotes any instant of the transmission.

It will be convenient to introduce three designations for time, namely T, t_i and t, which will be associated, respectively, with moments of transmission, of incidence at the obstacle and of echo reception. All three time scales are identical and have the same origin, but T, t_i and t are related by $T = t_\mathrm{i} - (r/c) = t - (2\,r/c)$, where c represents the wave velocity.

Assuming the scattering body to be at a sufficiently large distance from the transmitter for the inverse range law to be tenable, the pressure, at time t_i, of the wave incident on an element of body surface, $\mathrm{d}S$, at a distance r from the source and in the direction (θ, φ) is given by

$$p_\mathrm{i} = V P D_P(r, \theta) \exp[\mathrm{i}(\omega t_\mathrm{i} - k r)]/r,$$

where k is the wave number, $(2\pi/\lambda)$, and λ is the wave length in the fluid medium.

4. Boundary conditions and the resultant scattering

The particle velocity, u_i, of the wave incident at $\mathrm{d}S$ is given by

$$u_\mathrm{i} = \frac{p_\mathrm{i}}{\varrho c}\left(1 - \frac{\mathrm{i}}{k r}\right),$$

where ϱ is the density of the medium.

As the net normal particle velocity at the boundary is zero, the element $\mathrm{d}S$ acts as a point source of strength $-u_\mathrm{i} \cos\psi\, \mathrm{d}S$ [2], ψ denoting the angle of incidence at $\mathrm{d}S$.

The Kirchhoff approximation is now introduced. This assumes that the radiation scattered by the element is distributed evenly over a solid angle of 2π, secondary diffraction around the body or around curves onto other parts of the body being neglected. This amounts virtually to treating each element of surface as if it were embedded in an infinite plane baffle. Providing the radii of curvature and the dimensions of the body are at least a few wavelengths long, the error introduced by this approximation will be reasonably small.

The pressure back at the transducers at time t due to scattering from an element of surface, $\mathrm{d}S$, is then given by

$$\Delta p_s = -\mathrm{i}\frac{V P}{\lambda} D_P(r, \theta)\left(1 - \frac{\mathrm{i}}{k r}\right) \times \\ \times \frac{\exp[\mathrm{i}(\omega t - 2 k r)]}{r^2} \cos\psi\, \mathrm{d}S.$$

Since the scattering surface is defined by the equation $\varphi = F_3(r, \theta)$, $\mathrm{d}S$ and ψ are fully defined in terms of r and θ; thus for the projection of $\mathrm{d}S$ towards the origin we may write $\mathrm{d}S \cos\psi =$

[2] For a free boundary the same expression holds, except for a phase difference of π. Thus apart from a phase reversal, the rest of the treatment applies to "pressure release" bodies also.

$F_4(r,\theta)\,\mathrm{d}r\,\mathrm{d}\theta$. Hence,

$$\Delta p_s = -\mathrm{i}\,\frac{V P}{\lambda}\left(1-\frac{\mathrm{i}}{k r}\right)\frac{\exp[\mathrm{i}(\omega t - 2 k r)]}{r^2}\times D_P(r,\theta)\,F_4(r,\theta)\,\mathrm{d}r\,\mathrm{d}\theta .$$

Assuming that the range, r_1, of the nearest part of the scattering body is such that $k r_1 \gg 1$, by ignoring terms of the order of $1/k r$, we have

$$\Delta p_s = -\mathrm{i}\,\frac{V P}{\lambda}\,\frac{\exp[\mathrm{i}(\omega t - 2 k r)]}{r^2}\times D_P(r,\theta)\,F_4(r,\theta)\,\mathrm{d}r\,\mathrm{d}\theta . \qquad (1)$$

5. The receiver voltage

The voltage, ΔE, at the receiver at time t due to scattering from $\mathrm{d}S$ is given by

$$\Delta E = H\,D_H(r,\theta)\,\Delta p_s = B\,\frac{\exp(-\mathrm{i}\,2 k r)}{r^2}\,D_P(r,\theta)\,D_H(r,\theta)\,F_4(r,\theta)\,\mathrm{d}r\,\mathrm{d}\theta ,$$

where $B = -(\mathrm{i}\,V P H/\lambda)\exp(\mathrm{i}\,\omega t)$.

The voltage, $\mathrm{d}E$, at the receiver at time t due to scattering from the whole irradiated strip on the body between ranges r and $r+\mathrm{d}r$ is obtained by integrating over all the appropriate values of θ. The limits of θ where the body is illuminated are functions of range. Thus,

$$\mathrm{d}E = B\exp(-\mathrm{i}\,2 k r)\,\mathrm{d}r \times \int\limits_{r=\mathrm{const.}} \frac{D_P(r,\theta)\,D_H(r,\theta)\,F_4(r,\theta)}{r^2}\,\mathrm{d}\theta ,$$

the integral being taken over the above-mentioned strip. This integral, being a function of its limits, is a function of range.

The total instantaneous receiver voltage is obtained by integrating over all possible ranges and is given by

$$E = B\int\limits_0^{\infty}\int\limits_{r=\mathrm{const.}} \frac{D_P(r,\theta)\,D_H(r,\theta)\,F_4(r,\theta)}{r^2}\times\exp(-\mathrm{i}\,2 k r)\,\mathrm{d}\theta\,\mathrm{d}r .$$

Weighting each element of surface of the scattering body by the product of transmitter and receiver directivities appropriate to the position of that element, the directivity-weighted solid angle, $W_{\mathrm{w}}(r)$, subtended at the transducers by those parts of the body surface within range r is given by

$$W_{\mathrm{w}}(\mathrm{r}) = \int\limits_0^{r}\int\limits_{r=\mathrm{const.}} \frac{D_P(r,\theta)\,D_H(r,\theta)\,F_4(r,\theta)}{r^2}\,\mathrm{d}\theta\,\mathrm{d}r .$$

Hence,

$$E = B\int\limits_0^{\infty}\frac{\mathrm{d}W_{\mathrm{w}}(r)}{\mathrm{d}r}\exp(-\mathrm{i}\,2 k r)\,\mathrm{d}r . \qquad (2)$$

For the case where the transducers are non-directional, this integral is frequently solved approximately by the FRESNEL zone method. That method gives little insight into the mechanism of echo formation, so we proceed differently.

6. Integration of $[W_{\mathrm{W}}^{(1)}(r)\exp(-\mathrm{i}\,2 k r)]$ over a point where $W_{\mathrm{W}}(r)$ is finitely discontinuous

Let the range to the nearest part of the scattering body be denoted by r_1 and the range to the furthest part of the body receiving direct irradiation (i. e. excluding those parts which are in geometrical shadow) be denoted by r_{f}. The directivity-weighted, total solid angle, $W_{\mathrm{w}}(r)$, will be zero up to range r_1 and will, in general, increase continuously to a maximum value at range r_{f} and then remain constant at this value up to range infinity (Fig. 3 a).

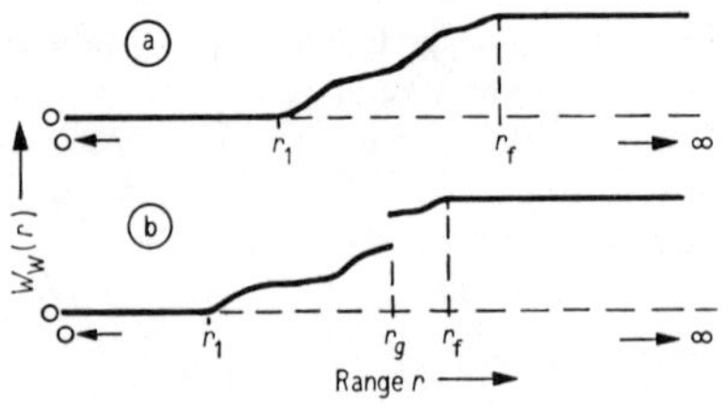

Fig. 3. Typical modes of variation of the directivity-weighted, total solid angle; for (a) and (b) see text.

However, wherever, at range r_g between r_1 and r_{f}, a region of the body's surface coincides with the incident wavefront, $W_{\mathrm{w}}(r)$ will exhibit a discontinuous, finite increase (Fig. 3 b). One example of such a situation is where part of the directly irradiated surface forms part of a sphere whose centre coincides with the transducers (Fig. 4 a).

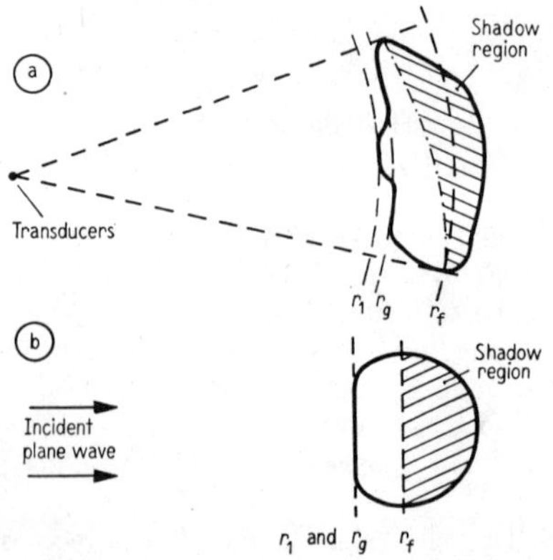

Fig. 4. Examples of bodies exhibiting discontinuities in $W(r)$ or $A(r)$; for (a) and (b) see the text.

Another example is where part of the surface is flat and a plane wave is incident normally on that part (Fig. 4 b). (In the latter example it is however more appropriate to deal in terms of projected area, rather than in terms of solid angle, as the plane wave implies that $r \to \infty$.)

It may therefore be necessary to perform the integration of eq. (2) over a domain which includes a point (or points) where $W_w(r)$ is finitely discontinuous, and the relevant procedure for this will now be established.

Assume that, from range 0 up to range r_g, the continuous quantity $W_w(r)$ may be denoted by $W_{wg-}(r)$. Assume also that, after a finite discontinuity at range r_g, $W_w(r)$ is again continuous up to range infinity and that in this second region it may be denoted by $W_{wg+}(r)$. Let the size of the discontinuity, $W_{wg-}(r_g) - W_{wg+}(r_g)$, be denoted by D.

Thus

$$W_w(r) = \begin{bmatrix} W_{wg-}(r) & 0<r<r_g \\ W_{wg+}(r) & r_g<r<\infty \end{bmatrix} = \begin{bmatrix} W_{wg-}(r) & 0<r<r_g \\ W_{wg+}(r)+D & r_g<r<\infty \end{bmatrix} - D\,H(r-r_g),$$

where $H(r-r_g)$ represents a Heaviside step function.

$$\frac{dW_w(r)}{dr} = \begin{bmatrix} dW_{wg-}(r)/dr & 0<r<r_g \\ dW_{wg+}(r)/dr & r_g<r<\infty \end{bmatrix} - D\,\partial(r-r_g),$$

where $\partial(r-r_g)$ is the Dirac delta function.

Hence, if the quantity $[W_w^{(1)}(r)\exp(-i\,2\,k\,r)]$, (where $W_w^{(n)}(r) \equiv d^n W_w(r)/dr^n$), is integrated over the range 0 to ∞, the integral will consist of the sum of the integrals on either side of r_g over which the integrand is continuous, plus the contribution due to the ∂ function at r_g.

As,

$$\int_0^\infty D\,\partial(r-r_g)\exp(-i\,2\,k\,r)\,dr = [W_{wg-}(r_g) - W_{wg+}(r_g)]\exp(-i\,2\,k\,r_g)$$

and as $dW_w(r)/dr$ is zero from ranges 0 to r_1 and from r_f to ∞,

$$\int_0^\infty \frac{dW_w(r)}{dr}\exp(-i\,2\,k\,r)\,dr = \tag{3}$$

$$= \lim_{\varepsilon\to 0}\int_{r_1}^{r_g-\varepsilon}\frac{dW_w(r)}{dr}\exp(-i\,2\,k\,r)\,dr + \lim_{\varepsilon\to 0}\int_{r_g+\varepsilon}^{r_f}\frac{dW_w(r)}{dr}\exp(-i\,2\,k\,r)\,dr - [W_{wg-}(r_g) - W_{wg+}(r_g)]\exp(-i\,2\,k\,r_g).$$

This equation is still formally correct even when there is no discontinuity in $W_w(r)$ at r_g. It is also applicable when there are discontinuities at r_g in derivatives of $W_w(r)$.

7. The effect of discontinuities in $W_W(r)$ or its derivatives

Let us write the indefinite integral over any range for which the integrand of eq. (2) is continuous as

$$\int \frac{dW_w(r)}{dr}\exp(-i\,2\,k\,r)\,dr = I(r).$$

Then, if the integrand is continuous for $r_1 \leqq r \leqq r_f$,

$$E = -B[I(r_1) - I(r_f)]. \tag{4}$$

However, if at any range r, the function $W_w(r)$ or one of its derivatives, with respect to range, contains a discontinuity, eq. (4) does not apply, and we must make use of (3).

Denoting $I_{g-}(r)$ and $I_{g+}(r)$ as the values of $I(r)$ corresponding to $W_{wg-}(r)$ and $W_{wg+}(r)$ respectively, eq. (4) is replaced by

$$E = -B\{[I_{g-}(r_1) - I_{g-}(r_g)] + [I_{g+}(r_g) - I_{g+}(r_f)] + [W_{wg-}(r_g) - W_{wg+}(r_g)]\exp(-i\,2\,k\,r_g)\}. \tag{5}$$

Comparing eq. (4) and eq. (5), we see that the discontinuity at r_g in $W_w(r)$ or in one of its derivatives, with respect to range, has introduced into the expression for the receiver voltage, E, the additional term

$$B\{[I_{g-}(r_g) - I_{g+}(r_g)] - [W_{wg-}(r_g) - W_{wg+}(r_g)]\exp(-i\,2\,k\,r_g)\}.$$

Similarly, any further discontinuities in $W_w(r)$ at other ranges will necessitate the splitting of the domain of integration at the ranges of the discontinuities, and each such discontinuity will introduce into E a term of the form just derived for the discontinuity at r_g.

Such discontinuities in $W_w(r)$ or in its derivatives, with respect to range, will also occur at the near and far limits of the directly irradiated part of the body in addition to their possible occurrence at intermediate ranges over the body.

If discontinuities in $W_w(r)$ or in its derivatives, with respect to range, occur at ranges r_1, r_2, r_3, ..., r_f, then the receiver voltage, E, at time t is given by

$$E = B\sum_{g=1}^{f}\{[I_{g-}(r_g) - I_{g+}(r_g)] - [W_{wg-}(r_g) - W_{wg+}(r_g)]\exp(-i\,2\,k\,r_g)\}. \tag{6}$$

8. An asymptotic expansion of the general integral and its application to the receiver voltage

The typical integral which gives the $I(r)$ terms in the above series is of the form

$$\lim_{\varepsilon \to 0} \int_{r_g+\varepsilon}^{r_{g+1}-\varepsilon} \frac{\mathrm{d}W_{\mathrm{w}}(r)}{\mathrm{d}r} \exp(-\mathrm{i}\,2\,k\,r)\,\mathrm{d}r\,.$$

This is of the type $\int^{r_a} f(r) \exp(-\mathrm{i}\,2\,k\,r)\,\mathrm{d}r$. A suitable form of solution of this indefinite integral can be obtained by integrating iteratively by parts, giving

$$\int^{r_a} f(r) \exp(-\mathrm{i}\,2\,k\,r)\,\mathrm{d}r = \qquad (7)$$
$$= -\exp(-\mathrm{i}\,2\,k\,r_a) \sum_{n=1}^{m} \left(\frac{-\mathrm{i}}{2\,k}\right)^n f^{(n-1)}(r_a) + R_m\,,$$

where $R_m = \dfrac{1}{(\mathrm{i}\,2\,k)^m} \displaystyle\int^{r_a} f^{(m)}(r) \exp(-\mathrm{i}\,2\,k\,r)\,\mathrm{d}r\,.$

But

$$|R_m| \leq \frac{1}{(2\,k)^m} \int^{r_a} |f^{(m)}(r)|\,\mathrm{d}r = \pm \frac{f^{(m-1)}(r_a)}{(2\,k)^m}\,.$$

Thus, for convergence of the series of eq. (7), $|f^{m-1}(r_a)|$ must be of smaller order than $(2\,k)^m$. However, although, in general, R_m does not tend to zero as m tends to infinity, convergence is not essential for useful application of the series, and eq. (7) gives the following asymptotic expansion:

$$\int^{r_a} f(r) \exp(-\mathrm{i}\,2\,k\,r)\,\mathrm{d}r =$$
$$= \frac{\mathrm{i}}{2\,k} \exp(-\mathrm{i}\,2\,k\,r_a) \sum_{n=0}^{\infty} \frac{f^{(n)}(r_a)}{(\mathrm{i}\,2\,k)^n}\,.$$

For large $k\,r$, such an asymptotic series can give results of considerable accuracy.

For our typical integral, the asymptotic expansion becomes

$$\int \frac{\mathrm{d}W_{\mathrm{w}}(r)}{\mathrm{d}r} \exp(-\mathrm{i}\,2\,k\,r)\,\mathrm{d}r = \qquad (8)$$
$$= -\exp(-\mathrm{i}\,2\,k\,r) \sum_{n=1}^{\infty} \frac{W_{\mathrm{w}}^{(n)}(r)}{(\mathrm{i}\,2\,k)^n}\,.$$

A typical term of the series of eq. (6) is then given by

$$[I_{g-}(r_g) - I_{g+}(r_g)] - [W_{\mathrm{w}g-}(r_g) - W_{\mathrm{w}g+}(r_g)] \times$$
$$\times \exp(-\mathrm{i}\,2\,k\,r_g) =$$
$$= -\exp(-\mathrm{i}\,2k\,r_g) \sum_{n=0}^{\infty} \frac{[W_{\mathrm{w}g-}^{(n)}(r_g) - W_{\mathrm{w}g+}^{(n)}(r_g)]}{(\mathrm{i}\,2\,k)^n}\,.$$

We note the special conditions at the near and far limits of irradiation, namely that

$$W_{\mathrm{w}1-}^{(n)}(r_1) = 0 \qquad (n = 0, 1, 2, \ldots),$$
$$W_{\mathrm{w}f+}^{(n)}(r_f) = 0 \qquad (n = 1, 2, 3, \ldots).$$

The quantity $[W_{\mathrm{w}g-}^{(n)}(r_g) - W_{\mathrm{w}g+}^{(n)}(r_g)]$ represents the magnitude of the discontinuity in the nth derivative, with respect to range, of $W_{\mathrm{w}}(r)$ at range r_g, so we denote it by $D(W_{\mathrm{w}}, g, n)$. Eq. (6) for the receiver voltage now becomes

$$E = \sum_{g=1}^{f} E_g\,, \quad \text{where}$$

$$E_g = \mathrm{i}\,\frac{M}{\lambda} \exp[-\mathrm{i}\,2\,k(r_g - r_1)] \sum_{n=0}^{\infty} \frac{D(W_{\mathrm{w}}, g, n)}{(\mathrm{i}\,2\,k)^n}$$
$$= \sum_{n=0}^{\infty} E(g, n)\,, \qquad (9)$$

$$M = V\,P\,H \exp[\mathrm{i}(\omega\,t - 2\,k\,r_1)] \qquad (10)$$

and

$$E(g, n) = \mathrm{i}\,\frac{M}{\lambda} \exp[-\mathrm{i}\,2\,k(r_g - r_1)]\,\frac{D(W_{\mathrm{w}}, g, n)}{(\mathrm{i}\,2\,k)^n}\,. \qquad (11)$$

E_g is the contribution to the total receiver signal due to all discontinuities occurring in different orders of the derivative at range r_g, while $E(g, n)$ is the contribution of an individual discontinuity in the nth derivative.

When r is so large that the change of $1/r^2$ over the scattering body may be neclected, then if the mean range to the body is r_{m}, $W_{\mathrm{w}}(r) = A_{\mathrm{w}}(r)/r_{\mathrm{m}}^2$, where $A_{\mathrm{w}}(r)$ is the total, directivity-weighted projection towards the transducers of that part of the scattering body within range r. Eq. (8) then becomes

$$E_g = \mathrm{i}\,\frac{M}{\lambda\,r_{\mathrm{m}}^2} \exp[\mathrm{i}\,2\,k(r_g - r_1)] \sum_{n=0}^{\infty} \frac{D(A_{\mathrm{w}}, g, n)}{(\mathrm{i}\,2\,k)^n}\,. \qquad (12)$$

Where the transducer directivities are uniform over the whole of the scattering body, in which case no directivity weighting is called for, the quantities ing unweighted quantities, $W(r)$ and $A(r)$. Cor- $W_{\mathrm{w}}(r)$ and $A_{\mathrm{w}}(r)$ are replaced by the correspond- respondingly, $D(W_{\mathrm{w}}, g, n)$ and $D(A_{\mathrm{w}}, g, n)$ in the above equations are replaced by $D(W, g, n)$ and $D(A, g, n)$, respectively.

Each term of the series of eq. (9) (or of eq. (12)) must be dimensionally the same. We recall that $D(W_{\mathrm{w}}, g, n)$ is based on the nth differentiation of $W_{\mathrm{w}}(r)$, with respect to r. Due to the quantity $(\mathrm{i}\,2\,k)^{-n}$, successive terms in the above series rise by one power of λ, so to keep matters dimensionally correct each differentiation implicit in $D(W_{\mathrm{w}}, g, n)$ must involve a drop in one power of a length. The lengths that might be involved are range, radii of curvature and dimensions of the scattering body. Thus successive terms in eq. (9) are of one order of λ/L different in magnitude from the preceding terms, where L is a linear combination of the body's range, curvature and dimensions. At it has been stipulated that all these latter quantities are large in terms of the wavelength, the implication is that the series of eq. (9) is generally likely to converge rapidly, and probably only one or two terms will be significant.

It should however be noted that the expression of eq. (8) does not apply where the discontinuities are not finite, and there are exceptional circumstances in which this method of expansion breaks down, as for example, when the function $W_w(r)$ equals $\exp[-a/(r-r_1)]$, which has not a TAYLOR series expansion about r_1.

9. Generalisation of the receiver response to the case where the transmission is not monotonic

On a time scale whose origin coincides with the beginning of transmission, let a voltage signal of the form $V_1(T)$ be applied to the transmitter, where $V_1(T)$ may be complex. Let the amplitude of the spectrum component of $V_1(T)$ over a narrow band, $d\omega$, around the angular frequency ω be given $v(\omega)$. $v(\omega)$ is the FOURIER transform of $V_1(T)$, and is given in complex form by

$$v(\omega) = \frac{1}{2\pi}\int_{-\infty}^{\infty} V_1(T)\exp(-\mathrm{i}\,\omega\,T)\,dT\,.$$

Correspondingly

$$V_1(T) = \int_{-\infty}^{\infty} v(\omega)\exp(\mathrm{i}\,\omega\,T)\,d\omega\,. \qquad (13)$$

Allowing for the frequency dependence of the transmitter and receiver sensitivities and directivities, the previous quantities $P, H, D_P(r,\theta)$, $D_H(r,\theta)$, $W_w(r)$ and $D(W_w, g, n)$ will be denoted by $P(\omega)$, $H(\omega)$, $D_P(r,\theta,\omega)$, $D_H(r,\theta,\omega)$, $W_w(r,\omega)$ and $D(W_w, g, n, \omega)$.

Considering the echo contribution due to a discontinuity in the nth derivative at range r_g, let the receiver response due to that part of the transmitted spectrum within the narrow band $d\omega$ around ω be denoted by $dE(g,n)$. Then from eq. (11) and eq. (10),

$$dE(g,n) = \frac{\mathrm{i}\,\omega}{2\pi c}\, v(\omega)P(\omega)H(\omega)\,\times$$
$$\times \exp\left[\mathrm{i}\,\omega\left(t-\frac{2r_g}{c}\right)\right]\left(\frac{c}{\mathrm{i}\,2\,\omega}\right)^n D(W_w, g, n, \omega)\,d\omega\,,$$

and the total receiver response due to a discontinuity in the nth derivative at range r_g is given by

$$E(g,n) = \frac{\mathrm{i}}{2\pi c}\left(\frac{c}{\mathrm{i}\,2}\right)^n \int_{-\infty}^{\infty} \omega^{1-n}\, v(\omega)P(\omega)H(\omega)\,\times \qquad (14)$$
$$\times \exp\left[\mathrm{i}\,\omega\left(t-\frac{2r_g}{c}\right)\right] D(W_w, g, n, \omega)\,d\omega\,.$$

10. An approximate solution for each received echo component

We now examine the form of eq. (14) when certain restricting conditions are applied.

Assume that transmitter and receiver form an idealised system of bandwidth $\Delta\omega$, such that $P(\omega)H(\omega) = P(\omega_0)H(\omega_0)$ when

$$\omega_0 - (\Delta\omega/2) < \omega < \omega_0 + (\Delta\omega/2)$$

and is zero everywhere else. Then

$$E(g,n) = \mathrm{i}\,\frac{P(\omega_0)H(\omega_0)}{2\pi c}\left(\frac{c}{\mathrm{i}\,2}\right)^n \int_{\omega_0-(\Delta\omega/2)}^{\omega_0+(\Delta\omega/2)} \omega^{1-n}\,\times \qquad (15)$$
$$\times D(W_w, g, n, \omega)\, v(\omega)\exp\left[\mathrm{i}\,\omega\left(t-\frac{2r_g}{c}\right)\right] d\omega\,.$$

If $[\omega^{1-n} D(W_w, g, n, \omega)]$ is a slowly varying function over the band $\omega_0 - (\Delta\omega/2)$ to $\omega_0 + (\Delta\omega/2)$,

$$E(g,n) \approx \mathrm{i}\,\frac{P(\omega_0)H(\omega_0)}{\lambda_0}\,\frac{D(W_w, g, n, \omega)}{(\mathrm{i}\,2\,k_0)^n}\,\times$$
$$\times \int_{\omega_0-(\Delta\omega/2)}^{\omega_0+(\Delta\omega/2)} v(\omega)\exp\left[\mathrm{i}\,\omega\left(t-\frac{2r_g}{c}\right)\right] d\omega\,,$$

where $2\pi f_0 = \omega_0 = k_0\,c$.

Providing $V_1(T)$ is such that most of its spectrum is concentrated within the limits $\omega_0 - (\Delta\omega/2)$ and $\omega_0 + (\Delta\omega/2)$, eq. (13) may be written

$$V_1(T) \approx \int_{\omega_0-(\Delta\omega/2)}^{\omega_0+(\Delta\omega/2)} v(\omega)\exp\left[\mathrm{i}\,\omega\left(t-\frac{2r_g}{c}\right)\right] d\omega\,.$$

Also, as the spectrum of $V_1(T)$ has been assumed to be centred around ω_0, it is convenient to rewrite $V_1(T)$ as

$$V_1(\mathrm{T}) = V(\mathrm{T})\,\exp(\mathrm{i}\,\omega_0\,T) =$$
$$= V\left(t-\frac{2r_g}{c}\right)\exp\left[\mathrm{i}\,\omega_0\left(t-\frac{2r_g}{c}\right)\right].$$

So

$$E(g,n) \approx \mathrm{i}\,V\left(\mathrm{t}-\frac{2r_g}{c}\right)\frac{P(\omega_0)H(\omega_0)}{\lambda_0}\,\times$$
$$\times \exp[\mathrm{i}(\omega_0 t - 2k_0 r_g)]\,\frac{D(W_w, g, n\,\omega_0)}{(\mathrm{i}\,2\,k_0)^n}$$

and

$$E_g \approx \mathrm{i}\,V\left(t-\frac{2r_g}{c}\right)\frac{P(\omega_0)H(\omega_0)}{\lambda_0}\,\times \qquad (16)$$
$$\times \exp[\mathrm{i}(\omega_0 t - 2k_0 r_g)]\sum_{n=0}^{\infty}\frac{D(W_w, g, n, \omega_0)}{(\mathrm{i}\,2\,k_0)^n}$$

When the simplification justifying eq. (12) applies,

$$E_g \approx \mathrm{i}\, V\left(t - \frac{2\, r_g}{c}\right) \frac{P(\omega_0) H(\omega_0)}{\lambda_0\, r_m^2} \times \qquad (17)$$

$$\times \exp[\mathrm{i}(\omega_0\, t - 2\, k_0\, r_g)] \sum_{n=0}^{\infty} \frac{D(A_\mathrm{w}, g, n, \omega_0)}{(\mathrm{i}\, 2\, k_0)^n}$$

Subject to the approximations of the solution and to the restrictions on $V(T)$ which are implied, it has been established that the equations for $E(g, n)$ and for E_g are of the same form whether the transmission is monotonic or not. It is thus convenient to drop the use of special nomenclature to denote that quantities such as P, H, λ, ω, k, $D(W_\mathrm{w}, \mathrm{g}, \mathrm{n})$ and $D(A_\mathrm{w}, g, n)$ are associated with the centre frequency of the transmission spectrum. Unless otherwise stated, it will be henceforward understood that such quantities are those applicable at the mean frequency. Eqs. (16) and (17), respectively, therefore become

$$E_g \approx i\, \frac{M_g}{\lambda} \exp[-\mathrm{i}\, 2\, k(r_g - r_1)] \sum_{n=0}^{\infty} \frac{D(W_\mathrm{w}, g, n)}{(\mathrm{i}\, 2\, k)^n} \qquad (18)$$

and

$$E_g \approx \mathrm{i}\, \frac{M_g}{\lambda\, r_m^2} \exp[-\mathrm{i}\, 2\, k(r_g - r_1)] \sum_{n=0}^{\infty} \frac{D(A_\mathrm{w}, g, n)}{(\mathrm{i}\, 2\, k)^n}, \qquad (19)$$

where

$$M_g = V\left(t - \frac{2\, r_g}{c}\right) P\, H \exp[\mathrm{i}(\omega\, t - 2\, k\, r_1)]. \qquad (20)$$

The correspondence with eqs. (9) and (12) is now very obvious.

E_g can be seen to correspond to a signal identical in envelope to the transmitted signal.

As before,

$$E = \sum_{g=1}^{f} E_g, \qquad (21)$$

and the total echo thus consists of the vectorial addition of a number of signals, each identical in envelope to the transmitted signal.

11. The implications of and departures from the approximate solution

Although the treatment assumed that no phase shift was introduced by the transmitter-receiver system over the passband $\Delta\omega$, a more realistic approximation would have been to assume a phase shift $\omega\,\tau_c + 2\, m\, \pi$, ($m$ = integer), to take place over the band. This would simply have resulted in each echo component being delayed by a time τ_c.

The further assumptions made in the derivation of eq. (18) will now be examined.

The assumption that the product of transmitter and receiver responses is uniform over a given band and zero outside it, coupled with the assumption that the significant portion of the spectrum of the transmitted signal lies within this band, has the following implications. Providing that $V(T)$ represents a reasonably behaved envelope (for instance, a rectangular envelope) of duration τ, then if τ is equal to or greater than the inverse of the bandwidth $\Delta\omega$, no serious distortion of the transmitted signal will occur. If τ is shorter than this, the echo component, $E(g, n)$, will consist of a distorted version of the transmitted signal, the rise and decay times of the envelope being lengthened the more τ is shortened. In fact, the phenomena normally associated with the use of equipment of inadequate bandwidth occur.

The next assumption was that ω^{1-n} does not vary very much over the relevant band. For the common case where the discontinuity is in the first derivative, ω^{1-n} equals unity and no restriction is implied. When the discontinuity is in the zero or the second derivative, the implication is that $\Delta\omega$ is small compared with ω_0. Where this is not the case, then, for $n = 0$, the components within the integral of eq. (15) will be more emphasised at the high frequency end than at the low, and for $n = 2$ the reverse will apply. This will tend to produce distortion somewhat akin to that met in a conventional circuit when the bandwidth is narrowed and the centre frequency is shifted off resonance.

The remaining assumption was that $D(W_\mathrm{w}, g, n, \omega)$ does not vary much over the relevant band. This is tantamount to saying that $W_{wg-}^{(n)}(r, \omega)$ and $W_{wg+}^{(n)}(r, \omega)$ do not vary much over this band. We note that the frequency dependence of $W_\mathrm{w}(r, \omega)$ is due to the directivity factors $D_P(r, \theta, \omega)$ and $D_H(r, \theta, \omega)$.

The influence of the directivity terms may be better appreciated by thinking for the moment of the transmission only. The width of the transmitter's primary lobe will increase gradually as the angular frequency decreases from $\omega_0 + (\Delta\omega/2)$ to $\omega_0 - (\Delta\omega/2)$. Thus, other than along the acoustic axis, the directivity term for any direction falling within the main lobe at the upper frequency will increase gradually as the frequency changes from $\omega_0 + (\Delta\omega/2)$ to $\omega_0 - (\Delta\omega/2)$. At angles beyond this main lobe, the variation of directivity with direction will be more complex. For angles within the above limit, the variation will be smaller, the closer the direction is to the acoustic axis and the smaller is $\Delta\omega/\omega_0$. Thus the pulse transmitted along non-axial directions will tend to be a distorted version of that transmitted along the axis, although the distortion will be small if $\Delta\omega/\omega_0$ is small and the bearing is not further removed from the acoustic axis than, say, the 6 dB point on the directivity curve at ω_0. The argument may straightforwardly be extended from the transmitted pulse to the received echo pulse, the product of the two transducer directivities being substituted for the transmitter directivity.

Yet another assumption is implicit in eq. (15). This is that the lower frequency limit, $\omega_0 - (\Delta\omega/2)$, is large enough for the KIRCHHOFF approximation to apply. When this assumption is not justified, there may be some distortion due to a falling off in the low frequency components. Possibly, there may then also be some effects due to creeping waves.

Despite all the various complications indicated above, the important point to note is that, even if the assumptions made in obtaining a solution of eq. (14) are not valid, but providing there are no significant low frequency contents in the incident wave packet which may give rise to creeping waves, the contribution to the echo caused by a discontinuity in $D(W_w, g, n)$ consists of a discrete component, and this will tend to have the form of a distorted version of the transmitted pulse. This form is likely to be a slowly varying function of direction in space, the least distortion being obtained on the acoustic axis.

12. Extension of the treatment to shorter ranges

The preceding treatment applies to the case where $k r_1 \gg 1$. A better approximation for closer ranges is obtained by taking into account the $1/\mathrm{i}\, k r$ term which was dropped in eq. (1). The lower range limit down to which the resulting formulae may be applied cannot be definitely stated here. However, it is probable that, due to the KIRCHHOFF approximation, the treatment will not be generally applicable at ranges less than the maximum dimension of the scattering body. But in the case of surfaces having very large radii of curvature or of large plane surfaces, the errors introduced by the KIRCHHOFF approximation become small and the treatment may be applicable down to smaller ranges than surmised above. Indeed, in the case of an infinite plane scattering surface, the assumptions of the KIRCHHOFF treatment are exactly fulfilled and consequently, the present extension of the treatment will then be valid right dowr to zero range.

In general, the ranges at which this extension is likely to prove useful are in the region where the assumption of transducers behaving as point transducers with superposed directivity patterns becomes questionable. We therefore drop the assumption of directional transducers, letting $D_P(r,\theta) = D_H(r,\theta) = 1$.

In place of eq. (2), retention of the $[1-(\mathrm{i}/k r)]$ term in eq. (1) leads to

$$E = B \int_0^\infty \frac{\mathrm{d}U(r)}{\mathrm{d}r} \exp(-\mathrm{i}\, 2 k r)\, \mathrm{d}r,$$

where $$\frac{\mathrm{d}U(r)}{\mathrm{d}r} = \left(1 - \frac{\mathrm{i}}{k r}\right) \frac{\mathrm{d}W(r)}{\mathrm{d}r}.$$

Eq. (9) is now replaced by

$$E_g = \mathrm{i}\, \frac{M}{\lambda} \exp[-\mathrm{i}\, 2 k (r_g - r_1)] \sum_{n=0}^{\infty} \frac{D(U, g\, n)}{(\mathrm{i}\, 2 k)^n},$$

where $D(U, g, n) = U_{g-}^{(n)}(r_g) - U_{g+}^{(n)}(r_g)$.

As $$U^{(n)}(r) = W^{(n)}(r) + \frac{1}{\mathrm{i}\, k} \sum_{m=1}^{n} \frac{(n-1)!}{(m-1)!} \frac{(-1)^{n-m}}{r^{n-m+1}} W^{(m)}(r),$$

so $$D(U,g,n) = D(W,g,n) + \frac{1}{\mathrm{i}\, k} \sum_{m=1}^{n} \frac{(n-1)!}{(m-1)!} \frac{(-1)^{n-m}}{r^{n-m+1}} D(W,g,m).$$

Coupled with the approximate solution for the pulse case, this yields

$$E_g = \mathrm{i}\, \frac{M_g}{\lambda} \exp[-\mathrm{i}\, 2 k (r_g - r_1)] \times \sum_{n=0}^{\infty} \frac{D(W,g,n)}{(\mathrm{i}\, 2 k)^n} [1 - K(n)], \tag{22}$$

where

$$\left.\begin{aligned} K(n) &= 0 \qquad (n=0), \\ &= \frac{2}{(n-1)!} \sum_{b=0}^{\infty} \frac{(n+b-1)!}{(-\mathrm{i}\, 2 k r_g)^{b+1}} \\ &\qquad (n = 1, 2, 3, \ldots). \end{aligned}\right\} \tag{23}$$

13. Descriptive account of the echo formation mechanism

The interpretation of eqs. (18) to (23) will now be described with the help of Fig. 5, which is a purely qualitative illustration.

On the top right of the figure is shown a relatively smooth, generally convex, rigid body, while on the left are the two transducers, which will be assumed non-directional in the first instance. A plot of the solid angle, $W(r)$, subtended at the transducers by that part of the body within range r is shown at Fig. 5 b. This quantity increases with range until all the directly irradiated part of the object is encompassed.

A plot of $\mathrm{d}W(r)/\mathrm{d}r$, shown at Fig. 5 c, exhibits two discontinuities, $D(W,1,1)$ and $D(W,3,1)$, corresponding to sudden changes of slope in the curve of $W(r)$ versus r. The ranges r_1 and r_3 at which these occur are associated respectively with the nearest point of the body and with some assumed prominent feature which has been indicated schematically in Fig. 5 a. Associated with another lesser feature, a change of slope of $\mathrm{d}W(r)/\mathrm{d}r$ has been shown at range r_2, together with a further change of slope at the limit, r_4, of the directly irradiated part of the body. The curve of $\mathrm{d}^2W(r)/\mathrm{d}r^2$ thus exhibits four discontinuities. Similar curves could be drawn for the n t h derivative of $W(r)$.

ditionally, there is an associated phase shift of $(1-n)\pi/2$ when the generating discontinuity is positive and of $(3-n)\pi/2$ when it is negative.

For the transmitted pulse and the scattering body shown at the top of Fig. 5, (which pulse has been drawn on an echo range scale to facilitate comparison with the echo), the envelope of the echo from the scattering body of Fig. 5 a would be ap-

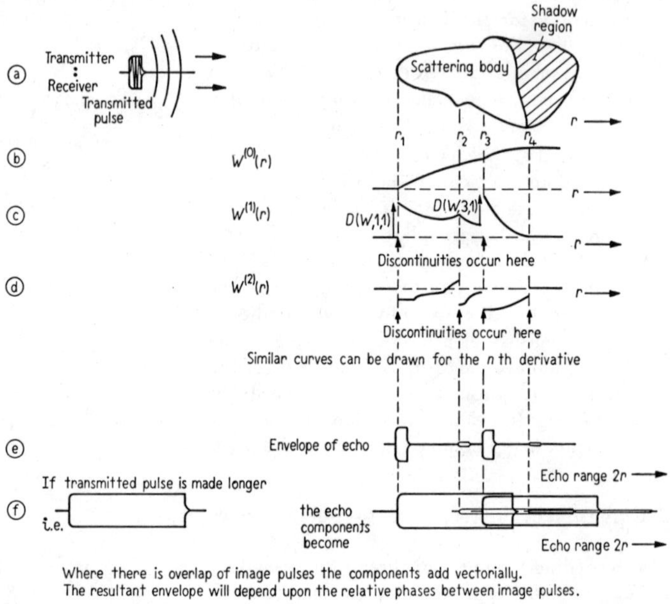

Fig. 5. Principles of echo formation for convex, rigid bodies whose dimensions are large in terms of wavelengths; for (a) to (f) see the text.

The interpretation of the theory is that if a pulse, such as that shown at Fig. 5 a, is transmitted towards the object, each discontinuity in $\mathrm{d}^nW(r)/\mathrm{d}r^n$ $(n=0, 1, 2, \ldots)$ engenders a component towards the echo, which component has the same envelope as the transmission pulse. The resultant of all components created at the same range by discontinuities in various derivatives of $W(r)$, which resultant is the quantity that is physically observable, also has an envelope identical with that of the transmitted pulse and is therefore termed an "image pulse".

The magnitude of each echo component is proportional to the size of its generating discontinuity, and also generally decreases fairly rapidly with increase in the order, n, of derivative in which the discontinuity occurs. Thus, for a given image pulse, usually only the first one or two orders of derivative in which discontinuities occur need to be considered. The magnitude of each echo component is also proportional to $(\text{frequency})^{1-n}$, so that, except for $n=0$ or 1, it falls with increase of frequency. The phase within each echo component depends upon the range of its generating discontinuity and, ad-

proximately as shown at Fig. 5 e. The image pulses at ranges r_1 and r_3 have been drawn proportional to $D(W,1,1)$ and $D(W,3,1)$, respectively, it being assumed that the contributions at these ranges due to discontinuities in $\mathrm{d}^2W(r)/\mathrm{d}r^2$ are small compared with those in $\mathrm{d}W(r)/\mathrm{d}r$. The amplitudes of the image pulses at ranges r_2 and r_4 have been drawn proportional to the discontinuities in $\mathrm{d}^2W(r)/\mathrm{d}r^2$ at these ranges, but on a smaller scale than the image pulses at ranges r_1 and r_3.

That echo contributions should be received only from certain discrete ranges and not from the whole of the scattering surface may be seen to be plausible by the following argument. As an element of the transmission pulse moves outwards over the body, the phase, $(\omega t - 2kr)$, associated with its echo contribution remains constant, range and time increasing in step. Thus, if there is no sudden change in solid angle between an element of surface at range r and one at range $r+\mathrm{d}r$, there will be no A. C. contribution to the echo. A sudden change will however leave a net A. C. resultant of magnitude both proportional to this change and to the amplitude of the element of incident pulse. Contributions of the successive elements of the transmission pulse will thus trace out an echo component having the same envelope as the transmission.

If the transmission pulse is arbitrarily lengthened as in Fig. 5 f, the image pulses will overlap. Where there is overlap, the components add vectorially and the resultant envelope will depend upon the relative phases between image pulses. Examples of the type of envelope that may arise from the vectorial addition of just two image pulses, assumed to have exponentially rising and decaying envelopes, are shown in Fig. 6. If one varies the aspect angle at

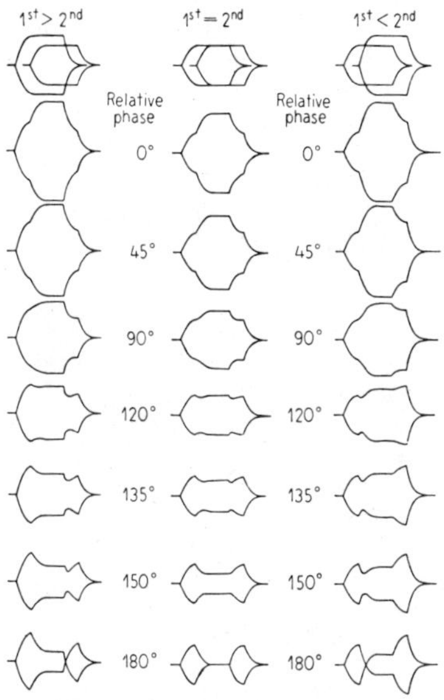

Fig. 6. Vectorial addition of two exponential image pulses.

which a body is viewed, the positions at which projection discontinuities occur will, in general, alter and, hence, so also will the range differences between the sources of image pulses. Should the body be very large, in terms of wavelengths, even quite small changes of aspect may cause changes of separation of the order of a wavelength. As a change of separation between two discontinuities of only $\lambda/4$ will produce a change of 180° in relative phase, it is evident that even small lateral relative movements of transducers and scattering body may cause large changes in the echo envelope when there is overlap of image pulses, and this affords a mechanism of fading.

Providing the transmission length is not large compared with the range extent of the obstacle and where the spacing of discontinuities is small relative to the transmission length and is such that many image pulses are contributing to the echo at any one time, then due to the random phases of the components, the shape of the echo envelope effectively loses its relationship to the transmission pulse envelope and is governed primarily by statistical considerations.

Thus, apart from the very long transmission case, there are three distinct types of echo envelope, depending upon the degree of overlap of image pulses. Where all image pulses are resolved, the echo envelope consists of a series of components whose envelopes are identical, except in size. Where there are few simultaneously overlapping pulses, the envelope shape is governed by laws of vectorial addition of pulse shapes. Where there are many simultaneously overlapping image pulses, the shape of the envelope is governed by probability laws.

When the transmission length is much greater than the range extent of the scattering body, a fourth type of envelope emerges. The echo now consists, in the main, of a central region in which all the image pulses are contributing simultaneously, flanked at beginning and end, respectively, by transient periods of growth or decrease of the number of contributing image pulses. These transients are of the relatively short duration corresponding to the spacing between the nearest and furthest discontinuities. Relative to the pulse length, the time delays between onset of the various image pulses are small and, as an approximation, their envelopes can all be considered to start at the same time, although the phase differences between the image pulses must not be ignored. Apart from during the short transients at beginning and end, the echo will therefore have approximately the same shape as the transmitted pulse. The magnitude of the echo envelope can be forecast where the magnitudes and relative phases of the image pulses are known, but when the scattering body contains an appreciable number of randomly spaced discontinuities, the magnitude of the echo envelope will be subject to probability considerations.

As the phase within one image pulse is, in general, different from that within another, there will be changes of phase within the echo (additional to the implicit sinusoidal variation) whenever the relative magnitudes of overlapping image pulses vary, as at the start or end of one of these overlapping image pulses. Apart from the very long transmission case, two types of phase behaviour can be discerned. For small numbers of simultaneously overlapping image pulses, the resultant phase will vary according to laws of vectorial addition, and as an example of this the phases associated with the several envelopes of Fig. 6 are illustrated in Fig. 7. For large numbers of simultaneously overlapping image pulses, the phase within the echo will vary randomly.

When the transmission length is much greater than the range extent of the scattering body, then over the main central echo region in which all the image pulses are contributing simultaneously, the phase will stay relatively constant, albeit with slow variations if the echo envelope is not flat.

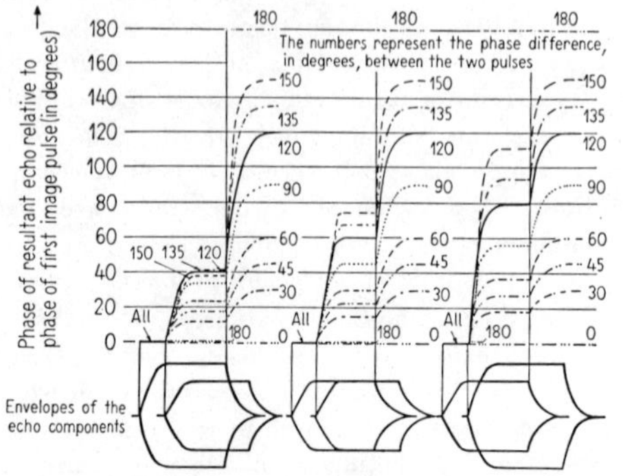

Fig. 7. Phases of echoes consisting of two overlapping exponential image pulses.

If the transmission is made long enough for a central region of the echo to occur in which all image pulses overlap at constant level, this region corresponds to the echo for C. W. transmission and will, of course, have both constant amplitude and phase.

All the foregoing description may be extended quite simply to the case where the transducer directivities are not uniform over the scattering body. The contribution of each element of body surface to the total solid angle must be weighted by the transmitter and receiver directivities appropriate to the position of that surface element. The total solid angle, $W(r)$, is replaced by the directivity-weighted total solid angle, $W_w(r)$, the principles remaining otherwise the same. The shape of the curve of $W_w(r)$ versus range will differ from that of $W(r)$ versus range, but the discontinuities will occur at the same ranges, the transducer directivities being themselves continuous. The magnitudes of the discontinuities will however be altered.

For a body whose complete echo or its major component arises from a single point, as occurs in the case of an infinite plane, or of a large sphere, the echo level is independent, or almost independent, of the shape of the product directivity curve. It will depend only on the transducer sensitivities in the direction of the echo-producing point. A correctly aimed, highly directional 'product beam' [3] of given peak sensitivity will give the same echo level as an omnidirectional 'product beam' of the same sensitivity.

When the change in range over the scattering body is small compared with the total range, the unweighted or weighted solid angle, $W(r)$ or $W_w(r)$, used in the above description may be replaced respectively by the unweighted or weighted areal projection, $A(r)$ or $A_w(r)$, towards the transducers of that part of the scattering body lying within range r.

Examples showing the application of the foregoing theory to some bodies of simple shape will be given in a further paper.

Acknowledgements

This paper is published by permission of the British Admiralty. The author is grateful to Mr. M. J. Daintith and to several other colleagues for helpful discussions on various points of this work.

(Received October 16th, 1961.)

References

[1] Fock, V. A. Journal of Physics, U.S.S.R. **10** [1946], 130.
[2] Franz, W. and Deppermann, K., Ann. Phys. **10** [1952], 361.
[3] Deppermann, K. and Franz, W. Ann. Phys. **14** [1954], 253.
[4] Van de Hulst, H. C., I. R. E. Trans. AP-4 [1959], 195.
[5] Wu, T. T. Phys. Rev. **104** [1956], 1201.
[6] Van de Hulst, H. C., Light scattering by small particles. Chapman and Hall, Ltd., London 1957, p. 333.
[7] Meecham, W. C., J. Rational Mech. Anal. **5** [1956], 323.

[3] The 'beam' obtained by taking the product of the transmitter and receiver directivities.

ACUSTICA
S. HIRZEL VERLAG · STUTTGART
22
Vol. 12 1962 No. 2

THE HIGH FREQUENCY ECHO STRUCTURE OF SOME SIMPLE BODY SHAPES

by A. Freedman

Admiralty Underwater Weapons Establishment, Portland, England

Summary

The echo structure and the backscattering cross section are calculated for each of several rigid bodies of simple shape. The calculations are based on the results of a companion paper which elucidated the primary echo formation mechanism applicable when high frequency acoustic waves (under C. W. or pulse conditions) are scattered by a rigid body immersed in an ideal, fluid medium.

Sommaire

On a calculé la structure de l'écho et le profil de la dispersion réfléchie dans le cas de plusieurs solides rigides et de formes simples. Les calculs utilisent les résultats d'un article déjà paru dans lequel on expliquait le mécanisme de la formation d'un écho primaire, mécanisme qui s'applique lorsque des ondes acoustiques de haute fréquence (sous forme d'ondes continues ou encore présentant les caractères d'impulsion) sonst dispersées par un solide rigide immergé dans un milieu parfaitement fluide.

Zusammenfassung

Die Struktur der Echos und der Rückstreuquerschnitt werden für eine Reihe von festen Körpern mit einfacher Form berechnet. Die Rechnungen beruhen auf den Ergebnissen einer gleichzeitigen Untersuchung, die den Mechanismus der Echobildung bei der Streuung kurzer akustischer Welllen an in einem idealen fluiden Medium befindlichen festen Körper aufgeklärt hat.

1. Introduction

If a rigid body, immersed in a ideal fluid medium, is irradiated by a pulse of small-amplitude, acoustic waves, the echo may be shown [1] to be composed of a number of discrete pulses. On the assumption of certain bandwidth limitations, the envelope of each of these pulses is (to a close degree of approximation) a replica of that of the transmission pulse. Hence, these echo pulses are termed 'image pulses'. An image pulse is generated whenever there is a discontinuity, with respect to range r, in $\mathrm{d}^n W(r)/\mathrm{d}r^n$, where $W(r)$ is the solid angle subtended at the transducers by that part of the scattering body within range r. n may be zero or any positive integer.

The results of [1] are here used to determine the echo structure and the backscattering cross section of several bodies of simple shape. To comply with the assumptions underlying the formulae abstracted from the previous paper, the spectrum of the transmission signal will be assumed not to contain appreciable components outside a bandwidth which is small compared with the mean transmission frequency. Apart from this restriction, the transmission will be deemed to be an amplitude-modulated pulse of arbitrary envelope.

2. Summary of principal general formulae

The relevant formulae from [1] will first be presented. For the reader's convenience, the former equation numbers will also be given.

Let r_1 denote the range of the nearest part of the scattering body and let k denote (2π/mean transmission wavelength). At ranges such that $k r_1 \gg 1$, the solution for the receiver voltage, E, is

$$E = \sum_{g=1}^{f} E_g, \qquad (1) \; ((21) \text{ in } [1])$$

where E_g represents an image pulse and is given by

$$E_g = \mathrm{i}\frac{M_g}{\lambda}\exp[-\mathrm{i}\,2\,k(r_g - r_1)] \sum_{n=0}^{\infty} \frac{D(W_{\mathrm{w}}, g, n)}{(\mathrm{i}\,2\,k)^n}, \qquad (2) \; ((18) \text{ in } [1])$$

in which

$$M_g = V\left(t - \frac{2\,r_g}{c}\right) P\,H \exp[\mathrm{i}(\omega\,t - 2\,k\,r_1)]. \qquad (3) \; ((20) \text{ in } [1])$$

In the above, $V(t - 2\,r_g\,c^{-1})$ = envelope of voltage pulse applied to the transmitter, (on a time-delayed scale); P = transmitter sensitivity; H = receiver sensitivity; c = wave velocity in fluid medium; ω = mean angular frequency of transmission; λ = mean wavelength; t = time; r_g = a range at which finite discontinuities occur in one or more values of $d^n W_w(r)/dr^n$, $(n=0,1,2,\ldots)$; $W_w(r)$ = value of $W(r)$ when each element of solid angle is weighted by the product of the corresponding transmitter and receiver directivities; $D(W_w, g, n)$ = discontinuity in $W_w^{(n)}(r)$, i. e. in $d^n W_w(r)/dr^n$, at range r_g; r_f = range of furthest directly irradiated part of scattering body.

If the range extent of the directly irradiated part of the scattering body is much smaller than the mean range, r_m, to the transducers, eq. (2) simplifies to

$$E_g = i\frac{M_g}{\lambda r_m^2}\exp[-i\,2\,k(r_g-r_1)]\sum_{n=0}^{\infty}\frac{D(A_w,g,n)}{(i\,2\,k)^n}, \quad (4)\ ((19)\ \text{in}\ [1])$$

where $D(A_w, g, n)$ = discontinuity in $A_w^{(n)}(r)$ at range r_g; $A_w(r)$ = directivity-weighted value of $A(r)$; $A(r)$ = area of projection towards the transducers of those parts of the scattering body within range r.

At ranges which are shorter than comply with $k r_1 \gg 1$ and whose lower limit, whilst not yet established, is believed, intuitively, to be of the order of the maximum body dimensions, E_g becomes

$$E_g = i\frac{M_g}{\lambda}\exp[-i\,2\,k(r_g-r_1)]\times \sum_{n=0}^{\infty}\frac{D(W,g,n)}{(i\,2\,k)^n}[1-K(n)], \quad (5)\ ((22)\ \text{in}\ [1])$$

where

$$K(n)=0 \quad (n=0),$$
$$= \frac{2}{(n-1)!}\sum_{n=0}^{\infty}\frac{(n+b-1)!}{(-i\,2\,k\,r_g)^{b+1}} \quad (n=1,2,3,\ldots). \quad (6)\ ((23)\ \text{in}\ [1])$$

The dropping of the suffix from $D(W_w, g, n)$ or $D(A_w, g, n)$ means that no directivity-weighting is implied.

The quantities "target strength", (T. S.), and "backscattering cross section", σ, defined in the usual manner, can be expressed as T. S. $= 10\log|J|^2$ and $\sigma = 4\pi|J|^2$, where at long ranges

$$J=\frac{1}{\lambda}\sum_{g=1}^{f}\exp[-i\,2\,k(r_g-r_1)]\sum_{n=0}^{\infty}\frac{D(A,g,n)}{(i\,2\,k)^n}, \quad (7)$$

and at short ranges, subject to a restriction described in the appendix,

$$J=\frac{1}{\lambda}\sum_{g=1}^{f}r_g^2\exp[-i\,2\,k(r_g-r_1)]\times \sum_{n=0}^{\infty}\frac{D(W,g,n)}{(i\,2\,k)^n}[1-K(n)]. \quad (8)$$

For certain body shapes some of the discontinuities, $D(W, g, n)$ or $D(A, g, n)$, are infinite, and the above formulae are then not applicable. A disc at non-axial incidence provides an example of this behaviour.

3. Examples of echo structure

The theoretical analysis of [1] utilised the KIRCHHOFF approximation. The condition, implicit in that approximation, that radii of curvature shall be large in terms of wavelengths will not be strictly adhered to in some of the examples, as for instance at the apex and base of a cone. An indication that this need not introduce very serious errors is given by comparison of the exact and physical optics solutions for the semi-infinite cone which SIEGEL, CRISPIN and SCHENSTED [2] show to be in excellent accord for large and for small cone angles in the electromagnetic case and for large cone angles in the acoustic case. The KIRCHHOFF assumptions are also violated in the case of infinitely thin, flat bodies of finite extent, but in this instance recourse to the analogy of flat radiators vibrating without a baffle indicates that the results of our physical optics analysis will provide useful answers. Experimental measurements of the FRAUENHOFER fields of such radiators have shown good agreement with the forecasts of physical optics theory for directions which do not approach too closely to the plane of the radiator, and it can similarly be expected that our echo theory will give reasonable forecasts for thin flat bodies providing the angle of incidence is not close to grazing. However, at angles approaching grazing incidence, it is conceivable that serious errors may arise due to the formation of creeping waves which could significantly affect the field in the vicinity of the edge.

In all the examples, except the last, the transducers will be assumed to be non-directional. Except where specifically stated to the contrary, the analyses apply to the situation where the range is sufficiently large for the incident wavefront to be considered plane over the extent of the scattering body, and where, also, eq. (4) rather than eq. (2) can be employed.

3.1. The sphere and the axially sited spherical cap

The geometry and designations are as shown in Fig. 1 a and g. The values of $A^{(n)}(r)$ within the range $r_1 \leqq r \leqq r_2$ are shown in Table I and plots of $A^{(n)}(r)$ are illustrated in Fig. 1 b to 1 e and 1 h to 1 k. Discontinuities occur at the near and far limits of irradiation and their values are listed in Table I.

For the sphere, the echo consists of two image pulses formed, respectively, at the near point and

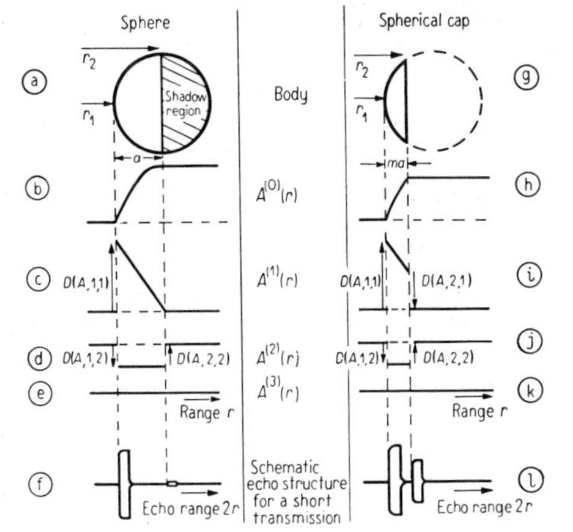

Fig. 1. Echoes from sphere and spherical cap; for (a) to (l) see the text.

Table I.

Discontinuity data for sphere and spherical cap.

n	$A^{(n)}(r)$	$D(A,1,n)$	$D(A,2,n)$ sphere	$D(A,2,n)$ spherical cap
0	$\pi a^2\left[1-\left(\frac{a-(r-r_1)}{a}\right)^2\right]$	0	0	0
1	$2\pi[a-(r-r_1)]$	$-2\pi a$	0	$2\pi a(1-m)$
2	-2π	2π	-2π	-2π
3	0	0	0	0

at the equator, and illustrated schematically in Fig. 1 f for an assumed short transmission pulse. The two image pulses are given by

$$E_1 = -\frac{M_1}{r_m^2}\frac{a}{2}\left(1+\frac{i}{2\,k\,a}\right) \qquad (9)$$

and

$$E_2 = i\,\frac{M_2}{r_m^2}\frac{1}{4\,k}\exp(-i\,2\,k\,a),$$

and the scattering cross section is

$$\sigma = \pi\,a^2\left[1-\frac{\sin 2\,k\,a}{k\,a}+\left(\frac{\sin k\,a}{k\,a}\right)^2\right]. \qquad (10)$$

The ratio of magnitudes of the two image pulses is approximately $2\,k\,a$, so that for sphere sizes large enough for the present theory to apply, the second component becomes quite small.

Weston [3] has treated the backscattering of an electro-magnetic pulse from a perfectly conducting sphere. His results can be interpreted as the formation of two image pulses, arising at the positions described above, of creeping wave contributions and of certain DC transients. Ignoring terms of the order of $(k\,a)^{-2}$ and beyond, and allowing for the use of the opposite sign convention for phase to that employed in this paper, the expression for Weston's first image pulse agress with eq. (9). The magnitude of his second image pulse, the derivation of which is based on exact theory, is stated to be of the order of $(k\,a)^{-1}$ of that of the first; this also agrees with our results and indicates that, even though the second image pulse is formed at parts of the surface bordering on grazing incidence; a treatment based on the Kirchhoff assumptions has given a magnitude which is, at worst, of the correct order.

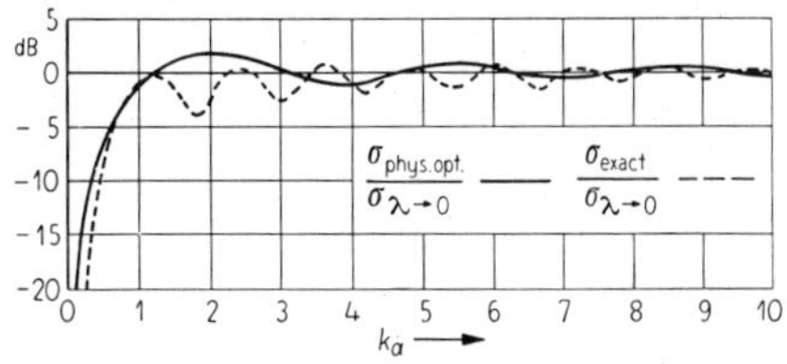

Fig. 2. Comparison of exact and physical optics solutions for the sphere.

Comparison in Fig. 2 of the curve of eq. (10) with the curve for the exact solution (Stenzel [4]) indicates that in the region of small $k\,a$ there are considerable differences, as would be expected, because a) the present treatment ignores the contribution of creeping waves, b) the formation of the latter at low values of $k\,a$ will particularly affect the value of the field near the equator and, hence, the magnitude of the second image pulse.

For the spherical cap at axial incidence, the first image pulse is the same as for the sphere, and the second image pulse, which is formed at the cap rim, (Fig. 1 g), is given by

$$E_2 = \frac{M_2}{r_m^2}\frac{a}{2}\left(1-m+\frac{i}{2\,k\,a}\right)\exp(-i\,2\,k\,m\,a).$$

Providing $m \not\approx 1$, this is of the same order of magnitude as the first image pulse. The scattering cross section is

$$\sigma = \pi\,a^2\left|4(1-m)\sin^2 k\,m\,a + m^2\left[1-\frac{\sin 2\,k\,m\,a}{k\,m\,a}+\left(\frac{\sin k\,m\,a}{k\,m\,a}\right)^2\right]\right|.$$

In the high frequency limit, under C. W. conditions,

$$E_{spherical\ cap}/E_{sphere} = 1-(1-m)\exp(-i\,2\,k\,m\,a).$$

The mode of variation of this quantity, with cap thickness, is illustrated in Fig. 3, which is plotted, for convenience, for $a = 5\,\lambda$. The initial part of the curve should be discounted as, in this region, the cap dimensions seriously violate the assumptions of the Kirchhoff approximation.

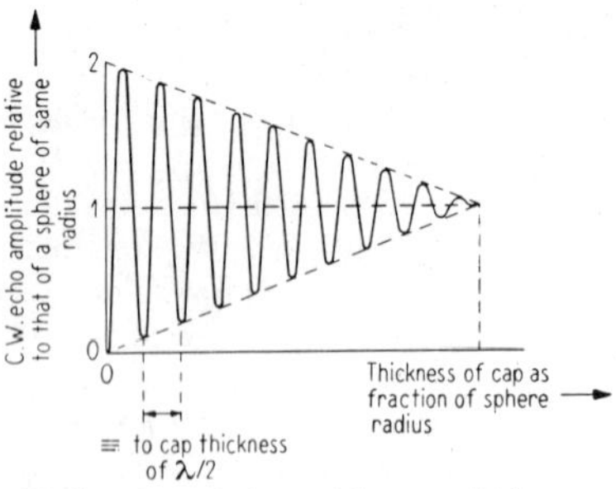

Fig. 3. Mode of variation with cap thickness of the C. W. echo amplitude from a spherical cap. (Plotted for the case where $\alpha = 5\,\lambda$.)

3.2. *The sphere at short range*

The geometrical parameters are as shown in Fig. 4 and the values of $W^{(n)}(r)$ and of the discontinuities occuring at the near and far limits of irradiation are listed in Table II.

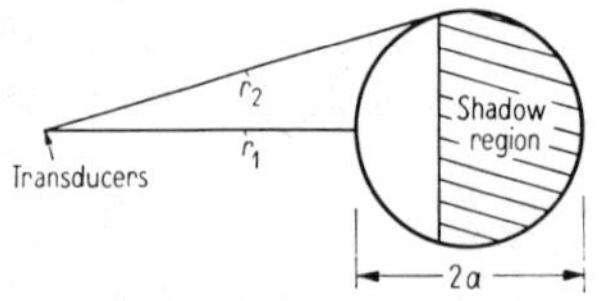

Fig. 4. Sphere when the incident wave is spherical.

Substitution in eq. (5) yields

$$E_1 = \mathrm{i}\, M_1 \frac{k}{(1 + a\, r_1^{-1})} \times \times \left[\frac{a\, r_1^{-1}}{(-\mathrm{i}\, 2\, k\, r_1)} + \sum_{n=2}^{\infty} \frac{(n-2)\,!}{(-\mathrm{i}\, 2\, k\, r_1)^n} \right]$$

and

$$E_2 = -\,\mathrm{i}\, M_2 \exp[-\mathrm{i}\, 2\, k (r_2 - r_1)] \times \times \frac{r_2}{r_1} \frac{k}{(1 + a\, r_1^{-1})} \sum_{n=2}^{\infty} \frac{(n-2)\,!}{(-\mathrm{i}\, 2\, k\, r_2)^n},$$

where $r_2 = r_1 (1 + 2\, a\, r_1)^{1/2}$.

Retaining only the first term in each of the above two series,

$$E_1 \approx -\frac{M_1}{r_1^2} \frac{a}{2} \left(1 + \frac{\mathrm{i}}{2\, k\, a}\right) \frac{1}{(1 + a\, r_1^{-1})},$$

$$E_2 \approx \mathrm{i} \frac{M_2}{r_1\, r_2} \frac{1}{4\, k} \exp[-\mathrm{i}\, 2\, k (r_2 - r_1)] \frac{1}{(1 + a\, r_1^{-1})}.$$

The magnitude of each of the two image pulses differs by the factor $1/(1 + a\, r_1^{-1})$ from the previously given long range values.

In the high frequency limit, $\sigma = \pi [a\, r_1 / (a + r_1)]^2$, which is the same result as derived by Senior and Siegel [5] from Weston's [3] analysis.

3.3 *The spheroid*

For incidence parallel to the axis of revolution (Fig. 5 a) and for incidence normal to this axis (Fig. 5 b), the values of $A^{(n)}(r)$ between r_1 and r_2, together with the relevant discontinuities, are given in Tables III and IV, respectively.

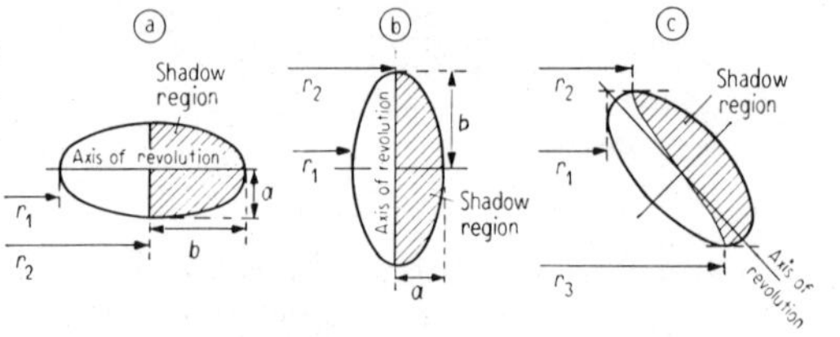

Fig. 5. The spheroid.
(a) Incidence along axis of revolution,
(b) Incidence normal to axis of revolution,
(c) Incidence at arbitrary angle.

Table III.
Discontinuity data for spheroid whose axis of revolution is parallel to the direction of incidence.

n	$A^{(n)}(r)$	$D(A, 1, n)$	$D(A, 2, n)$
0	$\pi (a/b)^2 [b^2 - (r_2 - r)^2]$	0	0
1	$2\,\pi (a/b)^2 (r_2 - r)$	$-2\,\pi\, a^2/b$	0
2	$-2\,\pi (a/b)^2$	$2\,\pi (a/b)^2$	$-2\,\pi (a/b)^2$
3	0	0	0

Table IV.
Discontinuity data for spheroid whose axis of revolution is normal to the direction of incidence.

n	$A^{(n)}(r)$	$D(A, 1, n)$	$D(A, 2, n)$
0	$\pi (b/a) [a^2 - (r_2 - r)^2]$	0	0
1	$2\,\pi (b/a) (r_2 - r)$	$-2\,\pi\, b$	0
2	$-2\,\pi (b/a)$	$2\,\pi (b/a)$	$-2\,\pi (b/a)$
3	0	0	0

Table II.
Discontinuity data for sphere at short range.

n	$W^{(n)}(r)$	$D(W, 1, n)$	$D(W, 2, n)$
0	$\frac{\pi\, a^2}{r (r_1 + a)} \left[1 - \left(\frac{a - (r - r_1)}{a}\right)^2\right]$	0	0
1	$\frac{\pi (r_1^2 + 2\, r_1\, a - r^2)}{(r_1 + a)\, r^2}$	$\frac{n!\, \pi (1 + 2\, a\, r_1^{-1})}{(1 + a\, r_1^{-1})(-r_1)^n} + \frac{\pi}{(1 + a\, r_1^{-1})\, r_1}$	0
2, 3, 4 ...	$\frac{n!\, \pi (r_1 + 2\, a)\, r_1}{(r_1 + a)(-r)^{n+1}}$	$\frac{n!\, \pi (1 + 2\, a\, r_1^{-1})}{(1 + a\, r_1^{-1})(-r_1)^n}$	$\frac{n!\, \pi}{(1 + a\, r_1^{-1})\, r_1 (-r_2)^{n-1}}$

For both directions of incidence, the echo consists of two image pulses.

For the "parallel to axis" direction,

$$E_1 = -\frac{M_1}{r_m^2}\frac{a^2}{2b}\left(1+\frac{i}{2kb}\right),$$

$$E_2 = i\frac{M_2}{r_m^2}\frac{a^2}{4kb^2}\exp(-i\,2kb),$$

$$\sigma = \frac{\pi a^4}{b^2}\left[1-\frac{\sin 2kb}{kb}+\left(\frac{\sin kb}{kb}\right)^2\right].$$

For the "normal to axis" direction,

$$E_1 = -\frac{M_1}{r_m^2}\frac{b}{2}\left(1+\frac{i}{2ka}\right),$$

$$E_2 = i\frac{M_2}{r_m^2}\frac{b}{4ka}\exp(-i\,2ka),$$

$$\sigma = \pi b^2\left[1-\frac{\sin 2ka}{ka}+\left(\frac{\sin ka}{ka}\right)^2\right].$$

At the high frequency limit, the above two values of σ correspond to the geometrical optics solution, $\sigma = \pi R_1 R_2$, where R_1 and R_2 are the principal radii of curvature at the nearest part of the scattering body.

When viewed along the axis of revolution, a spheroid for which $a = a_1$ and $b = m a_1$ will have identical echo structure with that of a spheroid viewed normally to the axis of revolution and for which $a = m a_1$ and $b = a_1/m$. The magnitude, phase and separation of the image pulses will be the same.

For angles of incidence other than along an axis of symmetry, the limits of irradiation will be of the type indicated in Fig. 5c. In addition to the main echo component formed at the range of the nearest part of the spheroid, there will be two further components, formed at ranges r_2 and r_3.

3.4. The cone and the truncated cone

For axial incidence on the curved surface, the designations for a complete right circular cone and for a cone truncated parallel to the base are given in Fig. 6 a and 6 g. The values of $A^{(n)}(r)$ between r_1 and r_2 are listed in Table V, and plots of $A^{(n)}(r)$ are shown in Fig. 6. Restricting the description to the most important features of these curves, there is a discontinuity in the second derivative at the cone apex, in the first derivative at the cone base and in the zero derivative at the surface of truncation. The values are presented in Table V and the echo structure is shown schematically in Fig. 6 f and 6 l.

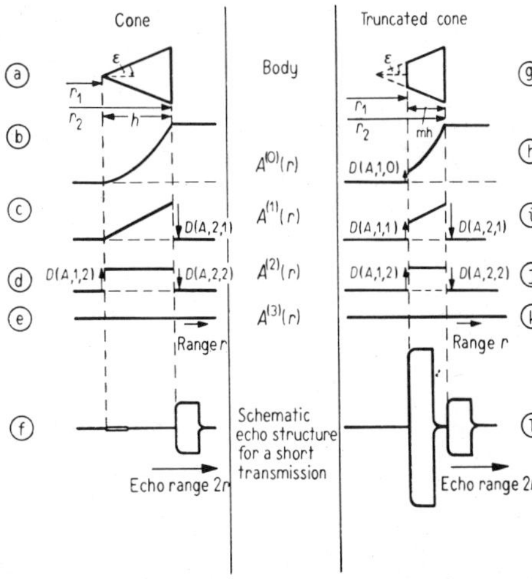

Fig. 6. Axial echoes from cone and truncated cone; for (a) to (l) see the text.

For the complete cone

$$E_1 = i\frac{M_1}{r_m^2}\frac{1}{4k}\tan^2\varepsilon,$$

$$E_2 = \frac{M_2}{r_m^2}\frac{h}{2}\left(1-\frac{i}{2kh}\right)\tan^2\varepsilon\exp(-i\,2kh),$$

$$\sigma = \pi h^2\tan^4\varepsilon\left[1-\frac{\sin 2kh}{kh}+\left(\frac{\sin kh}{kh}\right)^2\right].$$

As $h \to 0$, (E_1+E_2) can be shown to transform to the value of E corresponding to a disc.

At aspect angles, α, between 0° and 90°, (0° corresponding to axial incidence), image pulses are associated with the apex and with the nearest and furthest points on the base rim receiving direct irradiation. The sequence in which the image pulses arise is not only a function of aspect angle, but depends also upon whether the cone semiangle is less or greater than 45°. This is illustrated in Fig. 7. The special aspects, such as axial, normal to

Table V.
Discontinuity data for cone and truncated cone.

n	$A^{(n)}(r)$	$D(A,1,n)$ cone	$D(A,1,n)$ truncated cone	$D(A,2,n)$
0	$\pi[h-(r_2-r)]^2\tan^2\varepsilon$	0	$-\pi h^2(1-m)^2\tan^2\varepsilon$	0
1	$2\pi[h-(r_2-r)]\tan^2\varepsilon$	0	$-2\pi h(1-m)\tan^2\varepsilon$	$2\pi h\tan^2\varepsilon$
2	$2\pi\tan^2\varepsilon$	$-2\pi\tan^2\varepsilon$	$-2\pi\tan^2\varepsilon$	$2\pi\tan^2\varepsilon$
3	0	0	0	0

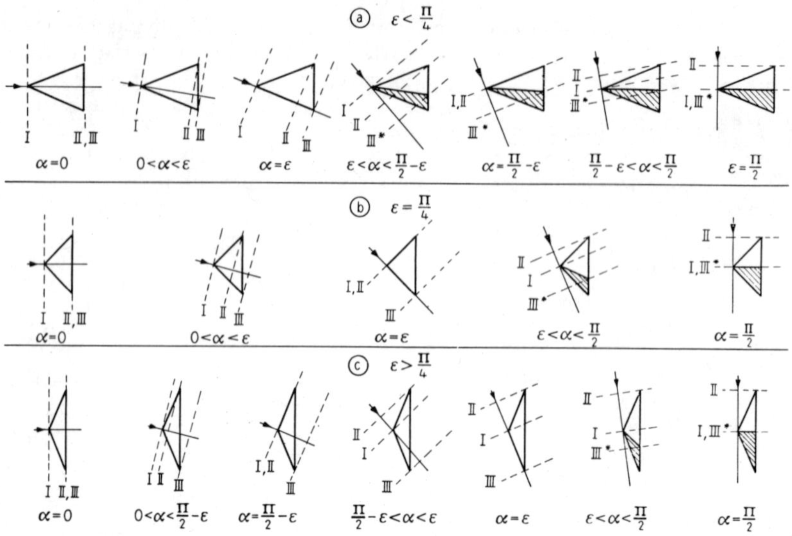

Fig. 7. The sequence of echo components from the curved surface of a cone.

Symbols:

ε Cone semiangle,
α Angle of incidence,
I Image pulse arising at apex,
II Image pulse arising at nearest point on base rim,
III Image pulse arising at furthest point on base rim,
III* Image pulse arising at furthest directly irradiated points on base rim.

axial and normal to the cone surface, are each associated with the coalescence of pairs of image pulses.

For the truncated cone at axial incidence

$$E_1 = -\mathrm{i}\,\frac{M_1}{r_\mathrm{m}^2}\,\frac{k\,h^2}{2}\,(1-m)^2 \times \left(1 - \frac{\mathrm{i}}{k\,h(1-m)} - \frac{1}{2\,k^2\,h^2(1-m)^2}\right)\tan^2\varepsilon\,,$$

$$E_2 = \frac{M_2}{r_\mathrm{m}^2}\,\frac{k\,h^2}{2}\left(\frac{1}{k\,h} - \frac{\mathrm{i}}{2\,k^2\,h^2}\right)\tan^2\varepsilon\,\exp(-\mathrm{i}\,2\,k\,m\,h)$$

and

$$\sigma = 4\,\pi\left(\frac{S}{\lambda}\right)^2\left[1 + \frac{2\sin 2\,k\,m\,h}{k\,h(1-m)^2} + \frac{1}{k^2\,h^2(1-m)^2} \times \left(\frac{1}{(1-m)^2} - \frac{2\cos 2\,k\,m\,h}{1-m} + \cos 2\,k\,m\,h\right)\right],$$

where S is the area of the surface of truncation and where terms of the order $(k\,h)^{-3}$ and beyond have been ignored.

3.5. *The semi-infinite cone*

The difference between this case and that of the finite cone is that, the base rim having been removed to infinity, the echo only consists of the image pulse from the apex. Thus, for axial incidence, we have

$$E = E_1 = \mathrm{i}\,\frac{M_1}{r_1^2}\,\frac{1}{4\,k}\,\tan^2\varepsilon\,, \quad \text{and}$$

$$\sigma = \frac{\lambda^2}{16\,\pi}\,\tan^4\varepsilon\,.$$

The latter is a well known physical optics result. Siegel, Crispin and Schensted [2] have shown that, to the degree of approximation used in their computations, there is complete agreement, for electromagnetic scattering, between this result and the exact solution when $\varepsilon \approx 90°$ and when $\varepsilon \approx 0°$. They have also shown that, for acoustic scattering, similar agreement with the exact solution exists when $\varepsilon \approx 90°$ but that when $\varepsilon \approx 0°$ the physical optics solution forecasts an echo amplitude which is twice that given by the exact solution. That a physical optics analysis should give too large an answer for very small cone angles is not altogether surprising, for at $\varepsilon \approx 0°$ one would intuitively expect the formation of surface waves, akin to creeping waves, and this would reduce the energy available for backscattering. What is surprising is that this does not appear to occur in the electro-magnetic case.

For aspect angles, α, such that $\tan\alpha$ is less than both $\tan\varepsilon$ and $\cot\varepsilon$, the values of $A^{(n)}(r)$ and of the discontinuity at the apex are shown in Table VI.

Table VI.
Discontinuity data for semi-infinite cone.

n	$A^{(n)}(r)$	$D(A, 1, n)$
1	$\dfrac{2\pi(r-r_1)\tan^2\varepsilon}{\cos^3\alpha(1-\tan^2\alpha\tan^2\varepsilon)^{3/2}}$	0
2	$\dfrac{2\pi\tan^2\varepsilon}{\cos^3\alpha(1-\tan^2\alpha\tan^2\varepsilon)^{3/2}}$	$\dfrac{-2\pi\tan^2\varepsilon}{\cos^3\alpha(1-\tan^2\alpha\tan^2\varepsilon)^{3/2}}$
3	0	0

Hence

$$E = E_1 = \mathrm{i}\,\frac{M_1}{r_1^2}\,\frac{\tan^2\varepsilon}{4k}\,\frac{1}{\cos^3\alpha(1-\tan^2\alpha\tan^2\varepsilon)^{3/2}},$$

$$\sigma = \frac{\lambda^2\tan^4\varepsilon}{16\pi}\,\frac{1}{\cos^6\alpha(1-\tan^2\alpha\tan^2\varepsilon)^3}.$$

Fig. 8 shows curves of $\sigma/\sigma_{\alpha=0}$ versus α for several values of ε. When $\alpha=\varepsilon$, the echo is associated with scattering from a cone generator of infinite length, and σ becomes infinite.

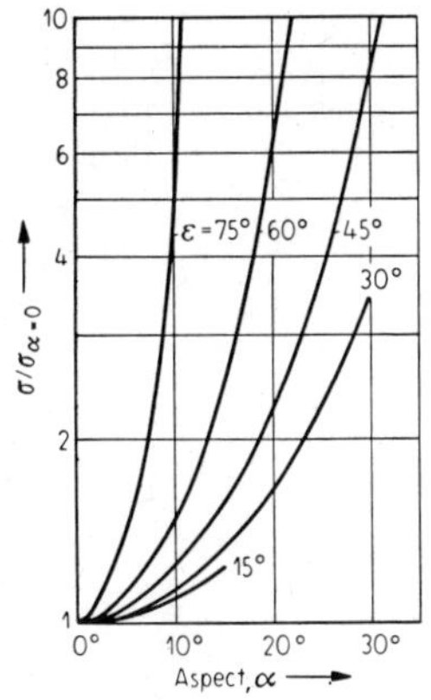

Fig. 8. Variation of backscattering cross section of semiinfinite cone with aspect.

3.6. *An arbitrary finite plane surface*

Consider a plane surface, of area S, whose outline is smooth but whose shape is otherwise arbitrary. Assuming normal incidence, the only discontinuity in $A^{(n)}(r)$ is in the zero derivative, being given by $D(A, 1, 0) = -S$. Hence

$$E = E_1 = -\mathrm{i}\,\frac{M_1}{r_1^2}\,\frac{S}{\lambda} \tag{11}$$

and
$$\sigma = 4\pi\left(\frac{S}{\lambda}\right)^2,$$

the latter being a well known result.

3.7. *The disc*

Assuming axially sited transducers and spherical incident waves, (Fig. 9), the values of $W^{(n)}(r)$ are shown in Table VII, together with the values of the discontinuities which occur at the centre and at the periphery of the disc.

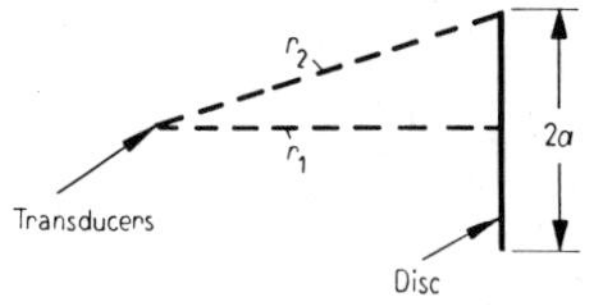

Fig. 9. Disc with axially sited transducers.

Table VII.
Discontinuity data for disc with transducers on axis.

n	$W^{(n)}(r)$	$D(W, 1, n)$	$D(W; 2, n)$
1, 2, 3, ...	$\dfrac{2\pi n!\,r_1}{(-r)^{n+1}}$	$\dfrac{2\pi n!}{(-r_1)^n}$	$\dfrac{2\pi n!\,r_1}{(-r_2)^{n+1}}$

Substitution in eq. (5) yields

$$E_1 = -\frac{M_1}{2r_1},$$

$$E_2 = M_2\,\frac{r_1}{2r_2^2}\exp\left[-\mathrm{i}\,2kr_1\left\{\left(1+\frac{a^2}{r_1^2}\right)^{1/2}-1\right\}\right].$$

As the range of the transducers increases till the incident wave front becomes plane, the two image pulses coalesce, giving

$$\lim_{r\to\infty} E = \lim_{r\to\infty}(E_1+E_2) = -\mathrm{i}\,\frac{M_1}{r_1^2}\,\frac{S}{\lambda},$$

in agreement with eq. (11).

For a plane incident wavefront, with the direction of incidence other than parallel to the disc axis, (Fig. 10 a), values of $A^{(n)}(r)$ are listed in Table VIII and plotted in Fig. 10 c to 10 e. The quantity, L, appearing there denotes

$$(r-r_1)/(r_2-r_1).$$

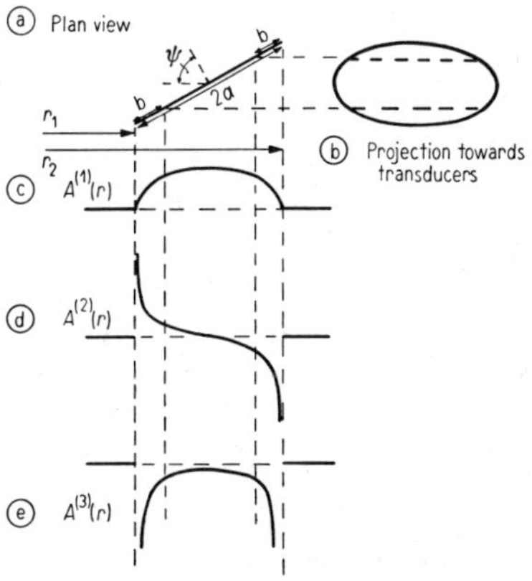

Fig. 10. The disc at non-axial incidence.

Infinite discontinuities occur at the near and far ranges of the disc, as listed in Table VIII, and the series solution of eq. (4) breaks down.

Table VIII.
Discontinuity data for disc at non-axial incidence.

n	$A^{(n)}(r)$	$D(A, 1, n)$	$D(A, 2, n)$
1	$2(r_2-r_1)\,(L-L^2)^{1/2}\cot\psi/\sin\psi$	0	0
2	$(1-2L)\,(L-L^2)^{-1/2}\cot\psi/\sin\psi$	$-\infty$	$-\infty$
3	$-[2(r_2-r_1)]^{-1}(L-L^2)^{-3/2}\cot\psi/\sin\psi$	∞	$-\infty$

This does not mean that image pulses do not occur at the near and far ranges of the disc. It is clear that there are sudden changes of projection at these ranges and, based on physical arguments put forward in Section 13 of [1], image pulses can be expected to arise at the limiting ranges.

This expectation is further supported by considering the echo from a disc which is doubly truncated so as to leave only the central region between the broken lines of Fig. 10 b. It is evident from Fig. 10 that, for $0<b<a$, the truncated disc will give rise to finite discontinuities at its near and far ranges. Whilst taking this truncated surface to the limit where $b\to 0$ does not appear to yield an expansion of convenient form for evaluating the echo components from a complete disc, the process indicates definitely that the latter surface will give rise to image pulses at its nearest and furthest points.

3.8. The infinite plane

The infinite plane can be considered as a disc whose radius has been made infinitely large. The former treatment for the disc with spherical incident waves applies, the sole difference being that, the disc rim having been removed to infinite range, the second image pulse does not occur. A single image pulse is generated at the pole of the transducer, and

$$E=E_1=-\frac{M_1}{2\,r_1}, \qquad \sigma=\pi\,r_1^{\,2}.$$

As would be expected, our physical optics approach has given the exact solution in this case.

3.9. The right circular cylinder

For a plane incident wavefront, with the direction of incidence other than parallel to or normal to the cylinder axis, the parts of the body which are

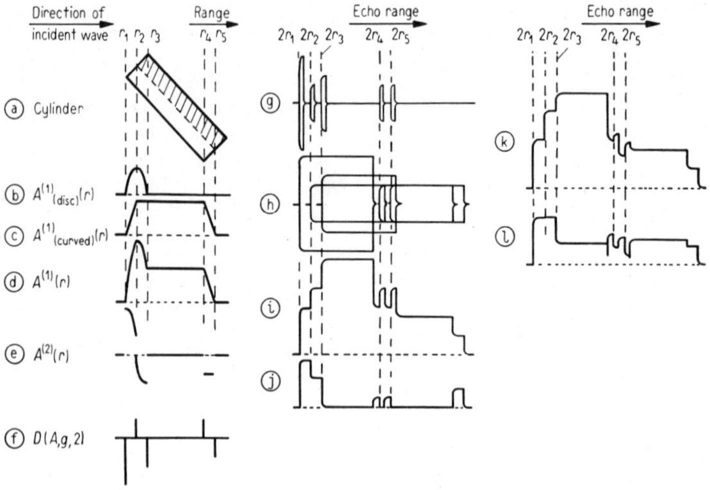

Fig. 11. Echo from right circular cylinder.

Explanation for (g) to (l):

(g) Envelope of receiver signal for a transmitted pulse short enough to resolve individual discontinuities,
(h) Image pulses produced by a longer transmission pulse,
(i) Upper limit of envelope (only one half shown) of receiver signal due to image pulses at (h) being in phase,
(j) Lower limit of envelope of receiver signal due to image pulses at (h).
(k) Envelope of receiver signal due to image pulses at (h) having relative phases 0°, 165°, 330°, 80°, 245°,
(l) Envelope of receiver signal due to image pulses at (h) having relative phases 0°, 80°, 160°, 230°, 310°.

directly irradiated consist of the nearest end-disc and of one half of the curved surface (Fig. 11 a).

The series expansion of eq. (4) has already been shown to break down in the case of the disc, and it breaks down in similar manner for the curved portion of the cylinder surface. Thus, a quantitative analysis cannot be proceeded with here; but a qualitative insight into the echo structure may be obtained by means of simplifying assumptions regarding the projection of the cylinder.

For the irradiated end-disc, the true shape of $A^{(1)}(r)$ over the range extent of the disc is a half ellipse, but this will be approximated by the shape shown in Fig. 11 b, which is not vertical at its near and far limits.

Remembering that $A^{(1)}(r)$ is proportional to the projected area of that part of a surface lying within an elementary range interval dr, we note that strips between ranges r and $r+dr$ on the irradiated part of the cylindrical surface increase from zero length at the nearest range, r_1, of the cylinder to a maximum at range r_2 (see Fig. 11 a). Between ranges r_2 and r_4, the length of strip remains constant, and between r_4 and r_5 it drops down towards zero again. The mode of increase between r_1 and r_2 and that of decrease between r_4 and r_5 are functions of aspect angle, but for simplicity a linear increase and a linear decrease will be assumed. The curve of Fig. 11 c results.

Addition of this curve to that for the end-disc gives the total $A^{(1)}(r)$ (Fig. 11 d). $A^{(2)}(r)$, illustrated in Fig. 11 e, exhibits five discontinuities, whose relative magnitudes are shown in Fig. 11 f.

Ignoring the contribution of discontinuities in higher order derivatives of $A(r)$, the echo structure for a short transmission pulse will be of the type illustrated in Fig. 11 g. Fig. 11 h shows the echo components resulting from an arbitrary lengthening of the transmission pulse. There is now overlap of image pulses, and the resulting, envelope will depend on their relative phases. These are a function of frequency and they may also vary appreciably for small variations of aspect angle. Maximal and minimal curves of envelope height, deduced by simple vector reasoning, are shown in Fig. 11 i and 11 j, and envelope curves for two arbitrarily chosen combinations of relative phase are presented in Fig. 11 k and 11 l.

We now examine the effect of transducer directivity on the echo. Assume that the transducers are non-directional in elevation, i. e. normal to the plane of Fig. 11 a, but that in azimuth the product of transmitter and receiver directivities is as shown in Fig. 12 a, in which the beam peak is arbitrarily directed towards the far end of the cylinder.

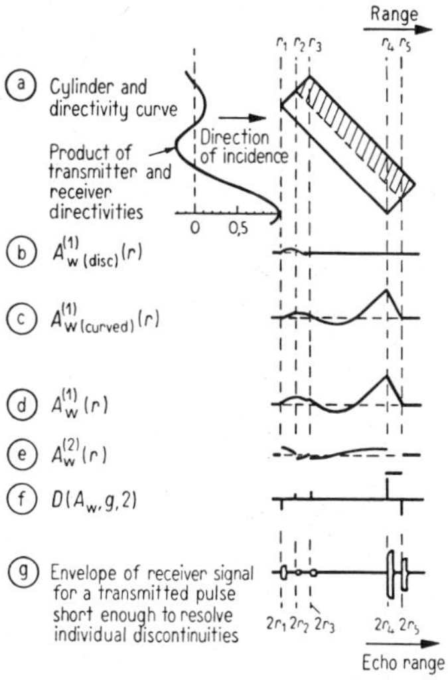

Fig. 12. Echo from right circular cylinder when the transducers are directional.

A computation in which each surface element is weighted approximately by the corresponding directivity value modifies the former curves of Fig. 11 b and 11 c to the shapes shown in Fig. 12 b and 12 c. Then proceeding as before, one arrives at the echo structure of Fig. 12 g, in which the relative amplitudes of the image pulses differ markedly from those shown in Fig. 11 g. Not only are these amplitudes a function of the shape of the 'product directivity' curve, but they also depend upon the beam centre bearing relative to the cylinder.

4. Experimental evidence

A considerable amount of experimental evidence has been gathered which gives qualitative support to the echo formation mechanism propounded in [1] and in this paper. A limited amount of quantitative evidence has also been obtained and there is an encouraging degree of agreement with the theoretical predictions. Not all the results agree with these predictions, and it is believed that this is due to the non-rigidity of the scattering bodies used. The experimental data will be presented in a later paper.

Acknowledgement

This paper is published by permission of the British Admiralty.

Appendix

Target strengths and backscattering cross sections at short ranges

Backscattering cross section, σ, and target strength, T. S., are normally defined as $\sigma = 4\pi\,|J|^2$

and T. S. = $10 \log |J|^2$, where $|J|^2 = r^2 \times$ back-scattered power density/incident power density.

Power density is a macroscopic concept applicable to a simple field such as a plane or spherical wave field. At large ranges from the scattering body, the scattered field is of this simple type, and the above definitions of σ and T. S. are valid. Eq. (7) and, of greater accuracy, eq. (8) are solutions for this case. At short ranges, the scattered field consists, in general, of the superposition of a number of wave trains, not necessarily spherical, which arise at different regions of discontinuity on the scattering body. The differences in range to these various generating sources may be of similar order to their actual range. The local scattered field is likely to be very complex, and power density concepts do not apply. If however one source of image pulses predominates sufficiently for the other contributions to be negligible, a simpler field results. For ranges at which this scattered field can be deemed to be behaving as a spherical wave in the vicinity of the transducers, eq. (8) may be used.

(Received October 16th, 1961.)

References

[1] Freedman, A., Acustica **12** [1962], 10.
[2] Siegel, K. M., Crispin, J. W. and Schensted, C. E., J. Appl. Phys. **26** [1955], 309.
[3] Weston, V. H., Exact near field and far field solution for the backscattering of a pulse from a perfectly conducting sphere. Univ. of Michigan Radiation Lab. Rept. No. 2778-4-T. 1959.
[4] Stenzel, H., Elektr. Nachr.-Techn. **15** [1938], 71.
[5] Senior, T. B. A. and Siegel, K. M., J. Res. Nat. Bur. Stand. – D. Radio Propagation **64 D** [1960], 217.

The following paper reports on an experimental study of the scattering of acoustic energy from solid and air-filled cylinders in water.

G. R. Barnard is a physicist. He received the M.A. degree in physics from the University of Texas in 1960. He is author of several papers in underwater acoustics research and has made significant contributions to the development of several successful sonar systems. He has been a member of the staff of the Defense Research Laboratory (now Applied Research Laboratories), University of Texas, since 1956 and has been Assistant Director since 1971.

Chester M. McKinney received the Ph.D. degree in physics from the University of Texas in 1950. He is the author of a number of papers in underwater acoustics and microwave radar. He has been a member of the staff of the Defense Research Laboratory (now Applied Research Laboratories), University of Texas, since 1946 and has been Director since 1965.

This paper is reprinted with the permission of the authors and the Acoustical Society of America.

Reprinted from THE JOURNAL OF THE ACOUSTICAL SOCIETY OF AMERICA, Vol. 33, No. 2, 226–238, February, 1961

23 Scattering of Acoustic Energy by Solid and Air-Filled Cylinders in Water*

G. R. BARNARD AND C. M. MCKINNEY
Defense Research Laboratory, The University of Texas, Austin, Texas
(Received August 8, 1960)

This paper presents the results of an experimental investigation of the scattering of acoustic energy from finite circular cylinders in water. Photographs are shown of the pulses scattered in the backward direction in the plane normal to the longitudinal axis and in the plane containing the longitudinal axis of a cylinder. The echo structure exhibits characteristics dependent upon the corners of the cylinder where there is a sharp change in the slope of the projected area and discontinuities between air, brass, and water. The photographs of pulses returned from the cylinders in the plane normal to the longitudinal axes imply the existence of surface waves which travel around the circumferences of the cylinders. Distribution-in-angle diagrams of acoustic energy scattered in the plane perpendicular to the longitudinal axes of solid and hollow air-filled cylinders were taken for changes in cylinder length, changes in pulse length, and changes in acoustic radii. Length modes of vibration do not affect significantly the energy scattered perpendicular to the longitudinal axis. When pulses, which are long compared to the diameter of the cylinder, are used, the scattered energy distribution is dependent upon the composition and the acoustic radius of the cylinder.

INTRODUCTION

THE theory of the scattering of sound by infinitely long cylinders is well known.[1–3] Some experimental studies of scattering by cylinders have been made but long cylinders were used and the measurements were restricted to the plane normal to the longitudinal axes of the cylinders.[4–10]

This paper presents the results of an experimental investigation of the scattering of acoustic energy from finite circular cylinders in water in the plane normal to the longitudinal axis and in the plane containing the longitudinal axis. Both solid and hollow air-filled brass cylinders were studied.

Presentation and discussion of the data are divided into two parts: general scattering (both backward and bistatic) and echo structure.

The purpose of the first study was to determine the changes brought about by using cylinders of different finite lengths. Polar diagrams were made of the distribution in angle of acoustic energy scattered in the plane normal to the longitudinal axis of the cylinders. Solid cylinders of the same diameter but of different lengths were used, while the frequency was held constant. Diagrams were also made of the general scattering from solid and hollow cylinders for several different pulse lengths.

The purpose of the second part of this study was to determine the distribution in angle of scattered acoustic energy from solid finite cylinders as a function of frequency. A range of ka from 0.33 to 24.4 was chosen, where a is the radius of the cylinder, k is $2\pi/\lambda$, and λ is the wavelength of sound. The scattered energy was measured at large distances from the cylinder and recorded on a polar diagram.

The second study was designed to determine the detailed nature of the echo structure of pulses scattered from finite cylinders and to correlate the echo structure with the physical characteristics of the cylinder. Pulse photographs of the echoes are shown for different scattering angles in the plane normal to the longitudinal axis and in the plane containing the longitudinal axis of solid and hollow cylinders.

EQUIPMENT AND METHODS

The measurements were made at the Defense Research Laboratory Lake Travis Test Station. Figure 1 describes the orientation of the cylinders in terms of angles θ and ϕ. This complication is necessary due to multiple aspects studied in this investigation. The angle ϕ is the angle between z axis and the longitudinal axis of the cylinder and θ is between the y axis and the axis

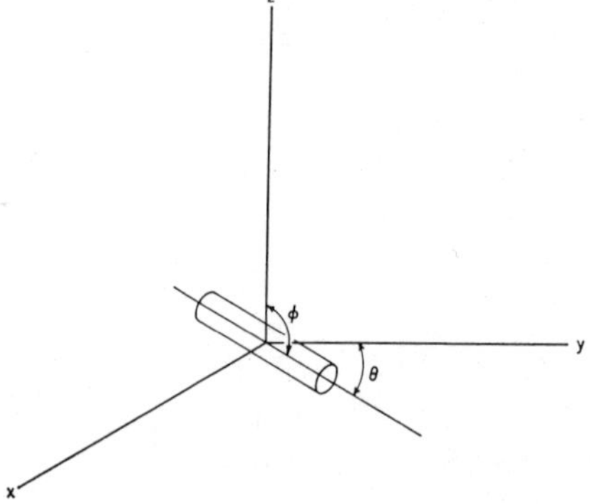

FIG. 1. Orientation of cylinder in coordinate system.

* This work was done under contract with the Bureau of Ships and Office of Naval Research.
[1] Lord Rayleigh, *The Theory of Sound* (Dover Publications, New York, 1945), Vol. II, p. 328.
[2] P. M. Morse, *Vibration and Sound* (McGraw-Hill Book Company, Inc., New York, 1936 and 1948), 2nd ed., p. 354.
[3] M. Lax and H. Feshback, J. Acoust. Soc. Am. **20**, 108 (1948).
[4] R. K. Cook and P. Chrzanowski, J. Acoust. Soc. Am. **17**, 315 (1946).
[5] F. M. Wiener, J. Acoust. Soc. Am. **19**, 4444 (1947).
[6] F. M. Wiener, J. Acoust. Soc. Am. **20**, 367 (1948).
[7] L. M. Lyamshev, Akust. Beih. XH. **5**, 122 (1959).
[8] L. Bauer, P. Tamarkin, and R. B. Lindsay, J. Acoust. Soc. Am. **20**, 583 (1948).
[9] L. Bauer, P. Tamarkin, and R. B. Lindsay, J. Acoust. Soc. Am. **20**, 858 (1948).
[10] J. J. Faran, Jr., J. Acoust. Soc. Am. **23**, 405 (1951).

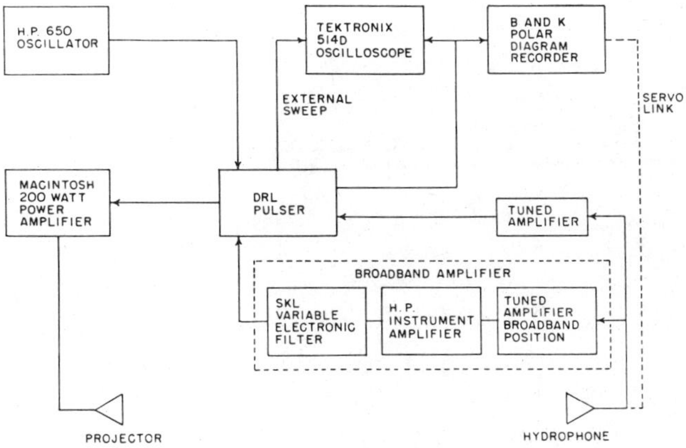

FIG. 2. System block diagram.

of the cylinder. In all cases the cylinder was insonified from the x direction and the hydrophone was rotated in the xy plane for the distribution-in-angle diagrams. The receiver (hydrophone) was held at a fixed distance from the cylinder by means of a boom and moved through 360° in the xy plane to probe the field in that plane. The axis of rotation of the boom coincided with the z axis of the coordinate system. The hydrophone boom was 10 ft long and was driven by an electric motor at a rate of one revolution every 10 min. The projector was supported independently at a distance from the cylinder only slightly greater than the length of the boom. The boom was adjusted to rotate in a horizontal plane, and the projector was adjusted to the same depth as the hydrophone. The projector and the hydrophone were placed at a depth of 12 ft below the surface of the water and approximately 40 ft above the bottom of the lake. No surface or bottom interference effects were encountered because of the directivity of the transducers.

"Line and cone" transducer[11] units of 7 in. aperture were used for the projector and hydrophone. The beamwidths of the individual units vary from 16° at 50 kc to 5.7° at 150 kc. They provided a linear beamwidth of 2.8 ft at 50 kc and 1 ft at 150 kc for a 10-ft test distance. The 10-ft test distance was adequate to ensure "far-field" conditions even at the highest frequency used (150 kc).

Figure 2 shows a block diagram of the electronic equipment used in the experiments. Pulses long compared to the diameters of the cylinders were used for the general scattering polar diagrams so that the measurements made were essentially the same as those for continuous transmission. Short pulses were used for the echo structure studies. The pulse repetition rate of 160 pulses/sec permitted two-way transmission in water for a maximum range of 15 ft.

[11] R. E. Mueser, J. Acoust. Soc. Am. 19, 952 (1947).

The cylinders used in the experiments varied from $\frac{1}{16}$-in. to 5-in. diam. The 1-in. and smaller cylinders were suspended by 0.039-in. diam monofilament nylon while the larger cylinders were hung with double strands of 0.058-in. diam monofilament nylon. The acoustic scattering from this nylon was found to be negligible.

PRESENTATION AND DISCUSSION OF DATA

General Scattering

Data of general scattering from cylinders are presented in this section. General scattering refers to the energy scattered at every angle, while back scattering refers to energy scattered back to the hydrophone where

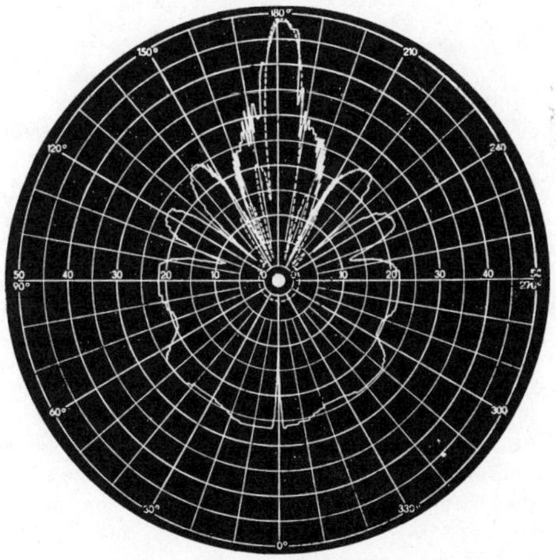

FIG. 3. Distribution in angle of scattered acoustic energy from a finite, circular, solid brass cylinder. Frequency: 150 kc; cylinder diameter: 1 in.; pulse length: 500 μsec; cylinder length: 6 in.; radial scale: 5 db per division; $\phi=0°$. Cylinder in position:———; cylinder removed: – – – .

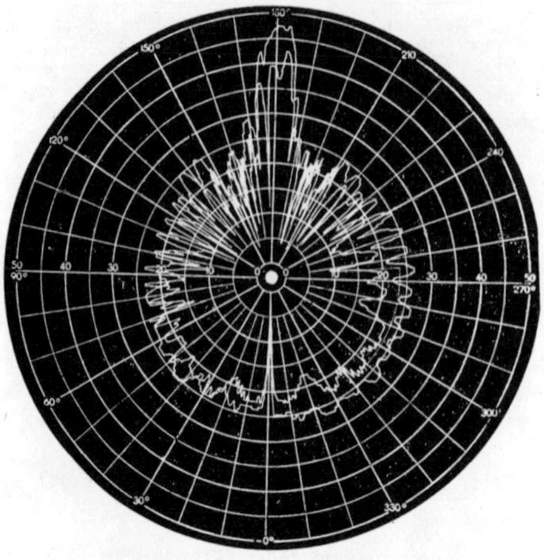

FIG. 4. Distribution in angle of scattered acoustic energy from a 3 in. diam solid brass cylinder superimposed on the pattern for energy scattered from a 3 in. diam hollow brass cylinder. Frequency: 150 kc; cylinder length: 6 in.; pulse length: 500 μsec; radial scale: 5 db per division; $\phi=0°$. Outside curve—solid cylinder; inside curve—hollow cylinder.

the projector and hydrophone are situated adjacent to each other. In this study the general scattering data were restricted to that obtained by rotating the hydrophone through 360° in the xy plane.

Figure 3 shows a typical polar diagram of the distribution in angle in the plane normal to the longitudinal axis of the cylinder of scattered acoustic energy from a 1 in. diam, 6 in. long cylinder taken at a frequency of 150 kc with 500 μsec pulses. The 500 μsec pulse corresponds to an extension in range of 14.4 in. The sound source is at 000°, and the notch appearing there is due to the hydrophone masking the cylinder from the projector as the hydrophone passed between the projector and cylinder. The radial scale is logarithmic and graduated at 5 db per division. The outside (solid) curve was taken with the cylinder in the water and the broken curve, with the cylinder removed.

The curve, with the cylinder removed, remains about 2 db above the zero level of the recorder until 130°, and then the minor lobes of the projector begin to appear. Two such lobes are seen at 150° and 210°. This inside curve traces out the projector beam pattern with the major lobe shown at 180°. Here the major lobe of the projector is in a direct line with the major lobe of the hydrophone and the direct path travel time coincides with the delay of the receiver gate. The broken curve is not a true beam pattern of the projector because at most of the hydrophone positions, the direct path (minimum range) from the projector to the hydrophone is less than the range corresponding to the delay time between the pulsing of the projector and the activation of the receiver. The receiver was time-gated to receive signals which had traveled from the projector through the water to the hydrophone over distances from 18.8 to 21.2 ft. Distortion is also present because the hydrophone does not revolve about an axis through the projector but about an axis midway between the projector and hydrophone.

The solid curve showing the disturbance due to the cylinder is relatively smooth (indicating omnidirectional scattering in the plane of measurement) until the projected beam (direct path) comes into the receiver gate. When the direct path contribution becomes significant, rapid fluctuations begin to appear due to constructive and destructive interference between the direct and scattered signals. The relative rate of change in phase due to boom position explains the rate of fluctuation. The gate width and the directivity of the transducers determine the angle at which interference due to the direct path transmission takes place. The presence of the cylinder did not produce an increase in the level of the forward projected signal during any phase of the studies.

It was found that for sufficiently long pulses the distribution in angle of scattered acoustic energy was the same for several finite length cylinders of the same diameter. The only variation from pattern to pattern was in target strength or magnitude of the signal level. The target strength increased by 6db when the cylinder was doubled in length from 2 to 4 in. and increased by 5 db when the cylinder length was changed from 6 to 12 in. These data indicate that target strength varies as the logarithm of the length squared which is in agreement with the classical theoretical prediction.

Polar diagrams were taken to indicate the change in the distribution of scattered energy as a function of pulse length for solid and hollow cylinders of the same size. For both types of cylinders the scattered energy was uniform in angular distribution for the short pulses where all reflections except the front surface reflection of the cylinders were gated out and disregarded by the recording instruments.

For the case where the pulses were long compared to the diameter of the cylinders there was a marked difference in the distribution of scattered energy for the two types due to the presence of compressional and shear waves in the solid cylinders. A comparison between the energy scattered from a 3 in. diam, solid brass cylinder and a 3 in. diam, hollow, brass-wall cylinder is shown by the polar diagram in Fig. 4. The data were taken using 500-μsec pulses at 150 kc. The energy scattered from the hollow cylinder is shown by the inside curve. The result indicates that there is a considerable difference in the distribution in angle of scattered energy for the two cases. This difference implies that the scattering is a function of the composition of the cylinder. Faran[10] and Hampton[12] have shown also that for

[12] L. D. Hampton, "Experimental study of scattering of acoustic energy from solid metal spheres in water," M. A. thesis, Department of Physics, the University of Texas, January, 1959 (unpublished).

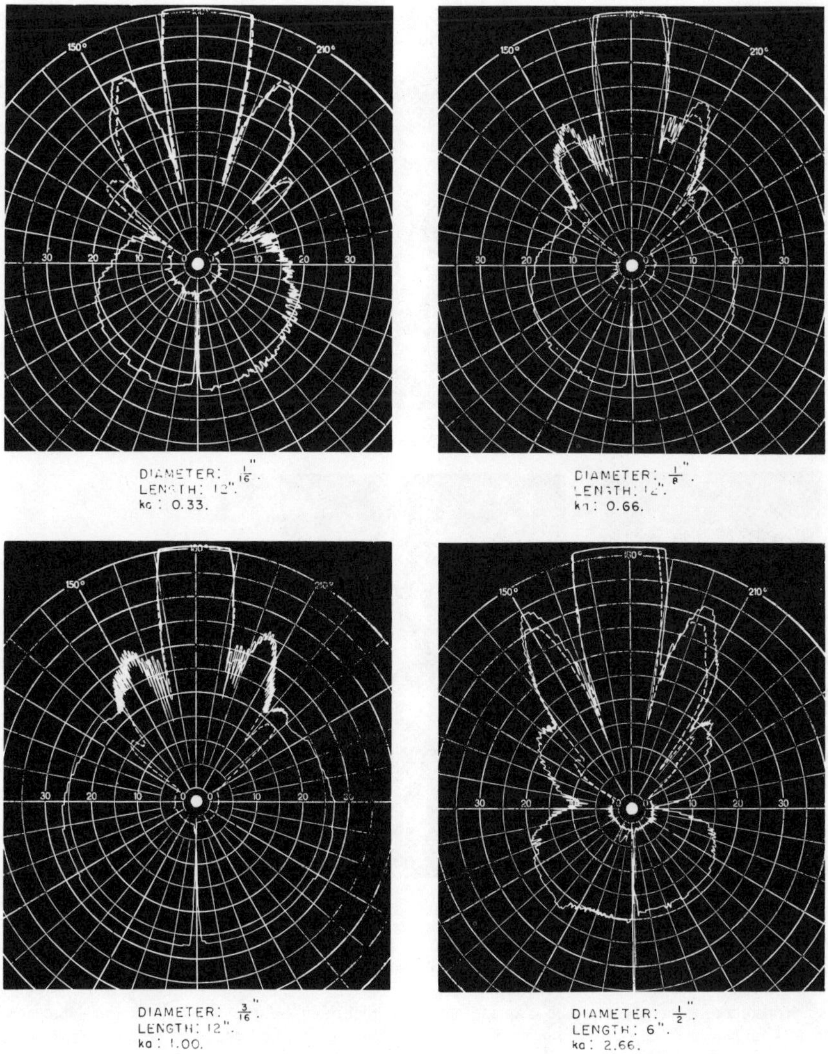

FIG. 5. Distribution in angle of scattered acoustic energy from different size solid brass cylinders. Frequency: 100 kc; pulse length: 500 μsec; $\phi=0°$; radial scale; 5 db per division.

cylinders and spheres the distribution in angle of scattered acoustic energy is dependent on the composition of the targets.

The dependence of scattered energy on acoustic radius is shown in Figs. 5–7. Solid brass rods with diameters of $\frac{1}{16}$, $\frac{1}{8}$, $\frac{3}{16}$, and 1/2 in., corresponding to acoustic radii *ka* of 0.33, 0.67, 1.00, and 2.66, respectively, at 100 kc, were used to obtain the polar diagrams in Fig. 5. Figure 6 shows similar data for a range of *ka* from 2.66 to 7.98, obtained by using a 1.0 in. diam cylinder at frequencies from 50 to 150 kc. A 3-in. diam rod, at the same frequencies, was used to obtain data for the range of *ka* from 7.98 to 23.94, shown in Fig. 7.

Because of the lower signal level from the small rods, greater electronic gain was needed to present curves comparable in size to those for the larger cylinders.

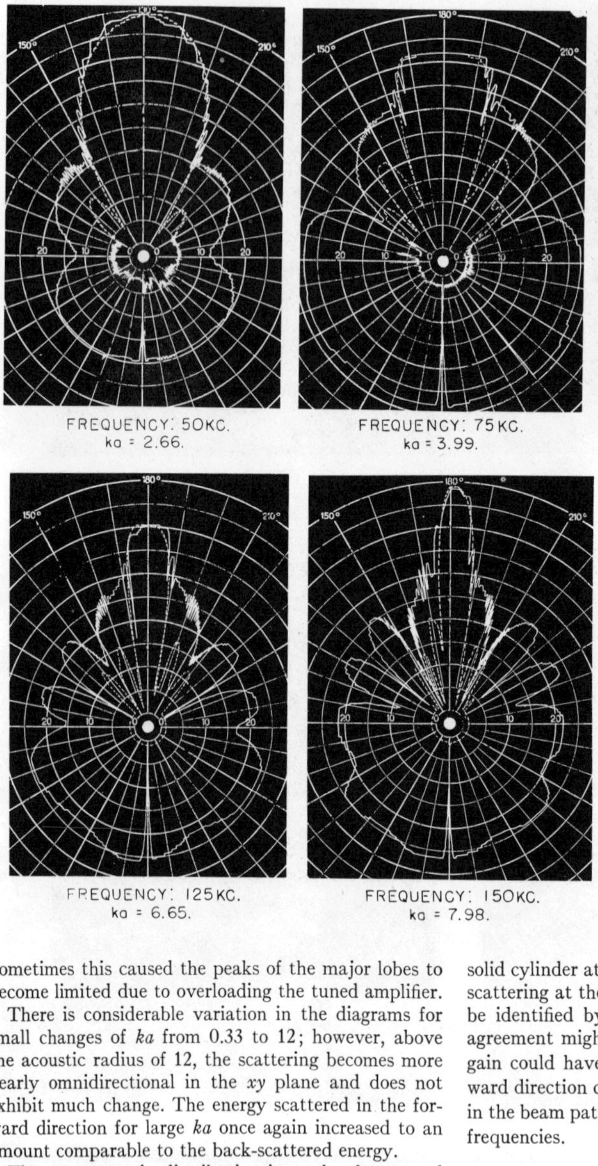

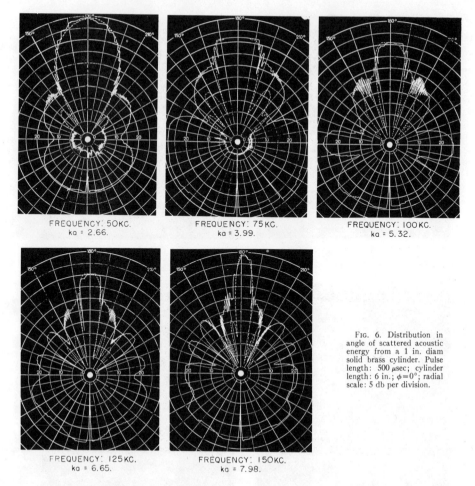

FIG. 6. Distribution in angle of scattered acoustic energy from a 1 in. diam solid brass cylinder. Pulse length: 500 μsec; cylinder length: 6 in.; $\phi=0°$; radial scale: 5 db per division.

Sometimes this caused the peaks of the major lobes to become limited due to overloading the tuned amplifier.

There is considerable variation in the diagrams for small changes of *ka* from 0.33 to 12; however, above the acoustic radius of 12, the scattering becomes more nearly omnidirectional in the *xy* plane and does not exhibit much change. The energy scattered in the forward direction for large *ka* once again increased to an amount comparable to the back-scattered energy.

The agreement in distribution in angle of scattered energy from different size targets representing the same *ka* is shown in Fig. 8. The polar diagram on the left indicates the agreement between scattering from the $\frac{1}{2}$ in. diam, solid rod at 100 kc and the 1 in. diam, solid rod at 50 kc. The second diagram is of the 1 in. diam, solid rod at 150 kc, superimposed on the 3 in. diam, solid cylinder at 50 kc. The curves corresponding to the scattering at the higher frequency on each diagram can be identified by the narrower major lobe curve. The agreement might have been better if exactly the same gain could have been obtained. Agreement in the forward direction cannot be expected due to the difference in the beam patterns of the transducers for the different frequencies.

Echo Structure

In this section are presented data relating to the detailed echo structure of sound scattered by cylinders. Photographs are shown of the pulses reflected from different size cylinders at various aspects as a function of pulse length and frequency. All oscilloscope photo-

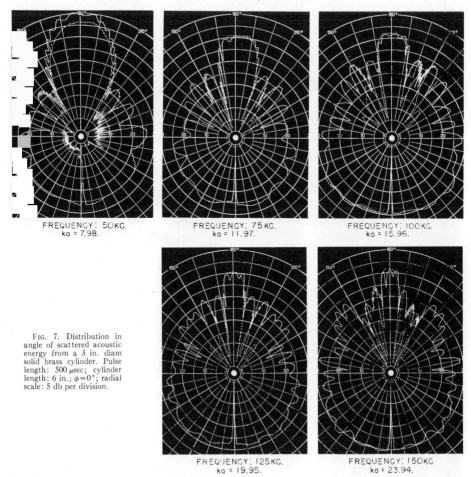

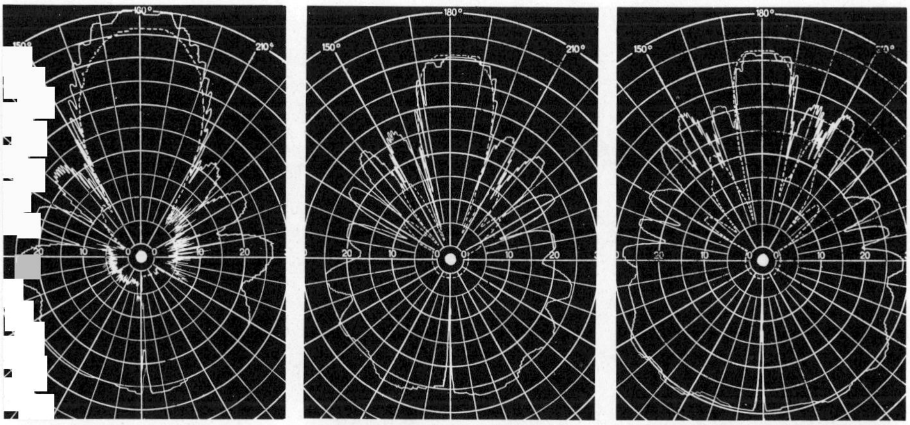

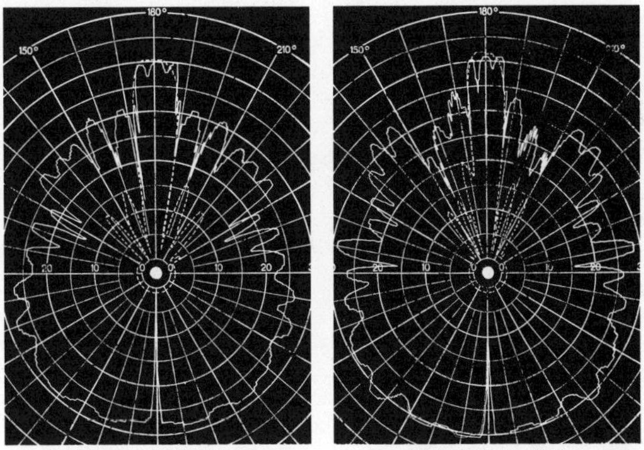

FIG. 7. Distribution in angle of scattered acoustic energy from a 3 in. diam solid brass cylinder. Pulse length: 500 μsec; cylinder length: 6 in.; $\phi=0°$; radial scale: 5 db per division.

graphs are shown with the sweep from left to right, that is, time is measured as increasing to the right.

The back-scattered pulses from a 5 in. diam, 10 in. long, hollow brass-wall cylinder are shown as a function of frequency in Fig. 9 and as a function of aspect in Fig. 10. The longitudinal axis of the cylinder lies in the plane of rotation. The first pulse shown on each photograph is the first pulse reflected from the cylinder. The projector and hydrophone were located side by side in the plane containing the longitudinal axis of the cylinder.

It can be seen from Fig. 9 that the pulse structure is independent of frequency for pulses as short as 50 μsec. As aspect of the cylinder is changed, a definite change in distance between pulses is noticed in Fig. 10. The pulses for the 45° aspect are measured to be 125 μsec apart. Using a velocity of sound in water of 4800 ft/sec, this corresponds to a two-way path of 7.14 in. between pulses or very nearly a spacing of 3.5 in. between points on the cylinder reflecting the pulses. For this 5 in. diam, 10 in. long cylinder at 45°, the origin of these pulses corresponds very closely to the four sharp corners of the cylinder. Using this criterion for a suspected echo, structure the values were computed for the distance between pulses for a 30° aspect. The second pulse should be 4.33 in. behind the first, the third should be 5 in. away from the first, and the last should be 9.33 in. from the first. The experimental values were measured to be 4.32, 5.19, and 9.78 in., respectively. Similar good agreement was found for the 40° and 50° aspects. These results suggest that a pulse is received from each point where there is a sharp change in the slope of projected area. Presented beside the photographs in Figs. 9 and 10 are drawings

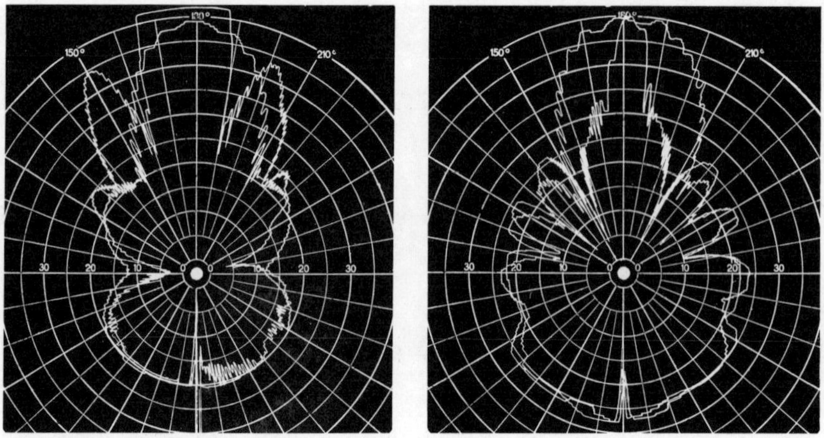

ka: 2.66 ka: 7.98

FIG. 8. The distribution in angle of scattered acoustic energy from different size cylinders representing the same ka. Pulse length: 500 μsec; cylinder length: 6 in.; $\phi=0°$; radial scale: 5 db per division.

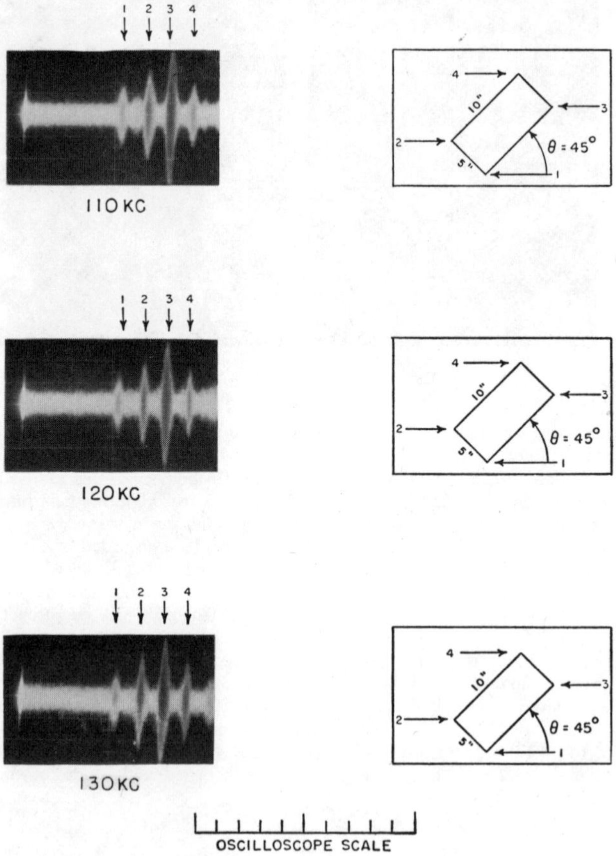

FIG. 9. Back scattered pulses from a brass wall hollow cylinder. Pulse length: 50 μsec; sweep time: 100 μsec per division; cylinder diameter 5 in.; cylinder length: 10 in.; $\theta=45°$; $\theta=90°$.

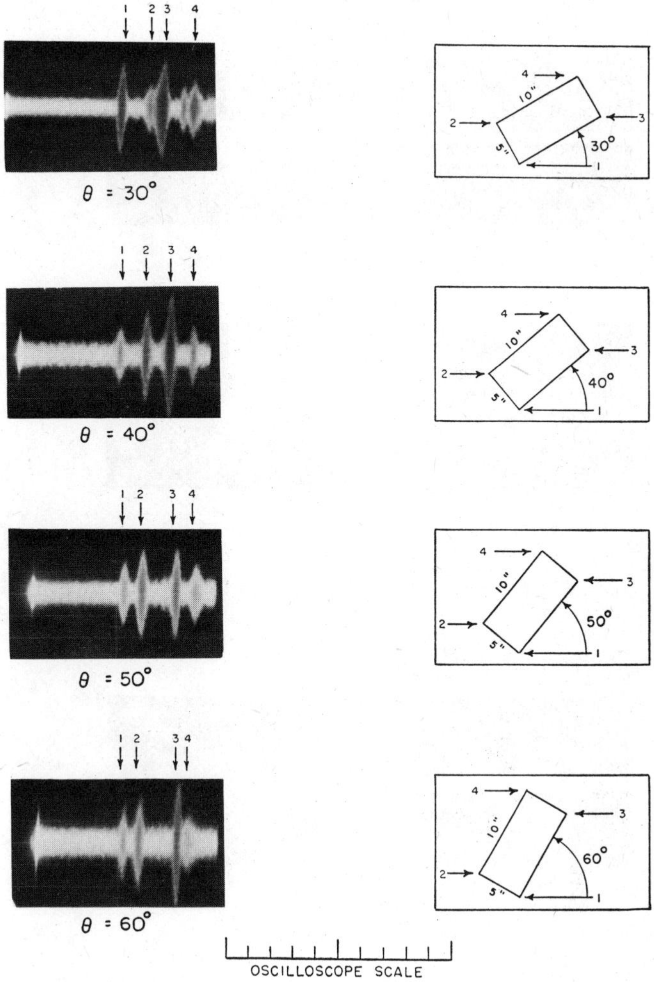

FIG. 10. Back scattered pulses from various aspects of a hollow brass wall cylinder. Frequency: 120 kc; pulse length: 50 μsec; cylinder diameter: 5 in.; cylinder length: 10 in.; sweep time: 100 μsec per division; $\phi = 90°$.

of the cylinders showing the sharp corners which apparently scatter the pulses. The right angle corners formed by the flat circular ends create these sharp changes in slope. In each case the largest pulse corresponds to the most distant edge after the long front surface and the smallest is usually the pulse from the rear discontinuity which is actually in a geometrical shadow zone. Two pulses appear in the 30° aspect corresponding to the longitudinal axis. This is the only case where they appear and is possibly peculiar to this particular angle of incidence.

The echo structure of this same hollow cylinder at its vertical and horizontal beam aspects and at an end-on aspect is presented in Fig. 11. In the four top photographs the longitudinal axis of the cylinder lies in the xy plane. Two photographs using different gains are presented at each aspect. The photographs on the left, using low gain, show the large surface reflection while the ones on right are magnified to show clearly the much smaller pulses appearing later in time. The lower two photographs were taken of the back-scattered pulses from the beam aspect in a plane perpendicular to the longitudinal axis of the cylinder.

The pulse reflections encountered may be due to a

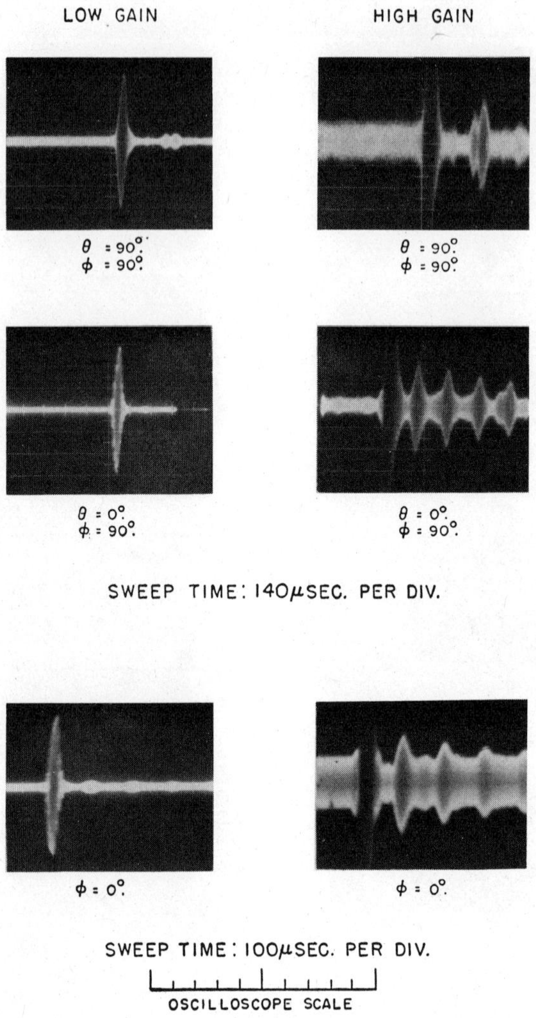

FIG. 11. Back scattered pulses from beam and end-on aspects of a hollow brass wall cylinder. Frequency: 120 kc; pulse length: 50 μsec; cylinder diameter: 5 in.; cylinder length: 10 in.

surface wave traveling ground the circumference of the cylinder. This concept is discussed by Franz,[13] and is termed a "creeping wave." Basically, the idea is that the path of the wave is from the projector to a point tangential to the cylinder, then around the surface of the cylinder until it reaches a point tangential to the receiver, then on to the receiver. In the case presented in this paper it appears that the multiple pulses are due to the wave making successive trips around the cylinder, which would explain the amplitude decay of successive pulses. The horizontal beam aspect case, where $\theta=0°$ and $\phi=90°$, indicates the same pulse structure as the vertical beam aspect case of $\phi=0°$.

In both cases, however, the five major pulses are spaced 190 μsec apart, corresponding to a water path length of 11 in. between pulses. The circumference of the cylinder is 15.7 in. The path around the back surface of the cylinder must then be 7.8 in. Assuming that the pulse will travel in water 5 in., it is implied that the path of 7.8 in. around the surface of the cylinder is equivalent to a 6 in. path in water. A suitable constant is then computed to convert the path around the circumference to an equivalent water path. This value is 0.764.

[13] W. Franz, Naturforsch., **9a**, 705 (1954).

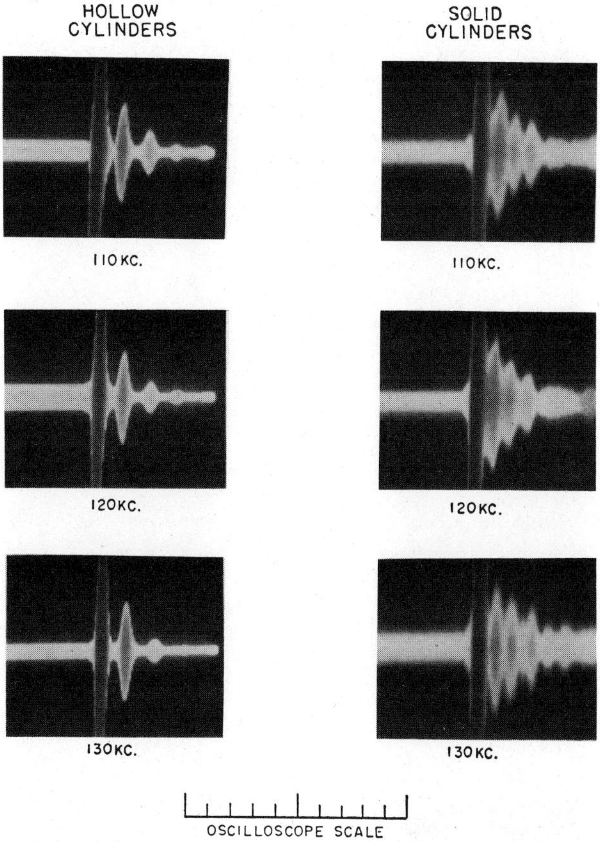

FIG. 12. Back scattered pulses from solid and hollow cylinders as a function of frequency. Pulse length: 50 μsec; $\phi=0°$; cylinder diameter: 3 in.; cylinder length: 6 in.; sweep time: 100 μsec per division.

The 15.7 in. circumference becomes an equivalent to 12 in. in water. If this conversion is valid, then the successive pulses, when computed with this factor, should be in agreement with successive revolutions around the cylinder. This is the case for each pulse and the error is in all cases less than an inch. This error can easily be explained due to inaccuracy in reading the spacing between pulses, the nonlinearity of the oscilloscope, and inaccuracy in using 4800 ft/sec as the exact velocity in water.

For the case of the horizontal beam aspect, a small pulse occurs 728 μsec after the large surface pulse and appears to be due to reflection from the back surface of the cylinder. Using a velocity of 1100 ft/sec (since the cylinder is air filled), this time delay corresponds to a range difference of 4.8 in. which is in reasonable agreement with the diameter of 5 in.

In the case of the end-on aspect two pulses appear 330 and 370 μsec behind the larger surface return. These correspond to water path distances of 18.96 and 21.1 in. The longest path here is, within experimental error, twice the length of the cylinder and may be the reflection from the discontinuity in the rear. If we consider a wave traveling around the surface, it must travel 25 in. Converting this to an equivalent water path yields 19.1 in., or very nearly the measured value of the other pulse.

A comparison of the echo structure between solid and hollow cylinders is shown in Figs. 12 and 13. The targets are 3 in. diam, 6 in. long, solid and thin-wall brass cylinders. Figure 12 shows the back-scattered pulses as a function of frequency, and Fig. 13 shows the scattered pulses as a function of pulse length. The pulse structure is seen to be independent of frequency.

The difference between the pulses reflected back and those reflected at an angle of 90° from a 3 in. diam, 6 in. long, hollow cylinder, suspended vertically, is depicted in Fig. 14. The photographs at the top were made using

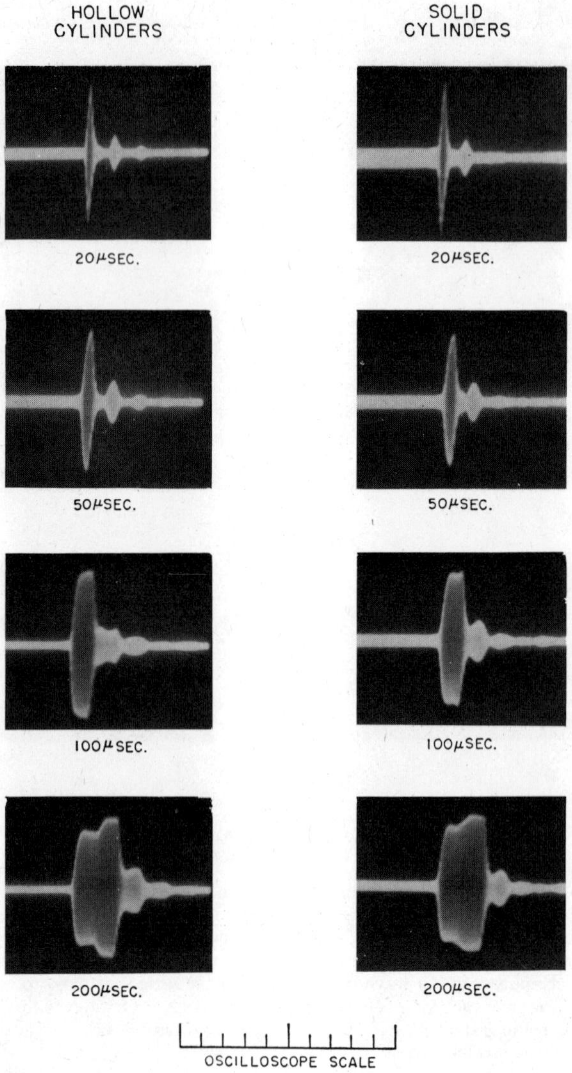

FIG. 13. Back scattered pulses from solid and hollow cylinders as a function of pulse length. Frequency: 120 kc; $\phi=0°$; cylinder diameter: 3 in.; cylinder length: 6 in.; sweep time: 100 μsec per division.

low gain to show the strong front reflection in each case. The gain was then increased on the oscilloscope to show the structure of the smaller pulses (lower photos).

For the back-scattering case the second, third, and fourth pulses are separated from the first in time by 120, 240, and 360 μsec corresponding to water paths of 6.91, 13.82, and 20.74 in., respectively. Converting the 9.42 in. circumference to an equivalent water path (using the factor 0.764) and calculating the distance between successive revolutions around the surface of the cylinders yields distances of 6.59, 13.78, and 20.97 in. Once again there is very good agreement between the measured and calculated values.

The next step in investigating this surface wave was to measure the pulse scattered back by a solid brass cylinder, which had the same dimensions as the small brass-wall hollow cylinder, and compare the spacing with calculated values. The measured distances from the

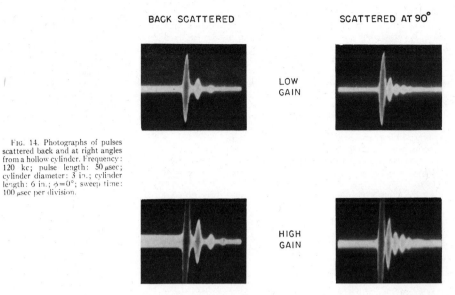

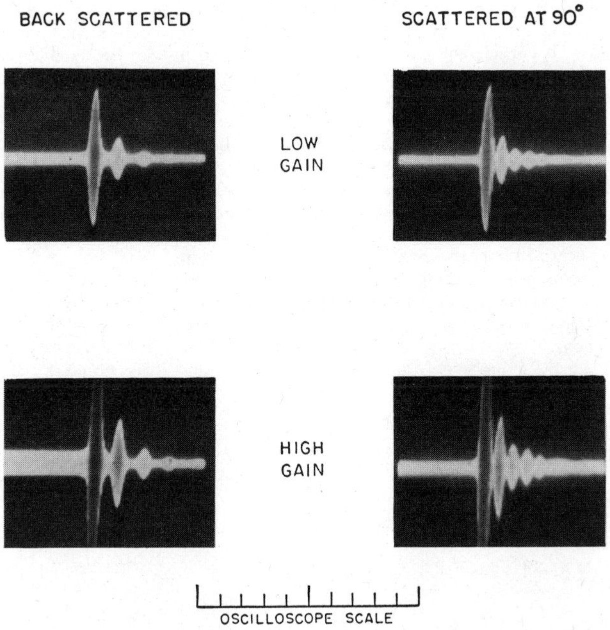

FIG. 14. Photographs of pulses scattered back and at right angles from a hollow cylinder. Frequency: 120 kc; pulse length: 50 μsec; cylinder diameter: 3 in.; cylinder length: 6 in.; $\phi=0°$; sweep time: 100 μsec per division.

large front surface scattered pulse, using the sound velocity in water, were 4.92, 8.64, and 13.2 in. Using the same procedure as before, a conversion factor for the solid brass cylinder was determined to be 0.41 as compared to 0.764 for the hollow cylinder. Using this conversion factor to change the cylinder circumference to an equivalent water path, the distances corresponding to the third and fourth pulses were calculated to be 8.78 and 12.64 in., or once again, very close to the measured values.

A final check on the agreement between measured and calculated values of pulse spacing was made on the 3 in. diam, hollow brass-wall cylinder for pulses scattered 90° from the angle of incidence. This case was chosen because there should be present those waves traveling around $\frac{1}{4}$ the circumference and subsequent revolutions of the circumference, and also around $\frac{3}{4}$ the circumference and subsequent revolutions around the circumference.

The pulses were measured at 85, 150, 200, and 260 μsec in time after the large front surface echo. These values correspond to water paths of 4.89 and 12.1 in. for the first and third pulses and 8.64 and 15.55 in. for the second and fourth pulses. The calculated values of path distance using the conversion factor of 0.764 for the path on the circumference were 4.8 and 12.0 in. for the case of the wave traveling $\frac{1}{4}$ and $1\frac{1}{4}$ times around the circumference. In the case where the wave traveled $\frac{3}{4}$ and $1\frac{3}{4}$ times around the circumference, the distances for total path were computed to be 8.4 and 15.6 in., again in good agreement with the measured values.

It can be seen from Fig. 12 that the echo structure of pulses scattered back by the hollow and solid 3 in. diam cylinders were independent of frequency. From Fig. 13 it can be seen that short pulses are necessary in order to achieve the resolution needed to study the echo structure. The use of longer pulses, where the range extent is much greater than the diameter of the cylinders, results in a grouping of the individual pulses. For pulses as long as 200 μsec no significant difference is seen between the echo structure of the hollow cylinder and that of the solid cylinder, except the spacing between subsequent echoes which is explainable in terms of surface waves traveling faster around a solid cylinder. The velocity of sound in brass is approximately four times the velocity of sound in water; therefore, if the surface waves penetrate the brass very deeply, it becomes reasonable to assume that this surface velocity may be greater for the solid than for the hollow cylinder.

SUMMARY AND CONCLUSIONS

The data presented in this paper show that there is no significant change in the distribution of scattered energy in the plane normal to and bisecting the longitudinal axis by cylinders immersed in water due to changes in finite length. This indicates that no length mode of vibration is present that affects the energy scattered perpendicular to the longitudinal axis.

There is considerable change in the distribution in angle of scattered energy for changes in pulse length and frequency. For the case of pulses which are long compared to the diameter of the cylinders, the compressional and shear waves present in the solid cylinders cause considerable change in the scattered energy with changes in frequency. This characteristic is very noticeable for values of acoustic radii *ka* less than 12.

The structure of echoes from cylinders is very aspect dependent. It appears that the major reflections are from points where there is a sharp change in the slope of the projected area and a definite discontinuity. The right circular cylinder is very representative of this situation.

When the cylinder is irradiated in a beam position, the multiple echoes exhibit a spacing structure that is dependent upon the water path to and from the points on the cylinder which are tangent to the projector and hydrophone, and upon the path around the circumference of the cylinder. The velocity of the sound around the circumference is greater than that in water, but less than the bulk velocity in brass. The velocity of this "creeping wave" is greater around the surface of solid brass than around the surface of a thin wall of brass, indicating that the path taken by the sound is partially within the brass.

ACKNOWLEDGMENT

The authors would like to thank Dr. C. W. Horton for his suggestion of the creeping wave as a means of describing the complex echo formation.

During World War II, the scattering of sound from the sea surface was studied under the designation "surface reverberation." This theoretical study was carried out to help in determining the geometry and kinematics of the sea surface from sound scattering data.

Carl Eckart received the Ph.D. degree in physics from Washington University in 1925. He has published many papers in the field of hydrodynamics of oceans and atmospheres. He was Assistant Director, Associate Director, and Director of the University of California Division of War Research from 1942–1946. He was Director of the Marine Physical Laboratory from 1946–1952, and Director and Professor of Physics and Oceanography of the Scripps Institution of Oceanography until his retirement in 1971. He is a member of the American Academy of Arts and Sciences; National Academy of Sciences, where he is Chairman of the Section of Geophysics; American Physical Society; American Association for the Advancement of Science; Acoustical Society of America; and the American Geophysical Union. He was awarded the Alexander Agassiz Medal for contributions to oceanography by the National Academy of Sciences in 1966, and the American Academy of Achievement Gold Plate Award in recognition of the Agassiz Medal in 1968.

This paper is reprinted with the permission of the author and the Acoustical Society of America.

Reprinted from THE JOURNAL OF THE ACOUSTICAL SOCIETY OF AMERICA, Vol. 25, No. 3, 566–570, May, 1953

24

The Scattering of Sound from the Sea Surface

CARL ECKART*
Institute for Advanced Study, Princeton, New Jersey
(Received January 5, 1953)

It is assumed that the equation of the sea surface is $z=\zeta(xyt)$, and that the time average $\Phi=\langle\zeta(x'y't)\zeta(x''y''t)\rangle$ is a function only of $\xi=x''-x'$, $\eta=y''-y'$. The scattering coefficient of long-wave sound is calculated and shown to be

$$\sigma=(c^2k^2/4\pi)^2\int\int\Phi(\xi\eta)\exp[-ik(a\xi+b\eta)]d\xi d\eta,$$

where $2\pi/k$ is the wavelength of the sound, and a, b, c are, respectively, the sum of the x, y, z direction cosines of the incident and scattered rays. For short-wave sound, the formula is more complicated, but independent of the wavelength of the sound, as it should be. However, it is shown that experiments with short waves yield much less information about the sea surface than do those with long waves. This is unfortunate, since the latter experiments are much more difficult than the former.

INTRODUCTION

THE scattering of underwater sound by the sea surface was studied, under the designation "surface reverberation," during the war.[1] It has also been noted that the study of scattered sound or other radiation can furnish information concerning the geometry and kinematics of the sea surface.[2,3] In order that such information may be obtained from the experimental data, a theoretical analysis of the problem is needed. This analysis is difficult, and many approximations are required to reach simple formulas. Despite this, it is hoped that the Eqs. (21) and (36) below will be useful for the purpose indicated.

The Sound Field

The acoustic pressure will be the sum of the incident pressure p_0 and secondary p_1 which is in turn the sum of the reflected and scattered sound. The incident sound will be given, to a sufficient approximation, by

$$p_0(xyz)=P[\exp(ikr_0)]/r_0 \tag{1}$$

where r_0 is the distance from the center of the source to the point xyz and P depends only on the direction of r_0. In order that all integrals in the following equations converge, it is necessary to assume that the source is quite directional and illuminates only a finite area of the otherwise infinite sea surface; P is therefore not constant.

The secondary (reflected plus scattered) sound will then be a solution of the wave equation and will consist only of outgoing waves. Helmholtz's theorem[4] is therefore applicable in the form

$$4\pi p_1(A)=\int\int_S\left\{\frac{\partial p_1}{\partial\nu}[\exp(ikr_1)/r_1]-p_1\frac{\partial}{\partial\nu}[\exp(ikr_1)/r_1]\right\}dS, \tag{2}$$

where

S is the sea surface, $z=\zeta(x, y, t)$;
ν is the normal to S, positive toward the source of sound;
A is any point on the same side of S as the source of sound; and
r_1 is the distance from dS to A.

This formula expresses the value of p_1 at a general point in terms of its boundary values and those of its normal derivative on the sea surface.

If the source of sound is underwater, the surface S is a "pressure release," and, to a very good approximation, one may set

$$p_0+p_1=0 \quad \text{on } S, \tag{3}$$

which determines the boundary values of the secondary radiation in terms of those of the (known) incident radiation. It is not so easy to obtain a good approximation for the normal derivative; one may, however, note that if S were a plane rather than an irregular surface, one would find

$$\partial p_0/\partial\nu=\partial p_1/\partial\nu \quad \text{on } S. \tag{4}$$

If S is not so irregular that some parts of it are shadowed by others, one may assume that Eq. (4) still holds; however, in deep shadows it would be more reasonable to suppose that $(\partial p_0/\partial\nu)+(\partial p_1/\partial\nu)=0$, so that some caution must be exercised in applying results obtained from Eq. (4). With this warning, it will be assumed that both

* On sabbatical leave from the Scripps Institution of Oceanography, La Jolla, California. This is S.I.O. Contribution No. 6?4 (New Series).

[1] "Physics of Sound in the Sea," NDRC Summary Technical Reports, Division 6, Vol. 8, Part II, chap. 12. "Principles of Underwater Sound," NDRC Summary Technical Reports, Division 6 Vol. 7, chap. 5. The reports have been reprinted and distributed by the National Research Council.

[2] V. V. Shuleikin, *Fisika Moria* (Physics of the Sea) (Izdatelstvo Akademii Nauk., Moscow, 1941), 833 pp.

[3] C. Cox and W. H. Munk, *The Measurement of the Roughness of the Sea Surface from Photographs of the Sun's Glitter*. Part I. The Method. (University of California, Scripps Institution of Oceanography, Reference No. 52-61, 1952). Contains bibliography.

[4] B. B. Baker and E. T. Copson, *The Mathematical Theory of Huyghens Principle* (Oxford University Press, London, 1939), pp. 23–28.

Eqs. (3) and (4) are valid, so that Eq. (2) may be written

$$4\pi p_1(A)=\iint_S (\partial/\partial\nu)\{[p_0 \exp(ikr_1)]/r_1\}dS. \quad (5)$$

If the source of sound is in the air, the sea surface is "hard" to a very good approximation, and the boundary condition $(\partial p_0/\partial\nu)+(\partial p_1/\partial\nu)=0$ replaces Eq. (3). The boundary values of p_1 are now difficult to determine with the same precision as $(\partial p_1/\partial\nu)$, but the approximation $p_0=p_1$ on S is valid to the same extent as Eq. (4). This results in an equation identical with Eq. (5) except for a negative sign on the right side.

If the source of sound is underwater, and p_1 is due to a hard, rough sea floor, this same equation applies. Clearly, intermediate cases are conceivable, but will not be considered here. Except for the sign, which is usually immaterial, Eq. (5) thus applies to all three cases. If r_0 and r_1 are everywhere much greater than the wavelength, this may be further simplified to

$$4\pi p_1(A)=\iint (P/r_0r_1)(\partial/\partial\nu)\{\exp[ik(r_0+r_1)]\}dS.$$

If the source is sufficiently directional, the denominator r_0r_1 may be replaced by its value at the origin of coordinates $r_{00}r_{10}$, while in the exponential

$$r_1+r_0=r_{10}+r_{00}+(ax+by+cz)$$

is used. The quantities a, b, c are, respectively, the sums of the x, y, z direction cosines of r_{10} and r_{00}. The replacement of z by ζ in the exponential, $\partial/\partial\nu$ by $\partial/\partial z$ and dS by $dxdy$ finally reduces this to

$$4\pi p_1(A)=\pm\{ikc[\exp(ikr_{10})]/r_{10}\}$$
$$\times\iint \bar{p} \exp[ik(ax+by+c\zeta)]dxdy, \quad (6)$$

where

$$\bar{p}=P[\exp(ikr_{00})]/r_{00}.$$

For future reference, it is to be noted that if ζ is identically zero, p_1 will consist entirely of the specularly reflected radiation. Hence, to the present approximation,

$$-4\pi p_0(A')=\{ikc[\exp(ikr_{10})]/r_{10}\}$$
$$\times\iint \bar{p} \exp[ik(ax+by)]dxdy, \quad (7)$$

where A' is the image of A in the xy plane. The notation

$$p_r(A)=\mp p_0(A')$$

will be used.

Thus far, it has been tacitly assumed that the function ζ is independent of the time t. Presumably, if the sea surface does not change too much during the time $2\pi/k$ so that ζ is a slowly varying function of t, the Eq. (6) will remain valid to the degree of approximation here achieved.

STATISTICAL THEORY FOR LONG-WAVE RADIATION

The Eq. (6) can be further simplified if the wavelength, $2\pi/k$ is much greater than the largest values of $|\zeta|$. Then

$$\exp(ikc\zeta)=1+ikc\zeta \text{ approx.}$$

The first term results in specularly reflected waves, and the scattered radiation is given by

$$p_s(A)=p_1(A)-p_r(A) \quad (8)$$
$$=\mp\{k^2c^2 \exp(ikr_{10})/4\pi r_{10}\}$$
$$\times\iint \bar{p}\zeta \exp[ik(ax+by)]dxdy. \quad (9)$$

The intensity of the scattered sound, per unit solid angle, will be

$$I_s=|p_s|^2r_{10}^2.$$

Because the sea surface is not constant, I_s will vary with time; only its mean value $\langle I_s\rangle$ will be calculated here. This is

$$\langle I_s\rangle=(k^2c^2/4\pi)^2\iiiint \bar{p}(x'y')\bar{p}^*(x''y'')$$
$$\times\langle\zeta(x'y't)\zeta(x''y''t)\rangle$$
$$\cdot\exp\{ik[a(x'-x'')+b(y'-y'')]\}$$
$$\times dx'dy'dx''dy'', \quad (11)$$

since only ζ depends on t. The function

$$\Phi=\langle\zeta(x'y't)\zeta(x''y''t)\rangle \quad (12)$$

therefore contains all the information about the sea surface that is required to determine $\langle I_s\rangle$; all other quantities in the integral depend only on the incident radiation. Being a time average, Φ does not depend on t, but may depend on the four coordinates of the two points x', y' and x'', y''. However, if the sea surface is homogeneous, ζ will depend only on the two differences $\xi=x''-x'$, $\eta=y''-y'$; throughout the following, it will be supposed that this is the case. Using ξ, η as one pair of variables, and introducing the abbreviation

$$J(\xi\eta)=\iint \bar{p}(x'v')\bar{p}^*(x'+\xi, y'+\eta)dx'dy', \quad (13)$$

Eq. (11) simplifies to

$$\langle I_s\rangle=(k^2c^2/4\pi)^2\iint J(\xi\eta)\Phi(\xi\eta)$$
$$\times\exp[-ik(a\xi+b\eta)]d\xi d\eta. \quad (14)$$

To evaluate this integral, one needs to know more about the function $\Phi(\xi, \eta)$; from Eq. (12), it readily follows that

$$\Phi(\xi, \eta) = \Phi(-\xi, -\eta), \tag{15}$$

and that

$$-\Phi(00) \leq \Phi(\xi\eta) \leq \Phi(00) = h^2, \tag{16}$$

where h is the rms value of ζ, the displacement of the sea surface from the plane $z=0$. (It is no loss of generality to assume that the arithmetic mean, $\langle\zeta\rangle$, is zero.) Any further properties of Φ must be found by observation, but it may be assumed that $\Phi(\xi, \eta) \rightarrow 0$ when $(\xi^2+\eta^2)^{\frac{1}{2}}$ becomes much larger than the length of the average train of surface waves. The function $J(\xi, \eta)$ will also approach zero for large ξ and η, but only when $(\xi^2+\eta^2)^{\frac{1}{2}}$ becomes greater than the dimensions of the area of the ocean that is insonified by the beam of incident radiation. If this dimension is much greater than the length of the trains of surface waves, one may approximate Eq. (14) by replacing $J(\xi, \eta)$ with

$$J(0, 0) = \int\int |\bar{p}|^2 dxdy; \tag{17}$$

then

$$\langle I_s \rangle = J(0, 0)\sigma, \tag{18}$$

where

$$\sigma = (k^2c^2)^2/(4\pi)^2 \int\int \Phi(\xi\eta) \exp[-ik(a\xi+b\eta)]d\xi d\eta \tag{19}$$

is a dimensionless quantity that may be called the scattering coefficient, or more descriptively, the scattering cross section for unit solid angle per unit area of sea surface.

Introducing the function

$$S(k_x k_y) = \int\int \Phi(\xi\eta) \exp[-i(k_x\xi+k_y\eta)]d\xi d\eta, \tag{20}$$

Eq. (19) simplifies to

$$\sigma = (k^2c^2/4\pi)^2 S(ka, kb). \tag{21}$$

The function defined by Eq. (20) may be called the spacial spectrum of the surface waves; it depends on the two components, k_x and k_y, of the wave number vector, and thus depends on both the direction and wavelength of the sinusoidal components of the surface waves.

It is therefore possible, by measuring σ for long wave radiation, to obtain values of the spacial spectrum. Unfortunately, it is necessary to vary the directional parameters a and b as well as the frequency of the incident radiation. This may be difficult in practice. The only measurements that are easily made are for scattering back to the source (reverberation), and these are usually made only at very small glancing angles. The experimental difficulty of measuring S in this way is thus very great.

One point deserves further notice. If k is so small that the spectrum function S can be replaced by $S(0, 0)$ in Eq. (21), the scattering coefficient is proportional to k^4, or inversely proportional to the fourth power of the wavelength of the radiation. This is the analog of the Rayleigh law for visible light scattered by a molecular medium.

STATISTICAL THEORY FOR SHORT-WAVE RADIATION

The difficulties in arranging the geometry of source and receiver are relatively small in one case. If, instead of sound, light is used, the sun[3] can be employed as source, and as the receiver, a camera in an airplane. In this way, the angular parameters are readily varied within wide limits. In this section, the formulas of the preceding section will be employed to calculate σ for very short-wave sound, no attempt being made to introduce the modifications required by the electromagnetic theory of light.

The simplifications of the preceding section are then no longer possible, and the full statistical theory of the sea surface must be invoked. Let $W(z)$ be the distribution function of ζ, defined by

$$\langle F[\zeta(xyt)]\rangle = \int F(z)W(z)dz, \tag{22}$$

which is to hold for every function F such that the time average of $F(\zeta)$ exists. In general, W will also depend on x, y, but if the surface is statistically homogeneous, W will depend only on z. An important special case of Eq. (22), for this as well as other applications, is

$$\langle \exp[-ik\zeta(xyt)]\rangle = \int \exp(-ikz)W(z)dz = Q(k); \tag{23}$$

Q is known as the characteristic function associated to W and ζ.

On taking the time average of both sides of Eq. (6), the simple result

$$4\pi\langle p_1(A)\rangle = \mp 4\pi p_0(A')Q(kc) \tag{24}$$

is readily obtained. It is reasonable[5] to identify $\langle p_1 \rangle$ with $p_r(A)$, the coherently reflected radiation. Then $Q(kc)$ is recognized as the (amplitude) reflection coefficient of the surface. Since $p_0(A')$ is very small except when the angles of incidence and reflection are equal, say to θ, the amplitude reflection coefficient is very approximately

$$R = \mp Q(2k \cos\theta). \tag{25}$$

It is readily seen from Eqs. (12) and (23) that $|R| \rightarrow 1$ for very long-wave radiation ($k \rightarrow 0$), and $R \rightarrow 0$ for very short-wave radiation ($k \rightarrow \infty$).

An important special case is the Gaussian distribu-

[5] This, and other considerations that follow, are entirely analogous to the refraction and scattering of light by three-dimensional molecular media; see, M. Born *Optik* (Berlin, 1933), Sections 74, 75, and 81.

tion, for which

$$W=[1/(2\pi)^{\frac{1}{2}}h]\exp(-z^2/2h^2),\quad Q=\exp(-k^2h^2/2),$$

h being the rms value of z. Then

$$R=\exp(-2k^2h^2\cos^2\theta), \tag{26}$$

and it is possible to estimate the reflection coefficient in special cases. Thus, if the frequency of the sound is 25 kc, $k=1$ cm^{-1}; if the sea is quite smooth, one may take $h=100$ cm, so that it is seen that $|R|$ will be very small unless $\cos\theta \leqslant 0.01$.

When $\cos\theta=0.01$, $|R|=0.135$, so that even at such small glancing angles and for such smooth seas, the loss on reflection will be more than 15 db. Possibly this description is unusual: in ordinary terminology, it indicates that the image of the source will be very distorted, and probably broken up into many images or parts of images. This is the idea on which the discussions of Shuleikin and Munk and Cox are based.

The energy that is thus lost from the coherently reflected beam will be found in the incoherently scattered radiation, $p_s=p_1-p_r$. Again defining I_s by Eq. (10), Eq. (6) yields

$$\langle I_s\rangle+|p_r|^2r_{10}^2$$
$$=(kc/4\pi)^2\int\int\int\int \bar{p}(x'y')\bar{p}^*(x''y'')$$
$$\cdot\langle\exp ik c[\zeta(x'y't)-\zeta(x''y''t)]\rangle$$
$$\cdot\exp\{ik[a(x'-x'')+b(y'-y'')]\}dx'dy'dx''dy'', \tag{27}$$

which may be compared with Eq. (11). To transform this integral, it is necessary to introduce the second distribution function of ζ, defined by

$$\langle F[\zeta(x'y't),\zeta(x''y''t)]\rangle$$
$$=\int\int F(z'z'')W_2(z'z'')dz'dz''. \tag{28}$$

The function W_2 is not independent of the x, y coordinates, even when the surface is statistically homogeneous, but then depends only on the differences $\xi=x''-x'$, $\eta=y''-y'$. Introducing the second characteristic function

$$Q_2(k',k'',\xi,\eta)$$
$$=\int\int \exp[-i(k'z'+k''z'')]W_2(z'z'')dz'dz'', \tag{29}$$

Eq. (27) becomes

$$\langle I_s\rangle+|p_r|^2r_{10}^2=\int\int J(\xi\eta)Q_2(-kc,kc,\xi,\eta)$$
$$\cdot\exp[-ik(a\xi+b\eta)]d\xi d\eta. \tag{30}$$

This integral cannot be evaluated without more specific information about Q_2. If W_2 is the bivariate Gaussian distribution function, then

$$Q_2=\exp\{-\tfrac{1}{2}[k'^2h^2+2k'k''\Phi(\xi\eta)+k''^2h^2]\}, \tag{31}$$

h and $\Phi(\xi,\eta)$ having the same significance as above. There is little evidence that the sea surface has the statistical properties implied by Eq. (31), but this assumption makes it possible to evaluate the integral of Eq. (30) approximately. One then has

$$\langle I_s\rangle+|p_r|^2r_{10}^2$$
$$=(kc/4\pi)^2\int\int J(\xi,\eta)\exp\{-k^2c^2[h^2-\Phi(\xi,\eta)]\}$$
$$\cdot\exp[-ik(a\xi+b\eta)]d\xi d\eta, \tag{32}$$

and the same assumptions that previously led to Eqs. (18) and (19) will now justify the approximation

$$J(\xi,\eta)\exp\{-k^2c^2[h^2-\Phi(\xi\eta)]\}$$
$$=J(\xi,\eta)\exp[-k^2h^2c^2]$$
$$+J(0,0)\exp\{-k^2c^2[h^2-\Phi(\xi,\eta)]\}. \tag{33}$$

It can be shown that when this is substituted into Eq. (32), the first term yields the value $|p_r|^2r_{10}^2$, so that one obtains the scattering coefficient as

$$\sigma=(kc/4\pi)^2\int\int \exp\{-k^2c^2[h^2-\Phi(\xi\eta)]\}$$
$$\times\exp[-ik(a\xi+b\eta)]d\xi d\eta, \tag{34}$$

which is to be compared with Eq. (19). This integral can now be evaluated by the method of steepest descent, which consists essentially in using the approximation

$$\Phi(\xi\eta)=h^2-\tfrac{1}{2}[\alpha^2\xi^2+\beta^2\eta^2+2\gamma^2\xi\eta]. \tag{35}$$

By suitably orienting the x and y axes, one may insure that $\gamma=0$; the integrals of Eq. (34) then reduce to known forms, and the final result is that

$$\sigma=(1/8\pi\alpha\beta)\exp\{-\tfrac{1}{2}[(a/\alpha c)^2+(b/\beta c)^2]\}. \tag{36}$$

The significant fact about Eq. (36) is the disappearance of the wave number k from the expression—the scattering coefficient is independent of the wavelength of the radiation. This is characteristic of the approximation of ray optics, which should, of course, be valid for short-wave radiation.[6] There is also some experimental evidence for this independence in the case of ultrasonic radiation scattered from the sea surface.

It remains to discuss the quantities α and β, which are the only parameters of the sea surface that remain in the

[6] Cox (private communication) has compared Eq. (36) with formulas derived entirely from ray optics. He finds that Eq. (36) should contain the additional factor $\cos^3(\mu/2)$, where μ is the angle between the reflected (scattered) ray and the vertical. This presumably indicates the inherent error of the present calculation, since Eqs. (4) and (6) could justifiably be modified so as to introduce such factors.

expression for σ. Clearly, they are given by

$$\alpha^2 = -\Phi_{\xi\xi}(0, 0), \quad \beta^2 = -\Phi_{\eta\eta}(0, 0);$$

the special orientation of the axes is such that

$$\gamma^2 = -\Phi_{\xi\eta}(00) = 0.$$

Now

$$\Phi_\xi(\xi, \eta) = (\partial/\partial\xi)\langle\zeta(xyt)\zeta(x+\xi, y+\eta t)\rangle$$
$$= \langle\zeta(xyt)\zeta_x(x+\xi, y+\eta, t)\rangle;$$

replacing x by $x-\xi$, y by $y-\eta$, this leads to

$$\Phi_\xi(\xi\eta) = \langle\zeta(x-\xi, y-\eta, t)\zeta_x(x, y, t)\rangle,$$

and hence

$$-\Phi_{\xi\xi}(\xi\eta) = \langle\zeta_x(x-\xi, y-\eta, t)\zeta_x(x, y, t)\rangle,$$

so that finally

$$\alpha^2 = \langle[\zeta_x(xyt)]^2\rangle, \tag{37}$$

which is to say that α is the rms value of $\partial\zeta(x, y, t)/\partial x$. In the same way, β is seen to be the rms value of the slope of the sea surface in the y direction.

One is thus lead to the disappointing conclusion that short-wave radiation will give much less information about the sea surface than long-wave radiation. The former yields only the values of α and β, and the orientation of the special axes here used; the latter yields the more interesting function $\Phi(\xi, \eta)$. It is true that, by introducing ray optics from the beginning, somewhat more information can be obtained, but it is apparently not possible to find Φ nor any information about W_2 from experiments of this sort with short waves. This is almost the reverse of the situation encountered in atomic and molecular physics, where the scattering of radiation whose wavelength is less than the dimensions of the scattering structure yielded much more information than did the longer waves. The reason is to be found in the way the rms height of the waves enters the formulas. The (slightly) analogous parameter in crystallographic work is the rms thermal displacement of the molecules, and this is small compared to the usual x-ray wavelengths employed. Thus, the two situations are not fundamentally different, but the quantitative relations between the parameters lead to the differing conclusions.

The following is an adaptation of the Eckart theory to reflections from nonperiodic pressure relief surfaces with experimental measurements to check the results of the theory.

The paper is taken, in part, from the dissertation by J. M. Proud, Jr., submitted in partial fulfillment of the requirements for the degree of Doctor of Philosophy at Brown University.

Dr. Beyer is a physicist. He has contributed in the fields of underwater sound, acoustic relaxation times, ultrasonic absorption in liquids and solids, and nonlinear acoustics. He is now Professor of Physics and Chairman of the Physics Department at Brown University, Providence, Rhode Island.

Dr. Tamarkin is a physicist. He has contributed in the fields of shock waves in air; finite amplitude waves; ultrasonics; underwater sound; solid-state, low temperature, ionospheric, and space physics; military systems research; and management of government research. He is now with the Advanced Research Projects Agency, Office of the Secretary of Defense.

This paper is reprinted with the permission of Dr. Beyer and the Acoustical Society of America.

Reprinted from JOURNAL OF APPLIED PHYSICS, Vol. 31, No. 3, 543–552, March, 1960

Printed in U. S. A.

25

Reflection of Sound from Randomly Rough Surfaces*

J. M. PROUD, JR.,†‡ R. T. BEYER, AND PAUL TAMARKIN§
Department of Physics, Brown University, Providence, Rhode Island
(Received September 23, 1959)

A study of the reflection of underwater sound from nonperiodic, pressure release surfaces is reported. The Eckart theory for wave reflection from rough surfaces has been adapted (with some modifications) to the experimental work. A portion of the investigation was directed toward a study of the dependence of the intensity reflected in the specular direction on angle of incidence, radiation wave number and the statistics of the reflecting surface. Secondly, a method is illustrated for determination of the rms amplitude and the correlation function of the reflecting surface from an analysis of the reflected intensity distribution. The radiation wavelength, in this case, must be much larger than the rms roughness amplitude.

I. INTRODUCTION

THE problem of wave reflection has been studied extensively in connection with diffraction grating phenomena in optics and acoustics. Current need for an understanding of the problem arises in the transmission of radar and radio waves over rough terrain, and in underwater sound propagation, where one must take account of reflections from the sea surface.

The study reported here is closely related to the latter situation, inasmuch as techniques of underwater sound were employed. Analogous results are to be expected, however, in the case of polarized electromagnetic radiation when the electric vector is parallel to the grooves of a perfectly reflecting rough surface. Two previous studies[1,2] have been reported which were similar experimentally to the investigation described here. Stationary periodic surfaces were used in both these cases to test the validity of several theories of rough surface reflection. This paper extends the investigation to the case in which the reflecting surfaces are time dependent and randomly rough. In the present experiment, a "randomly rough" surface is one whose nonperiodic profile is specified in terms of statistical quantities. The approach is such that one is able to form a description of the reflecting surface solely through an interpretation of the acoustic measurements. In particular, the description includes a specification of the correlation function and of the rms roughness amplitude about the average amplitude.

The investigation was concentrated on two aspects of the rough surface reflection problem. The first of these was a study of the dependence of the intensity reflected in the spectral direction upon such quantities as the angle of incidence, radiation wave number, and the statistics of the surface. In the second case, investigations were carried out which permitted a calculation of the rms amplitude and the correlation of a reflecting surface from an analysis of the reflected intensity pattern. It is shown that this can be done for a fixed radiation wavelength, which is much larger than the rms roughness amplitude.

II. THEORY

No existing theory of rough surface reflection solves the problem exactly. However, a large number of approximate theories have been developed.[3–11] The Eckart theory for sea surface reflection holds the advantage over most of the others in that it can be manipulated to apply to a variety of rough surface situations. It has been found to yield over-all agreement with experiments in the case of stationary sinusoidal surfaces.[1] Furthermore, it accounts explicitly for a directional incident beam shape, the type most often used experimentally. For these reasons, the Eckart theory has been adapted to the present problem.

The physical situation considered is illustrated in Fig. 1. An underwater sound source irradiates a perfectly reflecting (pressure release) surface whose displacement amplitude about the plane, $Z=0$, is specified by $\zeta=\zeta(x,t)$, where ζ is a slowly varying function of time. The incident pressure is given by

$$p=(P/r_0)\exp(ikr_0)$$

where P is a directional function for the source, r_0 is the distance from the source to any point on the surface, and k is the radiation wave number.

The solution of the problem consists in a determination of the acoustic pressure, p_1, at each point

* This work was supported in part by the Office of Naval Research.
† This paper is taken in part from a dissertation presented to Brown University Graduate School in partial fulfillment of the requirements for the degree of Doctor of Philosophy.
‡ Now at Allied Research Associates, 43 Leon Street, Boston, Massachusetts.
§ Now at RAND Corporation, Santa Monica, California.
[1] E. O. LaCasce, Jr., and P. Tamarkin, J. Appl. Phys. **27**, 138 (1956).
[2] Proud, Jr., Tamarkin, and Meecham, J. Appl. Phys. **28**, 1298 (1957).
[3] Lord Rayleigh, *Theory of Sound* (Dover Publications, New York, 1945), Vol. 2, pp. 89–96.
[4] H. S. Heaps, J. Acoust. Soc. Am. **27**, 698 (1955).
[5] J. G. Parker, J. Acoust. Soc. Am. **28**, 672 (1956).
[6] V. Twersky, J. Acoust. Soc. Am. **29**, 209 (1957).
[7] J. W. Miles, J. Acoust. Soc. Am. **26**, 191 (1954).
[8] W. C. Meecham, J. Appl. Phys. **27**, 361 (1956).
[9] L. M. Brekhovskikh, J. Exptl. Theoret. Phys. U.S.S.R. **23**, 275, 289 (1952).
[10] M. A. Isakovich, J. Exptl. Theoret. Phys. U.S.S.R. **23**, 305 (1952).
[11] C. Eckart, J. Acoust. Soc. Am. **25**, 566 (1953).

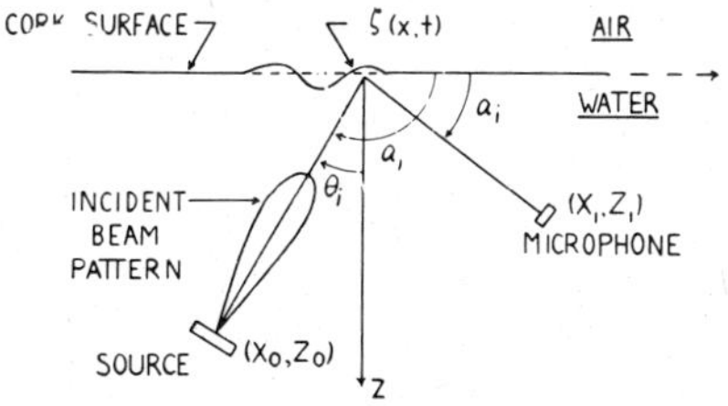

FIG. 1. Schematic diagram of the experimental configuration.

(x_1,z_1) in the half-plane $z\geq 0$. The Eckart solution makes use of the Helmholtz theorem and the approximation that the surface is nearly plane. The reflected pressure field is then given approximately by

$$p_1(x_1,z_1)=\frac{ikc}{r_{10}}\exp(ikcr_{10})\int D(x)$$
$$\times\exp[-ik\{ax+c\zeta(x,t)\}]dx, \quad (1)$$

where

$$D(x)=1/(4\pi)\int \bar{p}(x,y)\exp(-ikby)dy,$$

and

$\bar{p}=(P/r_{00})\exp(ikr_{00})$

r_{00} = the distance to the source from the center of coordinates.

r_{10} = the distance to the point (x_1,z_1) from the center of coordinates.

a, b, c are the sums of the x, y, z direction cosines for r_{10} and r_{00}.

The reflected intensity is proportional to $p_1p_1^*$, where p_1^* is the complex conjugate of p_1. Thus,

$$p_1p_1^*=(k^2c^2/r_{10}^2)\int\int D(x')D^*(x'')$$
$$\times\exp\{ikc[\zeta(x'',t)-\zeta(x',t)]\}$$
$$\times\exp[ika(x''-x')]dx'dx''. \quad (2)$$

By using the transformation: $x''=x+\xi$, $x'=x$, and averaging over time, we get

$$p_1p_1^*=(k^2c^2/r_{10}^2)\int\int D(x)D^*(x+\xi)$$
$$\times\langle\exp\{ikc[\zeta(x+\xi,t)-\zeta(x,t)]\}\rangle_t$$
$$\cdot\exp(ik\xi)dxd\xi. \quad (3)$$

The surface is assumed to possess the following properties:

$$\text{(a)}\quad \langle\zeta(x,t)\rangle_t=\langle\zeta(x+\xi,t)\rangle_t=0$$
$$\text{(b)}\quad \langle\zeta^2(x,t)\rangle_t=\langle\zeta^2(x+\xi,t)\rangle_t=h^2 \quad (4)$$
$$\text{(c)}\quad \langle\zeta(x,t)\zeta(x+\xi,t)\rangle_t=h^2\Phi(\xi).$$

The rms surface amplitude is h, and $\Phi(\xi)$ is the correlation function. The time averages in Eq. (4) express the homogeneity of the surface roughness in that the results of the averages are independent of the orgin of coordinates.

Results for a Gaussian Surface

In order to determine the time average in the integrand of Eq. 3 without further approximation, it is necessary to know more about the statistics of the surface. We shall assume here that the surface is Gaussian, i.e., that all of the basic distributions are Gaussian. The desired time average can then be performed with the aid of the bivariate Gaussian distribution. Thus,

$$\langle\exp\{ikc[\zeta(x+\xi,t)-\zeta(x,t)]\}\rangle_t$$
$$=(2\pi h^2)^{-1}[1-\Phi(\xi)]\int\int\exp[ikc(\zeta_1-\zeta_2)]$$
$$\times\exp\frac{-[\zeta_1^2+\zeta_2^2+2\Phi(\xi)\zeta_1\zeta_2]}{2h^2[1-\Phi(\xi)]}d\zeta_1d\zeta_2$$
$$=\exp\{-(chk)^2[1-\Phi(\xi)]\}. \quad (5)$$

By using this result, Eckart has obtained a geometric-optics approximate solution. In his calculation, the radiation wavelength is assumed to be much smaller than the surface roughness dimensions. We present here a solution valid for all wavelengths. By using Eq. (5), we can write Eq. (3) as

$$\langle p_1p_1^*\rangle_t=\frac{k^2c^2}{r_{10}^2}\Big\{\exp[-(chk)^2]\int\int D(x)D^*(x+\xi)$$
$$\times\exp[(chk)^2\Phi(\xi)-1]\cdot\exp(ika\xi)d\xi dx$$
$$+\exp[-(chk)^2]\int\int D(x)D^*(x+\xi)$$
$$\times\exp(ika\xi)d\xi dx\Big\}. \quad (6)$$

Now $\Phi(\xi)$ is equal to unity at $\xi=0$, and usually vanishes rapidly with increasing ξ. We therefore employ a Fourier integral representation as follows:

$$\exp[(chk)^2\Phi(\xi)]-1=(2\pi)^{-\frac{1}{2}}\int G(q)\exp(iq\xi)d\xi. \quad (7)$$

We then set

$$K(ka)=\int\int D(x)D^*(x+\xi)\exp(ika\xi)d\xi dx. \quad (8)$$

Hence,

$$(r_{10}^2/k^2c^2)\langle p_1p_1^*\rangle_t=K(ka)\exp[-(chh)^2]$$
$$+(2\pi)^{-\frac{1}{2}}\exp[-(chk)^2]\int G(q)K(q+ka)dq.$$

When the surface is plane ($h=0$) we obtain, in the specular direction ($a=0$, $c=c_s$),

$$(r_{10}^2/k^2c_s^2)\langle p_1p_1^*\rangle_t=K(0).$$

The reflected acoustic intensity relative to that reflected in the specular direction from a plane surface is then

$$I_1=(c/c_s)\exp[-(chk)^2]\cdot[K_n(ka)+F(ka)], \quad (9)$$

where

$$F(ka)=(2\pi)^{-\frac{1}{2}}\int G(q)K_n(q+ka)dq, \quad (10)$$

and

$$K_n(ka)=K(ka)/K(0) \quad (10a)$$

As a final result, a solution of the integral Eq. (10) yields an expression for the correlation function in terms of measurable acoustical quantities. The result is

$$\Phi(\xi)=(kch)^{-2}\ln\left[1+\frac{\bar{F}(\xi)}{\bar{K}_n(\xi)}\right], \quad (11)$$

where $\bar{F}(\xi)$ and $\bar{K}_n(\xi)$ are the inverse Fourier transforms of $F(ka)$ and $K_n(ka)$, respectively.

Thus, provided the rough surface is known to be Gaussian, one can (at least in principle) determine the correlation function by:

(1) knowing $K_n(ka)$; this is essentially equivalent to knowing the source beam pattern;
(2) determining $F(ka)$ by measurement of the reflected intensity distribution and applying Eq. (9).
(3) carrying out the operations indicated in Eq. (11).
(4) estimating the rms amplitude h by some auxiliary means since it cannot be determined from the acoustic data alone.

We have thus illustrated that it is possible to obtain the surface correlation function from acoustic measurements when the rms amplitude, h, and the surface statistics are known. Unfortunately, the difficulty in establishing these surface characteristics renders Eq. (11) of little use in practical situations. Indeed, it may not even be possible to obtain analytical solutions for $\Phi(\xi)$, when the surface statistics are other than Gaussian, and there is no evidence that the sea surface is ever describable by Gaussian statistics.[12] The statistical nature of the sea surface is discussed later.

Specular Results for Truncated $\Phi(\xi)$

When the surface is approximately purely random, that is, when there is little average correlation between neighboring points of the surface, the correlation function is nonvanishing only over a narrow region near the origin ($\xi=0$). It is not unreasonable, then, to expect $\Phi(\xi)$ to resemble

$$\Phi(\xi)=1, \quad -\epsilon_0\leq\xi\leq\epsilon_0$$
$$=0, \quad \text{otherwise.} \quad (12)$$

By using Eq. (11) we obtain

$$\bar{F}(\xi)=\bar{K}_n(\xi)[\exp(chk)^2-1], \quad -\epsilon_0\leq\xi\leq\epsilon_0$$
$$=0, \quad \text{otherwise.} \quad (13)$$

If the surface is "illuminated" uniformly over an area of half-width, ϵ,

$$K_n(ka)=\frac{\sin^2 ka\epsilon}{(ka\epsilon)^2}, \quad (14)$$

then

$$\bar{K}_n(\xi)=(2\pi)^{\frac{1}{2}}\epsilon^{-1}(1-\xi/2\epsilon), \quad -2\epsilon\leq\xi\leq 2\epsilon$$
$$=0, \quad \text{otherwise.} \quad (15)$$

Now, if $\epsilon_0\ll\epsilon$, $\bar{K}_n(\xi)$ can be replaced by its value at the origin. Equation 13 then becomes

$$\bar{F}(\xi)=(2\pi)^{\frac{1}{2}}\epsilon^{-1}[\exp(chk)^2-1]=\bar{F}(0). \quad (16)$$

On inverting Eq. 16 and substituting in Eq. (9), we get for the intensity in the specular direction (where $a=0$, $c=c_s$),

$$I^0=\exp[-(c_shk)^2]+(2\epsilon_0/\epsilon)\{1-\exp[-(c_shk)^2]\}. \quad (17)$$

Thus, if ϵ_0/ϵ is vanishingly small, the specular reflection decreases approximately as $\exp[-(c_shk)^2]$. For large hk, the specular intensity approaches $2\epsilon_0/\epsilon$. The dependence on c_shk is, of course, a consequence of the assumed Gaussian statistics. It will be shown later that in most cases, the specular intensity falls off as $1-(c_shk)^2+O(c_shk)^4$, independently of the surface statistics. Thus, by measurement of I^0 for small values of c_shk, one has the possibility of estimating the rms surface amplitude h. Somewhat more information can be obtained if the reflecting surface is highly random (small ϵ_0) in that it is reasonable to assume Gaussian statistics in such a case. Then, Eq. (17) may be used to determine ϵ_0 at short wavelengths.

Long Wave Theory

The above results demonstrate the difficulty in obtaining information about a reflecting surface from strictly wave reflection measurement. A technique is now given whereby the surface correlation function is determinable from long wave measurements for the case in which a knowledge of the complete surface statistics is lacking.

In the approximation that

$$kc\zeta_{\max}\ll 1, \quad (18)$$

[12] H. R. Seiwell and G. P. Wadsworth, Science **109**, 271 (1949).

Equation 2 can be written

$$p_1p_1^* \doteq (k^2c^2/r_{10}^2)\int\int D(x')D(x'')$$
$$\times\{1-ikc[\zeta(x',t)-\zeta(x'',t)]$$
$$-(k^2c^2/2)[\zeta^2(x',t)+\zeta^2(x'',t)+\zeta(x',t)\zeta(x'',t)]\}$$
$$\times\exp[-ika(x'-x'')]dx'dx'' \quad (19)$$

to second order in $kc\zeta$.

In the original Eckart theory, the scattering was described in terms of a scattered intensity proportional to the square of the magnitude of the difference in pressure reflected from the rough surface and that reflected from a plane surface replacing the rough one. This procedure dictates that one know both the amplitude and phase of these pressures in an experimental determination of the scattered intensity. The procedure we have adopted here leads to the experimentally simpler operation of forming the difference between plane and rough surface reflected *intensity*. No consideration of phase is then necessary. This, it is felt, is very important in that phase is usually a difficult experimental quantity. On employing Eq. (4), it is seen that

$$\langle p_1p_1^*\rangle_t = \frac{k^2c^2}{r_{10}^2}\Big\{[1-(kch)^2]\int\int D(x)D^*(x+\xi)\exp(ika\xi)d\xi dx$$
$$+(kch)^2\int\int\Phi(\xi)D(x)D^*(x+\xi)$$
$$\times\exp(ika\xi)d\xi dx\Big\}. \quad (20)$$

On following the general techniques applied in the theory for all wavelengths, we set

$$\Phi(\xi)=(2\pi)^{-\frac{1}{2}}\int g(q)\exp(iq\xi)dq. \quad (21)$$

Then it can be readily shown that the reflected intensity relative to that reflected in the specular direction from a plane surface is given by

$$I_1=(c/c_s)^2\{[1-(chk)^2]K_n(ka)+(chk)^2f(ka)\}, \quad (22)$$

where

$$f(ka)=(2\pi)^{-\frac{1}{2}}\int g(q)K_n(q+ka)dq. \quad (23)$$

As before, a knowledge of the beam shape specifies $K_n(ka)$. The function $f(ka)$ is then a measure of the scattering and may be found empirically by determining I_1 for the rough surface. The same procedures which led to Eq. 11 now yield

$$\Phi(\xi)=\bar{f}(\xi)/\bar{K}_n(\xi), \quad (24)$$

where $\bar{f}(\xi)$ and $\bar{K}_n(\xi)$ are Fourier transforms of $f(ka)$ and $K_n(ka)$, respectively.

Some insight into the effect of various correlation functions on the scattering can be gained from the following qualitative considerations.

Let us define a half-width, Δ_ξ, for $\bar{f}(\xi)$ and, similarly, Δ_{ka} for $f(ka)$. From the Fourier integral theorem it is known that $\Delta_\xi\cdot\Delta_{ka}\approx 1$. Now the half-width Δ_ξ is roughly determined by the half-width of the narrowest of $\Phi(\xi)$ or $\bar{K}_n(\xi)$. Thus, for example, if $\Phi(\xi)$ is very narrow (as for a highly random surface), Δ_ξ must also be small. In that case, Δ_{ka} must be large and the scattering is highly diffuse. It is important to note, however, that although the reflected radiation is diffuse in this case, it is not described by Lambert's law. Diffuse reflection obeying Lambert's law arises when the incident wavelength is considerably less than the surface roughness dimensions.[13] The results here are applicable only to long-wave radiation.

Special Cases

The function $K_n(ka)$ for the source used in the experiments described below can be represented quite accurately by

$$K_n(ka)=\exp[-\sigma^2(ka)^2/2] \quad (25)$$

where σ is a constant. The correlation function for the surface used in the long wave experiments is, to a good approximation

$$\Phi(\xi)=\cos(q\xi)\exp(-\xi^2/2b^2). \quad (26)$$

Then, from Eq. (24),

$$\bar{f}(\xi)=\sigma^{-1}\cos(q\xi)\exp(\xi^2/2\ell^2), \quad (27)$$

where

$$\ell^{-2}=b^{-2}+\sigma^{-2}. \quad (28)$$

By inverting Eq. (27), one obtains

$$f(ka)=(chk)^2(\ell/2\sigma)\{\exp[-\tfrac{1}{2}(q-ka)^2\ell^2]$$
$$+\exp[-\tfrac{1}{2}(q+ka)^2\ell^2]\}. \quad (29)$$

With the known values of σ, b, and q, together with an estimate of h, one can predict the intensity distribution, I_1. Conversely, when the surface correlation function is unknown [except that it is of the type Eq. (26)], the values of b and q can be found empirically by using Eq. (29). Of particular interest is the problem of finding the correlation function of the sea surface from acoustic measurements. Usually, $\Phi(\xi)$ is not too unlike Eq. (26). In the general situation, $\Phi(\xi)$ must be determined directly from the ratio of the two empirical functions in Eq. (24). The transforms $\bar{f}(\xi)$ and $\bar{K}_n(\xi)$ must be found by numerical methods.

By way of illustration, let us consider the case of a purely sinusoidal surface. The correlation function is then given by Eq. (26) with $b=\infty$. Then, Eqs. (29)

[13] E. Skudrzyk, Akustische Z. 4, 189 (1939).

and (22) yield

$$I_1 = (c/c_s)^2\{[1-(chk)^2]K_n(ka) + [(chk)^2/2][K_n(q-ka)+K_n(q+ka)]\}. \quad (30)$$

The first term on the right is that part of the reflected intensity which is just the plane surface specular reflection reduced by the factor $1-(kch)^2$. The second term represents the scattered intensity. The shape of these maxima is identical with that reflected in the specular direction, but they are located by the grating equation for first order.

$$q = \pm ka. \quad (31)$$

The effect of a damped harmonic correlation function (b finite) is to decrease the intensity at these maxima and to broaden them.

When the surface correlation is not oscillatory [e.g., $q=0$ in Eq. (26)] there are no first-order grating orders. Instead, the effect of scattering is merely to confuse the specular reflection.

A consideration of the grating equation together with the assumption of long wave radiation shows that the Eckart theory is applicable for wavelengths satisfying

$$h \ll \lambda < \Lambda, \quad (32)$$

where Λ is the wavelength of any periodic reflecting surface. It is interesting to note that this is essentially a statement of the conditions for which the Kirchhoff approximation is valid.

III. APPARATUS AND PROCEDURE

A. Apparatus

A pulsed underwater sound source irradiated the rough surfaces under study. The surfaces were made of cork (approximately pressure release) formed with the desired roughness profiles, and were floated on the surface of the water in a tank. The reflected pulses were observed by means of a scanning microphone. Details of the apparatus have been given previously.[1]

The basic construction of the surfaces used in the investigation was similar to that of periodic surfaces of previous studies.[1,2] A $\frac{3}{4}$-in. plywood backing is used to support sheet cork molded to give the desired roughness profile. The width of each surface is 18 in., whereas the lengths are more than twice this dimension. Since the beam typically irradiates an approximately circular region of 1 ft diam, one is assured that the entire main lobe of the beam makes contact with the rough surface. The surface lengths are such that the surfaces can be moved (along their lengths) with respect to the beam through wide limits. With reference to Fig. 1, the x and y directions run along the surface lengths and widths, respectively. The roughness variation is then a perturbation of the plane $z=0$. There is no y dependence of this variation so that the amplitude of variation can be specified by a function of x alone.

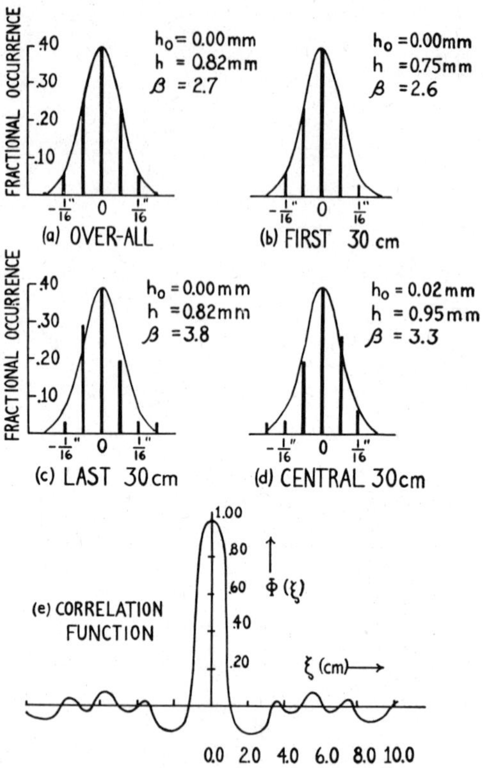

FIG. 2. Step-amplitude distributions and correlation function of surface I.

Surface I

Surface I was designed for use in investigating the dependence of the specular reflected intensity on the acoustic wave number, angle of incidence and surface roughness. The surface represents an attempt to establish a roughness whose constituent amplitudes are distributed according to a Gaussian law. In order to simplify construction, the roughness was composed of one hundred step-amplitude variations of equal width (1 cm). These consisted of flat cork strips cemented to the plywood base perpendicular to its length. The smallest step variation was limited to 1/32 in., the minimum thickness of available sheet cork.

The discontinuous distribution of amplitudes is shown in Fig. 2(a) and compared with a smooth Gaussian envelope. The ordinate represents the normalized frequency of occurrence of each amplitude shown along the abscissa. The hundred step amplitudes so distributed are arranged along the surface base in such a manner that any segment whose length is about equal

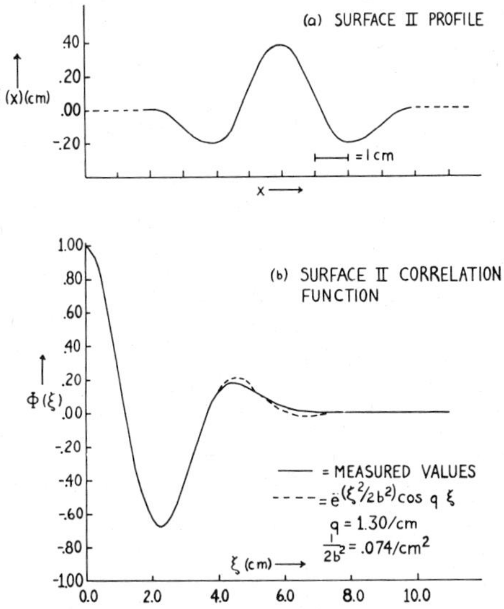

FIG. 3. Profile and correlation function of surface II.

to the diameter of the irradiated surface area (30 cm) has approximately:

(1) an average amplitude of zero;
(2) the same rms amplitude as the over-all surface;

and

(3) the same amplitude distribution as the over-all surface.

The degree to which these conditions were met is illustrated in Fig. 2(b,c,d). The distribution is shown for three 30 cm segments of surface; the first and last 30 cm of the roughness and a similar segment centered at the middle of the surface. Beside each figure is noted the average amplitude, h_0, the rms amplitude, h, and β, the ratio of the average fourth power of the amplitude to h^4. For a true Gaussian distribution, $\beta=3$.

The correlation function for Surface I is illustrated in Fig. 2(e). It demonstrates that there is little correlation on the average even between nearest neighbor step-amplitudes. The oscillations about zero indicate only slight irregular correlation for larger correlation distances. Note that while the surface profile is discontinuous, the correlation function is continuous.

Surface II

Surface II was constructed in order to investigate the dependence of the reflected intensity pattern on the surface correlation function. It was designed to yield an analytical correlation function that could be manipulated easily in the theoretical expressions.

To provide a smooth water-tight surface, a single $\frac{1}{16}$ in. cork sheet was bonded to a preformed cork mold attached to the plywood base. The surface is mostly planar with a brief spatial wave train impressed near the center. The profile of this surface "pulse" is shown in Fig. 3(a). The average amplitude is zero, while the rms amplitude h is 1.89 mm when the average is taken only over the rough portion of the surface. The plane portion of the surface is, of course, at $z=0$.

The correlation function (for positive argument) for Surface II is illustrated in Fig. 3(b) and is compared to the analytical function which approximates it. The adjustable parameters used in obtaining the fit are noted in the figure.

The correlation functions and other averages for Surfaces I and II have been computed as spatial averages over the lengths of the surfaces. Since the profiles themselves are fixed this method of averaging is completely equivalent to time averaging by a fixed observer observing the surface in uniform motion past a point in observer space. We note that the sea surface, for example, is not simple in this respect as the sea roughness profile itself changes in time.

B. Procedure

The portion of the investigation using Surface I was concerned with the dependence of the specular reflection on the quantity, c_shk. The reflected intensity was observed for fixed angles of incidence (fixed c_s) as the acoustic frequency (and therefore, k) was varied over the available range. This was done for four angles of incidence: 0°, 30°, 45°, and 60°. The value of h was such that the available range of c_shk was then approximately: 0.25 to 2.00. A calibration procedure was followed whereby all data were taken relative to reflections from the still water surface.

The time average of the (relative) reflected intensity was determined at each frequency as the rough surface passed through the area insonified by the incident beam. In order to obviate the difficulties of an actual time average of the received pulse amplitudes, a sampling technique was employed. That is, the surface was moved in 5 cm steps along the x axis and held at rest while the reflected pulse amplitude was recorded. Usually, ten "snapshot" readings of this kind were made. The average square of the ten readings was then proportional to the average reflected intensity. A more detailed sampling technique was determined to be unnecessary. This result, indicated by experiment, is to be expected in view of the uniformity of the surface, together with the fact that the beam irradiates a surface area of about 30 cm.

A correction factor was applied to the observed values of the average reflected intensity to account for the lack of perfect reflectivity of cork at higher frequencies. This was done by noting the reflection from a plane cork surface when the latter replaced the rough one. All of the observed average intensities were then adjusted by dividing them by the square of the reflection coefficient of the plane cork surface. This procedure is not strictly justifiable, since absence of perfect reflectivity indicates that the surface is not pressure release, as assumed in the theory. On the other hand, the discrepancy is only slight and occurs only at the highest frequencies used.

The long-wave theory was investigated experimentally by use of Surface II, with the intention of testing the theory and illustrating some of its limitations. Three angles of incidence were employed (24.5°, 30°, and 45°) while the frequency in each case was fixed at about 80 kc.

The function $K_n(ka)$ was defined in the theory by the intensity pattern observed when the sound beam is reflected from a plane, perfectly reflecting surface. $K_n(ka)$ was found experimentally by noting the form of the intensity distribution when the sound beam was reflected from the still water surface. All intensities observed in 1° to 2° steps throughout the angle describing the beam width were taken relative to: max $K_n(ka)=K_n(0)\equiv 1$.

The average reflected intensity I_1 was found by a technique similar to that just described for the determination of $K_n(ka)$. For each position of the receiving microphone, the average intensity was observed as Surface II moved through the incident sound beam in steps of 2 cm. About twenty readings were taken as the "rough" portion of the surface passed entirely through the incident beam. As in the case of Surface I, the average square of these readings was taken as the average intensity. In the neighborhood of the spectral direction, where the magnitude and slope of I_1 are large, observations were made at intervals of 2°. At angles well removed from the spectral direction, where I_1 varies slowly, sufficient detail was obtained by rotating the receiving microphone in steps of 5°. The reflected sound field was scanned in this way throughout the 180° in the water below the reflecting surface.

In specifying the rms amplitude, h, of Surface II we have given the value obtained by considering only the rough portion of the surface. Any value for h smaller than this is obtainable by including a part of the flat portion of the surface (amplitude=0) in the averaging procedure. This same flexibility arises in the acoustic measurements since the observed average intensity depends upon the number of reflection readings included which correspond to the reflection from the plane portion of the surface. Thus, the value of h to be used when applying the theory is somewhat elusive. The procedure used to obtain a value for h involved the use of the expression $I^0=1-(c_shk)^2$, which is valid provided $f(ka)$ is very small near $ka=0$.

IV. RESULTS

A. Spectral Reflections

The results of the specular reflection from Surface I are summarized in Fig. 4. Here, the average intensity,

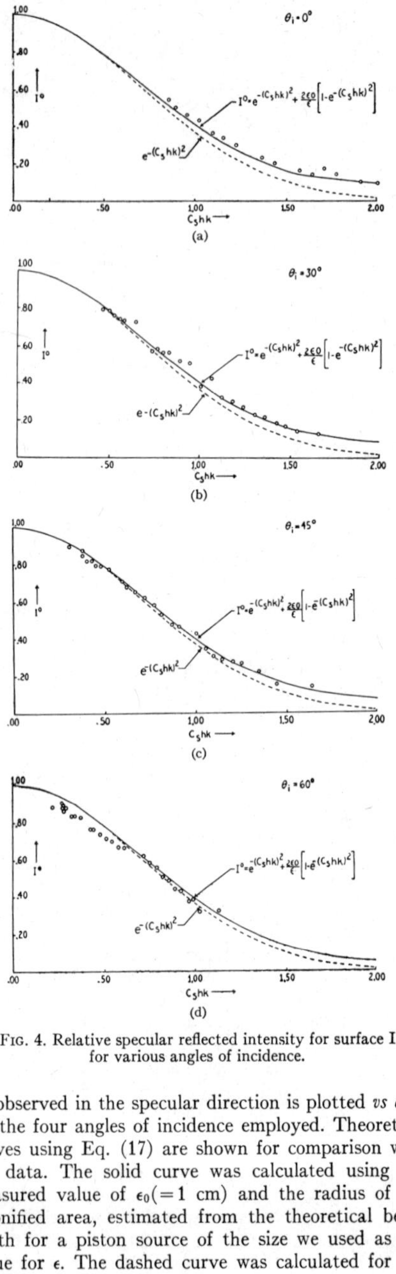

FIG. 4. Relative specular reflected intensity for surface I for various angles of incidence.

I^0, observed in the specular direction is plotted *vs* c_shk for the four angles of incidence employed. Theoretical curves using Eq. (17) are shown for comparison with the data. The solid curve was calculated using the measured value of $\epsilon_0(=1\text{ cm})$ and the radius of the insonified area, estimated from the theoretical beam width for a piston source of the size we used as the value for ϵ. The dashed curve was calculated for the case where ϵ_0 is vanishingly small, as for a purely random surface.

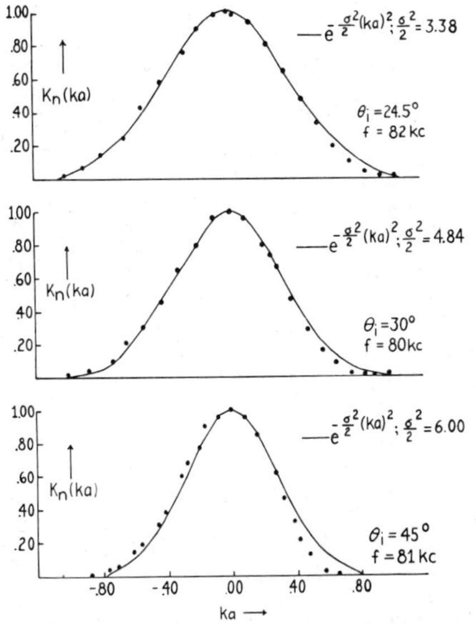

FIG. 5. Approximate Gaussian beam forms, $K_n(ka)$.

B. Distributions of Reflected Intensity

The distribution of reflected intensity and its relation to the correlation function was investigated at three angles of incidence. The results of an empirical determination of $K_n(ka)$ are illustrated in Fig. 5. The Gaussian function [Eq. (25)] was fitted to the data. The width parameter, σ, was then set by this procedure. The larger values of σ associated with greater angles of incidence are due to the decreasing depth of the source transducer.

The data obtained at $\theta_i=30°$ were used in a direct comparison with the theoretical expression, Eq. (22), where $f(ka)$ was calculated from Eq. (29). The Gaussian approximation for $K_n(ka)$ was also employed. The theoretical and experimental values of I_1 were matched at $ka=0$, so that the rms amplitude h is treated as a fitting parameter. The comparison is illustrated in Fig. 6 where theory is indicated by the solid curve. The cutoff value of ka, noted in the figure, is that value of ka for which the reflections are at the grazing angle, $\alpha_1=0$, in the forward direction.

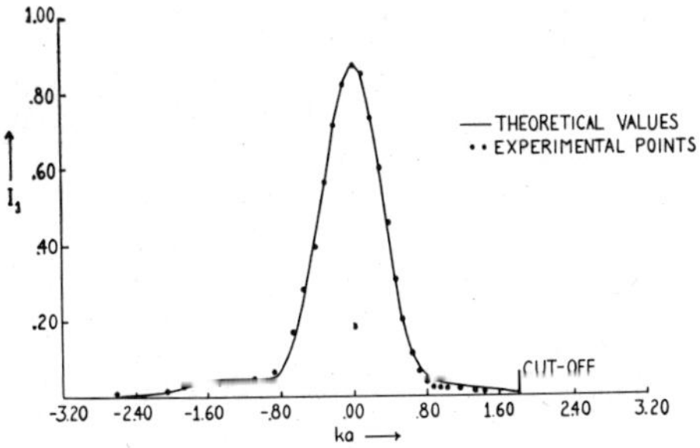

FIG. 6. Average reflected intensity distribution, $\theta_i=30°$.

An empirical evaluation of $f(ka)$ was determined from the same data using the back scattered reflections. The experimental points representing $f(ka)$ for negative ka are plotted in Fig. 7(a). The solid line again represents theoretical values determined from the directly measured surface correlation function. The parameters in the theoretical expression were then readjusted to determine $f(ka)$ empirically. The result is indicated by the dashed curve. In adjusting the parameters it is noted that the value of q is equal to the value of ka at which

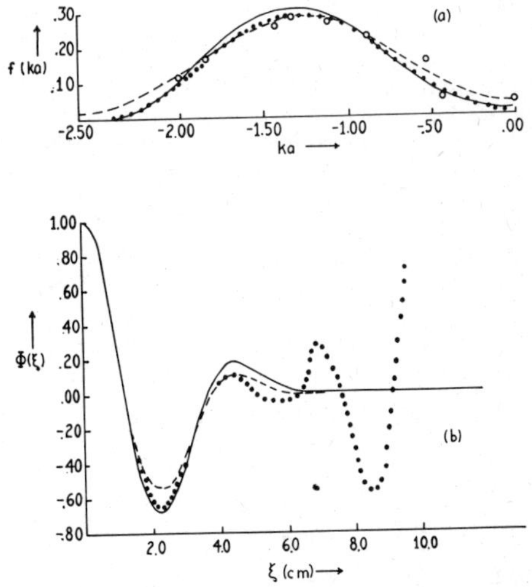

FIG. 7. Empirical determinations of correlation function, $\theta_i=30°$.

the maximum in $f(ka)$ occurs. The value of ℓ to be used in Eq. (28) can be found from the maximum value of $f(ka)$ or by the width as indicated by Eq. (29). The parameters b and q found in this manner were inserted into the assumed functional form of the correlation function. The results are indicated in Fig. 7(b) by the dashed curve. The solid curve represents the correlation function determined directly from measurements of the surface profile.

In the calculation just described, the choice of the analytical function used to fit the experimental data for $f(ka)$ was influenced by the expected theoretical form, Eq. (29). To the accuracy of the data, a number of other functional forms of $f(ka)$ could be assumed for obtaining an empirical evaluation. By way of illustration, the correlation function was calculated with the assumption that $f(ka)$ could be fitted with the following form:

$$f(ka)=A\,\sin^2 pka, \quad -\pi/(p)<ka<\pi/(p)$$
$$=0, \text{ otherwise.}$$

A comparison with experiment is denoted by the dotted curve in Fig. 7(a). From Eq. (24), one then obtains

$$\Phi(\xi)=[A\sigma/(2\pi)^{\frac{1}{2}}][1/(\xi)+\xi/(4p^2-\xi^2)] \times \sin\pi\xi/(p)\,\exp(\xi^2/2\sigma^2).$$

The dotted curve in Fig. 7(b) was determined from this expression. Because of the positive exponential term, the function diverges rapidly beyond $\xi=6.5$, approximately.

The above measurements indicated a distortion of the forward scattering due to the proximity of the cutoff to the maximum of the function, $f(ka)$. In order to clarify this effect, a second set of data was taken with $\theta_i=45°$. For this angle of incidence, most of the forward scattering is absent due to cutoff.

The assumption is made in the analysis of the data that the functional form of $f(ka)$ is given by Eq. (29). In determining the correlation function, the problem is

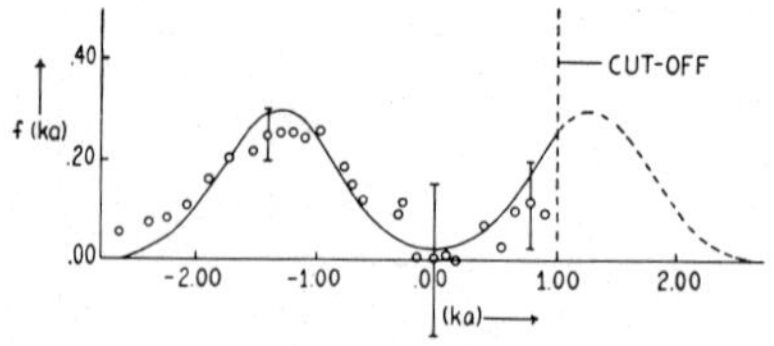

FIG. 8. $f(ka)$, experimental and theoretical, $\theta_i=45°$.

then one of evaluating the parameters b and q. The experimental values representing $f(ka)$ were calculated from smoothed values of $K_n(ka)$ and I_1. The results are shown in Fig. 8. The curve shown in the figure was calculated from the known values of b, q and σ. Estimates of the maximum possible experimental error are indicated. The large error near $ka=0$ is due to effect of determining experimentally a small difference between two large values.

The back scattering is illustrated again in Fig. 9(a). The experimental values are compared with the theoretical curve. The dashed curve represents the function $f(ka)$ when the parameters are adjusted to fit the data more closely. The resulting correlation function is illustrated in Fig. 9(b) where the dashed curve is the empirical result. The solid curve is the known function determined from the measured surface profile.

One further empirical determination of $\Phi(\xi)$ was carried out for the case where the forward scattering is nearly unaffected by the cutoff. The data and the fitted function, $f(ka)$, are shown in Fig. 10. By the methods outlined above, the correlation function was then determined and is illustrated in Fig. 11. Here, the dashed curve is the empirical result.

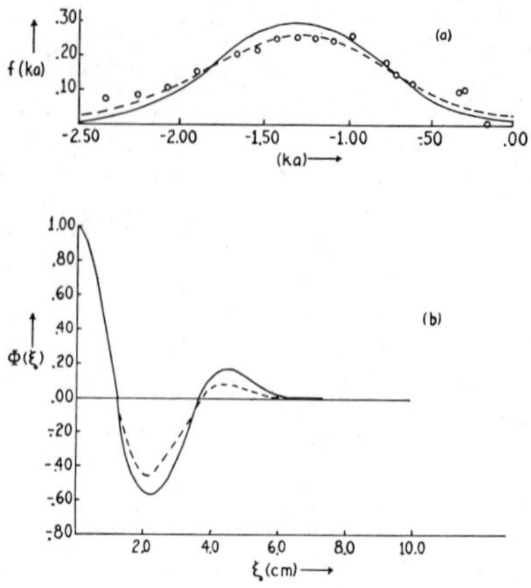

FIG. 9. Empirical determinations of correlation function, $\theta_i=45°$.

V. SUMMARY AND CONCLUSIONS

A. Specular Reflection

The results of the experiment and theory for the reflection from Surface I are illustrated in Fig. 4. The agreement is somewhat surprising since Surface I is only step-wise continuous. It thereby presents a clear violation of the basic condition required by the Eckart theory approximation that the surface slope is everywhere small.

The reason for the good agreement is probably due to the fact that the infinite slope portions of the surface reflect only a small portion of the incident sound, particularly at smaller angles of incidence. At the largest angle of incidence used (60°), there is definite disagreement with theory for the longer wavelengths. This may be related in some way to the fact that more reflection is occuring from the steep portions of the surface. The reason for the improving agreement with increasing $(c_s hk)$ may then be fortuitous.

The results of the spectral reflection study suggest that one might expect to determine the surface statistics by noting the dependence of I^0 on $(c_s hk)$. There is no available procedure to do this theoretically, however. Secondly, the spectral intensity is independent of the full surface statistics to second order in $(c_s hk)$. Furthermore, one can show that the fourth order term in $(c_s hk)$

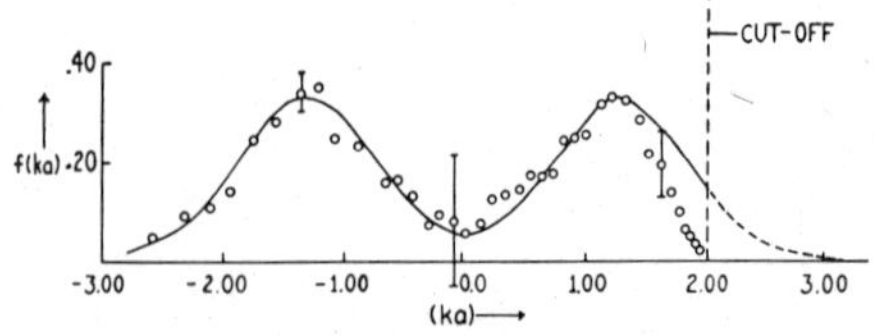

FIG. 10. $f(ka)$, experimental and theoretical, $\theta_i=24.5°$.

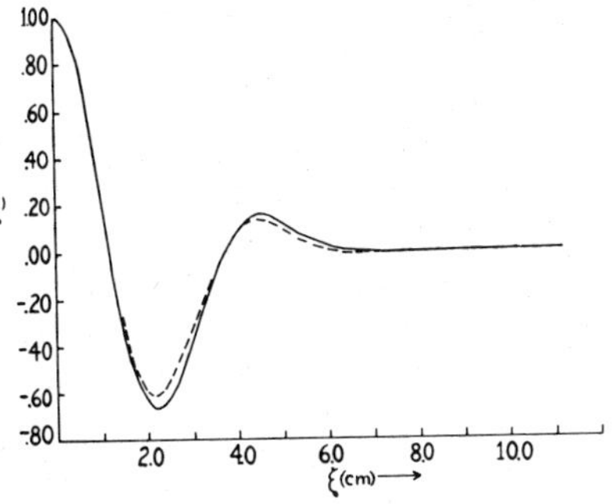

FIG. 11. Empirical determination of correlation function, $\theta_i=24.5°$.

depends only weakly on the statistics. For shorter wavelengths, where the higher order terms assume greater importance, one violates the condition on radiation wavelength and minimum radius of curvature which must hold in order that the Kirchhoff approximation is valid.[14]

In the case of a high degree of randomness such as we have constructed in Surface I, it is reasonable to assume Gaussian statistics. Then, one has the possibility of estimating the range of the correlation function by observing the value of I^0 for large (c_shk). It may be seen that I^0 approaches $2\epsilon_0/\epsilon$ in this limit. The experiment shows this effect quite clearly in Fig. 4.

B. Determinations of Correlation Function

We have shown that in theory one can form an estimate of the reflecting surface correlation function from acoustic measurements alone. It was shown, furthermore, that all the information about the surface is contained in the back scattering. This fact is particularly important in situations where the forward scattering is cut off as illustrated by the experimental situation we have considered with $\theta_i=45°$.

The correlation function provided by the profile of Surface II is quite general in that it is typical (though idealized) of many processes such as the sea surface roughness. The sea bottom is frequently characterized by a similar correlation function inasmuch as it is usually somewhat periodic. The acoustic measurements involve time averages which may be calculated by some sampling technique such as we have described. The procedure is apt to be difficult or even impossible when applied to the sea surface, however, in that the basic character of the sea roughness may change in time. That is, the sea surface roughness may not be a stationary random process. The reason is that the mechanisms generating the surface waves (local winds and large scale air mass motion) themselves change with time. For the type of acoustic measurements which need to be made, there appears to be only slight chance of obtaining the required data in a time that is short compared to the times characteristic of these changes.

When the radiation wavelength is very large compared to the dimensions of the surface irregularities, the reflected radiation is not greatly different from that reflected from a perfectly flat surface. That is, the scattered radiation which must be measured in order to determine the surface correlation function is not very evident. Great experimental accuracy is required, therefore, to obtain a representation of $f(ka)$. In fact, an uncertainty of more than one percent in the intensity measurements can lead to many times this uncertainty in the calculated values of $f(ka)$. This is particularly true in the vicinity of the spectral direction, where $f(ka)$ is determined from the small difference between two large quantities. In the experiment reported here it is seen that the correlation function is determined with approximately the same over-all accuracy as the function $f(ka)$.

When the surface correlation is not representable by some simple analytical form, the analysis leading to its computation must be performed by numerical methods. In particular, the Fourier transforms in Eq. (24) must be evaluated by numerical integration. This is not a real limitation inasmuch as the integration may be done readily by most electronic computers. In some instances it may suffice to assume that $f(ka)$ has some simple analytical form as we have done here. One then obtains an approximate correlation function which may be quite similar to the actual unknown function.

Further research into some of the basic problems of rough surface reflection is still necessary. All of the theories thus far developed are limited to surfaces which are nearly plane. A theory which is not so limited would have much greater applicability to real situations. Experimental studies of reflection from rough surfaces which have large slope and large amplitude compared to radiation wavelengths might provide considerable insight into the basic problem and might thus suggest theoretical techniques to be tried.

[14] B. B. Baker and E. T. Copson, *The Mathematical Theory of Huygens' Principle* (Oxford University Press, London, 1950) 2nd Edition.

The author of the following paper computes the reflection coefficients for the reflection of plane sound waves from a sinusoidal surface on the basis of an exact formulation and compares the theoretical result with experimental data. A more detailed treatment of the scattering from periodic surfaces is given by the author in a later publication.[1]

Dr. Uretsky is a physicist. He has contributed in the fields of atomic physics, acoustics, nuclear structure, field theory, and elementary particles. He is now on the staff of the Argonne National Laboratory.

This paper is reprinted with the permission of the author and the Acoustical Society of America.

[1]J. L. Uretsky, The scattering of plane waves from periodic surfaces, *Ann. Phys.* **33**, 400–427 (1965).

Reprinted from THE JOURNAL OF THE ACOUSTICAL SOCIETY OF AMERICA, Vol. 35, No. 8, 1293–1294, August, 1963
Copyright, 1963 by the Acoustical Society of America.
Printed in U. S. A.

26 Reflection of a Plane Sound Wave from a Sinusodial Surface

JACK L. URETSKY

Argonne National Laboratory, Argonne, Illinois
and
University of California, San Diego, Marine Physical Laboratory of the Scripps Institute of Oceanography
(Received 28 March 1963)

Reflection coefficients have been computed, on the basis of an exact formulation, for the reflection of a plane sound wave from a sinusoidal pressure-release surface. Comparison with experiment and some approximate formulations is given.

WE consider the problem of a plane acoustic wave reflected from a two-dimensional, sinusoidally shaped surface on which the pressure vanishes.[1] The following paragraphs contain a brief description of the technique of solution, and the accompanying figures compare the results with the experiment of La Casce and Tamarkin.[2] A more complete discussion of this marvelously complex problem will be published elsewhere.

Formally, we are concerned with the solution of the integral equation obtained from[3]

$$\psi(x,y) = \psi_i(x,y) + \tfrac{1}{4}i \int_{-\infty}^{\infty} dx' H_0(k|\mathbf{r}-\mathbf{r}'|)\phi(x') \quad (1)$$

by putting the y coordinate on the bounding surface where the left-hand side vanishes. The quantity $H_0(x)$ is the usual Hankel function of the first kind. The bounding surface is

$$y = \eta(x) = b\cos px, \quad (2)$$

and $\mathbf{r}'$ refers to an arbitrary point on the surface with coordinates $[x',\eta(x')]$. The quantity $\phi(x')$ is related to the normal derivative of the velocity potential $\psi(x,y)$ evaluated at the reflecting surface, and the incident wave is given by

$$\psi_i(x,y) = \exp ik[\lambda_0 x - \mu_0 y]. \quad (3)$$

This defines the notation for the wavenumber k and the direction cosines λ_0 and μ_0.

The crucial step in the present formulation of the problem is to recognize that $\phi(x)$ admits of a Fourier series representation:

$$\phi(x) = k\sum_{j=-\infty}^{\infty} i^{-j}A_j \exp[ik(\lambda+jp/k)x] \equiv k\sum_j i^{-j}A_j \exp(ik\lambda_j x), \quad (4)$$

so that the integral equation obtained from Eq. (1) can be transformed into an infinite set of linear equations:

$$\sum_{j=-\infty}^{\infty} M_{nj}A_j = (-1)^n J_n(bk\mu_0). \quad (5)$$

The (complex) matrix elements M_{nj} are given by

$$M_{nj} = -(2\pi)^{-1}\sum_{l=-\infty}^{\infty} (-1)^l \int_{-\infty}^{\infty} dt (t^2-\mu_l^2)^{-1} J_{n-l}(bkt) J_{l-j}(bkt), \quad (6)$$

where the $J_l(x)$ are the usual Bessel functions, and μ_l is the sine of the grazing angle of the lth-order reflection coefficient and is given by

$$\mu_l = (1-\lambda_l^2)^{\frac{1}{2}} = +i(\lambda_l^2-1)^{\frac{1}{2}}. \quad (7)$$

The λ_l were defined in Eq. (4).

Finally, the physically interesting quantities—the reflection coefficients for the various orders of reflection—are determined from the boundary coefficients A_j by use of the relation

$$R_l = (2\mu_l)^{-1}\sum_{j=-\infty}^{\infty} A_j J_{l-j}(bk\mu_l). \quad (8)$$

The major complication of the problem (other than the usual difficulties associated with inverting infinite matrices) is in the

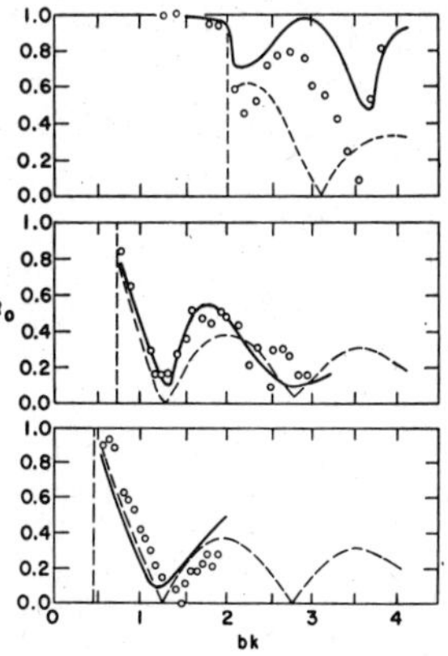

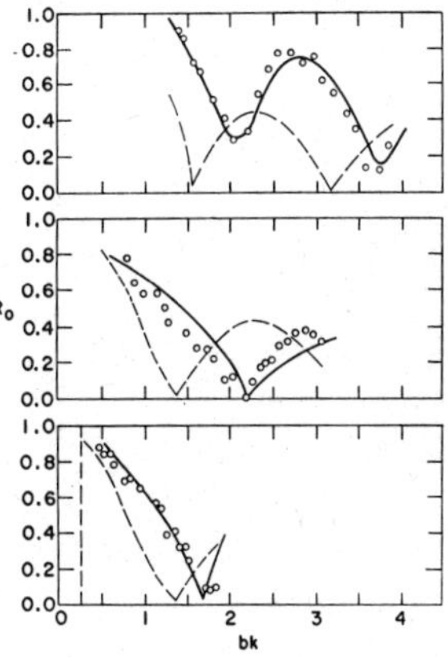

FIG. 2. The same as Fig. 1 except that the angle of incidence is 40° from normal.

FIG. 1. The reflection coefficient R_0 for normal incidence. The parameters of the reflecting surface are, from top to bottom, $b = 0.32$, 0.24, and 0.15 cm and $p = 6.64$, 3.12, and 3.08 cm^{-1}. The dashed curve represents the predictions of Rayleigh's theory and is taken, along with the experimental points, from reference 2. The solid curve shows the present calculation.

calculation of the matrix elements M_{nj}. For fixed l in the summation of Eq. (6), the integral may be shown to be

$$-i(2\mu_l)^{-1}J_{n-l}(bk\mu_l)J_{l-j}(bk\mu_l)+(bk)R_{n,j}{}^l(bk\mu_l). \quad (9)$$

The function $R_{nj}{}^l(x)$ is a generalization of the hypergeometric function. Most previous attempts to solve this problem have involved approximations that ignore the contribution of the real function $R_{nj}{}^l$. It was interesting to learn that if one attempted to evaluate this contribution to M_{nj} as a power series in p/k (Rayleigh's approximation), then every term of the expansion vanished identically.

The set of Eqs. (5) was solved numerically with the aid of Argonne's IBM-704 and the University of California (San Diego) CDC-1604 computers. The technique was the standard one of starting with a 1×1 matrix (the central element), and then increasing the size step by step until the values of the low-order reflection coefficients had "converged." A "goodness of solution" criterion was provided by computing the value of

$$\sum_l R_l\mu_l/\mu_0. \quad (10)$$

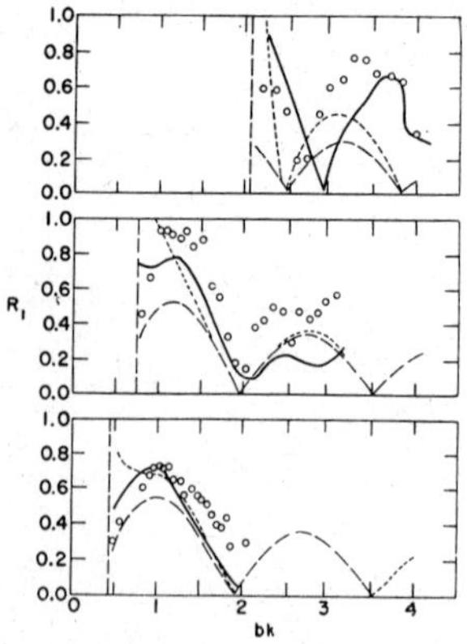

FIG. 3. The reflection coefficient R_1 for normal incidence. The curves for the different values of parameters b and p are in the same order as in Fig. 1. The short dashes are the predictions of Brekhovskikh, the long dashes those of Eckart. Both are copied from reference 2.

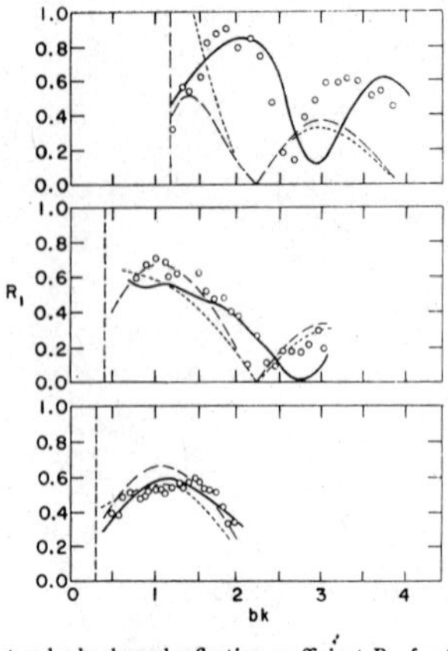

FIG. 4. The first-order backward reflection coefficient R_{-1} for 40° incidence. Otherwise the caption of Fig. 3 applies.

Conservation of flux requires that this quantity have the value unity.

The accompanying figures show the comparison between the experimental results of La Casce and Tamarkin[2] and the calculation described here. I feel that the general trend of agreement is encouraging. Unfortunately, the nature and the description of the experiment are such that a detailed analysis of the extent of the agreement is not possible since there is no basis currently available for discussing the precision or the necessity of corrections to the experiment.

I am indebted to Dr. B. A. Lippman for helpful discussions, to Dr. F. N. Spiess for the repeated hospitality of the Marine Physical Laboratory, and to R. Morell of Argonne for some of the programming.

This work was supported in part by the Office of Naval Research and in part by the U. S. Atomic Energy Commission.

[1] This problem was stated, and incorrectly treated, by Lord Rayleigh, *Theory of Sound* (Dover Publications, Inc., New York, 1945), Sec. 272a. A good bibliography of subsequent work prior to 1958 is contained in the review article by Iu. P. Lysanov, Soviet. Phys.—Acoustics **4**, 1 (1958).

[2] E. O. La Casce, Jr., and P. Tamarkin, J. Appl. Phys. **27**, 138 (1956).

[3] P. M. Morse and H. Feshbach, *Methods of Theoretical Physics* (McGraw-Hill Book Company, Inc., New York, 1953), pp. 811 *ff*.

The authors of the following paper measured reflection coefficients of bottoms in a number of locations. They have found that pressure release bottoms are fairly common.

Dr. Jones is an electrical engineer. He has contributed in the fields of acoustics, circuit theory, electromagnetic theory, and shock and vibration. He is Professor of Electrical Engineering at Bradley University, Peoria, Illinois.

Charles B. Leslie is an electrical engineer. He has contributed in the fields of underwater acoustics, shallow water transmission, antisubmarine warfare, and data processing. He is an Electronic Scientist at the U.S. Naval Ordnance Laboratory.

L. E. Barton is a radio engineer. He is a part-time consultant in radio engineering at the U.S. Naval Research Laboratory.

This paper is reprinted with the permission of the senior author and the Acoustical Society of America.

Reprinted from THE JOURNAL OF THE ACOUSTICAL SOCIETY OF AMERICA, Vol. 36, No. 1, 154–157, January, 1964

27

Acoustic Characteristics of Underwater Bottoms

J. L. JONES,* C. B. LESLIE, AND L. E. BARTON

U. S. Naval Ordnance Laboratory, White Oak, Silver Spring, Maryland
(Received 16 October 1963)

Acoustic bottom-reflection coefficients have been measured at five ocean, lake, and river locations. A correlation technique was employed, using octave-band random noise from 200 to 3200 cps as a source. Values obtained ranged from −0.92 to +0.55. Only at Lake Travis, Texas, did the coefficient change substantially with frequency. Angle of incidence made little or no difference up to 45°. Further evidence is presented that pressure-release bottoms are fairly common.

INTRODUCTION

IN 1958, results were published[1] of an investigation of the bottom characteristics of Triadelphia Lake, Brighton, Maryland, where the U. S. Naval Ordnance Laboratory (NOL) Acoustical Facility is located. Measurements of the bottom-reflection coefficient over the frequency range from 200–3200 cps and at various angles of incidence were accomplished by using a correlation technique. It is the purpose of this article to present the results of similar measurements that have subsequently been carried out at various other locations.

METHOD

The measurements were made in areas having level bottoms with water depths between 40 and 60 ft. An asesmbly consisting of two identical, pressure-sensitive hydrophones was lowered to a location near the bottom so that the effect of surface reflections was negligible. As indicated by Fig. 1, hydrophone 1 was positioned very close to the projector so that bottom reflections received by it were negligible as compared to the direct arrival. Several feet away from the projector was located hydrophone 2, which, therefore, received an appreciable bottom-reflected signal in addition to the direct arrival. If a simple sinusoidal signal were transmitted, there would be no way to separate the arrivals over these two paths, since they would simply add to a resultant sinusoid of the same frequency but usually different in both amplitude and phase than either constituent. However, a band-limited Gaussian random signal was projected and the signals received by hydrophones 1 and 2 were crosscorrelated.

The resulting correlogram has two regions of interest, one corresponding to the direct arrival and the other due to the bottom reflection. If the frequency band of the projected random signal is relatively high and the geometry is such that the path difference between the direct and bottom-reflected arrivals is significant, the two regions of the crosscorrelogram are quite distinct, as indicated by Fig. 2; otherwise, they coalesce and cannot be distinguished by inspection.

In any case, it is possible to add to the signal received by hydrophone 2 two hydrophone-1 signals, each of which is appropriately phased and then delayed by means of an electrical delay line and adjusted in magnitude so as to cancel as completely as possible both portions of the correlogram. The reflection coefficient is then determined from a knowledge of the ratio of the canceling-signal magnitudes and the ratio of the direct and reflected pathlengths. The latter ratio is most accurately determined from the electrical delays required for cancellation.

Field measurements were made at the following locations:

(1) in the Patuxent River near the NOL Test Facility, Solomons, Maryland;

(2) in the harbor near the NOL Test Facility, Ft Monroe, Virginia;

(3) in the Atlantic Ocean near the NOL Test Facility, Ft. Lauderdale, Florida;

---

* Present address: Electrical Engineering Dept., Bradley University, Peoria, Ill.

[1] J. L. Jones, C. B. Leslie, and L. E. Barton, J. Acoust. Soc. Am. **30**, 142 (1958).

(4) in the Potomac River near Indian Head, Maryland;

(5) In Lake Travis, Texas, near Mansfield Dam. The field measurements were easily accomplished since the underwater components indicated on Fig. 1 were light and the assembly was easily rigged. Auxiliary equipment consisted of a noise generator, an adjustable band pass filter, a power amplifier, and a dual-channel magnetic-tape recorder to record the hydrophone signals.

On playback in the laboratory, the tapes were analyzed, using a polarity-coincidence correlator.[2,3] This instrument displays an oscillogram showing the fraction of time that the two hydrophones have the same instantaneous polarity, minus the fraction of time that they have opposite polarities, as a function of τ, the time that the signal from hydrophone 1 is delayed. For a Gaussian process, the relationship between the polarity-coincidence correlation function $r(\tau)$ and the true (normalized) correlation function $\rho(\tau)$ is $r(\tau)=(2/\pi)$ arc sin $\rho(\tau)$.[4]

The use of this normalized correlator is necessary in this application to obtain good accuracy. An indicator of this type takes advantage of the fact that the random signal and ambient background are uncorrelated. Furthermore, because of the normalization, cancellation of one portion of the correlogram makes the remaining portion more prominent, as illustrated in Fig. 3. Consequently, by monitoring with this correlator, it is possible to establish the ratio of the canceling-signal magnitudes in the presence of background noise much more precisely than would be possible if another type of indicator (such as an ordinary voltmeter) were used.

In areas where probe hydrophones could be inserted into the bottom, the acoustic velocity in the bottom just beneath the interface was measured by techniques described in the previous article.[1]

RESULTS

The ocean was rather rough when the measurements were made at Fort Lauderdale. Consequently, it was necessary to make 3-sec loops of specially selected portions of the magnetic tapes in order to realize sufficient positional stability for the correlation analysis. The results obtained from the best tape loops are presented in Table I. As expected for these solid-coral and hard-packed coral-sand bottoms, the reflection coefficient is positive and relatively large.

At the locations listed in Table II, the reflection coefficients were found to be independent of frequency between 200 and 3200 cps. No variation was found in the reflection coefficient with angle of incidence at these locations. Measurements were made from 0° to 45°, except at Solomons, Maryland, where the measurements were restricted to a 0° to 30° range.

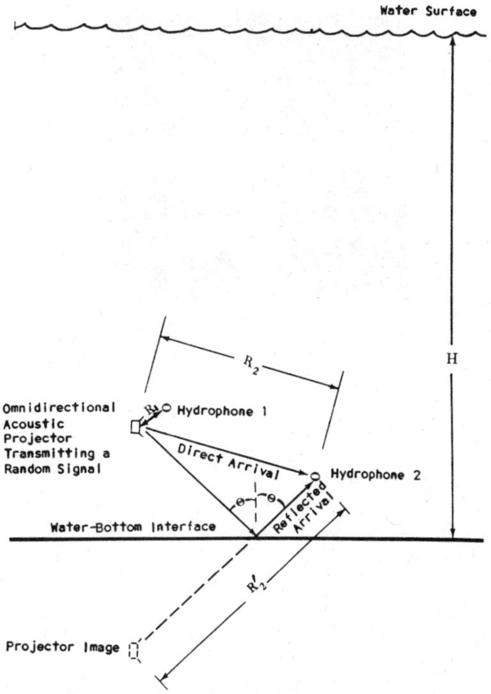

FIG. 1. Arrangement for bottom-reflection studies by correlation techniques.

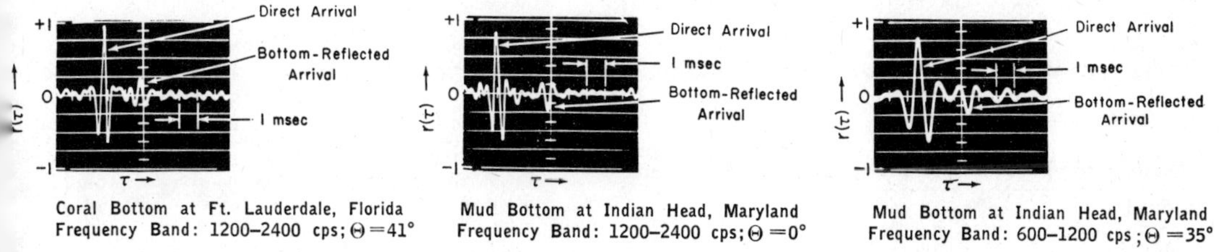

FIG. 2. Crosscorrelograms illustrating the direct and bottom-reflected arrivals.

[2] V. C. Anderson, Harvard Univ. Acoust. Res. Lab. Tech. Mem. No. 37 (Jan. 1956).
[3] J. C. Munson and L. E. Barton, U. S. Naval Ord. Res. & Develop. Rept. 4244 (Sept. 1956).
[4] J. H. Van Vleck, Mass. Inst. Technol. Radiation Res. Lab. Rept. 51 (July 1943).

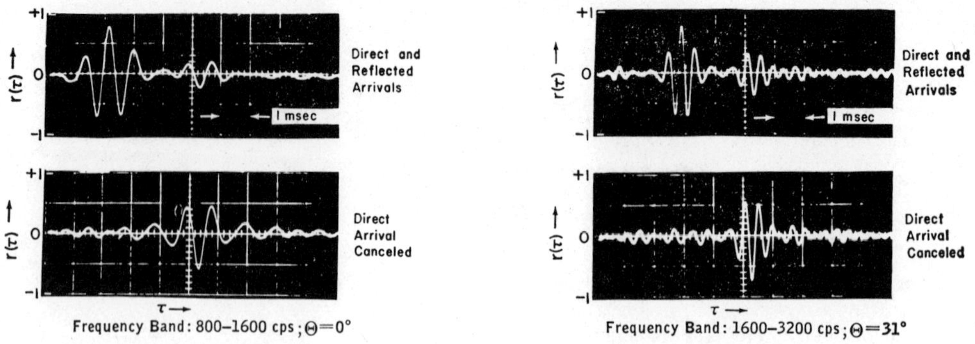

FIG. 3. Correlograms obtained at Lake Travis over the 800- to 1600- and 1600- to 3200-cps frequency bands.

The results from Lake Travis, Texas, showed that the reflection coefficient there did vary with frequency. Table III presents the results obtained for an angle of incidence of 31°. To within the accuracy of the measurements, the same results were obtained at normal incidence.

Table IV shows the results of the bottom-velocity measurements.

DISCUSSION

An accuracy significantly better than 10% in determining the reflection coefficient was relatively easily attained when its magnitude exceeded 0.7. As the magnitude decreases, the percentage error, of course, increases. It was not possible to distinguish reliably between a reflection coefficient of 0.1 and 0.15, although the difference between 0.1 and 0.2 could be distinguished. The use of a higher-powered projector to increase the signal-to-noise ratio would have resulted in improved accuracy.

The bottom-velocity results are consistent with the measured values of reflection coefficient. In areas where the bottom velocity is near that of water, the reflection coefficient is small. As was found previously at Triadelphia Lake,[1] the bottom at Indian Head, Maryland, is a lossy medium of low acoustic velocity. The loss increases rapidly as the frequency is raised. The presence of entrapped gas produces these characteristics, which in turn create a pressure-release interface. Since similar acoustic behavior has been observed in Lake Gem Mary near Orlando, Florida,[5] in the Sea of Moscow[6] and in Lake Maracaibo, Venezulea,[7] it appears that pressure-release bottoms are fairly common.

The measurements at Lake Travis, Texas, were undertaken with the hope of demonstrating the existence of bottoms that are harder acoustically than those found at Ft. Lauderdale, Florida. This seemed reasonable on the basis of knowledge that solid rock underlies Lake Travis in the region of Mansfield Dam. It was known, of course, that several inches of silt covered the rock, but it was hoped that the acoustical properties of this silt would not be too different from those of water.

The results obtained at Lake Travis, as compiled in Table III, are much different than had originally been expected. Nevertheless, they are quite understandable if the silt is assumed to possess acoustical properties similar to those at Triadelphia Lake and Indian Head, Maryland. At the higher frequencies, the attenuation in the silt is so high that the thin silt layer can be considered acoustically to be infinitely thick. Consequently, at relatively high frequencies, a very effective pressure-release interface can be created by a few inches of silt, irrespective of the characteristics of the underlying medium. At lower frequencies, the attenuation is reduced to the point where reflections from the silt–rock interface are returned to the water–silt interface with

TABLE I. Results of reflection measurements at Ft. Lauderdale, Florida.

Bottom material	Frequency band	Angle of incidence	Reflection coefficient
Solid coral	300–600 cps	0°	+0.4
Solid coral	600–1200 cps	0°	+0.4
Coral sand	600–1200 cps	0°	+0.4
Solid coral	600–1200 cps	41°	+0.55
Solid coral	1200–2400 cps	41°	+0.50
Coral sand	600–1200 cps	41°	+0.50
Coral sand	1200–2400 cps	41°	+0.45

TABLE II. Reflection coefficients at various locations.

Site	Bottom material	Reflection coefficient
Indian Head, Md.	Soft mud	−0.85
Ft. Monroe, Va.	Soft mud	+0.15
Solomons, Md.	Mud	+0.2
Solomons, Md.	Sand	+0.3

[5] R. J. Bobber, J. Acoust. Soc. Am. **31**, 250 (1959).
[6] N. A. Grubnik, Akust. Zh. **6**, 446 (1960) [English transl Soviet Phys.—Acoust. **6**, 447 (1961)].
[7] F. K. Levin, Geophysics **27**, 35 (1962).

TABLE III. Results of reflection measurements at Lake Travis, Texas.

Frequency band	$\Theta=31°$ Reflection coefficient
200–400 cps	−0.55
300–600 cps	−0.64
400–800 cps	−0.72
600–1200 cps	−0.83
800–1600 cps	−0.92
1600–3200 cps	−0.94

sufficient amplitude to modify the impedance seen at this upper interface. This, in turn, reduces the magnitude of the reflection coefficient. A quantitative analysis of this situation can be carried through by using the formula for the reflection coefficient for three layers.[8]

TABLE IV. Bottom sound velocities at various sites.

Site	Bottom Material	Sound Velocity
Indian Head, Md.	Soft mud	150–400 ft/sec
Ft. Monroe, Va.	Soft mud	5100–5400 ft/sec
Solomons, Md.	Mud	4700–5000 ft/sec
Solomons, Md.	Sand	5400–5800 ft/sec

This analysis, as well as a much more detailed description of this entire study, is available.[9] The results of this analysis agree quite well with the experimental data.

ACKNOWLEDGMENT

The authors gratefully acknowledge the use of facilities made available by The Defense Research Laboratory of the University of Texas during the measurements at Lake Travis.

[8] C. B. Officer, *Introduction to the Theory of Sound Transmission* (McGraw-Hill Book Co., Inc., New York, 1958), p. 220.

[9] J. L. Jones, U. S. Naval Ord. Lab. Tech. Rept. 62-196 (Nov. 1962).

Ocean-bottom reflectivity was measured using explosive sound sources and the results were compared with calculations based on the assumption of a multilayered model of the bottom. The overall agreement between the measured and calculated coefficients is good.

Robert S. Winokur has been on the staff of the U.S. Naval Oceanographic Office since 1961. His research has involved studies of the acoustic properties of bottom sediments, bottom reflectivity, bottom reverberation, and volume scattering. He has directed various programs in underwater acoustics and oceanographic research with the objective of developing improved prediction models and measurement techniques. He is currently Director of the Acoustical Oceanography Division of the U.S. Naval Oceanographic Office.

This paper is reprinted with the permission of the senior author and the Acoustical Society of America.

Reprinted from THE JOURNAL OF THE ACOUSTICAL SOCIETY OF AMERICA, Vol. 44, No. 4, 1130–1138, October 1968

28

Sound Reflection from a Low-Velocity Bottom*

ROBERT S. WINOKUR AND JOYCE C. BOHN

U. S. Naval Oceanographic Office, Washington, D. C. 20390

Measurements of ocean-bottom reflectivity were made using explosive sound sources and bottom reflection coefficients determined for grazing angles between 2° and 85°. Theoretical reflection coefficients were computed using a multilayered model of the ocean bottom, in which the lowermost layer is considered to be semi-infinite and solid. The layers are assumed to have plane-parallel interfaces and to be absorbing. Sediment sound-speed measurements made on cores collected in the area, and used in the theoretical computations, indicate the bottom in this region to be a low-velocity sediment interspersed with thin high-velocity layers. Oscillograph records made of the bottom-reflected signals show a phase reversal, with respect to the incident wave at the angles theoretically predicted. Comparisons of theoretical and measured reflection coefficients are made at 1, 2, and 4 kHz, and, in general, the over-all agreement is good. An unusual increase in measured reflection coefficient with frequency is observed at the angle of intromission and can, in part, be explained by the presence of subbottom reflectors.

INTRODUCTION

DURING studies of sound propagation in the ocean, the effect of bottom reflection has been found to be highly variable with changes in area, grazing angle, and frequency. During the past few years, various investigators[1–5] have conducted studies using theoretical models of the ocean bottom to compute changes in bottom reflection coefficients with grazing angle and to compare the results with measured reflection coefficients. This paper presents the results of a deep-water study of this type. Theoretical predictions of bottom reflection coefficients for a low-velocity bottom are compared with experimentally measured values at three frequencies; 1, 2, and 4 kHz.

* A shorter version of this paper was presented at the 71st Meeting of the Acoustical Society of America [J. Acoust. Soc. Amer. **39**, 1241(A) (1966)].

[1] G. R. Barnard, J. L. Bardin, and W. B. Hempkins, "Underwater Sound Reflection from Layered Media," J. Acoust. Soc. Amer. **36**, 2119–2123 (1964).

[2] F. R. Menotti, S. R. Santaniello, and W. R. Schumacher, "Studies of Observed and Predicted Values of Bottom Reflectivity as a Function of Incident Angle," J. Acoust. Soc. Amer. **38**, 707–714 (1965).

[3] M. V. Brown and J. H. Rickard, "Interference Pattern Observed in Reflections from the Ocean Bottom," J. Acoust. Soc. Amer. **37**, 1033–1036 (1965).

[4] H. P. Bucker, J. A. Whitney, G. S. Yee, and R. R. Gardner, "Reflection of Low-Frequency Sonar Signals from a Smooth Ocean Bottom," J. Acoust. Soc. Amer. **37**, 1037–1051 (1965).

[5] R. S. Winokur, "Theoretical Computations of Sound Reflection from a Layered Ocean Bottom," U. S. Naval Oceanogr. Office IR No. 0-33-65 (Oct. 1965).

I. DESCRIPTION OF FIELD EXPERIMENT

The experimental data were obtained in an area having a water depth of about 2400 fathoms (f) in the Yucatán Basin of the Caribbean Sea. The bottom in this area appeared to be very smooth, as observed from 12-kHz echo-sounder records collected during the measurements. Two ships and explosive charges containing 1.8 lb of TNT were used to make the measurements. One ship was the receiving ship and was drifting, while the second ship was the source ship and proceeded to open range-dropping charges at predetermined ranges. The grazing angle of the bottom-reflected signals was varied between 2° and 85° by changing the detonation depth of the explosive sources as well as the range between the source and receiver. Several shots were used at each range and the range between the ships was monitored by radar. The experimental geometry is illustrated in Fig. 1. A source depth of about 1800 m was used for the steep and intermediate grazing angles, while deeper charges were used to obtain the shallow grazing angles at the longer ranges.

The direct and reflected signals were received by a calibrated omnidirectional hydrophone located at a depth of about 90 m and having a flat frequency response in the range of interest. The output of the hydrophone was paralleled into three separate channels of amplification and then recorded broad band on magnetic tape.

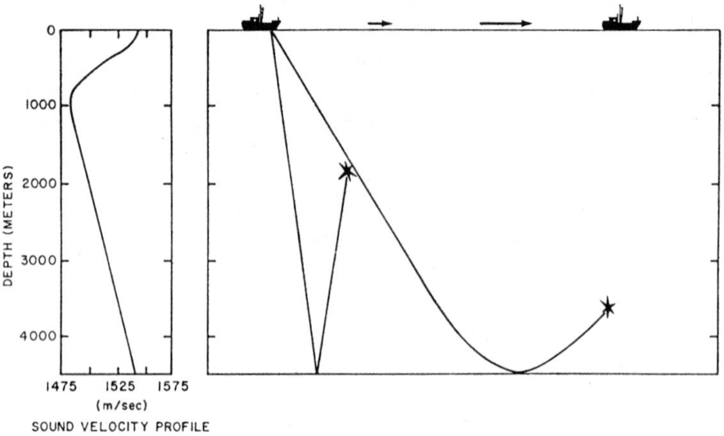

FIG. 1. Experimental geometry and sound-velocity profile.

II. DATA REDUCTION

The recorded signals were played back through octave-band filters having center frequencies of 1, 2, and 4 kHz. The signals were displayed on a high-speed oscillographic recorder, and the peak amplitude in the first few cycles of the bottom-reflected signal was determined. An average source level was computed by making source-level measurements at the steep grazing angles. The source level was arrived at by correcting the peak amplitude of the direct arrival for spherical spreading and sea water absorption along the ray path, so as to yield a level for a reference distance from the source. The grazing angle and path length for the bottom-reflected signals were determined by converting temperature and salinity data collected at the location of the measurements into a sound-speed profile by using Wilson's equation[6] and by means of a ray-tracing computer program. The sound-speed profile for this area is also shown in Fig. 1.

In this paper, the bottom reflection coefficient is defined as the ratio of the peak-amplitude reflected signal to the peak-amplitude incident signal at the ocean-bottom interface. This ratio was determined by utilizing the measured source level and by correcting the bottom-reflected signals for sea-water absorption and spherical spreading along the determined ray path. The sea water absorption was calculated by using the low-frequency attenuation coefficient reported by Thorp.[7,8] The bottom reflection coefficients were obtained by converting the measured octave-band levels to spectrum levels and by correcting for the frequency slope across the band.

[6] W. D. Wilson, "Speed of Sound in Sea Water as a Function of Temperature, Pressure, and Salinity," J. Acoust. Soc. Amer. **32**, 641–644, 1357(L) (1960).

[7] W. H. Thorp, "Deep-Ocean Sound Attenuation in the Sub- and Low-Kilocycle-per-Second Region," J. Acoust. Soc. Amer. **38**, 648–654 (1965).

[8] W. H. Thorp, "Analytic Description of the Low-Frequency Attenuation Coefficient," J. Acoust. Soc. Amer. **42**, 270(L) (1967).

III. DESCRIPTION OF THE THEORETICAL MODEL

The theoretical ocean-bottom model used is described in Refs. 2, 9, and 10 and assumes a multilayered absorbing bottom in which the sediment layer thickness, sound speed, density, absorption, and number of reflecting layers are assigned arbitrary values. The ocean is considered as a semi-infinite fluid overlying a series of fluid sedimentary layers of finite thickness. The lowermost layer in the sedimentary sequence is considered to be a semi-infinite solid. That is, both a longitudinal and a shear wave are propagated in this layer. The layers are further assumed to be absorbing and to have plane-parallel interfaces. The pressure and particle velocity are continuous across the interface between layers. The bottom model was machine programmed to compute the steady-state peak-amplitude reflection coefficient and phase shift for a single-frequency plane wave. The geometry of the bottom reflection model and the input parameters required are illustrated in Fig. 2.

IV. DESCRIPTION OF ACOUSTIC AND PHYSICAL PROPERTIES

The input parameters for the theoretical computations were determined from a 13-ft core taken in the immediate vicinity of the reflection measurements. Sediment sound-speed measurements were made at 10-cm intervals along the length of the core by means of a pulse technique.[11] The sediment density, porosity grain size, and other physical and engineering properties

[9] M. C. Karamargin, "A Treatment of Acoustic Plane Wav Reflection from an Absorbing Multi-Layered Liquid and Soli Bottom," U. S. Navy Underwater Sound Lab. Tech. Memo 913-91-62 (16 Jul. 1962).

[10] M. C. Karamargin and B. J. Klein, "Some Theoretica Computations of Distortions at Reflection from an Absorbin Multi-Layered Liquid and Solid Bottom," U. S. Navy Underwate Sound Lab. Tech. Memo. 910-169-63 (27 Aug. 1963).

[11] R. S. Winokur and S. Chanesman, "A Pulse Method for Soun Speed Measurements in Cored Ocean Bottom Sediments," U. Naval Oceanogr. Office IM No. 66-5 (Aug. 1966).

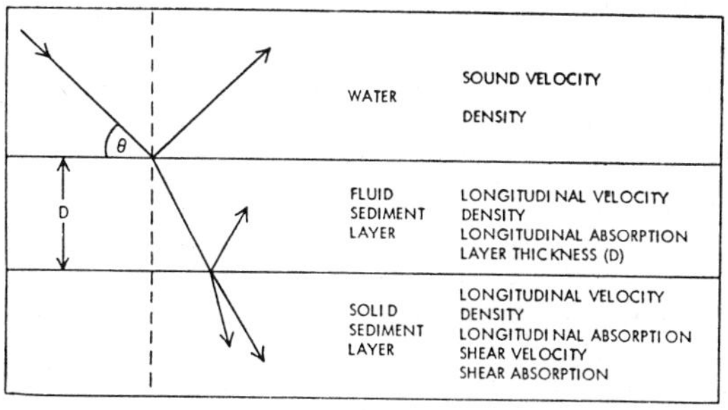

FIG. 2. Geometry and input parameters for theoretical model.

were measured at the same intervals. The laboratory sound-speed measurements were corrected to estimated *in situ* values by applying corrections for differences in temperature and pressure between measurement conditions and *in situ* conditions.[12] Wilson's tables for the speed of sound in sea water[6,13] were used to make the corrections.

The corrected laboratory (*in situ*) sediment sound speed and sediment density versus depth below the top of the core are illustrated in Fig. 3. The sound speed of the bottom water is included for comparison. It can be observed that the core represents a low-velocity bottom interspersed with thin high-velocity layers at depths of 90, 290, and 350–390 cm. In general, the bottom in this area is a high-porosity calcareous sediment having an average porosity of about 70% and a mean calcium carbonate content of about 70%. The low-velocity nature of high-porosity sediments has been established by others[12,14,15] and is consequently not surprising.

It is apparent that there are many combinations of sediment layering to choose from when selecting the proper input parameters for the theoretical model. During this judgment process, the following general guidelines were taken into consideration: (1) to determine whether it is possible to use a simple model to satisfactorily predict observed reflection coefficients, (2) to determine whether one sedimentary model could be used to predict the observed reflectivity for the three frequencies utilized, and (3) to determine whether shallow sediments can be used to predict bottom reflectivity for signals that are assumed to represent reflection from a shallow penetration of the bottom. A requirement for selecting the proper sedimentary model is that the bottom structure should take into account the effective penetration of the signal into the bottom.

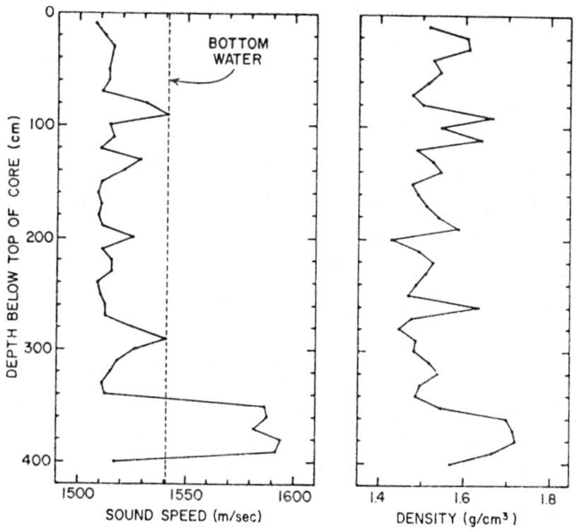

FIG. 3. Sediment sound speed and density versus depth below top of core.

Considering the foregoing guidelines for the sedimentary model, the number of cycles of the reflected signal used, and the resolution corresponding to the octave-bandwidth filters it was necessary to compromise in the selection of the number of layers utilized in the prediction model. For example, the effective penetration at 1 kHz is greater than at 4 kHz, and it becomes necessary to consider deeper layering. The following factors were also taken into account in the determination of the layering selected: (1) the first significant increase in sound speed does not occur until 80–90 cm; (2) the peak amplitude of only the first few cycles of the short pulse explosive signal was utilized, and the experimental data, consequently, represent the reflec-

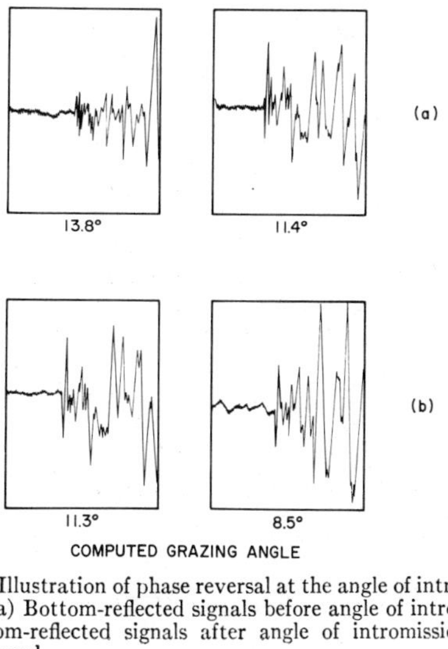

FIG. 4. Illustration of phase reversal at the angle of intromission (11.3°). (a) Bottom-reflected signals before angle of intromission. (b) Bottom-reflected signals after angle of intromission—180° phase reversal.

---

[12] E. L. Hamilton, "Sediment Sound Velocity Measurements Made In Situ from Bathyscaph Trieste," J. Geophys. Res. 68, 5991–5998 (1963).

[13] W. D. Wilson, "Tables for the Speed of Sound in Distilled Water and in Sea Water," U. S. Naval Ordnance Lab. Rep. 6747 (1959).

[14] S. Katz and M. Ewing, "Seismic Refraction Measurements in the Atlantic Ocean Part VII: Atlantic Ocean Basin, West of Bermuda," Bull. Geol. Soc. Amer. 67, 475–509 (1956).

[15] J. C. Fry and R. W. Raitt, "Sound Velocities at the Surface of Deep-Sea Sediments," J. Geophys. Res. 66, 589–597 (1961).

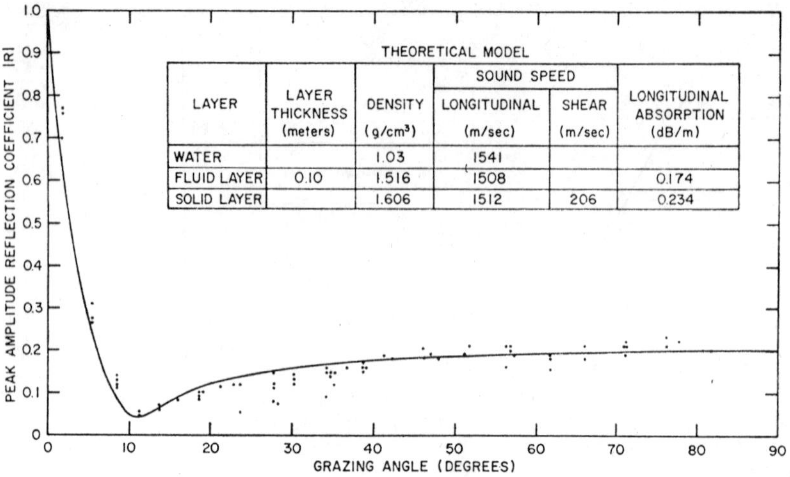

LAYER	LAYER THICKNESS (meters)	DENSITY (g/cm³)	SOUND SPEED LONGITUDINAL (m/sec)	SOUND SPEED SHEAR (m/sec)	LONGITUDINAL ABSORPTION (dB/m)
WATER		1.03	1541		
FLUID LAYER	0.10	1.516	1508		0.174
SOLID LAYER		1.606	1512	206	0.234

FIG. 5. Comparison of predicted and measured reflection coefficients at 1 kHz. ——: Predicted. ····: Measured.

tion from a shallow penetration of the bottom; (3) the higher-velocity layer at 80–90 cm probably does not contribute significantly to the measured reflectivity at 2 and 4 kHz because of its depth, and (4) the 80–90-cm layer probably does not contribute to the 1-kHz reflectivity at the shallow grazing angles because of the increased travel time to the layer.

For the reasons given above, only the upper 30 cm of the core were utilized for the theoretical predictions. The layering used represents a combination of the measured density and sound speed, and observed lithology. The layer boundaries correspond to measured density–sound-speed (acoustic-impedance) changes, although they were observed to coincide with changes in geologic structure. The theoretical computations were made using the model and parameters presented in Figs. 2 and 3. The sedimentary model selected is assumed to consist of a thin fluid sediment layer 10 cm thick, overlying a semi-infinite homogeneous elastic layer. The upper 10 cm of the core represents a calcareous silt containing scattered *Globigerina* and the sediment from 10 to 30 cm, which forms the basis of the semi-infinite layer, is also a calcareous silt, but containing thin lenses of fine calcareous sand.

The values used for longitudinal absorption were extrapolated from Shumway's[16] high-frequency data relating absorption to porosity, and by using an assumed first-power frequency dependence. The shear absorption in the solid layer was assumed to be zero. The shear sound speed in the solid layer was calculated by means of the aggregate theory, which has been used for suspensions of mineral particles within a liquid,[17,18] and by using the well-known equations for longitudinal and shear velocities in elastic media.[19] For this computation, the aggregate bulk modulus was computed by using the proportional precentages of the compressibility of a sediment that is composed of predominately calcium carbonate and the compressibility of water. The sediment compressibility was determined from Ref. 19, while the water compressibility was computed using the sound velocity and density of the bottom water. The aggregate bulk modulus, measured sediment sound velocity, and sediment density were then used to compute a shear modulus, which, in turn, was used in the final determination of a shear velocity. The computed shear velocity was found to be about 206 m/sec or about 0.14 of the longitudinal sound velocity. This value is considerably lower than the 0.5 relationship observed in Laughton's[20] data, but is in agreement with values calculated by Hamilton.[18]

V. REFLECTION FROM A LOW-VELOCITY BOTTOM

The longitudinal sound speed measured in the upper 30 cm of the core is about 2% lower than the sound speed for the adjacent bottom water. For such a low velocity bottom, acoustic theory predicts that the reflection coefficient decreases from its value at normal incidence to zero at an angle referred to as the angle of intromission.[21] Theory also predicts a 180° phase reversal in the direction of the first motion of the reflected signal, with respect to the incident signal, as the incident acoustic energy becomes more grazing than the angle of intromission. The theoretically predicted phase reversal was observed to occur on the oscillograp

[16] G. Shumway, "Sound Speed and Absorption Studies of Marine Sediments by a Resonance Method—Parts I and II," Geophysics 25, 451–467 (I), 659–682 (II) (1960).

[17] R. J. Urick, "A Sound Velocity Method for Determining the Compressibility of Finely Divided Substances," J. Appl. Phys. 18, 983–987 (1947).

[18] E. L. Hamilton, "Geoacoustic Models of the Sea Floor. 1. Shallow Bearing Sea; 2. Mohole (Guadalupe Site)," U. S. Navy Electron. Lab. Rep. 1283 (19 Apr. 1965).

[19] L. H. Adams, "Elastic Properties of Materials of the Earth Crust," in *Internal Constitution of the Earth*, B. Gutenberg, E (Dover Publications, Inc., New York, 1951), 2nd ed., Chap. I pp. 50–80.

[20] A. S. Laughton, "Sound Propagation in Compacted Ocea Sediments," Geophysics 22, 233–260 (1957).

[21] J. W. Strutt Lord Rayleigh, *The Theory of Sound* (Dov Publications, Inc., New York, 1945), 2nd ed., Vol. 2, pp. 78–8

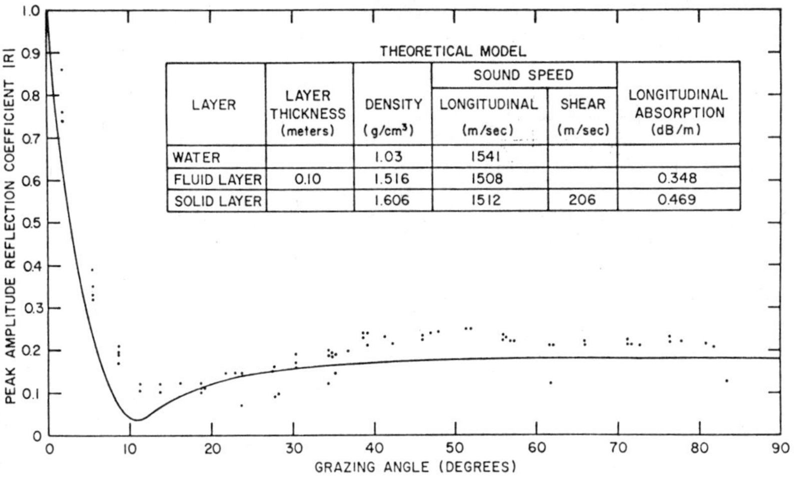

FIG. 6. Comparison of predicted and measured reflection coefficients at 2 kHz. ——: Predicted. · · · · : Measured.

records between 11.3° and 11.4°. The phase reversal and decrease in amplitude of first motion that is typically found in the region of the angle of intromission is illustrated in Fig. 4. Based on the measured sediment sound speed and density and on the water sound speed and density, the theoretically predicted angle of intromission is 11°, which agrees extremely well with the observed angle of about 11.3°. Since the theoretical calculation is dependent on the core data, this angle also serves as a check on the validity of the laboratory-measured and -corrected sediment sound speed.

VI. COMPARISON OF EXPERIMENTAL AND THEORETICAL RESULTS

Comparisons of the experimentally measured and theoretically generated reflection coefficients at 1, 2, and 4 kHz are presented in Figs. 5–7. Note that the absolute value of the peak amplitude reflection coefficient is plotted. The comparison at 1 kHz is illustrated in Fig. 5. The over-all agreement is very good, and the theoretical curve is either generally within the range of scatter for those angles where several measurements were made or within about 0.03 of the measured data. Figure 6 presents the results of the comparison at 2 kHz. In general, the predicted values form the lower limit for the measured reflection coefficients, with the worst agreement occurring between about 40° and 57° and for grazing angles less than about 11°. In terms of bottom reflection loss, the theoretical prediction tends to predict a higher loss at 2 kHz than is actually measured, although the disagreement at the higher grazing angles generally corresponds to a reflection loss difference of less than 3 dB. In comparison with Fig. 5, it can be observed that a discrepancy is forming in the region

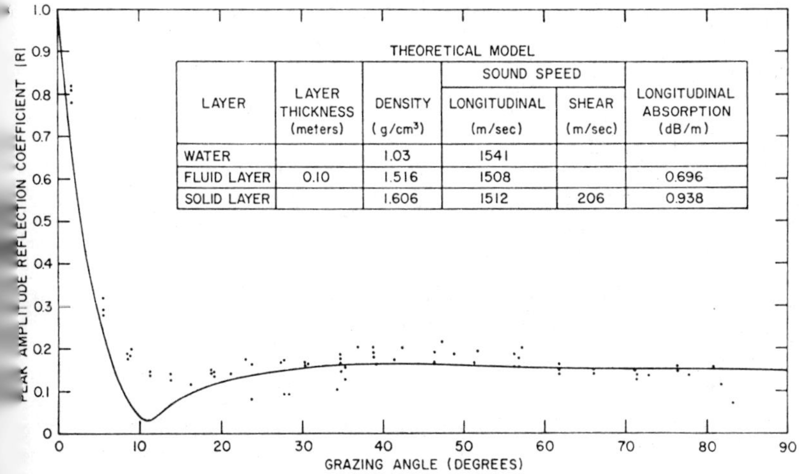

FIG. 7. Comparison of predicted and measured reflection coefficients at 4 kHz. ——: Predicted. · · · · : Measured.

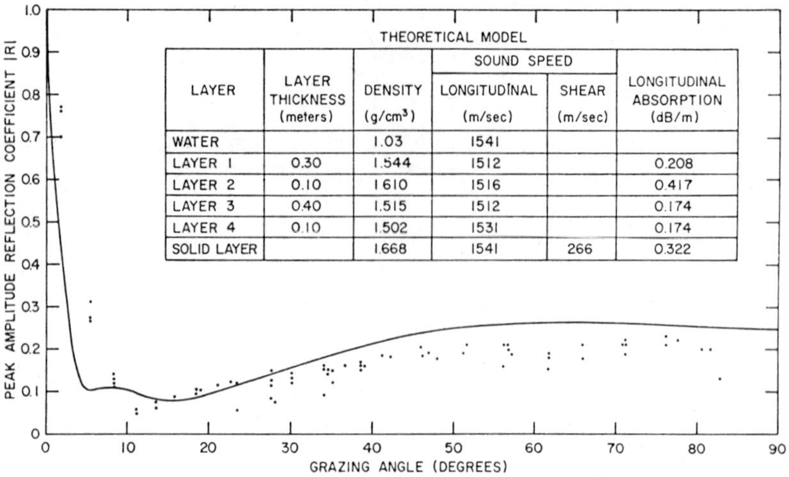

LAYER	LAYER THICKNESS (meters)	DENSITY (g/cm³)	SOUND SPEED LONGITUDINAL (m/sec)	SOUND SPEED SHEAR (m/sec)	LONGITUDINAL ABSORPTION (dB/m)
WATER		1.03	1541		
LAYER 1	0.30	1.544	1512		0.208
LAYER 2	0.10	1.610	1516		0.417
LAYER 3	0.40	1.515	1512		0.174
LAYER 4	0.10	1.502	1531		0.174
SOLID LAYER		1.668	1541	266	0.322

FIG. 8. Comparison of predicted and measured reflection coefficients at 1 kHz using a six-layer model. ——: Predicted. ····: Measured.

of angles around the angle of intromission at 11.3°. This discrepancy occurs in two ways. The measured reflection coefficients are beginning to deviate from the predicted values and they are increasing with frequency.

The comparison between the measured and predicted reflection coefficients at 4 kHz is presented in Fig. 7. The over-all agreement is good, except for the angles in the region of the angle of intromission. The frequency discrepancy is again obvious, particularly at the angle of intromission. At this angle, the measured reflection coefficients increase with frequency. This effect is not in agreement with other studies that show that bottom loss increases or reflection coefficients decrease with increasing frequency[22,23] and, for the purposes of this study, can be referred to as a frequency inversion or reversal.

VII. DISCUSSION

A number of possible explanations exist for the observed frequency anomaly in the region around the angle of intromission. One contribution to the increase in reflection coefficient with frequency may be the use of octave-band analysis for comparison with predictions based on a single-frequency model. The wide bandwidths require that corrections be made across the band for the frequency slope of (1) the source, (2) the differential sea-water absorption, and (3) the bottom reflectivity or sediment attenuation. These corrections take on increasing significance as the frequency and range increase. The measured bottom reflection coefficients were corrected for (1) and (2) above, but were not corrected for the effect of sediment attenuation across the band. Since the over-all comparison between the measured and predicted curves is generally very good, it appears that the use of octave bands did not severely limit the data.

It is possible to show that, as a second contributing factor, at the angle of intromission the increase in reflection coefficient with frequency may result from the presence of thin subbottom layers. Since the measured reflection coefficients never become zero at the angle of intromission, it may be assumed that the measured peak amplitude signal was reflected by a shallow subbottom reflector. Such a reflector may exist at the velocity increase observed in Fig. 3 at 80–90 cm. In view of the relatively good agreement between the measured and predicted bottom reflectivity, it appears that the initial sedimentary model chosen justifies the assumption that reflections from deeper layers should not affect the first few cycles of the reflected signals, although deeper layers may be important at the angle of intromission. The assumption to neglect the deeper layers appears to be justified further by the absence of interference patterns in the measured data, which would reflect the contribution from the subbottom; however, it is possible that for the octave-band analysis used, fewer fluctuations result compared to interference patterns observed for single frequency data.

In order to determine whether a more detailed sedimentary model that included possible subbottom reflectors could explain the unusual frequency effect observed at the angle of intromission, a six-layer model was formulated based on the measured sediment density and sound speed in the upper 90 cm of the core. Figures 8–10 illustrate the theoretical predictions resulting from this model as well as the comparison with the measured values. It can be seen that, in general, there is little agreement between the measured and predicted reflection coefficients, except for those angles in the region around the angle of intromission. The oscillations in the

[22] H. W. Marsh, T. G. Bell, and C. W. Horton, *Reflection and Scattering of Sound by the Sea Bottom*, AVCO Marine Electron. Office, Ed. (AVCO Marine Electronics Office, New London, Conn., 1965), Part II, pp. 29–40.

[23] B. F. Cole, "Marine Sediment Attenuation and Ocean-Bottom-Reflected Sound," J. Acoust. Soc. Amer. **38**, 291–297 (1965).

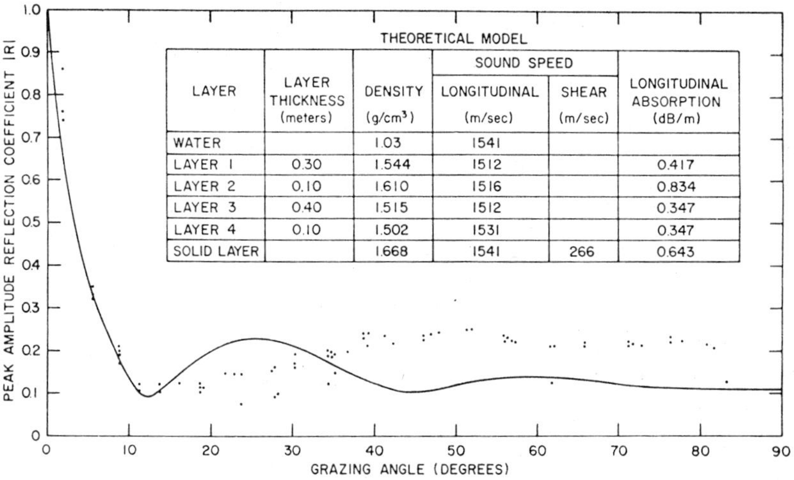

LAYER	LAYER THICKNESS (meters)	DENSITY (g/cm³)	SOUND SPEED LONGITUDINAL (m/sec)	SOUND SPEED SHEAR (m/sec)	LONGITUDINAL ABSORPTION (dB/m)
WATER		1.03	1541		
LAYER 1	0.30	1.544	1512		0.417
LAYER 2	0.10	1.610	1516		0.834
LAYER 3	0.40	1.515	1512		0.347
LAYER 4	0.10	1.502	1531		0.347
SOLID LAYER		1.668	1541	266	0.643

FIG. 9. Comparison of predicted and measured reflection coefficients at 2 kHz using a six-layer model. ——: Predicted. ····: Measured.

theoretical curves result from the interference effects expected for a layered model and are in contrast to the smooth curves in Figs. 5–7, where the 10-cm layer had only a slight influence on the theoretical curves. The lack of over-all agreement indicates that the detailed six-layer model does not provide an adequate description of the observed reflectivity; however, it is useful at the angle of intromission.

Figure 11 presents a comparison of the measured and predicted reflection coefficients as a function of frequency at the angle of intromission. As previously discussed, the measured values increase with frequency. The values predicted by the three-layer model decrease slightly with frequency, while the reflection coefficients predicted by the six-layer model increase and show excellent agreement with the measured values as 2 and 4 kHz. It appears that the frequency inversion at the angle of intromission can, in part, be explained by the presence of subbottom reflectors and the resulting interference effects.

A possible contributing factor to the disagreement between the measured and predicted reflection coefficients at the angle of intromission could be the frequency-dependent absorption. In this region, the reflection coefficients would have a frequency-dependent effect related to the absorption. It was found that if the absorption was increased as the two-thirds power of frequency instead of the first power, there would be better agreement using the three-layer model, but the absorption values then would not be consistent with the absorption constant derived by the first-power extrapolation. The two-thirds power dependence is also not in agreement with other studies. Cole[23] found that a first-power frequency dependence was predominant for a frequency range of 100–900 Hz for bottom-loss data for a low-velocity bottom in the Gulf of Alaska. Wood

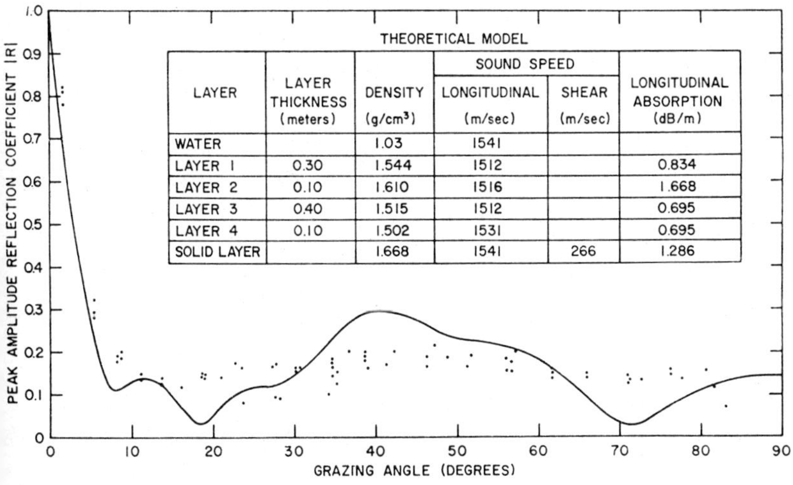

LAYER	LAYER THICKNESS (meters)	DENSITY (g/cm³)	SOUND SPEED LONGITUDINAL (m/sec)	SOUND SPEED SHEAR (m/sec)	LONGITUDINAL ABSORPTION (dB/m)
WATER		1.03	1541		
LAYER 1	0.30	1.544	1512		0.834
LAYER 2	0.10	1.610	1516		1.668
LAYER 3	0.40	1.515	1512		0.695
LAYER 4	0.10	1.502	1531		0.695
SOLID LAYER		1.668	1541	266	1.286

FIG. 10. Comparison of predicted and measured reflection coefficients at 4 kHz using a six-layer model. ——: Predicted. ····: Measured.

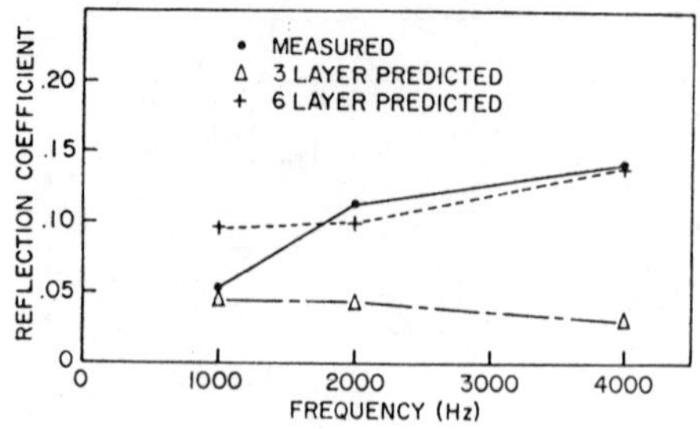

FIG. 11. Comparison of predicted and measured reflection coefficients at the angle of intromission.

and Weston[24] observed a first-power relation in the frequency range from 4 to 50 kHz. Although there is no simple theory to explain the first power sediment absorption frequency dependence, if it is valid for all sediments it would be expected to hold between 1 and 4 kHz as well.

A final contributing factor to the divergence between the observed and predicted results could be the combined effects of the choice of layering used for the predictions, the proximity of the core to the insonified area, and the lateral variability of bottom sediments. The choice of layering and input parameters is usually based on a combination of measured and interpreted data, and is left to the discretion and judgment of the individual investigator. The acoustic and physical properties of the bottom often change gradually, while the theoretical model requires exact layer thicknesses and number of layers. A slight change in the sedimentary conditions can sometimes bring about a frequency-dependent variation in the predicted curves that would be present at all grazing angles.

A possible contributing factor is that the core is not from the precise location or area of bottom insonified. The core represents a discrete point, while the reflection measurements were made as the point of reflection on the bottom changed as the source and receiving ship separated. A related factor is that the ocean bottom may be structurally complex and variable over relatively short distances. Results presented in Refs. 2 and 25 indicate that complex acoustic layering exists over small lateral distances and that exact predictions based on a single core may be difficult. The relatively good over-all agreement between the measured and predicted reflection coefficients using the initial three-layer model suggests that in this area the surficial sediments may not be too variable, and that variations in sediment layering and properties probably do not contribute significantly to the observed differences around the angle of intromission.

[24] A. B. Wood and D. E. Weston, "The Propagation of Sound in Mud," Acustica 14, 156–162 (1964).

[25] J. J. Gallagher, "Variability in Derived Sediment Sound Velocity as a Function of Core Analysis and its Effect in Determining Theoretical Bottom Reflectivity," U. S. Navy Underwater Sound Lab. Tech. Memo. 2213-78-68 (20 Feb. 1968).

VIII. SUMMARY AND CONCLUSIONS

In summary, a set of experimentally measured and theoretically predicted bottom reflection coefficients were compared as a function of grazing angle and frequency for a low-velocity bottom. The theoretical reflection coefficients were obtained by using the measured sediment sound speed and physical properties of a core in conjunction with a multilayered mathematical model of the ocean bottom. Since only the first few cycles of the bottom-reflected signals were used in the data analysis, it was assumed that the observed reflection coefficients resulted from a shallow penetration of the bottom. Consequently, only the upper 30 cm of the core were utilized for the theoretical predictions.

A comparison between the experimental and theoretical results at 1, 2, and 4 kHz shows good over-all agreement. In particular, there is excellent agreement between the observed and predicted angle of intromission and associated phase reversal. This agreement further supports Hamilton's[12] results that laboratory measurements of sediment sound speed can be corrected to *in situ* conditions by applying corrections for temperature and pressure. An unusual frequency dependence, which resulted in an increase in reflection coefficient with frequency, was observed to exist at the angle of intromission. This effect, referred to as a frequency inversion, may be explained at 2 and 4 kHz by assuming that the measured reflection coefficients at this angle resulted from interference effects associated with the presence of subbottom reflectors.

The good agreement between the measured and predicted results for the limited number of data presented in this paper does not permit wide-spread conclusions to be made. The results show that a simple three-layer model that considers only the upper 30 cm of the bottom can be used to predict bottom reflectivity for a particular bottom at 1, 2, and 4 kHz for signals that are assumed to represent a shallow penetration of the bottom. A detailed six-layer model that considered the upper 90 cm of the bottom was tested, with the results showing a general lack of agreement with the measured reflection coefficients. This model was, however, useful for predicting the frequency inversion at the angle of intromission.

ACKNOWLEDGMENTS

The authors gratefully acknowledge the support of the U. S. Navy Underwater Sound Laboratory for permitting them to use the basic theoretical model modified and programmed by M. C. Karamargin. This model was used in conjunction with the computer model developed at the Naval Oceanographic Office and was extremely valuable for this investigation. The authors are indebted to John E. Allen, Benjamin A. Watrous, James Cole, Paul M. Dunlap, Laurence C. Breaker, and Wayne E. Renshaw, and to the other members of the field party for their respective assistance in collecting and analyzing the data.

Underwater Transducers and Calibration

VI

In the following paper, an exact derivation is given for the low-frequency acoustic sensitivity of barium titanate cylindrical tubes.

Robert A. Langevin is a mathematical physicist. He has contributed in the fields of computer applications, digital simulation, and operations research. He is Associate Director of the Auerbach Corporation, Arlington, Virginia.

This paper is reprinted with the permission of the author and the Acoustical Society of America.

THE JOURNAL OF THE ACOUSTICAL SOCIETY OF AMERICA VOLUME 26, NUMBER 3 MAY, 1954

The Electro-Acoustic Sensitivity of Cylindrical Ceramic Tubes

29

R. A. LANGEVIN*
Transistor Products, Incorporated, Boston 35, Massachusetts
(Received October 16, 1953)

Increasing use is being made of cylindrical barium titanate tubes as underwater acoustic transducers. An exact derivation is given for the low-frequency acoustic sensitivity of such tubes for a variety of boundary and polarization conditions.

SUMMARY OF SYMBOLS

ν = Poisson ratio of ceramic,
N = an even integer,
R, θ, Z = axes in cylindrical coordinate system,
l = length of ceramic tubes, in meters,
a = inner radius of ceramic tube in meters,
b = outer radius of ceramic tube, in meters,
P_0 = external pressure in newtons/meter²,
g_P, g_T, d_P, d_T = electromechanical constants of ceramic material,
ϵ = absolute dielectric constant of ceramic in farads/meter,
T_R, T_θ, T_Z = stresses in cylindrical coordinate system, newtons/meter²,
$dR, Rd\theta, dZ$ = line elements in cylindrical coordinates,
E_R, E_θ, E_Z = components of electric-field in volts/meter, produced by T_R, T_θ, T_Z,
V = open circuit output voltage developed by element,
ρ = ratio of inner to outer radius,
dQ_R, dQ_θ, dQ_Z = components of charge, in coulombs, produced by T_R, T_θ, T_Z in an elementary volume,
C = capacity of element in farads,
$\mathbf{D}$ = electric displacement vector, coulombs/meter²,
$\mathbf{dS}$ = vector element of surface area,
φ = two-dimensional stress function,
A, B, C, D, E, F, G = numerical constants.

INTRODUCTION

RECENT developments in the utilization of cylindrical ceramic tubes as electro-acoustic transducers has focused attention on the need for acoustic sensitivity data for tubes used in such applications. This paper presents a derivation of the voltage/pressure sensitivity of such tubes for a variety of types of polarization. In addition, account is taken of several varieties of boundary conditions which are of practical importance. The necessary elastic theory is developed in Appendix *A*, while the derivation of the sensitivity for nine specific cases is carried out in the body of the paper. It must be understood at the outset that the values of sensitivity calculated herein are valid only for frequencies below and reasonably well removed from the lowest lying resonant frequency of the structures considered. In this range the response is "stiffness controlled" and the stresses determined from elastic theory can be used alone to calculate the sensitivity.

POLARIZATION

We are in all cases dealing with a piezoelectrically active, cylindrical, ceramic tube. Such a structure with dimensions and an attached system of cylindrical coordinates is illustrated in Fig. 1. Three different types of electroding and polarization are contemplated, as follows:

Radial

The inside and outside lateral surfaces of the tube are both fully electroded and the polarization is everywhere radial, i.e., parallel to *R*.

Longitudinal

Both ring shaped ends of the tube (at $Z=0$ and $Z=l$) are fully electroded and the polarization is everywhere longitudinal, i.e., parallel to *Z*.

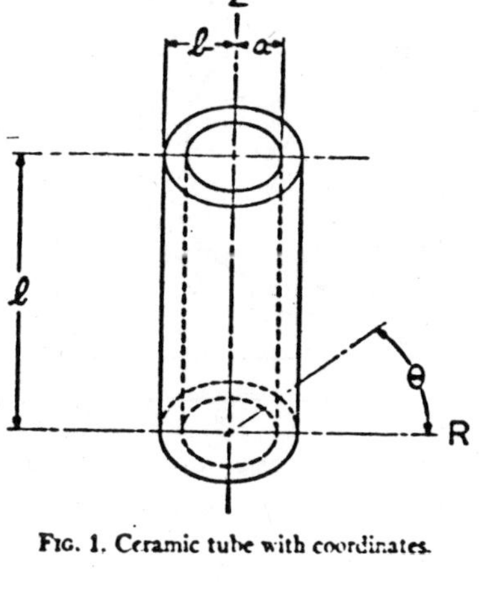

FIG. 1. Ceramic tube with coordinates.

* Formerly with Clevite-Brush Development Company Cleveland, Ohio.

Tangential

The tube is supposed divided into an even number of curved plates by planes $\theta=0$, $\theta=2\pi/N$, $\cdots$, $\theta=(N-1)\cdot 2\pi/N$. Each plate is then bounded by surfaces defined by the intersection of the tube with adjacent diametral planes. These surfaces are supposed fully electroded so that the polarization is everywhere tangential, i.e., perpendicular to R and Z. In operation, all of the plates are considered to be connected electrically in parallel. The connection and direction of polarization are illustrated in Fig. 2 for the special case $N=4$.

BOUNDARY CONDITIONS

In all cases it is supposed that the inside lateral surface of the tube is completely shielded from radiation while the outside lateral surface is exposed to a uniform radiation field, P_0 newtons/meter². With respect to the ends of the tube at $Z=0$ and $Z=l$, there are three possibilities of practical importance:

Shielded ends.—Both ends are completely shielded from the radiation field.

Exposed ends.—Both ends are exposed to the radiation field, P_0 newtons/meter².

Capped ends.—Both ends of the tube are closed by caps so that the cross section between $R=a$ and $R=b$ is subjected to a radiation field $b^2P_0/(b^2-a^2)$ newtons/meter².

CASES CONSIDERED

The cases considered in the following derivations of voltage/pressure sensitivity are the nine obtained by considering each set of boundary conditions for tubes with each type of polarization. They will be found tabulated in Fig. 3.

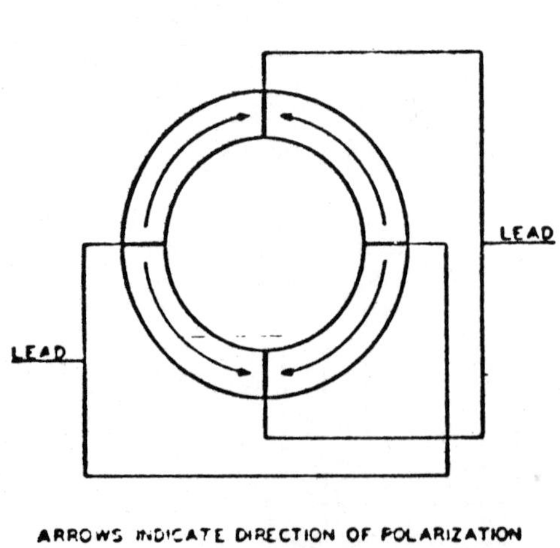

FIG. 2. Electrode arrangement for tangential polarization.

	Polarization	Ends
Case I	Radial	Shielded
Case II	Radial	Exposed
Case III	Radial	Capped
Case IV	Longitudinal	Shielded
Case V	Longitudinal	Exposed
Case VI	Longitudinal	Capped
Case VII	Tangential	Shielded
Case VIII	Tangential	Exposed
Case IX	Tangential	Capped

FIG. 3. Polarization and boundary conditions.

SENSITIVITY CALCULATIONS

The ceramic material with which we are dealing is typified by barium titanate or by barium titanate with lead or other additions in small proportions. The electro-mechanical activity of representative samples of such materials is characterized by parallel and transverse constants:

$$g_P=1.26\times 10^{-2}\ \text{volt/meter//newton/meter}^2,$$
$$g_T=-5.2\times 10^{-3}\ \text{volt/meter//newton/meter}^2,$$
$$d_P=190\times 10^{-12}\ \text{coulomb/meter}^2\text{//newton/meter}^2,$$
$$d_T=-78.5\times 10^{-12}\ \text{coulomb/meter}^2\text{//newton/meter}^2.$$

If ϵ is the absolute dielectric constant of the material parallel to the direction of polarization, the g and d constants are related by a simple expression, $g\epsilon=d$.

For current materials, Fig. 4 shows representative relative dielectric constant values as a function of temperature.

In response to the external forces to which the tube is subjected, there will result three stresses, T_R, T_θ, and T_Z, the first two parallel and perpendicular to R respectively in the $R\theta$ plane, while T_Z is parallel to Z. From the results obtained in the Appendix, these stresses are found to be,

for shielded ends,

$$T_R=\frac{a^2b^2P_0}{b^2-a^2}\cdot\left\{\frac{1}{R^2}-\frac{1}{a^2}\right\}, \tag{1}$$

$$T_\theta=\frac{a^2b^2P_0}{b^2-a^2}\cdot\left\{-\frac{1}{R^2}-\frac{1}{a^2}\right\}, \tag{2}$$

$$T_Z=0. \tag{3}$$

for exposed ends,

$$T_R=\frac{a^2b^2P_0}{b^2-a^2}\cdot\left\{\frac{1}{R^2}-\frac{1}{a^2}\right\}, \tag{4}$$

$$T_\theta=\frac{a^2b^2P_0}{b^2-a^2}\cdot\left\{-\frac{1}{R^2}-\frac{1}{a^2}\right\}, \tag{5}$$

$$T_Z=-P_0. \tag{6}$$

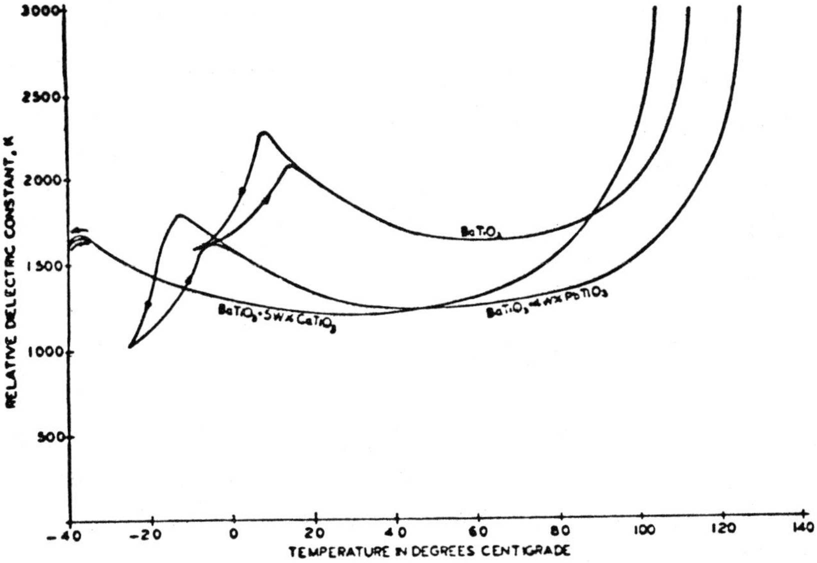

FIG. 4. Dielectric constant of various materials.

for capped ends,

$$T_R=\frac{a^2b^2P_0}{b^2-a^2}\cdot\left\{\frac{1}{R^2}-\frac{1}{a^2}\right\}, \tag{7}$$

$$T_\theta=\frac{a^2b^2P_0}{b^2-a^2}\cdot\left\{-\frac{1}{R^2}-\frac{1}{a^2}\right\}, \tag{8}$$

$$T_Z=\frac{-b^2P_0}{b^2-a^2}. \tag{9}$$

Our problem is then to find expressions for the total potential developed between the electroded surfaces of the tube as a function of the external stress, P_0. (Note: As will be observed from Eqs. (1)–(9) above, we have chosen to represent the external stress as $-P_0$ in the calculations).

Radial Polarization

Choosing a volume element with edges dR, $Rd\theta$, dZ we observe that the stresses T_R, T_θ, and T_Z will produce electric fields given respectively by

$$E_R=\frac{\partial V_R}{\partial R}=g_PT_R \text{ volts/meter},$$

$$E_\theta=\frac{\partial V_\theta}{\partial R}=g_TT_\theta \text{ volts/meter},$$

$$E_Z=\frac{\partial V_Z}{\partial R}=g_TT_Z \text{ volts/meter},$$

so that the total voltage developed between the outer and inner surfaces of the tube will be simply

$$V=\int_a^b\{g_PT_R+g_T(T_\theta+T_Z)\}dR.$$

We can then obtain the voltage developed by the tube in response to the applied stress by substituting the appropriate stress functions from the preceding Eqs. (1)–(9), and carrying out the indicated integration. There result

Shielded ends

$$V/P_0=b\{g_P(1-\rho/1+\rho)+g_T\}, \tag{10}$$

where

$$\rho=a/b.$$

Exposed ends

$$V/P_0=b\{g_P(1-\rho/1+\rho)+g_T(2-\rho)\}. \tag{11}$$

Capped ends

$$V/P_0=b\{g_P(1-\rho/1+\rho)+g_T(2+\rho/1+\rho)\}. \tag{12}$$

Longitudinal Polarization

For the case of longitudinal polarization, we can obtain an expression for the voltage developed between the electroded ends by the following line of reasoning.

As before, we choose a volume element with edges dR, $Rd\theta$, and dZ. The charge, dQ, developed between the faces at Z, and $Z+dZ$ is then related to the stresses by the equations:

$$\frac{dQ_R}{Rd\theta dR}=d_TT_R,$$

$$\frac{dQ_\theta}{Rd\theta dR}=d_TT_\theta,$$

$$\frac{dQ_Z}{Rd\theta dR}=d_PT_Z.$$

Hence, the total charge developed between the electroded faces will be given by:

$$\int dQ = \int_0^{2\pi}\int_a^b \{d_P T_Z + d_T(T_R + T_\theta)\} R\,dR\,d\theta.$$

Now the capacity of this end electroded tube will be simply

$$C = \frac{\epsilon\pi(b^2 - a^2)}{l},$$

and since $d = g\epsilon$ and $Q = CV$, we have that

$$V = \frac{Q}{C} = \frac{l}{\pi(b^2 - a^2)} \int_0^{2\pi}\int_a^b \{g_P T_Z + g_T(T_R + T_\theta)\} R\,dR\,d\theta.$$

As before, substitution of the appropriate expressions for the stress components permits the immediate integration of the foregoing to obtain

Shielded ends

$$\frac{V}{P_0} = \frac{2lg_T}{(1-\rho^2)}. \tag{13}$$

Exposed ends

$$\frac{V}{P_0} = l\left\{\frac{2g_T}{(1-\rho^2)} + g_P\right\}. \tag{14}$$

Capped ends

$$\frac{V}{P_0} = l\left\{\frac{2g_T + g_P}{(1-\rho^2)}\right\}. \tag{15}$$

Tangential Polarization

We consider again a volume element with edge[8] dR, $Rd\theta$, dZ, and note that the charge, dQ, developed by each of the stress components will be given by

$$\frac{dQ_R}{dRdZ} = d_T T_R,$$

$$\frac{dQ_\theta}{dRdZ} = d_P T_\theta,$$

$$\frac{dQ_Z}{dRdZ} = d_T T_Z,$$

so that the total charge developed between the electroded surfaces of a curved plate bounded by the planes $\theta = q\cdot 2\pi/N$ and $\theta = (q+1)\cdot 2\pi/N$ will be simply

$$Q = \int_0^l\int_a^b \{d_P T_\theta + d_T(T_R + T_Z)\} dR\,dZ. \tag{16}$$

Now if we knew the capacity of this curved plate, we could, making use of the relation $Q = CV$, find the voltage developed between the electroded faces. This capacity not being readily available, we digress briefly to calculate it.

Consider then a curved plate bounded by the surfaces:

$$R = a,\quad R = b,$$
$$\theta = 0,\quad \theta = 2\pi/N,$$
$$Z = 0,\quad Z = l.$$

The potential, V, between the faces $\theta = 0$, and $\theta = 2\pi/N$ must then satisfy Laplace's equation, $\Delta V = 0$, subject to the boundary conditions.

$$V = 0,\quad \theta = 0 \text{ for all } R,$$
$$V = V_0,\quad \theta = 2\pi/N \text{ for all } R.$$

From symmetry considerations we may expect a solution independent of R and Z. This leads at once to the simplified form of Laplace's equation

$$\partial^2 V/\partial\theta^2 = 0,$$

so that $V = A\theta + B$.

$\theta = 0$, $V = 0$ gives at once $B = 0$, while

$\theta = 2\pi/N$, $V = V_0$ gives $A = V_0 N/2\pi$, so that

$$V = V_0 N\theta/2\pi.$$

Now if $\mathbf{D}$ is the electric displacement vector within the plate, and $d\mathbf{S}$ is an elementary element of area, we have that

$$\oint \mathbf{D}\cdot d\mathbf{S} = Q,$$

where the integration is extended over the surface of the plate.

However,

$$\mathbf{D} = \epsilon\mathbf{E} = -\epsilon\,\mathrm{grad}\,V = -\frac{\epsilon V_0 N}{2\pi}\cdot\frac{1}{R},$$

so that

$$\oint \mathbf{D}\cdot d\mathbf{S} = \int_0^l\int_a^b \frac{\epsilon V_0 N}{2\pi}\cdot\frac{1}{R} dR\,dZ = Q,$$

and

$$Q = \frac{\epsilon V_0 N l}{2\pi}\cdot\log_e\frac{b}{a} = CV_0.$$

Hence,

$$C = \frac{\epsilon N l}{2\pi}\log_e\frac{b}{a}.$$

Using this result with Eq. (16) above, we then find that

$$V = \frac{1}{N}\cdot\frac{2\pi}{l\log_e(b/a)}\int_0^l\int_a^b \{g_P T_\theta + g_T(T_R + T_Z)\} dR\,dZ.$$

Substituting appropriate expressions for the stress components from Eqs. (1)-(9), and carrying out the indicated integrations, we easily find the following:

For shielded ends

$$\frac{V}{P_0}=\frac{1}{N}\cdot\frac{2\pi b}{\log_e(1/\rho)}\left\{g_T\left(\frac{1-\rho}{1+\rho}\right)+g_P\right\}. \quad (17)$$

For exposed ends

$$\frac{V}{P_0}=\frac{1}{N}\cdot\frac{2\pi b}{\log_e(1/\rho)}\left\{g_T\left(\frac{1-\rho}{1+\rho}\right)(2+\rho)+g_P\right\}. \quad (18)$$

For capped ends

$$\frac{V}{P_0}=\frac{1}{N}\cdot\frac{2\pi b}{\log_e(1/\rho)}\left\{g_T\left(\frac{2-\rho}{1+\rho}\right)+g_P\right\}. \quad (19)$$

Data Presentation

In order to present the information implicit in Eqs. (10)–(19) in usable form, Figs. 5–7 have been prepared

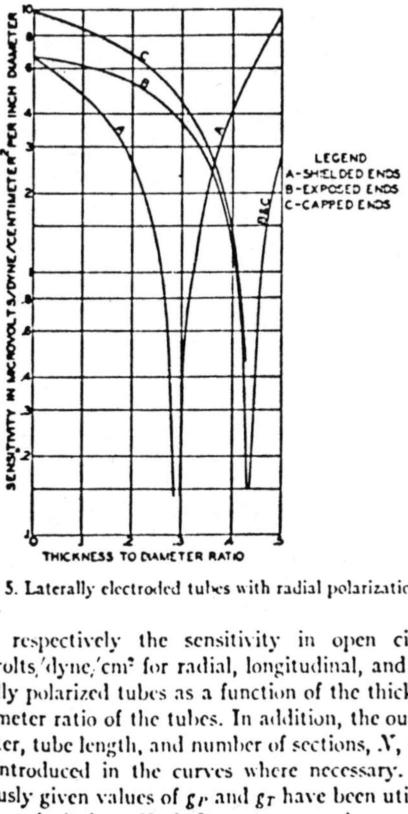

FIG. 5. Laterally electroded tubes with radial polarization.

giving respectively the sensitivity in open circuit microvolts/dyne/cm² for radial, longitudinal, and tangentially polarized tubes as a function of the thickness to diameter ratio of the tubes. In addition, the outside diameter, tube length, and number of sections, N, have been introduced in the curves where necessary. The previously given values of g_P and g_T have been utilized in these calculations. Each figure presents three curves corresponding to the three sets of boundary conditions treated in this report, *viz.*: shielded, exposed, or capped ends.

COMMENTS

1. It should be noted that at all frequencies sufficiently low that a wavelength in the medium is long

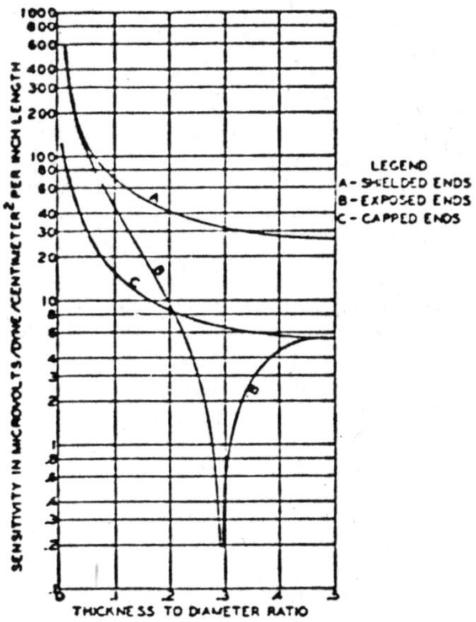

FIG. 6. End electroded tubes with longitudinal polarization.

compared to the greatest dimension of the tube, the foregoing theory is exact. Therefore, any discrepancy between theory and experiment must be a consequence of lack of fulfillment of the assumptions herein utilized. The most likely sources of difficulty would appear to be the following:

(a) Deviations of the tube from the form of a right circular cylinder.

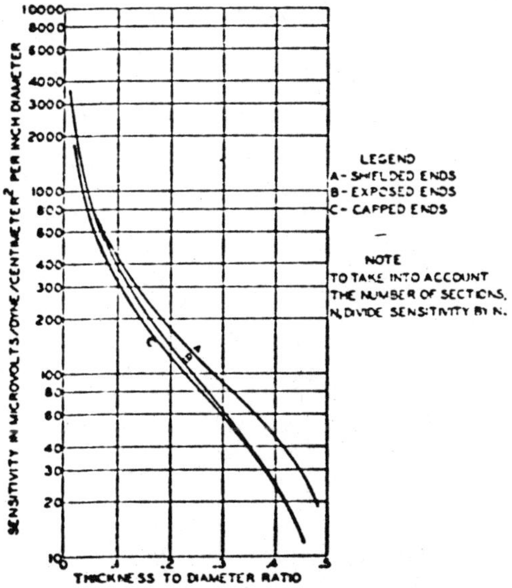

FIG. 7. Longitudinally electroded tubes with tangential polarization N sections (N even) in parallel.

(b) Inhomogeneity leading to locally different values of the g_r and g_T constants of the material.

(c) Introduction of radial constraints (clamping) at the ends of the tube.

(d) Imperfect pressure release at the inside of the tube; i.e. at $R=a$, $P\neq 0$.

(b) may be a particular source of trouble with the transverse polarized tubes since, as has been seen, the electric field in a single curved section is inversely proportional to the radius vector, R. This may mean that it will be impossible to obtain full polarization of the outer layers of the cross section before the electrical breakdown field of the inner layers is reached. It is, of course, obvious that this effect will become less and less important as the thickness to diameter ratio is decreased.

APPENDIX A

Calculation of stresses produced in a right circular cylinder whose exterior surfaces are exposed to a uniform hydrostatic pressure.

We contemplate a right circular cylinder of length, l, and inner and outer radii, a, and b, respectively. We suppose that the inside lateral surface is shielded from pressure, while the outside of the cylinder is exposed to a uniform hydrostatic pressure, P_0, newtons/meter². We wish to calculate the stresses resulting from this pressure field.

For this purpose it is convenient to utilize cylindrical coordinates, R, θ, Z, and we shall then expect from the symmetry of the problem that our solutions will be independent of the angular variable, θ.

The solution is most easily obtained[1] by introducing a stress function, ϕ, which then must satisfy the fourth order partial differential equation

$$\left(\frac{\partial^2}{\partial R^2}+\frac{1}{R}\frac{\partial}{\partial R}+\frac{\partial^2}{\partial Z^2}\right)^2\phi=\nabla^2\nabla^2\phi=0. \tag{1}$$

The stress components T_R, T_θ, and T_Z will then be given by

$$T_R=\frac{\partial}{\partial Z}\left(\nu\nabla^2\phi-\frac{\partial^2\phi}{\partial R^2}\right), \tag{2}$$

$$T_\theta=\frac{\partial}{\partial Z}\left(\nu\nabla^2\phi-\frac{1}{R}\frac{\partial\phi}{\partial R}\right), \tag{3}$$

$$T_Z=\frac{\partial}{\partial Z}\left([2-\nu]\nabla^2\phi-\frac{\partial^2\phi}{\partial Z^2}\right), \tag{4}$$

$$T_{RZ}=\frac{\partial}{\partial R}\left([1-\nu]\nabla^2\phi-\frac{\partial^2\phi}{\partial Z^2}\right), \tag{5}$$

where ν is the Poisson ratio for the material.

Three different sets of boundary conditions are appropriate to the problem, *viz.*,

[1] See Timoshenko, *Theory of Elasticity* (McGraw-Hill Book Company, Inc., New York, 1934), p. 309.

Shielded ends.—If the ends of the tube are shielded from the radiation, the boundary conditions become:

$$\begin{aligned} &R=a, & &T_R=0, & &T_{RZ}=0 \text{ (all } Z).\\ &R=b, & &T_R=-P_0, & &T_{RZ}=0 \text{ (all } Z),\\ &Z=l, & &T_Z=0 \text{ (all } R), & &\\ &Z=0, & &T_Z=0 \text{ (all } R). & & \end{aligned} \qquad \text{I}$$

Exposed ends.—If, on the other hand, the ends of the tube are exposed to the pressure, P_0, we must alter the boundary conditions to obtain:

$$\begin{aligned} &R=a, & &T_R=0, & &T_{RZ}=0 \text{ (all } Z),\\ &R=b, & &T_R=-P_0, & &T_{RZ}=0 \text{ (all } Z),\\ &Z=l, & &T_Z=-P_0 \text{ (all } R), & &\\ &Z=0, & &T_Z=-P_0 \text{ (all } R). & & \end{aligned} \qquad \text{II}$$

Capped ends.—Finally another practically useful structure is one in which the inside of the tube is shielded from the pressure by "capping" the ends of the tube. When this is done, the resultant pressure on the ends of the tube is greater than $-P_0$ by a factor $b^2/(b^2-a^2)$ leading to the boundary conditions,

$$\begin{aligned} &R=a, & &T_R=0, & &T_{RZ}=0 \text{ (all } Z),\\ &R=b, & &T_R=-P_0, & &T_{RZ}=0 \text{ (all } Z),\\ &Z=l, & &T_Z=\frac{-P_0b^2}{b^2-a^2} \text{ (all } R), & &\\ &Z=0, & &T_Z=\frac{-P_0b^2}{b^2-a^2} \text{ (all } R). & & \end{aligned} \qquad \text{III}$$

From a study of Eqs. (2)–(5), together with the boundary conditions, it is not too difficult to discover that a satisfactory solution of Eq. (1) for the stress function will be

$$\phi=Z(A\log_e R+BR^2+C)+DZ^3. \tag{6}$$

We may then hope that the constants in this function can be chosen so that the stress components will satisfy the requisite boundary conditions. This is indeed the case. To proceed directly, (6) can be used in (2)–(5) to calculate the stress components. There results:

$$T_R=\frac{A}{R^2}+(4\nu-2)B+6\nu D, \tag{7}$$

$$T_\theta=-\frac{A}{R^2}+(4\nu-2)B+6\nu D, \tag{8}$$

$$T_Z=(8-4\nu)B+6(1-\nu)D, \tag{9}$$

$$T_{RZ}\equiv 0.$$

The algebra involved in evaluating these constants from the boundary conditions can be much simplified by the introduction of a new constant, $E=4B+6D$. With this substitution the preceding equations can be simply written:

$$T_R=\frac{A}{R^2}+\nu E-2B, \tag{7}$$

$$T_\theta=-\frac{A}{R^2}+\nu E-2B, \tag{8}$$

$$T_Z=(1-\nu)E+4B. \tag{9}$$

Introduction of the boundary condition, I, into (7)-(9) gives

$$A=\frac{a^2b^2P_0}{b^2-a^2},$$

$$B=\frac{a^2b^2P_0}{b^2-a^2}\cdot\frac{1}{2a^2}\left(\frac{1-\nu}{1+\nu}\right),$$

$$E=\frac{a^2b^2P_0}{b^2-a^2}\cdot\left(\frac{-4}{a^2}\right)\cdot\frac{1}{2(1+\nu)}.$$

From this we find:

for shielded ends

$$T_R=\frac{a^2b^2P_0}{b^2-a^2}\left\{\frac{1}{R^2}-\frac{1}{a^2}\right\}, \tag{10}$$

$$T_\theta=\frac{a^2b^2P_0}{b^2-a^2}\left\{-\frac{1}{R^2}-\frac{1}{a^2}\right\}, \tag{11}$$

$$T_Z=0. \tag{12}$$

Using boundary condition II in (7)-(9) we find:

$$A=\frac{a^2b^2P_0}{b^2-a^2},$$

$$B=\frac{a^2b^2P_0}{b^2-a^2}\left\{\frac{\dfrac{1-2\nu}{a^2}+\dfrac{\nu}{b^2}}{2(1+\nu)}\right\},$$

$$C=\frac{a^2b^2P_0}{b^2-a^2}\left\{\frac{\dfrac{2}{b^2}-\dfrac{6}{a^2}}{2(1+\nu)}\right\},$$

which gives us

for exposed ends

$$T_R=\frac{a^2b^2P_0}{b^2-a^2}\left\{\frac{1}{R^2}-\frac{1}{a^2}\right\}, \tag{13}$$

$$T_\theta=\frac{a^2b^2P_0}{b^2-a^2}\left\{-\frac{1}{R^2}-\frac{1}{a^2}\right\}, \tag{14}$$

$$T_Z=-P_0. \tag{15}$$

Finally, employing the boundary conditions III in Eqs. (7)-(9), we obtain

$$A=\frac{a^2b^2P_0}{b^2-a^2},$$

$$B=\frac{a^2b^2P_0}{b^2-a^2}\left\{\frac{\dfrac{1-(1+\gamma)\nu}{a^2}+\dfrac{\gamma\nu}{b^2}}{2(1+\nu)}\right\},$$

$$C=\frac{a^2b^2P_0}{b^2-a^2}\left\{\frac{\dfrac{-2(2+\gamma)}{a^2}+\dfrac{2\gamma}{b^2}}{2(1+\nu)}\right\},$$

where

$$\gamma=b^2/(b^2-a^2).$$

Substitution of these constants in the Eqs. (7)-(9) then yields for the stress components:

$$T_R=\frac{a^2b^2P_0}{b^2-a^2}\left\{\frac{1}{R^2}-\frac{1}{a^2}\right\}, \tag{16}$$

$$T_\theta=\frac{a^2b^2P_0}{b^2-a^2}\left\{-\frac{1}{R^2}-\frac{1}{a^2}\right\}, \tag{17}$$

$$T_Z=-\frac{b^2P_0}{b^2-a^2}. \tag{18}$$

The Eqs. (10) through (18) constitute, then, the solution of the stated problem.

Remarks

In retrospect it is evident that these results could have been obtained much more simply: For consider; Eqs. (7)-(9) could have been replaced at once by

$$T_R=(A/R^2)+F, \tag{19}$$

$$T_\theta=-(A/R^2)+F, \tag{20}$$

$$T_Z=G, \tag{21}$$

where F and G are new constants independent of each other. This particularly simple form of the stress components could, no doubt, have been obtained directly by a somewhat different choice of stress function. We have, however, made no attempt to construct this function, being satisfied for the present with the admittedly "brute force" approach herein employed.

The following paper considers spherical and cylindrical ceramic transducers and introduces the concept of the transformation coefficient of mechanical stresses in order to increase the sensitivity of the transducers as receivers of acoustic energy.

A. A. Anan'eva is a member of the staff of the Acoustics Institute of the Academy of Sciences of the USSR, Moscow. He is the author of several papers on barium titanate and piezoelectric transducers. He has also studied the effect of various additives to the composition of the barium titanate and other ceramic transducer materials.

Reprinted from the AIP translation of Soviet Physics Acoustics, 1956, V. 2, p. 8–24

30

Acoustic Nondirectional Ceramic Receiving Transducers

A. A. ANAN'EVA

ACOUSTIC NONDIRECTIONAL CERAMIC RECEIVING TRANSDUCERS

A. A. Anan'eva

The paper examines piezoelectric acoustic receiving transducers with spherical and cylindrical directivity, which represent thin shells of barium titanate ceramic. The concept of the transformation coefficient of mechanical stresses is introduced, and it is shown that when the character of the transformation is properly selected and the corresponding arrangement of electrodes properly executed, we may obtain a high sensitivity for the receiving transducers when the directional characteristics are good.

It is known that barium titanate ceramic, used as a material for acoustic receiving transducers, has a low sensitivity in comparison to other piezoelectric substances. For a medium-quality barium titanate ceramic, the piezoelectric modulus in the direction of polarization is $d_{33} = 3.6 \cdot 10^{-6}$ CGSE, the modulus in a direction perpendicular to polarization is $d_{32} = -1.5 \cdot 10^{-6}$ CGSE and the dielectric constant is $\epsilon = 1{,}000$ to 1,200. The ratio of the piezoelectric moduli to the dielectric constant (this ratio determines the sensitivity of the acoustic receiving transducers in the no-load regime [1]) is small and in the usual transducer assemblies with plain or package transducing elements the utilization of barium titanate proves impractical. In these cases, preference may be given to barium titanate solely because of its greater strength.

However, there is one field of acoustic instrumentation where the utilization of polarized barium titanate ceramic considerably simplifies the solution of the problem posed — that is in the field of designing nondirectional acoustic receivers, for example, nondirectional radio-broadcasting microphones, nondirectional acoustical and hydroacoustical measuring receivers, etc.

Spherical nondirectional measuring hydrophones of barium titanate ceramic, and later cylindrical receiving transducers as well, had been produced in the acoustics laboratory of the Phys. Inst. Acad. Sci. as early as 1950.

For receivers with a spherical directional characteristic, hollow spheres of barium titanate glued together out of two halves (Figure 1a) were used in the capacity of a detecting element, and they were polarized radially; for receivers with a circular directional characteristic, radially polarized hollow cylinders were used (Figure 1b).

We should note that it is difficult to obtain an ideal spherical directional characteristic, since the unavoidable structural elements serving as output leads and as cable connectors disrupt the spherical symmetry. As an example, the directional characteristics of a spherical receiver made of barium titanate are shown in an equatorial plane (Figure 2) and in a plane containing an output lead (Figure 3).** In the given case the output lead from the inner electrode was passed through a metal tube which was soldered to the outer metal surface of the sphere. As we can see, the directional characteristics in the equatorial

---

* Presented at the Second All-Union Conference on Piezoelectricity, May 26, 1955

** The polar characteristics given in this paper were obtained under conditions of an unbounded medium by means of an automatic pulse arrangement for obtaining the directional characteristics for the hydroacoustic emitters and receivers.

plane were satisfactory, where they are substantially distorted in the plane containing the output lead.

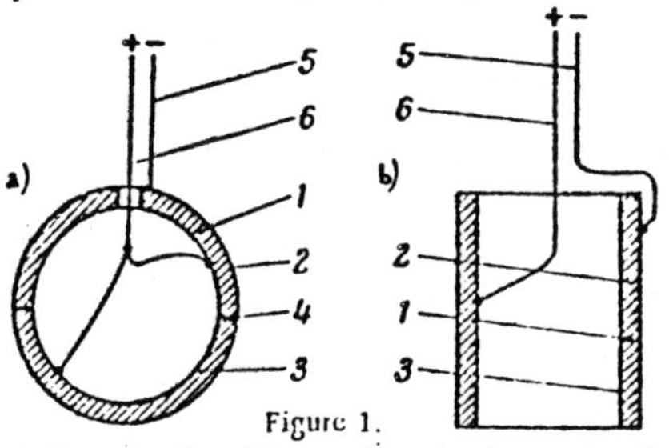

Figure 1.

1) Ceramic. 2 and 3) metallization layer; 4) splice of two hemispheres along equatorial plane; 5 and 6) outlets.

In a cylindrical receiver the problem of constructing the output lead is not as sharply posed, since the cylindrical output lead, naturally, does not disrupt cylindrical symmetry. However the presence of the output lead still affects the directional characteristics in the axial planes, which is easy to see from Figure 6. Let us note that the directional characteristics of cylindrical receivers in the plane perpendicular to the axis (Figures 4 and 5) are, generally speaking, better than the characteristics of spherical receivers in an equatorial plane. The presence of the glued joint in the great-circle of the spherical receivers does not produce a noticeable effect upon the directional characteristics; this was verified with an experimental receiver in which the joint was located in the axial plane of the output lead.

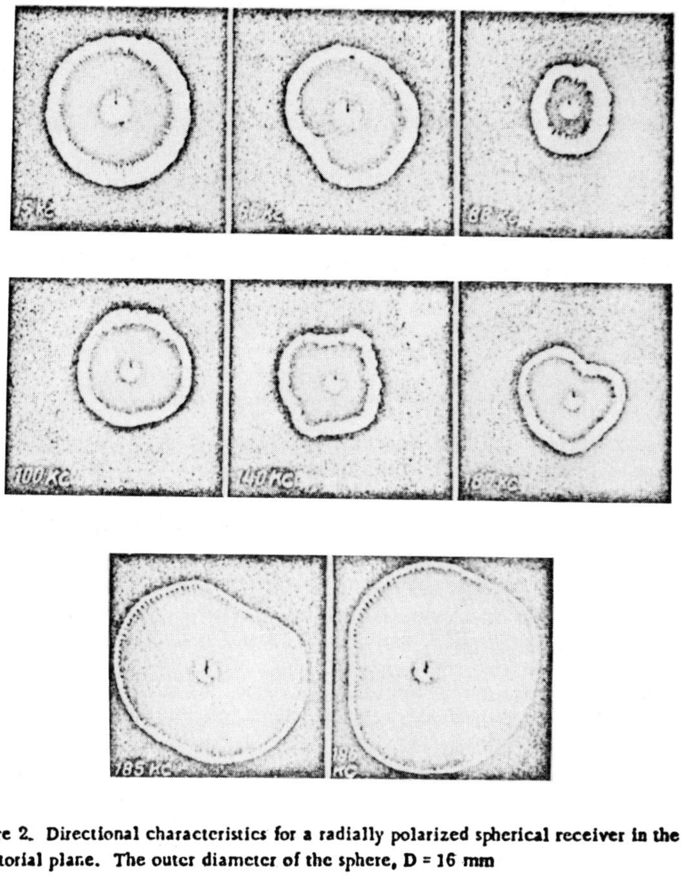

Figure 2. Directional characteristics for a radially polarized spherical receiver in the equatorial plane. The outer diameter of the sphere, D = 16 mm

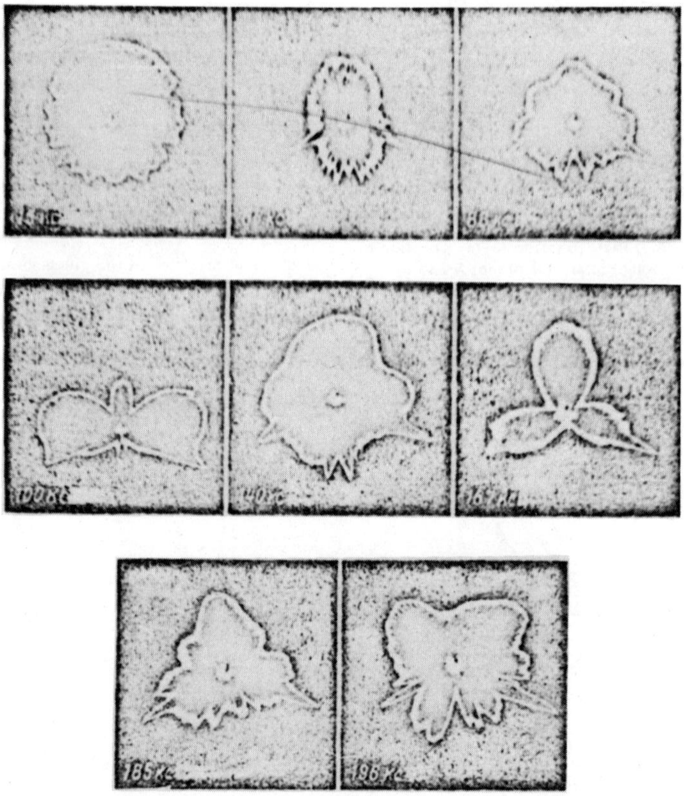

Figure 3. Directional characteristics of a spherical receiver with radial polarization in the plane containing the output lead. The external diameter of the sphere D = 16 mm.

Wide-band nondirectional acoustic receivers are the ones most often required for practical purposes. The working frequency-band, in which the sensitivity is constant, is limited at the upper end by the frequencies of natural vibration of the ceramic or the resonant frequencies of the structural elements serving to lead in the cable and hermetically seal it. As an example, Figure 7a shows the frequency response for the sensitivity of a cylindrical receiving transducer with rubber end caps and a soft cable input lead; in this case the resonances of the cable attachments are absent and the sensitivity undergoes little change right up to the frequency for which longitudinal resonance occurs in the cylinder. In the same figure (b), the frequency responses are shown for a cylinder of the same type having a brass cap at one end and a cap with a brass input tube at the other end. The presence of these elements leads to a sharp drop in the frequency near 15 kc.

We should note that the piezoelement in the form of the cylindrical tube has three fundamental types of internal resonances: 1) that determined by longitudinal vibrations of the cylinder (first resonance for $\lambda = 2h$, where λ is the wave length in the ceramic, h is the height of the cylinder); 2) that related to the diameter of the cylinder (first resonance for $\lambda \approx \pi D$, where D is the diameter of the cylinder); and 3) that related to the wall thickness (first resonance for $\lambda \approx 2a$, where a is the wall thickness). Figure 8 gives

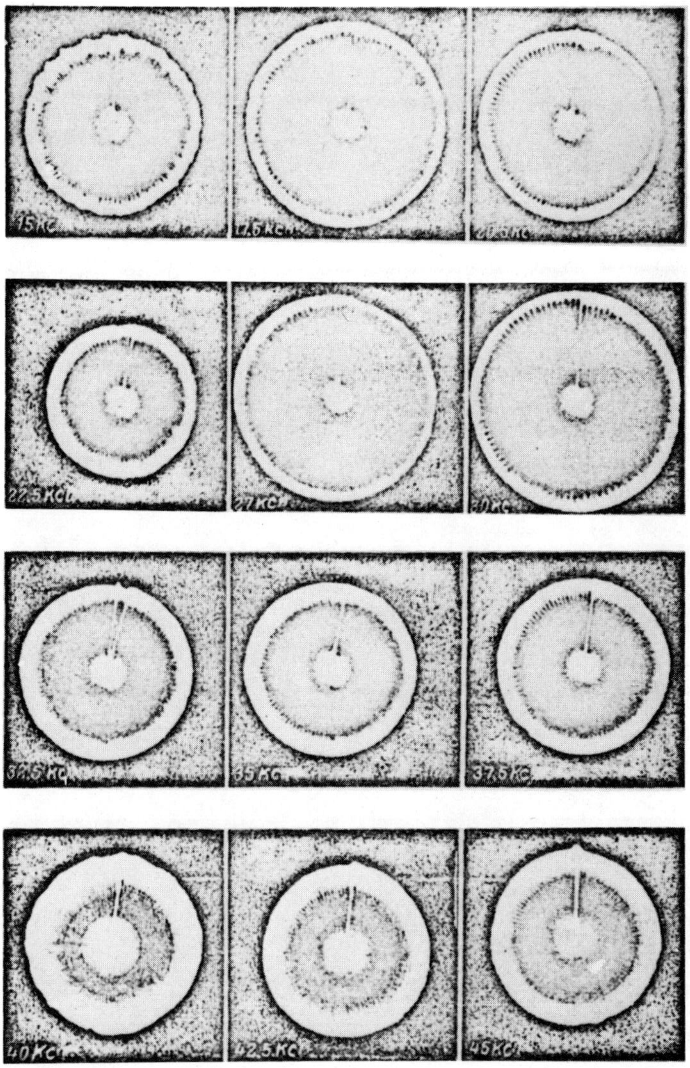

Figure 4. Directional characteristics for a cylindrical receiver with radial polarization in a plane perpendicular to the axis of the cylinder in the frequency range 15–45 kc. Outer diameter of the cylinder D 32 mm.

the frequency characteristic of the electrical impedance of the cylindrical detecting element, and the enumerated resonances are clearly seen in it. For a piezoelement in the form of a hollow sphere, the resonant frequencies may be calculated; however, when a joint is present, generally speaking, which does

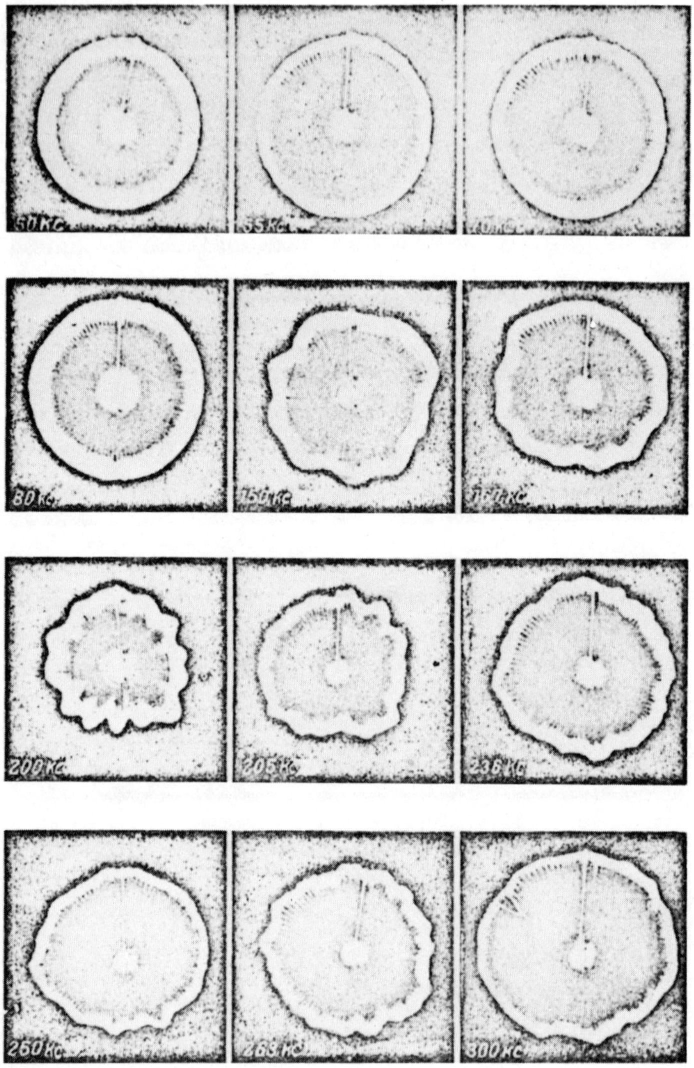

Figure 5. Directional characteristics of a cylindrical receiver with radial polarization in a plane perpendicular to the axis of the cylinder in the frequency range 50-300 kc. Outer diameter of the cylinder D = 32 mm.

not possess a sufficiently high rigidity, the hemispheres vibrate "freely" to a considerable degree and the experimental resonant frequencies do not coincide with the ones calculated for a hollow sphere without a joint. It has been experimentally discovered that the following two types of resonances take place: 1) those which depend on the diameter $\left(\lambda \approx \frac{\pi D}{2}\right)$ and 2) those which depend upon the wall-thickness ($\lambda \approx 2a$).

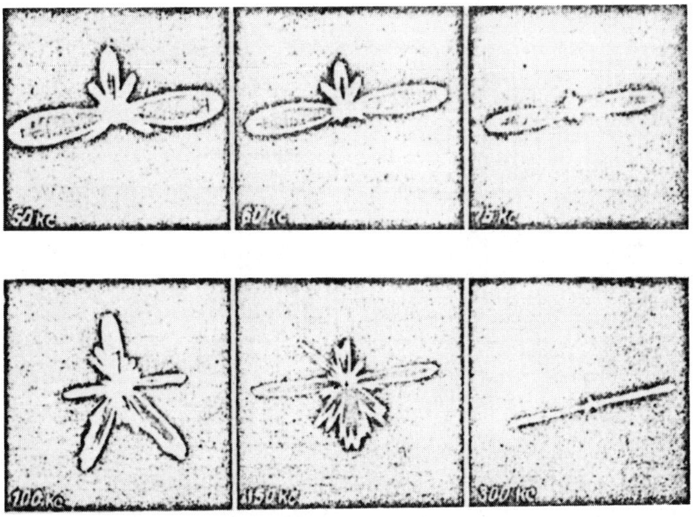

Figure 6. Directional characteristic in an axial plane of a cylindrical receiver with radial polarization. Outer diameter of the cylinder D=32 mm.

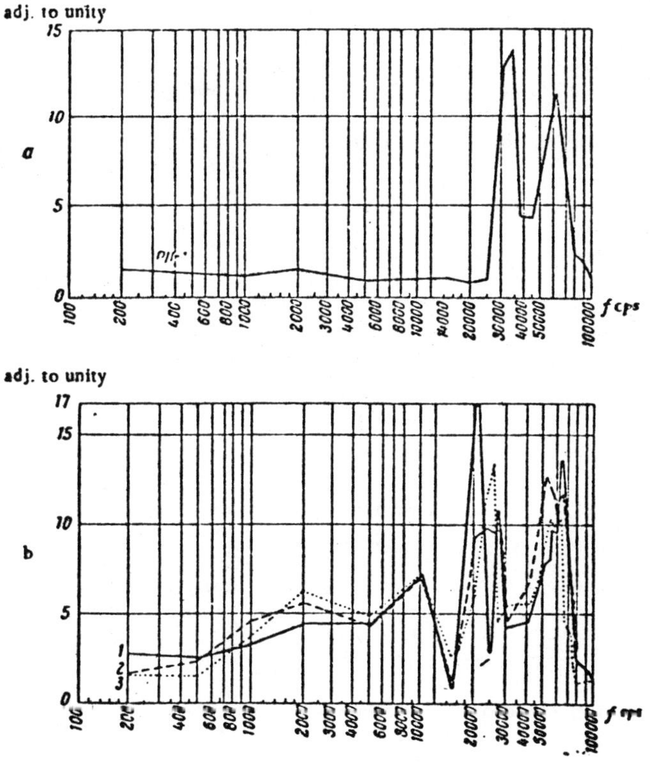

Figure 7. Examples of the frequency responses for the sensitivity of cylindrical receiving transducers with radial polarization.
a) Cylinder ends capped by a rubber layer; b) the cylinders possess metal membrane end-caps (1, 2, 3 are the characteristics of three receiving transducers with identical geometric dimensions).

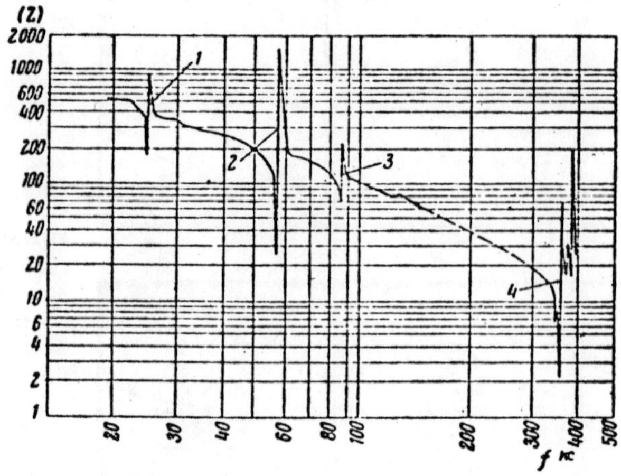

Figure 8. Spectrum of the internal vibrations in the cylindrical receiving transducer. 1) Fundamental longitudinal resonance of the cylinder; 2) fundamental radial resonance of the cylinder; 3) first harmonic radial oscillations; 4) resonance of thickness vibrations in the cylinder.

The above discussion shows that the most simply constructed cylindrical and spherical receiving transducers enable us to obtain the required frequency and polar responses with proper design; however, their low sensitivity (of the order of units of microvolts/bar) is their great shortcoming. Therefore, we pose the problem of raising the sensitivity of ceramic receiving transducers while retaining a circular or spherical directional characteristic.

Let us examine the ways of raising the sensitivity of nondirectional receiving transducers which are constructed as shown in Figure 1. For the static sensitivity of the receiving transducer built in the form of a hollow sphere with radial polarization, we obtain the formula

$$\frac{E}{p_0} = R\frac{(1-2\sigma)^2}{(3-6\sigma+4\sigma^2)}\left[g_{32}(3-5\sigma)+g_{33}\sigma\right], \quad (1)$$

where E is the emf in the no-load regime, p_0 is the acoustic pressure on the surface of the sphere, R is the external radius, σ is the ratio of the wall thickness to the external diameter, $g_{33} = \frac{4\pi d_{33}}{\epsilon}$ and $g_{32} = \frac{4\pi d_{32}}{\epsilon}$.

For a small wall thickness ($\sigma \ll 1$), we obtain approximately

$$\frac{E}{p_0} \doteq g_{32}R. \quad (2)$$

It is interesting to note that for $\sigma = \frac{1}{2}$, i.e., for a point-contact inner electrode at the center of a solid ceramic sphere, a sensitivity equal to zero is obtained.

For a cylindrical receiving transducer with radial polarization, the sensitivity is determined in the following manner [2]:

$$\frac{E}{p_0} = R\left(g_{33}\frac{\sigma}{1-\sigma} + g_{32}\right) \quad (3)$$

with the same notation as in formula (1). Let us note that as a consequence of the opposite sign in the moduli d_{33} and d_{32}, the sensitivity of a cylindrical receiving transducer is equal to zero for the condition

$$\sigma = \frac{d_{32}}{d_{32} - d_{33}}.$$

As in the case of the sphere, the sensitivity is approximately equal to Rg_{32}, when the wall is thin.

On the basis of the above it is plain that only two fundamental methods exist for increasing the sensitivity of such receiving transducers: direct increase of the piezo modulus d_{32}, and increase of the external diameter of the sphere or cylinder. The first method as yet shows little promise. The second method may be used, but only until diffractional limitations begin to apply. Therefore, we selected a different method for increasing the sensitivity of ceramic receiving transducers with spherical or cylindrical symmetry. Actually, the problem is one of utilizing a piezoelectric material which has an excessively large dielectric constant in the most practical manner. For that purpose, evidently, it is necessary to utilize the maximum piezomodulus inherent in the given material on the one hand, and on the other hand, to increase the effective voltage in the directions corresponding to this piezomodulus. Simultaneously, it is permissible, in view of the excessively large dielectric constant of barium titanate, to increase the effective size of the detecting element in the direction of greatest piezomodulus. We shall define as mechanical stress transformers those mechanical arrangements which simultaneously resolve all of these three problems.

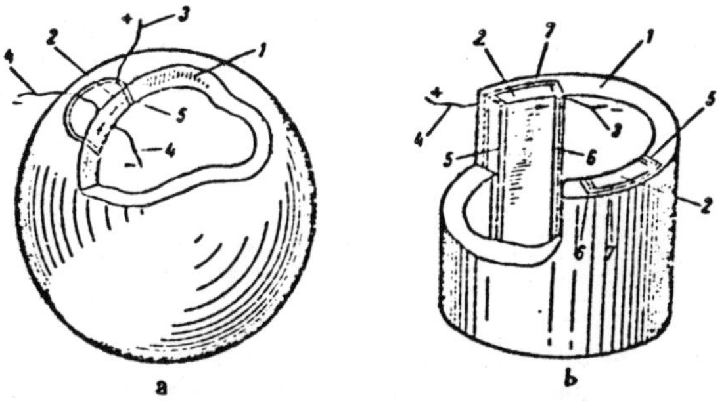

Figure 9. Construction of receiving transducers with spherical and cylindrical receiving surfaces.
1) Cylindrical shell (metallic), spherical (glass) shell; 2) piezoelectric ceramic element; 3 and 4) output leads; 5 and 6) metallic coating; 7) insulator.

The concept of a mechanical stress transformer may be illustrated very clearly by the example of the first receiving transducers with mechanical transformations which we built. We first thought it practical to utilize cylindrical or spherical shells with built-in small piezoelements (Figure 9) in the capacity of stress transformers, in order to retain a spherical or circular directional characteristic. A thin-walled sphere or cylinder will transform external radially directed forces into tangential forces which are applied to the piezoelements it is natural that both the polarization and the arrangement of electrodes must be so selected that the maximum output emf is produced, as is shown in Figure 9.

If the pressure acts upon the surface of a thin-walled sphere (or on the lateral surface of a cylinder) and the thickness of the wall is small enough, we may neglect normal stresses in the material (i.e., we may assume that the external pressure causes chiefly tangential stresses at each point in the cross-section of the cylinder or sphere). The ratio of the mechanical stress upon the cross-section to the acoustic pressure at the surface of the sphere $\frac{\sigma_T}{p_0}$ shall be defined as the transformation coefficient K. We shall calculate the transformation coefficient for two cases: 1) for a thin-walled spherical shell, 2) for a thin-walled cylindrical shell.

Thin-Walled Spherical Shell

The pressure, p_0, is specified on the outer surface of a sphere whose radius is equal to R. Inside the sphere p = 0. At each point in the cross-section of the sphere equal mechanical stresses will act in two directions perpendicular to each other $\sigma_T = \sigma_\phi$: let us compute them. We shall mentally take a cross-section of the sphere through a great-circle. The sum of the normal forces acting upon the circular cross-section (Figure 10a) with an area $S = 2\pi R^2$, is equal to $F = \int_0^{\pi/2} p\,ds$.

where

$$p = p_0 \sin \vartheta, \quad dS = 2\pi\rho R d\vartheta = 2\pi R^2 \cos \vartheta d\vartheta.$$

Therefore,

$$F = 2\pi R^2 p_0 \int_0^{\pi/2} \sin 2\vartheta d\vartheta = p_0 \pi R^2,$$

the stress in the direction τ is

$$\sigma_\tau = \frac{F}{S} p_0 \frac{R}{2a}.$$

From this,

$$K_\tau = \frac{R}{2a},$$

and since $\sigma_\tau = \sigma_\phi$,

$$K_\phi = \frac{R}{2a}.$$

Thin-Walled Cylinder

Let a pressure p_0 be specified on the outer lateral surface of a cylinder, with height h, radius R and wall thickness a. Inside the shell the pressure $p = 0$. The faces of the cylinder are free, i.e., shielded from the action of acoustic pressure; therefore, at each point in the cross-section tangential stresses σ_τ will act in a direction perpendicular to the generant of the cylinder; stresses in a direction parallel to the generant of the cylinder are equal to zero.

Let us mentally take a cross-section of the cylinder through a generant (Figure 10b) and set aside one-half of the cylinder; we shall replace its effect upon the other half of the cylinder by equivalent forces which are normal to the plane of the cross-section. The sum of the projections of all forces acting on two rectangular cross-sections with an overall area $S = 2ha$, is equal to

$$F = h \int_0^\pi p dS,$$

where

$$p = p_0 \sin \vartheta,$$
$$dS = R d\vartheta.$$

Therefore,

$$F = h \int_0^\pi p_0 R \sin \vartheta d\vartheta = 2hRp_0;$$

$$\sigma_\tau = \frac{F}{S} = p_0 \frac{R}{a}.$$

From this, the transformation coefficient is

$$K_\tau = \frac{R}{a}.$$

When the dimensions of the sphere or cylinder are sufficiently great, and the wall is sufficiently thin, large values of the transformation coefficient may be obtained.

It is evident that the mechanical transformation may be utilized independently of how the polarization of the detecting element is oriented, and even independently of what type of element is selected.

In this particular case, however, we may fully utilize the advantages of barium titanate only by tangential polarization of the detector element. When this is done we make use of the largest piezomodulus d_{33}, and it is possible to select a sufficiently large effective length for the detector element (along the arc of

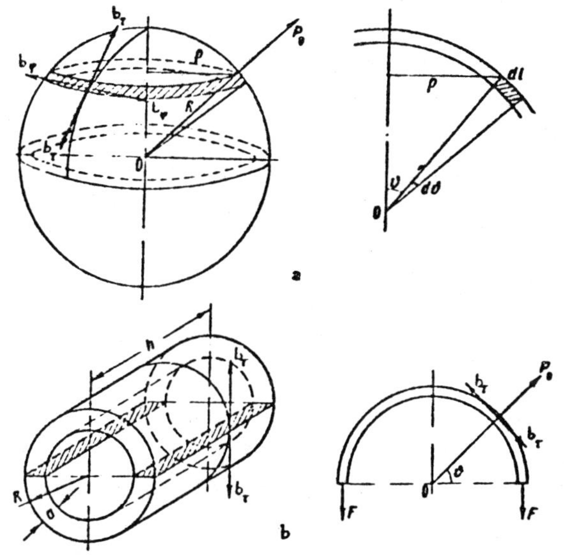

Figure 10

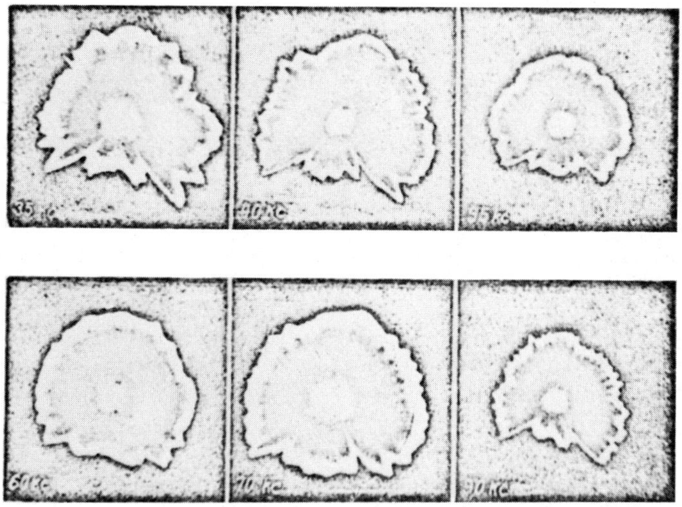

Figure 11. Directional characteristics of a receiving transducer with a spherical receiving surface (diameter of the sphere 29 mm)

a great-circle for a sphere and along the circumference for a cylinder). Under these conditions the large dielectric constant of ceramic barium titanate is not a disadvantage, but on the contrary is an advantage.

We prepared cylindrical and spherical receivers according to the diagrams in Figure 9. However, the directional characteristics of such receivers proved to be unsatisfactory (Figures 11 and 12), and therefore we decided to utilize the very same piezoelectric material in the capacity of a material for the mechanical transformer. Thus, we arrived at a construction in which the piezoelement is a thin-walled shell with a special electrode arrangement.

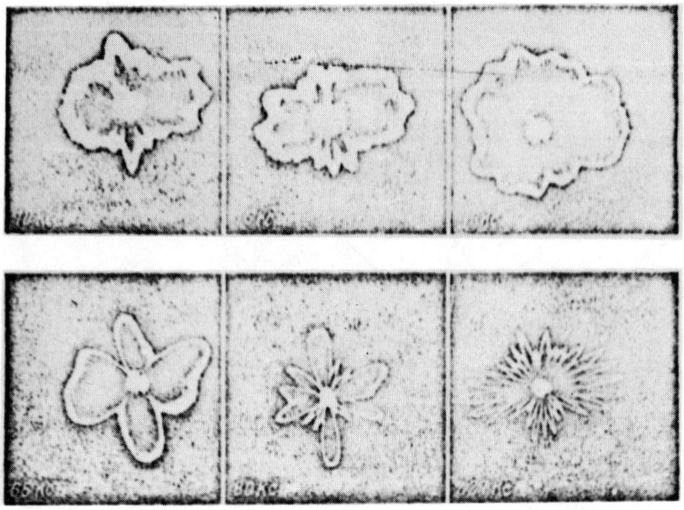

Figure 12. Directional characteristic for a receiving transducer with a cylindrical receiving surface (diameter of the cylinder 38 mm).

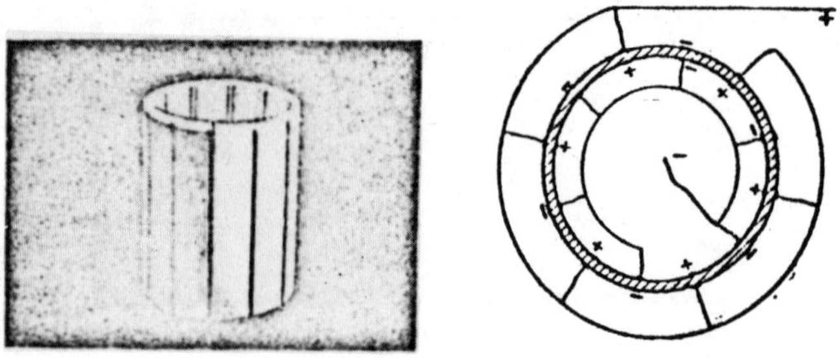

a b

Figure 13. a) External view of a ceramic detecting element in the form of a continuous cylinder with electrodes fired onto its lateral surfaces; b) circuit for combining the electrodes when polarizing the ceramic and when the acoustic receivier is in operation.

When the construction is cylindrical, the large magnitude of the dielectric permeability of the ceramic made it possible to utilize non-sectioned ceramic thin-walled cylinders with electrodes fired onto the lateral surfaces of the cylinder. With an electrode distribution over the cylindrical surfaces which is shown in Figures 13, a and b, tangential polarization of the ceramic is effected. In utilizing the method of surface application of the electrodes, it is also possible to effect the tangential polarization of a thin-walled ceramic sphere, as is shown in Figure 14.

In a case when a cylindrical or spherical acoustical receiver possessing surface applied electrodes is utilized in the capacity of an air microphone, there is no necessity for using any shielding devices whatsoever, except, perhaps, a shielding grid. For hydrophones it is necessary to utilize an external shielding layer of a dielectric with a dielectric constant which is substantially lower than that of barium titanate.

Various plastics may be utilized as material for such a layer; in particular, we utilized a layer made of rubber.

The external view of the latest models of our highly-sensitive cylindrical hydrophones is shown in Figure 15. An example of the directional characteristics of cylindrical hydrophones with tangential polarization is shown in Figures 16 and 17. The band of frequencies in which the polar characteristics remain circular is narrowed in comparison to that for a receiver with a homogeneous radial polarization, since we changed the character of the piezoelement's symmetry. The internal resistance of hydrophones of the specified type is considerably higher than that of hydrophones having the same dimensions but possessing radially-polarized elements; however, the internal resistance remains within permissible limits for engineering application, due to the large dielectric permeability of the ceramic.

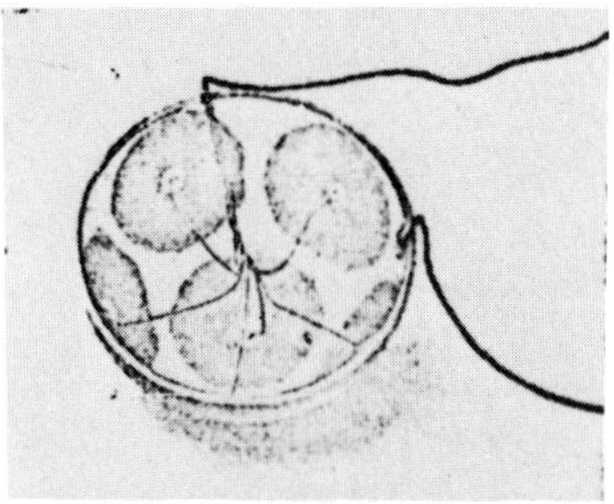

Figure 14. Element of a spherical acoustical receiver-hemisphere with surface electrodes, allowing tangential polarization of the ceramic to be effected.

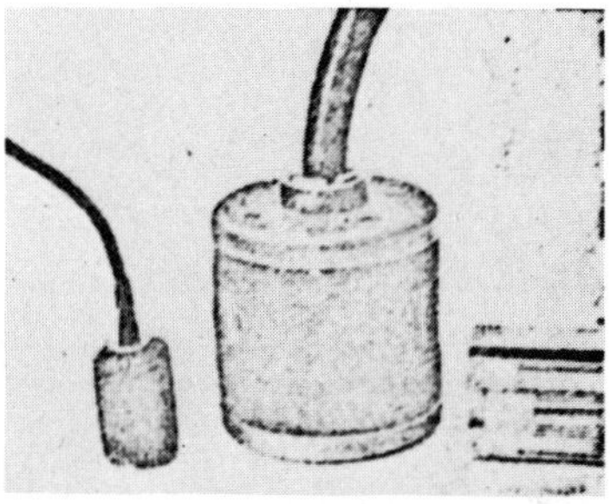

Figure 15. Overall view of hydrophones with detecting elements in the form of a radially-polarized ceramic cylinder.

TABLE 1

Polarization	Shapes of receiving surface	
	Sphere	Cylinder (with free faces)
Radial	$R g_{32} = 4.7 \cdot R$	$R g_{32} = 4.7R$
Longitudinal	—	$R g_{32} \frac{1}{a} = 4.7 R \frac{1}{a}$
Tangential	$\frac{2\pi R^2}{an} g_{33} = 71.0 \frac{R^2}{an}$	$\frac{2\pi R^2}{an} g_{33} = 71.0 \frac{R^2}{an}$

Notation in the table: a is the wall-thickness of the ceramic shell in cm; l is the distance between electrodes in cm; R is the radius of the cylinder or sphere in cm; n is the number of cells (on a great-circle of the sphere or on a diameter of the cylinder).

Let us compute the static sensitivity of thin-walled spherical and cylindrical receivers of barium titanate ceramic, utilizing the definition for the coefficient of mechanical transformation which was introduced earlier. The sensitivity of the piezoelectric receiver of the simplest type, in a no-load regime, is determined by the formula

$$\frac{E}{p_0} = gl$$

where E is the emf developed across the coatings of the piezoelement for a specified acoustic pressure upon its surface, g is the corresponding piezoconstant, l is the effective length of the piezoelement (the distance between electrodes).

In utilizing systems with mechanical transformation we gain an additional advantage in receiver sensitivity:

$$\frac{E}{p_0} = Kgl,$$

where K is the transformation coefficient of mechanical stress. For the cases of a thin-walled sphere and a thin-walled cylinder with free ends, the expressions for K were given above.

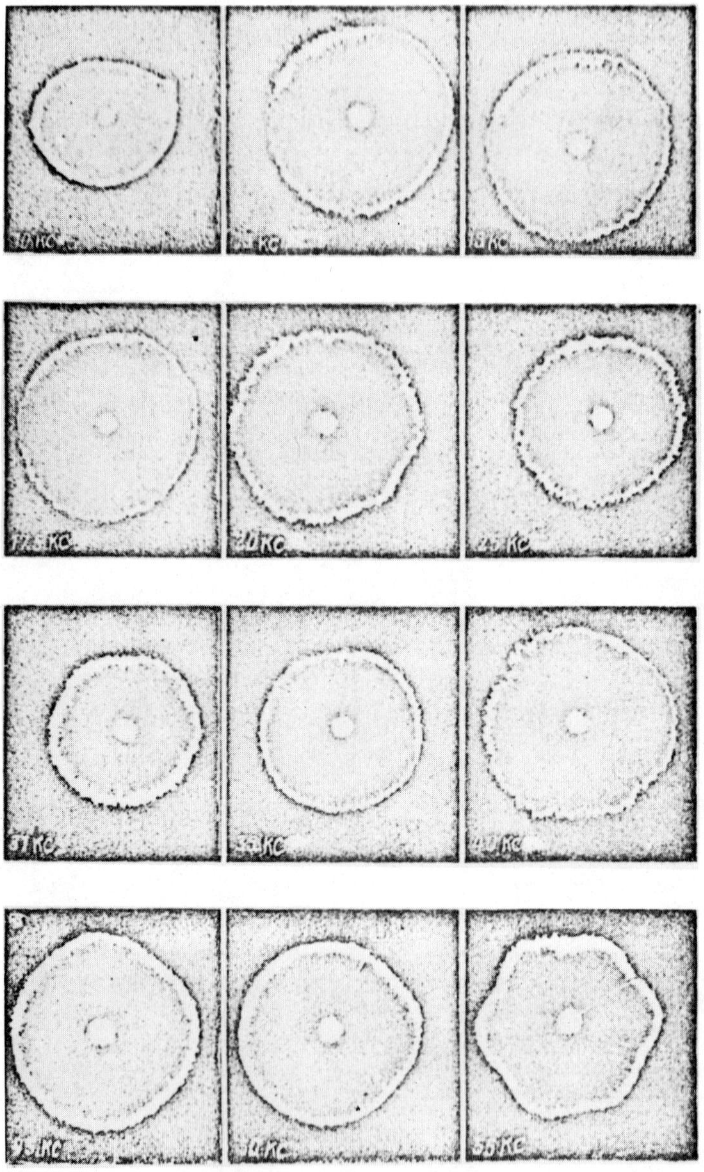

Figure 16. Directional characteristics of a ceramic acoustic receiver with tangential polarization in the frequency range 10-55 kc. Diameter of the cylinder D = 32 mm; height h = = 40 mm.

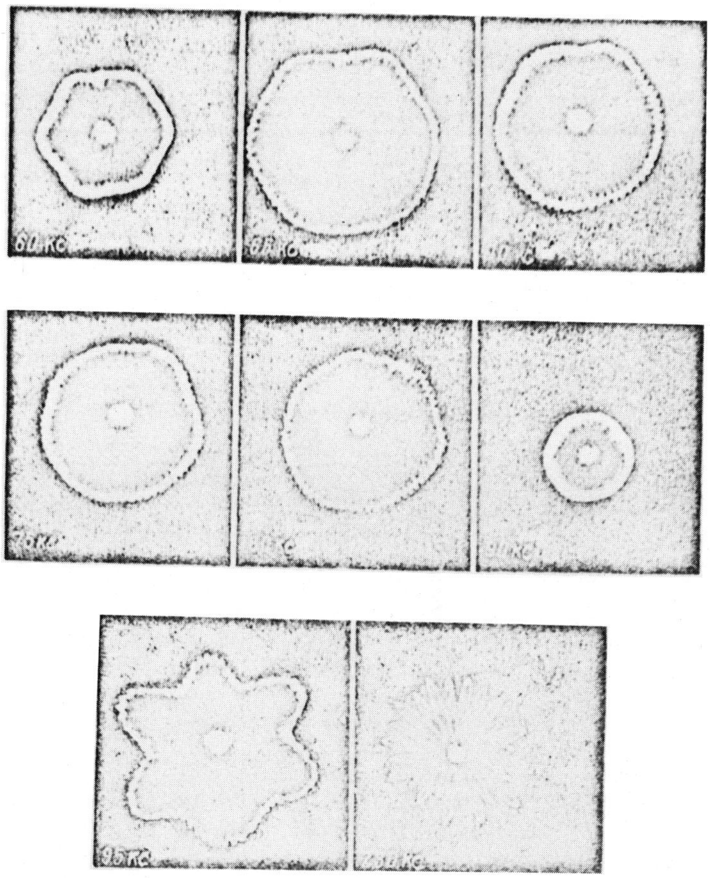

Figure 17. Directional characteristics of a ceramic acoustic receiver with tangential polarization in the frequency range of 60-239 kc. The diameter of the cylinder D = 32 mm; height h = 40 mm.

In Table 1, formulas are given for calculating the sensitivity in the no-load regime in μv/bar, for ordinary barium titanate ceramic with a dielectric permeability $\epsilon = 1200$ and a piezoelectric modulus $d_{33} = 3.6 \times 10^{-6}$ CGSE.

In Table 2, data is given for cylindrical piezoelements for hydrophones with tangential polarization. For comparison, the maximum sensitivity which may be obtained for radial polarization of cylinders with the same dimensions are given (cylinders with rigid diaphragms capping the ends, which increase the longitudinal amplification in the shell for a specified sound pressure). The table indicates two values for the calculated sensitivity: the lesser value is calculated for the case when the portion of the surface occupied by the electrodes is taken into account, and the larger value for the case when the electrodes are infinitely narrow.

The design of cylindrical tubes with a finite wall thickness when various methods are used for capping the ends [2] results in the formulas provided in Table 3.

TABLE 2

Specified piezoele-ment No.	Dimensions of cylinders: Height (mm)	Dimensions of cylinders: Diameter (mm)	Wall-thickness, (mm)	Capacity ($\mu\mu f$)	Static sensitivity, (μv/bar): Calculated values	Static sensitivity, (μv/bar): Average measured values	Sensitivity for radial polarization (μv /bar)	Working frequency band
1	40	52	1.5	500	264-238	234	17	up to 20 kc
2	40	52	1.5	500	264-238	250	17	up to 20 kc
3	40	52	1.5	500	264-238	240	17	up to 20 kc
4	40	52	3.0	900	132-110	135	16.2	up to 20 kc
5	40	52	3.0	900	132-110	106.5	16.2	up to 20 kc
6	40	52	3.0	900	132-110	101.5	16.2	up to 20 kc
7	40	30.9	0.9	690	127.5-115	101	10.9	up to 50 kc
8	40	32.0	1.5	—	73	64.4	10.7	up to 50 kc
9	23.6	14.6	1.0	—	19	—	4.75	up to 100 kc

TABLE 3

	Method of securing ends	Method of polarization: I, radial polarization	Method of polarization: II, longitudinal polarization	Method of polarization: III, tangential polarization
1	Ends shielded from the pressure	$R\left(g_{33}\frac{\sigma}{1-\sigma}+g_{32}\right)$	$h\frac{g_{33}}{2(\sigma-\sigma^2)}$	$A\left(\frac{g_{32}\sigma}{1-\sigma}+g_{33}\right)$
2	Ends subjected to the pressure	$R\left[g_{33}\frac{\sigma}{1-\sigma}+g_{32}(1-2\sigma)\right]$	$h\left[\frac{g_{32}}{2(\sigma-\sigma^2)}+g_{33}\right]$	$A\left[g_{32}\frac{\sigma(3-2\sigma)}{1-\sigma}+g_{33}\right]$
3.	Ends closed by solid diaphragm (pressure on the ends is amplified).	$R\left(g_{33}\frac{\sigma}{1-\sigma}+g_{32}\times\frac{3-2\sigma}{2-2\sigma}\right)$	$h\left[\frac{2g_{32}+g_{33}}{4(\sigma-\sigma^2)}\right]$	$A\left[g_{32}\frac{1+2\sigma}{2(1-\sigma)}+g_{33}\right]$

In Table 3, the parameter A has the form

$$A=\frac{1}{h}\frac{2\pi R}{\ln\frac{2}{1-2\sigma}},$$

σ is the ratio of wall thickness to the external diameter of the cylinder, h is the height of the cylinder; the remaining definitions are the same as before.

When we place $\sigma = 0$, it is easy to prove that the approximate expressions for the static sensitivities of thin-walled spherical and cylindrical shells given in Table 1 are the limiting cases of the generalized calculation of the sensitivities of a cylinder and sphere with finite wall-thicknesses.

It is of interest to clarify to what degree the approximate formula is usable for practical computation in a case which differs from the limiting case, i.e., for finite values of σ. It is evident that it is im-

• For a cylinder with rigid diaphragms as end caps. For a cylinder with free ends, when it is radially polarized the sensitivity would be 1.5 times smaller.

practical to select large values of σ, even if we do not approach the region near $\sigma = 0$ where the sensitivity decreases rapidly, since for a thick wall the sensitivity falls sharply. Therefore, we selected values of σ which were no greater than 0.05 – 0.06.

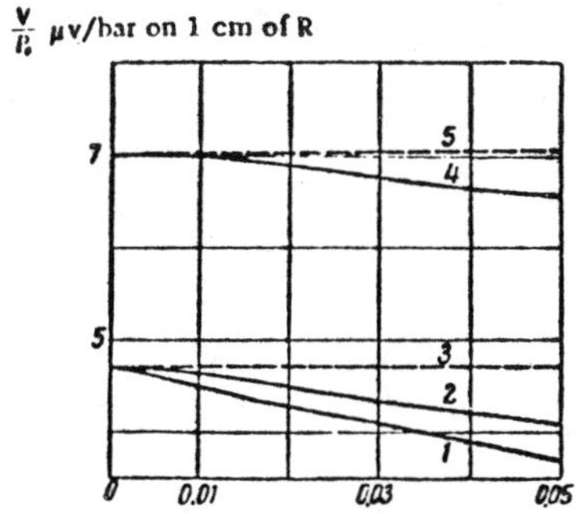

Figure 18. Calculation of static sensitivity of ceramic cylindrical and spherical acoustic receivers with radial polarization
1 and 3 (Sphere); 2 and 3 (cylinder) (ends are shielded from pressure); 4 and 5 (cylinder) (ends are closed by rigid membranes which amplify the pressure upon them).

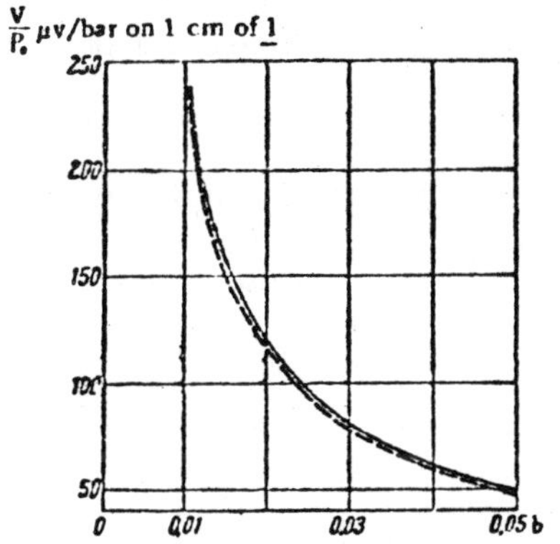

Figure 19. Calculation of the static sensitivity of a cylinder with longitudinal polarization (ends of the cylinder are shielded from pressure).

Figure 18 provides values for the sensitivity of acoustic receivers, calculated per 1 cm external radius for radial polarization. The dotted horizontal straight lines depict the sensitivity calculated by the approximate formulas, the solid curves depict calculation by the precise formulas. As we can see, for $\sigma = 0.05$ the discrepancies between the results of calculation by the approximate and precise formulas are 2.1 db for a sphere, 1.2 db for a cylinder with free ends, and 0.73 db for a cylinder with diaphragms capping its ends.

A similar comparison is given in Figure 19 for the case of a cylinder with longitudinal polarization and free ends. Here the advantage of small σ is clearly seen; the discrepancy between the results of computation by the precise formula (solid curve) and by the approximate formula (dotted curve) is extremely small and for $\sigma = 0.05$ is only 0.46 db. Finally, Figure 20 shows a comparison with the case of a cylinder with tangential polarization. Here the error in calculation by the approximate formula for $\sigma = 0.05$ is 1.16 db. Thus, it is always possible in practice to calculate the static sensitivity of an acoustic receiver by the approximate formulas.

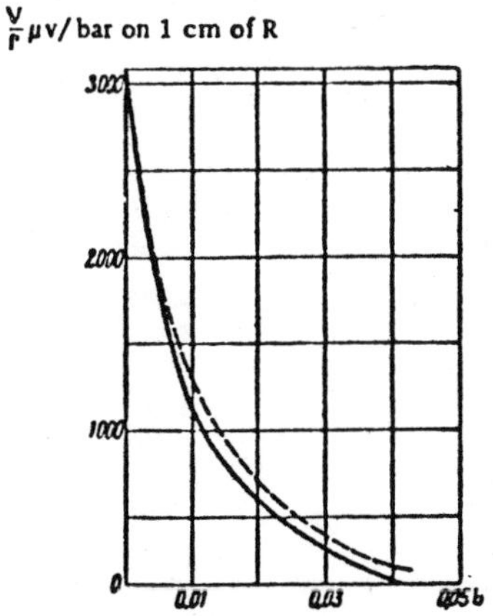

Figure 20. Calculation of the static sensitivity of a cylinder with tangential polarization (the ends are shielded from pressure).

Let us note that for cylindrical receivers with tangential polarization the limit of the sensitivity when σ is decreased is determined in practice by the conditions imposed by the piezoelement strength and by the necessity of obtaining an electrical impedance which is not too high.

Summarizing, it may be said that piezoelectric acoustic receivers in the form of thin shells of barium titanate ceramic with surface-attached electrodes permit us to obtain good directional characteristics with high sensitivity and an internal electrical impedance which is not too great. It should be said that the above statement applies not only to piezoelements in the form of closed shells, but also to dome-shaped and even flat shells; the latter may be utilized in case a high axial concentration coefficient is required.

In conclusion the author would like to take this opportunity to express his appreciation to N. N. Andreev and V. S. Grigor'ev for their consistent attention and help in this work.

LITERATURE CITED

[1] N. N. Andreev, "Piezoelectrical Crystals and Their Application," Electricity 2, 5-13 (1947).

[2] R. A. Landevin, "The Electro-Acoustic Sensitivity of Cylindrical Ceramic," J. Acoust. Soc. Am., 26, 3, 421-427 (1954).

Received July 29, 1955

Acoustical Institute
Academy of Sciences of the USSR
Moscow

Transducer materials such as ceramics have high mechanical strength in compression but low strength in tension. Mechanical bias permits the application of a compression force through a spring so that when the transducer is excited, it does not experience a tension load. Although mechanical bias has a great amount of application in practice, it is not generally realized that the idea was originally invented by Harry B. Miller.

Mr. Miller is an acoustics and communication engineer. He has done research on the near-field behavior of transducer arrays including the influence on premature cavitation, mutual interaction effects of the elements in large transducer arrays, reverberation in real and simulated rooms, subjective effects in binaural listening, magnetic recording head design; he was the first to reduce pulsed frequency modulation to practice.

He was Head of the Acoustic Engineering Department of the Brush Development Company for 14 years, and is now Senior Project Engineer at the New London Laboratory of the Navy Underwater Systems Center.

This paper is reprinted with the permission of the author and the Acoustical Society of America.

Reprinted from THE JOURNAL OF THE ACOUSTICAL SOCIETY OF AMERICA, Vol. 35, No. 9, 1455, September 1963

31

Origin of Mechanical Bias for Transducers

HARRY B. MILLER
General Dynamics/Electronics, Rochester, New York
(Received 22 May 1963)

The people who were familiar with the birth of this invention have agreed to speak for the inventor before the origin gets hopelessly forgotten.

IN a recent book,[1] and in innumerable articles during the past few years, the term "mechanical bias" is used and referred to as if the concept has always been with us. It has not. It had to be invented. The time was 1954 and the inventor was Harry B. Miller. The invention was published in March 1960 as U. S. Patent No. 2,930,912. When the invention was explained to Dr. Harold L. Saxton of NRL in 1954 shortly after its reduction to practice, he greeted the invention with enthusiasm and predicted that it would be widely used.

Briefly, the invention resides in the recognition that a very soft spring can provide a very large dc force and still retain its low ac stiffness. For example, in a typical transducer a soft spring is stretched almost to the breaking point, where it delivers 9000 lb of precompression force; yet the stiffness of the spring is less than 10% of the stiffness of the "sandwich" transducer that it is compressing. The importance of this low stiffness is that the coupling coefficient k is preserved almost intact, and, hence, the over-all tuned bandwidth of the transducer is preserved. In addition, since the "sandwich" is only allowed to go into *relative* tension (compared to the bias reference), it is even possible to eliminate the cement at the joints. Illustrations from the patent are shown in Fig. 1.

It should be pointed out that, if the teachings of the prestressed-concrete art had been followed, the bandwidth of the transducer would have suffered tremendously.

The late Fritz Pordes of USL recommended universal adoption of the invention and publicly commended the author. The late P. N. Arnold of NRL considered the invention an important contribution to the field of underwater acoustics. The author has received permission from H. L. Saxton of NRL, C. L. Buchanan of NRL, and E. J. Parssinen of USL to use them for corroboration if the origin of mechanical bias again becomes forgotten.

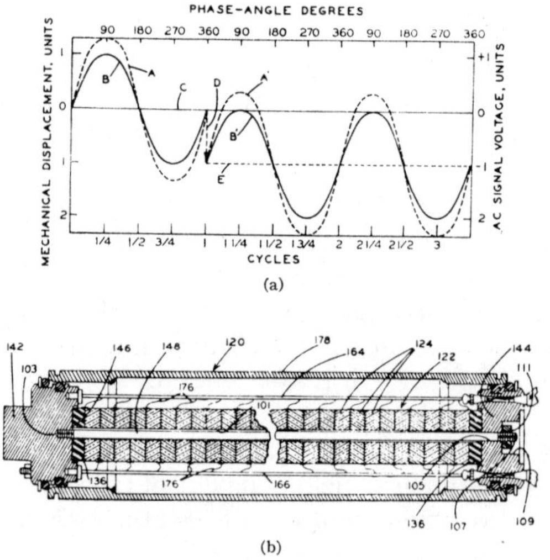

FIG. 1. Illustrations from the patent disclosing "mechanical bias."

[1] V. M. Albers, *Underwater Acoustics* (Plenum Press, Inc., New York, 1963).

The following paper describes the application of the principles of the Luneberg lens to spherical structures used for focusing underwater sound.

Dr. Toulis is a physicist. He has contributed in the fields of the acoustical properties of matter, irradiation physics of water, microwave circuits, antenna theory, sonar transducers, sonics and ultrasonics, underwater sound, sonar systems, and oceanographic instrumentation. He is on the staff of the Autonetics Division of North American Aviation, Inc.

This paper is reprinted with the permission of the author and the Acoustical Society of America.

Reprinted from THE JOURNAL OF THE ACOUSTICAL SOCIETY OF AMERICA, Vol. 35, No. 3, 286–292, March, 1963

32

Acoustic Focusing with Spherical Structures

W. J. TOULIS

Electric Division, Daystrom, Inc., Poughkeepsie, New York
(Received 23 November 1962)

A variety of structures in the form of spheres can provide sharp focusing and directivity gains comparable to those from a circular piston with the same diameter. With an isotropic refractive sphere, spherical aberration does not impair focusing except for angular resolutions of the order of a few degrees and less. Luneberg lenses with compliant tubes or refractive liquid packets eliminate spherical aberration, but the focal distance may change as a result of dispersion in the resultant index of refraction. The outer shell of a Luneberg lens alone, and without the inner core, provides an angular resolution which is comparable to that for the full lens, but with a decrease in the directivity gain which is not as great as the reduction in the amount of refractive matter. Primary feeds with considerable directivity are essential for all types of lenses except those with the focus near the outer surface of the lens.

INTRODUCTION

THE focusing of sound waves by an acoustic lens was recognized long ago to have characteristics similar to those of an optical lens. Inasmuch as thin lenses do not offer any superior advantage over pistons, reflectors, and other single-beam-forming structures, their use has been limited to specialized applications. Thick lenses, and especially lenses in the form of a sphere, have been neglected because of the presence of spherical aberration. Such a neglect is fully justifiable for optical applications where requirements for extremely high resolving powers are commonplace, but not in the field of underwater sound where directivity indices greater than 30–35 dB are quite rare. Furthermore, spherical aberration may be overcome in a spherical acoustic lens when the index of refraction is designed to vary with the radius in the manner suggested by Luneberg's equations.[1]

The superiority of the spherical over the thin lens is not in improved focusing, but rather in its ability to provide many directional beams both simultaneously and independently. Arrays of transducers can also be designed with the aid of delay lines and phasing networks to provide multiple-beam-forming capabilities, provided a large number, if not all, of the transducers in the array are utilized at any one moment to form either one or many beams. On the other hand, the operational versatility of the spherical lens is much greater in that only one small transducer is necessary to form one beam. Consequently, the multiple beams with a single spherical lens may be operated independently in terms of both time and frequency of operation and for either receiving or transmitting acoustic signals with a minimum amount of electronic equipment. Otherwise, a receiving array might have an advantage in terms of

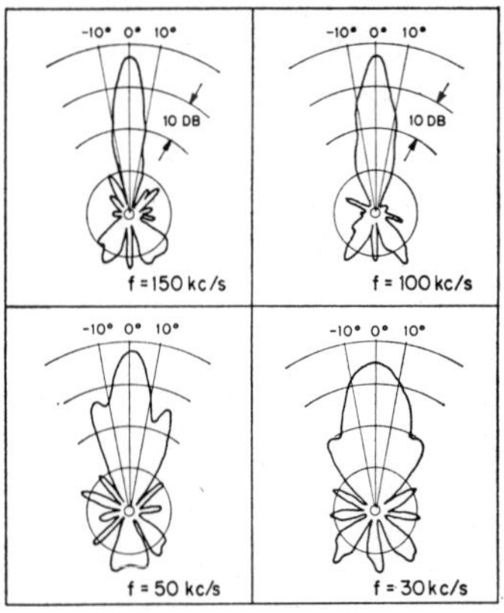

FIG. 1. The four polar patterns at 150, 100, 50, and 30 kc were recorded with a spherical liquid lens 6 in. in diam. The liquid enclosed in lens was flurolube, a fluorine compound. The measurements were recorded at NEL's Calibration Station at Sweetwater Lake.

[1] R. K. Luneberg, "Mathematical Theory of Optics," Brown University Lecture Notes (1944), pp. 208–213.

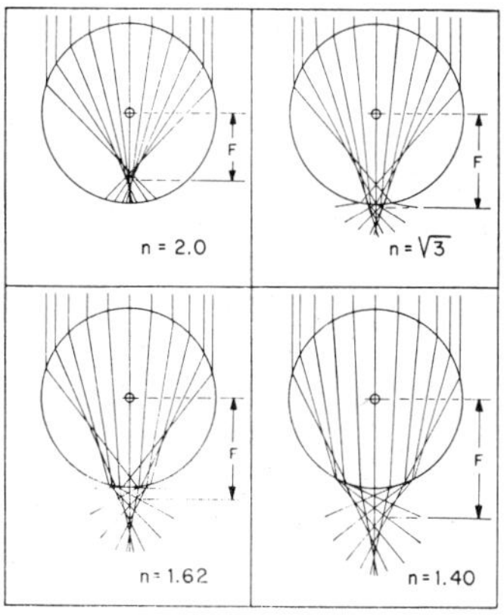

FIG. 2. The ray-theory focusing with a liquid spherical lens is shown for four different indices of refraction: $n=2.00$, 1.73, 1.62, and 1.40. The focal distance F is defined to be the position where the 30° and 60° rays converge.

weight in air, but this is quite unlikely when the array is used to project as well as to receive the sound.

The basic aim of this paper is to demonstrate that a simple spherical acoustic lens with a uniform index of refraction or a more sophisticated version thereof will provide, in conjunction with appropriate sensors, very attractive focusing and frequency-response characteristics. Luneberg's equations, compliant tubes, and widely spaced refractive scatterers represent potential variables that may influence greatly the ultimate weight and beam-forming characteristics of a spherical lens.

SIMPLE LIQUID LENS

Figure 1 illustrates four polar patterns that were recorded with a spherical acoustic lens. The latter was in the form of a thin spherical shell of aluminum, 6 in. in diameter, that was filled with the liquid fluorolube. The polar patterns are not unlike those that may be expected from a circular piston with a diameter comparable to that of the lens. Spherical aberration does not seem to be an appreciable factor, and the reason for this becomes apparent when the paths of the rays are traced in accordance with Snell's law.

The ray diagrams in Fig. 2 indicate that spherical aberration seems to increase as the index of refraction decreases. Actually such a conclusion is only partly correct in that the minimum circle of confusion is less when the focal position is inside rather than outside the lens; for focusing outside the lens, the aberration increases with the focal length, but its angular magnitude remains essentially constant so that the maximum resolving power at the onset of spherical aberration is independent of the index of refraction.

The precise focusing behavior of the liquid spherical lens may be considerably more complicated than is depicted by the ray tracings in Fig. 2. Although the angular spacing between the drawn rays within the lens is uniform as assumed, the acoustic energy content varies from ray to ray so that the relative concentration of rays near the focal region is not a true measure of the signal level. In addition, a considerable portion of the incident energy may be reflected[2] at the surface of the lens as a result of the difference in the speed of sound and density in the two media. Rho-c's close to that for water are to be preferred because reflection losses are at a minimum. An additional factor that influences the shape of the polar pattern is the size of the hydrophone or transducer at the focus. A detector whose dimensions are small compared to the wavelength of sound is required. If the dimensions are half-wave or more, then the width of the main lobe becomes progressively greater. For example, the polar pattern for 150 kc in Fig. 1 is considerably broader than it would have been because the dimensions of the detector corresponded to a 10° arc on the surface of the lens.

The significance of spherical aberration on the ultimate resolving power of the liquid lens can be demonstrated by considering a ray-theory study of the acoustic energy distribution and the relative path difference in the vicinity of a focal position. The results of such a study are shown in Figs. 3 and 4 for the specific case where the probe is on the outer surface of the lens. Although the energy concentration, in Fig. 3, increases with the index of refraction, the side-lobe level rises also so that the net improvement is very gradual. On the other hand, the difference in path length in Fig. 4 among the different rays suggests that an index of refraction near 1.8 ought to represent the optimum value

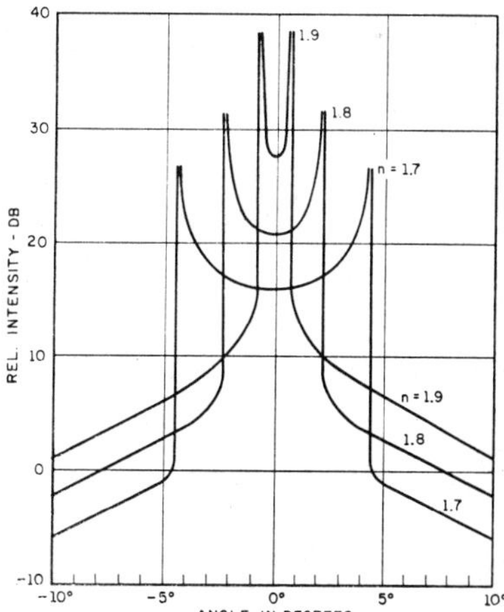

FIG. 3. The ray-theory intensity concentration at the outer surface of a simple spherical lens relative to the incident intensity is shown as a function of the angle measured from the center of the lens. The angular dispersion is a measure of the spherical aberration for three different indices of refraction: 1.7, 1.8, and 1.9.

[2] K. V. Mackenzie, J. Acoust. Soc. Am. **32**, 221 (1960).

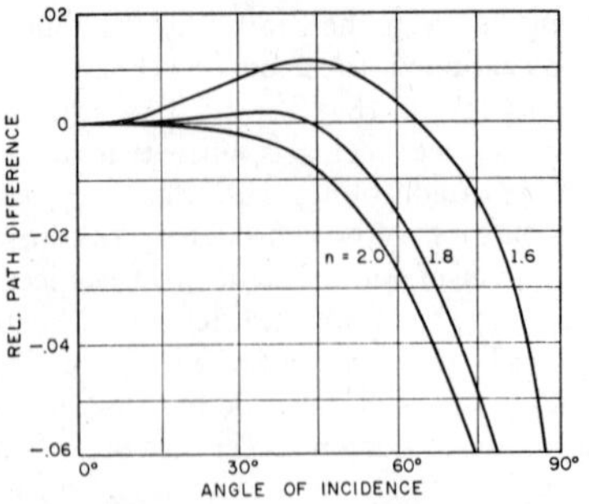

FIG. 4. The difference in path length relative to that of the center ray is shown for the rays with different angles of incidence on the lens when the rays terminate on the opposite surface of the lens with three different indices of refraction: 2.0, 1.8, and 1.6.

for a minimum amount of destructive interference between the different rays when the wavelength of the acoustic wave is comparable to the minimum circle of confusion.

VARIABLE INDEX OF REFRACTION LENSES

Ray-theory analysis indicates that focusing with a simple liquid lens suffers in terms of differences in phase as well as insufficient convergence among the rays. These difficulties may be resolved either by a continuously variable index of refraction or a combination of fixed and variable indices of refraction. Luneberg's equations define certain distributions in the index of refraction[1,3] for a spherical lens wherein all the rays focus at a point. These are characterized in general by an index of refraction which rises gradually from unity at the edge to a maximum value at the center. Of these distributions, the simplest and best known is

$$n^2 = 2 - r^2, \quad (1)$$

where n is the index of refraction at radius r when r is defined to have a value of unity at the outer surface of the lens. For a focal point well beyond the lens surface, the index of refraction may be stated quite accurately for most situations that do not require an extreme amount of resolving power as

$$n^2 \doteq n_c^2 - r^2(n_c^2 - 1) \quad (2)$$

wherein the appropriate value for n_c is that determined by Morgan[3] for the exact distribution of the index of refraction. Figure 5 shows the reliance of n_c on the focal length.

A somewhat different method for focusing all rays at a point involves a combination of a partially constant and partially variable index of refraction. Figure 6 illustrates two such arrangements.[3,4] The principal advantage of the combination lens in comparison with the Luneberg lens is in the reduction of constructional detail. This is possible because the index of refraction does not have to be continuously variable all the way to the outer surface.

Inasmuch as a continuously variable index of refraction is impractical with liquids and very difficult with solid structures, then a stepped index must be considered as a practical alternative. The simplest conceptual arrangement involves uniform-size scatterers which are distributed in a prescribed manner. Kock and Harvey[5] demonstrated that rigid obstacles in air effectively increase the density of the displaced air by 50%, and, therefore, the speed of sound decreases in the limit of packing in the mixture to 82% of that in the fluid. However, the width of the obstacles has to be considerably less than the wavelength of sound in the enveloping fluid if the apparent index of refraction is to be independent of frequency.

An alternative method of changing the speed of sound through a mixture occurs when the rigid obstacles or scatterers are replaced by objects or liquids that are more compressible than the enveloping fluid. In water this is a practical method because there are a few liquids whose compressibility is 2–3 times greater. On assuming that packets of a compressible liquid are small compared to the wavelength of sound, then for a uniform distribution of such packets in an enveloping liquid, the index of refraction n_m for the mixture is

$$n_m^2 = \{1 + p[(\rho c^2/\rho_x c_x^2) - 1]\}\{1 + p[(\rho_x/\rho) - 1]\}, \quad (3)$$

where ρ and c are the density and speed of sound for the enveloping liquid, respectively; ρ_x and c_x are the corresponding values for the compressible liquid in the packets; and p is the volume-packing factor or propor-

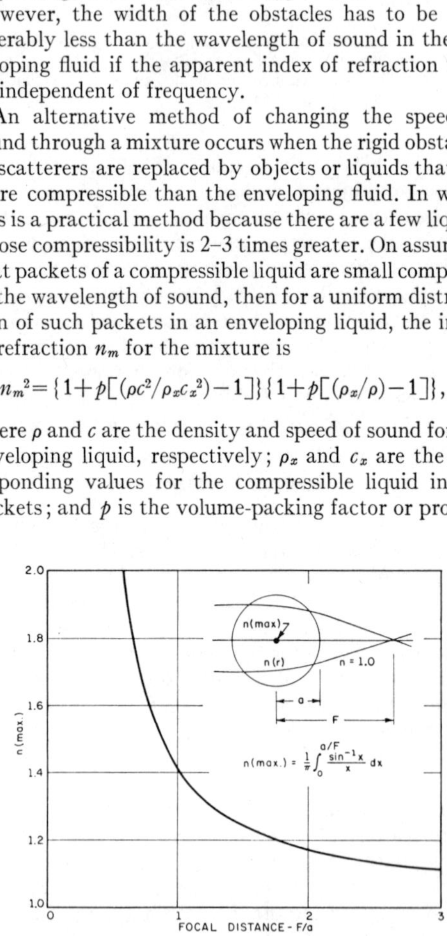

FIG. 5. The maximum index of refraction at the center of a Luneberg lens is shown as a function of the resultant focal distance when the index of refraction is distributed in accordance with Luneberg's equations.[3]

[3] J. P. Morgan, J. Appl. Phys. 29, 1358 (1958).
[4] R. B. Saunders, "Modification of the Luneberg Lens," Technical Rept. NR384–206, Contract Nonr 1286(01) (1955).
[5] W. E. Kock and F. K. Harvey, J. Acoust. Soc. Am. 21, 47 (1949).

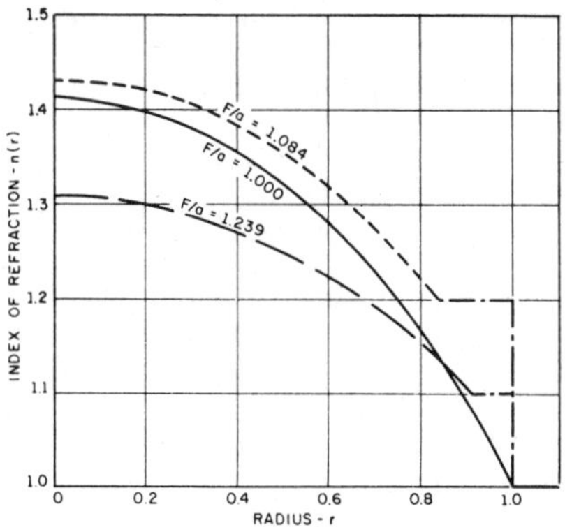

FIG. 6. The distribution in the index of refraction as a function of the radius is shown for two types of lenses: the original Luneberg solution with $F/a=1.0$, and the combination or hybrid type of distribution[4] with $F/a=1.084$ and 1.239.

tion of a unit volume of the mixture that is occupied by the liquid in the packets. Equation (3) suggests that the packet liquid should have a high density as well as a high compressibility if the index of refraction of the mixture is to be large and if the volume of packets is to be small. For a Luneberg lens with the focal point on its outer surface, the total volume occupied by the packets is approximately 40% when a liquid is selected with an index of refraction equal to $2^{1/2}$. For focal distances well beyond the lens boundary, the total packet volume should decrease further and in proportion to (n_c^2-1), where n_c is the index of refraction at the lens center.

COMPLIANT TUBE LENS

The difficulties that might be encountered in supporting a series of liquid packets within a Luneberg type of lens may be overcome to a great extent by a comparatively rigid structure of metallic compliant tubes. The acoustic properties of such tubes have already been described by Toulis,[6,7] and consequently only the lens design and resultant characteristics need to be discussed in detail. Although compliant tubes may be highly compressible, the acoustic compressibility of the mixture of a liquid and compliant tubes and the resultant index of refraction are not free from dispersion, and especially at frequencies near the resonance frequency for the pulsating mode of vibration. In general, the index of refraction n for the mixture[7] may be stated in the following form:

$$(n^2-1)=H_qC_0/C, \tag{4}$$

where C is the compressibility of the enveloping liquid, C_0 is the apparent compressibility of the mixture with compliant tubes at zero frequency, and H_q represents the dispersion factor, which for frequencies well below the dipole mode of vibration may be expressed simply as

$$H_q=(1-\omega^2/\omega_1^2)^{-1}. \tag{5}$$

In Eq. (5), ω is the angular frequency of the acoustic wave, while ω_1 corresponds to the resonance frequency of the compliant tubes in the pulsating mode of vibration.

Since the index of refraction might be appreciably dispersive if the compliant tubes were used near mechanical resonance, then any one of Luneberg's solutions, such as Eq. (1), can be satisfied at only one frequency. However, it follows from Eq. (2) that satisfactory focusing is attainable at other frequencies but with different focal lengths. The latter does not represent a serious limitation over a broad range of frequencies if the resolving power is not very great or if the operational frequencies are well below the resonance frequency of the compliant tubes; that is, defocusing is not likely to be serious because the effective focal volume has a minimal dimension of at least one-third the wavelength of the acoustic wave, and, furthermore, the focal volume becomes greater at the lower freqquencies. Thus the operational focal length of the lens should be designed to satisfy Luneberg's equation at a frequency near the upper end of the band if it is to be in a suitable position to respond to all the frequencies in a band.

From the design point of view, it is convenient to state the index of refraction in terms of the center-to-center spacing s between adjacent compliant tubes:

$$n^2=1+(s_e/s)^2, \tag{6}$$

where s_e is that spacing for a uniformly spaced volume array which contributes a compressibility equal to that for the enveloping fluid at any frequency of interest. On equating Eqs. (2) and (6), the relative spacing s between the compliant tubes at any radius within the lens is established:

$$s/s_e \doteq \left[(n_c^2-1)(1-r^2)\right]^{-1/2} \tag{7}$$

However, this equation assumes that a continuous distribution of compliant tubes exists, whereas a stepwise distribution is necessary because of the much greater compressibility of compliant tubes in comparison with an equal volume of water, for example.

In order to satisfy one of Luneberg's equations, the simplest distribution is a series of concentric spherical shells, wherein the compliant tubes are in the form of hoops with a uniform spacing which is characteristic of that shell. On assuming that Eq. (7) is used to establish approximately, if not exactly, the radial position of concentric shells, then for the nth shell the total tube length L_n can be calculated by assuming that each compliant tube influences a cross-sectional area equal to s^2; that is, in a shell with an infinitely small thickness dr, the length of tubing

$$dL_n=4\pi(r^2/s^2)dr. \tag{8}$$

[6] W. J. Toulis, J. Acoust. Soc. Am. **29**, 1021 (1957).
[7] W. J. Toulis, J. Acoust. Soc. Am. **29**, 1027 (1957).

With the aid of Eq. (7) for s^2, then

$$L_n = \frac{4\pi}{s_e^2} \int_{r_n-1/2}^{r_n+1/2} (n_c^2 - 1)(1 - r^2) r^2 \, dr \tag{9}$$

wherein the limits of integration correspond to the inner and outer radial boundaries of the nth shell:

$$r_{n+\frac{1}{2}} = (r_n + r_{n+1})/2$$

and

$$r_{n-\frac{1}{2}} = (r_n + r_{n-1})/2. \tag{10}$$

Inasmuch as the thickness of the shell is $s_n = r_{n+\frac{1}{2}} - r_{n-\frac{1}{2}}$ and the spacing between adjacent hoops of compliant tubes is also s_n, the effective volume of the nth shell is

$$V_n = s_n^2 L_n = 4\pi r^2 s_n.$$

Consequently, the resultant spacing for the nth shell may be expressed as

$$s_n = 4\pi r_n^2 / L_n. \tag{11}$$

For the specific case where $n_c^2 = 2$, and the spacing at the center $s_c = s_e$, Eq. (9) yields

$$L_n = 4\pi / s_c^2 \times [(r_{n+\frac{1}{2}}^3)(1 - \tfrac{3}{5} r^2_{n+\frac{1}{2}}) - (r^3_{n-\frac{1}{2}})(1 - \tfrac{3}{5} r^2_{n-\frac{1}{2}})], \tag{12}$$

and the exact-spacing calculations can be determined with the aid of Eq. (11). Actually, the above expression for L_n may be greatly simplified so that the spacing for all but the outermost shell may be expressed simply as

$$s_n / s_c \doteq (s_c / t_n)(1 - r_n^2)^{-1}, \tag{13}$$

where t_n is the thickness of the nth spherical shell if it is different from s_n.

In practice, a series of concentric spherical shells of compliant tubes might be tedious to fabricate, as different diameters are required for all the compliant tube hoops. On the other hand, a series of concentric cylindrical shells allows many hoops to have a given diameter, but the calculations for the exact spacings between compliant tubes become very tedious. The latter problem may be overcome by determining the tube positions in terms of the spherical shell theory and then transposing the tubes to the nearest cylindrical shell. Thus the total length of tubing should average out to be about the same for both the spherical and cylindrical distributions in a lens.

It would appear that the varied spacings between the tubes, close together at the center of the lens and far apart near the outer surface, would cause the resonance frequency and compressibility of the compliant tubes to shift as a result of mutual coupling. Although such behavior was observed with plane arrays,[6] it is not expected to be a significant factor in volume arrays because the phase of the mutual interaction differs randomly from tube to tube. Thus the frequency-response characteristics of a single compliant tube should be the same whether it is measured in free space or in a uniform volume array of compliant tubes, at least for tube-to-tube spacings which are considerably less than the wavelength of sound in the mixture.

Figure 7 shows the distribution of compliant tubes in a Luneberg lens, 6 ft in diameter, that was fabricated and then employed to record the polar patterns in Fig. 8. All the patterns were recorded with a hydrophone in conjunction with a compliant-tube reflector behind the hydrophone—both of which were fixed relative to the lens for the measurements in Fig. 8. The corresponding directive gain is shown in Fig. 9 and suggests that the effective aperture of the Luneberg lens is practically indistinguishable from that of a circular piston of the same diameter in a plane rigid baffle of infinite extent.

PRIMARY FEEDS AND SIDE LOBES

Inasmuch as a lens is a passive device whose main purpose is to gather and concentrate rather than amplify a signal, a primary feed or probe must be employed at the focus to project or detect a signal. The ideal feed is one that directs all of the acoustic energy toward the lens and radiates none in other directions. Moreover, the illumination or insonification of the lens by the feed should be tapered so that it is a maximum through the center of the lens and decreases gradually to zero at the edges of the lens for side lobes comparable to those from a circular piston in a plane rigid baffle. Thus for a focal point on the outer surface of the lens, a small omnidirectional transducer with a pressure-release plane re-

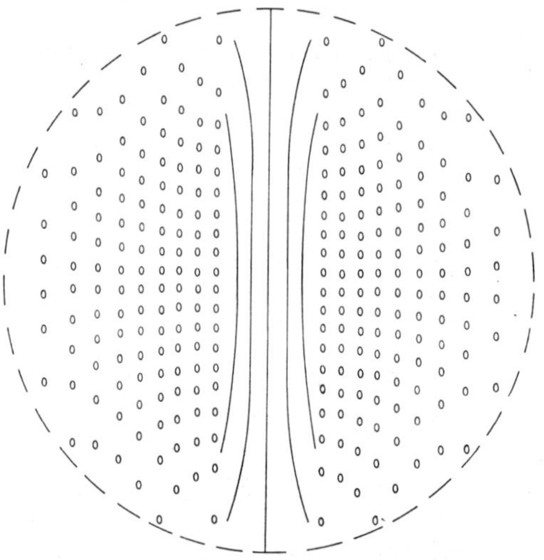

FIG. 7. The drawing depicts the location of compliant tubes at a plane through the center of the 6-ft lens. The distribution of the tubes is consistent with Luneberg's equation in all radial directions. The central core consists of vertical tubes. All other tubes are arranged in circular hoops of concentric cylindrical shells.

flector behind it ought to be an efficient and satisfactory primary feed. Without the pressure-release plane reflector behind it, the transducer–lens combination would have a directive gain that might be as much as 7.8 dB down as a result of inefficient illumination and concentration of the acoustic signal on the lens. This loss in directive gain, which appears as an increase in the side-lobe level, can be reduced to only 3 dB if the omnidirectional transducer is replaced with a dipole type of transducer that has a figure-of-eight polar pattern. A rigid reflector is better than an omnidirectional transducer, but it may be down by as much as 4.8 dB when compared to a pressure–release reflector.

Side lobes may be reduced beyond those to be expected from a circular piston by providing additional tapered illumination from adjacent beam-forming feeds. Since the first side lobe might be only 15–18 dB down

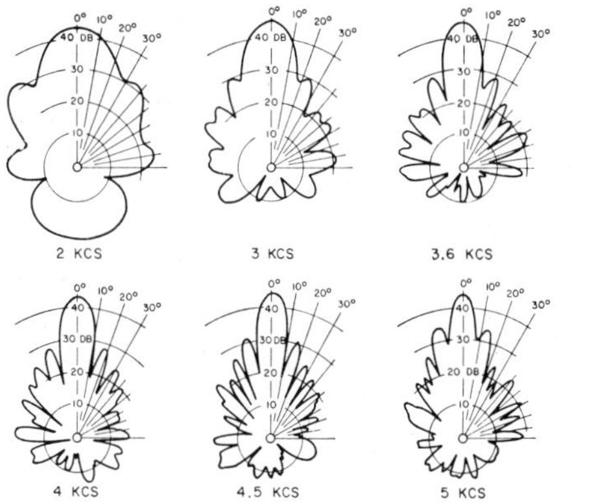

FIG. 8. The six polar patterns at 2, 3, 3.6, 4, 4.5, and 5 kc were recorded with a complaint tube lens 6 ft in diam and with the prime feed fixed at a single distance near the lens' outer surface. The prime feed consisted of a hydrophone and a compliant tube reflector whose reflectivity was poor at 2 kc and lower frequencies. The measurements were recorded at NEL's Calibration Station at Sweetwater Lake.

relative to the main lobe, then a network should be provided that permits signal levels of this order of magnitude to be exchanged between adjacent feeds. An alternative arrangement that may assist in the reduction of side-lobe level requires that the focal point be well inside the lens in order to enhance the illumination taper across the lens aperture. A short focal length requires a large index of refraction for the lens design, and, therefore, the amount of refractive material might be unduly excessive.

For a focal point well beyond the outer surface of the lens, the simple dipole or simple plane-reflector arrangement is not capable of concentrating all of the projected energy on the lens, inasmuch as the capture angle presented by the lens will be less than 180° by a factor inversely proportional to the focal length in terms of the outer radius of the lens. A sufficiently large horn or an

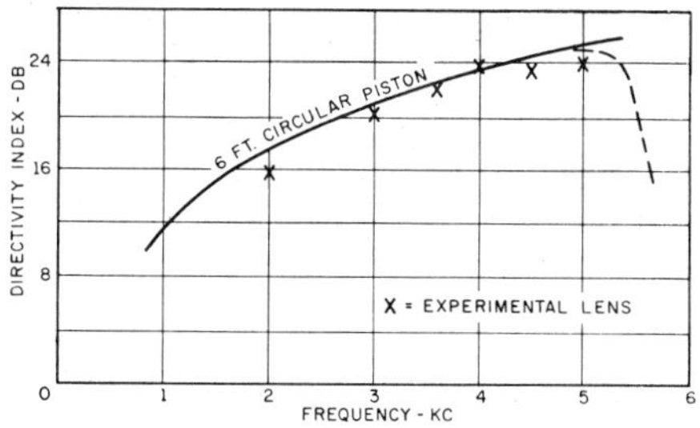

FIG. 9. The directivity index for a compliant tube lens and circular piston, both 6 ft in diam, are compared as a function of frequency. The values for the lens are calculated from the polar patterns in Fig. 7. The directivity index decreases rapidly above 5 kc because of resonance in the compliant tubes at 8.5 kc, and because the prime feed was fixed and not refocused at each frequency.

array of transducers can be designed to provide the appropriate angular coverage; the resultant polar patterns and directive gains should then be comparable to those from a baffled circular piston with a diameter equal to that of the lens itself.

FOCAL LENGTHS AND NARROW BEAMS

The ray theory of focusing with a Luneberg type of lens indicates that the focus is a point, and, therefore, extrapolation in terms of wave theory suggests that the focal position should correspond in the limit to the volumetric dimensions of a quantum for a specific spectral line; that is, the effective diameter of a quantum of wave energy is roughly one-half of the spectral-line wavelength. Specifically, this condition is fulfilled by the polar patterns shown in Fig. 8 for a Luneberg lens that was designed in accordance with Eq. (1), which is the simplest of the family with the focus on the outer surface of the lens.

It would appear from the above analysis that a focal length well beyond the outer lens surface should result in increasingly narrower beams if the focal volume corresponds always to that of a quantum. Actually, the effective volume may be considerably greater. Diffraction theory indicates that the angular beamwidth is independent of focal length, and directive gain may deteriorate rapidly unless a primary feed with a sufficiently large aperture is employed to coincide with the effective dimensions of the focal region. Since the directive gain and the resolving power of a lens with a long focus are, at best, the equal of a spherical lens with the focus on the lens' outer surface, and since such a lens requires an extended type of primary feed, then there is no justification for its use except possibly to minimize the amount of refractive matter; that is, the volume of the latter decreases inversely with the focal length at long focal positions.

An alternative and more effective method for conserving the amount of refractive medium in a lens involves a Luneberg lens with most of the inner core of the lens left inactive; that is, in accordance with Eqs. (1) and (2), the index of refraction approaches unity at the outer shell, and, therefore, a minimum of refractive material may be required in the shell for a given directive gain. Consider, for example, the design of an outer shell lens in which the distribution of the index of refraction is specified by Eq. (1). The length of compliant tubing in the outer shell may be calculated with the aid of Eq. (12):

$$L=\tfrac{2}{5}(V/s_c^2)[1-(r^3/2)(5-3r^2)], \qquad (14)$$

where s_c is the center-to-center spacing between compliant tubes at the center of the lens, $V(=4\pi a^3/3)$ is the physical volume of the lens sphere, a is the outer radius of the sphere, and r is the inner radius (in proportion to a) of the outer shell. The proportion of the tubing in a thin outer shell in comparison with that for a full lens is

$$\frac{L\text{ (in shell)}}{L\text{ (in full lens)}} \doteq (15/2)(1-r^2). \qquad (15)$$

In contrast, the radiation-capture area of the thin shell relative to that of the lens may be expressed as

$$\frac{A\text{ (shell)}}{A\text{ (full lens)}} \doteq 2(1-r). \qquad (16)$$

Thus the length of tubing decreases as the square of the thickness, $(1-r)$, of the shell, whereas the effective capture area of the shell decreases only with the thickness. Consequently, a shell-type lens may yield appreciably more directive gain than a full, but physically smaller, lens with the same amount of compliant tubing.

The potential reduction in refractive material or the increase in directive gain for a given amount of refractive material in a shell type of lens is counterbalanced, in part, by the requirement of a pie-shaped polar pattern for the primary feed and, in part, by the prominence of side lobes near the main lobe. The pie-shaped pattern is readily attainable by a linear array of transducers whose total length is inversely proportional to the thickness of the lens shell. Thus the number of hydrophones or transducers may not be small if the lens is to be employed at its maximum focusing capacity. Although the first side lobe might be only 10 dB down from the main lobe when the shell is very thin, the successive side lobes decrease rapidly. Specifically and ideally the polar pattern of a thin-shell lens is similar to that of a ring transducer which is describable in terms of a Bessel function of the zeroth order.

Perhaps the most interesting feature of the shell-type lens is that it can provide sharp main lobes which are even sharper than might be expected from a full lens of the same diameter. The directive gain may not be as high, but the amount of refractive material may be much less. Side lobes also may be reduced to a considerable extent with appropriate amounts of coupling from adjacent primary feeds.

ACKNOWLEDGMENTS

The experimental and theoretical aspects of this paper reflect a portion of the author's research effort while he was employed at the U. S. Navy Electronics Laboratory, San Diego (1951–1959) and also at Convair, a division of General Dynamics, San Diego (1959–61).

In addition to studying the propagation of displacement waves in the sea, the following authors discuss the theory of displacement reception and the design and construction of sensitive displacement hydrophones of both the variable capacitance and variable reluctance types.

Professor Lieberman has conducted research in ultrasonics, underwater sound, hydrodynamics, properties of liquids, electromagnetic propagation, and solid state. He is Professor of Physics at the University of California, La Jolla, California.

Dr. Robert A. Rasmussen has conducted research in underwater and physical acoustics. He is a member of the staff of the Marine Physics Laboratory of the Scripps Institution of Oceanography, San Diego, California.

This paper is reprinted with the permission of the senior author and the Acoustical Society of America.

THE JOURNAL OF THE ACOUSTICAL SOCIETY OF AMERICA VOLUME 36, NUMBER 5 MAY 1964

33

Propagation of Displacement Waves in the Sea

L. N. LIEBERMANN AND R. A. RASMUSSEN

University of California, San Diego, La Jolla, California 92037

(Received 18 December 1963)

In the past, underwater acoustic data have been obtained with pressure-sensitive hydrophones, which neglect the particle displacement in an acoustic wave. Displacement reception possesses certain advantages, among which are the possibilities of direction finding and also of acoustic observations at the sea surface. The theory of operation of a displacement hydrophone is discussed and the technique is given for constructing instruments sensitive to a displacement of 0.1 Å at 1000 cps. The propagation of displacement observed at sea is compared with theoretical predictions. The vertical component of displacement is found to be maximum on the sea surface as predicted; the horizontal component vanishes at the sea surface, but at greater depths can be used to observe the wave-propagation direction. At short ranges, the acoustic-displacement field is closely approximated by a dipole source; at longer ranges, the displacement field diminishes less rapidly ($1/r$ dependence) than is predicted by the dipole theory.

INTRODUCTION

HYDROPHONES utilized for conventional undersea acoustic-propagation studies measure acoustic pressure rather than particle displacement. The analytical relationship between these two quantities, although familiar, is by no means trivial: At the sea surface, the underwater acoustic pressure is essentially zero, whereas the vertical component of the acoustic displacement is a maximum; a conventional (pressure-response) hydrophone yields no signal near the sea surface although a strong sound field may be present. In addition, pressure is a scalar quantity that ignores the directionality information inherent in the vector properties of a propagating displacement wave.

Acoustic-displacement receivers are common in air (e.g., the ribbon microphone), but hitherto have had little use under water. The reason is immediately apparent from the farfield relationship between pressure p and particle velocity $\dot{\xi}$, given by $p=\rho c\dot{\xi}$. For a given acoustic pressure, the particle velocity in air is nearly 1000 times greater than in water because of the density difference. A practical detectable sound pressure at sea is 1 dyn/cm², corresponding to a displacement of 0.1 Å (at 1000 cps), or far less than a molecular dimension!

In spite of these limitations, undersea displacement detectors of high sensitivity have been constructed and utilized to explore the properties of acoustic-displacement propagation at or near the sea surface and at depth. In the ensuing sections, the elementary theory of propagation of displacement is given, construction principles of displacement hydrophones are discussed, and, finally, the theory is compared with actual observations at sea.

I. THEORY OF DISPLACEMENT RECEPTION

The exact theory of underwater sound propagation has been treated by a number of authors for the idealized case of the plane sea surface and sea bottom.[1] The extension of this theory to displacement reception is straightforward. However, the theory of a rough sea surface and rough bottom with variable reflectivity is difficult and thus far has not been given. In spite of this simplification, it is useful to begin the discussion with the idealized plane sea surface and bottom.

A perfectly plane sea surface will be a nearly perfect specular reflector for acoustic waves. In this ideal case, the acoustic field is readily calculated by replacing the reflecting surface by an identical image source located the same distance above the surface, as shown in Fig. 1. For a sinusoidal wave of frequency $f=\omega/2\pi$, the velocity potential is

$$\Phi=(A/r_1)\exp i(\omega t-kr_1)-(A/r_2)\exp i(\omega t-kr_2). \quad (1)$$

It is seen from the geometry that if d and h are the

[1] For example, see L. M. Brekhovskikh, *Waves in Layered Media* (Academic Press Inc., New York, 1960).

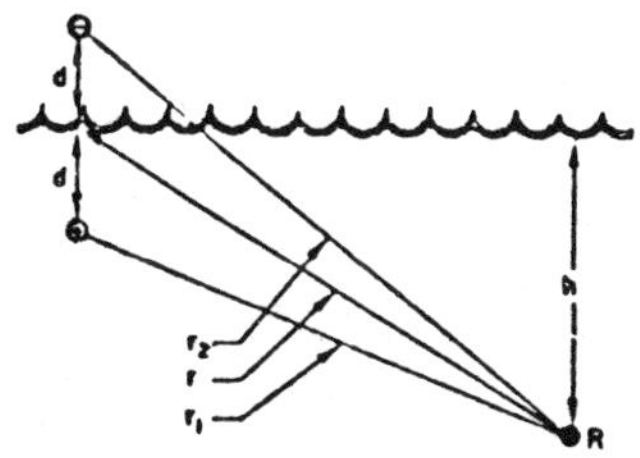

FIG. 1. Image-source model of smooth surface.

source and receiver depths then $r_2-r_1 \simeq 2hd/r$. Hence, for long distances ($r \gg d$), Eq. (1) becomes

$$\Phi=(A/r)[(\exp ihdk/r)-(\exp -ihdk/r)][\exp i(\omega t-kr)].$$

The vertically received displacement is given by

$$\xi_v=-(1/\omega)(\partial\xi/\partial h)=-(A_0dk/\omega r^2)\cos(hdk/r), \quad (2)$$

where the time dependence has been omitted.

Note that as the depth of the receiver decreases ($h\rightarrow 0$) vertical displacement reception improves, reaching a maximum on the surface. This is not true for the sonrce depth; on the average, increasing the source depth increases reception. It is also seen that attenuation in the vicinity of the sea surface is higher than for an unbounded medium, increasing as the square of the range. Square-law attenuation is also experienced by acoustic pressure at longer ranges ($r \gg 2hdk/\pi$) and is commonly referred to as "Lloyd-mirror attenuation."

The horizontal component of displacement[2] is of interest because it permits acoustic direction location with a point receiver, as is discussed more fully below. Horizontal displacement is given by

$$\xi_h=\frac{1}{\omega}\frac{\partial\Phi}{\partial r}=\frac{A_0k}{\omega}\left[\frac{i}{r}\sin\frac{hdk}{r}-\frac{1}{r^3}\cos\frac{hdk}{r}+\frac{1}{r^2}\sin\frac{hdk}{r}\right]. \quad (3)$$

The first term is the only term of importance in the far field (long range), inasmuch as the other terms exhibit r^{-2} and r^{-3} dependence. Note that at long ranges ($\sin hdk/r\rightarrow hdk/r$) the predicted amplitude dependence of horizontal displacement is inverse square with range, just as with vertical displacement.

Horizontal displacement vanishes at the sea surface, as seen from Eq. (3) with $h=0$. The maximum horizontal displacement will be received when the receiver is located at the depth $h=nr\lambda/4d$, where n is an integer. For example, when the range is 1000 m and $f=1000$ cps, the first reception maximum occurs at 40 m depth when the source is located at 10 m. However, some horizontal reception is predicted at shallower depths and, indeed, horizontal reception in the actual ocean was observed even at 10 cm.

A. Effect of Bottom Reflection

An idealized ocean is now considered in which the bottom as well as the surface are plane reflectors. The bottom is considered to be "hard," producing perfect specular reflection but without phase reversal as at the surface.

Consider, initially, a single reflection from the bottom: the source and its surface image, constituting a dipole, are reflected in the bottom, as shown in Fig. 2(a). Note that reception from the bottom images occurs as if the receiver were located at a new depth h'. The magnitude of this virtual receiver depth will be approximately twice the bottom depth. At first glance, this relatively large value might be expected to affect the received vertical and horizontal displacements as given by Eqs. (2) and (3). However, Eq. (3), giving the horizontal displacement, no longer vanishes on the surface, thus violating the surface boundary condition. This difficulty is removed at the surface by inserting still another image dipole above the surface. But this system of images will be asymmetrical about the bottom, violating the boundary condition there. The final solution is found by constructing an array of images as shown in Fig. 2(b). Each additional source is farther away from the boundary, diminishing its contribution to the field and yielding a finite limiting value, which satisfies the boundary condition. The velocity potential is then of the form

$$\Phi=\sum_{l=0}^{\infty}(-1)^l\frac{\exp ikr_l}{r_l},$$

which satisfies the wave equation. This field can also be represented as a set of normal modes, each of which propagates along the layer with its own velocity; the latter representation is more usual in dealing with waveguides and, of course, gives equivalent results.

II. DISPLACEMENT HYDROPHONE

Although modern hydrophones are of the pressure-response type, the displacement hydrophone has not been completely overlooked in the past. The possibilities of utilizing displacement hydrophones were explored during World War I,[3] but applications were not

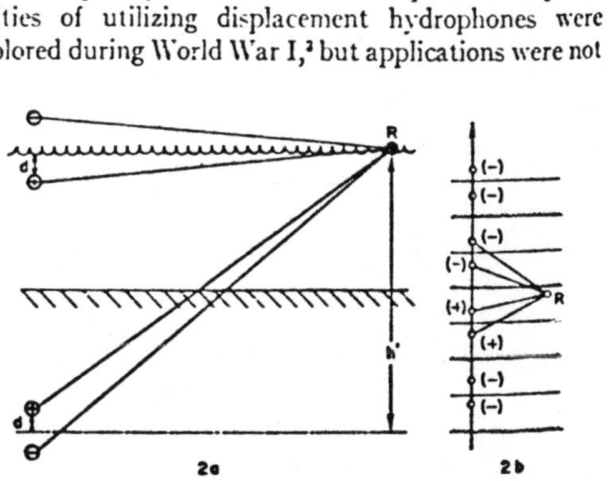

FIG. 2. (a) Image-source model of smooth surface and bottom. (b) Dipole array of smooth surface and bottom.

[2] Strictly speaking, the radial component is calculated, but this effectively horizontal at reasonable ranges.

[3] A. B. Wood, *Sound* (G. Bell & Sons Ltd., London, 1957), pp. 346, 457.

extensive, undoubtedly owing to the lack of sufficient sensitivity with then available techniques.

All techniques for measuring acoustic displacement can be reduced to either of two categories: (1) the outer case of the detector partakes of the acoustical-particle displacement, with the electromechanical transduction wholly internal; or (2) the motion of the case is nearly zero, with displacement (relative to the housing) being imparted to a flexible diaphragm or ribbon. The familiar accelerometer is an example of the first category; the ribbon microphone is of the second. In denser media (e.g., water), it is difficult to suppress motion of the housing, making transducers in the second category less attractive.

Before discussing implementation of these principles, it is necessary to examine the mechanism by which an acoustic wave sets the hydrophone into motion.

Hydrophone Motion Induced by Sound

An object upon which a sound wave is impinging experiences a vibrational motion. This motion arises because of the acoustic "shadow" produced on the far side of the object; the unequal sound pressure on both sides results in a net driving force. The phenomenon persists even for objects far smaller than a wavelength, as is shown in the following discussion.

For simplicity, consider a plane sound wave of the form $\xi=\xi_0 \exp i(\omega t-kr)$, where ξ is the particle velocity. A spherical hydrophone will perturb the plane wave by producing a scattered wave; the resultant field is the sum of the scattered and incident waves. The velocity potential of the scattered waves (at long ranges) is given by[4]

$$\Phi'=\frac{-\xi_0 ka^3}{3r}\left\{\frac{\beta_1-\beta}{\beta_1}+\frac{3(\rho_1-\rho)}{\rho+2\rho_1}\cos\theta\right\}\exp i(\omega t-kr), \quad (4)$$

where the hydrophone has a small radius a (and $a\ll\lambda$) and acoustic properties specified by compressibility β_1 and density ρ_1; β and ρ are properties of the medium (seawater). Only the second term giving an angular dependence of scattering is of interest here, inasmuch as it leads to asymmetry in the scattered wave, forming the "shadow." It is to be noted that a "shadow" arises from a discontinuity in density but not from a discontinuity in compressibility. Thus, the compressibility of the displacement hydrophone has no effect on its receiving sensitivity.

It is clear from the above discussion that *asymmetrical* scattering and vibrational motion of the hydrophone are merely different manifestations of a single phenomenon. Hence, the acoustic velocity potential of a vibrating sphere should be identical with the asymmetrical part of the scattering given in Eq. (4). The velocity potential (far field) of a vibrating sphere is[5]

$$\Phi'=-[(ka^3)/(2r)](\xi_1-\xi_0)\cos\theta \exp i(\omega t-kr), \quad (5)$$

where ξ_1 refers to the hydrophone motion; thus, $(\xi_1-\xi_0)$ is the velocity of the sphere *relative* to the surrounding water. Equating Eq. (4) and Eq. (5) yields

$$\xi_1/\xi_0=3\rho/(2\rho_1+\rho). \quad (6)$$

Inasmuch as the density of seawater is nearly unity, $\xi_1/\xi_0 \simeq 3/(2\rho_1+1)$.

It is seen that the hydrophone displacement or velocity can be increased by diminishing its average density; the maximum gain is the factor 3 compared to the undisturbed particle displacement. In the present work, the hydrophone density was approximately unity, with no special effort directed toward weight reduction.

Principles of Hydrophone Design

Any one of several electromechanical transducers, e.g., moving coil, variable reluctance, variable capacity, etc., may be employed as a displacement or velocity sensor. Mechanical elements comprising the transducer assembly and the corresponding equivalent electrical circuit are sketched in Fig. 3, wherein M is the moving mass of the hydrophone housing, m is the mass of the internal inertial element, K is the stiffness of the spring coupling the two masses, and R is the resistance offered to the motion of the inertial mass relative to the housing. The quantities ξ_1 and x denote the small displacements of the masses with respect to a fixed reference frame. If, for example, a variable reluctance transducer is utilized, m is the mass of the armature, the spring stiffness is that of the armature suspension, and R includes the viscous drag of the fluid in the gap separating the armature and the pole pieces.

In the variable-reluctance device, the differential velocity v is of interest, since the open-circuit voltage appearing across the coil terminals will be proportional to this quantity. In designing a variable capacity transducer, the differential displacement $s=(\xi_1-x)$ is of primary interest.

Analysis of the equivalent circuit based on the assumption of a constant, sinusoidal housing displace-

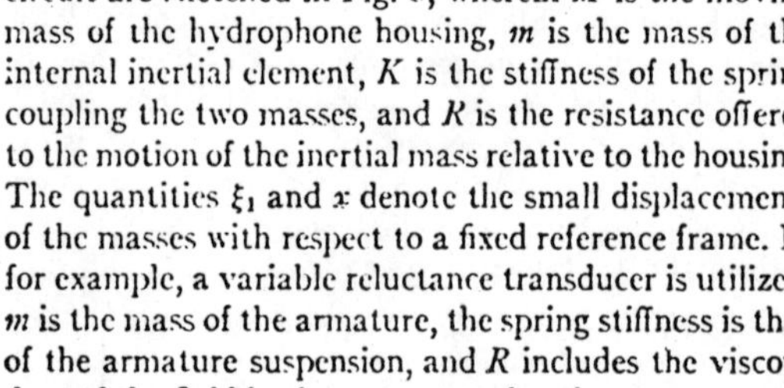

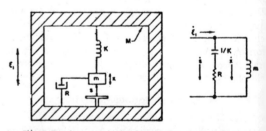

FIG. 3. Displacement hydrophone and equivalent circuit.

[4] Lord Rayleigh, *Theory of Sound* (Dover Publications, Inc., New York, 1945), Sec. 335.

[5] Reference 4, Sec. 325.

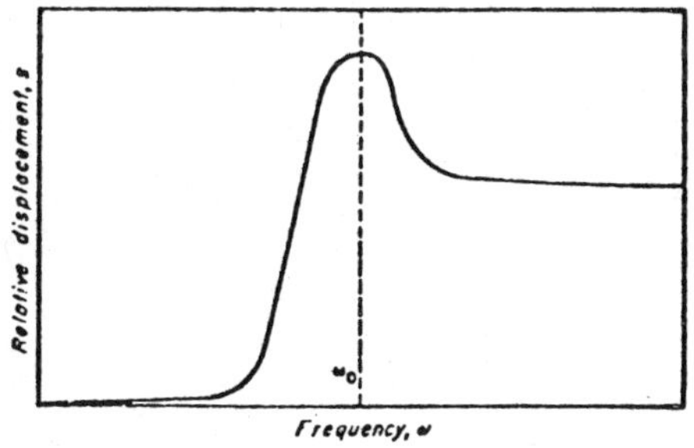

FIG. 4. Frequency response of displacement hydrophone.

ment yields the result

$$s=\frac{v}{j\omega}=[\xi_1\cdot j\omega m]\Big/\left[R+j\left(\omega m-\frac{K}{\omega}\right)\right]. \quad (7)$$

The relative displacement s as a function of frequency has the approximate appearance shown in Fig. 4. It will be noted that the ratios v/ξ_1 and s/ξ_1 are identical; hence, selection of the particular type of transducer to be utilized can be based on considerations other than mechanical gain, e.g., ruggedness and data-transmission-system requirements.

Mechanical methods for improving the gain and frequency-response characteristics of the different transducers can be applied with varying degrees of complication. As an example, a simple method for improving the gain is shown in Fig. 5. Accurate alignment of the transducer within its housing is essential to good directional characteristics; in this respect, the variable-capacity transducer appears to be the more desirable.

Two transducer types were employed in the present investigations. The basic construction of the variable-reluctance unit used to study displacement at the sea surface is shown in Fig. 6(a), and details of the variable-capacity unit used in more-general studies may be seen in Fig. 6(b). While the variable-reluctance unit proved to be more sensitive, permitting detection of particle displacements as small as 0.1 Å, the variable-capacity transducer was more simply instrumented for use in a free-floating, telemetered instrumentation system. The photograph in Fig. 7 shows the variable-capacitance transducer assembly. The variable capacitor consists of the inertial cube attached by means of a leaf spring to the case, and either of the two circular disks, the choice depending upon the displacement component (horizontal or vertical) to be measured. The capacitance variation is used to modulate the frequency of a miniature, radiofrequency oscillator, which may also be seen in the photograph.

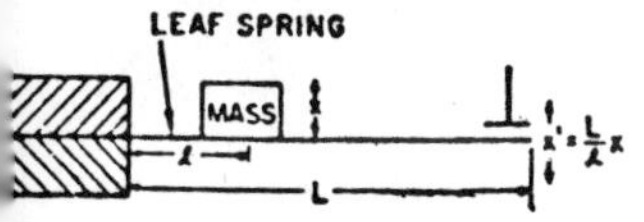

FIG. 5. Mechanical multiplication of displacement.

An expression for the relative sensitivity of this transducer, i.e., its frequency deviation as a function of acoustic displacement, can be obtained as follows. For small displacements ξ_1, the frequency deviation Δf is proportional to the changes in capacity or to changes in plate·separation L. Hence, $\Delta f/f=\Delta C/C=s/L$. Substituting from Eqs. (6) and (7) yields

$$\frac{\Delta f}{f}\propto\frac{\rho\xi_0}{L(2\rho_1+\rho)}\cdot\frac{j\omega m}{R+j[\omega m-(K/\omega)]}. \quad (8)$$

The sensitivity of the variable-capacity transducer may be defined as the fractional frequency change per unit particle displacement ξ. From measurements involving direct microscopic observation of the hydrophone displacement, as well as those based upon conversion of sound pressure to particle displacement, the sensitivity was determined to be $1.6\pm0.3\cdot10^{-4}$ Å^{-1} at

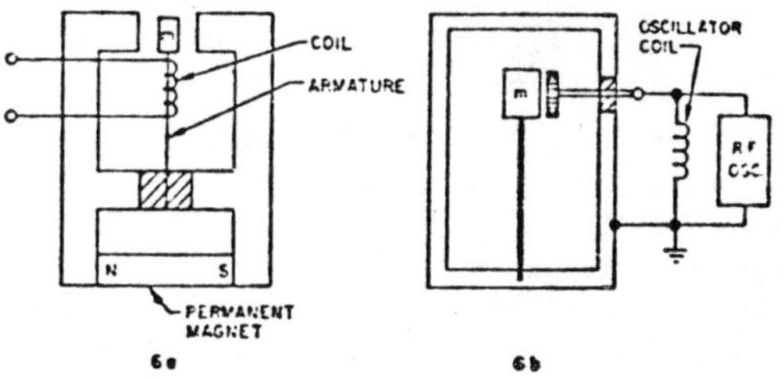

FIG. 6. Variable-reluctance and variable-capacity displacement hydrophones.

the resonant frequency of 880 cps. System noise in the capacitative device limited detection to particle displacements greater than 0.6 Å or, equivalently, to pressure levels exceeding 14 dB (*re* 1 μbar) at the resonant frequency. The limiting sensitivity in the reluctance device corresponds to a pressure of approximately 0 dB (*re* 1 μbar) at frequencies well above resonance.

A horizontal directivity pattern for the variable-capacity hydrophone measured in the field under optimum conditions, i.e., still water, controlled rotation, etc., is shown in Fig. 8.

III. OBSERVATIONS

A. General Procedure

A photograph showing the telemetering float and hydrophone appears in Fig. 9. The frequency-modulated radio signal from the hydrophone is coupled to the transmitter on the sea surface via a miniature coaxial cable. Vibrational isolation of the transducer from the float and the cable is provided by a chain of rubber bands shunting a short length of the cable. A similar

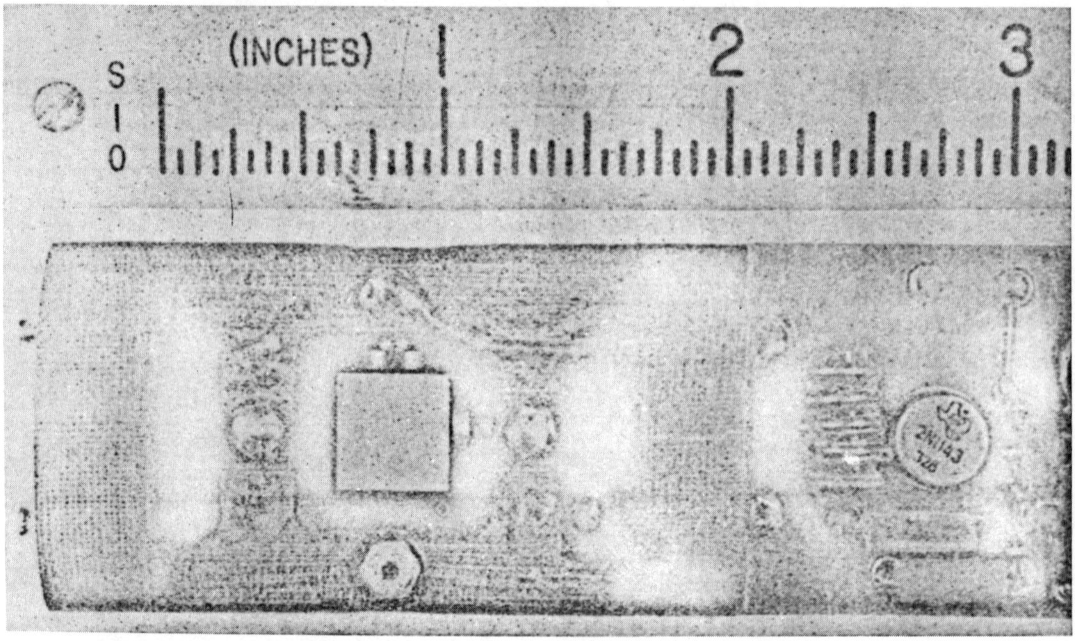

FIG. 7. An experimental, variable-capacity displacement hydrophone.

compliance decouples the vertical-aligning weight suspended from the base of the hydrophone.

The float consists of an inflated rubber torus mounted to a Plexiglas® rim. A circular aluminum plate attached to the upper edge of this rim serves as an antenna ground plane and as a mounting platform for the transmitter, the air propellers, and batteries.

Slow rotation of the float by means of air propellers permitted continuous monitoring of the vertical or horizontal hydrophone alignment. In monitoring the horizontal component of displacement, two maxima and two minima are observed during each revolution of the float, differences between the extreme displacement amplitudes agreeing with those predicted by the directivity pattern; any disagreement indicates misalignment of the transducer. On the other hand, observations of the vertical component of displacement should reveal no fluctuations correlated with transducer rotation.

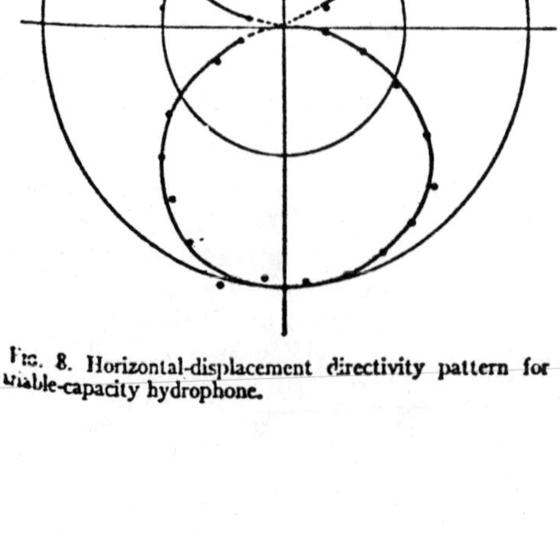

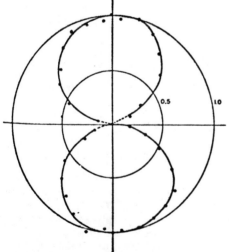

FIG. 8. Horizontal-displacement directivity pattern for the variable-capacity hydrophone.

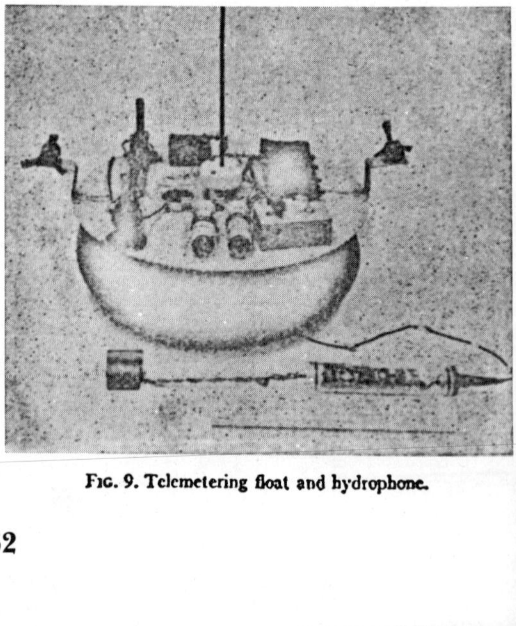

FIG. 9. Telemetering float and hydrophone.

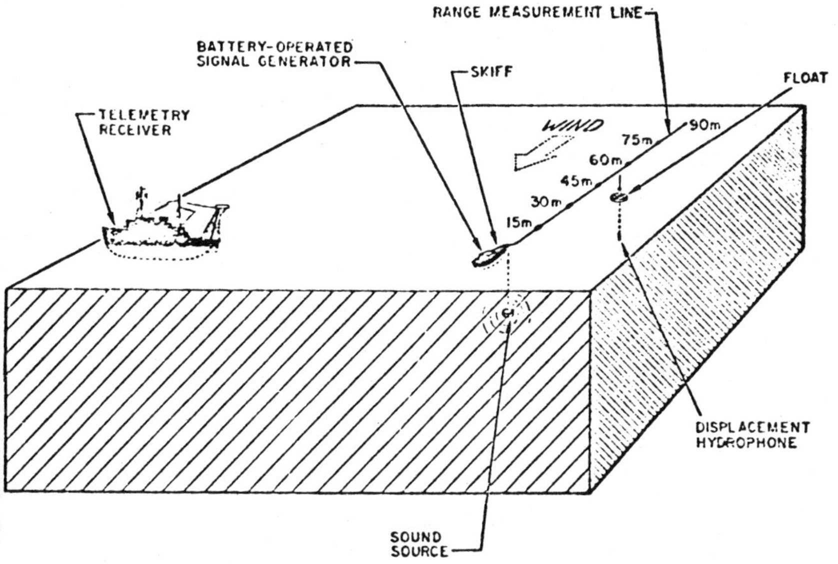

FIG. 10. Arrangement of measurement apparatus.

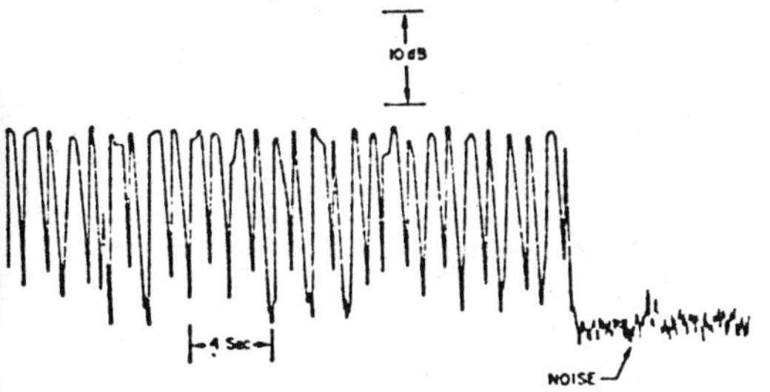

FIG. 11. Part of a typical record obtained in horizontal-displacement studies.

Measurement procedures may be described generally with the aid of Fig. 10. An omnidirectional sound source was lowered to the desired depth below a small skiff containing the electronic equipment necessary for driving the source. The range-measurement line consisted of colored floats attached at 15-m intervals to a light nylon line extending upwind from the skiff. Receiving and recording apparatus were located aboard a larger vessel well-removed from the skiff. Horizontal-range information was communicated between the skiff and the larger vessel via a "walkie-talkie" radio link, the range marks corresponding to passage of the float past

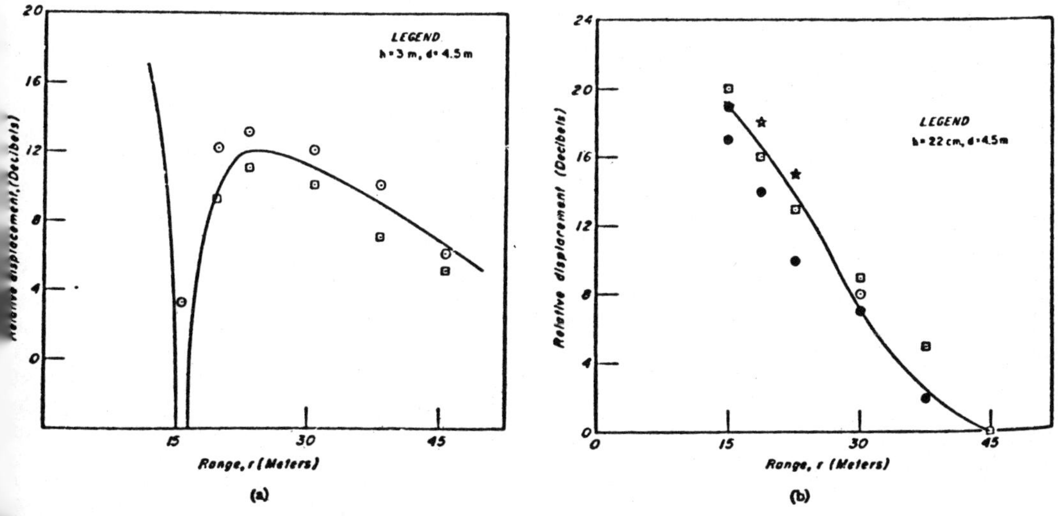

FIG. 12. Comparison of measured and predicted variation of horizontal displacement with range. (a) Receiver depth 3 m; (b) receiver depth 22 cm.

the range markers. The water depth exceeded 100 f (fathoms) at all times.

B. Transmission at Short Ranges

Part of a typical record obtained in experimental studies of horizontal displacement at close ranges may be seen in Fig. 11. It is to be noted that maximum and minimum levels are in excellent agreement with those predicted by the measured directivity pattern (Fig. 8), indicating good vertical alignment with the maximum levels providing accurate measurements of horizontal particle displacement.

The graphs appearing in Figs. 12 and 13 demonstrate close agreement between idealized theory and measured variations of horizontal displacement with range and depth. The solid curves in these plots were calculated, using Eq. (3). It is interesting to note that, although this theory is based on an ideal plane, perfectly reflecting sea surface, it is in agreement with short-range observations taken with 2-ft maximum waveheights.

Equally definitive results were obtained in the studies of vertical displacement; the interesting result that displacement increases as the hydrophone is brought nearer the sea surface is verified. Fluctuations in the hydrophone output due to rotation were less than 2 dB when observing vertical displacement.

C. Transmission at Increased Ranges

The above discussion illustrated the agreement between observation and the elementary dipole theory at ranges less than a few hundred feet. At longer ranges, and particularly at ranges considerably larger than the bottom depth, elementary dipole theory no longer suffices. The theoretical discussion given in an earlier section described how multiple reflection between surface and bottom leads to an infinite array of dipoles. However, quantitative predictions are difficult because the reflectivity of the bottom is not only uncertain, but may exhibit spatial and angular variations. Generally,

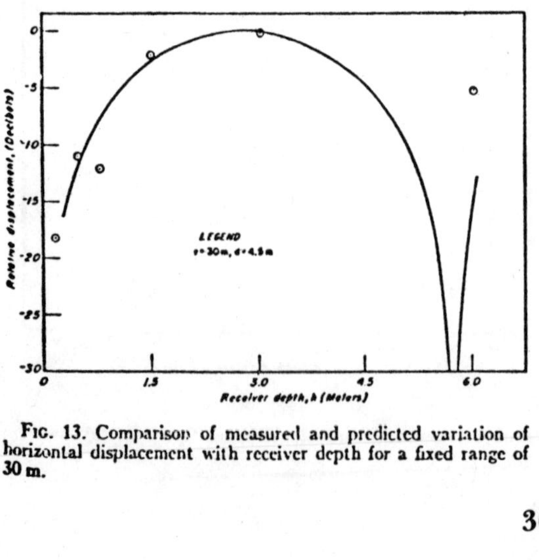

FIG. 13. Comparison of measured and predicted variation of horizontal displacement with receiver depth for a fixed range of 30 m.

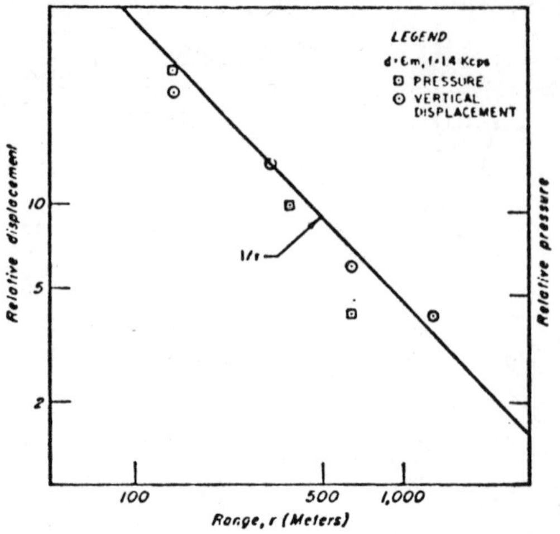

FIG. 14. Variation of displacement at longer ranges. The receiver is floating at the surface ($h \cong 2$ cm). The water depth was 100 f. Simultaneous reception with a pressure hydrophone at 6-m depth is included for comparison.

multiple bottom-surface reflections will tend to diminish, rather than enhance, interference; hence, the $1/r^2$ attenuation observed at short ranges may be expected to be reduced, possibly to $1/r$ attenuation. On similar grounds, it may be argued that the surface and bottom tend to confine acoustical energy in one dimension, permitting spreading loss only in the horizontal plane. This argument again tends toward $1/r$ attenuation. Observations made in 100-f water confirm this expectation: Fig. 14 suggests that at least to ranges of 2000 m the attenuation is $1/r$. (Ranges were limited to less than 2000 m because of the low source level.)

The single-dipole theory also predicts image interference at these ranges with a pressure hydrophone (at 6 m) as well, leading to the expectation of $1/r^2$ dependence. However, observations with a pressure hydrophone, shown in Fig. 14, also demonstrate that image interference has been suppressed, as in displacement observations; indeed, *acoustic pressure exhibits comparable attenuation with vertical displacement at the surface.* It appears that, in practical applications, displacement reception will suffer no greater attenuation than is presently observed with pressure hydrophones.

ACKNOWLEDGMENTS

A portion of these experiments was conducted at the SACLANT ASW Research Center by one of the authors (LNL). Use of the Center's ships and technical facilities is gratefully acknowledged. Considerable technical assistance was also contributed by Norman Head of the Marine Physical Laboratory.

This paper represents results of research sponsored by the U. S. Office of Naval Research. This is a contribution from the Scripps Institution of Oceanography, University of California, San Diego.

In the following paper, the reciprocity method of calibrating acoustic transducers was first developed. This method is universally used for calibrating transducers used as primary standards in calibration facilities.

Dr. W. R. MacLean (deceased, 1964) conducted research in measurement of sound, electromagnetic theory, and electrical instruments. He was Professor of Electrical Engineering at the Polytechnic Institute of Brooklyn.

This paper is reprinted with the permission of the Acoustical Society of America.

Absolute Measurement of Sound Without a Primary Standard

34

W. R. MacLean
101 Lafayette Avenue, Brooklyn, New York
(Received April 23, 1940)

An application of the principle of reciprocity applied to electroacoustic transducers led W. Schottky some time ago to derive a universal relationship between the action of a reversible unit as a microphone and as a loudspeaker.[1] His results, properly used, can lead to the absolute measurement of sound by means of purely electrical observations. Such absolute measurement in its practical form amounts to the calibration of a microphone. (Or in its less practical form, to the calibration of a loud-speaker.) It is the purpose of this paper to outline the theoretical basis for this type of measurement. The microphone undergoing calibration does not itself have to be reversible.

A first application of Schottky's results leads to a free field calibration in terms of either open-circuit voltage or short circuit current. A slightly different application, carried out in a chamber instead of in free space, leads to a pressure calibration in terms of the same units. These two methods are sketched out below.

First Consequences of Reciprocity

The results of this section, in part the same as those obtained by Schottky, are based upon the assumption of the validity for this case of one of the laws of reciprocity, which may be stated as follows: In a passive, linear 4-pole, (1) the two short circuit transfer impedances are equal, (2) the two open-circuit transfer admittances are equal, and (3) the open-circuit voltage ratio in one direction is equal to the short circuit current ratio in the other direction. For both electrical and mechanical cases involving a finite number of degrees of freedom, this follows from the symmetry of the system determinant. However, two reversible electroacoustic transducers coupled to the same medium and accessible only through the two pairs of electrical terminals form a 4-pole which is neither wholly mechanical nor electrical. A fully general and satisfactory proof of the validity of the law has not, to the knowledge of the writer, been given for this case. For that reason, its applicability is herein assumed; however, in an appendix, the scope of the assumption is reduced by proving the theorem from a more elementary assumption. In the above, the two transducers and the surroundings have been taken as linear and passive, but otherwise, quite general.

This generality may now be restricted. The surroundings are assumed to be free space, one of the transducers, U, may be general, but the other, U', is a pulsating sphere connected through a conservative mechanism to an electro-mechanical transducer of one degree of freedom on each side. The mechanical system is assumed to be highly damped, and the radius of the sphere always negligibly small compared with the wave-length. Two points, O and D, separated a distance d, are chosen in free space. U' is placed at D. A point on, in, or near U, but fixed with respect to it, is chosen, as well as a direction also fixed with respect to this unit. The point is called the center, and the direction is called the axis of U, although they are arbitrarily chosen. This transducer is placed with its center coincident with O and with its axis pointing toward D.

With this configuration fixed, the following quantities are defined for U:

M_o: Free field sensitivity as a microphone in open-circuit abvolts per bar
M_s: Free field sensitivity as a microphone in short circuit abamperes per bar
S_o: Free space calibration as a speaker in bars at distance d, per abampere driving current (1
S_s: Free space calibration as a speaker in bars at distance d, per terminal abvolt driving.

S is in terms of pressure on the (arbitraril[y] chosen) axis at distance d, from the (arbitraril[y] chosen) center. M is in terms of f.f. pressure a[t] the center for waves arriving along the axi[s]

[1] Schottky, "Tiefempfangsgesetz," Zeits. f. Physik 36, 689 ff (1926).

having at the center a radius of curvature d. Similar primed quantities apply to the transducer U', except that the center is the center of the sphere, and the question of an axis is eliminated by symmetry.

If U is now driven with 1 abampere, the f.f. pressure at D will be S_o. The voltage generated in U' by this field will therefore be $S_o M_o'$.

If the 4-pole is driven in the other direction with the same driving current, the same generated voltage will result. Similarly by using all four of the laws of reciprocity, the following relations are found:

$$\begin{cases} S_o M_o' = S_o' M_o, \\ S_s M_s' = S_s' M_s, \\ S_o M_s' = S_s' M_o, \\ S_s M_o' = S_o' M_s. \end{cases} \qquad (2)$$

These reduce to the three equations:

$$\frac{S_0}{M_0} = \frac{S_0'}{M_0'} = \frac{S_s}{M_s} = \frac{S_s'}{M_s'}, \qquad (3)$$

which means that for any unit the relation:

$$\frac{M_0}{S_0} = \frac{M_s}{S_s} = H \qquad (3')$$

holds, and that this ratio is independent of design. It remains only to determine the value of H. Since it is independent of design, the ratio can be obtained by computing it for the idealized spherical transducer.

For this purpose, the mechanism of U' is assumed to be devoid of mass, stiffness, and friction, with the exception of a large mechanical impedance Z_m, imposed upon the one degree of freedom coupling with the electrical side. The mechano-electrical transducer is likewise free of impedances, both mechanical and electrical. Electrodynamic coupling is chosen with a force factor, k. The mechanical advantage of the mechanism is so chosen that:

$$Q = \dot{x}, \qquad (4)$$

where Q is the volume current of the sphere, and $\dot{x}$ is the velocity of the mechanical coordinate. The mechanism as a 4-pole connecting volume current and linear velocity acts as an ideal transformer with unit turns ratio, i.e., as a direct connection. The impedance Z_m is then in series on the mechanical side. The electromechanical coupler itself has a circuit matrix:

$$\begin{vmatrix} 0 & k \\ -k & 0 \end{vmatrix}. \qquad (5)$$

The radiation impedance of a pulsating sphere in terms of volume current and pressure is known to be:

$$Z_r = \frac{r}{4\pi a^2 - j2a\lambda} \doteq j\frac{r}{2a\lambda} \cdots 2\pi a \ll \lambda, \qquad (6)$$

where r is the characteristic resistance of the medium, a is the radius of the sphere, and λ is the wave-length. Putting (4), (5) and (6) together, it is seen that the equivalent circuit is as shown in Fig. 1.

The box marked k is the electromechanical transducer. No equivalent circuit can be shown for this since its matrix is not symmetrical. The matrix of the entire 4-pole shown in Fig. 1 is:

$$\begin{vmatrix} (Z_r + Z_m) & k \\ -k & 0 \end{vmatrix} \qquad (7)$$

wherein the first diagonal element applies to the acoustical end, and the second to the electrical. The applied pressure at the acoustical end is the free field pressure. The microphone sensitivity in terms of short circuit current is seen to be the reciprocal of the short circuit transfer impedance of Fig. 1 when driven acoustically. This amounts to:

$$M_s' = \frac{k}{k^2} = \frac{1}{k}. \qquad (8)$$

The reciprocal of the s.c. transfer impedance when driven electrically is the volume current of the sphere for unit applied voltage, which is numerically the negative of the above value. Multiplication by the radiation resistance of the sphere gives the pressure at the surface of the sphere:

$$P_0 = -\frac{1}{k} Z_r. \qquad (9)$$

(9) reduced for divergence and phase at distance d is the free-space voltage calibration of the unit as a speaker:

$$S_s' = -\frac{1}{k} Z_r \frac{a}{d} e^{-j((2\pi/\lambda)d)} \qquad (10)$$

Whence the value for H becomes:

$$H=\frac{M_s'}{S_s'}=-\frac{1}{Z_r}\frac{d}{a}e^{j((2\pi/\lambda)d)}=\frac{2d\lambda}{r}e^{j((2\pi/\lambda)d+\pi/2)}. \quad (11)$$

Ignoring the phase factor:

$$H=\frac{2d\lambda}{r}. \quad (12)$$

By means of (12) the characteristic of a unit as a microphone may be computed in absolute units if its characteristic as a speaker is known, and vice versa. It shows that if any unit is flat as a speaker operated by $\frac{\text{const. current}}{\text{const. voltage}}$ it rises on the low end as an $\frac{\text{o.c. voltage}}{\text{s.c. current}}$ free field microphone at the rate of 20 db per decade (approx. 6 db per octave). This law was called by Schottky the "Tiefempfangsgesetz."

Equation (12) is exact; however, the conditions under which M and S are defined must be kept in mind. The result is independent of the choice of center and axis, but each coordinated pair of M and S are associated with *one* choice. Let the greatest linear physical dimension of U be called l (the arbitrary center is made part of the physical configuration for the purpose of determining l). (12) is independent of the relative magnitudes of l, d and λ. However, if $l \ll d$, then for a *pressure* microphone, M will be practically the same as the plane wave f.f. calibration. For a velocity microphone, this will also be true so long as $\lambda \ll d$, but as soon as λ becomes comparable with d, the calibration in terms of pressure for waves of radius d (which is M) is appreciably larger than the plane wave value. The two values are related by the factor $(1+(c/\omega d)^2)^{\frac{1}{2}}$ where c = speed of propagation and $\omega = 2\pi$ frequency; hence the plane wave calibration can be obtained from M by dividing by this factor.

Analogous but different results are obtained if the procedure is carried out in a small confined chamber instead of in free space. In this case, the chamber is assumed to be so small with respect to the wave-length that the pressure is the same everywhere within. In this space are placed an arbitrary reversible unit and also the

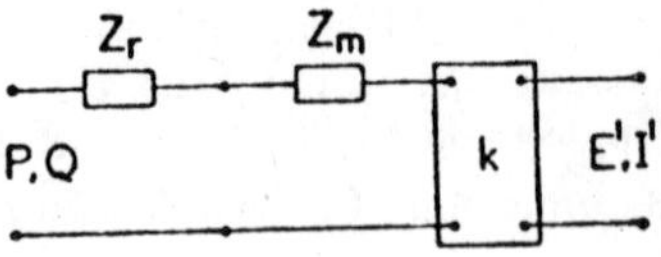

FIG. 1. Equivalent circuit of idealized spherical transducer.

same special spherical unit as before. Fig. 1 is now changed in that the free space radiation impedance of the sphere is replaced by the impedance of the chamber. An equivalent circuit of the entire system is shown in Fig. 2. U is again the arbitrary transducer. Quantities analogous to (1) are defined, but using the symbols T instead of M, and R instead of S. T is the calibration of the unit as a transmitter in terms of pressure *on* the diaphragm. (In this case, only pressure type units can be used.) R is the calibration of the unit as a receiver in terms of volume displacement when operating into zero external impedance. The arbitrary impedance added to the mechanical side of U' is made so large that the volume admittance of the sphere vanishes compared to that of the chamber. Z_o and Z_s are the acoustical impedances of the unit U with electrical terminals open and shorted. By examining Fig. 2, it is seen that a set of equations analogous to (2) can be written down. The first would be:

$$j\omega R_o\frac{Z_oZ_c}{Z_o+Z_c}T_o'=j\omega R_o'\frac{Z_oZ_c}{Z_o+Z_c}T_o.$$

It is seen that in all four,

$$\text{either } j\omega\frac{Z_oZ_c}{Z_o+Z_c} \text{ or } j\omega\frac{Z_sZ_c}{Z_s+Z_c} \quad (13$$

enter on both sides as factors and cancel out Analogies to (3) and (3′) now follow, the latte being:

$$\frac{T_o}{R_o}=\frac{T_s}{R_s}=G. \quad (14$$

The ratio G, independent of design, is found before by computing it from U', the idealize spherical transducer; the equivalent circuit is t same as in Fig. 1 without the Z_r, so that i matrix is (7), also without the Z_r. T_s is the r ciprocal of the s.c. transfer impedance wh driven acoustically, and $j\omega R_s$ of the s.c. trans impedance when driven electrically. These tv

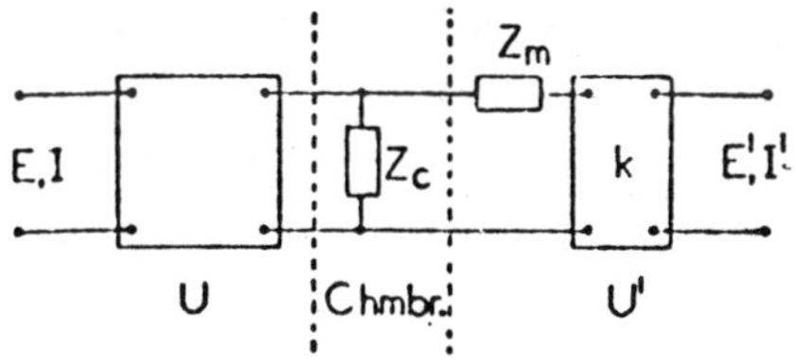

FIG. 2. Equivalent circuit of idealized and general transducers coupled to a chamber.

values are the negative of each other, hence:

$$G = -j\omega = \omega e^{-j(\pi/2)}. \quad (15)$$

Or, dropping the phase factor:

$$G = \omega. \quad (16)$$

From (16) the absolute calibrations of a unit as a receiver (R) and as a pressure microphone (T) can be computed from one another.

Taking R as a measure of the pressure sensitivity of the unit as a receiver—which will actually be the case in terms of relative pressure generated if it operates into a pure compliance soft compared with the diaphragm (the ear cavity) the following law is deduced from (14) and (16): If a unit is flat as a receiver operated by $\frac{\text{constant current}}{\text{constant voltage}}$, it rises on the high end as an $\frac{\text{o.c. voltage}}{\text{s.c. current}}$ pressure microphone at the rate of 20 db per decade (approx. 6 db per octave). If there is a "Tiefempfangsgesetz" for free space, as Schottky found, there is also a "Hochempfangsgesetz" for a chamber.[2]

THE ABSOLUTE MEASUREMENT OF SOUND—FREE FIELD

The results given by (12) and (16) can be used to obtain the absolute calibration of a microphone in terms of free field or applied pressure. The pressure method suffers from some of the difficulties common to all chamber measurements, such as cooling of the walls, which, however, has been compensated for in the past in work on the thermophone and pistonphone. The free field problem which will be treated first suffers from the difficulty of simulating free space conditions; however these are likewise met with in Rayleigh disk measurements.

In the experiment leading to the f.f. calibration of the microphone there enter: (a) the microphone to be calibrated, U, (b) a reversible unit, U_x, and (c) a sound generator, U_0. Since it is desired ultimately to obtain the plane wave f.f. calibration in terms of pressure, and since the curvature of the wave affects the calibration of a velocity microphone at low frequencies, a somewhat more strictly specified experiment is needed in this case than for a pressure microphone. This more restricted experiment will be described since it includes both cases.

The sound generator U_0 should be a zero order radiator, i.e., one which behaves as a point source at sufficiently low frequencies. On, in, or near U, U_x, and U_0 arbitrary centers are chosen. Similarly, three axes are chosen. Let l be the greatest physical dimension of any of the three devices (the centers are included as part of physical configuration).

U and U_x are now placed one after the other at the same spot in the sound field of U_0. More specifically, U is placed co-axially with U_0, and their centers are separated by a distance d. This distance is chosen so that $l \ll d$. Without changing U_0 in any way, U_x is substituted for U. The voltages generated by each, E and E_x, are recorded. If M_0 and M_0^x are the o.c. free field calibrations of U and U_x for waves of radius d, then since the sound field was the same in both cases:

$$\frac{E}{M_0} = \frac{E_x}{M_0^x} \quad \text{whence:} \quad \frac{M_0}{M_0^x} = \frac{E}{E_x}. \quad (17)$$

The use in (17) of M (which is for waves of radius d) must be justified, since U_0 does not necessarily emit spherical waves at all frequencies. For high frequencies with $\lambda \ll d$, the values for response of U and U_x in plane, spherical, or actual field would be nearly identical since $l \ll d$. For a velocity microphone, the curvature effect becomes apparent for $d \sim \lambda$ or $d \ll \lambda$. But then we have also $l \ll \lambda$ and therefore U_0 behaves as a spherical radiator. Consequently the use of M is justified.

U_0 is now removed, and the two units placed in juxtaposition with their axes coincident but

[2] The words "receiver" and "transmitter" have opposite meanings in the German and English literature.

opposed, and their centers separated distance d. U_x is driven with a current I'. The f.f. pressure at U's center is $I'S_o^x$, where, as before, S_o^x is the calibration of U_x as a speaker. If U now puts out the o.c. voltage E', then:

$$E' = I'S_o^x M_o, \tag{18}$$

where the use of M is similarly justified. But now from (12) we have the value for the ratio of M_o^x to S_o^x. So from (18) and (12):

$$M_o M_o^x = \frac{E'}{I'} \frac{2d\lambda}{r}, \tag{19}$$

which together with (17) gives:

$$M_o = \left(\frac{E'E}{I'E_x} \frac{2d\lambda}{r}\right). \tag{20}$$

(20) is the absolute calibration of the given microphone in o.c. abvolts per f.f. bar for waves of radius d. Quite similar results are obtained for s.c. abamperes per f.f. bar, and analogous results for the calibration of a unit as a speaker, by applying (12) to the above, for instance.

If the microphone being tested is one of the pressure type, M is practically equal to the plane wave value; if the microphone is a velocity type, and the plane wave value is desired, the correction factor $\{1+(c/\omega d)^2\}^{-\frac{1}{2}}$ must be applied.

THE ABSOLUTE MEASUREMENT OF SOUND—CHAMBER PRESSURE

To apply the method to obtain the pressure calibration of a given microphone, there enter the same three devices as before, except that this time they would want to be of such physical size and otherwise so chosen as to be suitable for work in a chamber. Coupling all three to a small chamber and driving U_o, the analog of (17) is obtained:

$$\frac{T_o}{T_o^x} = \frac{E}{E_x}. \tag{17'}$$

For the second experiment, U_o may or may not be left in the chamber. If it is, its diaphragm impedance must be added in parallel to the chamber impedance, the former taken with the electrical connections in the condition in which they are left. Calling Z_c, Z_o, Z_o^x the impedance of the chamber, the o.c. diaphragm volume impedance of U, and the same for U_x, the pressure generated in the chamber by virtue of a current I' in U_x is:

$$P = I'R_o^x j\omega Z, \tag{21}$$

where Z is the parallel combination of the chamber and the two diaphragm impedances. If E' is now the o.c. voltage in U, then the analog of (18) is:

$$E' = j\omega I'R_o^x T_o Z, \tag{18'}$$

which together with (16) gives:

$$T_o T_o^x = \frac{-jE'}{I'Z}. \tag{19'}$$

Whence (19') and (17') lead to:

$$T_o = \left(\frac{E'E}{I'E_x} \frac{1}{Z}\right)^{\frac{1}{2}} \tag{20'}$$

after dropping the phase factor. This is the absolute calibration of the given microphone in o.c. abvolts per bar applied pressure. Instead the s.c. calibration could have been obtained or even the calibration of a unit as a receiver. The application of this formula requires a knowledge of the impedance of the chamber as modified by wall cooling and by the unit coupled to it, these taken with their electrical terminals o.c. in the given case, or s.c. for the s.c. case.

The derivation of the principle results given by (12), (16), (20) and (20') was carried out with c.g.s. electromagnetic units. This was done to facilitate the computation of the idealized spherical transducer which used magnetic coupling. However, electrostatic coupling could have been used and the work carried through with e. units. Nevertheless, the four principle equations would remain unaltered but giving M_o, for instance, in statvolts per bar rather than abvolts per bar.

It is customary to express M and S in terms of bars on the one hand and volts or amperes on the other. It is therefore neither c.g.s. nor m.k.s., but mixed. If d, λ, ω, Z, and r are in c.g.s. units, and E and I in international units then the four equations become:

$$H = 10^{-7}\frac{2d\lambda}{r}, \qquad (12a)$$

$$G = 10^{-7}\omega, \qquad (16a)$$

$$M_o = (10^{-7})^{\frac{1}{2}}\left(\frac{E'E}{I'E_x}\,\frac{2d\lambda}{r}\right)^{\frac{1}{2}}, \qquad (20a)$$

$$T_o = (10^{-7})^{\frac{1}{2}}\left(\frac{E'E}{I'E_x}\,\frac{1}{Z}\right)^{\frac{1}{2}} \qquad (20'a)$$

with M, S, T, R in bars and volts or amperes. (H and G are ratios of these quantities.)

In all methods previously employed to obtain absolute measurements, a primary standard was necessary. This primary standard was, in a word, a sound detecting or generating device whose construction was so simple that it could be reliably computed. In other words, a computable standard was used. It would seem impossible to obtain absolute measurements without such a standard. In the present case, such a computable standard, namely the spherical transducer, also enters into the picture, but it remains only hypothetical, i.e., it has to be used in the theoretical work, but it does not have to be built.

Appendix

To prove the law of reciprocity as it has been used, it suffices to assume that the 4-pole formed by two acoustically coupled transducers can be represented as the limiting case of an infinite sequence of electromechanical systems of a finite but indefinitely increasing number of degrees of freedom. Such a proof carries considerable weight, although it is not as satisfying as a real wave solution.

The proof is carried through for the case of electromagnetic and/or electrostatic couplers. A similar result can no doubt be obtained for piezoelectric couplers. Fig. 3 represents one of the systems in the infinite sequence of systems with increasing numbers of degrees of freedom whose limit is supposed to be the acoustical case. If reciprocity holds for them all, it presumably holds for the limit as well. The logic of this is not too precise.

N_m is a mechanical system with an arbitrary number of coordinates available on the left-hand side, and another arbitrary number available on the right-hand side. W_1 and W_2 are electromechanical couplers devoid of unnecessary mechanical or electrical impedances. Each side is homogeneous, that is, only one kind of coupling, static or magnetic is present on any one side of N_m. N_1 and N_2 are two general electrical networks. N_{12} represents eventual direct electric coupling between N_1 and N_2. 1 and n are the input and output meshes. It is noted that a single

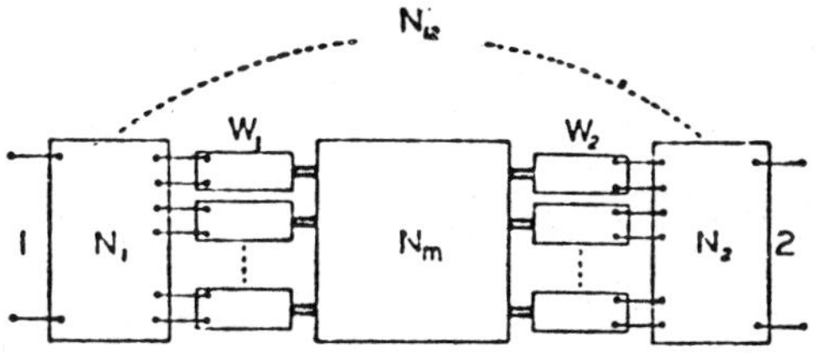

FIG. 3. Equivalent circuit of one out of the infinite sequence of 4-poles whose limit is the acoustically coupled 4-pole.

coupler on each side would not be sufficient, for that would imply that a single mechanical coordinate was imposed between the acoustical and electrical systems. Such might be the case with a cone speaker, but it is not the case for a velocity microphone with an undulating ribbon or with a condenser microphone with a flexible diaphragm. The entire system is assumed to be passive and linear.

For Fig. 3, a system matrix Δ can be set up. The entire principle of reciprocity rests merely upon the relation between the two minors Δ_{1n} and Δ_{n1} of this matrix. If their determinants are equal, reciprocity applies, otherwise not. They will certainly be equal if Δ is symmetrical. Since the two-rowed matrix of a simple electrostatic coupler, formed, say by a rigid piston facing a back plate, is symmetrical, one concludes that reciprocity applies if all couplers are electrostatic, for then the entire system matrix will also be symmetrical. The two-rowed matrix representing a simple magnetic coupler is, however, skew symmetric as shown in (5), so that it is necessary to demonstrate the validity of reciprocity in this case.

Both sides magnetic

If the couplers on both sides are magnetic, the system matrix would appear:

$$\Delta = \left|\begin{array}{c|c|c} N_1 & W_1 & N_{12} \\ \hline -W_1^T & N_m & W_2 \\ \hline N_{12}^T & -W_2^T & N_2 \end{array}\right| \qquad (22)$$

Δ is broken up into several component oblongs. The three on the main diagonal are square, and each would be the matrix of that part of the circuit alone which is indicated therein if all its accessible meshes were shorted (unconstrained). W_1 is not, in general, square. It represents half the coupling terms due to the like-named couplers. Were it not for the peculiar characteristic of magnetic coupling, the middle left oblong would be merely the transform of W_1. As it is, every term there has a minus sign. A similar situation obtains for W_2. N_{12} is also not square in general. This represents direct electric coupling. The lower left oblong is the transform of this with plus sign.

The two minors are derived from Δ as shown below:

$(-)^n|\Delta_{1n}| =$, $(-)^n|\Delta_{n1}| =$. (23)

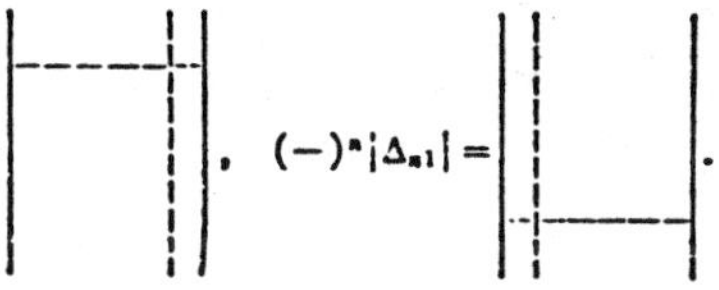

In each case, the determinant within the dotted line is evaluated to find the minor. In the first case, the first row and last column are stripped from Δ, and in the second case, the first column and last row are stripped. The second one can now be rotated on its main diagonal without changing its value. Then in both cases it will be the first row and last column that are stripped, but in the first case from Δ, and in the second case from Δ^T, i.e., from the transform of Δ. Δ and Δ^T differ only in that the sign of every element in the W sections is different. If in the second case every row and column contained in N_m is multiplied through by -1, the value of the minor determinant will not be changed, since an even number of negatives has been used. But now it is identical with the first minor, for every element in the W sections has changed back again, and the elements of N_m have all been multiplied twice and hence are unchanged Therefore if W_1 and W_2 are both magnetic, the law of reciprocity applies.

W_1 magnetic and W_2 static

This case differs from the above in that only the W_1 sections have opposite signs. In this case we proceed as before, except that instead of multiplying only those rows and columns contained in N_m by -1, we include those contained in N_2 as well. As a result of this, the W_1 sections change their signs back again; N_m, N_2, and the W_2 sections get multiplied twice and remain as they were, but the two N_{12} sections are multiplied only once and hence change signs. The two minor determinants are now identical, except that in the second one, the N_{12} sections are negative. Unless all the elements of N_{12} are zero, the two are definitely unequal, and the law of reciprocity does not hold. Since the last column was multiplied by -1 but removed to form the minor, the minor determinant has been multiplied by an odd number of negatives; so if all elements of N_{12} vanish, reciprocity holds, but with *phase reversal.* Hence for magnetic coupling on one side and static on the other, reciprocity holds only when there is no direct electric coupling between the two electrical sides of the system. In the calibration problem, this was not the case, and hence the assumption was justified, but the example of such cases where reciprocity does not apply means that a certain amount of care must be observed in using it.

The following paper describes a calibrator which can be used to calibrate gradient hydrophones at low frequency in the laboratory.

Mr. Bauer is recognized for many significant contributions in research and for numerous acoustical developments. These include microphones for acoustical measurements, pistonphones, heartbeat measurements, hearing aids, sound ranging, military and civilian communications, and public address systems. He has also conducted psychoacoustic research related to directional hearing in the air and underwater, and the measurement of loudness. He has authored numerous papers, he holds more than 50 patents and has lectured widely on acoustics. He has been a Visiting Professor of Engineering Acoustics at The Pennsylvania State University.

Mr. Bauer is a Fellow of the Institute of Electrical and Electronics Engineers, a Fellow of the Acoustical Society of America, and an Associate Editor of its Journal. He is a Fellow of the Audio Engineering Society, its Executive Vice President in 1967–1968, President in 1968–1969 and a recipient of its Potts Memorial Award. He is Vice President of the CBS Laboratories in charge of acoustics and magnetics.

This paper is reprinted with the permission of the author and the Acoustical Society of America.

Reprinted from The Journal of Acoustical Society of America, Vol. 39, No. 3, 585–586, March 1966

35

Laboratory Calibrator for Gradient Hydrophones

B. B. Bauer

CBS Laboratories, Stamford, Connecticut 06905

A calibrator has been developed for gradient hydrophones consisting essentially of a water-filled drum driven axially by two transducer motors. To the first approximation, the sound pressure and velocity functions within the drum are those that one encounters at the pressure node and velocity loop in a standing wave.

Measurement of gradient (or "velocity") hydrophones has always been a difficult accomplishment, especially at low frequencies where the long wavelengths virtually ensure the presence of reflections from nearby surfaces. Being confronted with the task of developing low-frequency gradient hydrophones, we found the conventional methods of measurement to be unwieldy, time-consuming, and expensive. A reliable laboratory device for generating a theoretically predictable sound velocity in water became a practical necessity.

The gradient hydrophone calibrator (shown in Fig. 1) developed at CBS Laboratories consists essentially of a heavy, water-filled, stainless-steel drum with stiffened ends, supported horizontally in an I-beam frame by means of a vibration-isolation suspension that minimizes floor and frame vibrations. A port in the uppermost surface of the drum admits the hydrophones to be measured, and a corresponding opening at the top of the frame holds an azimuth scale and pointer for investigation of directional response. The port can be closed by means of a suitable porthole cover equipped with a gland for passing the hydrophone cable. When closed, the drum can be pressurized safely to more than 100 lb/sq in by means of a hydraulic pump.

The drum is vibrated in the axial mode by two transducer motors placed at its ends and fastened to the frame. To the first approximation, the fluid moves in phase with the drum, producing the equivalent of a standing wave that is eminently suited for calibrating gradient hydrophones.

Fig. 1. Calibrator for gradient hydrophones.

The proximate theory of operation is extremely simple. Referring to Fig. 2, consider two waves of equal amplitude and wavelength λ traveling in opposite directions in a rigid pipe.[1] The resulting standing wave can be expressed in terms of root-mean-square sound pressure p, and velocity u as a function of the maximum root-mean-square pressure p_{max} and the distance x from the pressure node 0, as follows:

$$p = p_{max} \sin(2\pi x/\lambda), \tag{1}$$

and

$$u = -j(p_{max}/\rho c) \cos(2\pi x/\lambda), \tag{2}$$

where ρ is the density of the fluid and c is the velocity of sound.

We select two planes equidistant by $L/2$ from the pressure node, and find that the particle velocities are of equal sign and have equal root-mean-square values u_e. But this is precisely the condition occurring at the end walls of a rigid drum of length L oscillating with a root-mean-square velocity u_e where,

$$u_e = -j(p_{max}/\rho c) \cos(\pi L/\lambda), \tag{3}$$

and, since u_{max}, by definition, is equal to $-jp_{max}/\rho c$, and u_e is measurable with a velocity pickup attached to the end of the drum, u_{max} is clearly expressable as $u_e/\cos(\pi L/\lambda)$ and p_{max} as $j(u_e \rho c)/\cos(\pi L/\lambda)$. Therefore, we can conveniently rewrite the equations for pressure and velocity within the drum as follows:

$$p = j[u_e \rho c/\cos(\pi L/\lambda)] \sin(2\pi x/\lambda), \tag{4}$$

and,

$$u = [u_e/\cos(\pi L/\lambda)] \cos(2\pi x/\lambda). \tag{5}$$

The length of the drum L (as well as, incidentally, the diameter) of the calibrator is 12 in. The wavelength, at, say, 500 cps, is 120 in. Thus, u does not differ from u_e by a calculable factor greater than $1/\cos(\pi 12/120) = 1.05$. Calibration can be checked by deriving the velocity function from pressure measurements obtained with a pressure hydrophone placed on the axis off the center of the drum. Call this off-center displacement x_1; then the pressure reading is,

$$p = j[u_e \rho c/\cos(\pi L/\lambda)] \sin(2\pi x_1/\lambda); \tag{6}$$

and for long wavelengths this reduces to

$$p = j(u_e \rho c) 2\pi x_1/\lambda = j\omega x_1 \rho u_e, \tag{7}$$

where ω is the angular frequency $2\pi f$. With constant driving force, throughout the range of frequency where the tank–driver system is mass controlled, u_e will be roughly inversely proportional to frequency and the pressure-versus-frequency relationship of the offset hydrophone will be approximately a constant whereas the velocity will vary inversely with the frequency.

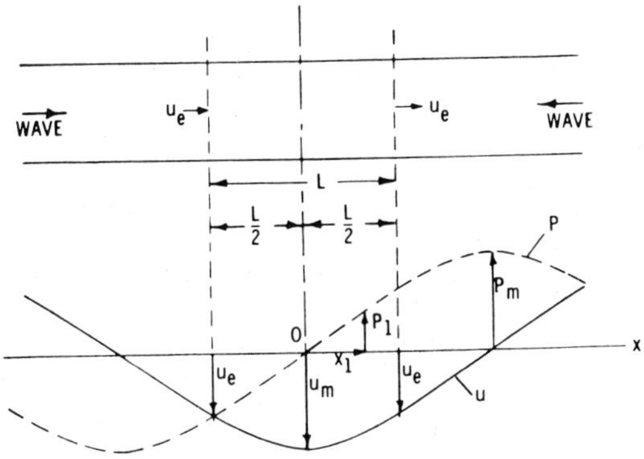

FIG. 2. Pressure and velocity relationships within the calibrator drum.

The forces needed to drive the drum are modest, as is seen from the following example. Assume that the desired root-mean-square velocity u_{max} is that corresponding to a maximum root-mean-square sound pressure of 100 dyn/sq cm, or $u_{max} = 100/154\,000 = 65 \times 10^{-5}$ cm/sec rms. If the drum and the water weigh 400 lbs (or have a mass of 182 000 g), the drum impedance at 500 cps is $6.28 \times 500 \times 182\,000 = 5.7 \times 10^8$ mechanical ohms. The force required to drive the drum at the desired velocity is $65 \times 10^{-5} \times 5.7 \times 10^8 = 370 \times 10^3$ dyn rms, equivalent to less than 1 lb. Thus, moderate-size drivers can produce root-mean-square particle velocity corresponding to relatively large sound-pressure level.

Obviously, it is an oversimplification to assume that the tank behaves as a rigid body; the tank, frame, driving system, and fluid enter into a variety of resonant conditions that must be minimized by proper design and assembly. George Myers, of our Laboratories, has observed that the tendency to resonate is diminished by filling the drum somewhat less than full. With the drum full, some resonances appear to stem from trapped air bubbles. When adjusted properly, the calibrator is well-behaved and permits the study of frequency response and the polar patterns of gradient hydrophones over the complete range of frequencies from the infrasonic up to several hundred cycles per second. Because the wave is plane, no corrections are required for wave sphericity.

***Acknowledgment*:** The author is grateful to Alfred L. DiMattia and Robert L. Townsend, who contributed to the design and analysis of the calibrator, and to Louis A. Abbagnaro and George C. Myers, who developed the operating technique to produce results in good agreement with the theoretical predictions.

[1] For example, see H. F. Olson, *Acoustical Engineering* (D. Van Nostrand Co., Inc., Princeton, N. J., 1957), p. 14.

Large transducers, such as those used in sonar systems, require large volumes of water for normal far-field calibrations. This paper describes a system for calibration of large transducers from near-field measurements, making it possible to calibrate large transducers in relatively small volumes of water.

While Mr. Trott was with the U.S. Navy Underwater Sound Reference Laboratory, he developed measurement techniques and calibration procedures for underwater sound. He spent the year 1966–1967 as a liaison scientist for acoustics at the London Branch of the Office of Naval Research. He is now a Research Physicist in the Transducer Branch of the Acoustics Division of the Naval Research Laboratory.

This paper is reprinted with the permission of the author and the Acoustical Society of America.

Reprinted from The Journal of the Acoustical Society of America, Vol. 36, No. 8, 1557–1568, August 1964

36

Underwater-Sound-Transducer Calibration from Nearfield Data

W. James Trott

U. S. Navy Underwater Sound Reference Laboratory, Orlando, Florida 32806
(Received 28 April 1964)

Farfield transmitting response, freefield sensitivity, and directivity of large audiofrequency transducers can be obtained directly from measurements made in the near field (Fresnel zone) of a transducer array especially designed to radiate a plane wave of constant amplitude within a portion of the Fresnel zone. The theory of the measurements and of the design of the special measuring array are supported by experimental data. Design data for building suitable measuring arrays are given.

INTRODUCTION

Many problems in the calibration of underwater-sound transducers are caused by inadequate dimensions and nonideal boundaries of the environment. Although various measures can be taken to improve the conditions, limitations still remain, especially for large transducers and low frequencies. Acoustic reflections from the surface, bottom, and shore of lakes can be reduced by using baffles, directional measuring transducers, and pulse techniques; sound absorbers and pulse techniques can be used in tanks; but all these corrective measures become progressively less effective as the wavelength of sound increases. Response and sensitivity data obtained by standard procedures[1] can be corrected to some extent for interfering reflections and less-than-adequate separation distances, but no such corrections are available for farfield measurement of directional characteristics—valid directivity measurements can be made only in the correct environment.

The great distances and tremendous volumes of water needed for large present-day transducers resonant at frequencies below 10 kc/sec are not easily obtained. If some new technique would permit the measurements to be made with an extremely short distance between the source and the receiver, then either the greater distance loss of the unwanted signal (reflection) or the greater delay between it and the direct signal would extend the abilities of existing facilities to the calibration of large low-frequency transducers. For this reason, there has been much recent interest in trying to determine farfield characteristics from data obtained in the near field or Fresnel zone of a source. Hanish,[2] Horton and Innis,[3] Pachner,[4] and others have analyzed the situation as a boundary-value problem.

The nearfield and the farfield measurements are related to each other in this paper by means of the reciprocity parameters appropriate to the mode of wave propagation in the two regions. This derivation is somewhat specialized in its application, however, so the same equations will then be derived by means of Huygens' principle and shown to have broader application. It is believed that the two derivations will provide a better understanding of the nearfield technique than would either one alone. The third part of this paper deals with the theory, instrumentation, and procedure for developing a Huygens' wavefront by means of a shaded array of point sources. Because the reciprocity principle applies to the arrangement, the array can also be used to integrate the near field of an unknown source transducer.

[1] *American Standard Procedures for Calibration of Electroacoustic Transducers, Particularly Those for Use in Water Z24.24–1957* (American Standards Association, New York, 1957).

[2] S. Hanish, "Notes on Computation of Far-Field Pressure from Near-Field Data," U. S. Naval Res. Lab. Sound Div. (Mar. 1962; unpublished).

[3] C. W. Horton and G. S. Innis, Jr., "The Computation of Far-Field Radiation Patterns from Measurements Made near the Source," J. Acoust. Soc. Am. **33**, 877–880 (1961).

[4] J. Pachner, "Investigation of Scalar Wave Fields by Means of Instantaneous Directivity Patterns," J. Acoust. Soc. Am. **28**, 90–92 (1956).

I. THEORY

A. Derivation by Reciprocity Parameters

If a transducer is effectively a point on a spherical wave, a line on a cylindrical wave (line parallel to the axis of the cylinder), or a plane on a plane wave, the freefield receiving sensitivity will be the same in each case.

The transmitting response will depend on the mode and extent of the wave propagation. Normally, the transmitting current response S_s is defined in terms of the sound pressure in a spherical wave in the far field.[1] The ratio of the freefield voltage sensitivity to this transmitting current response is the reciprocity parameter calculated by MacLean,[5] Cook,[6] and Foldy and Primakoff.[7]

Consider, as a source of sound, a piston transducer having an active face of dimensions equal to two wavelengths or more. Within the near field, the sound wave is essentially plane and the average pressure is constant over an aperture normal to the beam axis at positions along the axis. Stenzel[8] has shown that the radiation impedance of a circular or square piston having a radius or half-width a for $ka>5(k=2\pi/\lambda)$ is essentially resistive, and the ratio of the average pressure to the piston velocity is the plane-wave or characteristic impedance ρc of the medium.

Simmons and Urick[9] have demonstrated experimentally, and Williams[10] has shown theoretically, that the average pressure in the aperture at any position in the near field of a piston is constant. Seki, Granato, and Truell[11] have calculated the phase across the aperture of a piston transducer at various positions along the beam axis and shown that the wave is plane to a high degree of approximation.

We define a nearfield (plane-wave) transmitting current response S_p as the average pressure in the near field over an effective area A for a unit driving current (Fig. 1). In a physical sense, all reciprocity parameters are the ratio of the volume velocity emanating from a transducer to the sound pressure at a specified reference point. Consequently, with a fixed volume velocity from one piston transducer, the nearfield average plane-wave pressure is related to the farfield spherical-wave pressure by the ratio J_s/J_p of the spherical-wave reciprocity parameter to the plane-wave reciprocity parameter. It follows then that the freefield voltage sensitivity M is given by

$$M=S_pJ_p=S_sJ_s, \quad (1)$$

where $J_p=2A/\rho c$ (from Ref. 9), $J_s=2D\lambda/\rho c$ (from Ref. 5), A is the effective area of the piston transducer, D is the reference distance for the farfield transmitting current response, λ is the wavelength of sound, and ρc is the characteristic impedance of the medium.

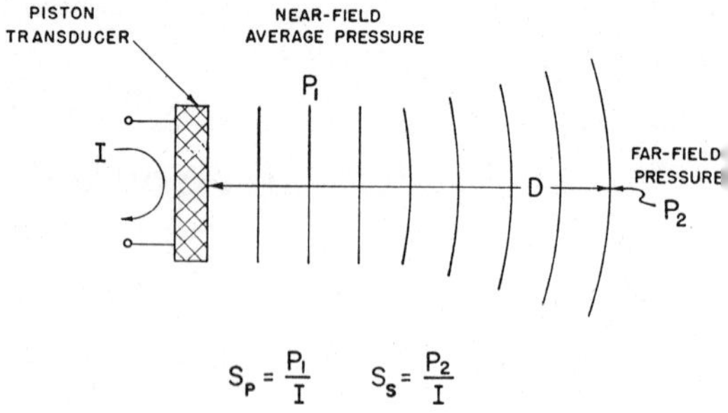

FIG. 1. Near and far field of piston transducer.

The transmitting current response in the near field S_p can be expressed in terms of the voltage output E_T of a probe hydrophone scanning a plane aperture in the near field, the probe's freefield voltage sensitivity M_T and the current I driving the piston transducer:

$$S_p=\int E_T dA/IM_TA. \quad (2)$$

From Eqs. (1) and (2) is obtained the farfield transmitting current response S_s of this piston transducer

$$S_s=S_pJ_p/J_s=\frac{\int E_T dA}{IM_TA}\frac{A}{D\lambda}=\frac{\int E_T dA}{IM_TD\lambda}. \quad (3)$$

The voltage E_T must be determined point by point in magnitude and phase. The ratio of E_T to I is the transfer impedance between the piston transducer and the probe hydrophone. It is the same in the reverse direction; that is, it is the same when the probe is used as a source, except for a possible change in sign because of different electromechanical coupling in the transducer and hydrophone. For this condition, the current driving the probe is I_T and the output voltage of the piston transducer is E. Thus, by equating E/I_T to E_T/I and multiplying both sides of Eq. (3) by the spherical-wave reciprocity parameter, the freefield voltage sensitivity M of the piston transducer is obtained:

$$M=\int E dA/I_TS_TD\lambda. \quad (4)$$

[5] W. R. MacLean, "Absolute Measurement of Sound without a Primary Standard," J. Acoust. Soc. Am. **12**, 140–146 (1940).

[6] R. K. Cook, "Absolute Pressure Calibration of Microphones," J. Acoust. Soc. Am. **12**, 415–420 (1941).

[7] L. L. Foldy and H. Primakoff, "A General Theory of Passive Linear Electroacoustic Transducers and the Electroacoustic Reciprocity Theorem," J. Acoust. Soc. Am. **17**, 109–120 (1945); **19**, 50–58 (1947).

[8] H. Stenzel, "Die akustische Strahlung der rechteckigen Kolbenmembran," Acustica **2**, 263–281 (1952).

[9] B. D. Simmons and R. J. Urick, "The Plane Wave Reciprocity Parameter and Its Application to the Calibration of Electroacoustic Transducers at Close Distances," J. Acoust. Soc. Am. **21**, 633–635 (1949).

[10] A. O. Williams, Jr., "The Piston Source at High Frequencies," J. Acoust. Soc. Am. **23**, 1–6 (1951).

[11] H. Seki, A. Granato, and R. Truell, "Diffraction Effects in the Ultrasonic Field of a Piston Source and Their Importance in the Accurate Measurement of Attenuation," J. Acoust. Soc. Am. **28**, 230–238 (1956).

In a similar fashion, we can calibrate a line transducer in the near field by defining a transmitting current response S_c as the average pressure along a line aperture in the cylindrical wave radiated within the near field. The farfield transmitting current response S_s of the line transducer is defined in the same manner as before (Fig. 2). The freefield voltage sensitivity M of this line transducer is related to the nearfield transmitting current response S_c and the cylindrical-wave reciprocity parameter derived by Bobber and Sabin:[12]

$$M = S_c J_c = S_s J_s, \tag{5}$$

where $J_c = 2L(a\lambda)^{\frac{1}{2}}/\rho c$, L is the length of line transducer, a is the reference distance for S_c such that $1 < ka < 2\pi(L/\lambda)^2$.

The nearfield transmitting current response S_c of a line transducer can be expressed in terms of the voltage E_T received by a probe hydrophone scanning a line parallel to the source line at distance a. Thus,

$$S_c = \int_0^L E_T dx / I M_T L. \tag{6}$$

From Eqs. (5) and (6), we obtain the farfield transmitting current response S_s of the line transducer:

$$S_s = S_c J_c / J_s = \left(\int_0^L E_T dx / I M_T \right) [(a\lambda)^{\frac{1}{2}} / D\lambda]. \tag{7}$$

The nearfield separation distance a must be measured and inserted in the equation because the cylindrical wave is diverging.

The freefield-voltage sensitivity is obtained as before from the farfield transmitting current response and the spherical-wave reciprocity parameter:

$$M = \left(\int_0^L E dx / I_T S_T \right) [(a\lambda)^{\frac{1}{2}} / D\lambda]. \tag{8}$$

The voltage is measured in magnitude and phase.

The line transducer can be calibrated by means of nearfield measurements and used to integrate the near sound field of a piston transducer along a coordinate in the direction of its length. By scanning the sound field in a direction normal to the length of the line, the average pressure can be determined, and the farfield transmitting current response S_s of the piston transducer can then be computed from

$$S_s = \int_0^y E_L I dy / I M_L D\lambda, \tag{9}$$

where E_L is the open-circuit voltage of the line transducer and M_L is the freefield voltage sensitivity of the line transducer.

[12] R. J. Bobber and G. A. Sabin, "Cylindrical Wave Reciprocity Parameter," J. Acoust. Soc. Am. 33, 446–451 (1961).

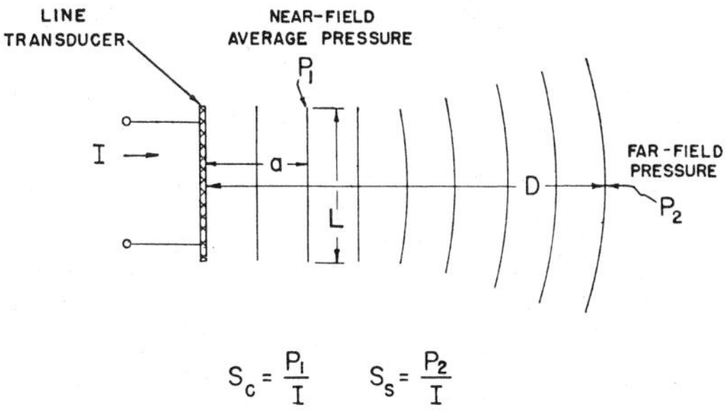

FIG. 2. Near and far field of line transducer.

The freefield voltage sensitivity M of the piston transducer is

$$M = \int_0^y E L dy / I_L S_L D\lambda, \tag{10}$$

where S_L is the farfield transmitting current response of the line transducer and I_L is the current driving the line transducer.

In Eqs. (3), (4), and (7)–(10), the integration can be extended to include all contributions. In Eqs. (9) and (10), the length L of the line transducer and the length of scan should be greater than the dimensions of the measured transducer.

The plane-wave reciprocity parameter is justified on the basis of the nondivergence of the near field of a piston transducer. The cylindrical-wave parameter is justified on the basis of the cylindrical-wave distance loss in the near field of a line transducer. The equations derived here have been applied to nearfield probing measurements to obtain accurate results for both the transmitting current response and the freefield voltage sensitivity of transducers.

The limit $1 < ka$ for Eqs. (7) and (8) is for a thin-line transducer. For a thick-line transducer (diameter of the line larger than the wavelength), measurements have shown that the minimum distance a from the center must be greater than twice the diameter.

B. Derivation by Huygens' Principle

Equations (3) and (4) can be derived by means of Huygens' principle with broader applicability than is indicated by the derivation using the reciprocity parameters. Imagine an unknown transducer receiving a plane wave and a plane surface as the source of this plane wave (Fig. 3). To the receiver (hydrophone), the plane wave is effectively of infinite extent. The output voltage E of the hydrophone is measured in magnitude and phase for each position of the measuring transducer, which is a point source on the plane surface. The voltage output of the hydrophone is represented by the product of the freefield voltage sensitivity M of the hydrophone, the transmitting current response S_T of the measuring

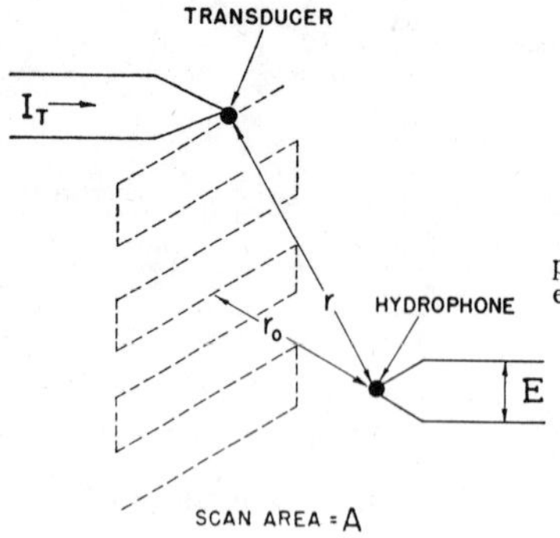

FIG. 3. Plane wave produced by Huygens' principle.

(source) transducer, the current I_T driving the source, and the ratio of the reference distance D for S_T to the measuring distance r:

$$E = MS_TI_T(D/r). \tag{11}$$

Integrating the voltage over the scanned area A produces

$$\int EdA = MS_TI_TD\int (1/r)dA. \tag{12}$$

Huygens' principle[13] states that a plane wave of infinite extent can be mathematically replaced by the area of the first half-wave Fresnel zone ($\pi\lambda r_0$) with a source amplitude $1/\pi$ times the infinite plane-wave amplitude, insofar as the effect on the hydrophone is concerned. In this first half-wave zone $r \approx r_0$, so

$$\int EdA = (MS_TI_TD/\pi)(\pi\lambda r_0/r_0). \tag{13}$$

From this expression, we can obtain the freefield voltage sensitivity of the hydrophone:

$$M = \int EdA/S_TI_TD\lambda, \tag{14}$$

which is the same as Eq. (4) that was derived from the reciprocity parameter. This equation now is not restricted to the plane normal to the beam axis. The equation applies regardless of the directivity of the hydrophone. The equation also applies to the sum of an array of point hydrophones for any orientation.

The transmitting current response, Eq. (3), is obtained by reversing the previous procedure and again equating the transfer impedance in one direction to its equal in the opposite direction.

The directivity of the unknown and the rapid change of phase as the scanned area is increased restrict the area that must be scanned.

A more rigorous derivation by Huygens' principle[14] r quires that the probe have the directional characterist $(1+\cos\theta)$, where θ is the angular direction with respe to a normal to the scanned plane. Two probes alor the normal connected in opposite phase and a hal wavelength apart will produce the desired directivi from 0° to 60°—a sufficient range to cover the region the unknown. If a line is used to scan the plane, tl desired directivity can be obtained by using two lines phase, $\frac{1}{4}$ wavelength apart in the plane. Baker ar Copson[14] show that this derivation requires $1/r_0$ to l negligible in comparison to k, or $kr_0 > 1$. This is also tl limitation for line-scanning.[12] Directivity of the prol has very little effect on the accuracy when the plar scanned is the plane normal to the beam axis of tl unknown transducer, but from measurements it appea to be important for other planes scanned.

In some of the earlier work[3], the farfield directivi was computed by means of the Kirchhoff–Helmhol formula from measurements of pressure amplitude ar phase over a plane aperture in the near field. Point-b point equality between the pressure gradient and tl product of ik times the pressure was assumed ($i = \sqrt{-1}$ Patterns computed from such measurements agree wit the measured farfield patterns only in angular positior close to the beam axis. The derivation in this paper c the basis of Huygens' principle for the receiving cond tion shows the reason. The computed directivity wi agree with the measured directivity only if the diffra tion constant, the ratio of the blocked to the freefiel pressure, is independent of the orientation of tl measured transducer, which would be true for only few cases. Consider the unknown hydrophone to be directional piston transducer enclosed in a pillbo shaped housing. It may be subject to twice the freefiel plane-wave pressure for sound incident in the directio of maximum sensitivity and to only a fraction of tl freefield pressure at other directions. This variation i the diffraction constant from 2 to <1 is not accounte for in calculating directivity from measurements in or plane by the Kirchhoff–Helmholtz formula as it wa used by Horton and Innis.[3] It would appear from th derivation in this paper, however, that valid directivit results could be obtained by scanning several planes an restricting the computation to ±20° or ±30°.

II. DESIGN OF A MEASURING ARRAY

Instead of scanning the plane area point by point, a array can be used as the source of the plane wave. Thi array must be a special kind of transducer. It must b acoustically transparent to eliminate standing wave between it and any transducer to be measured in an orientation. The transparency can be achieved by con structing the array of many piezoelectric transducer each small with respect to the wavelength, widel

---

[13] Lord Rayleigh, *Theory of Sound* (Macmillan & Co. Ltd., London, 1926), 2nd ed., Sec. 283.

[14] B. B. Baker and E. T. Copson, *The Mathematical Theory* *Huygens' Principle* (Oxford University Press, London, 1950 2nd ed., Sec. 4.6.

spaced, and operating well below resonance. The array must have a near sound field different from that of any transducer discussed so far. It must produce a plane wave of constant level within the region to be occupied by the unknown transducer.

The plane wave of infinite extent was divided into annular Fresnel half-period zones in the derivation of Eq. (13). The effect of the wave at point P, the position of the hydrophone in Fig. 3, was mathematically expressed as the contribution from an area of the wavefront equal to the first zone $\pi\lambda r_0$, with a strength $1/\pi$ times the amplitude of the wave. If this zone were divided into a large number of rings, the vector-summation diagram of the contributions from these rings to the sound pressure at P would approach the shape of a semicircle, the diameter being twice the plane-wave pressure expressed in Eq. (13). Suppose that the wave were represented as emanating from a circular piston source of area equal to the first zone for the position of point P. The distance r_0 would then be at the outer limit of the near field of this source. At some shorter distance, the area would be equal to two half-period zones and the vector-summation diagram would take the shape of a circle; the resultant pressure at this new point would be zero. Thus, on the axis in the near field of a circular-piston source, the sound pressure will vary between zero and twice the sound pressure in the simulated plane wave. That is, the sound pressure will vary between 0 and 2 unless the source strengths of successive Fresnel zones are progressively reduced so that the vector summation takes the form of a spiral, the resultant pressure being always one. Shading of this form has been used to reduce the level of the minor lobes in the farfield directivity of a source. From this nebulous relationship, it has been suggested that the same shading that would eliminate the minor lobes in the farfield directivity would also eliminate the undulations in the nearfield sound pressure.

von Haselberg and Krautkrämer[15] showed mathematically and experimentally that a circular source to which the excitation is applied in the form of a Gaussian function (exp $-r^2$) has a constant pressure along the axis in the near field. This type of shading is known to produce a farfield directivity without minor lobes. This shading will not serve our purpose, however, since the pressure level falls off rapidly for positions in the aperture off the beam axis. Linhardt[16] showed that a bell-shaped amplitude distribution smooths the near field.

Consider the unshaded piston as a source of sound of increasing frequency. In the manner discussed by Stenzel,[17] the radius of the source is a, and we shall analyze it for increasing values of ka, where $k=2\pi/\lambda$. For $ka=\pi$, a region of pressure maximum originates at the center of the piston, and a null forms at 90° to the beam axis in the farfield directivity. As ka increases, successive maxima and minima appear at the center of the piston and move out radially across the face of the piston and out along the beam axis within the near field. Concurrently, successive minor lobes in the farfield directivity appear in the plane of the piston source and rotate inward toward the beam axis.

From knowledge of shading to suppress minor lobes in the farfield directivity and the relationship between the position of the undulations in the near field to the position of the successive minor lobes, we can now attempt to find a shading function that will produce the desired constant-level near field—that is, a constant-level, three-dimensional sound field, not just along the beam axis. Instead of suppressing all minor lobes as was done by Gaussian shading, we shall produce an array in which the first minor lobe is attenuated slightly, but minor lobes of higher orders are suppressed by increasingly greater amounts. Our special measuring array will thus consist of many small piezoelectric transducers, widely spaced, and shaded in source strength for suppression of the higher-order minor lobes, thus reducing the undulations in the near sound field that exist close to the center of an unshaded array.

A family of plane-area arrays with varying numbers of elements but having similar characteristics in farfield directivity and constancy of pressure in the near field can be designed from a family of shaded-element line arrays. The general equation for the directivity of the family of shaded-element line arrays can be expressed as

$$p=(\sin m\phi/m\sin\phi)\cos^n\phi, \qquad (15)$$

where $\phi=(\pi d/\lambda)\sin\theta$, d is the element spacing, λ is the wavelength, and θ is the angle in the plane of the line between the normal to the line and the direction of a distant point. The cosine term is the directivity of a line array of equally spaced elements whose source strengths are proportioned to the coefficients of a binomial series having the power n. This unit is replicated m times with a center-to-center spacing of d. For example, if the relative amplitudes in a 3-element line are 1,2,1 (coefficients of the series for $(a+b)^2$), the directivity in the plane of the line is $\cos^2\phi$. Replication of four of these elemental arrays, thus

$$\begin{array}{cccccc} 1 & 2 & 1 & & & \\ & 1 & 2 & 1 & & \\ & & 1 & 2 & 1 & \\ & & & 1 & 2 & 1 \end{array} \quad = \quad 1\ \ 3\ \ 4\ \ 4\ \ 3\ \ 1,$$

produces a line array whose directivity is $p=(\sin 4\phi/4\sin\phi)\cos^2\phi$.

The directivity of this array contains only the first minor lobe at the level -24 dB *re* the level of the main beam as shown by Stenzel[17] in his Fig. 11. Compare this

[15] K. von Haselberg and J. Krautkrämer, "Ein Ultraschall-Strahler für die Werkstoffprüfung mit verbessertem Nahfeld," Acustica **9**, 359–364 (1959).
[16] F. Linhardt, "Über den Einfluss der Schallfeldinterferenzen auf den Fehlernachweis in Festkörpern," Metall **12**, 1085–1092 (1958); **13**, 1133–1138 (1959).
[17] H. Stenzel, *Leitfaden zur Berechnung von Schallvorgängen* (Julius-Springer Verlag, Berlin, 1939).

to an array shaded for no minor lobes; 1,5,10,10,5,1. This should increase the level of the off-axis sound pressure above that achieved with a Gaussian-shaded source.[15]

By selecting the values of m and n in Eq. (15), it is possible to retain only the first minor lobe and suppress all subsequent minor lobes. Concurrently, in the near field of a plane-area array designed from this line array, there will be one maximum on the beam axis at the far end of the near field and a maximum at the periphery of the array to maintain the pressure level over a greater portion of the aperture. This condition exists when a plane-area array is designed from a line array having the directivity represented by Eq. (15) if $m=9$ and $n=5$. The shading function is obtained by normalizing 9 replications of a row of 6 elements whose source strengths are proportioned to the coefficients of a binomial series having the power 5. The shading function is 0.03, 0.19, 0.5, 0.81, 0.97, 1.0, 1.0, 1.0, 1.0, 0.97, 0.81, 0.5, 0.19, 0.03. If $m=10$ and $n=6$, the shading function is 0.02, 0.11, 0.34, 0.66, 0.89, 0.98, 1.0, 1.0, 1.0, 1.0, 0.98, 0.89, 0.66, 0.34, 0.11, 0.02.

A shading function computed in this manner can be used to produce a 2-dimensional plane array. Mathematically, it can be treated as a plane-surface source on an infinite rigid baffle. The sound pressure p in the far field is given by Stenzel[17] as

$$p\sim\int_{\text{area}} F(x,y)e^{-ikr}dA,$$

where $F(x,y)$ is the shading function. If $F(x,y)=g(x)h(y)$,

$$p\sim\int_x g(x)e^{-ikx\cos\alpha}dx\int_y h(y)e^{-iky\sin\beta}dy.$$

This is the product of the directivity of a line equivalent of the array along x and a line equivalent of the array along y. If $g(x)$ and $h(y)$ are identical shading functions, the condition described by this equation is shown in Fig. 4(a) for a 10-element line. This relationship has been referred to as the second product theorem by Pennell and coworkers,[18] and has been shown to produce

ae	be	ce	de	ee
ad	bd	cd	dd	ed
ac	bc	cc	dc	ec
ab	bb	cb	db	eb
aa	ba	ca	da	ea

(a) Circular symmetry

e'	e'	e'	e'	e'
d'	d'	d'	d'	e'
c'	c'	c'	d'	e'
b'	b'	c'	d'	e'
a'	b'	c'	d'	e'

(b) Square symmetry

FIG. 4. Application of line-shading to area arrays having circular and square symmetry.

[18] W. O. Pennell *et al.*, "Directivity Patterns of Sound Sources," NDRC Rept. C4-sr287-089, HUSL [PB L 86108] (Apr. 1942).

approximately circular symmetry. Rows and columns of elements in the plane array are spaced a distance d.

Square symmetry is represented by one quadrant in Fig. 4(b). The shading terms are primed to show that new shading factors must be obtained by normalizing the coefficients obtained by solving the set of equations $5e'=e$, $4d'+e'=d$, $3c'+d'+e'=c$, etc., so that Eq. (15) is also the farfield directivity of the square-symmetry array in the plane containing the beam axis and parallel to a side.

For design purposes, it is desirable to be able to compute the sound-pressure variations within the near field. The task is extremely difficult for an unshaded array and even more so for a shaded array. However, since undulations in the magnitude of the sound pressure originate at the center of an array and move out along the beam axis with increasing frequency of the sound wave, some measure of the near field can be obtained by computing the nearfield pressure at the center of the array as a function of frequency. For circular symmetry,[17] the amplitude of the undulations is constant with frequency if the array is unshaded. Across the face of the array at any one frequency, the amplitude of the undulations gradually diminishes from the center of the array toward the edge. For a square array,[8] the undulations at the center of the array gradually diminish in amplitude with increasing frequency; across the face of the array at any one frequency, they retain a constant amplitude. A computation of the pressure versus frequency at the center of a circular-symmetry array will show the greatest amplitude variation that will exist for the same basic shading, whether the array has circular or square symmetry.

Equation (111) in Stenzel[17] describes the pressure in the near field along the beam axis of a circular unshaded piston as a function of kR, where R is the radius of the piston and $k=2\pi/\lambda$. At the center of the piston [$Z=0$ in Stenzel's Eq. (111)], the normalized or relative pressure is

$$p=2\sin\tfrac{1}{2}kR\exp[i(\omega t+\tfrac{1}{2}\pi-\tfrac{1}{2}kR)]. \qquad (16)$$

This equation shows that the pressure varies between 0 and 2—the same conclusion as was reached by application of Fresnel-zone theory.

Note that in the shading function $m=9$, $n=5$ there exists a symmetry in the coefficients about the 0.5 coefficient; that is, $0.19+0.81=0.03+0.97=1$. The circular-symmetry array can be considered to be a piston of unity source strength having a radius equal to the distance from the center to the element of 0.5 source strength. The shading can be superimposed on this piston by the addition of ring sources. An inner ring (No. 1) of source strength -0.03 and width d has an average radius equal to the distance from the center to the element of source strength 0.97. Ring No. 2 of source strength -0.19 and width d has an average radius equal to the distance from the center to the element of source

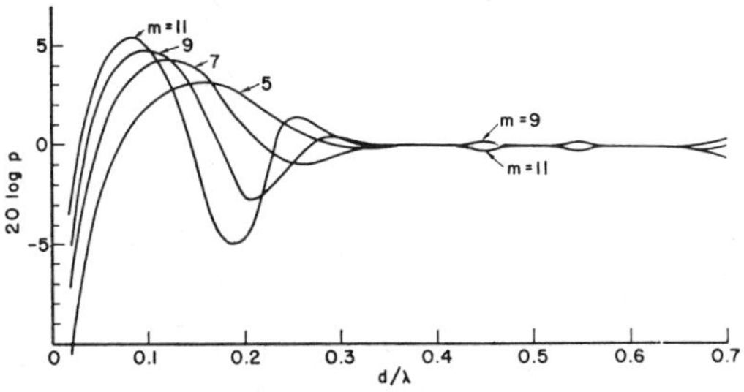

m	SHADING FUNCTIONS	
5	0.03, 0.19, 0.5, 0.81, 0.97	0.97, 0.81, 0.5, 0.19, 0.03
7	0.03, 0.19, 0.5, 0.81, 0.97, 1	1, 0.97, 0.81, 0.5, 0.19, 0.03
9	0.03, 0.19, 0.5, 0.81, 0.97, 1, 1	1, 1, 0.97, 0.81, 0.5, 0.19, 0.03
11	0.03, 0.19, 0.5, 0.81, 0.97, 1, 1, 1	1, 1, 1, 0.97, 0.81, 0.5, 0.19, 0.03

℄

FIG. 5. Relative sound-pressure level at the center of shaded circular-symmetry area arrays, computed from Eq. (17).

strength 0.81. Ring No. 3 of source strength 0.19 and width d has an average radius equal to the distance from the center to the element of 0.19 source strength. The outer ring (No. 4) has a source strength 0.03.

Stenzel also derives an expression for the pressure on the beam axis in the near field of a ring source. Again, for $Z=0$ in Stenzel's Eq. 114, the normalized pressure at the center of a ring is $p=2w_i \sin \frac{1}{2}kd \times \exp[i(\omega t+\frac{1}{2}\pi-kR_i)]$, where w_i is the source strength density of the ring and R_i is the average ring radius.

From these two equations, the pressure at the center of this family of shaded arrays has been derived by superposition of rings of source-strength density $\frac{1}{2}$, $\pm w_1$, $\pm w_2$, $\pm w_3$, upon the piston of source-strength density one. For half of the family of arrays, n odd, there will be a ring of source strength $\frac{1}{2}$ with a radius R. For these arrays, the sound pressure at the center is given by

$$p=\sin kR+i(\cos \tfrac{1}{2}kd - \cos kR - 4w_1 \sin \tfrac{1}{2}kd \sin kd - 4w_2 \sin \tfrac{1}{2}kd \sin 2kd - 4w_3 \sin \tfrac{1}{2}kd \sin 3kd - \cdots). \quad (17)$$

In the example $m=9$, $n=5$: $0.19=w_1$, $0.81=1-w_1$; $0.03=w_2$, $0.97=1-w_2$. The width of the rings is the element spacing d.

In the remaining half of the family of arrays, n even, there is no ring of half source-strength density, and the radius R is the radius to a point half-way between a ring with density greater than a half and a ring of density less than a half. The pressure at the center of these arrays is given by

$$p=\sin kR+i[1-\cos kR-4w_1 \sin \tfrac{1}{2}kd \sin \tfrac{1}{2}kd - 4w_2 \sin \tfrac{1}{2}kd \sin 3(\tfrac{1}{2}kd) - 4w_3 \sin \tfrac{1}{2}kd \sin 5(\tfrac{1}{2}kd) - \cdots]. \quad (18)$$

Equations (17) and (18) are based on a stepped-density distribution rather than a point-distribution array. This stepped density is valid for the arrays considered since the actual array of points is a square configuration and circular symmetry has been computed into the shading. In either case—stepped-density or point distribution—when the distance d is greater than 0.8 wavelength, as is later shown, the near field is no longer constant and a second-order beam is forming in the far field in the plane of the array.

Equation (17) was used to compute the relative sound pressure versus frequency at the center of a circular-symmetry array derived from the shaded-line example $m=9$, $n=5$. Additional examples were also computed with $n=5$ and $m=5$, 7, and 11. The computed curves and the line shading functions are shown in Fig. 5. All four curves approach the 0-dB level at a point between $d/\lambda=0.3$ and 0.35. The plot is symmetrical about the point $d/\lambda=0.5$. That is, the value for $d/\lambda=0.8$ is the same as for $d/\lambda=0.2$. The line shading $m=9$, $n=5$ was chosen for constructing two measuring arrays, one having circular symmetry and one having square symmetry.

III. EXPERIMENT

A. Measuring Array

Figure 6 shows the circular-symmetry measuring array. A 12×12 array of elements with corner elements missing (140 elements in all) was constructed on the basis of the shaded line represented by $m=9$, $n=5$ in Fig. 5. The strength of the end elements shaded to 0.03 of the center elements was considered small enough to neglect. This was a mistake, as will be seen later.

The elements are PZT-4 piezoelectric ceramic cylin-

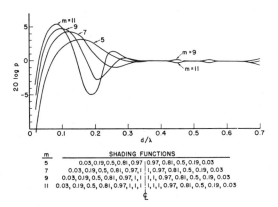

FIG. 6. Nearfield measuring array.

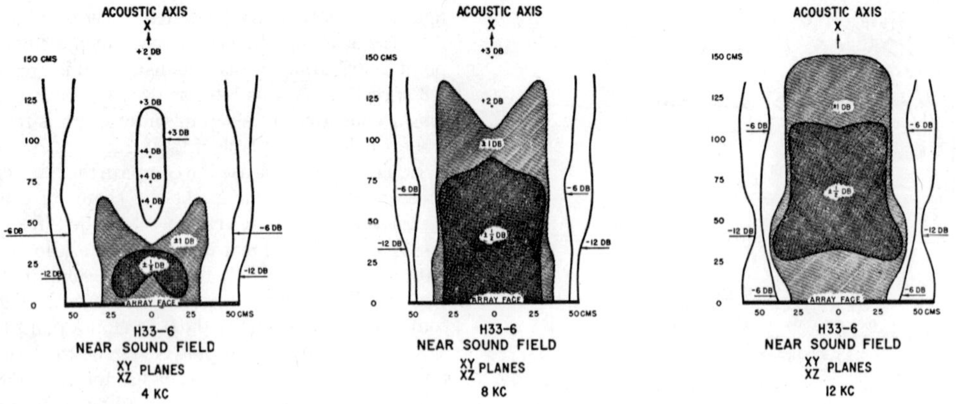

FIG. 7. Near sound field of the measuring array at 4, 8, and 12 kc/sec.

ders, $\frac{1}{2}$ in. diam, $\frac{1}{2}$ in. long, and $\frac{1}{8}$ in. wall thickness. The cylinders, capped on each end with glass-to-metal seals, are air-filled, connected in parallel, and spaced 10 cm apart in oil-filled tygon plastic tubes also spaced 10 cm apart. Element shading is obtained by the use of series-connected capacitors. The elements resonate at 70 kc/sec, well above the frequency range 4–12 kc/sec where the array operates. The freefield voltage sensitivity of this acoustically transparent and acoustically stiff array with a diffraction constant of unity is constant from a few cycles per second up to 15 kc/sec or more. The array is 112 cm square. It is mounted between wire screens for electrical shielding.

The near field of this array was measured from 1 to 15 kc/sec at 80 points, grid-spaced every 15 cm from 15 to 150 cm along the beam axis and at 0, 15, 30, 37, 45, and 50 cm vertically and horizontally from the beam axis. Representative points were also measured diagonally to the vertical and horizontal axes. Figure 7 shows the uniformity of the sound pressure obtained in the near field of this array at 4, 8, and 12 kc/sec. The plane of the array is represented by the line at the bottom of each plot; the acoustic-beam axis is shown in the vertical direction along the X axis. The near field at intermediate frequencies was also uniform, as was shown by the smoothness of the recorded sweep-frequency data. The sound pressure in the shaded areas of the plots is constant within $\pm\frac{1}{2}$ and ±1 dB as shown. The contours for -6 and -12 dB *re* the average sound-pressure level within the shaded areas are also shown. The maxima on the beam axis for 4 and 8 kc/sec are related to the maximum shown in Fig. 5.

Stenzel[17] shows in his Fig. 53 the variation in sound pressure at the edge of a circular unshaded piston source. The amplitude varies between 0.35 and 0.7 times the average pressure at the center. Measurements of this shaded array at the distance R, Eq. (17), off the acoustic axis show a variation between 0.48 and 0.63 for a total spread of about $2\frac{1}{2}$ dB. This variation would have been reduced if the shading function for $n=7$ had been used; it probably would have been reduced if the elements shaded to 0.03 had been included in the array.

Figure 8 shows the measured and computed nearfield transmitting current response of this measuring array. The levels of the curves computed from Eq. (17) have have been adjusted to agree with that of the measured transmitting current response. The computed curves are for a position at the center of the array; the measured curves are for the distances 15 and 30 cm from the array on the beam axis. Omitting the elements shaded to 0.03 did produce some pressure variation. A few values of the sound pressure were computed by summing the contributions from the point sources to a point 15 cm from the array on the beam axis. They are shown on the lower curve.

B. Calibration of the Measuring Array

Measurements can be made on an unknown transducer within this constant-sound-pressure, plane-wave region just as if the source were far removed. The nearfield transmitting current response S_p and the ratio of the freefield voltage sensitivity M to the effective area A of the array must be measured.

Within the useful frequency range (4–12 kc/sec) for the experimental array, the freefield voltage sensitivity is the element sensitivity. The nearfield transmitting current response S_p can be measured with a calibrated hydrophone:

$$S_p = E_H / M_H I, \tag{19}$$

where E_H is the open-circuit output voltage of the standard hydrophone, M_H is the freefield voltage sensitivity of the standard, and I is the current input to the array.

The ratio of the freefield voltage sensitivity of the array to its effective area can be measured in a comparison calibration by means of a small projector and the same calibrated hydrophone. The projector is placed close to the array for one measurement. Then, the calibrated hydrophone is placed in the far field of the small source. The separation distance can be the same as that used in the first measurement.

Although only two measurements are made, the three arrangements shown in Fig. 9 are considered in deriving the equations. In arrangement 1, the small source is shown as the receiver (hydrophone) in the near field of the array. These two transducers are assumed to be reversible for the derivation of the equation. The nearfield transmitting current response of the array is

$$S_p = E_s/M_sI, \tag{20}$$

where M_s is the freefield voltage sensitivity of the transducer referred to as the small source.

The equation for arrangement 2 is obtained by applying the reciprocity parameters to Eq. (20). The transfer impedance is the same in arrangements 1 and 2: $S_p = M/J_p = M(\rho c/2A)$, where M is the freefield voltage sensitivity of the array.

In arrangements 2 and 3, the only measurements actually made, the measurements are made in the far field of the small source, so that $M_s = S_sJ_s = S_s(2D\lambda/\rho c)$, where D is the distance between the small source and the hydrophone in arrangement 3, λ is the wavelength of sound, and ρc is the characteristic wave impedance. When these values for S_p and M_s are substituted in Eq. (20) and E_s/I is replaced by its equivalent E/I_s, the equation for arrangement 2 becomes

$$M/A = (E/I_sS_s)(1/D\lambda). \tag{21}$$

The sound pressure denoted by I_sS_s is measured with a calibrated hydrophone in arrangement 3:

$$I_sS_s = E_H/M_H. \tag{22}$$

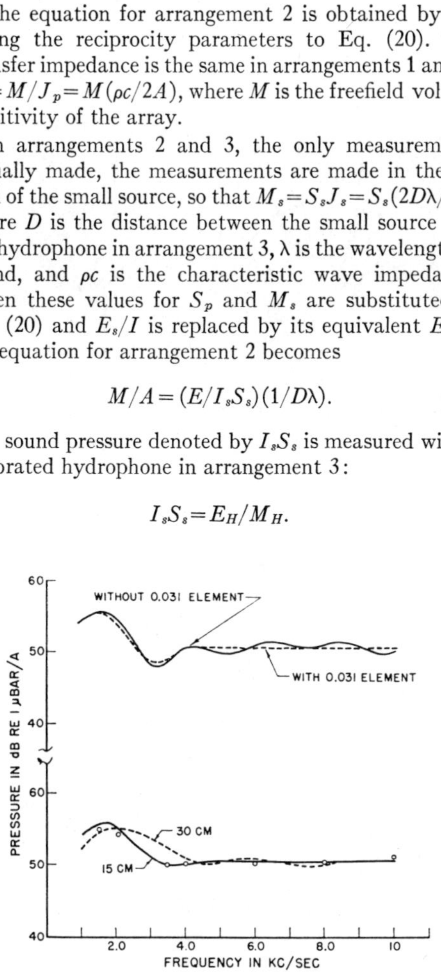

FIG. 8. Nearfield transmitting current response of the measuring array. *Top*: Pressure at center of array computed from Eq. (17). *Bottom*: Curves from data obtained at 15 and 30 cm from array on beam axis; o points computed for array of point sources 15 cm from array on beam axis.

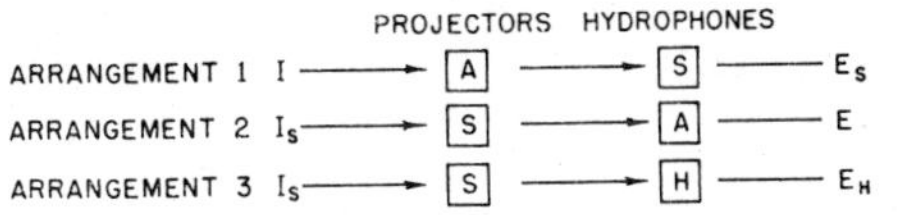

FIG. 9. Transducer arrangements for deriving array-calibration equations.

Equation (21) can then be written

$$(M/A) = (E/E_H)(M_H/D\lambda). \tag{23}$$

The values obtained by Eqs. (20) and (23) can be averaged for several positions in the near field of the array and at several frequencies within the operating range. The data so obtained are sufficient to permit use of the array to calibrate large transducers.

C. Calibration with Nearfield Array

The calibrated measuring array can be used to calibrate an unknown transducer so long as the volume of unknown transducer is not larger than the constant-sound-pressure region in the near field of the array. The freefield voltage sensitivity M_x of an unknown transducer is

$$M_x = E_x/S_pI, \tag{24}$$

where E_x is the open-circuit voltage measured at the terminals of the unknown and I is the current into the measuring array.

If the unknown is measured as a source, its nearfield transmitting current response is the average sound pressure p in its plane-wave near field divided by its driving current: $S_{px} = p/I_x$. This average sound pressure p exists over an effective area A_x, which is less than the effective area A of the array. Thus, the pressure measured by the array is $p = (E/M)(A/A_x)$. Therefore, $S_{px} = (E/MI_x)(A/A_x)$.

The farfield transmitting current response S_{sx} of the unknown is related to the nearfield transmitting current response by the ratio of the plane-wave reciprocity parameter J_p to the spherical-wave reciprocity parameter $J_s = 2D\lambda/\rho c$, where D is the reference distance (generally 1 m) and λ is the wavelength of sound: $S_{sx}/S_{px} = A_x/D\lambda$.

The farfield transmitting current response of the unknown, as obtained by measurements in the near field with this measuring array, is

$$S_{sx} = EA/MI_xD\lambda. \tag{25}$$

Equation (25) is the same as Eq. (3) derived for probe-scanning by means of the reciprocity parameters and Huygens' principle. The measuring array performs the integration of the sound field. Equation (24) can be expressed in the form of Eqs. (4) and (14), but it is unnecessary.

A cylindrical transducer having a delay-line beam-forming network was calibrated by using the measuring

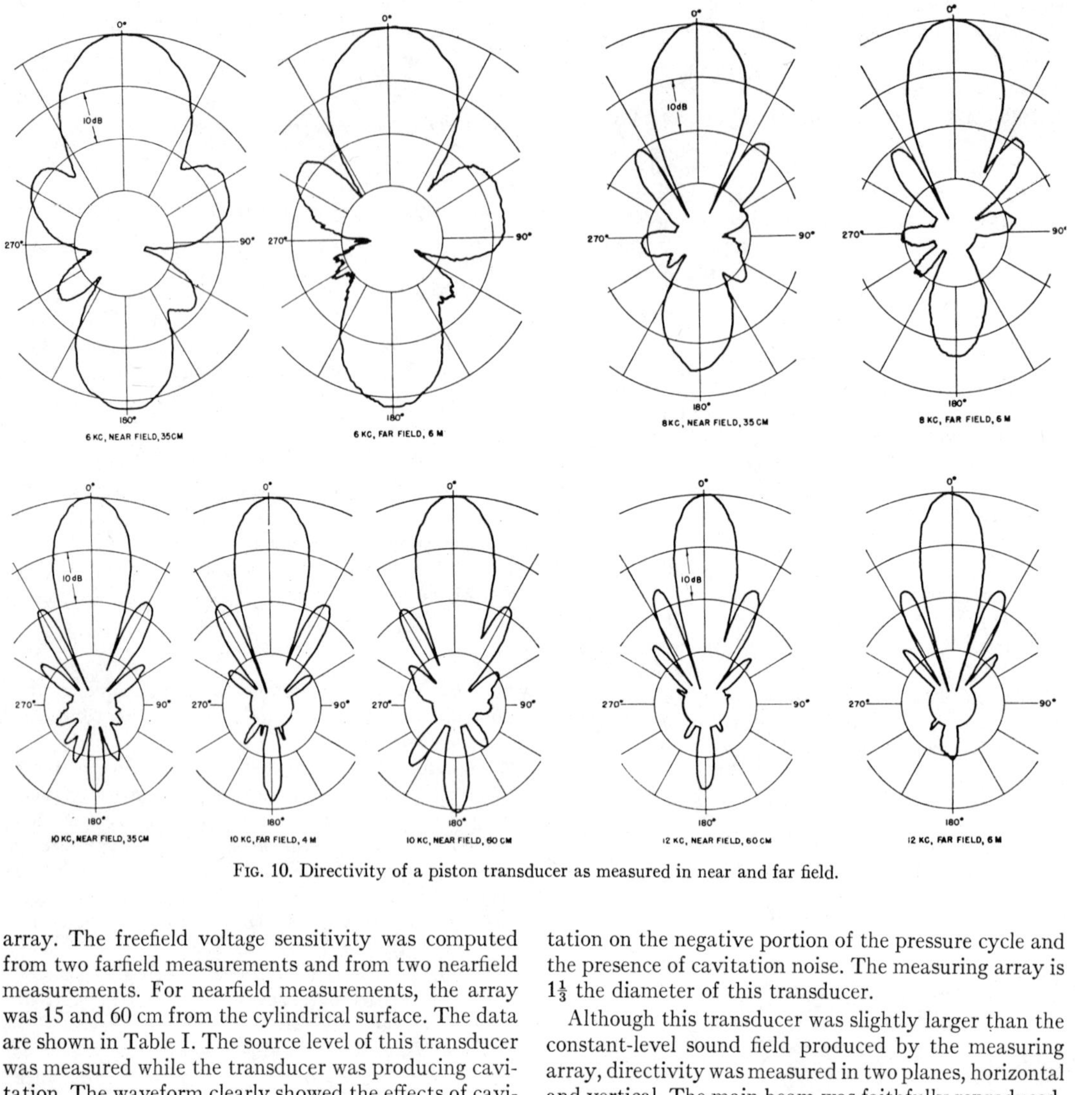

FIG. 10. Directivity of a piston transducer as measured in near and far field.

array. The freefield voltage sensitivity was computed from two farfield measurements and from two nearfield measurements. For nearfield measurements, the array was 15 and 60 cm from the cylindrical surface. The data are shown in Table I. The source level of this transducer was measured while the transducer was producing cavitation. The waveform clearly showed the effects of cavitation on the negative portion of the pressure cycle and the presence of cavitation noise. The measuring array is $1\frac{1}{3}$ the diameter of this transducer.

TABLE I. Freefield voltage sensitivity (in dB *re* 1 V/μbar) of cylindrical transducer measured through beam-forming network.

Freq. (kc/sec)	Measured in far field Meas. 1	Measured in far field Meas. 2	Measured in near field Meas. 1	Measured in near field Meas. 2
6	−120.5	−119.7	−121.1	−121.0
7	−117.1	−115.8	−116.9	−116.9
8	−112.0	−110.8	−111.8	−112.0
9	−106.9	−105.3	−106.3	−106.5
10	−101.1	− 99.6	−100.6	−101.1
11	− 95.0	− 93.2	− 94.4	− 95.3
12	− 89.1	− 87.6	− 88.9	− 89.7
13	− 95.7	− 94.7	− 97.0	− 97.2

Although this transducer was slightly larger than the constant-level sound field produced by the measuring array, directivity was measured in two planes, horizontal and vertical. The main beam was faithfully reproduced, but some distortion was apparent in the minor lobes. The measurement of directivity requires greater precision in design of the measuring array than is required for measurement of response or sensitivity. In the measurement of response, the output from the array is never more than 6 dB below the output of a single array element measured separately. In the measurement of directivity, the ratio may be 20 dB or more.

The freefield voltage sensitivity of a circular-piston transducer 55 cm in diameter, or half the size of the measuring array, was measured with the same reliability as shown in Table I for the cylindrical transducer. Directivity was measured at the distance 35 cm, which was as close as the transducer could be placed and still be rotated through 360°. Patterns at 6, 8, 10, and 12 kc/sec are shown in Fig. 10. Directivity was measured at the same frequencies in the far field at the distance 6 m, and at 10 kc/sec in the near field at 60 cm for comparison.

IV. MEASURING-ARRAY DESIGN DATA

Experimental data have shown that a particular shading function can be used to produce a multielement, plane-area array that has a constant-amplitude sound pressure over a region of its near field. One shading function was used to produce two arrays, one of circular symmetry and one of square symmetry. The calibrations described in this paper were obtained with both arrays. The basis for producing the area shading is a line shading function. The farfield directivity of the shaded line array is given in general terms by Eq. (15). The sound pressure at the center of a circular-symmetry area array is given as a function of frequency by Eqs. (17) and (18). On the basis that this point is the origin of variations in the level of the near sound field and the point at which the variations are a maximum, we establish the useful calibrating range for an array design.

The line shading function can be represented by two numbers, the values of m and n in Eq. (15) representing the farfield directivity. Examples of the sound pressure at the area-array center for four shading functions are shown in Fig. 5. In the example $m=5$, $n=5$, the line shading function is produced by five replications of the fifth-power binomial-series coefficients. The rth coefficient of the binomial series can be expressed as

$$\binom{n}{r}=n!/r!(n-r)!,\quad r=0,\,1,\,2,\,\cdots,\,n.$$

If $m=n$ as in the example $m=5$, $n=5$ in Fig. 5, the first shading coefficient to the right of center can be expressed in general terms by

$$\sum_{r=1}^{n}\binom{n}{r}\Big/\sum_{r=0}^{n}\binom{n}{r}=\sum_{r=1}^{n}\binom{n}{r}(0.5)^{n}.$$

The second shading coefficient to the right of center can be expressed by

$$\sum_{r=2}^{n}\binom{n}{r}(0.5)^{n},$$

and so on.

These coefficients can be recognized as the probability of at least r occurrences in n independent trials, when the probability in any single trial is 0.5. The coefficients are available in published Tables[19] for values of n from 2 to 49.

As is apparent from Fig. 5, the shading function for a line when $m\neq n$ is obtained by adding elements of unity

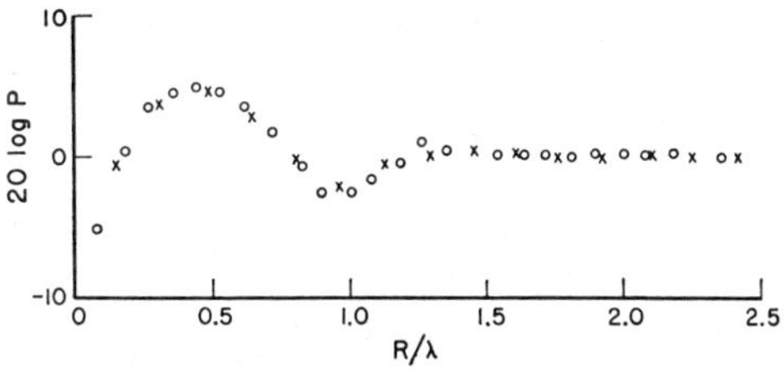

FIG. 11. Relative sound-pressure level at the center of a circular-symmetry measuring array: ○ Computed from Eq. (17); $R=9d/2$; $m=9$, $n=5$; $(\sin 9\phi/9\sin\phi)\cos^5\phi$; shading function (half line length) = 0.031, 0.19, 0.5, 0.81, 0.97, 1, 1. × Computed from Eq. (18); $R=16d/2$; $m=16$, $n=20$; $(\sin 16\phi/16\sin\phi)\cos^{20}\phi$; shading function (half line length) = 0.02, 0.06, 0.13, 0.25, 0.41, 0.59, 0.75, 0.87, 0.94, 0.98, 1, 1, 1.

source strength in the center region of the array in number equal to $(m-n)$. The sound pressure computed by means of Eq. (17) for $m=9$, $n=5$ has been replotted in Fig. 11 as a function of the piston radius R divided by the wavelength of sound. A shading function for $n=20$ was selected from the Tables, and a value of $m=16$ was found for which Eq. (18) gave the same curve.

The low-frequency limit for constant sound pressure at the center can be taken from Fig. 11 as $R/\lambda=1.2$. The upper frequency limit for the array has been stated as $d/\lambda=0.8$, a limit determined by the element spacing. Thus, for the array that was constructed, the lower limit is $R/\lambda=9d/2\lambda=1.2$, or $d/\lambda=2.4/9$; and the upper limit is $d/\lambda=0.8$. The ratio of upper frequency limit to lower limit is 3.

Thus, for a given array design, we are limited at the lower frequency by R/λ and at the upper frequency by d/λ. For the second example in Fig. 10, the upper limit would be equal to $5\frac{1}{3}$ times the lower frequency limit. One could guess from these examples that a shading function in the range $n=10$ to 14 would have $m=n$. Equations (17) or (18) could then be used for verification.

For design purposes, the freefield voltage sensitivity of the elements can be computed approximately from the piezoelectric constants. This is the sensitivity of the array. The effective area of the array, A in Eq. (1) for the plane-wave reciprocity parameter, is the area of the piston of radius R in Eq. (17) or (18), if the array is designed for circular symmetry. With the freefield voltage sensitivity and the plane-wave reciprocity parameter, the nearfield transmitting current response, which is the 0-dB level (Fig. 11) for sound pressure per ampere, can be computed. Measurements show that the diameter of the aperture for constant sound pressure is the distance between the 80% shading coefficients in the line shading function, or about half the width of the array. The axial depth of the constant-sound-pressure near field is approximately R^2/λ.

An array can be constructed as shown in Fig. 6, or a number of identical shaded lines can be constructed with the elements connected in parallel and the shading obtained by means of capacitors in series with each ele-

[19] National Bureau of Standards, Applied Mathematics Series No. 6, *Tables of the Binomial Probability Distribution* (U. S. Government Printing Office, Washington, D. C., 1950).

ment. The number of lines may be equal to the number of elements in the line. Circular symmetry would then be obtained by connecting the lines in parallel with capacitors in series with each line to adjust the lines to the shading function. Array shapes other than square could be obtained by changing the line spacing or the number of lines. The correct nearfield transmitting current response and the ratio of the freefield voltage sensitivity to the effective area can be determined by nearfield measurements, as previously described.

V. APPLICATIONS AND CONCLUSIONS

Existing facilities for the calibration of underwater-sound transducers can extend their capabilities by making measurements in the near sound field. A measuring array can be designed and built to produce a plane-wave, constant-level sound field for calibration of transducers of any size up to one-half the dimensions of the array. The instrumentation, procedures, and calculations are similar to those used to calibrate a transducer in the far field of a source. The method requires no special training. Measurements can be obtained by using continuous sound, pulsed sound, or a band of noise.

In this array concept of a nearfield calibration, the measurements are within the near field of the measuring array. Thus, the unknown transducer may be omnidirectional with no plane-wave near field, it may be directional with a plane-wave near field, or it may even be an isophase spherical or cylindrical surface. By reciprocity, the measuring array correctly integrates the sound field of any of these unknown transducers. Thus, from nearfield measurements, the freefield voltage sensitivity of a transducer is obtained from a voltage, a current, and the computed or measured nearfield transmitting current response of the measuring array, as in Eq. (24). The farfield transmitting current response of a transducer is obtained from a voltage, a current, the freefield voltage sensitivity, the effective area of the measuring array, and the wavelength of sound, as in Eq. (25). The only knowledge about the transducer that is required is its physical size in relation to the size of the measuring array or, more specifically, in relation to the size of the uniform plane-wave region in the near field of the measuring array when it is used as a source. The transducer can be rotated within this plane-wave near field to obtain its directivity pattern directly, without any computation. Because the reciprocity principle applies to this experiment, it is possible to measure the directivity of the unknown transducer, using it as the source. Other acoustic parameters such as source level, transmitting voltage response, and transmitting power response can be derived from similar equations.

If the number of calibrations will not justify the construction of a measuring array, then probe- or line-scanning can be used, and the farfield characteristics can be determined by means of Eqs. (3), (4), (9), and (10). The Kirchhoff–Helmholtz formula can be used to compute directivity within the limitations specified. Electrical coupling between the measuring and the unknown transducer is easier to identify in array measurements than in probe measurements.

Small transducers can be calibrated at frequencies below 10 kc/sec in a lake in the near field of a line array to eliminate surface and bottom reflections—a more economical method for this purpose than the construction of an area array. The shading function represented by $m=5$, $n=5$ (Fig.5) produces an ideal cylindrical wave for this type of measurement. The transition from spherical to cylindrical divergence occurs without a sound-pressure maximum as is common to the area arrays. An absolute calibration can be performed with two lines of length almost equal to the water depth, and by the application of the cylindrical-wave reciprocity parameter.[12] The effective length of the line L, in the reciprocity parameter, can be determined from the design (the distance between 0.5 shading coefficients), or can be obtained by a reciprocity calibration using a calibrated hydrophone.

ACKNOWLEDGMENTS

This project was possible only through the cooperation and technical competence of the personnel in the Transducer and Calibration Divisions of the U. S. Navy Underwater Sound Reference Laboratory. They constructed the arrays and obtained the experimental data. R. J. Bobber's interest and criticism throughout this research and in the preparation of this paper are also gratefully acknowledged.

Due to the requirements for operation of transducers at great depths in the ocean, it is necessary to calibrate them at correspondingly high pressures. The high-pressure tube provides such a means of calibration.

Louis G. Beatty has been primarily responsible for the development of low-sonic and infrasonic controlled-environment calibration techniques and systems.

Robert J. Bobber is the Technical Director of the Naval Research Laboratory, Underwater Sound Reference Division at which the other two authors are staff members.

This paper is reprinted with the permission of the senior author and the Acoustical Society of America.

Reprinted from THE JOURNAL OF ACOUSTICAL SOCIETY OF AMERICA, Vol. 39, No. 1, 48-54, January 1966

37

Sonar Transducer Calibration in a High-Pressure Tube

LOUIS G. BEATTY, ROBERT J. BOBBER, AND DAVID L. PHILLIPS

U. S. Navy Underwater Sound Reference Laboratory, Orlando, Florida 32806

A unique tube facility for the calibration of sonar transducers under hydrostatic pressures to 8500 psi is described. The pressure vessel consists of the modified liner from a 16-in. gun. The liner is 50 ft long and has an inside diameter of 15 in. An active-impedance principle is applied to establish plane progressive waves within the tube. Resonant transducers can be calibrated in the frequency range 100–1500 Hz; nonresonant, hard transducers can be calibrated in the range 40–1500 Hz.

INTRODUCTION

THE use of sonar and other underwater-sound transducers at ever-increasing ocean depths has led to the need for tanks in which acoustic calibration and evaluation measurements can be made in a hydrostatic-pressure-controlled environment. At high-audio and ultrasonic frequencies, this need can be met with an anechoic tank in which conventional free-field methods are used. The sound field in the tank approximates a free field only when the length, width, and depth of the tank are large as compared with the wavelength of sound in the water. This fact sets the low-frequency limit, which, in the Navy Underwater Sound Reference Laboratory's (USRL) 9500-gal anechoic tank,[1] is 2–10 kHz, depending on the type of measurement, the transducer size, and the Q of the transducer. The frequency limit would be lower for a larger tank, but, because high hydrostatic pressure is involved, both engineering problems and tank cost multiply rapidly as tank size increases. The replacement cost of the USRL's present 1000-psi tank facility would be several hundred-thousand dollars. It would cost several million dollars to build a substantially larger facility for lower frequencies and higher pressures.

Several hydrophone calibration techniques are available for low-audio and infrasonic frequencies.[2–4] These techniques are limited, however, by either or both of the requirements that (1) the hydrophone being calibrated must have a high acoustic impedance and (2) the tank dimensions must be small as compared with the wavelength of sound in water.

Thus, there has been no feasible method or facility for calibrating at pressure greater than 1000 psi a sonar transducer that is, for example, a 1-ft cube resonant at 500 Hz. A large anechoic tank for operation at high pressure would be very costly; present standard codes are inadequate for the size and pressure required so that the design and fabrication of such a tank would itself constitute a research project. The size of the transducer and its low acoustic impedance at resonance eliminate all of the low-frequency techniques.

I. MEASUREMENTS IN A TUBE

Transducers like the example just cited can be calibrated in a tube whose diameter is less than a half-wavelength and whose length is several wavelengths. The wavemotion can be controlled in such a tube, and the cost and engineering problems are manageable.

To calibrate sonar transducers under high static pressures, USRL has a tube facility (Fig. 1) in which the pressure vessel consists of the liner from a 16-in. naval gun. The tube is 50 ft long and 15 in. i.d., with a wall thickness of 2.3 in. The tube is fitted with closing plugs on each end and small holes along its length to allow insertion of instruments, particularly probe hydrophones, through the walls. The operating high-pressure limit is 8500 psi, equivalent to the pressure at an ocean depth of almost 20 000 ft.

A unique method is applied to establish sound waves of known characteristics in the USRL tube. An electroacoustic transducer at one end of the tube produces

[1] C. L. Darner, "An Anechoic Tank for Underwater Sound Measurements under High Hydrostatic Pressures," J. Acoust. Soc. Am. **26**, 221–222 (1954).
[2] L. L. Beranek, *Acoustic Measurements* (John Wiley & Sons, Inc., New York, 1949).
[3] W. J. Trott and E. N. Lide, "Two-Projector Null Method for Calibration of Hydrophones at Low Audio and Infrasonic Frequencies," J. Acoust. Soc. Am. **27**, 951–955 (1955).
[4] C. C. Sims and T. A. Henriquez, "Reciprocity Calibration of a Standard Hydrophone at 16 000 psi," J. Acoust. Soc. Am. **36**, 1704–1707 (1964).

FIG. 1. Tube Facility at U. S. Navy Underwater Sound Reference Laboratory.

sound waves that travel as plane progressive waves down the length of the tube and are absorbed by a second active electroacoustic transducer at the other end.

It has been demonstrated[5–7] that an active transducer can be used as a characteristic terminal impedance for an acoustic transmission line, and that the standing-wave ratio in the line and the radiation load impedance on the source transducer can be controlled. In practice, the transducers at the two ends of the tube are driven by the same electronic amplifier, but the amplitude and phase of the signal fed to one transducer are adjusted relative to those fed to the other one. The system is shown schematically in Figs. 2 and 3 for two different calibration arrangements. Probe hydrophones at several

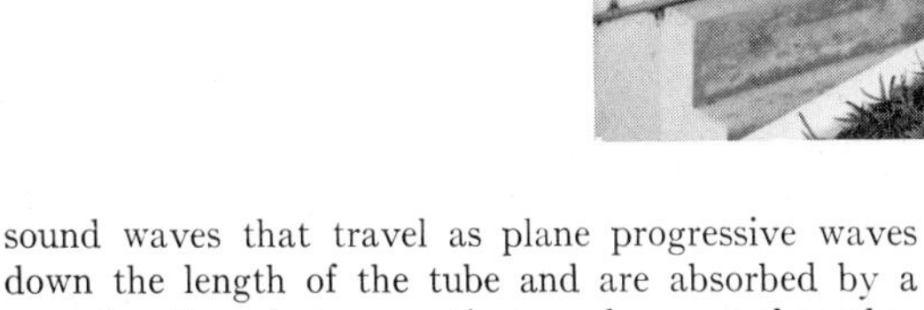

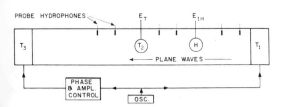

FIG. 2. Single-termination arrangement; T_1 and T_3 are terminal transducers, T_2 and H are unknown or standard transducers; E_T and E_{1H} are open-circuit voltages.

[5] R. J. Bobber and L. G. Beatty, "Impedance Tube for Underwater Sound Transducer Evaluation," J. Acoust. Soc. Am. **31**, 832(A) (1959).

[6] R. J. Bobber, "Active Load Impedance," J. Acoust. Soc. Am. **34**, 282–288 (1962).

[7] L. G. Beatty, "Acoustic Impedance in a Rigid-Walled Cylindrical Sound Channel Terminated at Both Ends with Active Transducers," J. Acoust. Soc. Am. **36**, 1081–1089 (1964).

locations sample the acoustic pressure distribution along the tube and facilitate adjustment of the standing- and progressive-wave conditions in the tube. If the transducer being calibrated nearly fills the cross section and prevents propagation of plane progressive waves the full length of the tube, it is still possible to establish the plane-wave condition in that part of the tube between the sound source and the transducer being calibrated.

The frequency of the first transverse mode in the tube, for which the sound pressure no longer is uniform across a diameter, is about 2 kHz. Simple plane progressive waves can be obtained at frequencies up to 1.5 kHz.

The primary function of the facility is to measure how much the sensitivity of a transducer is changed by variation of the hydrostatic pressure. Although relative measurement of sensitivity versus pressure is the most important use of the tube, transducers are calibrated routinely at any desired static pressure to 8500 psi by comparison with a small standard hydrophone that has been calibrated previously by a coupler-reciprocity technique,[4] or by a tube-reciprocity technique.[8,9]

The radiation impedance of the transducer under measurement in the tube is not the same as it would be in a free field. For small, hard, nonresonant transducers, this difference is negligible and the calibration is equivalent to a free-field calibration. If the ratio of radiation impedance to mechanical impedance is *not*

[8] L. G. Beatty, "Reciprocity Calibration within a Rigid-Walled Water-Filled Tube," J. Acoust. Soc. Am. **35**, 810(A) (1963).

[9] L. G. Beatty, "Reciprocity Calibration in a Tube with an Active Impedance Termination," J. Acoust. Soc. Am. **39**, 40–47 (1966).

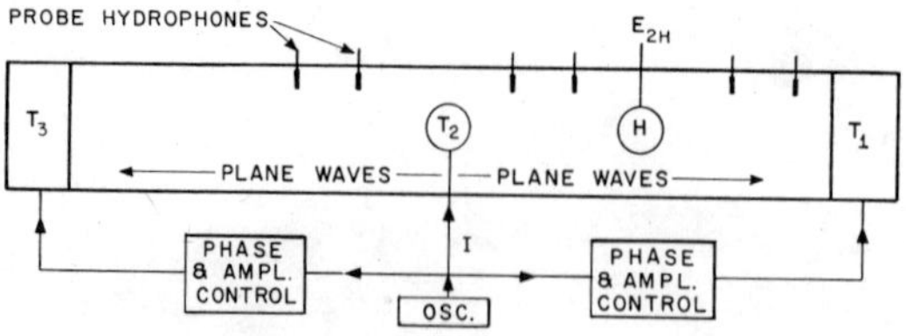

FIG. 3. Double-termination arrangement; T_2 is the reciprocal transducer and H is the hydrophone in a reciprocity calibration.

small, the transducer is calibrated first in a free field and then in the tube at the same low pressure. The two calibration curves will be different, but the difference, for all practical purposes, is a function of the boundary conditions imposed by the tube wall and is independent of the hydrostatic pressure in the tube. The difference, therefore, can be treated as a correction to calibrations in the tube at other pressures. The correction for resonant transducers, along with the theoretical detail pertaining to the tube-reciprocity calibration method and the equivalence of tube and free-field calibrations, are presented in companion paper.[9] Inasmuch as the boundary conditions in the tube are similar to those for an element in a large planar array, the sensitivity or response measured in the tube is approximately the same as that of an element in an array.

II. COMPARISON CALIBRATION

A comparison calibration, or calibration by comparison of the sensitivity of a transducer with that of a standard hydrophone, is made with the single-termination arrangement shown in Fig. 2, where T_2 is the unknown and H is the standard hydrophone. When the phase and amplitude of T_1 are properly adjusted with respect to those of T_3, plane progressive sound waves produced by T_1 impinge on both H and T_2; the sound energy not absorbed by H or T_2 continues down the tube to be absorbed by T_3.

There being negligible dissipation in the tube, H and T_2 are subjected to the same plane-wave pressure. The free-field voltage sensitivity M_{T2} of the unknown transducer T_2, then, is given by

$$M_{T2}=M_H E_T/E_{1H}, \tag{1}$$

where M_H is the free-field voltage sensitivity of the standard hydrophone.

For this measurement, the plane-wave region must extend to T_2 from a point between T_1 and H. Transducers T_1 and T_3 are piston radiators that do not quite fill the cross section of the tube; consequently, there is a short region where the sound field diverges near both T_1 and T_3. Hydrophone H is small and does not significantly disturb the plane waves. The unknown T_2 may be as long as 11 in. in the direction of a diameter of the tube. When T_2 is large, the region between T_3 and T_2 contains standing waves as well as progressive waves.

The standard hydrophone can be used also to measure the tube transmitting response of the unknown transducer. For this purpose, the double-termination arrangement shown in Fig. 3 is used. The source T_2 radiates plane waves in both directions; both T_1 and T_3 act as absorbers. The sound pressure is measured by the standard hydrophone H. The tube transmitting current response of T_2 is then

$$S_{T2}=E_{2H}/(M_H I), \tag{2}$$

where I is the current driving transducer T_2.

III. RECIPROCITY CALIBRATION

A reciprocity calibration requires both of the arrangements shown in Figs. 2 and 3. Consider T_1, T_2, and H to be the projector, reciprocal transducer, and hydrophone, respectively, in a reciprocity calibration. Then, the measurements described for comparison calibrations constitute the three necessary steps of a reciprocity calibration:

$$\begin{aligned} T_1 &\longrightarrow H \dashrightarrow E_{1H}, \\ T_1 &\longrightarrow T_2 \dashrightarrow E_T, \\ I \dashrightarrow T_2 &\longrightarrow H \dashrightarrow E_{2H}. \end{aligned} \tag{3}$$

The sensitivity of the hydrophone is then

$$M_H=[E_{1H}E_{2H}J_t/(E_T I)]^{\frac{1}{2}}, \tag{4}$$

where J_t is the reciprocity parameter for the tube boundary condition. Beatty shows[9] that, if the mechanical impedance of the reciprocal transducer is high over the frequency range of the calibration, $J_t=2A/\rho c$, where A is the cross-sectional area of the tube and ρc is the specific acoustic impedance of the water in the tube. The speed of sound c in the water in the tube is 4700 ft/sec at 0 psi; it increases to 5000 ft/sec at 8500 psi. With M_H determined, the transmitting or receiving sensitivity of T_2 can be computed also.

If the mechanical impedance of the unknown transducer is high over the frequency range of the calibration, it can serve as the reciprocal transducer T_2 in the reciprocity calibration; if the unknown transducer does not satisfy this condition, it must be used as the hydrophone, and a transducer that has been developed for the purpose is used as the reciprocal transducer.

IV. MEASUREMENT DATA

The measured sensitivities of transducers calibrated in the tube by the comparison method agree with those obtained by the reciprocity method. The reciprocity method requires more measurements than the comparison method, but it is independent of a standard hydrophone that has to be calibrated elsewhere. The choice between the two methods is determined by convenience.

The free-field voltage sensitivities of small nonresonant hydrophones measured in the tube have been the

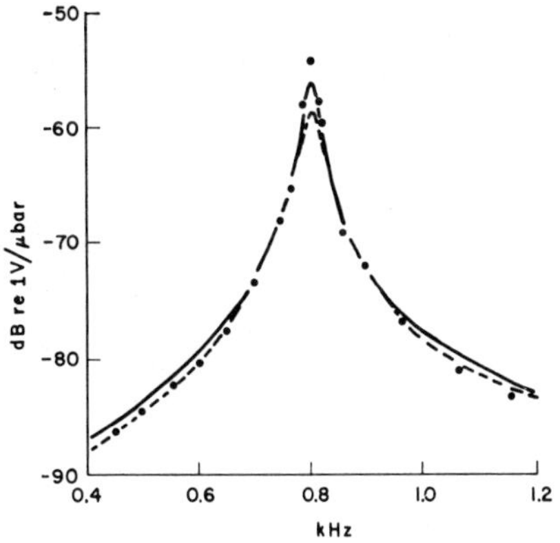

FIG. 4. Free-field voltage sensitivity of a variable-reluctance transducer. The tube values include the frequency adjustment $\Delta f_0=35$ Hz. ——: Free field No. 1. - - -: Free field No. 2. ●: Tube.

same as those measured by conventional free-field methods.

Calibrating resonant transducers at low frequencies always is difficult, even when high hydrostatic pressure is not a parameter. Such transducers usually operate on an electromagnetic transduction principle; they tend to be nonlinear and to have very high values for Q. Figure 4 shows the free-field voltage sensitivity of a variable-reluctance transducer of $Q=26$; Fig. 5 shows that of a moving-coil transducer of $Q=22$. The tube data have been plotted against an adjusted frequency scale $f'=f\Delta f_0$, where f is the measured frequency and Δf_0 is the ratio of the resonant frequency in the free field to the resonant frequency in the tube. For example, tube data shown at the resonance frequency are plotted 20 Hz lower than the frequency at which they actually occurred in the tube.

As can be seen from Figs. 4 and 5, the agreement between tube and free-field measurements at off-resonance frequencies is good to excellent. The two free-field measurements were made in different facilities; the disagreement between the tube calibration and the free-field calibration is not worse than that between the two free-field calibrations. The disagreement at the resonance frequency (Fig. 4) is greater than the ± 1 dB at other frequencies, and is attributed partially to the high Q of the transducer, which makes measurement of the peak value very sensitive to small variations in the calibration conditions. There was no evidence of nonlinearity in the variable-reluctance transducer; it was stable, and measurements could be repeated at all frequencies. Changes in sensitivity due to changes in hydrostatic pressure could be detected and measured. The moving-coil transducer was nonlinear at resonance and a stable measurement condition could not be obtained in the tube at that frequency. As yet, transducers cannot be calibrated in the tube at signal levels and frequencies at which they are nonlinear.

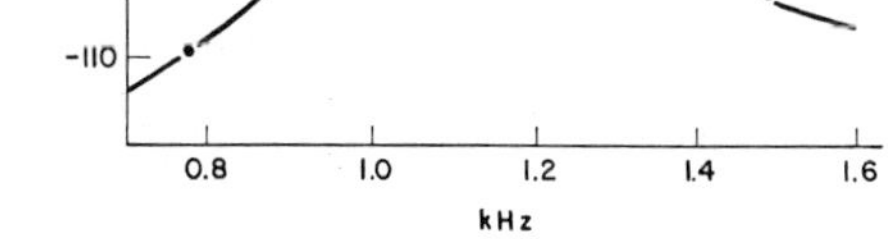

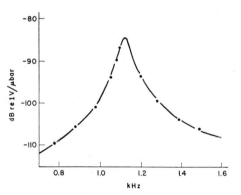

FIG. 5. Free-field voltage sensitivity of a moving-coil transducer. The tube values include the frequency adjustment $\Delta f_0=20$ Hz. ——: Free field. ●: Tube.

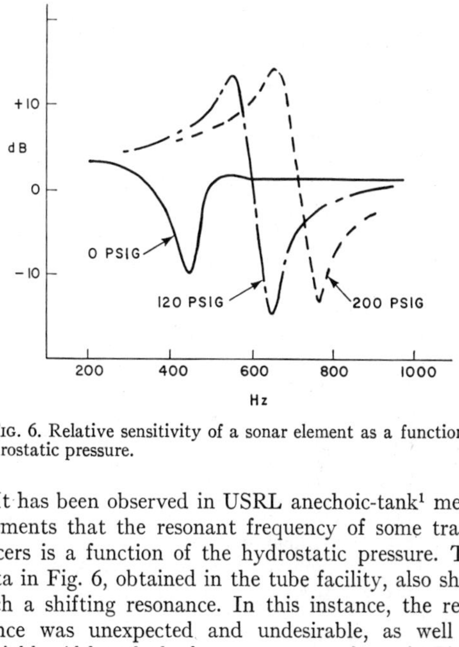

FIG. 6. Relative sensitivity of a sonar element as a function of hydrostatic pressure.

It has been observed in USRL anechoic-tank[1] measurements that the resonant frequency of some transducers is a function of the hydrostatic pressure. The data in Fig. 6, obtained in the tube facility, also show such a shifting resonance. In this instance, the resonance was unexpected and undesirable, as well as variable. Although the frequency range shown in Fig. 6 is below the nominal low-frequency limit of the USRL anechoic-tank facility, qualitative measurements in the anechoic tank verified the shifting resonance of the sonar element.

Line hydrophones of considerable length can be calibrated in the tube. In Fig. 7, the results of endfire measurements obtained for the pressure range 0–200 psi on the 8-ft line of Fig. 8 are compared with theory. Only the endfire sensitivity can be measured; but, if it can be assumed that the hydrophone is a uniform line, then the directivity is known and the broadside sensitivity can be computed. Figure 7 also shows the comparison between the broadside sensitivity computed

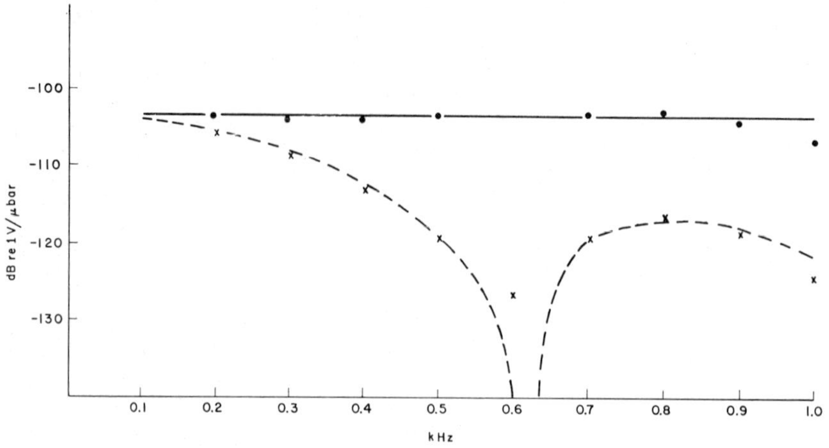

FIG. 7. Free-field voltage sensitivity of an 8-ft line hydrophone. *Broadside*: ——: Free field. •: Computed from tube endfire data and theoretical directivity. *Endfire*: - - -: Theory. x: Tube values.

from measured endfire values and the broadside sensitivity actually measured in a free-field calibration. The agreement is good except near 0.6 kHz, where the directivity pattern shows a null in the endfire direction.

A unique capability of the tube facility is measurement of the transmitting response of elements of large arrays under baffle conditions similar to those that exist in the array. One element calibrated in the tube was a doublet transducer; that is, its front and back radiations were 180° out of phase. In a closely packed array, the radiation load of such an element may approach that of a large piston (ρcA mechanical ohms). The radiation load of a doublet alone in a free field is predominantly a mass reactance and is very small. This element almost filled the tube cross section; where spaces remained between the element and the tube walls, baffles were provided. Thus, measurements were made under almost full-load conditions.

Calibration measurements have been made up to the maximum hydrostatic pressure 8500 psi and throughout the frequency range 40–1500 Hz. The low-frequency limit is set by the capability of the piezoceramic terminal transducers to produce measurable sound pressures in the tube when they are driven by the existing electronic equipment. Plane progressive wave techniques are used to the low-frequency limit 100 Hz. From 40 to 100 Hz, conventional low-frequency techniques are used wherein uniformity of sound pressure is assumed in the region of the tube where T_2 and H of Fig. 2 are located. Because of this assumption, neither T_2 nor H can be resonant at frequencies below 200 Hz when measurements are made below 100 Hz.

V. IMPEDANCE MEASUREMENTS

In addition to measurement of sensitivity and response, impedance measurements can be made at the terminals of the unknown while it is driving the system. The radiation impedance consists of three components. If the cross section of the transducer is smaller than that of the tube, the sound will diverge near the transducer, and will give rise to a radiation mass reactance. In the tube region where plane waves are present, there may be any combination of progressive or standing waves. The plane progressive waves react on the transducer as a radiation resistance, and the plane standing waves as a radiation reactance.

The authors have shown[6,7] that the radiation load

FIG. 8. Carriages for rigging transducers in the tube facility.

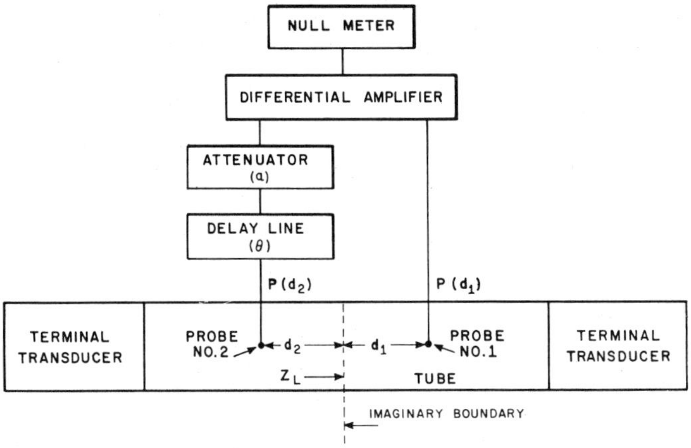

FIG. 9. Two-probe system for measuring the wave impedance Z_L at an imaginary boundary in a tube.

due to the plane waves can be adjusted to virtually any desired value. This adjustment then permits loading the transducer with any arbitrary radiation impedance.

VI. ESTABLISHMENT OF PLANE PROGRESSIVE WAVES

The key to achieving propagation of plane waves from one end of a tube to the other is a practical method for monitoring the standing-wave or progressive-wave condition in the tube while the relative amplitude and phase of the source and the terminal transducer are adjusted. Two methods have been used.

In the first, the output voltages of three probe hydrophones spaced along the tube are monitored simultaneously. The distances between probes are so chosen that no pair of probes is separated by an integral number of half-wavelengths. When equal sound pressures are measured at the three probe positions, the standing-wave ratio is 1—that is, only plane progressive waves are present. This method is feasible, but tedious.

The second and preferred method is a variation of a two-probe technique developed by Bouyoucous[10] for measuring the terminal impedance in a tube. By this method, the wave impedance and therefore any combination of standing and progressive plane waves can be identified. If plane progressive or standing waves are propagated in a rigid-walled tube with a source transducer at one end and a load or terminal plane of impedance Z_L at the other end, the sound pressure $P(d)$ in a plane at a distance d from the terminal plane is given by

$$P(d)=P_L \cos kd + jZ_0U_L \sin kd, \tag{5}$$

where P_L and U_L are the pressure and the velocity at the load impedance $Z_L=P_L/U_L$; Z_0 is the characteristic impedance of the medium; k is the wavenumber. There is nothing in the derivation of Eq. 5 that limits its application to terminal impedance, so the wave impedance at any imaginary plane boundary can be found. The pressure at a point on each side of the boundary shown in Fig. 9 is given by

$$P(d_1)=P_L \cos kd_1 + jZ_0U_L \sin kd_1, \tag{6}$$

$$P(d_2)=P_L \cos kd_2 + jZ_0U_L \sin kd_2. \tag{7}$$

The magnitude and phase of the output of probe No. 2 can be adjusted until they are equal to those of probe No. 1. Then, if the sensitivities of the probes are equal,

$$(ae^{j\theta})P(d_2)-P(d_1)=0, \tag{8}$$

where a and θ are the magnitude and phase adjustments that make the outputs of the probes equal.

If the imaginary boundary is taken equidistant from each probe, $d_2=-d_1$; then, substituting Eqs. 6 and 7 into Eq. 8 and solving for Z_L/Z_0 produces

$$Z_L/Z_0=b(1+ae^{j\theta})/(1-ae^{j\theta}), \tag{9}$$

where $b=(-j \cot kd_2)^{-1}$ is a constant for a given frequency. Solutions to Eq. 9 are found easily with the aid of the familiar Smith chart.[11] The normalized impedance Z_L/Z_0 can be determined from a and θ, or the adjustments a and θ necessary to achieve a desired impedance can be found. These adjustments to the output of probe No. 2 are made with the attenuator and the delay line of Fig. 9. The signals to the two terminal transducers are varied until the amplitudes and phases of the two probe output signals are matched, as determined by the differential amplifier and the null meter.

In addition to effecting more complete identification of the wave conditions in the tube, the two-probe method and Eq. 9 are useful for the impedance measurements previously discussed. For the special case of plane progressive waves, $a=1$, $\theta=-2kd$; then, for

[10] J. V. Bouyoucous, "A Two-Probe Method for Measuring Acoustical Impedance," Harvard Univ. Acoust. Res. Lab. Internal Mem. (8 Dec. 1955).

[11] P. H. Smith, "An Improved Transmission Line Calculator," Electronics 17, No. 1, 130–133, 318, 320, 322, 324, 325 (1944).

waves moving from probe No. 2 toward probe No. 1, Eq. 9 reduces to $Z_L/Z_0=+1$. For waves moving from probe No. 1 toward probe No. 2, Eq. 9 reduces to $Z_L/Z_0=-1$.

The relative magnitude and phase of the probe voltages are measured in a modification of the coupler[4] that is used to calibrate the standard hydrophone. The magnitudes of the sensitivities of the probes differ by only 0.3–0.5 dB throughout the whole frequency and hydrostatic pressure ranges of the facility, and this difference is accounted for by inserting a correction into the measuring system. All of the probes have virtually the same difference in phase between output and input signals.

VII. TRANSDUCER SIZE

Although the inside diameter of the tube is 15 in., a sealing ring at each end reduces the diameter of the access opening to 14 in. Transducers that can fit through that opening can be accommodated in the tube. A transducer that radiates in two opposite directions, as T_2 does in Fig. 3, can be of the size of the maximum access opening. Transducers such as T_2 in Fig. 2 must be small enough to allow sound energy to pass by or around them and to allow the terminal transducer (T_3) at the rear to influence the wave conditions in front of T_2. The maximum size that will allow sound to pass by or around a transducer is not known; however, no problems connected with transducer size have arisen in the calibration of transducers up to 11 in. in diameter and 6 in. long.

VIII. TRANSDUCER RIGGING

Normally, if both receiving and transmitting measurements are desired, the transducer to be calibrated is rigged in the center of the test chamber as illustrated by T_2 in Figs. 2 and 3. This arrangement can be used for either comparison or reciprocity measurements. When receiving sensitivity only is required, the rigging arrangement of H in Figs. 2 and 3 may be used.

Transducers are rigged in the tube on specially constructed carriages such as those shown holding the line hydrophone in Fig. 8. These carriages conform to the walls of the tube and are supported on their convex surfaces by countersunk ball-bearing rollers to facilitate movement inside the chamber. The tube is slanted 15° from the horizontal, and the carriages roll downhill into the tube. Carriages are held in place in the chamber by attaching them to a wire cable whose upper end is tied to an expandable clamping ring fitted into the upper end of the chamber.

The probe hydrophones in the top of the tube or at radial positions 120° in either direction from the top extend only 4 in. into the tube. Electrical cables from transducers T_2 or H are brought out through an unused probe hole. There are 11 probe holes; usually, only 10 probe hydrophones are used at any one time.

IX. TUBE

The gun-barrel liner was obtained from the now defunct Naval Gun Factory. The Gun Factory also designed the closures and the stuffing glands for the probe holes. Machine work was done at the Charleston Naval Shipyard.

The end plugs are large continuous-thread screws. "Mushroom" pads, used in large naval guns to seal the high dynamic pressures, serve as seals for the 8500-psi static pressure.

The tube was installed with the axis slanted 15° from the horizontal because the arrangement was economical and also because it allows the tube to be opened after only a small amount of water has been drained. The slant also facilitates the removal of entrapped air by conventional circulation procedures.

Air bubbles also are removed by exposing the water to vacuum during filling and draining, and by storing the water under vacuum. Temperature control is being planned.

The facility has been in use for research projects for about two years. It is currently in routine use as one of eight facilities at the Underwater Sound Reference Laboratory that serve the sonar-calibration needs of the Navy.

In the following paper, the author developed the scaling law for the sound signal developed by explosive charges. This scaling law is now accepted by all workers in the field.

Dr. Weston is a physicist. He received the D.Sc. degree in physics from London University. He has worked in the Royal Naval Scientific Service since 1951 and is a Senior Principal Scientific Officer at the Admiralty Research Laboratory. His principal research interest has been in underwater acoustics, particularly sound propagation and oceanography. He has also worked with underwater explosives as acoustic sources and in signal processing.

This paper is reprinted with the permission of the author and the Institute of Physics (London).

REPRINTED FROM THE
PROCEEDINGS OF THE PHYSICAL SOCIETY, VOL. LXXVI, p. 233, 1960

38

Underwater Explosions as Acoustic Sources

BY D. E. WESTON
Admiralty Research Laboratory, Teddington, Middlesex

MS. received 19*th November* 1959, *in revised form* 30*th March* 1960

Abstract. The manner in which underwater explosions differ from low-amplitude point sources of sound is considered theoretically, especially effects due to cavitation near the sea surface. Some measured differences between the acoustic source levels of various size charges and some absolute charge source levels are given. These experimental results are presented for charge sizes between 0·002 and 50 lb, for charge depths from 7 to 60 fathoms, and for frequencies from 25 c/s to 6·4 kc/s. The results at a given depth are shown to obey a simple scaling law. Theoretical source levels are calculated by Fourier analysis of shot pressure–time curves reported by Arons. At high frequencies the theoretical spectral energies of the shock wave and the bubble pulses are simply added together, but at low frequencies it is necessary to take account of phase. In general there is very good agreement between the experimental and theoretical levels, and certain small discrepancies are explained in terms of bubble migration and related effects.

§ 1. INTRODUCTION

UNDERWATER explosions have been used as acoustic or seismic sources in many connections: for example in geophysical prospecting, in radio-acoustic and pure acoustic navigation, for signalling including distress signalling, and for the investigation of underwater sound propagation. Explosions provide a pulse source of great power, so that measurements at long ranges may be made without costly generating equipment. In addition they contain energy in a wide range of frequencies and are not restricted in depth. They have some disadvantages as acoustic sources in research work, mainly due to finite-amplitude effects.

In the course of various sea trials organized by the Admiralty Research Laboratory, many explosive charges have been fired, and much experimental information accumulated. From this it has been possible to deduce absolute source levels for the various charge sizes, which are needed if transmission loss is to be found from measurements of received signal level. In addition there have been several special experiments to measure the energy differences between charges. A knowledge of these differences is necessary so that measurements of various workers with various charges may be reduced to a common standard. The present results are mainly for 1-lb charges of T.N.T. but some data are given for other charges from detonators up to 50 lb—a range of 1 : 25 000 in weight of explosive.

This paper considers first the special nature of the underwater explosion as an acoustic source, particularly effects due to surface cavitation, so that the experimental results may be processed having regard to these effects. The measured charge differences and free-field source levels are then reported, and discussed first in the light of the simple theoretical scaling laws. The theoretical Fourier spectra for the shot pressure–time curve reported by Arons *et al.* (Arons 1948, 1954, Arons, Slifko and Carter 1948, Arons and Yennie 1948) are calculated and compared with experiment, and the modification due to such effects as bubble migration are examined. A generalization of some of the theoretical laws at low frequencies, together with its application to underground explosions and other disturbances, is given elsewhere (Weston 1960).

§ 2. The Underwater Explosion

The general character of underwater explosions is well established, and Cole (1948) may be given as a general reference. When a high explosive is detonated the detonation wave propagates through it and it is converted to incandescent gas at a very high pressure. A spherically symmetrical shock wave is radiated into the water. The pressure rise in the shock wave is practically instantaneous, and is followed by a decay which is initially exponential with time constant typically a fraction of a millisecond. By the time the gas bubble has expanded to its maximum radius the radiated pressure has become slightly negative. Due to the water inertia the bubble overshoots its equilibrium radius, so it begins next to contract, and then undergoes a damped radial oscillation. Observations of up to 10 oscillations have been reported, the actual number depending on how soon the bubble breaks up into smaller parts. At each bubble minimum a pressure pulse is radiated, of smaller amplitude but longer duration than the shock wave. The first bubble pulse has an impulse comparable with that of the shock wave. The radiated pressure–time curve is quantitatively described in § 7.

When the shock wave reaches the surface it is reflected as a tension wave, which may cause cavitation. This can sometimes be seen on cine records as a white flash just below the surface. If the water surface is initially fairly smooth a slight roughening of the same extent as the flash is visible, which is usually described as the 'black ring'. In addition, with a strong shock wave the area above the charge may rise into a dome of broken water, this spray dome having a diameter less than that of the black ring. The precise mechanism of the cavitation phenomenon is not certain, but it can best be understood in terms of the multiple layer theory of Wood, Lakey and Butterworth (1924). The reflected tension wave causes a fracture just below the water surface, and the momentum associated with the first part of the shock is trapped in the region above the fracture. The next part of the shock wave is reflected at the new surface and a second fracture formed. A series of layers will result, stopping when the amplitude in the tail of the shock is insufficient to cause cavitation. These layers will collide and break up to form the spray dome. The region with one layer only may be identified with the black ring. This picture is of course idealized, and in practice will be blurred by such effects as surface roughness. In addition the cavitation regions will probably not be voids with sharp boundaries, but more like layers of bubbles—so producing their white appearance.

The most spectacular surface effect is the plume, which may appear as a bush-like eruption or as a thin spout. The plume occurs later than the spray dome and is due to the oscillating bubble moving upwards under gravity and arriving at the surface. For the deeper charges the bubble breaks up before reaching the surface, and the plume is replaced by an upwelling of frothy water.

§ 3. Differences from a Low-amplitude Point Source

3.1

The underwater explosion differs from a fixed low-amplitude acoustic source both because of the high pressure generated, and because the source may move whilst it is still radiating. The most direct effect of the finite amplitudes is that the different parts of the shock wave travel with different velocities, and there is a Riemann broadening plus a continual sharpening of the pulse. The latter keeps the viscous energy loss at the shock front at a high level by continuously transferring some of the spectral energy to the higher frequencies. These effects are less important beyond about 100 yards and are discussed further in § 6. Very near the charge the velocity of propagation of the shock wave is supersonic, but the saving in transit time is typically only a small fraction of a millisecond.

3.2

If there is cavitation the tension wave reflected from the water surface will not be of full strength, and the signal propagated to a distance may be affected. This occurs mainly for the shock wave at the sea surface, but bubble pulses may also cause cavitation. Theoretically cavitation may still occur after an intermediate shock wave reflection at the sea bottom, or even at the bottom itself if there is sufficient gas there to make the boundary behave as a free surface.

The study of cavitation in this connection is too large a subject to be treated adequately in this paper, and therefore the brief account below concentrates on presenting the main results. It applies to shock wave cavitation at the sea surface, and assumes the charge is not so shallow that the explosion gases blow out through the surface on their first expansion.

There has been much controversy on the effective tensile strength of water, even when discussion is limited to results on cavitation above underwater explosions. The author has measured $7\frac{1}{2}$ fathoms as the critical depth for the appearance of the corrugations (or black ring) above a 1-lb charge, corresponding to 290 lb in^{-2} for the sum of water breaking tension and the 15 lb in^{-2} atmospheric pressure. The breaking tension is likely to vary with the shock wave duration and therefore with the charge size, but will here be assumed constant. Similarly the critical depth for the spray dome was measured as $4\frac{1}{2}$ fathoms, in close agreement with the results of Kolsky *et al.* (1949).

For shallower charges the theoretical horizontal extent of the surface cavitation is usually given simply by the slant range at which the shock pressure has fallen to the above figure (290 lb in^{-2}). In extreme cases there may be corrections due to the hydrostatic pressure gradient, the curvature of the cavitation surface, refraction, surface roughness etc. The ' cavitation angle ' is a useful quantity, defined as the angle to the horizontal made by a line joining the source to the edge of the cavitated area. The cavitation effects are most likely to be serious for large shallow charges, e.g. 50 lb at 7 fathoms, which has a calculated cavitation angle of 14°.

To investigate the effect on a distant signal it is necessary to compare the cavitation angle with the various angles important in describing the undisturbed propagation. At high frequencies only the rays with small grazing angles at the surface are normally effective for propagation, since for larger angles there is a change over from specular to diffuse reflection. Typically the cavitation angle is greater than the angles of the rays carrying the high-frequency energy, so cavitation is unimportant for the high frequencies. At low frequencies rays with angles greater than 14° may be effective, so that cavitation effects will sometimes be significant. This is especially true at very low frequencies, where the direct and surface reflected rays will normally tend to cancel one another out and produce a dipole spreading law. However it is shown later that the major contribution to the low-frequency energy comes from the first bubble pulse, so that overall the low-frequency effects are seldom important. For all the depths, sizes and frequencies discussed in this paper only marginal effects due to cavitation are to be expected.

3.3

Some of the energy lost on surface reflection is re-radiated as cavitation noise, and as noise due to the water in the spray dome falling back. There is also the falling-back of the water in the plume. Some of these noises may be observed at close range, but it is thought unlikely that they are significant at long range.

3.4

When a strong shock wave is reflected at a very small glancing angle from a free surface the reflection is irregular, and the separate pressure and tension waves are replaced by a single pulse of reduced magnitude, as discussed theoretically by Penney and Pike (1950) etc. In idealized conditions this irregular reflection will eventually occur near the surface for any charge weight and depth, and for a 50-lb charge at 7 fathoms the calculated range is 4600 ft. However, refraction and rough surface effects make its occurrence uncertain, and it is probably not important at all in the present connection.

3.5

Energy is radiated for as long as the bubble continues to oscillate; for the larger charges and shallower depths the bubble rises appreciably during one oscillation and some of the energy is radiated from a reduced source depth, so that surface image interference may be more serious. The worst case considered here is that of a 50-lb charge at 7 fathoms, where the vertical rise due to gravity in one oscillation is one quarter of the original source depth. This leads at very low frequencies to a decrease of $2\frac{1}{2}$ dB in the contribution of the first bubble pulse, the shock wave of course being unaffected. In addition it is with these very conditions of large shallow charges that migration makes the bubble pulse energy output relatively unimportant. Thus it appears that reduced radiation depth effects may usually be ignored in the present connection. Bubble migration is more important in its effect on the energy spectrum of the bubble pulse, as discussed in § 8.2, which also presents more background information on migration generally.

All the effects considered in this section are marginal, and not too serious for the data presented here. A few other data have been rejected because of these effects. Note also that with a slight change of circumstances the effects could lead to considerable error, and it is necessary to bear the limits of all of them in mind.

§ 4. MEASURED DIFFERENCES BETWEEN CHARGES

4.1

There are surprisingly few charge comparisons in the literature, and **most** workers have corrected their results assuming energy is proportional to W (charge weight) or $W^{2/3}$, though neither is correct in general. Worzel and Ewing (1948) report some comparisons of pressure measured at a distance for various charges, and O'Brien (1960) gives a similar comparison for Admiralty Research Laboratory records of the low-frequency signals from 1-lb and 50-lb charges fired on the sea bed at about 20 fathoms (averaging 23 dB at about 10 c/s). All these results agree roughly with those presented below. However the main need is for comparisons at one depth and in a series of reasonably narrow frequency bands, preferably for charges fired off the bottom and preferably of energy rather than pressure.

4.2

The measurements here and in § 5 are of the *energy flux density*, since this is the quantity for pulses which obeys the same transmission laws as *intensity* for continuous-wave sources. In practice the quantities actually measured in the two cases are $\int p^2\, dt$ and p^2, where p is acoustic pressure and t is time. The comparisons presented in table 1 were made on different occasions, typically with the explosions several miles distant and with the different sizes of charges fired in succession at the same position. The signals were received on a barium titanate hydrophone, amplified, and fed to a series of eight octave-filters (covering the range 25 c/s to 6·4 kc/s) each followed by a small analogue computer.

Table 1. Measured Differences from a 1-lb Charge

Measured octave-band energy differences in dB from a charge demolition S.C. 1 lb (T.N.T.)

Frequency (c/s)	35	70	140	280	560	1100	2200	4500
Depth (fathoms)								
Detonator electric No. 79 Mark 1.N, corresponding to about 0·002 lb T.N.T.								
5	−34·0	−31·0	−30·7	−30·0	−28·0	−25·5	−23·5	−22·7
10	−35·0	−30·3	−30·0	−30·3	−27·3	−25·0	−20·0	−18·0
20	−35·0	−33·0	−31·6	−33·4	−31·2	−28·6	−22·0	−24·0
30	−46·0	−41·5	−35·5	−36·0	−30·9	−30·0	−28·7	−27·0
Charge demolition $1\frac{1}{4}$ lb T.N.T.								
7	+0·2	+0·3	+0·5	+0·1	+0·2	+0·0	−0·1	+0·1
Bundle $3 \times 1\frac{1}{4}$ lb T.N.T.								
7	+0·3	+1·5	+2·5	+1·7	+2·1	+2·1	+1·7	+1·5
Bundle 4 × 1 lb S.C. (T.N.T.)								
20	+4·5	+4·6	+4·7	+4·7	+4·7	+4·8	+4·7	—
Charge demolition 50 lb Mark 3 T.N.T.								
5	—	+19·0	+16·0	+15·0	+15·0	+15·0	+12·0	+11·0
7	+12·5	+14·6	+15·2	+16·2	+14·8	+13·7	+13·0	+12·5
20	—	+18·2	+16·5	+15·0	+14·5	+14·0	+14·0	+10·0
30	+14·5	+17·0	+17·5	+16·9	+15·4	+13·4	+10·0	+9·0
40	+14·6	+16·2	+16·2	+16·1	+14·8	+12·9	—	—

The computers or octave-energy units were specially designed six-valve assemblies consisting of a calibrated potentiometer, two-stage feedback amplifier, two-valve square-law device, Miller integrator, and output valve feeding a pen recorder. The square-law device was based on that of Tucker (1952), and comprised a phase-splitter feeding a double triode with output taken from the common cathode. The units could accept a wide range of input levels, frequencies and pulse durations. It was necessary that they should be simple and physically small since several had to be used simultaneously, sometimes on board ship, and they have been operated successfully over a number of years. In all work with charges the possibility of the shock wave causing overload has to be watched carefully, so that the wide-band level plus all the octave-energy unit levels were monitored for overload using thyratron triggers and neon indicators.

4.3

In addition to the results in table 1, comparisons of 5-lb charges (5 lb 60/40 R.D.X./T.N.T.) with four 1-lb charges have shown that the mean

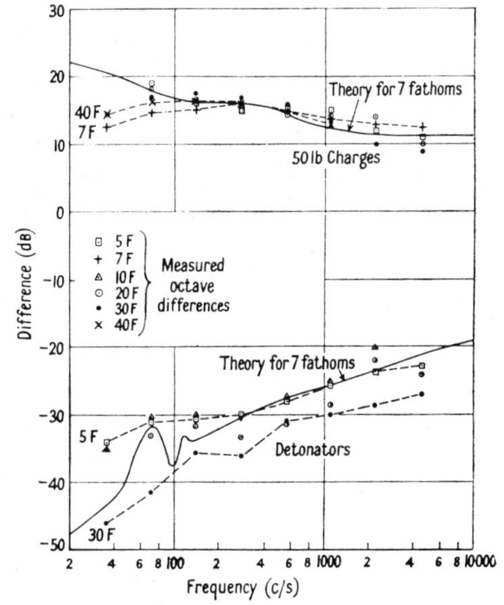

Figure 1. Differences from a 1-lb charge.

difference is less than 0·3 dB for frequencies between 113 c/s and 1740 c/s. The general accuracy of the table 1 comparisons is to about ± 1 dB, in a few cases it is a little worse but in several cases it is to about $\pm 0{\cdot}3$ dB. In particular the $1\frac{1}{4}$-lb figures and the 50-lb figures at 7 and 40 fathoms are reliable, being derived from a very large number of observations. Thus the small changes in the low-frequency figures for 50-lb charges between 7 and 40 fathoms are real, and the systematic low-frequency variation in the detonator figures is also real. The detonator and 50-lb results are illustrated in figure 1, in which the theoretical lines should be temporarily ignored since the actual values of the differences will be discussed in § 8. Many of the table 1 results are unexpected, for example the differences for the $1\frac{1}{4}$-lb, $3 \times 1\frac{1}{4}$-lb and some other charges are significantly smaller than expected on any reasonable theory. This may be connected with the manner of detonation, and shows that it is unwise to rely on theoretical differences. The table 1 results were taken in water depths varying from a few tens of fathoms to several hundred fathoms, and there is no reliable evidence for any variation with measurement site.

4.4

During the firing of a long series of charges it is possible to obtain information on their constancy of acoustic output, provided there is no random variation due to the propagation through the medium. For 1-lb charges the scatter in the octave-energy levels is less than $\pm 0{\cdot}5$ dB, being greatest at the extreme high and low frequencies. The scatter for 50-lb charges is greater than this, possibly due to variable bubble migration, and for detonators greater still. The detonator variability at 35 c/s, of magnitude several decibels, is explained in § 7 by the uncertain cancellation effect.

§ 5. Measured Charge Source Levels

5.1

From the shots fired at closer ranges it has been possible to deduce the absolute free-field source levels presented in table 2: The energy flux spectrum levels have been obtained by measuring $(1/\rho c)\int p^2\,dt$ for an octave band as described in § 4, and dividing by the bandwidth. Here ρc is the acoustic impedance, and strictly the term 'energy flux' may only be applied if there is a plane wave front, e.g. near a free surface the true energy flux is not measured. There are some other difficulties in strictly defining charge acoustic source levels because of finite amplitude effects, but these effects are small beyond about 100 yards. The choice of 100 yards as the reference distance is also convenient because it is typical for the range from charge to first boundary reflection. Shots at distances between about 100 yards and several miles have been used, though the more distant shots are only suitable if the water is deep. At small ranges it may be assumed that attenuation and absorption are insignificant at the lower frequencies, so that there is cylindrical spreading for ranges greater than a few water depths. The transmission loss may then be calculated from a simplified theory based on

---

Table 2. Absolute Source Levels for a 1-lb Charge

Frequency (c/s)	35	70	140	280	560	1100	2200	4500
Free-field source spectrum levels of energy flux density for 60-fathom 1-lb charge, in dB w.r.t. 1 erg · cm^{-2} $cycle^{-1}$ at 100 yards	+17·2	+12·8	+9·9	+5·4	+2·8	−0·5	−5·0	−13·9
Differences with charge depth								
60F–45F :	+5·1	−2·0	Expected to be small from theory and experiment, assumed zero					
60F–40F :	+4·4	+1·0						
60F–7F :	+0·4	0·0						

Weston (1959) if certain assumptions about the bottom reflection coefficient are made. At high frequencies there is spherical spreading out to appreciable ranges. The correction for transmission loss finally used is thus frequency dependent and takes account of many different factors, but it may be pointed out that at such close ranges the correction depends but little on the propagation mechanisms assumed. In addition for some shots separate measurements have been made of the energies of the arrivals following the different ray paths. Unfortunately the subject of propagation loss is too large to be discussed properly here, even with the restriction to close ranges.

5.2

Very large numbers of results have been used to obtain table 2 but the measurement of absolute levels is intrinsically more difficult than that of the table 1 differences. Thus at, and above, 140 c/s the results for all depths have been lumped together, which has both theoretical and experimental justification. In this frequency range the accuracy relative to neighbouring spectrum levels is to about ± 1 dB, but the absolute accuracy depends on additional factors such as hydrophone calibrations and is perhaps a little worse. At, and below, 70 c/s the variation with depth is not in general accurately known, but the best estimate is entered. However the large variation from 60 to 45 fathoms is known to within $\pm 0{\cdot}5$ dB, so for the 35 c/s octave the enormous ratio of about 3:1 in energy is real, and presumably due to the changing bubble pulse time (see § 8.3).

5.3

To find the best source level values for any charge size and depth it is necessary to start with the 1-lb 60-fathom figures in table 2, convert to the required depth,

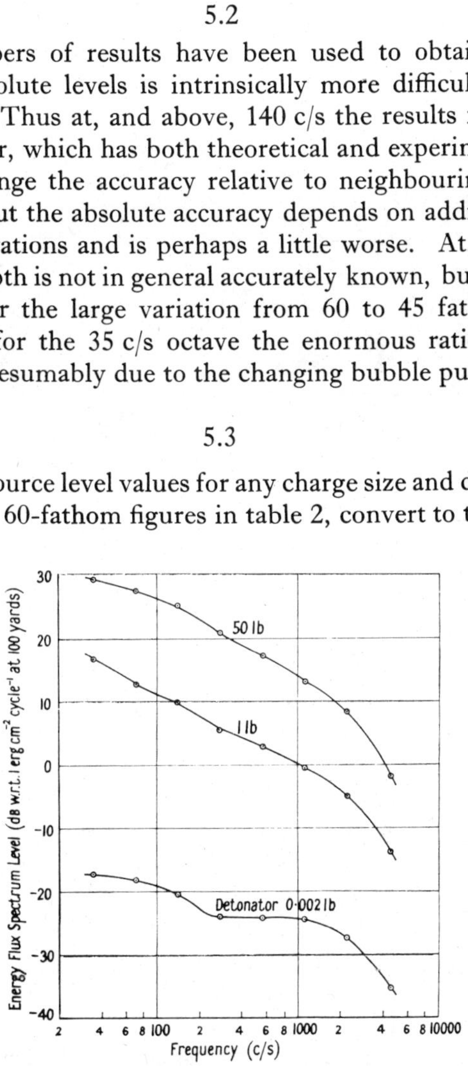

Figure 2. Measured absolute source levels for 7-fathom charges.

and then convert to the required size using smoothed table 1 differences. The results of this procedure are illustrated in figure 2 for various-sized charges at 7 fathoms.

§ 6. Scaling Law

For charges of different sizes fired at the *same depth* there is a simple scaling law. Identical pressure curves are obtained when plotted against a reduced time $t/W^{1/3}$ and measured at corresponding values of reduced range $r/W^{1/3}$. This applies rigorously to the shock and to the bubble pulses, but only if effects such as bubble migration and surface cavitation may be neglected. The various energy spectra may therefore be reduced to 1-lb charge equivalents, and this is illustrated in figure 3 for the 7-fathom results from figure 2. The reduction is carried out with the simplifying approximation that pressure is inversely proportional to range, and the errors arising are discussed at the end of §6.

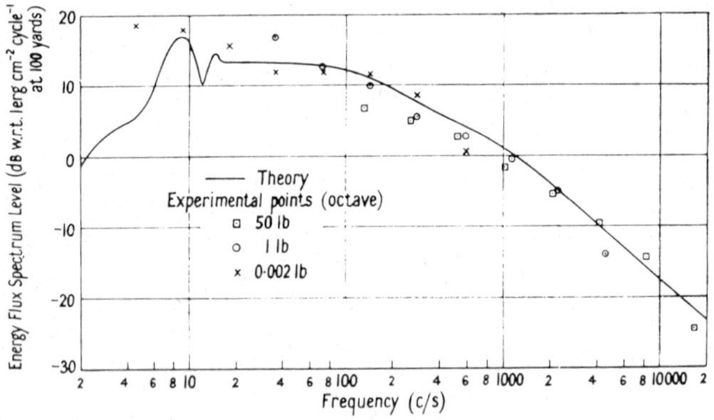

Figure 3. Source levels for 7-fathom charges scaled to 1 lb.

Table 3. Factors for Reduction of Spectrum Level to a 1-lb Charge Equivalent

W (lb)	0·002	50
$W^{1/3}$ or frequency factor	0·126	3·68
$W^{-4/3}$ (dB or level change)	+36	−22·7

The first step is multiplication of the frequency by $W^{1/3}$. Secondly, consider that the total charge energy is proportional to W (assuming spreading without attenuation), and also that the radiated energy per octave is proportional to W. Due to the frequency shift the octave bandwidth is proportional to $W^{-1/3}$, so that the spectrum level or energy per cycle is proportional to $W^{4/3}$. Thus the second step in the reduction is multiplication of spectrum level by $W^{-4/3}$, which may be deduced alternatively from equations (5), (9) and (14) of § 7. The conversion factors are given in table 3.

Disregarding temporarily the theoretical curve in figure 3, it may be seen that the experimental points lie fairly well on a single line. This is evidence for the correctness of both the source level spectrum shapes and the measured differences between charges. Consideration of the experimental spreading laws quoted in § 7 shows that the rate of shock wave pressure decay is a little faster than the spherical spreading assumed above. However this means the scaling laws need slight modification only at the higher frequencies, such that the 50-lb results for reduced frequencies above 1 kc/s should be decreased an extra 1·5 dB.

The effect of bubble migration in lowering the 50-lb results in the middle frequencies is discussed in § 8.2, after the presentation of the Fourier analysis predictions.

§ 7. Fourier Analysis

7.1

The best data for the explosion pressure–time curves have been given by Arons *et al.*, both for bubble pulses (Arons, Slifko and Carter 1948, Arons 1948, Arons and Yennie 1948) and recently for the shock wave at long ranges (Arons 1954). Experimental curves for charges at 500 feet depth are given by Arons and Yennie, and using these Raitt (1952) has already carried out a numerical Fourier analysis. However it is desired here to obtain analytical expressions for the spectrum level, so that variation with charge size, depth etc. may be easily investigated. Thus a different approach has been followed, the spectrum being split into two parts and certain simplifying assumptions made. At the high frequencies the shock wave and bubble pulses have been analysed separately, and the resulting spectrum energies added. At low frequencies the shock wave and bubble pulses have been replaced by impulses and account taken of phase. In this section the results are illustrated by the numerical values for a 1-lb T.N.T. charge at 20 fathoms, as measured 100 yards away, and its theoretical spectrum is shown in figure 4.

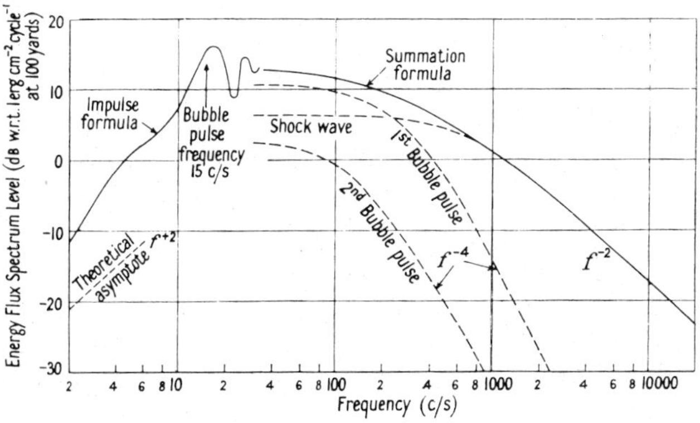

Figure 4. Theoretical source spectrum for 20-fathom 1-lb charge.

7.2

According to Arons (1954) the shock wave values are:

Peak pressure

$$P_0 = 2{\cdot}16 \times 10^4 (W^{1/3}/r)^{1{\cdot}13}\,\text{lb in}^{-2} \quad (W \text{ in lb, } r \text{ in ft})$$
$$= 2{\cdot}35 \times 10^6\,\text{dyn cm}^{-2} \text{ for 1 lb at 100 yards.} \qquad \ldots\ldots(1)$$

Positive impulse

$$I_0 = 1{\cdot}78\, W^{1/3} (W^{1/3}/r)^{0{\cdot}94}\,\text{lb sec in}^{-2}$$
$$= 576\,\text{dyn sec cm}^{-2}. \qquad \ldots\ldots(2)$$

Time constant

$$t_0 = 58\, W^{1/3} (W^{1/3}/r)^{-0{\cdot}22}\,\mu\text{sec}$$
$$= 2{\cdot}03 \times 10^{-4}\,\text{sec}. \qquad \ldots\ldots(3)$$

The above time constant describes the initial decay of the shock, after which the shock pressure drops more slowly. It is possible to define a slightly greater time constant which gives a better description for the lower frequency components:

$$t_0 = I_0/P_0 = 2{\cdot}44 \times 10^{-4} \text{ sec.} \qquad \ldots\ldots(4)$$

The Fourier analysis for the energy spectrum of an exponentially decaying pulse may easily be shown to be

$$E_0(f) = \frac{2P_0^2}{\rho c(1/t_0^2 + 4\pi^2 f^2)}, \qquad \ldots\ldots(5)$$

where f is frequency. This is a well-known result but in deriving it more than one worker has omitted the factor 2, a frequent and surprising cause of error in Fourier analysis using exponential functions being the neglect of the 'negative frequencies' (see also eqn (9)). In the plot in figure 4 the t_0 value from equation (4) is used, since it gives a better overall fit. At low frequencies the spectrum is flat, but changes over near $1/(2\pi t_0) = 653$ c/s (the '3-dB down point') to an f^{-2} or -6 dB per octave law. The rise time at the shock front is only a few microseconds, so that the -6-dB slope should hold up to a limit of several tens of kc/s (cf. Arons, Yennie and Cotter 1949, Arons and Wertheim, unpublished†).

7.3

Now the bubble pulses are fairly symmetrical, and their shape is well matched by an exponential rise followed by an exponential decay. From Arons (1948), for the first bubble,

Peak pressure $P_1 = 3450(W^{1/3}/r)$ lb in^{-2}

$$= 0{\cdot}793 \times 10^6 \text{ dyn cm}^{-2}. \qquad \ldots\ldots(6)$$

Positive impulse $I_1 = 9{\cdot}58(W^{1/3}/r)d_0^{-1/6}$ lb sec in^{-2}

$$= 953 \text{ dyn sec cm}^{-2}. \qquad \ldots\ldots(7)$$

Here d_0 is the depth in feet plus 33, and has been evaluated for a charge at 20 fathoms. A time constant for either side of the pulse may be defined in a manner similar to that of equation (4):

$$t_1 = I_1/2P_1 = 6{\cdot}01 \times 10^{-4} \text{ sec.} \qquad \ldots\ldots(8)$$

The energy spectrum for the double exponential pulse may easily be shown to be

$$E_1(f) = \frac{8}{\rho c}\left(\frac{P_1/t_1}{1/t_1^2 + 4\pi^2 f^2}\right)^2. \qquad \ldots\ldots(9)$$

This is also plotted in figure 4. At low frequencies the spectrum is flat like that for the shock waves, but changes over near $1/(2\pi t_1) = 265$ c/s ('6-dB point') to an f^{-4} or -12 dB per octave law. At high frequencies the predicted spectrum level will be too great, since the -12-dB spectrum slope corresponds to the discontinuity in pressure gradient at the peak of the simulated bubble pulse (see Weston 1960), which is not present in the real pulse. However at these frequencies the bubble contribution to the total spectrum energy is insignificant.

† 1950, Woods Hole Oceanographic Institution, ref. no. 50-32.

For the second bubble there are the approximate relationships

$$P_1/P_2 = 4{\cdot}72, \qquad P_2 = 0{\cdot}168 \times 10^6 \text{ dyn cm}^{-2} \qquad \ldots\ldots (10)$$

$$I_1/I_2 = 2{\cdot}47, \qquad I_2 = 386 \text{ dyn sec cm}^{-2} \qquad \ldots\ldots (11)$$

$$t_2 = I_2/2P_2 = 11{\cdot}5 \times 10^{-4} \text{ sec.} \qquad \ldots\ldots (12)$$

The spectrum follows from equation (9) and is also plotted in figure 4, with its '6-dB point' at 135 c/s. The contribution of further bubble pulses may be neglected with little error.

7.4

The total radiated energy spectrum shown in figure 4 is obtained by summing the energies due to the shock wave and the first two bubble pulses. It may be noted that according to Arons and Yennie (1948) rather more than half the total chemical energy of the charge is associated with the shock wave, but even including the bubble pulses rather less than half the total energy is eventually radiated as acoustic waves. Below about 230 c/s it may be seen that the major contribution to the spectrum energy comes from the first bubble pulse, and it is only above 230 c/s that the shock wave predominates. The summation spectrum shows no appreciable downward kink at this change over frequency, although it might under some conditions (the 7-fathom curve in figure 3 has such a tendency).

For bubbles far from a boundary the interval between the shock wave and the bubble pulse peak is

$$T_1 = 4{\cdot}36 W^{1/3} d_0^{-5/6} = 6{\cdot}59 \times 10^{-2} \text{ sec at 20 fathoms.} \qquad \ldots\ldots (13)$$

Thus the 'bubble pulse frequency' $1/T_1 = 15{\cdot}2$ c/s. The summation method of calculating spectrum level is valid only for frequencies much greater than $1/T_1$, typically for frequencies above, say 35 c/s.

7.5

At low frequencies the shock wave and bubble pulses may be replaced by their impulses I_0, I_1 and I_2 occurring at times 0, T_1, and $T_3 = T_1 + T_2$. The interval between the first and second bubble pulses $T_2 = 0{\cdot}72 T_1 = 4{\cdot}75 \times 10^{-2}$ sec. In addition, at low frequencies allowance must be made for the fact that the pressure goes negative between the pulses, so that the impulse over the whole pressure–time curve is zero. This is taken into account by adding a steady negative pressure of duration T_3. The energy spectrum for such a system may easily be shown to be

$$E_L(f) = (2/\rho c)[\{I_0 + I_1 \cos 2\pi f T_1 + I_2 \cos 2\pi f T_3 - N \sin 2\pi f T_3\}^2 + \{I_1 \sin 2\pi f T_1 + I_2 \sin 2\pi f T_3 - N(1 - \cos 2\pi f T_3)\}^2], \quad \ldots\ldots (14)$$

where $N = (I_0 + I_1 + I_2)/(2\pi f T_3)$. This impulse formula for the spectrum level is valid for frequencies much less than $1/2\pi t_1 = 265$ c/s, or say less than 140 c/s. Thus it overlaps the summation curve by two octaves, in a region where the spectrum level curve forms a plateau. The impulse spectrum has a main maximum near 15 c/s (the bubble pulse frequency), with minor oscillations above this frequency.

Now since the total impulse is zero the spectrum level at very low frequencies must be proportional to f^2 (Weston 1959b), and on expansion the asymptotic form of equation (14) is found to be

$$E_L(f) = (2\pi^2 f^2/\rho c)\{I_1(T_1 - T_2) - (I_0 - I_2)T_3\}^2. \qquad \ldots\ldots (15)$$

For the usual range of depths the particular values of I and T are such that there is almost complete cancellation, and the coefficient of f^2 is very small. For example at 20 fathoms the asymptotic level becomes 24·4 dB higher if one puts $I_1 = I_2 = 0$, leaving only the shock impulse. Thus the spectrum slope predicted by equation (14) in the 1–10 c/s region is greater than the +6 dB per octave of equation (15), and the spectrum curve falls off sharply just below the bubble pulse frequency. With real charges, however, the bubble impulse is likely to vary from shot to shot (due to migration, turbulence, etc.), and the cancellation may not be so effective. Thus the average spectrum level is likely to be higher than that given by equation (15), and would be expected to vary greatly. The scatter in the measured 35 c/s level for detonators is certainly larger than that for any other charge or frequency, and the mean level is higher than expected. It should be noted that cancellation does not affect the total energy output, but only changes the frequency distribution.

§ 8. Comparison of Measurements with Theory

8.1. *General Agreement*

A theoretical curve for 7 fathoms is shown in figure 3, again based on equations (5), (9) and (14). Different depths have deliberately been chosen for the theoretical curves discussed in § 7 and the present section, so that the theoretical change with depth may be seen. The major difference arises from the changed bubble pulse frequency, so that the 7-fathom curve of figure 3 shows a larger region with a flat spectrum than does the 20-fathom curve of figure 4. The 7-fathom depth was not chosen to illustrate the comparison of theory and experiment because agreement is particularly good there (it is in fact rather better at the other, greater, depths), but because it exhibits the special effects due to bubble migration. By applying the scaling laws of § 6 to the 7-fathom spectrum curve the theoretical 7-fathom difference curves shown in figure 1 may be derived.

The general agreement between measurements and theory in figures 1 and 3 is good, especially for the 1-lb results in figure 3. The existence of a maximum at the bubble pulse frequency is shown in the detonator results of both figures 1 and 3, though there is a smoothing out tendency since the measurements are all in octave bands. It may be noted that the humps in figures 1 and 3 may be shown to exist from independent measurements of detonator to 1-lb differences and by direct observations of the detonator source spectrum. The lowness of the theoretical prediction for detonators for frequencies below the bubble pulse frequency (figures 1 and 3) was explained at the end of § 7, and there is the possibility of a slight enhancement near the bubble pulse frequency due to the contribution of higher order pulses.

8.2. *Bubble Migration*

The only remaining significant discrepancy in figure 3 is the low experimental value for the low-frequency 50-lb results, amounting to about 5 dB at 35 c/s (or 130 c/s scaled). The sign and value of this source level discrepancy agree with those shown in figure 1 for the 50-lb–1-lb differences. It is thought to be due to bubble migration under gravity which, as already noted in § 3.5, is most important for large charges at shallow depths—so that in extreme cases there is virtually no bubble pulse at all. The phenomenon has been studied by Taylor

(1942), Bryant (1942), Kennard (1943), Friedman (1947) and others. Migration is also influenced by the nearness of boundaries, such that an oscillating bubble is repelled by a free surface and attracted towards a rigid surface, but this effect is only important for the smaller charges.

The explosion bubble moves very slowly when it is expanded, but may rise rapidly under gravity during its contracted phase. This is because the momentum is proportional to the time integral of the bubble buoyancy, which will produce a larger velocity when the associated bubble size is small. In considering migration effects it is useful to know the values of the maximum bubble radius a, and the vertical rise b between the initiation and the first bubble pulse. From, for example, Arons (1948),

$$a = 12{\cdot}6(W/d_0)^{1/3}\,\text{ft} = 3\,\text{ft for 1 lb at 7 fathoms.} \qquad \text{......(16)}$$

Kennard (1943) gives the approximate formula

$$b \sim 4\sqrt{W/P_a}\,\text{ft} \sim 1{\cdot}7\,\text{ft for 1 lb at 7 fathoms,} \qquad \text{......(17)}$$

where P_a is the hydrostatic pressure in atmospheres.

When there is migration some of the kinetic energy of radial motion is transferred upon contraction to kinetic energy of vertical motion. This factor rather than the gas pressure can then determine the minimum bubble radius, which is thereby increased. It may be noted that it is normally only about twice the original charge radius. The peak bubble pulse pressure may be much reduced, but in contrast the impulse is affected little. The main effect on the bubble pulse spectrum will be that of a bodily shift to lower frequencies, the '6 dB down' frequency being reduced in approximately the same ratio as the peak pressure. Migration also causes an increase in the bubble interval T_1.

According to Kennard the above reduction ratio depends on the value of b/a; for a 1-lb charge at 7 fathoms $b/a \sim 0{\cdot}6$ and the predicted reduction ratio is about 0·2. This should lead to an appreciable lowering of the spectrum level in the neighbourhood of 200 c/s, and though there is experimental support for this in figure 3 it is of doubtful significance, e.g. no reliable evidence of variation with depth was obtained for 140 c/s and above. However for a 50-lb charge at 7 fathoms the bubble radius a is 11 ft, and the calculated rise is about the same. Under these conditions the bubble pulse is of very little importance, and in fact the low-frequency 50-lb results are fitted extremely well by the shock wave theoretical spectrum alone. However, it might again be necessary to take account of the bubble pulse if the 50-lb measurements were extended to frequencies below about 10 c/s.

It is worthwhile calculating the depth at which such bubble migration effects become small. This may be arbitrarily defined as corresponding to $b/a = 0{\cdot}3$, with a predicted reduction ratio of about 0·7. From equations (16) and (17) for the critical depth

$$d_0 \sim 200\,W^{1/4}\,\text{ft.} \qquad \text{......(18)}$$

This corresponds to about 1 fathom for a detonator, 30 fathoms for a 1-lb charge, and 80 fathoms for a 50-lb charge.

Now the explosion bubble must eventually break up when it stops oscillating, and the break-up is likely to be accelerated when it is rising rapidly under gravity. It may be noted that a bubble is unstable in form when it is contracting (Taylor

1942), and the shape may be distorted at the minimum radius—where the speed of rise is a maximum. This suggests that the bubble pulse may be virtually eliminated when the migration is serious. Experiments by other workers have shown that the bubble distortion is indeed greatest for shallow charges (and also when the original charge is not spherically symmetrical), and that bubble pulses are missing for large shallow charges (quite apart from catastrophic effects such as blow-out at the surface). This also suggests the possibility of a systematic variation of source level with firing area etc., due to changes in the sea turbulence conditions. There is however no reliable evidence for this, e.g. no significant changes in the measured differences with site.

8.3. *Variation with Depth*

The variation of charge-source level with depth is obviously of general interest, because charge sources are used on various occasions at a variety of depths. There is also a particular interest because charges may be fired at different depths at the same position, perhaps to investigate mode excitation functions in shallow water. It is necessary to be able to separate variation due to the source from that due to propagation. There is little source variation at high frequencies since the shock wave corresponds to a high impedance source, but at low frequencies there is considerable variation.

Now the shock wave is unaffected by depth, as noted above. The bubble pulse peak pressures are also unaffected for the range of depths used (provided the bubble is stationary), and their impulses vary only very slowly—with $d_0^{-1/6}$. The only parameter changing rapidly with depth for non-migratory bubbles is the bubble interval T_1 proportional to $d_0^{-5/6}$. Thus, on going from 7 to 60 fathoms there is a four-fold increase in the bubble-pulse frequency. Some of these effects have already been seen when comparing figures 3 and 4 (see also figure 5), and there are also the special effects which do not obey the simple scaling laws —e.g. gravity migration.

Consider now the experimental results, and first the detonator to 1-lb differences in figure 1 where there are large real differences with depth. On going from 5 to 30 fathoms the calculated detonator bubble-pulse frequency moves up from about 60 to 160 c/s, and this is evident in the results. This moving up is also responsible for the decreasing (algebraic) difference at the lowest frequencies as depth is increased, but a quantitative theory is difficult because of the variable cancellation effect discussed in §7. Changes in the 1-lb spectrum should have a relatively small effect on these differences.

The 50-lb–1-lb differences in figure 1 show less depth variation, because the bubble-pulse frequencies tend to lie below the range of interest—though there should be marginal effects due to the 1-lb 40-fathom charges. The most obvious depth effects are for the lowest frequencies at 7 fathoms, due to the 50-lb charge migration already discussed in § 8.2. A small variation may be expected at the middle frequencies too, because of the $d_0^{-1/6}$ impulse law plus small bubble migration effects on both the 1-lb and 50-lb charges. However this is swamped by the experimental scatter, even though the latter is surprisingly small.

The absolute charge source levels shown in table 2 exhibit appreciable depth variations only for frequencies at, or below, about 70 c/s. The 35 c/s and 70 c/s values are plotted in figure 5 and compared with the theoretical curve for a single frequency at the octave centre. The experimental and theoretical levels agree in mean level, the magnitude of the depth variations is correct, but detailed agreement is completely lacking. This cannot be wholly ascribed to the width of the measurements band, and it may be yet another phenomenon in which bubble migration plays a part.

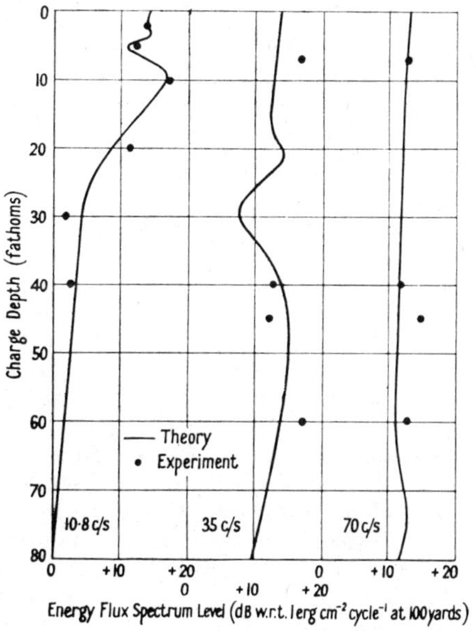

Figure 5. Dependence of 1-lb source levels on charge depth. Experimental points: arbitrary reference for 10·8 c/s, absolute for 35 c/s and 70 c/s.

Figure 5 also shows experimental data for the narrow bandwidth at 10·8 c/s, based on unpublished ground-wave measurements by W. W. Reay in the area off Perranporth, Cornwall, described by Merriweather (1958). The considerable depth dependence measured there was due both to the source variation and the mode excitation function, and it was necessary to correct for the latter by assuming a sinusoidal form with a maximum at 25 fathoms in water 45 fathoms deep. The general level has been adjusted to give the best fit to the theoretical line but is known to be approximately correct; the agreement in shape is excellent.

§ 9. Conclusions

Surface cavitation and the change of source position due to bubble migration are shown to have only a small effect on the acoustic signal received at a distance. The measured differences between charges (table 1) and the absolute charge source levels (table 2) obey the simple scaling laws, and agree well with the predictions from Fourier analysis. For the larger and shallower charges it is also necessary to allow for the effect of migration in reducing the contribution of the bubble pulses.

Many minor problems remain, especially some connected with the finite-amplitude propagation discussed in § 3. In general it is still desirable to use experimental rather than theoretical values for charge differences and source levels, especially when migration makes accurate prediction difficult. Thus the most important part of this paper is the presentation of the experimental results in tables 1 and 2, with accuracy generally to about ± 1 dB.

Acknowledgments

Thanks are due to the Captains and Ships' Companies of R.R.S. *Discovery II* and many of H.M. Ships for their patient help during arduous trials. The author is also grateful to many colleagues, some of whose results have been included, and especially to W. W. Reay, A. S. Merriweather and A. A. Horrigan.

References †

Arons, A. B., 1948, *J Acoust. Soc. Amer.*, **20**, 277 (U.E.R. II, 481).

—— 1954, *J. Acoust. Soc. Amer.*, **26**, 343.

Arons, A. B., Slifko, J. P., and Carter, A., 1948, *J. Acoust. Soc. Amer.*, **20**, 271 (U.E.R. II, 475).

Arons, A. B., and Yennie, D. R., 1948, *Rev. Mod. Phys.*, **20**, 519 (U.E.R. I, 1137).

Arons, A. B., Yennie, D. R., and Cotter, T. P., 1949, *Nav. Ord. Rep.*, 478 (U.E.R. I, 1473).

Bryant, A. R., 1942, *Road Research Laboratory Report*—Index 10 (U.E.R. II, 163).

Cole, R. H., 1948, *Underwater Explosions* (Princeton: University Press).

Friedman, B., 1947, *Inst. for Math. and Mech., N.Y.U.*, Rep. 166 (U.E.R. II, 329).

Kennard, E. H., 1943, *David Taylor Model Basin Report* (U.E.R. II, 377).

Kolsky, H., Lewis, J. P., Sampson, M. T., Shearman, A. C., and Snow, C. I., 1949, *Proc. Roy. Soc.* A, **196**, 379.

Merriweather, A. S., 1958, *Geophys. J.*, **1**, 73.

O'Brien, P. N. S., 1960, *Geophys. J.*, **3**, 29.

Penney, W. G., and Pike, H. H. M., 1950, *Rep. Progr. Phys.*, **13**, 46.

Raitt, R. W., 1952, *Oceanographic Instrumentation* (N.A.S.–N.R.C. Publ. 309), 70.

Taylor, G. I., 1942, *Min. Home Security*, S.W.19 (U.E.R. II, 131).

Tucker, M. J., 1952, *Electronic Engng*, **24**, 466.

Weston, D. E., 1959, *Proc. Phys. Soc.*, **73**, 365.

—— 1960, *Geophys. J.*, **3**, 191.

Wood, A. B., Lakey, R. H., and Butterworth, S., 1924, A.R.L./S/12 (U.E.R. I, 497).

Worzel, J. L., and Ewing, M., 1948, *Geol. Soc. Amer. Mem.*, 27.

† Many of the papers cited have been reprinted in a joint Anglo-American compendium of reports published in 1950 by the Office of Naval Research and entitled *Underwater Explosion Research* (shortened here to U.E.R.).

Finite Amplitude Effects

VII

Dr. Westervelt developed the theory of intermodulation of signals due to the nonlinearity of the medium. This is covered in the two abstracts of papers presented at the Acoustical Society and the more detailed paper published later in the *Journal of the Acoustical Society of America.*[1]

Dr. Westervelt is Professor of Physics at Brown University, Providence, Rhode Island. He was consultant to the Assistant Attaché for Research, U.S. Navy, at the American Embassy, London, in 1951–1952, and a member of the subcommittee on aircraft noise, NASA, 1954–1959. He is a Fellow of the Acoustical Society of America and he has contributed in the fields of physical effects of high amplitude sound waves, air acoustics, underwater sound, and high energy physics, as well as relativity and gravitation.

These papers are reprinted with the permission of Dr. Westervelt and the Acoustical Society of America.

---

[1]While Dr. Westervelt was stationed at the London, England branch office of the U.S. Office of Naval Research in 1951, he met the late Captain H. J. Round, English pioneer in the development of the superheterodyne receiver. Captain Round was carrying out experiments with an underwater magnetostriction projector in his private laboratory. The work was being done for Dr. Paul Vigoureux who was then at the Admiralty Research Laboratory. Captain Round happened to have an 18-kHz transducer operating in air and when Dr. Westervelt walked in front of the beam, he was startled to hear a loud low-frequency hum, rich in harmonics, but highly directive, coming from such a tiny projector. The fundamental he heard seemed about 100 Hz, while the emitter was not more than about six inches on a side. He immediately concluded that Round was supplying his RF driver either with an unfiltered power supply or at worst raw ac, and that the demodulation was occurring either in the air or in his own ears. It was at this moment that the concept of an end-fire array first occurred to him.

Journal of the Acoustical Society of America, vol. 32, 1960, p. 934(A)

39

Abstract from

Parametric End-Fire Array

PETER J. WESTERVELT

A high-power modulated sound beam acts like an end-fire directional array for the modulation frequency. This occurs because the nonlinear terms in the equations of motion cause such a beam to act like a distribution of sources for the modulation frequency. If the sound beam is unmodulated, it behaves precisely like a parametric amplifier for any sound traveling in the direction of the beam and thus can be used as a highly directional receiver. In order to test this concept, two carrier beams having approximately the same frequency $\omega_1 \approx \omega_2 \approx \omega$ were superimposed on each other. The modulation frequency is then the difference $\omega_s = |\omega_1 - \omega_2|$. If the pressure of each carrier is the same and equal to P_o, the modulation percentage will be 100. The expression for the radiated intensity I_s at a distance R_o far from the source is

$$I_s = \frac{\omega_s^4 P_o^4 S_o^2 \left[1 + \tfrac{1}{2} \rho_o c_o^{-2} (d^2P/d\rho^2)\right]^2}{2(8\pi)^2 \rho_o^3 c_o^9 R_o^2} \times \frac{1}{\alpha^2 + k^2 \sin^4(\theta/2)}$$

in which S_o is the cross-sectional area of the carrier beam, α is the pressure attenuation coefficient for the carrier beam, θ is the angle measured from the axis of the carrier beam, and $k = \omega_s/c_o$. Maximum radiation occurs at $\theta = 0$ in the forward direction of the carrier. Experiments, described in an accompanying paper, agree well with the theoretical expression which, incidentally, shows the same angular dependence as Rutherford scattering in atomic theory.

Supported by the Office of Naval Research.

Journal of the Acoustical Society of America, V. 32, 1960, p. 935(A)

40

Abstract from

Experimental Investigation of a Parametric End Array

J. L. S. BELLIN
PETER J. WESTERVELT,
and R. T. BEYER

To obtain experimental verification of the expression for an acoustic end-fire array derived by Westervelt, a 13.4 Mc crystal, mounted inside a water-filled tank, was driven simultaneously by 13 Mc and 14 Mc transmitters. A small barium titanate probe, free to rotate in the horizontal plane, was used to detect the 1 Mc difference frequency radiated by the interacting carrier beams. This transducer was connected in turn with a sensitive calibrated receiver. The entire system permits a determination of the directivity pattern and the absolute magnitude of the radiated 1 Mc signal. The results of experiments using this arrangement, as well as of trials with both higher and lower difference frequencies, are in good agreement with the theoretical expression mentioned above.

Reprinted from THE JOURNAL OF THE ACOUSTICAL SOCIETY OF AMERICA, Vol. 35, No. 4, 535–537, April, 1963

41

Parametric Acoustic Array

PETER J. WESTERVELT

Department of Physics, Brown University, Providence 12, Rhode Island
(Received 28 December 1962)

This paper presents the theory of highly directional receivers and transmitters that may be "constructed" with the nonlinearity of the equations of fluid motion.

GENERAL APPROACH

A THEORETICAL study of the scattering of sound by sound has been given by the author,[1] and Lighthill[2] has suggested that the author's method might be applied to a beam of sound. This paper deals with the consequences of Lighthill's suggestion.

It has long been known, both theoretically[3] and experimentally,[4] that two plane waves of differing frequencies generate, when traveling in the same direction, two new waves, one of which has a frequency equal to the sum of the original two frequencies and the other equal to the difference frequency. These "sum" and "difference" waves have an existence that is, in the following sense, independent of the existence of the primary generating waves: consider a semipermeable screen capable of totally absorbing the generating waves, yet freely transmitting the sum and difference waves; then these latter waves will be launched into an independent existence. (Recent claims[5,6] to the contrary notwithstanding, it is known[1,7] that when the two generating waves intersect at a nonzero angle, the scattered sum and difference waves will not exist outside the zones of interaction; that is, the scattered waves will not have an independent existence in the sense just described.)

This paper is concerned primarily with the launching of a difference frequency wave by two high-frequency, collimated beams of sound. The semipermeable screen is provided by the absorptive properties of the medium that attenuates the high-frequency waves at a rate that may be one, or more, orders of magnitude greater than the attenuation of the difference frequency wave. Barring relaxation phenomena, which might give rise to an absorption maximum for the generating waves, the sum wave is attenuated, along with the second harmonics of the primary waves, even more strongly than the primary waves themselves are attenuated.

Several simplifying assumptions and approximations adopted in this work are now listed with the realization that not all of them are justified, in view of the accuracy with which the results can be verified experimentally.

First, the equations of motion for an ideal fluid, devoid of viscosity and heat conduction, form the basis of the study; the effects of attenuation are introduced in an *ad hoc* way.

The second assumption concerns the choice of a pair of superimposed collimated beams as the primary generating waves. It is assumed that the beam is so narrow, and the collimation so perfect, that the volume distribution of sources may be represented adequately by the line distribution located along the axis of the primary beams.

Other approximations, not difficult to eliminate in a fuller treatment, include the following: there is no attenuation of the difference wave; the pressure attenuation coefficients for each of the two primary waves are equal, and one or more orders of magnitude less than the wavenumber of the difference wave, and they result essentially from linear processes—that is, nonlinear attenuation is negligible.

[1] P. J. Westervelt, J. Acoust. Soc. Am. **29**, 934 (1957); **29**, 199 (1957).
[2] M. J. Lighthill, Math. Revs. **19**, 915 (1958).
[3] H. Lamb, *Dynamical Theory of Sound* (Edward Arnold and Company, London, 1931).
[4] A. L. Thuras, R. T. Jenkins, and H. T. O'Neill, J. Acoust. Soc. Am. **6**, 173 (1935).
[5] K. U. Ingard and D. Pridmore-Brown, J. Acoust. Soc. Am. **28**, 367 (1956).
[6] L. W. Dean, III, J. Acoust. Soc. Am. **32**, 934(A) (1960).
[7] J. L. S. Bellin and R. T. Beyer, J. Acoust. Soc. Am. **32**, 339 (1960).

THEORETICAL DETAILS

Lighthill's exact equation for arbitrary fluid motion is well known by now to be

$$(\partial^2\rho/\partial t^2)-C_0^2\nabla^2\rho=-C_0^2\Box^2\rho=(\partial^2/\partial x_i\partial x_j)T_{ij}, \quad (1)$$

where

$$T_{ij}=\rho u_iu_j+p_{ij}-C_0^2\rho\delta_{ij}+D_{ij} \quad (2)$$

and D_{ij} represents the viscous stresses. Equation (1) can easily be written with the pressure p on the left so that, neglecting viscosity, it becomes

$$\Box^2p=(\partial^2/\partial t^2)(\rho-C_0^{-2}p)-(\partial^2/\partial x_i\partial x_j)\rho u_iu_j. \quad (3)$$

In order to calculate the difference frequency, it is only necessary to retain terms on the right of Eq. (3), which are quadratic in the field variables. To this approximation, and when $\nabla\times\mathbf{u}=0$, it is easy to show that

$$(\partial^2/\partial x_i\partial x_j)\rho u_iu_j\approx\rho_0[(\nabla\cdot\mathbf{u})^2+\mathbf{u}\cdot\nabla^2\mathbf{u}+\nabla^2\tfrac{1}{2}u^2]$$
$$=\rho_0^{-1}C_0^2[-\Box^2(\tfrac{1}{2}p^2+\tfrac{1}{2}\rho_0^2C_0^{-2}u^2)+\nabla^2\rho_0^2C_0^{-2}u^2]. \quad (4)$$

Also, to the same approximation,[1]

$$\rho-C_0^{-2}p\approx-\tfrac{1}{2}C_0^{-2}(d^2p/d\rho^2)_{\rho=\rho_0}\rho^2$$
$$=-\tfrac{1}{2}C_0^{-6}(d^2p/d\rho^2)_{\rho=\rho_0}p^2. \quad (5)$$

Now, terms like $\Box^2$ in the source contribute nothing to the radiation field so, dropping it in Eq. (4), and making use of Eqs. (4) and (5), we may rewrite Eq. (3) as follows:

$$\Box^2p_s=-\tfrac{1}{2}C_0^{-6}(d^2p/d\rho^2)_{\rho=\rho_0}(\partial^2/\partial t^2)p_i^2-\rho_0\nabla^2u_i^2. \quad (6)$$

The subscript s serves to label the scattered—that is, the difference frequency—wave, which is generated by the collimated primary waves; these primary waves are designated by the subscript i. The attenuation of the primary wave is small enough to allow the substitution $u_i=\rho_0^{-1}C_0^{-1}p_i$, which, along with the fact that

$$\nabla^2p_i^2=\Box^2p_i^2+C_0^{-2}(\partial^2/\partial t^2)p_i^2$$

and the fact that $\Box^2p_i^2$ contributes nothing, allows Eq. (6) to be written in the form

$$\Box^2p_s=-\rho_0(\partial q/\partial t), \quad (7)$$

where

$$q=\rho_0^{-2}C_0^{-4}[1+\tfrac{1}{2}\rho_0C_0^{-2}(d^2p/d\rho^2)_{\rho=\rho_0}](\partial/\partial t)p_i^2. \quad (8)$$

Now, q is the simple source strength density resulting from the primary waves p_i.

We assume the following form for the primary waves:

$$p_i=P_0e^{-\alpha x}[\cos(\omega_1t-k_1x)+\cos(\omega_2t-k_2x)], \quad (9)$$

where $|\omega_1-\omega_2|\ll\omega_1\approx\omega_2$, and $k_1=\omega_1/C_0$, and $k_2=\omega_2/C_0$. This represents two plane waves of equal amplitude at the origin, and both attenuated at the same rate. The waves are assumed to be collimated in a beam with a cross-sectional area S_0. It is then a simple matter to obtain the difference-frequency component of q by

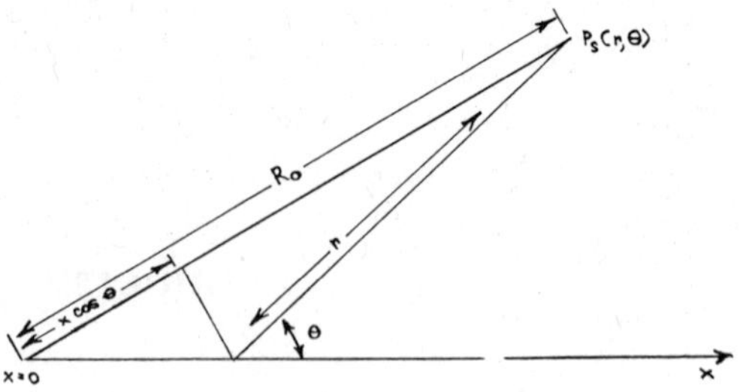

FIG. 1. Geometry of the problem.

inserting Eq. (9) into Eq. (8), which gives, in complex form,

$$q=\rho_0^{-2}C_0^{-4}[1+\tfrac{1}{2}\rho_0C_0^{-2}(d^2p/d\rho^2)_{\rho=\rho_0}]P_0^2(-i\omega_s)$$
$$\times e^{-i\omega_st+i(k_s+2i\alpha)x}, \quad (10)$$

in which $\omega_s=|\omega_1-\omega_2|$, and $k_s=\omega_s/C_0$. It might be added here that the medium is supposed not to be dispersive. Now, the general solution of Eq. (7) is given by the volume integral

$$p_s(\mathbf{r})=-\frac{i\omega_s\rho_0}{4\pi}\int\frac{qe^{ik_s|\mathbf{r}-\mathbf{r}'|}}{|\mathbf{r}-\mathbf{r}'|}dV',$$

which simplifies, using the geometry of Fig. 1, to the form

$$p_s=-\frac{i\omega_s\rho_0S_0}{4\pi}\int_0^l\frac{qe^{ik_sr}}{r}dx.$$

In the above equation, S_0dx represents the elementary volume of the source density, this source density being distributed along the x axis. The integration need only be carried out to a distance l such that q is attenuated to a negligible value. In practice, it is easier to let $l\to\infty$. If $R_0>\omega_s/(\alpha^2C)$, the integral may be put in the form

$$p_s=\frac{-i\omega_s\rho_0AS_0e^{ik_sR_0-i\omega_st}}{4\pi R_0}\int_0^l e^{i[k_s+2i\alpha-k_s\cos\theta]x}dx, \quad (11)$$

where

$$A=-i\omega_s\rho_0^{-2}C_0^{-4}[1+\tfrac{1}{2}\rho_0C_0^{-2}(d^2p/d\rho^2)_{\rho=\rho_0}]P_0^2.$$

This integral is elementary and yields, for $l\gg\alpha^{-1}$, the result

$$p_s=(\omega_s^2\rho_0A/8\pi R_0)e^{ik_sR_0-i\omega_st}/[i\alpha+k_s\sin^2(\theta/2)]. \quad (12)$$

Corresponding to the sound pressure, p_s is the intensity I_s given by

$$I_s=\frac{\omega_s^4P_0^4S_0^2}{2(8\pi)^2\rho_0^3C_0^9}\left[1+\tfrac{1}{2}\rho_0C_0^{-2}\left(\frac{d^2p}{d\rho^2}\right)_{\rho=\rho_0}\right]^2$$
$$\times\frac{1}{R_0^2[\alpha^2+k_s^2\sin^4(\theta/2)]}, \quad (13)$$

and the total power radiated W_s is

$$W_s=\frac{\omega_s^3 P_0^4 S_0^2}{64\rho_0^3 \alpha C_0^8}[1+\tfrac{1}{2}\rho_0 C_0^{-2}(d^2p/d\rho^2)_{\rho=\rho_0}]^2. \quad (14)$$

The angular dependence of Eq. (13) is the same as that for Rutherford scattering in atomic theory. The angle at which the intensity is reduced by one-half is given approximately by

$$\theta_{\frac{1}{2}}\approx 2(\alpha/k_s)^{\frac{1}{2}}. \quad (15)$$

In order to apply the above concepts to a discussion of highly directional receivers, we imagine that one of the primary waves is the signal that we wish to detect, and thus it emanates from behind the high-frequency carrier generator. The carrier is modulated by the incoming wave, and *vice versa*, leading to sideband waves that, in the absence of damping, increase linearly in amplitude with distance along the carrier beam. It is necessary to place a microphone in the carrier beam, and the output of the microphone must then be fed into a conventional radio set in order to demodulate the signal. Some interesting possibilities suggest themselves, such as, for example, suppressing the carrier with acoustic filters.

ACKNOWLEDGMENTS

This work has been supported by the U. S. Office of Naval Research, and it formed the substance of a paper (Abstract P9) presented at the Fifty-Ninth Meeting of the Acoustical Society of America, June 1960.

Note added in proof. An experimental investigation of the end-fire array discussed here has recently been published by J. L. S. Bellin and R. T. Beyer, J. Acoust. Soc. Am. **34**, 1051 (1962).

The author of the following paper shows that, although the Fay and Fubini solutions appear to be contradictory, each holds for a different region of the flow. The Fubini solution holds close to the source and the Fay solution holds in the region far from the source.

Dr. Blackstock was Associate Professor of Electrical Engineering at the University of Rochester. He is now Visiting Associate Professor of Electrical Engineering at the University of Texas, and Special Research Associate at the Applied Research Laboratories at the University of Texas. He has been concerned with basic research in acoustics and high intensity sound. He is a Fellow of the Acoustical Society of America and was Chairman of the Committee on Education in Acoustics of this Society.

This paper is reprinted with the permission of the author and the Acoustical Society of America.

Reprinted from THE JOURNAL OF THE ACOUSTICAL SOCIETY OF AMERICA, Vol. 39, No. 6, 1019–1026, June 1966

42

Connection between the Fay and Fubini Solutions for Plane Sound Waves of Finite Amplitude

DAVID T. BLACKSTOCK

Acoustical Physics Laboratory, Electrical Engineering Department
University of Rochester, Rochester, New York 14627

Plane, progressive, periodic sound waves of finite amplitude are considered. The well-known solutions of Fay and Fubini are reviewed. At first glance, the two solutions seem contradictory, but, actually, each holds in a different region of the flow, the Fubini solution close to the source and the Fay solution rather far from the source. In the intermediate, or *transition*, region, neither solution is valid. A more general solution is obtained by using a method commonly employed for waves containing weak shocks. For distances up to the shock-formation point, the general solution reduces exactly to the Fubini solution. For distances greater than about 3.5 shock-formation lengths, the general solution is practically indistinguishable from the sawtooth solution, which, in turn, is the limiting form of Fay's solution for strong waves. The form of the general solution shows clearly how, in the transition region, the Fubini solution gives way to the sawtooth solution. The problem of an isolated cycle of an originally sinusoidal wave is also considered. Finally, some limitations on the weak-shock method are discussed. In the periodic-wave problem, the general solution is found to be inaccurate for distances greater than $1/\alpha$, approximately, where α is the small-signal absorption coefficient. In Appendix A, a brief extension of the analysis to spherical and cylinrical waves is given.

INTRODUCTION

MODERN interest in finite-amplitude sound, at least from the point of view of acousticians, dates back to the 1930's. Some of the first work in this period is still the best, and, indeed, two early solutions are now regarded as classics; namely, those of Fay[1] (1931) and Fubini[2] (1935). Both solutions deal with the propagation of plane periodic waves in a semiinfinite fluid.

Though Fubini's solution came later, it is convenient to consider it first. Let the boundary condition at the origin $x=0$ be given by

$$u(0,t)=u_0 \sin\omega t, \qquad (1)$$

where u is particle velocity, ω frequency, and t time. For this boundary condition, a very good "first-order" solution of the exact equations of motion for lossless fluids is

$$u/u_0=\sum_n (2/n\sigma)J_n(n\sigma)\sin n(\omega t-kx), \qquad (2)$$

where $k-\omega/c_0$, c_0 is the small-signal sound speed, J_n is the ordinary Bessel function of order n, and σ is the distance in terms of shock-formation lengths (i.e., $\sigma=x/\bar{x}$, where $\bar{x}$ is the distance at which the wave first becomes discontinuous). Equation 2 has come to be known as the Fubini solution.[3]

Fay was not concerned with such a specific boundary condition as Eq. 1 but, instead, sought the most nearly stable periodic waveform in a viscous perfect gas. Adapting his solution to the case at hand, we have

$$\frac{u}{u_0}=\sum_{n=1}^{\infty}\frac{2/\Gamma}{\sinh[n(1+\sigma)/\Gamma]}\sin n(\omega t-kx), \qquad (3)$$

where Γ is a group of constants characterizing the importance of nonlinearity relative to that of dissipation.

[1] R. D. Fay, J. Acoust. Soc. Am. 3, 222–241 (1931).

[2] E. Fubini, Alta Frequenza 4, 530–581 (1935). Fubini's work lay largely unappreciated for many years. P. J. Westervelt first brought it to light in this country; see, for example, J. Acoust. Soc. Am. 22, 319–327 (1950).

[3] This is despite the fact that there are certain differences between Eq. 2 and the actual solution that Fubini gave. First, Fubini calculated a second-order as well as a first-order series. In terms of Eq. 2, the second-order series would be a cosine series whose over-all magnitude is smaller than the given sine series by a factor u_0/c_0 (incidentally, Fubini's second-order series should be corrected by multiplying by $-\frac{1}{2}$). Second, Fubini used Lagrangian rather than Eulerian coordinates (this implies, for example, a slightly different boundary condition than our Eq. 1). Within the "first-order" framework used in this paper, however, Eq. 2 is equivalent to the solution Fubini actually gave. See also Ref. 15. For a discussion of the limits and accuracy of Eq. 2, see D. T. Blackstock, J. Acoust. Soc. Am. 34, 9–30 (1962).

Equation 3 is commonly referred to as the "Fay solution."[4]

Despite the fact that the two solutions are highly regarded, they do not by themselves support each other. For example, though it might be supposed that Eq. 2 is simply the limiting form of Eq. 3 as the viscosity vanishes, such is not the case. The "lossless" limit of Eq. 3, found by letting $\Gamma \rightarrow \infty$, is

$$u/u_0 = \sum_n [2/n(1+\sigma)] \sin n(\omega t - kx), \qquad (4)$$

which is the series representation of a sawtooth wave and certainly does not agree[5] with Eq. 2. How, then, can both solutions be valid? The answer is that the respective ranges of validity of the two solutions are different. The Fubini solution holds near the source, in particular in the shockfree region $\sigma < 1$. The Fay solution is valid in the region where the shocks are fully formed and have begun to decay—that is, rather far from the source.

Yet there remains some mystery about the region of transition, where neither the Fubini solution nor the sawtooth (or Fay) solution is accurate. This region is relatively short, extending from $\sigma = 1$ to approximately $\sigma = 3.5$. Across this region, the functional form of the harmonic amplitudes changes from $2(n\sigma)^{-1} J_n(n\sigma)$ to $2[n(1+\sigma)]^{-1}$, as is clear from Eqs. 2 and 4. This is a rather drastic change and requires investigation.

The purpose of this paper is to show one way of bridging the gap between the Fubini and Fay solutions. To this end, a more-general solution, one that contains the Fubini and sawtooth solutions as limiting cases and also covers the transition region in between, is developed. The method of solution is one commonly employed in treating weak, nonuniform shock waves.[6,7] By nonuniform is meant unsteady—that is, growing or decaying—shocks. Unfortunately, use of the weak-shock method allows us to connect the Fubini solution directly with the sawtooth solution but not with the Fay solution. This is not deemed a serious impairment of our purpose, however, because the relation of the sawtooth solution to Fay's solution has already been shown.

The weak-shock method is not the only one that can be used to connect the two classic solutions, of course. The solution of Burgers' equation has also been exploited.[8] In some respects, the approach using Burgers' equation is preferable because the Fay solution itself, not just the sawtooth solution, emerges as one of the limiting cases. However, the solution of Burgers' equation is extremely complicated (this is the penalty that one pays for generality). The behavior in the transition region is particularly obscure. By contrast, the weak-shock solution shows very simply and directly how the Fubini solution gives way to the sawtooth solution across the transition region. In passing, we might add the unifying comment that, in general, for shocks in otherwise continuous waveforms, the weak-shock method has been shown to be a limiting form of the method based on Burgers' equation.[9]

I. ANALYSIS

As noted above, the method of solution used here is one commonly employed in dealing with waveforms that contain weak shocks. Lossless, progressive-wave relations are utilized to describe continuous sections of the waveform between shocks. To these relations are added the weak-shock formula that connects shock speed with particle velocity on either side of each shock. The resulting system of equations can then be solved to give waveform shape, shock paths, and shock amplitudes. The analytical sketch of the method given here is, necessarily, brief. For more general and complete treatments, see Friedrichs[6] and Whitham.[7] Our approach closely parallels that of Whitham.

A. General

The progressive-wave solution for continuous, forced wave motion is basically due to Earnshaw.[10] Let the particle velocity at some convenient point, which we take as the origin of our coordinate system, be specified as a function of time—that is,

$$u(0,t) = g(t). \qquad (5)$$

The outward-propagating wave can then be described in terms of a parameter ϕ as follows[11]:

$$u = g(\phi), \qquad (6)$$

$$t - \phi = (x/c_0)[1 + \beta c_0^{-1} g(\phi)]^{-1}, \qquad (7)$$

where β depends on the equation of state of the fluid.[12] Equation 7 shows that $\phi = t$ at $x = 0$. Thus ϕ represents

[4] Here again, some license is used. The solution that Fay actually gave is more general. It probably represents the limiting form for any wave that is periodic at the source, not just the originally sinusoidal wave. The form that Fay's result takes when Eq. 1 is the boundary condition is found by putting Fay's arbitrary constant α_0 equal to Γ^{-1}; see, for example, D. T. Blackstock, J. Acoust. Soc. Am. **36**, 534–542 (1964). Also, though Fay considered only viscous perfect gases, his solution is equally valid for viscous, heat-conducting fluids of arbitrary equation of state. Finally, Fay used Langrangian coordinates and put his result in terms of pressure, not particle velocity.

[5] Fubini himself was puzzled by this; see Art. 22 in Ref. 2, p. 575.

[6] K. O. Friedrichs, Commun. Pure Appl. Math. **1**, 211–245 (1948).

[7] G. B. Whitham, Commun. Pure Appl. Math. **5**, 301–348 (1952), especially the Appendix.

[8] See, for example, S. I. Soluyan and R. V. Khokhlov, Vestn. Mosk. Univ. Ser. III Fiz. Astron. 3, 52–61 (1961), or D. T. Blackstock, Ref. 4.

[9] M. J. Lighthill, *Surveys in Mechanics*, G. K. Batchelor and R. M. Davies, Eds. (Cambridge University Press, Cambridge, England, 1956), pp. 250–351.

[10] S. Earnshaw, Trans. Roy. Soc. (London) **150**, 133–148 (1860); see also J. W. Strutt Lord Rayleigh, Proc. Roy. Soc. (London) **84**, 247–284 (1910).

[11] See, for example, D. T. Blackstock, Ref. 3.

[12] For perfect gases $\beta = (\gamma+1)/2$, where γ is the ratio of specific heats; for other fluids $\beta = 1 + B/2A$, where $B/2A$ is the coefficient of the first nonlinear term in an isentropic equation of state. See, for example, R. T. Beyer, J. Acoust. Soc. Am. **32**, 719–721 (1960).

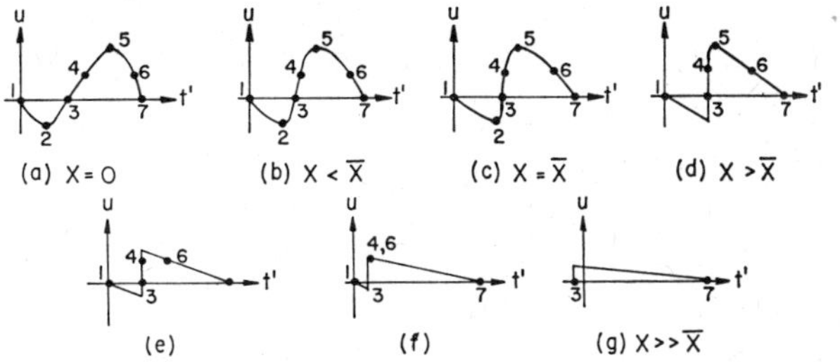

FIG. 1. Progressive distortion and (shock-induced) decay of a traveling wave. Numbered points identify specific wavelets (values of the particle velocity u); also, $t'=t-x/c_0$ and $\bar{x}$ is the shock formation distance.

the time that a given signal or "wavelet" (i.e., value of u) left the origin. It will be seen from Eq. 6 that each value of ϕ is to be associated with a particualr point on the waveform, for example, point 1, 2, 3, etc. in Fig. 1. If $\beta g/c_0=\beta u/c_0$ is small as compared with unity, Eq. 7 can be approximated as

$$t'=\phi-(\beta c_0^{-2}x)g(\phi), \tag{8}$$

where $t'=t-x/c_0$ is a retarded time based on the speed at which zeroes of the waveform travel. Equation 8 may be thought of as a first-order approximation of Eq. 7 and, hereinafter, is used in its place.[13]

Next, consider that this continuous-waveform solution gives birth to a shock at time $\bar{t}$ at distance $\bar{x}$ from the origin. The time of arrival t of the shock at any subsequent point is given by

$$t=\bar{t}+\int_{\bar{x}}^{x}(1/v)d\lambda, \tag{9}$$

where v is the shock-propagation speed. This speed depends on the value of the particle velocity on either side of the shock. Let u_a and u_b stand for particle velocity just ahead of and just behind the shock, respectively. The following equation is a first-order approximation of the exact shock relation that connects v with u_a and u_b:

$$v=c_0+\tfrac{1}{2}\beta(u_a+u_b) \tag{10}$$

(whence the theorem that weak shocks propagate with a speed that is the mean of the propagation speed of signals, or wavelets, just ahead and just behind the shock). To first order, therefore,

$$v^{-1}=c_0^{-1}[1-\tfrac{1}{2}\beta c_0^{-1}(u_a+u_b)], \tag{11}$$

and Eq. 9 becomes

$$t_s'=\bar{t}'-\tfrac{1}{2}\beta c_0^{-2}\int_{\bar{x}}^{x}(u_a+u_b)d\lambda. \tag{12}$$

Bars continue to indicate values at the instant of shock formation and primes retarded time. The subscript s has been used to signify the value associated with the shock. It is frequently convenient to convert Eq. 12 to the differential equation

$$dt_s'/dx=-\tfrac{1}{2}\beta c_0^{-2}(u_a+u_b). \tag{13}$$

Equations 12 (or 13), 6, and 8 are sufficient to determine the entire motion. Equations 6 and 8 give the waveform shape in the continuous regions between shocks. The path and amplitude of each shock are determined by Eq. 13 in conjunction with Eqs. 6 and 8 when the latter are evaluated at the point just behind the shock ($u=u_b$, $\phi=\phi_b$, and $t'=t_s'$) and at the point just ahead of the shock ($u=u_a$, $\phi=\phi_a$, and $t'=t_s'$). Whitham[7] chose to eliminate particle velocity and to work instead with what amounts to our quantities t_s', ϕ_a, and ϕ_b. This leads to an illuminating geometrical rule (the "equal lobes" rule) based on the $g(\phi)$ curve; the rule allows one to pick out proper values of ϕ_a and ϕ_b very easily. Our approach here, which is more convenient in certain cases, is to eliminate the parameter ϕ. In principle, Eqs. 6 and 8 can be combined to give

$$t'=g^{-1}(u)-\beta xc_0^{-2}u, \tag{14}$$

where g^{-1} is the inverse function corresponding to g; that is, $g[g^{-1}(u)]=u$. Thus, just ahead of the shock we have

$$t_s'=g^{-1}(u_a)-\beta xc_0^{-2}u_a, \tag{15a}$$

and just behind

$$t_s'=g^{-1}(u_b)-\beta xc_0^{-2}u_b. \tag{15b}$$

Equations 15a, 15b, and 13 (or 12) can then be solved simultaneously for u_a, u_b, and t_s'. A generalization of these formulas to cover certain classes of spherical and cylindrical waves is given in Appendix A.

[13] Our specification of boundary condition is different from Whitham's. Whitham couched his analysis in terms of the classical "piston problem," where the motion of the piston is given. In that case, the particle velocity is known not at a fixed point but rather at a moving point: namely, the face of the piston. In place of Eq. 5, one then has $u(\xi,t)=g(t)$, where $\xi=\int g dt$ is the piston displacement. The solution becomes slightly more complicated in that x must be replaced by $x-\xi$ in Eqs. 7 and 8. In many experiments, however, the source condition is poorly defined or impossible to measure directly. In this case, the measurement of the field nearest the source can constitute the "boundary condition." As far as the theoretician is concerned, the first measurement point serves as the origin, and Eq. 5 is the correct specification of the "source." Actually, the difference between this specification and Whitham's becomes negligible as long as the observer is not close to the piston. A detailed discussion of this point has been given previously by the author (see Ref. 3). Mathematically, the boundary condition we have used leads to a somewhat simpler analysis than Whitham's, but the difference is not of major importance.

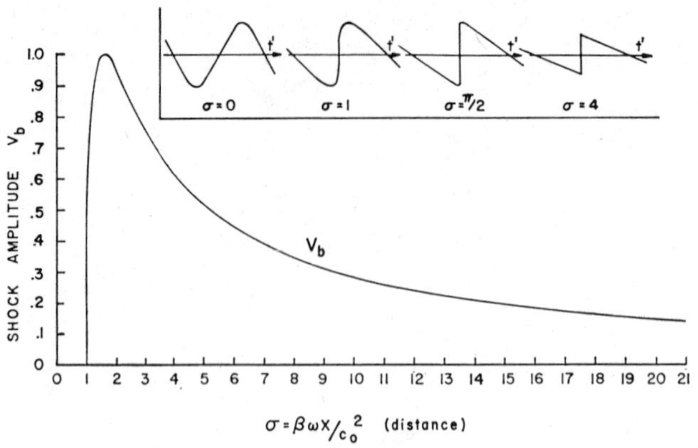

FIG. 2. Shock amplitude of an originally sinusoidal wave. Inset shows the waveform at various distances from the source.

B. Sawtooth Waves

The method sketched briefly above is now applied to the problem of an infinite wavetrain whose time variation is a pure sinusoid at $x=0$. Let the source excitation be

$$u(0,t)=u_0 \sin\omega t,\ t>-M, \quad (16)$$

where M is a large positive number.[14] The shock-formation distance in this case is[11] $\bar{x}=(\beta\epsilon k)^{-1}$, where $\epsilon=u_0/c_0$. If we choose dimensionless quantities $V=u/u_0$, $y=\omega t'=\omega t-kx$, $\sigma=x/\bar{x}$, and $\Phi=\omega\phi$, then Eqs. 6, 8, 14, and 13 are, respectively, for this case

$$V=\sin\Phi, \quad (16)$$

$$y=\Phi-\sigma\sin\Phi \quad (17)$$

$$=\sin^{-1}V-\sigma V, \quad (18)$$

$$dy_s/d\sigma=-\tfrac{1}{2}(V_a+V_b). \quad (19)$$

Furthermore, evaluation of Eq. 15 just ahead of and just behind the shock gives, respectively,

$$y_s=\sin^{-1}V_a-\sigma V_a, \quad (20a)$$

$$y_s=\sin^{-1}V_b-\sigma V_b. \quad (20b)$$

Various stages in the distortion of the waveform are shown in the insert in Fig. 2.

First, let us determine the relation between shock amplitude and distance. Any period of the wave can be considered, but it is simplest to focus attention on the one between $y=-\pi$ and $y=\pi$. Here, the shock forms at the origin $y=0$, and so $y_s=0$ initially. In the next increment of time Δt, the shock catches up with a point on the waveform just ahead and is overtaken by a point just behind. If the particle velocity of the point just behind is $V=\Delta$, then that just ahead is $V=-\Delta$. This is true because of the symmetry properties of the sine function (see Eq. 18). At the end of the increment Δt, therefore, the shock has attained an amplitude Δ (the peak-to-peak value will be 2Δ). As Eq. 19 shows, however, the shock remains stationary at $y=0$ because $V_b=\Delta$ and $V_a=-\Delta$. In succeeding increments of time, this process is continued. The bottom of the shock grows as rapidly as the top so that symmetry is always preserved. A partial solution of Eqs. 19–20b is, therefore,

$$y_s=0,\quad V_a=-V_b, \quad (21)$$

and we are left with the following equation for determining shock amplitude V_b:

$$V_b=\sin(\sigma V_b). \quad (22)$$

The only solution of this transcendental equation in the region $0\leqslant\sigma<1$ is $V_b=0$, in agreement with our expectations. For values of $\sigma>1$, another root appears, and this is the one that gives the shock amplitude. (As σ becomes progressively larger, other roots appear, but these have no physical significance here; jumping from one root to another would correspond to changing the shock amplitude discontinuously.) To solve Eq. 22 for V_b, first note that the equation may be expressed as $\sigma^{-1}=j_0(\sigma V_b)$, where $j_0(\theta)=(\sin\theta)/\theta$ is the spherical Bessel function of order zero. Then, if the inverse is taken,

$$V_b=\sigma^{-1}j_0^{-1}(\sigma^{-1})H(\sigma-1), \quad (23)$$

where $H(\sigma-1)$ is the Heaviside unit step function and is inserted to emphasize that $V_b=0$ when $\sigma<1$. Here, $j_0^{-1}(\sigma^{-1})$ is to be read the "quantity whose zero-order spherical Bessel function is $1/\sigma$." An asymptotic expression for V_b for large values of σ may be obtained as follows. A graphical sloution of Eq. 22 suggests that, as σ becomes large, σV_b approaches π. Putting $\sigma V_b=\pi-\delta$, where δ is small, we find $\sin\sigma V_b=\sin(\pi-\delta)=\sin\delta\doteq\delta=\pi-\sigma V_b$. Thus, for large values of σ,

$$V_b=\pi/(1+\sigma). \quad (24)$$

This relation and a table of zero-order spherical Bessel functions have been used to construct the graph of shock amplitude shown in Fig. 2. The accuracy of the

[14] Choosing the source excitation to begin at $t=-M$ rather than at $t=0$ relegates to the background a consideration of what happens at the very front of the wave. The problem of the wavefront is analyzed in the following section.

asymptotic formula, Eq. 24, exceeds 2% when σ is greater than about 3.6.

A novel result becomes evident when Eq. 24 is put in terms of dimensional quantities:

$$u_b = \pi u_0/(1+\beta u_0 kx/c_0). \quad (25)$$

As x becomes very large, this reduces to

$$u_b \sim \pi c_0/\beta kx, \quad (26)$$

which result was first obtained by Whitham,[7] who indeed derived it for the more general case of periodic (not just sinusoidal) piston displacement. Equation 26 is interesting because it implies that the wave amplitude at great distances is independent of the source amplitude u_0. This broad implication is correct, but the specific form of the result is wrong. At such distances, as is made clear later, the weak-shock method fails. Explicit account must be taken of the effect of dissipation over the entire waveform. When the dissipation is due to viscosity and heat conduction, the correct formula for wave amplitude at great distances is[8]

$$|u| = (4\alpha c_0/\beta k)e^{-\alpha x}, \quad (27)$$

where α is the small-signal attenuation coefficient for a sinusoidal wave of frequency ω.

From this point on, the analysis is very similar to that frequently used to obtain the Fubini solution.[11] The Fubini solution is merely a Fourier-series representation of the wave in the shockfree region, $0 \leqslant \sigma < 1$, where the entire waveform is continuous. As Bessel showed in an entirely different connection,[15] the harmonic amplitudes are obtained by shifting from an integration over $y(-\pi<y<\pi)$ to an integration over $\Phi(-\pi<\Phi<\pi)$. The shift in integration is straightforward for the region $0\leqslant\sigma<1$, because there, as inspection of Eq. 17 shows, Φ is a single-valued function of y. In physical terms, for a given value of σ in the range $0\leqslant\sigma<1$, there is one and only one value of Φ—and therefore one and only one value of the particle velocity V—for each value of y. Once the shock forms, however, several different values of Φ exist at one particular value of y: namely, at y_s. Thus, when $\sigma>1$, the one-to-one correspondence between Φ and y is destroyed. Nevertheless, there are still

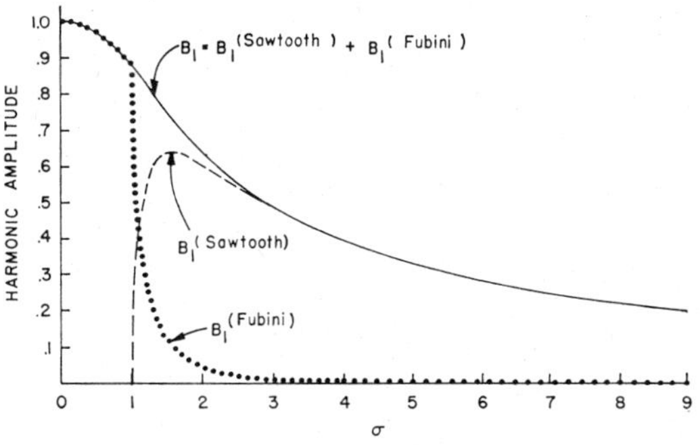

FIG. 3. Amplitude of the fundamental component B_1 in an originally sinusoidal wave. See Eqs. 29 and 30.

ranges of Φ in which Φ is a single-valued function of y: namely, the ranges $-\pi<\Phi<\Phi_a$ and $\Phi_b<\Phi<\pi$, where $\Phi_a=\sigma V_a$ and $\Phi_b=\sigma V_b$ (see Eqs. 16 and 20). This gives us a clue as to how the shift in integration can be accomplished when $\sigma>1$.

The explicit mathematical steps are as follows. Because of symmetry, the y integration need be carried out only over the range $0<y<\pi$. Let $V=\sum B_n \sin ny$; then,

$$B_n = (2/\pi)\int_0^\pi V \sin ny\,dy$$

$$= (2/\pi)\int_0^\pi \sin\Phi \sin ny\,dy \quad (28)$$

$$= (2/n\pi)\left\{-\sin\Phi \cos ny\Big|_{y=0}^{y=\pi} + \int_{y=0}^{y=\pi} \cos ny \cos\Phi\, d\Phi\right\},$$

where $\Phi=\pi$ when $y=\pi$, and $\Phi=\Phi_{\min}$ when $y=0$. The value of $\Phi_{\min}$ is zero for $0\leqslant\sigma<1$, but is Φ_b for $\sigma>1$. Since, by Eq. 17, $\cos\Phi d\Phi=\sigma^{-1}(d\Phi-dy)$, we have

$$B_n = (2/n\pi)\left\{\sin\Phi_{\min}+\sigma^{-1}\int_{y=0}^{y=\pi} \cos ny(d\Phi-dy)\right\}$$

$$= (2/n\pi)V_b + (2/n\pi\sigma)\int_{\Phi_{\min}}^{\pi} \cos n(\Phi-\sigma\sin\Phi)d\Phi \quad (29)$$

$$= B_n^{(\text{sawtooth})} + B_n^{(\text{Fubini})}, \quad (30)$$

where use has been made of the fact that $\sin\Phi_{\min}$ is the shock amplitude V_b.

The association of the sawtooth solution with the first term in Eq. 29 and the Fubini solution with the second term is justified as follows: If $0\leqslant\sigma<1$, then $V_b=0$ and $\Phi_{\min}=0$, in which case

$$B_n = (2/n\sigma)J_n(n\sigma), \quad (31)$$

giving us the classical Fubini solution, Eq. 2. On the other hand, if $\sigma\gg1$. then V_b is given by Eq. 24 and

[15] F. W. Bessel, Abhandl. Berliner Akad. Wis. Math. Cl. **1824**, 1–52 (1826), or *Abhandungen von Friedrich Wilhelm Bessel* (Verlag von W. Engelmann, Leipzig, Germany, 1875), Vol. 1, pp. 84–109. Bessel was concerned with Kepler's problem; Eq. 17 has exactly the same form as Kepler's second law of planetary motion. Because Bessel solved this transcendental equation by Fourier series long before Fubini (Ref. 2) did, and in a much simpler and more direct manner, I originally proposed (see Ref. 3) that Eq. 2 be called the Bessel–Fubini solution rather than the Fubini solution. F. V. Hunt has pointed out to me (personal communication), however, that Eq. 14 arises in quite a number of physical problems besides planetary motion and finite-amplitude sound waves. For example, the motion of a phonograph needle in a record groove or of a cam follower can also be couched in terms of Kepler's problem. See, for example, D. H. Cooper, J. Audio Eng. Soc. **12**, 2–7 (1964). Inasmuch as it would be terribly confusing to attach Bessel's name to solutions of all these diverse problems, it is undoubtedly preferable to stick to the older description "Fubini solution" in reference to Eq. 2, this despite the inaccuracies noted in Ref. 3.

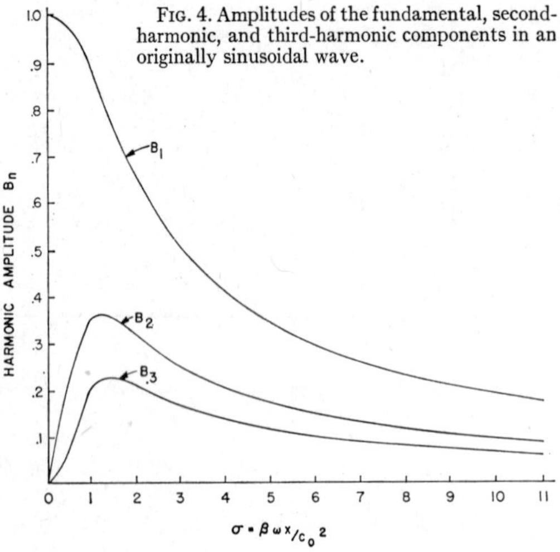

FIG. 4. Amplitudes of the fundamental, second-harmonic, and third-harmonic components in an originally sinusoidal wave.

$\phi_{\min} \to \pi$; in which case, Eq. 29 reduces to

$$B_n = 2/[n(1+\sigma)], \tag{32}$$

agreeing with the sawtooth solution, Eq. 4. It is now clear how the mathematical transition from the Fubini solution to the sawtooth solution occurs. The complete solution is simply the sum of two solutions. As one dies out, the other grows. The behavior is sketched in Fig. 3 for the fundamental B_1. Notice how rapidly the Fubini solution dies out for $\sigma > 1$. The Figure also makes clear that in the limiting regions, $0 \leqslant \sigma < 1$ and $\sigma \gg 1$, only one solution remains.

For practical purposes, as is easily seen from Fig. 3, the transition region extends from $\sigma = 1$ to $\sigma \doteq 3.5$. In this region, the complete solution, Eq. 29 with V_b given by Eq. 23, must be used. No analytical simplifications have as yet been obtained. However, the coefficients B_1, B_2, and B_3 have been tabulated by using the asymptotic formulas and a numerical evaluation of the integral in Eq. 29. The results are shown in Fig. 4.

C. Isolated Sine-Wave Cycle

In the previous problem, an infinite train of waves was considered. No particular attention was paid to the cycle at the head of the train. The first cycle may behave differently, however. For simplicity, consider a source that generates a single sine-wave cycle. If the excitation is

$$u(0,t) = u_0 \sin\omega t, \quad -\pi \leqslant \omega t \leqslant \pi, \tag{33}$$

then the wave behaves in exactly the same manner as any single cycle in an infinite train (see Fig. 2). The shock forms in the middle of the cycle, and, since $V_a = -V_b$, the shock and the zeroes of the wave travel at the same speed, a "one-tooth" wave develops, and the decay is as given by Eq. 23 (and ultimately by Eq. 24). But if the phase of the excitation is shifted 180°, i.e., if

$$u(0,t) = -u_0 \sin\omega t, \quad -\pi < \omega t < \pi, \tag{34}$$

the wave motion is much modified. Instead of a single shock in the middle of the cycle, two shocks form, one at the head and one at the tail. In the case of the head shock, $V_a = 0$, whereas for the tail shock $V_b = 0$. The shocks do not, therefore, move at the same speed as the zero; the head shock moves faster, the tail shock slower. Thus the wave elongates as it propagates and eventually resembles an N wave. The behavior is sketched in Fig. 5.

It is a simple matter to find the shock amplitude V_b for the wave pictured in Fig. 5. First of all, both shocks are born at $\sigma = 1$. But because of symmetry, only half the wave need be dealt with—say that between the head shock and the zero. Equations 12 and 15b for this case are, in terms of the dimensionless quantities introduced earlier,

$$y_s = -\pi - \tfrac{1}{2}\int_1^\sigma V_b d\sigma, \tag{35}$$
$$y_s = -\pi + \sin^{-1} V_b - \sigma V_b,$$

where we have put $\omega g^{-1}(V_b) = -\pi + \sin^{-1} V_b$ in order to deal only with the principal branch of the arcsine function. Elimination of y_s and differentiation with respect to σ yields the differential equation

$$[(1-V_b^2)^{-\frac{1}{2}} - \sigma] dV_b/d\sigma = \tfrac{1}{2} V_b, \tag{36}$$

or

$$V_b^2 d\sigma/dV_b + 2V_b\sigma = 2V_b(1-V_b^2)^{-\frac{1}{2}}. \tag{37}$$

The solution that satisfies the condition $V_b = 0$ at $\sigma = 1$, is

$$\sigma = 2V_b^{-2}[1-(1-V_b^2)^{\frac{1}{2}}]. \tag{38}$$

Solving for V_b, we get

$$V_b = 2\sigma^{-1}(\sigma-1)^{\frac{1}{2}} H(\sigma-1). \tag{39}$$

Thus, the shock amplitude is zero at $\sigma = 1$, increases to unity at $\sigma = 2$, and decays thereafter. As σ becomes very

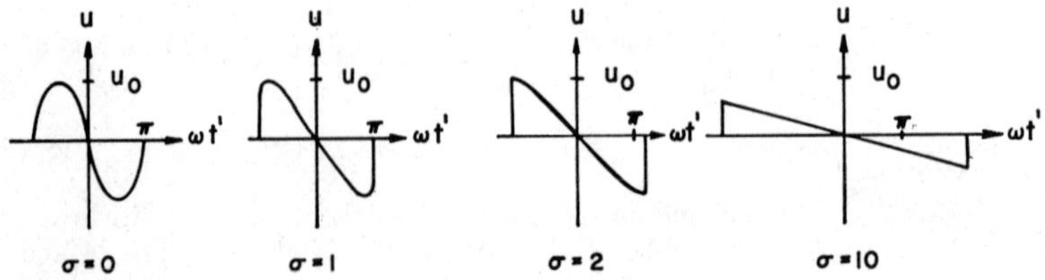

FIG. 5. Progressive distortion and subsequent decay and elongation of a one-cycle wave produced when the excitation at the origin is $u = -u_0 \sin\omega t$, $-\pi < \omega t < \pi$.

large, the wave decays as $1/(\sigma)^{\frac{1}{2}}$, or, in other words, like an N wave.[16]

D. Limitations of the Method

Consider again the case of the infinite wavetrain whose shape at the source is sinusoidal. As we have seen, weak-shock theory leads to the prediction that the wave amplitude (and each harmonic amplitude) will decay as $1/x$ as x becomes large. The basic assumptions in this theory are that (1) nonlinear effects are important, and (2) dissipative effects may be neglected (save at the shock fronts, where they are accounted for implicitly through use of the Rankine–Hugoniot relations). Suppose that these assumptions are now reversed. Let us neglect nonlinear effects and instead take explicit account of viscosity and heat conduction (or any other agencies of dissipation). In this case, there is only one harmonic, the fundamental, and, therefore, the wave amplitude decays as $e^{-\alpha x}$, where α is the small-signal attenuation coefficient. The question is, which type of decay is more important? Clearly, if αx is extremely small, the weak-shock decay rate will be more important. But no matter how small α is, there will always be a distance beyond which exponential decay is more rapid than inverse first-power decay. Thus at very great distances from the origin the formulas derived using weak-shock theory can be expected to be inaccurate. This is why Whitham's result, Eq. 26, cannot be relied on for indefinitely large distances.

Let us attempt to make a quantitative estimate based on these ideas. Consider the fundamental component in a sawtooth wave. The amplitude B_1 of this component at the point $\sigma=\sigma_m$ is given by Eqs. 29 and 24 as

$$B_1=2/(1+\sigma_m). \qquad (40)$$

Assume that the wave continues to behave as a sawtooth for a distance $\Delta\sigma$ beyond this point. Then, $B_1=2/(1+\sigma_m+\Delta\sigma)$, and the rate of decay is

$$-dB_1/d(\Delta\sigma)=2/(1+\sigma_m+\Delta\sigma)^2. \qquad (41)$$

On the other hand, if beyond this point the wave behaves as a small-signal wave subject to ordinary attenuation, we must write $B_1=[2/(1+\sigma_m)]e^{-\alpha\Delta x}$, where $\Delta x=\Delta\sigma/\beta\epsilon k$. In this case, the rate of decay is

$$-dB_1/d(\Delta\sigma)=(2\alpha/\beta\epsilon k)/(1+\sigma_m)e^{-\alpha\Delta x}. \qquad (42)$$

Now, equate the two rates of decay and let $\Delta\sigma\to 0$. The value of σ_m thus obtained, which we call $\sigma_{\max}$ and interpret as the maximum distance at which the sawtooth solution is valid, is given by

$$1+\sigma_{\max}=(\beta\epsilon k/\alpha), \qquad (43)$$

or, approximately,

$$x_{\max}=1/\alpha. \qquad (44)$$

This estimate agrees very nicely with an earlier one obtained using the solution of Burgers' equation and based on a comparison of shock thickness with wavelength.[17]

Lighthill[9] has given another way, and indeed a very general one, of estimating the point beyond which the weak-shock method gives inaccurate results. He defines a Reynolds number R such that when $R>1$, approximately, the shock-wave-decay formulas are expected to hold. Conversely, when $R<1$, approximately, linear (thermoviscous) attenuation theory should be used. For our sawtooth-wave problem, the Reynolds number is given by

$$R=(\beta\epsilon k/\alpha)\int_0^{\pi} Vdy, \qquad (45)$$

where V is the appropriate solution of Burgers' equation[8] (the constant Γ appearing in Eq. 3 is equal to $\beta\epsilon k/\alpha$). It turns out that R becomes unity in this equation when $x\doteq 2/\alpha$, a value that gives reasonable agreement with Eq. 44.

There is a corollary ot the limitation on distance prescribed by Eq. 44; namely, that the wave must be strong enough to begin with that $x_{\max}\gg\bar{x}$ or equivalently,

$$\beta\epsilon k\gg\alpha. \qquad (46)$$

If this is not the case, then whatever shocks that form will be so weak that there is no point in using the weak-shock method at all. The wave is so nearly in the infinitesimal-amplitude class that direct effects of dissipation over the entire waveform are at least as important as nonlinear effects. In such a case, the perturbation solution of Keck and Beyer[18] could well be used.

ACKNOWLEDGMENT

This work was supported by the U. S. Air Force Office of Scientific Research, Office of Aerospace Research. David M. Ryon carried out the numerical calculations that led to Fig. 4.

[16] For N-wave decay formulas, see, for example, L. D. Landau and E. M. Lifshitz, *Fluid Mechanics* (translated from the Russian by J. B. Sykes and W. H. Reid, Addison-Wesley Publ. Co., Inc., Reading, Mass., 1959), Art. 95.

[17] See, for example, D. T. Blackstock, Ref. 4.

[18] W. Keck and R. T. Beyer, Phys. Fluids 3, 346–352 (1960).

Appendix A. Spherical and Cylindrical Sawtooth Waves

The analysis for spherical and cylindrical waves is exceedingly simple provided the radial distance r is large compared with a wavelength. In a previous note,[A1] Fubini-like solutions were derived for spherical and cylindrical waves. A general sawtooth solution similar to Eq. 4 was also obtained, but not by the methods described in this paper. Here, we show how the two solu-

[A1] D. T. Blackstock, J. Acoust. Soc. Am. 36, 217–219 (L) (1964).

tions may be brought together by employing a generalization of the weak-shock method.[A2]

In the region where the waveform is continuous, the generalization of Eqs. 6 and 8 is[A1]

$$w=g(\phi), \tag{A1}$$

$$t'=\phi-(\beta c_0^{-2}h)g(\phi), \tag{A2}$$

where

$$w=(r/r_0)^a u, \tag{A3}$$

and a and h are as follows for the three different kinds of waves:

Plane	$a=0,$	$h=r,$	(A4)
Spherical	$a=1,$	$h=r_0 \ln(r/r_0),$	(A5)
Cylindrical	$a=1/2,$	$h=2(r_0)^{\frac{1}{2}}[(r)^{\frac{1}{2}}-(r_0)^{\frac{1}{2}}].$	(A6)

This solution corresponds to a given excitation or boundary condition at the reference point $r_0(h=0)$ of

$$w|_{h=0}=u(r_0,t)=g(t), \tag{A7}$$

where the definition of the retarded time t' has been generalized to $t'=t-(r-r_0)/c_0$ (for plane waves $r_0=0$).

The shock path is still found from Eq. 13 (or 12) because that equation was derived from the Rankine–Hugoniot equations, which are quite general. But because each of Eqs. (A4–A6) satisfies $dh/dr=(r_0/r)^a$, we have $dt_s'/dr=(dt_s'/dh)dh/dr=(r_0/r)^a(dt_s'/dh)$, so that Eq. 13 can be generalized to

$$dt_s'/dr=-\tfrac{1}{2}\beta c_0^{-2}(w_a+w_b). \tag{A8}$$

It is seen that Eqs. A1, A2, and A8 have exactly the same form as their plane-wave counterparts, Eqs. 6, 8, and 13. Consequently, all the formulas heretofore given for plane waves may be made to apply to spherical and cylindrical waves simply by replacing u with w and x with h.

For an originally sinusoidal wave, we take the excitation at $r=r_0$ to be $u=u_0 \sin\omega t$. The Fourier-series representation of the solution is

$$u=u_0(r_0/r)^a\sum B_n \sin n[\omega t-k(r-r_0)], \tag{A9}$$

where

$$B_n=(2/n\pi)W_b+(2/n\pi f)\int_{\Phi_{\min}}^{\pi} \cos n(\Phi-f\sin\Phi)d\Phi. \tag{A10}$$

Here $W_b=(r_0/r)^a(u_b/u_0)$ stands for the reduced shock amplitude analogous to V_b; i.e.,

$$W_b=(1/f)[j_0^{-1}(1/f)]H(f-1), \tag{A11}$$

and f is a dimensionless form of the corrdinate stretching function h—namely, $f=\beta\epsilon kh$—such that $f=1$ signifies shock formation. When $f<1$, the solution reduces to

$$u/u_0=(r_0/r)^a\sum(2/nf)J_n(nf)\sin ny, \tag{A12}$$

where, here, $y=\omega t-k(r-r_0)$. On the other hand, the sawtooth solution is obtained by letting f be large as compared with unity:

$$u/u_0=(r_0/r)^a\sum[2/n(1+f)]\sin ny. \tag{A13}$$

Finally, the distance limitation on the sawtooth solution may be estimated. Using the same criterion employed previously for plane waves, we find

$$r_{\max}=\beta\epsilon kr_0/\alpha[1+\beta\epsilon kr_0 \ln(r_{\max}/r_0)], \tag{A14}$$

for spherical waves and

$$r_{\max}=(2\alpha)^{-1}+\tfrac{1}{2}br_0\{1+[1+2(b\alpha r_0)^{-1}]^{\frac{1}{2}}\}, \tag{A15}$$

where $b=[1-(2\beta\epsilon kr_0)^{-1}]^2$, for cylindrical waves. In the limit as $r_0\to\infty$, both formulas give $r_{\max}-r_0=1/\alpha$ in agreement with the result for plane waves. For finite values of r_0, however, the sawtooth region ends sooner for spherical and cylindrical waves than it does for plane waves. This is to be expected. Nonlinear effects are sharply reduced in diverging nonplanar waves. Estimates of $r_{\max}$ for spherical and cylindrical waves have also been obtained by Naugol'nykh, Soluyan, and Khokhlov, who used the solution of a modified form of Burgers' equation.[A3] Their analysis leads to equations that agree with Eqs. A14 and A15, except that α is replaced by $2\alpha/\pi^2 \doteq 0.2\alpha$.

[A2] The method here is based on a direct solution of the equations of motion for spherical and cylindrical waves. An alternative approach would be to use Whitham's general method for nonplanar but quasi one-dimensional waves [J. Fluid Mech. **1**, 290–318 (1956)]. Whitham used ideas taken from ray theory and in fact obtained what might be called a ray theory for finite-amplitude waves. Our results may be obtained using the ray-theory method, providing, however, that one first modifies Whitham's analysis to take account of the fact that the flow ahead of each shock is disturbed, not quiescent.

[A3] K. A. Naugol'nykh, S. I. Soluyan, and R. V. Khokhlov, Akust. Zh. **9**, 54–60 (1963) [English transl.: Soviet Phys.—Acoust. **9**, 42–46 (1963)]; Vestn. Mosk. Univ. Fiz. Astron. **4**, 65–71 (1962).

The next paper is an experimental study of the intermodulation of two acoustic signals by the medium due to its nonlinearity at high sound levels. The results confirm Westervelt's theory.

Dr. Muir is a Research Scientist at the Applied Research Laboratories at the University of Texas. His research has been in the field of nonlinear acoustics. He is Supervisor of the Nonlinear Research Section of the Applied Research Laboratories.

Dr. Blue was employed at the U.S Navy Mine Defense Laboratory at Panama City, Florida, from 1961–1968. He has been a Research Scientist at the Applied Research Laboratories at the University of Texas since 1968.

This paper is reproduced with the permission of the authors and the Acoustical Society of America.

J. Acoust. Soc. Amer., 1969, V. 46, p. 227–232.
Copyright 1969 by The Acoustical Society of America

Received 11 February 1969 **13.10; 14.5**

43

Experiments on the Acoustic Modulation of Large-Amplitude Waves

T. G. Muir and J. E. Blue

Applied Research Laboratories, The University of Texas at Austin, Austin, Texas 78712*

A series of experiments were preformed to study the parametric relationship between two high-frequency carrier waves and the difference-frequency wave created by their nonlinear interaction. The measurements were made in a freshwater lake in the farfield of all radiations. A split-face piston projector, 2 in.×2 in., was driven by two pulsed power amplifiers, and the resulting acoustic fields were measured with three separate calibrated hydrophones. Before the experiments were conducted, the usefulness of a frequency selective acoustic filter (SOAB) in identifying pseudosound was established. In the first experiments, a 143 kHz difference-frequency wave having an extrapolated source level of 81 dB *re* 1 μbar at 1 yd was created from the nonlinear interaction of a 981- kHz and a 1124-kHz radiation whose source levels were 121 and 113 dB, respectively. The difference-frequency sound propagated according to the spherical spreading law, independently of the existence of the carriers, with a 1.7° half-power beamwidth. The difference-frequency SPL was found to be proportional to the product of carrier-wave amplitudes, except at high carrier-wave source levels, and proportional to the difference frequency to a power of 1.7. With few exceptions, the results showed good agreement with a mathematical model of the problem devised by Westervelt.

INTRODUCTION

The production of highly directional sound by the nonlinear interaction of two high-frequency primary waves has been the subject of a number of papers over the past few years and has raised hopes of obtaining highly directional, low-frequency sound beams with small apertures. This phenomenon can be thought of as a volumetric acoustic array with the zone of modulation forming its boundaries. The first theoretical analysis of the problem was made by Westervelt.[1] This work was augmented by an experimental model study made by Bellin and Beyer,[2] who measured the difference-frequency level and beam pattern in the nearfields of their radiations. Their results and the later experimental results of Hobaek[3] showed a greater difference-frequency directivity than predicted by the Westervelt model. The present paper presents the results of five experiments performed to illustrate the nature of nonlinear acoustic modulation and to explore the range of validity of the mathematical model.

* Formerly the Defense Research Laboratory.

[1] P. J. Westervelt, "Parametric Acoustic Array," J. Acoust. Soc. Amer. 35, 535–537 (1963).

[2] J. L. S. Bellin and R. T. Beyer, "Experimental Investigation of an End-Fire Array," J. Acoust. Soc. Amer. 34, 1051–1054 (1962).

[3] H. Hobaek, "Experimental Investigation of an Acoustical End Fired Array," J. Sound Vibration 6, 460–463 (1967).

I. THEORETICAL RESULTS

Westervelt[1] derived a relatively simple mathematical model to predict the sound-pressure amplitude of the difference-frequency waves as a function of range and angle from the axis of symmetry. Several simplifying assumptions were made in order to arrive at a solution. Some of these are: (1) that the equations of motion are those for an ideal fluid devoid of the effects of viscosity and heat conduction; (2) that the effects of attenuation can be introduced in an *ad hoc* manner; (3) that the primary generating waves are directive and perfectly collimated so that the volume distribution of sources produced by their interaction may be represented by a line distribution located along the axis of the primary beams; and (4) that the excess pressure, being a function only of the excess density, i.e.,

$$p=\sum_{n=1}^{\infty}\frac{1}{n!}\left(\frac{d^n p}{d\rho^n}\right)_{\rho=\rho_0}\rho^n, \tag{1}$$

may be approximated by retaining only the second-order terms so that

$$p\simeq c_0^2\rho+\tfrac{1}{2}(d^2p/d\rho^2)_{\rho=\rho_0}\rho^2. \tag{2}$$

With some minor modification, Westervelt's model

may be expressed as

$$P_s(R,\theta)=\frac{\omega_s^2 P_1 P_2 S}{8\pi R\rho_0 c_0^4}\left[1+\frac{\rho_0}{2c_0}\left(\frac{d^2P}{d\rho^2}\right)_{\rho=\rho_0}\right]\times\frac{1}{[\alpha^2+K_s^2\sin^4(\theta/2)]^{\frac{1}{2}}},\quad(3)$$

where ω_s is the angular difference frequency (equal to the difference of the angular frequencies ω_1 and ω_2 of the primary waves); P_1 and P_2 are the amplitudes of the primary waves; S is the cross-sectional area of the collimated zone; ρ_0 and c_0 are, respectively, the ambient density and sound velocity in the medium; R is the range to the measurement point from the projector; θ, the angle between the observation point and the acoustic axis; α, the mean absorption coefficient of the primary waves; and K_s, the wavenumber of the difference-frequency wave. The term

$$\frac{\rho_0}{c_0}\left(\frac{d^2P}{d\rho^2}\right)_{\rho=\rho_0}$$

is the parameter of nonlinearity, commonly called B/A. For the problem at hand, B/A is a result of the approximation of Eq. 2 and is thought to have a value of 5.2 for water at 30°C.[4]

The normalized angular dependence of the sound pressure at the difference frequency can be written from Eq. 3 as

$$C(\theta)=\frac{P_s(R,\theta)}{P_s(R,0)}=\left[\frac{\alpha^2}{\alpha^2+K_s^2\sin^4(\theta/2)}\right]^{\frac{1}{2}}.\quad(4)$$

This equation leads to a beam pattern with no side lobes. Naze and Tjøtta[5] modified Eq. 4 to account for the finite aperture of the primary beams of circular transducers to obtain

$$D(\theta)=C(\theta)|2J_1(K_s a\sin\theta)/K_s a\sin\theta|,\quad(5)$$

where a is the radius of the transducer. They also modified Eq. 2 to account for the divergence of the primary beams to obtain

$$E(\theta)=C(\theta)(2/\pi)[\arccos x-x(1-x^2)^{\frac{1}{2}}],\quad x<1,\quad(6)$$
$$=0\qquad x\geq 1,$$

where $x=(K_s a/4)\sin\theta$. For a rectangular transducer, one can show that Eq. 5 becomes

$$F(\theta)=C(\theta)|\sin(K_s a\sin\theta)/K_s a\sin\theta|,\quad(7)$$

where $2a$ is the width or length of the transducer, according to which directivity is being considered.

[4] R. T. Beyer, "Parameter of Nonlinearity in Fluids," J. Acoust. Soc. Amer. 32, 719–721 (1960).

[5] J. Naze and S. Tjøtta, "Nonlinear Interaction of Two Sound Beams," J. Acoust. Soc. Amer. 37, 174–175 (L) (1965).

In all of the theory developed for the production of the difference-frequency wave by the nonlinear interaction of two primary waves, the effect of the Fresnel zone, which is present in any experimental test of the theory, is omitted because the interaction there is very complex. This fact coupled with the assumptions discussed above renders the mathematical model approximate in light of the accuracy with which the experimental measurements may be made. The theory is nonetheless surprisingly accurate in many respects, as is shown later.

II. SOAB EXPERIMENT

Prior to beginning the full-scale measurements, a novel experiment was performed in the Applied Research Laboratories' model tank to study the effects of pseudosound or modulation in the measurement system. This experiment proved to be valuable in later measurements subject to the effects of pseudosound and is included here for this reason. Two ½-in. diam barium titanate circular pistons operating in the 3,3 mode were mounted side by side in an oil-filled housing and driven separately at frequencies of 1130 and 1015 kHz. A U. S. Navy Standard E-8 hydrophone was used at a range of 25 ft to measure the sound-pressure level (SPL) of each carrier; these were each set at 64 dB *re* 1 dyn/cm². The 115-kHz difference-frequency wave was measured at the output of a bandpass filter and its level was found to be 10 dB *re* 1 dyn/cm².

A legitimate question arises as to whether or not the source of the difference-frequency sound is actually a modulation in the water or merely a modulation on either the face of the hydrophone or in the hydrophone preamplifier. One way to answer this question would be to employ a semipermeable screen or acoustic filter that would suppress the carrier waves while passing the difference-frequency sound. A good approximation to such a filter is afforded by a material designed by Cramer[6] and marketed by the Goodrich Company under the trade name of SOAB. Sound absorption in this material is dependent upon the slight porosity caused by aluminum powder imbedded in Butyl rubber. Its acoustic impedance is relatively close to that of water so that the normal incidence reflection coefficient near 115 kHz is lower than −20 dB. The 2 ft×2 ft×¾ in. slab of SOAB used in our experiments had an attenuation slope of roughly 1.5 dB/100 kHz at normal incidence, between 100 kHz and 1 MHz.

The test for pseudosound involved placing the SOAB directly in front of the projector and directly in front of the hydrophone and measuring the levels of both carrier and difference-frequency sounds. When the SOAB was placed at the hydrophone, the megacycle frequency waves were both attenuated by 14 dB, while the 115-kHz difference-frequency sound was attenuated by 3 dB. When the SOAB was placed in

[6] Products Materials for Underwater Sound Applications, B. F. Goodrich Co. Product Brochure (1961), 2nd ed.

front of the projector, however, the carriers were again attenuated by 14 dB, but the amplitude of the difference-frequency sound was reduced to a level lower than system noise. In this case, the supply of carrier waves to the parametric acoustic array in front of the projector was significantly reduced causing a great inefficiency in the acoustic nonlinear interaction. The point of the experiment is that the modulation did, in this case, take place in the water between the projector and the hydrophone. This test can easily be made to help identify the source of difference-frequency sound in other experiments and is generally quite useful in avoiding the confusing effects of pseudosound.

III. PROPAGATION EXPERIMENT

Although the SOAB test can provide a necessary condition for the determination of pseudosound, it is not necessarily a sufficient condition. It could be argued, for example, that modulation takes place on the SOAB itself. The difficulties caused by pseudosound have arisen before in nonlinear acoustics to the degree that many of the published experimental results have not enjoyed widespread credibility. For this reason, it is probably worthwhile to overemphasize the elimination of pseudosound. In order to overcome all objections completely, one must resort to nature's own frequency-dependent filter—the medium itself. A propagation experiment was therefore designed to demonstrate the independent existence of difference-frequency sound at ranges concomitant with high absorption of the generating waves.

In this experiment, a split-face transducer was used in conjunction with a dual-channel transmitting system to launch the high-frequency carrier waves. A block diagram of the experiment is given in Fig. 1; all of the remaining experiments discussed in this paper were done with essentially the same apparatus. The transducer was made up of four slabs of lead zirconate-lead titanate ceramic, each of which was $\frac{1}{2}$ in.×2 in.×0.070 in., operating in the 3,3 mode. The slabs were mounted in an oil-filled housing so that they formed a 2 in.×2 in.-square piston. Alternating slabs were wired in parallel, resulting in a composite dual-channel transducer. Other experimenters have used a single-crystal projector, but

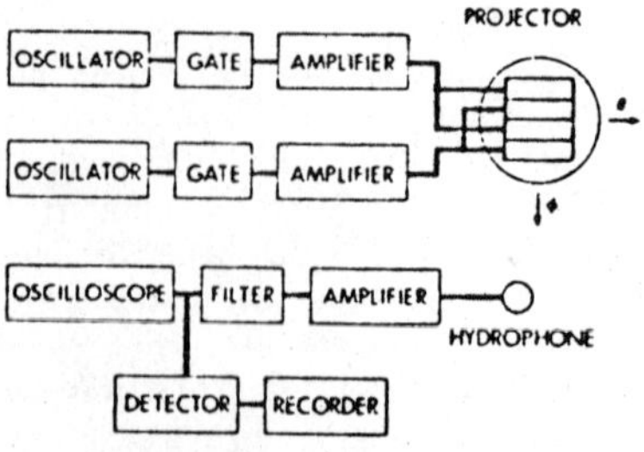

FIG. 1. Block diagram of equipment.

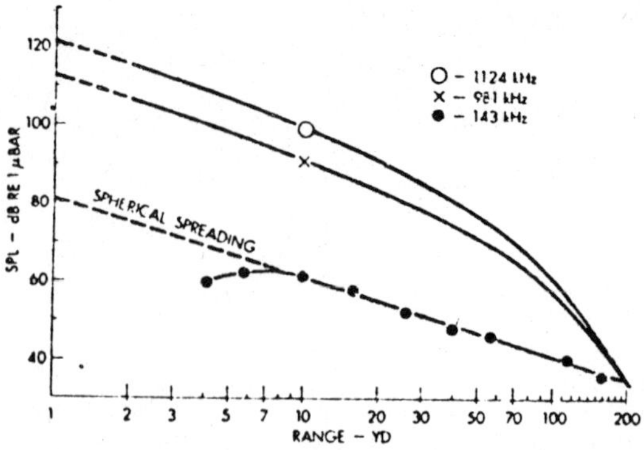

FIG. 2. Axial SPL versus range for the two high frequencies and the difference frequency.

this difference probably has little significance in the outcome of the experiment. Hydrophones used to measure the difference-frequency sound were omnidirectional probes having a high frequency rolloff above 200 kHz.

The measurements (and those to follow) were made at Applied Research Laboratories' Lake Travis Test Station, in isothermal fresh water having a temperature of 84°F. The projector was placed at a depth of 10 ft and axial SPLs were measured as a function of range. The results are shown in Fig. 2. The 1124- and 981-kHz carriers were projected as 1-msec pulses at source levels of 121 and 113 dB *re* 1 dyn cm² at 1 yd, respectively. Detailed measurements of the carrier-wave-propagation curve were not made in this particular experiment; however, similar measurements have been made many times, and there is no reason to suspect that this curve is anything but simple spherical spreading plus absorption. The curve shown in the Figure follows spreading and absorption for fresh water[7] and passes through the two points actually measured. The 143-kHz difference-frequency sound propagation curve, on the other hand, was accurately monitored at nine range stations out to a maximum range of 160 yd. At a range of 10 yd, the acoustic modulation appears to have been completed, and the propagation curve begins to follow the spherical-spreading law. At about 7 yd, the curve passes through a maximum and then decays with decreasing range, demonstrating a nearfield effect associated with the parametric acoustic array. The pseudosound problem began to be noticeable at ranges shorter than 4 yd so that it was not possible in this experiment to delineate completely the nearfield of the parametric array. One can extrapolate to a hypothetical 1-yd source level for the modulation product; in this experiment, a hypothetical source level of 81 dB *re* 1 dyn/cm² at 1 yd was obtained. This is 36 dB below the mean source level of the carriers and corresponds to an amplitude conversion

[7] J. J. Markham, R. T. Beyer, and R. B. Lindsay, "Absorption of Sound in Fluids," Rev. Mod. Phys. 23, 353–411 (1951).

efficiency of almost 2%. No pulse-length dependence was observed in this or any of our experiments. The difference-frequency signal appeared simply as a low Q, flat-topped CW pulse. In the farfield of the parametric array, the behavior of difference-frequency sound is similar to that of an ordinary linear radiation.

IV. BEAM-PATTERN MEASUREMENTS

The only significant difference between a linear radiation and a nonlinearly produced difference frequency radiation is the directional character of the latter. Beam patterns were taken in both principal planes of the projector, in the farfield of both carrier and difference frequency sources. The results are shown in Fig. 3. The horizontal beam patterns at the carrier-wave frequencies contain the normal −13-dB first minor lobes expected of a rectangular aperture. No such minor lobes appear in the corresponding difference-frequency beam pattern. Instead, a slowly decaying, rather broad skirted pattern is obtained down to a level of 50 dB below the lobe peak. The half-power beamwidth of the difference-frequency radiation is 1.7°. In the vertical plane, the carrier-frequency beam patterns display a trilobal structure that is characteristic of an element spacing greater than a wavelength. This feature is, of course, undesirable in most underwater acoustical systems, but it was deliberately designed into this experiment to shed some light on the creation of minor lobes in the difference-frequency beam pattern. Comparison of Fig. 3(b) with Fig. 3(a) shows that the ordinary −13-dB minor lobes associated with the unshaded rectangular aperture do not give rise to any lobal structure in the difference-frequency beam patterns, but the −4-dB repeat major lobes of the primary beams in the vertical plane do give rise to two ~−8-dB lobes in the difference-frequency beam pattern.

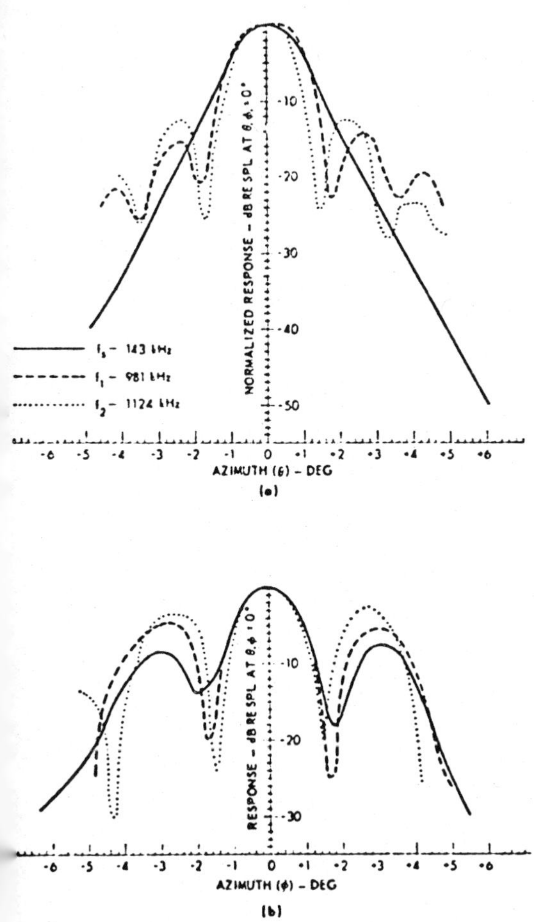

FIG. 3. θ and φ beam patterns at 10-yd range.

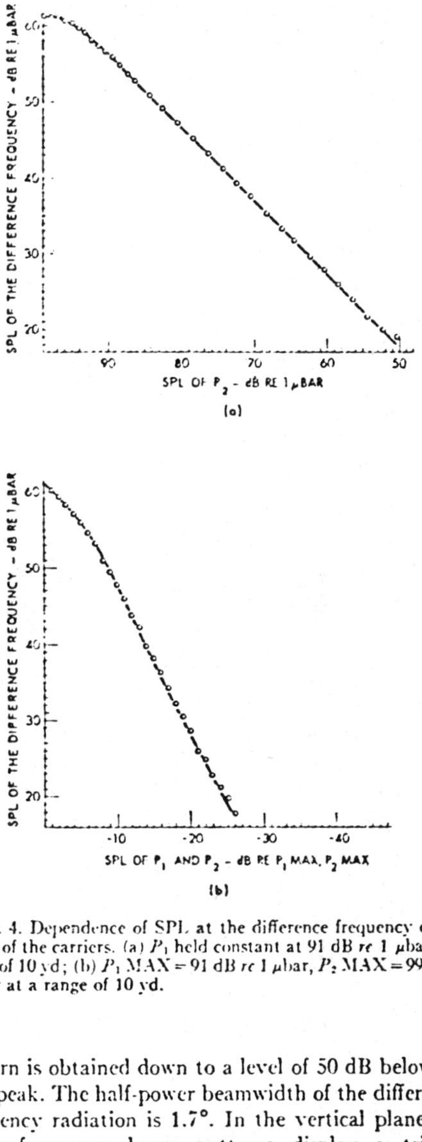

FIG. 4. Dependence of SPL at the difference frequency on the SPL's of the carriers. (a) P_1 held constant at 91 dB *re* 1 μbar at a range of 10 yd; (b) P_1 MAX = 91 dB *re* 1 μbar, P_2 MAX = 99 dB *re* 1 μbar at a range of 10 yd.

V. AMPLITUDE EXPERIMENT

The production of a respectable difference-frequency SPL is dependent on the levels of the carrier wave supplied to the parametric array. The dependence of difference-frequency amplitude on the amplitudes of the carrier waves was measured in the farfield of the parametric array in order to establish experimentally the functional relationship between these parameters. The results are shown in Fig. 4. In the experiment described by Fig. 4(a), the 981-kHz carrier was held constant at an amplitude of 91 dB *re* 1 dyn/cm² while the 1124-kHz carrier was decreased in amplitude, thus allowing a measurement of SPL at the difference frequency versus the SPL of one of the carriers. Throughout most of the range of measurement, the data shows a slope of 1, indicating a perfect linear dependence. As the level of the 1124-kHz carrier approaches 95 dB, the curve departs from linearity so that a 1-dB increase in carrier amplitude is no longer reflected as a 1-dB increase in difference-frequency amplitude. This fact is further demonstrated in Fig. 4(b), which is a plot of the difference-frequency SPL versus the level of both carriers. Again, a rolloff from the expected linear curve of slope 2 appears as the carriers reach their maximum attainable values. This departure raises some interesting questions regarding the efficiency of the modulation process at higher power levels and these are discussed after the last experiment is described.

VI. FREQUENCY-RESPONSE EXPERIMENT

The last measurement in this series was designed to explore the frequency dependence of the modulation product. In this experiment, both carrier-wave radiations were maintained at a level of 80 dB *re* 1 dyn/cm² at a range of 10 yd, while their frequency separation was varied. This allowed an experimental determination of difference-frequency SPL as a function of difference frequency. The results are shown in Fig. 5. The most important feature of the experimental curve is its slope.

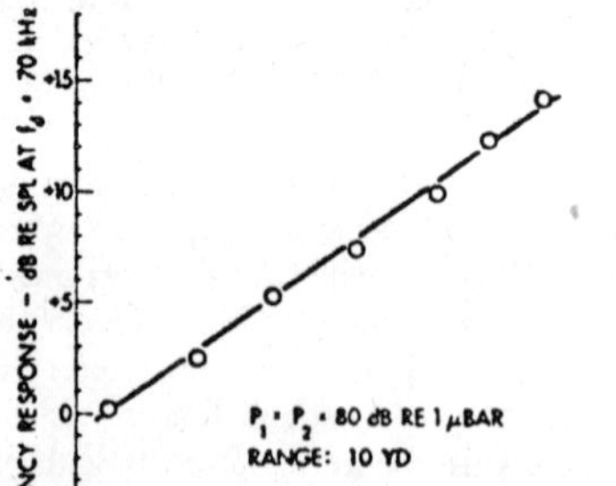

FIG. 5. Dependence of the SPL of the difference frequency on the difference frequency.

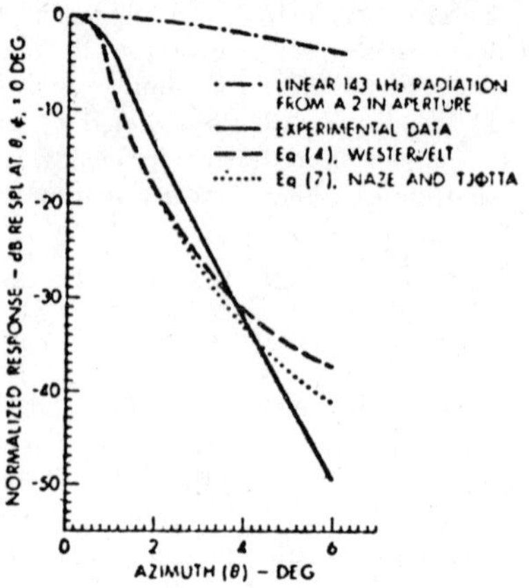

FIG. 6. Comparison of experimental and theoretical beam patterns.

It can be shown that the slope of the curve is approximately equal to 1.7, so that the difference-frequency amplitude was found to be proportional to the difference frequency to a power of 1.7. A discussion of the significance of this dependence and its agreement with the theoretical model of the problem is given below.

VII. DISCUSSION

The results of the SOAB and propagation experiments should remove any remaining doubts concerning the ability of a nonlinear interaction of collinear primary waves to produce a modulation product capable of independent existence. The utility of a frequency-selective filter in identifying pseudosound may be of interest to other experimenters.

The horizontal-directivity data acquired in these experiments are compared in Fig. 6 to the theoretical predictions of Eq. 4, Westervelt's result, and of Eq. 7, which contains the difference-frequency aperture factor, after Naze and Tjøtta. Although Eq. 7 does show better agreement at the larger azimuth angles, neither curve shows exact agreement with the experimental data. This should be expected from consideration of the simplifying assumptions embraced by the derivation; moreover, from this standpoint, the agreement is actually quite satisfying.

The results of Bellin and Beyer,[2] Hoback,[3] and those of Zverev and Kalachev,[8] show a narrower half-power beamwidth than that predicted by the mathematical model while the present data demonstrate a broader beamwidth. This discrepancy is probably due to differences in the experiments. The present measurements were made in the field in the Fraunhofer zone, but most

[8] V. A. Zverev and A. I. Kalachev, "Measurement of the Scattering of Sound by Sound in the Superposition of Parallel Beams," Sov. Phys.—Acoust. 14, 173–178 (1968).

of the earlier measurements were made in tanks at ranges within the Fresnel zones of either the carrier or difference-frequency radiations. Since the $K_r a$ values for most of the earlier experiments were higher than that for the present data, the increased significance of aperture factors like those contained in Eqs. 5 and 7 also helps to explain this discrepancy. Both the predicted and measured beam patterns show the absence of minor lobes in the θ direction, as can be seen from examination of Fig. 6. The predicted response of the projector driven at 143 kHz is also shown in the Figure. The half-power beamwidth is 11° as compared to 1.7° for the nonlinearly produced radiation, thus demonstrating a ~6:1 improvement in resolution at the half-power points.

The Φ direction beam patterns of Fig. 3(b) offer further insight into the process of beam formation at the difference frequency. $C(\theta)$ or $C(\Phi)$ from Eq. 7 is the controlling factor in both the θ and Φ directions since this factor decreases much more raipdly than does the directivity factor for the projector at the difference frequency for the parameters of this experiment. The trilobal structure in the Φ direction of the beam pattern at the difference frequency, nonetheless, demonstrates the importance of the beam patterns at the carrier frequencies in determining the beam pattern at the difference frequency. The pattern seems to be following the product of the amplitudes of the carriers. This effect is masked in the θ direction by the spread of the primary beam over the lower sidelobes.

Results of the amplitude measurements of Fig. 4 show good agreement with the model in that the difference-frequency amplitude follows a P_1P_2 dependence over most of the dynamic range measured. The rolloff from the quadratic law at high carrier-wave amplitudes has not been explained. It could be that the approximation of Eq. 2 is no longer valid and that more terms should be included in it, or that higher-order interactions are beginning to sap the conversion process.

The absolute levels of difference-frequency sound in the regions below the amplitude rolloff were compared to the theoretical predictions. In so doing, all parameters were taken into Eq. 3 as they were measured at the 10-yd range of the amplitude experiment. The cross-sectional area was taken to be that encompassed by the half-power projector beamwidth. Agreement between theory and experiment was surprisingly exact, more so than one would expect; for example, the discrepancy at carrier-wave levels of 91 dB *re* 1 μbar at 10 yd was only 1 dB.

The last experiment showed a difference-frequency amplitude dependence on frequency to the power of 1.7 rather than to the power of 2.0 as is implied from a cursory examination of the theoretical result of Eq. 3. When the frequency-dependent carrier-wave absorption coefficients are considered, however, the frequency response factor in Eq. 3 becomes

$$G(\omega)=\omega_d^2/(\omega_1^2+\omega_2^2). \tag{8}$$

Over the range of this experiment, the frequency response predicted by $G(\omega)$ yields a slope of 1.88 rather than the experimentally observed slope of 1.7. Further examination of the data shows that a total cumulative error of 1 dB in the measurement could have accounted for this small discrepancy between theory and experiment, thus making the difference insignificant.

The implications of possible practical applications for nonlinear acoustics have not been discussed here, but it should be obvious that the creation of highly directive sound, devoid of undesirable minor-lobe structures, may find application in some acoustical systems not limited by the inefficiency of the process. Berktay has published extensively on this topic.[9-11] The problem of high-resolution subbottom profiling in shallow water is an example of a possible application.

VIII. CONCLUSIONS

Good agreement between Westervelt's analysis of the nonlinear interaction of two high-frequency primary waves and experiment was found in the course of this investigation. The amplitude, frequency response, and azimuthal character of the nonlinearly produced difference-frequency radiation were found to be reasonably compatible with theory. Exceptions noted were a rolloff in the efficiency of the conversion process at high carrier-wave source levels and a slightly broader half-power beamwidth than predicted as well as measured in previous experiments in the nearfield. The independent existence of the difference-frequency sound in the farfield of all radiations was established to demonstrate its nature and its possible use in acoustical systems not limited by the inefficiency of the conversion process.

ACKNOWLEDGMENTS

The authors wish to thank J. E. Stockton and R. H. Wallace for suggesting some of these experiments. This work was supported by the U. S. Navy Office of Naval Research.

[9] H. O. Berktay, "Possible Exploitation of Non-Linear Acoustics in Underwater Transmitting Application," J. Sound Vibration 2, 435–461 (1965).

[10] H. O. Berktay, "Parametric Amplification by the Use of Acoustic Non-Linearities and Some Possible Applications," J. Sound Vibration 2, 462–470 (1965).

[11] H. O. Berktay, "Some Proposals for Underwater Transmitting Applications of Non-Linear Acoustics," J. Sound Vibration 6, 244–254 (1967).

Acoustic Cavitation in Liquids

VIII

The following paper demonstrates that cosmic rays provide nucleation for ultrasonic cavitation.

Dr. Sette is Professor of Physics at the University of Messina, Messina, Italy, and a Research Physicist at the Ultraacoustics Institute "O. M. Corbino" in Rome, Italy. His research has been concerned, primarily, with the development of nucleation in water by radiation.

F. Wanderlingh is a Research Scientist at the Ultraacoustics Institute "O. M. Corbino" in Rome, Italy.

This paper is reprinted with the permission of the senior author and the Editors of *Physical Review*.

Reprinted from THE PHYSICAL REVIEW, Vol. 125, No. 2, 409–417, January 15, 1962
Printed in U. S. A.

44

Nucleation by Cosmic Rays in Ultrasonic Cavitation

D. SETTE AND F. WANDERLINGH
Institute of Physics, University of Messina, Messina, Italy, and Ultracoustics Institute "O. M. Corbino," Rome, Italy
(Received July 3, 1961)

Nucleation in ultrasonic cavitation is due to cosmic rays. The neutron component seems sufficient to explain the experiment in water, through the creation of oxygen recoil nuclei which would act as nucleating agents; energy deposit and radiolytic effects produce small overheated and oversaturated regions in which cavities of radius larger than the critical one may originate. The microcavities would be created in the absence of a sound field and grow to visible size when an adequate sound field is applied. Enclosing the tank in which the liquid is contained with lead or paraffin screens results in threshold variations which are due to absorption and slowing down of primary neutrons and, probably, to nuclear reactions induced by the gamma and meson components of cosmic radiation in the screen material through the production of secondary neutrons.

1. INTRODUCTION

THE majority of experiments on sound-induced cavitation can be explained by assuming that the process consists mainly of the growth of pre-existing nuclei to visible size through evaporation during the negative part of the pressure cycle. The size of nuclei would be of the order of 0.1–10 μ, and to maintain them in the liquid, the presence of one or more stabilizing mechanisms has been suggested.[1]

The origin of the nuclei, however, has defied for a long time any attempt at investigation. One of the authors investigated the possibility that cosmic radiation could be responsible for nucleation and reported[2] some results obtained in water which support such an hypothesis. At the same time, Libermann and Rudnick[3] in trying to clarify how bubble chambers work performed an experiment in pentane and acetone and obtained a strong decrease of the sonic cavitation threshold in the presence of a neutron source. Their results point out that C or O recoil nuclei may be responsible for the creation of cavitation nuclei, i.e., gaseous bubbles having a radius larger than the critical one, as a consequence of energy deposit in small regions (thermal spikes) which subsequently explode with a mechanism similar to that suggested by Seitz[4] to explain the action of ionizing radiation in bubble chambers. The presence of sound would be necessary for the production of cavitation nuclei.

The present paper reports an investigation of the relation between penetrating radiation and ultrasonic cavitation in water, which supports evidence that cosmic radiation may be responsible for the origin of cavitation nuclei in ordinary water, and makes possible the identification of recoil oxygen nuclei as the nucleating agents. In addition, there has been developed a new method for the examination of the influence of radiation on the cavitation threshold which also appears of interest for the detection of nuclear particles.

2. EXPERIMENTAL METHODS

The cavitation threshold in commercial distilled water was measured at 1 Mc/sec in a tank (volume about 20 liters) with a procedure similar to that followed in previous research[2]: the voltage at the quartz source was increased by steps and applied for 30 sec, after a rest time of 30 sec; at the onset of cavitation the sound was quickly removed and a minimum of 5 min was allowed before a new run was started. The sound intensity for which cavitation was visually detected was taken as the cavitation threshold. All measurements were performed at room temperature (23°–25°C). The threshold found in aerated distilled water was about 1 w/cm². Figure 1 gives a general scheme of the experimental disposition. In order to avoid possible cavitation at the interface quartz-water, a focusing device[5] formed of conical and parabolic mirrors was used; the cavitation always starts in this region well separated from any solid walls. The distribution of sound energy in the field has been checked by direct measurement obtaining a calibration to pass from the voltage applied to the source to the sound density in the focal region. The equivalent average intensity used in the following refers to plane waves giving the same sound density. The onset of

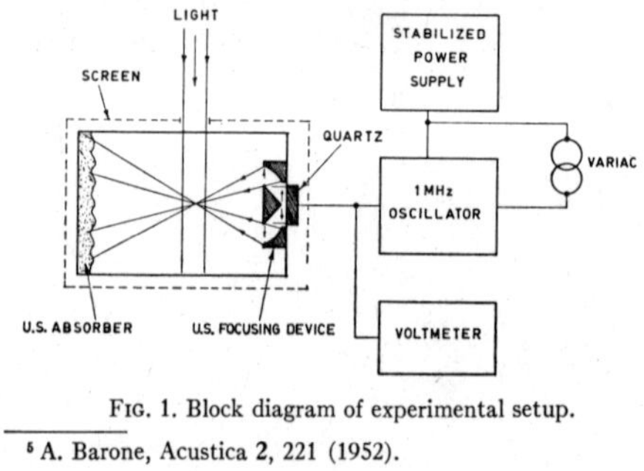

FIG. 1. Block diagram of experimental setup.

[1] M. Strasberg, J. Acoust. Soc. Am. 31, 163 (1959); F. E. Fox and K. F. Herzfeld, J. Acoust. Soc. Am. 26, 984, (1954); E. N. Harvey *et al.*, J. Appl. Phys. 18, 162 (1947); W. R. Turner, Vitro Laboratories, Silver Spring Laboratory, Maryland, Technical Note N 4329–12960, 1960 (unpublished).
[2] D. Sette, Third Meeting of the International Commission on Acoustics, Stuttgart, 1959 (unpublished).
[3] D. V. Liberman and I. Rudnick, Third Meeting of the International Commission on Acoustics, Stuttgart, 1959 (unpublished); D. V. Liberman, Phys. Fluids 2, 466 (1959).
[4] F. Seitz, Phys. Fluids 1, 2 (1958).
[5] A. Barone, Acustica 2, 221 (1952).

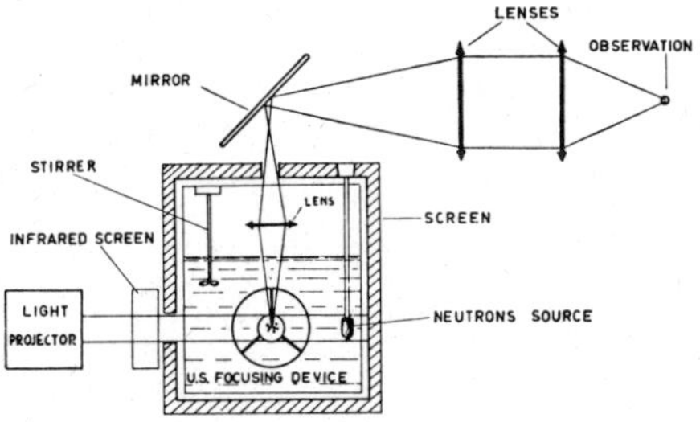

FIG. 2. Optical observation of cavitation.

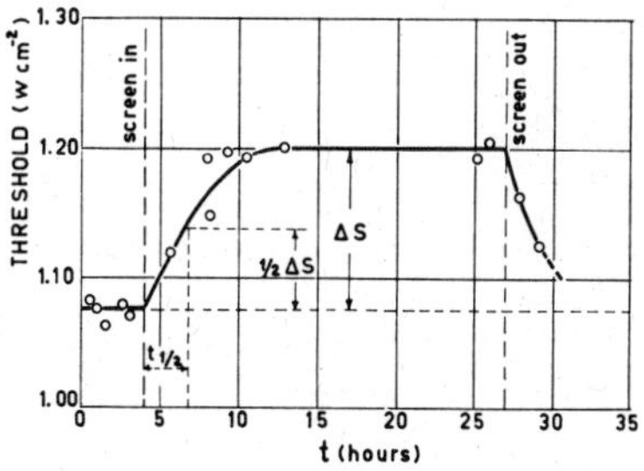

FIG. 4. Threshold in distilled water surrounded by a 5-mm lead screen.

cavitation has been observed visually by means of an optical disposition (Figs. 1 and 2) at a distance of about 2 m from the tank for safety purposes when a neutron source is used. A stirrer slowly (60 rpm) moves the water. Figures 1 and 2 show the position of a screen. A simple mechanical support has been devised to use screens of various thicknesses singly or in combinations.

The definition of cavitation threshold used by us is an operative one. There is, in fact, no sharply defined power density necessary for the cavitation onset; there is instead for each power density a probability that the phenomenon occurs during a fixed time interval. Such a probability is finite already at power densities well below the operative threshold and increases to a value of one for densities larger than the threshold. The operative threshold can be taken, in first approximation, as a weighted average of sound intensities at which the onset may occur, the weights being the probabilities that cavitation starts during the fixed time interval T. To clarify the definition of threshold we have performed some preliminary experiments trying to evaluate the average delay with which onset follows the application of the sound field; in this experiment the sound was applied at a fixed intensity and held until cavitation set in. In Fig. 3 (curve a) the reciprocal (F) of the average delay has been plotted as a function of sound intensity (I); curve b will be discussed later. The probability of observing the cavitation in the time T at a certain power can be put equal $F\times T$. The variation of F with power is due to (1) an increase of volume in our experimental setup in which the sound level may easily produce the growth of nuclei, and (2) the possibility of the growth of nuclei of smaller radii. If one calculates $\sum_i I_iF_iT/\sum_i F_iT$, using an adequate number of points of curve a (Fig. 3) between the intensity I_1 and I_2 for which, respectively, $F\times T$ is appreciably larger than zero and has value 1 ($F=0.033$, T being 30 sec), one obtains a sound intensity of 1.01 w/cm² which is practically coincident with the threshold measured in the operative way (1 w/cm²).

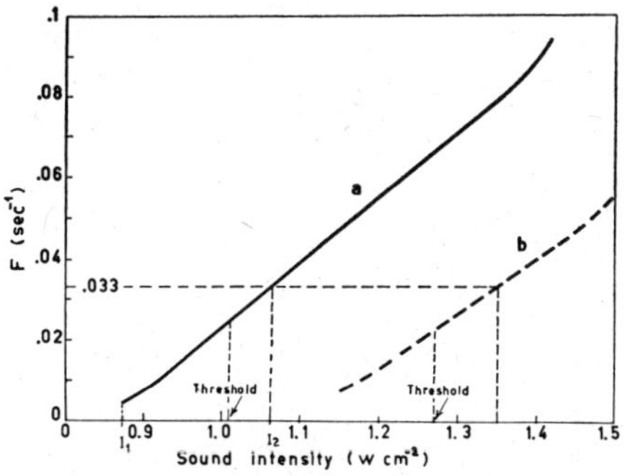

FIG. 3. Inverse of average delay of cavitation onset in distilled water; a sound field of constant intensity is applied to the liquid. (a) Unscreened tank; (b) 20-mm lead screen surrounding the tank.

3. EXPERIMENTAL RESULTS

(a) Lead Screen

Sette[2] has found that the cavitation threshold in ordinary as well as in aerated distilled water is considerably increased if the tank is surrounded by a lead screen of 15-mm thickness. In order to determine better the

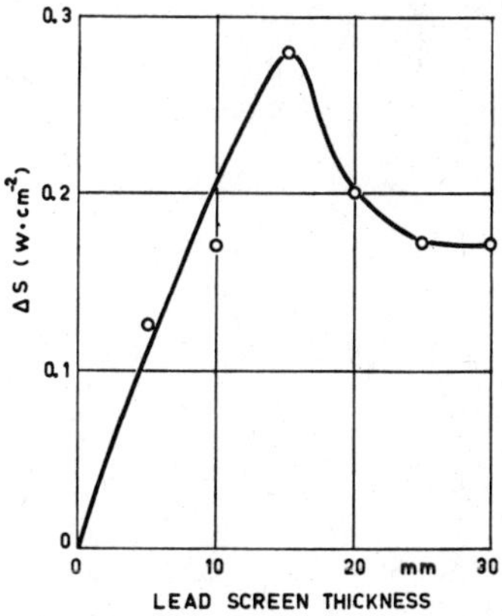

FIG. 5. Threshold variation in distilled water as a function of lead screen thickness.

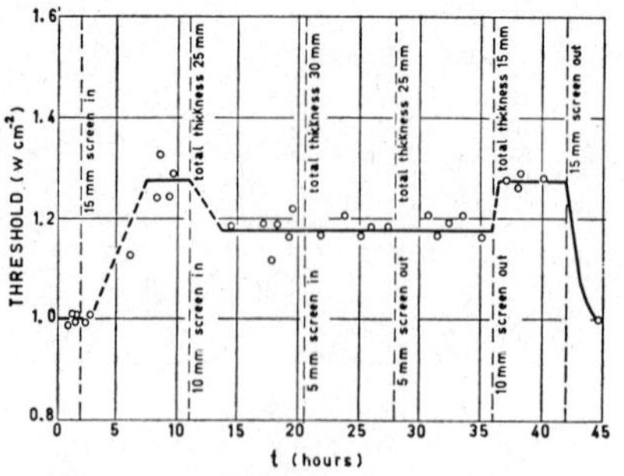

FIG. 6. Threshold in distilled water with lead screens of different thicknesses.

effect of lead screens, three parallelepiped enclosures of the tank have been prepared. They have removable sides and can operate in combination; the lead thicknesses of the enclosures are from the inner one 15, 10, and 5 mm. With this disposition it is possible to study the cavitation threshold when the lead screen surrounding the tank has a total thickness of 5, 10, 15, 20, 25, and 30 mm.

Figure 4 gives a typical run with a 5-mm lead screen. Each point is the average of five determinations. In each run new distilled water has been used; the cavitation threshold has been measured a few hours before screening the tank in order to ensure that no disturbing effects were present. The cavitation threshold increases as a consequence of the presence of the screen and returns to the starting value when sufficient time has elapsed after the removal of the enclosure.

From the observation of the various curves an approximate evaluation of the time required for $\frac{1}{2}$ variation of the threshold when the screen is inserted or removed has been obtained: $\tau_{\frac{1}{2}}=70$ min. This value is only to be taken as indicative of the order of magnitude, and as such it is valid for all thickness and for the increase and decrease of the threshold. Very probably, however, differences are present, but they cannot be exactly evaluated at present.

In Fig. 5 the variations of threshold obtained in runs with different lead thicknesses are collected. It has been already pointed out that the threshold can be associated with the average delays with which cavitation follows

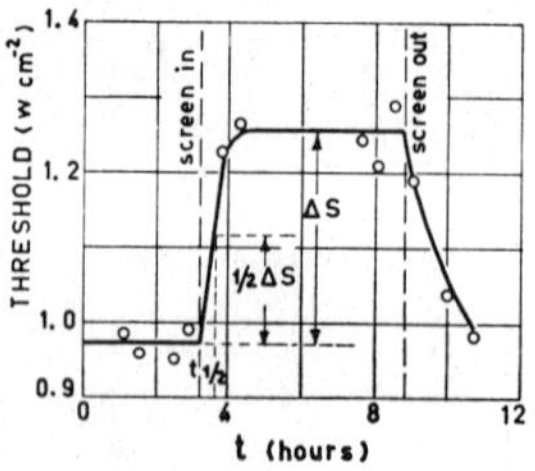

FIG. 7. Threshold in distilled water surrounded by a 15-mm paraffin screen.

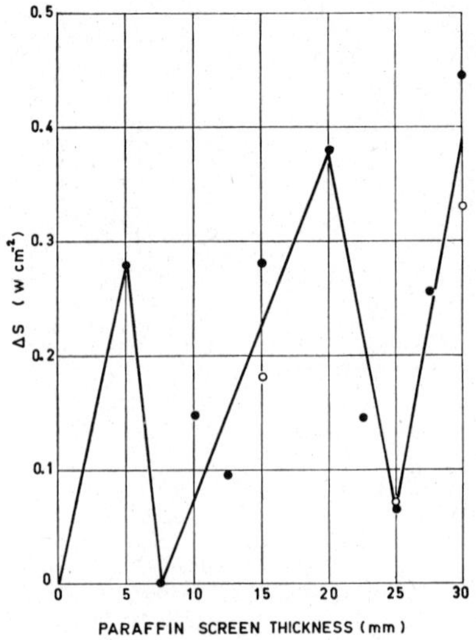

FIG. 8. Threshold variation in distilled water as a function of paraffin screen thickness.

the application of sound of different intensities (experiment of Fig. 3). At the threshold intensity an average delay of about 40 sec was found.

In order to confirm the existence of the maximum in the curve of Fig. 5 an experiment has been performed using the same water specimen and changing the screen

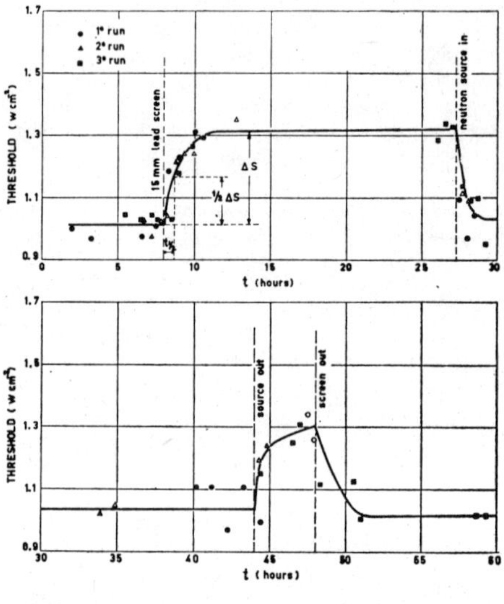

FIG. 9 Threshold in distilled water surrounded by a 15-mm lead screen and in the presence of a Ra–Be neutron source.

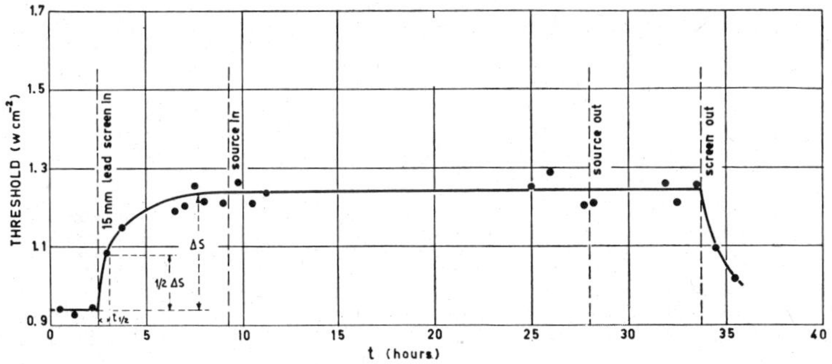

FIG. 10. Threshold in distilled water surrounded by a 15-mm lead screen and in the presence of a source of neutrons (in a paraffin container) with energies lower than 1 Mev.

thickness in the order 0, 15, 25, 30, 25, 15, and 0 mm (Fig. 6).

(b) Screens of Highly Hydrogenated Materials

Four enclosures of paraffin having 5, 7.5, 10, and 15 mm wall thickness have been built making possible experiments with the following screen thicknesses (mm): 5, 7.5, 10, 12.5, 15, 20, 22.5, 25, 27.5, and 30. Figure 7 gives the result for 15-mm screen. The time for $\frac{1}{2}$ variation of threshold is about 25 min. Figure 8 shows the results obtained with the paraffin screens: dots refer to experiments with different water specimens, circles to experiments with the same water filling. An experiment has also been performed with a 5-mm enclosure of perspex (polymethyl methacrylate) obtaining an increase of the threshold in water of about 19%, and a $\tau_{\frac{1}{2}}$ value of the same order than with paraffin screens.

(c) Neutron Irradiation

A 10-mC equivalent Ra–Be neutron source has been used: the flux is 1.5×10^5 neutrons/sec. The maximum neutron energy is about 10 Mev; the energy distribution is approximately Gaussian and it has a maximum around 4 Mev. The concomitant emission of γ rays has no consequence on cavitation as will be discussed later.

To find the influence of neutrons on the cavitation threshold the latter has been measured during the following steps: (1) no screen; (2) 15-mm lead screen; (3) neutron source in the liquid; (4) neutron source excluded; (5) lead screen removed. The source was fixed at one end of an aluminum rod which could enter into the tank through a hole in the lead screen (diameter 1 cm). Figure 9 gives the results obtained with various specimens.

Figure 10 refers instead to the case in which a paraffin cylinder, having walls and bottom of 25-mm thickness (Fig. 11), has been fixed into the tank at the beginning of the experiment and receives the neutron source. In this second experiment the energy of the neutrons reaching the liquid as reduced by a factor of 10. In fact, the mean-free path of neutrons having energies between 1 and 10 Mev in paraffin is 10 mm; this means that the neutrons emitted with energy E undergo on the average 2.5 collisions during their path through paraffin, reducing their energy (E') according to $\ln(E/E')=2.5\xi$, the slowing down power ξ for paraffin being 0.913. Thus the result is $E'=0.1\ E$.

It is evident that while the fast neutrons emitted from the source bring the threshold to the original value for unscreened water, the neutrons having energies smaller than 1 Mev are not able to produce any change in the cavitation threshold.

4. DISCUSSION OF NUCLEATION MECHANISM

The preceding results show that elementary particles as neutrons may, and cosmic radiation does, influence ultrasonic cavitation. The simplest hypothesis seems to be that penetrating radiation may lead to the formation of cavitation nuclei, i.e., of gaseous bubbles with very small radii.

Seitz[4] suggested a mechanism for nucleation in bubble chambers i.e., in a liquid near the critical point, which has been used also by Libermann and Rudnick[3] to furnish an interpretation of their results on ultrasonic cavitation in pentane and acetone. It would seem, therefore, interesting to see if Seitz' mechanism can explain the results of the experiments on ultrasonic cavitation in water and, if necessary, what changes may be reasonably introduced in the theory.

According to Seitz the primary penetrating particles produces, inside the liquid contained in a bubble chamber, small hot regions (thermal spikes) as a consequence of energy deposits either of the same primary particles or of secondary ionizing particles. The hot regions may explode in cavities of a radius larger than the critical one required for stability; these cavities can grow subsequently through evaporation of the liquid. In the case of hydrogen and propane bubble chambers in normal working conditions, Seitz's analysis indicated that the knocked-on electrons of about 1 kev produced by high-energy particles along their path are responsible for nucleation. Libermann and Rudnick, in the analysis of their results on sound cavitation in pentane and acetone at room temperature and in the presence of a neutron source, indicate C and O recoil nuclei as the nucleation agents.

When a thermal spike explodes into a nucleus, effects due to inertia, viscosity, and thermal conductivity of the liquid come into play and they are very difficult to be evaluated exactly. An approximate discription of the process may be obtained simply by considering the energy balance for the bubble formation without taking into account dynamic effects. By these calculations one obtains a lower evaluation of the energy that the particles have to deposit in a length equal to the diameter of cavities of critical size in order to produce a nucleus. Such calculations were done by Seitz for the bubble chambers situation and extended by Libermann and Rudnick to the case in which a sound wave is present, delivering energy for the cavity formation during the negative part of the pressure cycles. The formula that these authors find is

$$U=0.21(4\pi\sigma^2/P)\times[-1+2(10^{-6}n_vH-P_{gi})/P]\text{ Mev cm}^{-1}, \quad (1)$$

where $U=-dE/dx$ is the density of energy deposited by the ionizing particles, σ the surface tension (dynes cm^{-1}), $P=p_a+p_{go}+p_{gi}$ (bars), p_a the amplitude of acoustic pressure, p_{go} the pressure of gas above liquid, p_{gi} the pressure of gas inside cavity, n_v the number of moles of vapor per unit volume at temperature and pressure inside the cavity, and H the heat of vaporation per moles (erg $mole^{-1}$). Pressures are positive if they contract the bubble. On the right side of Eq. (1) there is the sum of three terms which represent, in order, the contributions to energy balance of surface, evaporation, and sound energies.

In the case of water[6] $\sigma=72.75$ dyne cm^{-1}; $p_{gi}=2.66\times10^{-2}$ bar; $H=44\times10^{10}$ erg $mole^{-1}$, $n_v=10^{-6}$ cm^{-3} (at pressure p_{gi} and $T=25°C$). Equation (1) yields for water:

$$U\simeq13.97\times10^3(0.80/P^2-1/P)\text{ Mev cm}^{-1}. \quad (2)$$

The maximum densities of energy deposit for possible primary particles are not larger than those for protons. The energy loss as a function of energy (E) for protons moving in water may be evaluated by means of the semi-empirical method suggested by Hirschfelder and Magee.[7] One finds a maximum for U which is about 800 Mev cm^{-1} at an energy between 0.05 and 0.1 Mev. Because the method is not reliable below proton energies of 0.1 Mev, the results are to be taken only as indicative: they show however, that the maximum energy deposit occurs at an energy lower than 0.1 Mev. The determination of U for O^{16} nuclei may be obtained starting from the value calculated for U in air according to Livesay[8] and converting it to the case of water using Evans' curves[9] of stopping power per electrons vs atomic number of the material. One can estimate for O^{16} nuclei in water a U_{max} of 6000–7000 Mev cm^{-1} for energies between 4 and 5 Mev.

In ordinary water other elements are present as impurities which may yield recoil nuclei when struck by high-energy particles. Since, however, the number of impurity atoms is extremely low in comparison with the number of water molecules, and since the cavitation threshold is usually not greatly influenced by a small variation of impurity content (the threshold is about the same for distilled and tap water), we assume that the presence of impurity atoms can be at this stage ignored, and that the maximum possible value of U is that experienced in the case of O^{16} nuclei (6000–7000 Mev cm^{-1}). Using this value in Eq. (2) the acoustic pressure needed for nucleation would be about 3.6 bars, which differs from the results of these experiments because the cavitation threshold under ordinary conditions in water is about 1.20 bars (1 w cm^{-2}). The disagreement between Seitz's theory and these experiments is more strongly emphasized if one remembers that by bringing into account dynamic effects the acoustics pressure needed for cavitation, according to this scheme, increases noticeably.

It is also to be observed that the mechanism now under examination would require the simultaneous presence of sound and ionizing event for the onset of cavitation. This does not seem to be in agreement with the experiment under our conditions of observation. First of all it would be difficult to explain the observed long time constants with which the threshold changes as a consequence either of screening or of the presence of the neutron source. In addition, cavitation should be a much rarer process than it is, if the nucleation is to be attributed, as it seems, to cosmic rays. In fact the total number of cosmic particles at sea level[10] being of the order of 2×10^{-2} cm^{-2} sec^{-1} and the section of our tank being 500 cm^2, and assuming that each particle entering the tank produces on average one cavity and that the events are equally distributed in the liquid mass (volume 2×10^4 cm^3), one can calculate an approximate value for the number of events happening in the liquid volume in which the acoustic pressure can produce cavitation (few cm^3). One gets 10^{-3} event per second, and therefore, one should observe on the average one cavitation process every 1000 sec while the observed delay of the onset of cavitation following the application of the sound field is about 40 sec at the cavitation threshold intensity. In addition, in the experiment with the neutron source inserted inside the lead screen (Fig. 9), the threshold returns to the original value of un-

[6] *Handbook of Chemistry and Physics*, edited by C. D. Hodgman (Chemical Rubber Publishing Company, Cleveland, Ohio, 1952).
[7] J. Hirschfelder and J. Magee, Phys. Rev. **73**, 107 (1948).
[8] D. Livesay, Can. J. Phys. **34**, 203 (1956).
[9] R. Evans, *The Atomic Nucleus* (McGraw-Hill Book Company, Inc., New York, 1955).
[10] B. Rossi, Revs. Modern Phys. **20**, 537 (1948); D. J. Montgomery: *Cosmic-Ray Physics* (Princeton University Press, Princeton, New Jersey, 1949).

screened water, although the number of neutrons active in the liquid is much higher than the number of high-energy particles of cosmic radiation reaching the tank (also the neutrons seem to be more effective particles). It could be that either only a few neutrons are emitted from the source with sufficient energy for the formation of nuclei, or that the equilibrium density of nuclei in the liquid is controlled by elements other than radiation, such as impurities present in the liquid.

It seems, therefore, plausible to assume that in the conditions under which we observe cavitation in water, the creation of nuclei occurs without requiring the presence of sound. Therefore Liberman and Rudnick's modification of Seitz's treatment does not apply, and one deduces from Seitz's formula

$$2rU = 4\pi\sigma r^2 + \tfrac{4}{3}\pi r^3 n_v H + \tfrac{4}{3}\pi r^3 P_{go}, \quad (3)$$

that to produce cavities of about 1 μ, densities of energy deposit of the order of 47 500 Mev cm^{-1} are necessary. If the dynamic effects were taken into account the value would be enlarged appreciably. Because there are no events produced by 10-Mev neutrons which can give such densities and while the experiment with the neutron source inside the tank enclosed with lead screen (Fig. 9) shows that nuclei are instead originated we conclude that Seitz's treatment in the way in which it was developed for the bubble chamber situation is inadequate to explain normal cavitation in water.

It is to be noted that the penetrating radiation produces in water intense radiolytic effects as well as local heating. When radiolysis[11] is caused by a particle having a mass equal to or larger than the proton mass, OH free radicals are formed along the path of the ionizing particle, while H free radicals are distributed at about 150 A from the same line (they are originated by δ rays). This situation leads to reactions among free radicals of the same kind with a copious production of H_2 and O_2 molecules.

The region in proximity to the path of particles is, therefore, in a very peculiar condition which is not properly described as simply overheated. The possibility of evolving a better theory is bound to an adequate description of the status of molecules in this region. It does not seem satisfactory to use parameters which are normally introduced to describe situations of a quite different nature. If, however, in the absence of such a theory, one wishes to approximately describe the situation, one should assume that the ionizing particles create in water small overheated and oversaturated regions which originate nuclei. The above analysis would indicate that in the case of water oversaturation plays a major role. This seems plausible because an increase of solved gases may help the formation of nuclei directly and drastically reduce the surface tension.[12]

It seems therefore to us that the experiment indicates the following conclusions on the nucleation mechanism in water:

(1) Primary particles of cosmic radiation as well as neutrons from a Ra–Be source can create nuclei directly or through secondaries; cavities originate in small regions in which very special conditions exist as a consequence of the heating and the radiolytic effects produced by ionizing paritcles.

(2) Nucleation does not require the presence of sound, and when nuclei are created they have a finite life in the liquid.

The secondary particles which may originate in water are recoil H or O nuclei. Both of them are produced by neutrons from the Ra–Be source in the experiment with the screened tank (Fig. 9); the maximum density of energy deposit for protons is about 800 Mev/cm and it occurs at an energy lower than 0.1 Mev; the maximum density of energy deposit for O recoil nuclei is 6000–7000 Mev/cm at about 4 Mev.

In order to determine if protons may act as nucleation agents we have performed the second experiment (Fig. 10) with the neutron source, surrounding the latter with a paraffin enclosure of 2.5 cm (the maximum energy of neutrons being reduced to 1 Mev). In this case protons of energy higher than 0.1 Mev (which therefore will pass during their flight into the liquid through the energy of maximum deposit density) are still produced in large number while the maximum value of energy for O recoil nuclei is 0.25 Mev which correspond to a maximum density of energy deposit of about 1000 Mev/cm. No variation in the cavitation threshold has been observed when neutrons with a maximum energy of 1 Mev are shot into the liquid, thus showing that protons, as well as ionizing particles which deposit energy into the liquid at a rate lower than 1000 Mev/cm are not able to produce nucleation.

The oxygen recoil nuclei are, therefore, the only secondary particles which may produce cavitation nuclei in water. To do so they muct have an energy which surely is higher than 0.25 Mev; energies somewhat lower than 2.5 Mev are, however, sufficient. The latter energy value[13] has been obtained from the results of the first experiment with the Ra–Be source when neutrons with maximum energy of 10 Mev produce nucleation.

---

[11] *Proceedings of the Second United Nations International Conference, on the Peaceful Uses of Atomic Energy, Geneva, 1955* (United Nations, Geneva, 1958), Vol. 7, Report on Session 12B, pp. 513–610; M. Haissinsky, *La Chimie Nucleaire et ses Applications* (Masson, Paris, 1957).

[12] Concentrations of the order of 5×10^{19} gas molecules dissolved per cm^3 would be easily produced (Haissinsky) to which a pressure of 120 atm corresponds. Assuming in first approximation the validity of C. G. Kuper and D. H. Travena [Proc. Phys. Soc. (London) **A65**, 46 (1952)] calculation on the influence of small gas solution on the surface tension of liquids also for high values of gas content, a reduction of about 50% would result in σ.

[13] The corresponding value of U is 5000 Mev/cm.

5. CREATION OF CAVITATION NUCLEI BY COSMIC RADIATION

According to the above discussion the creation of cavitation nuclei should be possible by ionizing particles which during their life in the liquid may deposit energy at a rate of the order of a few thousand Mev cm^{-1} (5000 Mev/cm is surely sufficient). The analysis has shown that 2.5-Mev oxygen recoil nuclei are in these conditions.

We wish now to examine if cosmic particles either directly (when charged) or through secondaries, may explain quantitatively the experiment in ordinary water. The direct nucleation by charged particles (proton, electrons, mesons) can be excluded. The maximum densities of energy deposit of protons (800 Mev/cm) are not sufficient for the creation of nuclei and the densities of energy deposit for electrons and mesons are much lower than for protons. Moreover, the free radicals created by these particles take part mainly in recombination reactions.[11]

Turning now to nucleation produced through secondaries, we observe that energetic electrons and mesons could produce recoil nuclei, but it can be easily seen that events of this kind are rare and so can be ignored. In fact the cross section for these collisions[14] is

$$\sigma = 8\pi a^2 Z^2 E_R^2 / Mc^2 E_D, \quad (4)$$

where a = Bohr radius, 5.3×10^{-9} cm, E_R = Rydberg energy, 13.6×10^{-6} Mev, Z = atomic number of the struck nucleus, E_D = minimum required for the energy given to the nucleus (2.5 Mev for O in our case), and M = rest mass of the nucleus. We obtain for the oxygen nuclei:

$$\sigma = 2\times10^{-4} \text{ barn}.$$

These conclusions are in agreement with the results of Sette[2] for water irradiated with gamma rays (Compton electrons of about 1 Mev) and of Libermann[3] in pentane irradiated with 20-Mev β^+; no effect on the cavitation threshold was found.

Cosmic neutrons with energy equal to or higher than 10 Mev can produce oxygen recoil nuclei with an energy higher than 2.5 Mev, and they are present in the cosmic radiation at sea level with a density of about 2×10^{-4} cm^{-2} sec^{-1}.[10] High-energy neutrons appear, therefore, as the particles in cosmic radiation which may lead to the formation of small cavities near the surface of the liquid and which diffusing into the volume, may act as cavitation nuclei.

An approximate calculation can be carried out to see if their density in the cosmic radiation reaching the tank may be sufficient to explain the experiment. The geometrical cross section of our liquid mass is about 5×10^2 cm^2, and we can assume that about 10^{-1} neutron having energy higher than 10 Mev reaches the liquid per second. We suppose also that each neutron produces at least one cavity which has a finite mean life. The experiments with lead and paraffin screens give different values for $\tau_{\frac{1}{2}}$. In order to make an orientative calculation we assume the intermediate value of $\tau_{\frac{1}{2}} = 3\times10^3$ sec, for the time required to reach the $\frac{1}{2}$ variation of threshold as a consequence of inserting the screen. Assuming an exponential law for the decrease of the number of nuclei, λ being a constant, we have

$$\tau_{\frac{1}{2}} = (\ln 2)/\lambda = 3\times10^3 \text{ sec}.$$

When the screen is removed the increase of the number of nuclei will proceed according to

$$dN = Kdt - \lambda N dt, \quad (5)$$

N being the total number of nuclei in the liquid (volume about 2×10^4 cm^3) at time t and K (sec^{-1}) the rate of creation of cavities.

From (5), we have

$$N = (K\tau_{\frac{1}{2}}/\ln 2)(1 - 2^{-t/\tau_{\frac{1}{2}}}).$$

As equilibrium value of N, we have

$$N = K\tau_{\frac{1}{2}}/\ln 2 = 3\times10^3 K/\ln 2. \quad (6)$$

Assuming that each fast neutron reaching the liquid produces at least one nucleus, we take for K the value 10^{-1}. The result is

$$N = 4.34\times10^2,$$

and per unit volume

$$n = 2.17\times10^{-2}. \quad (7)$$

In our experiment the volume in which the sound field was sufficient to produce cavitation (growth of nuclei) was about $V = 2$ cm^3. In addition, the fluid under sound radiation changes continuously as a consequence of motion due to the stirrer and to the gradient of radiation pressure. Velocities of the order of 0.5 cm/sec have been observed in the focal region; and one can calculate that the fluid in this zone is renewed every 2 sec. The volume of the fluid which passes through the region in which suitable conditions for the growth of nuclei exist in t seconds is $Vt/2 = t$ cm^3 and the resultant number of nuclei in it is

$$n^* = 2.17\times t\times10^{-2}. \quad (8)$$

The average delay of cavitation onset is obtained from (8) for $n^* = 1$ as $t = 46$ sec, which is quite in agreement with the observed value (40 sec).

In this calculation approximate values have been used. One has to consider, however, two circumstances which have not been taken into account although they may favor the formation of nuclei: (1) Neutrons with energy not much lower than 10 Mev are probably active; (2) neutrons with very high energy are present in cosmic radiation which could produce more than one oxygen recoil nuclei.

It seems, therefore, that the neutron component of cosmic radiation is sufficient to explain the observed

[14] G. J. Dienes and G. M. Vineyard, *Radiation Effects in Solid* (Interscience Publishers, Inc., New York, 1957).

cavitation in ordinary water, through the production of oxygen recoil nuclei which would be the agents creating cavities.

6. DISCUSSION OF SCREEN EXPERIMENTS

Figures 5 and 8 show that small thicknesses of lead or paraffin screens produce an increase of threshold and that one (15-mm lead) or two (7.5- and 25-mm paraffin) maxima are found as the thickness grows to 30 mm. The increase of threshold is connected with a decrease of efficient primaries in the radiation reaching the liquid and therefore in microcavity density, as a consequence of absorption and slowing down in the screen. In Fig. 3, curve b has been obtained by the same procedure as curve a, in this case enclosing the tank with a 20-mm lead screen. The corresponding value of the threshold is about 1.27 w/cm². The shift of the curve F vs power when the lead screen is inserted corresponds to an increase of threshold which is associated with a reduction of nuclei present in the liquid.

The maxima in the threshold screen thickness curves indicate the presence of processes which facilitate nucleation, and depend on screen thickness. Although further experimentation and more nuclear data are needed to clarify the process, a tentative suggestion is that mesons and γ rays of cosmic radiation could induce nuclear reactions in the screen material with the emission of secondary particles, as for instance, neutrons.

In paraffin one could have the reaction[15,16] $C^{12}(\gamma,n)$-C^{11}, whose differential cross section has a maximum for a γ energy of 22 Mev. The total cross section is 0.05 barn.

Similar reactions[15–17] induced by gamma rays in paraffin and lead are: $C^{12}(\gamma,p3n)B^8$, $C^{12}(\gamma,2n)C^{10}$, $Pb^{204}(\gamma,n)Pb^{203}$, $Pb^{206}(\gamma,n)Pb^{205}$, $Pb^{207}(\gamma,n)Pb^{206}$, and $Pb^{208}(\gamma,n)Pb^{207}$.

Moreover mesons may react with protons in nuclei according to

$$\mu^- + p \rightarrow n + \nu,$$

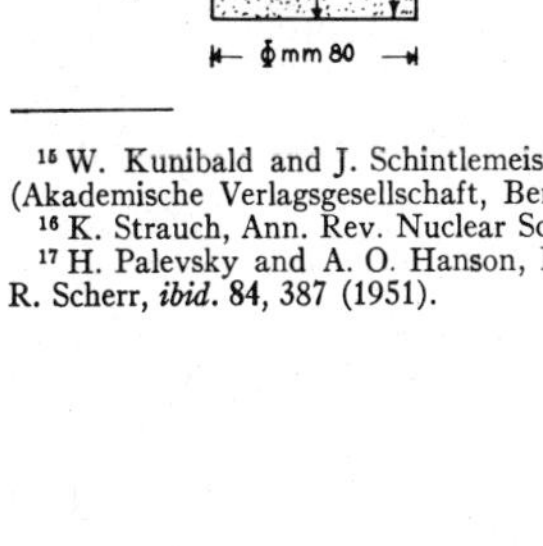

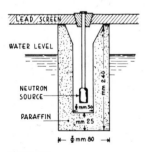

FIG. 11. Paraffin enclosure used for slowing down neutrons.

giving rise to neutrons having energies of 10–20 Mev; as a consequence, one or more particles, mainly neutrons, escape from the excited nucleus. Average emission of 1.6 neutrons per absorbed meson in lead and of 0.8 neutrons per meson in carbon with cross sections of 0.02 barn and 0.002 barn, respectively, have been reported.[18] From these data it seems that mesons are more effective in lead than in paraffin. It is to be noted that cross sections of the order of 10^{-2} barn assume interest in meson or gamma reactions because the meson component of cosmic radiation is about 10^2 times the proton and neutron component, and because the gamma component is still higher than the nuclear one by a factor of 30–50.

The importance of reactions of this kind in producing secondary neutrons reaching the liquid will depend on thickness, increasing with the thickness to a maximum for each reaction when the probability of reaction is maximum and the probability of subsequent absorption of secondaries in the screen is low. A process of this kind may compensate for and overcome the absorption of primary particles in the screen and qualitatively explain the observed threshold dependence with screen of lead and paraffin. In the latter case more than one reaction should be present. It is to be mentioned that (γ,n) reactions have significant cross sections only for narrow energy bands.

The more rapid increase of threshold for the paraffin screens than for the lead ones seems in agreement with the conclusion that neutrons are responsible for nucleation, taking into account the more effective slowing-down of neutrons in paraffin.

More research is in progress in order to increase the knowledge of the mechanism responsible for the screening action.

7. CONCLUSIONS

The experimental results presented in this paper confirm the suggestion given by one of the authors in a previous communication and establishes that the nucleation responsible for ultrasonic cavitation is connected with cosmic radiation.

More theoretical work is needed to fully clarify the mechanism of production of gaseous nuclei using an adequate description of the status of molecules in the small regions in which high-energy particles are absorbed. In order to explain the experiment in ordinary water it seems necessary to take into account heating and radiolytic effects which produce small overheated and H_2 and O_2 oversaturated regions along the path of radiation inside which cavities may be produced. In addition, cavities originated in the absence of sound seem to be stabilized with processes associated with impurities which eventually would control the density of nuclei in ordinary water.

[15] W. Kunibald and J. Schintlemeister, *Tabellen der Atomkerne* (Akademische Verlagsgesellschaft, Berlin, 1958), Vol. 1, Part I.
[16] K. Strauch, Ann. Rev. Nuclear Sci. 105, 2 (1953).
[17] H. Palevsky and A. O. Hanson, Phys. Rev. **79**, 242 (1950); R. Scherr, *ibid.* **84**, 387 (1951).
[18] R. D. Sard and M. F. Crouch, in *Progress in Cosmic-Ray Physics* (North-Holland Publishing Company, Amsterdam, 1954), Vol. II.

The analysis of possible nucleation processes indicates that oxygen recoil nuclei produced by the neutron component of cosmic rays act as nucleating agents.

The effects of lead or paraffin screens on the cavitation threshold are probably due to a decrease of the present neutron component reaching the liquid and to the production of secondaries in nuclear reactions induced by mesons and gamma rays in the screen material.

A cell formed by a liquid in an adequate sound field and suitably screened, constitutes a device which allows the examination of the influence of various penetrating particles entering the cell through a hole, on cavitation threshold. The device also allows one to distinguish the various types of particles: for example, its response to neutrons depends on their energy.

ACKNOWLEDGMENT

The authors express their thanks to Dr. M. Bertolotti for useful discussions.

The following experiment utilizes very small purified droplets of the test liquid, thereby minimizing the possibility of foreign nucleation sites. As a result, the measured strength of the liquid agrees very well with the strength predicted by theory.

Dr. Apfel has published several papers in the field of acoustic cavitation in liquids. He was Research Assistant in the Division of Engineering and Applied Physics at Harvard University from 1966–1970, and Research Fellow from 1970–1971. He is now Assistant Professor in the Department of Engineering and Applied Science at Yale University. In 1971, he was awarded the A. B. Wood Medal and Prize by the Institute of Physics (London) for his work on cavitation.

This paper is reprinted with the permission of the author and the Acoustical Society of America.

Reprinted from The Journal of the Acoustical Society of America, Vol. 49, No. 1, (Part 2), 145–155, January 1971

Printed in U. S. A.

Received 24 July 1970

45

10.2

A Novel Technique for Measuring the Strength of Liquids *

Robert E. Apfel

Harvard University, Cambridge, Massachusetts 02138

A novel technique has been effectively utilized in studying the properties of liquids under conditions that are not ordinarily accessible. This technique involves the use of an acoustic standing-wave field established in a column of one liquid in order to trap an immiscible droplet of another liquid. In the experiment reported here, a filtered ether droplet suspended in filtered glycerine was superheated and acoustically stressed until the combination produced an explosive liquid-to-vapor phase transition. The experiment was performed under atmospheric conditions at which the normal boiling point of ether is 35.6°C. The measured tensile strength varied linearly from 17 bars at 130°C to 0 bars at 146°C. Previous measurements of the tensile strength of liquids have not come within a factor of 2 or 3 of the theoretical predictions based on homogeneous nucleation theory, a theory which describes the vaporization conditions for pure liquids. The results reported here are in good agreement with that theory. This novel agreement is attributed to the ability to obtain pure liquid samples by utilizing small (0.5-mm-diam) filtered droplets.

INTRODUCTION

In this section we set forth some of the basic nomenclature used in discussing the liquid-to-vapor phase transition; we selectively review experimental and theoretical efforts relative to this phase transition; and we introduce a novel technique that has been effectively utilized in measuring the ultimate strength of liquids.

At a flat liquid–gas interface, the conditions under which the liquid-to-vapor transformation will occur are described by the vapor pressure-versus-temperature curve (P vs T) for a particular liquid. If the interface is not flat, as in the case of a vapor cavity in a liquid, it is possible to attain a state normally associated with the vapor phase without that phase developing. When a phase boundary (as described by the P–T diagram) is crossed without the new phase developing, the phase boundary is said to be transgressed.[1] When this transgression of the boundary occurs along a constant pressure path, the liquid is said to be superheated. When it occurs along a constant temperature path, the liquid remains a liquid at a pressure lower than the vapor pressure for that temperature; sometimes the pressure is negative, corresponding to a positive tensile stress. There is no word, such as "superheat," to express this condition, but it is clear that transgression of the phase boundary has occurred.

If transgression has occurred, the system is said to be in a metastable state. This term "metastable" refers to the fact that the entropy of the system is not at an absolute maximum, which implies stable equilibrium, but is at a relative maximum.[2]

A metastable liquid reaches its "limit of transgression" when a small increase in temperature or decrease in pressure will produce vapor cavity formation "in a reasonable period of time." The state of the liquid at this limiting point is often called the "cavitation threshold."

In discussing cavitation in a metastable liquid, it is often advantageous to distinguish between cavity nucleation (or inception) in a pure liquid and nucleation caused by something foreign to that pure liquid, such as some other pure substance, solid impurities, or some kind of radiation acting on the liquid. Nucleation within a pure liquid is called "homogeneous" nucleation, and "impure" nucleation is called "heterogeneous." Heterogeneous nucleation usually prevails in nature and prevails more often than not in the laboratory in which the experimenter is trying to observe homogeneous nucleation.

One basic obstacle to successful homogeneous nucleation experiments is the presence in the liquid sample of imperfectly wetted solids, such as container surfaces or suspended motes (dirt). (See Ref. 3, pp. 70–88.)

This obstacle has been overcome by Wakeshima and Takata in their measurements of the limit of superheat of various hydrocarbons.[4] In their experiment, a small droplet of the test liquid was heated as it rose in an immiscible host liquid until the droplet exploded into its vapor. The results of these experiments, and some by Kenrick, Gilbert, and Wismer,[5] and Briggs,[6] who superheated liquids in capillary tubes, are in good agreement with homogeneous nucleation theory.

In this theory it is assumed that density fluctuations cause the rapid formation and destruction of minute vapor cavities; the probability that one of these fluctuations develops into an observable vapor cavity in a reasonable period of time is then calculated. The fundamental work on this theory was performed by Döring.[7] Others, including Volmer,[1] Fisher,[8] and more recently, Apfel,[3] have refined Döring's work. They all agree on the general form of the transcendental nucleation expression that describes the pressure–temperature states for the onset of cavitation:

$$\frac{16\pi\sigma^3(T)/3[P(T)-P_0]^2}{kT}=\ln C(T). \tag{1}$$

The left-hand side of this equation represents the "energy barrier" to nucleation, W_m, divided by kT. W_m is found first by calculating the reversible isothermal work to form a vapor cavity of arbitrary size, and then by calculating that value of the size variable that makes the work an extremum. A vapor cavity of such size, which is called a "critical" cavity, is in unstable mechanical equilibrium and will grow to observable size if the temperature is infinitesimally increased or if the pressure in the liquid is infinitesimally decreased. In Eq. 1, T is the absolute temperature, k is Boltzmann's constant, $\sigma(T)$ is the liquid–vapor surface tension, and $[P(T)-P_0]$ is the pressure differential across the cavity interface, $P(T)$ being the pressure of the vapor on the interface and P_0 being the pressure in the liquid.

The complexities of the kinetics of the nucleation process reside in the omnibus variable $C(T)$. Precise evaluation of $C(T)$ would demand knowledge of evaporation and condensation coefficients, accommodation coefficients, and other information related to the liquid state. Approximate expressions for $C(T)$ can be found in the referenced work.[1,3,7,8] Although evaluation of these expressions for a given liquid may often differ by orders of magnitude, the spread in $\ln C(T)$ is less than 25%. Such a spread typically produces an "uncertainty" of about 2° or 3°C in the prediction of the limit of superheat of a liquid and of about 1 or 2 bars in the prediction of the tensile strength of a liquid. We further note that for a wide variety of liquids the cited expressions show little change. This leads to the approximation that $\ln C(T) \sim 45$-75.

Until recently, Eq. 1 had been unsuccessful in predicting the outcome of experiments designed to measure the tensile strength of liquids. Presumably, the liquids that were studied were not as pure as the experimenters thought, and hoped, they were.

Berthelot was one of the earliest of those who have set out to measure the tensile strength of liquids.[9] The method he used is described as follows: Deaerated water, which partially fills a closed tube, is heated until it does fill the tube; then it is cooled. Up to a certain degree of cooling, the water maintains the full volume, after which the vapor forms suddenly (and audibly), relieving the tensile stresses in the liquid. In this manner, Berthelot estimated the tensile strength of water to be between 30 and 50 bars.

A phenomenon similar to the one that Berthelot observed in the laboratory may occasionally occur under natural circumstances. Roedder has suggested that extremely high negative pressures occur in aqueous liquid- and vapor-filled microscopic inclusions in minerals[10]: "In some microscopic inclusions (consisting of aqueous liquid and vapor) in minerals, freezing eliminates the vapor phase because of greater volume occupied by the resulting ice. When vapor fails to nucleate again on partial melting, the resulting negative pressure (hydrostatic tension) inside the inclusions permits the existence of ice I crystals under reversible metastable equilibrium, at temperatures as high as +6.5°C and negative pressures possibly exceeding 1000 bars . . . ."

It is possible that these microscopic inclusions are so small (the largest dimension is of order 30 μ) that no foreign nucleation sites, such as motes, exist in the samples that register these high tensions. The actual site of the nucleation (i.e., within the liquid or at the liquid–solid interface) has not been observed.[10]

The highest statically measured value for the tensile strength of chemically pure water is about one-fifth of the lowest theoretical value. L. Briggs performed this experiment in which a tube containing "mote-free" water was spun at a speed fast enough to establish a tensile stress of up to 277 bars (at 10°C) in the liquid before cavitation.[11] Briggs made the important observation that, whereas mote-free water was essential in his tensile-strength tests, boiled distilled water sustained the same superheating as mote-free water. Presumably, motes in superheated water are either destroyed or more easily wetted than motes in water near room temperature.

Purity was also found to be essential in measurements by M. Greenspan and C. Tschiegg of the acoustic cavitation threshold of carefully filtered water.[12] They found that at room temperature such water could sustain acoustic-stress amplitudes of over 200 bars for a few seconds.

In hopes of overcoming some of the obstacles encountered in measuring the ultimate strength of liquids, we have designed an experiment that encompasses the following features: By preparing small filtered samples, we were able to increase our chances of obtaining mote-free liquid; by injecting this sample into a filtered immiscible "host" liquid, we were able to isolate the test sample from solid container surfaces; and by measuring the acoustic cavitation threshold at high temperatures (when the sample was superheated), our experimental apparatus was capable of producing acoustic pressure amplitudes comparable to the theoretical threshold predictions. Our results for the tensile strength of superheated ether are, in fact, in very good agreement with the predictions of homogeneous nucleation theory. This experiment is now described in detail.

I. APPARATUS AND PROCEDURE FOR INTRODUCING HOST AND TEST LIQUIDS

The apparatus for this experiment consists of four major sections: the central section that contains the host liquid and in which heating and acoustically driven systems operate, the two input systems for the introduction of the host and test liquids into the central system, the evacuation system, and thermal and acoustical sensing probes and related apparatus. A schematic of the whole system is shown in Fig. 1.

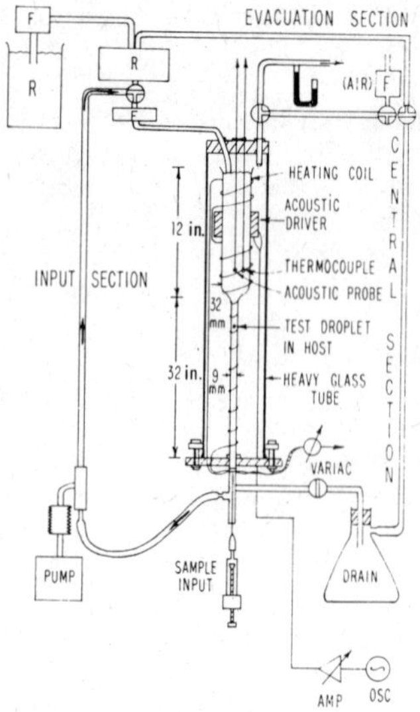

FIG. 1. Over-all diagram of apparatus. Reservoirs, R; filters, F.

A. The Evacuation System

The evacuation system is used to draw the host liquid through the filters into the central system, to degas the host liquid partially, and to speed up the drainage process. To accomplish these tasks, a simple aspirator unit was used.

B. The Input Systems

With the assistance of the evacuation system, the host liquid is drawn first through the prefilter into the intermediate reservoir and through the final filter into the inner cylinder of the central system. Once this central system is filled, the host liquid can be circulated and filtered in the closed loop shown by the arrows in Fig. 1. The micro-bellows pump (Research Appliance Co., Allison Park, Pennsylvania) forces the liquid around the loop and through a Millipore filter (Millipore Filter Co., Bedford, Massachusetts) of 1.2-μm pore size. This circulation system was used to clean the glycerine. The particulate contamination of the glycerine could be estimated by observing the light scattered from particles when a collimated beam of light was sent through the central tube in a direction perpendicular to the observation direction. After extended closed-loop filtering such "Tyndal-effect" scattering was considerably reduced although not absent.

The input system for the test liquid is also shown in Fig. 1. A gas-tight Teflon-tipped screw-type Hamilton syringe (Hamilton Co., Whittier, California) is used to force a small amount of ether through a 0.45-μm Millipore filter into the central system containing gylcerine.

C. The Central System

The central system is composed of thermal and acoustic subsystems.

1. *Thermal Subsystem*

Owing to buoyancy forces, the injected droplet rises in its host. A positive temperature gradient (up being positive) is established in the liquid-filled tube by a nichrome heating coil (0.13-in.-diam wire) wrapped around the tube. The electrical input is controlled by a variac. Small pieces of high-temperature glass-fiber tape keep the windings in place at the desired spacing.

The windings were adjusted so that the temperature in a region just above the point where the tube diameter increases was nearly constant. Henceforth, this region of nearly constant temperature is referred to as the "test region."

2. *Acoustic Subsystem*

The acoustic driving unit is a cylindrical PZT-4 lead zirconate transducer (Clevite Co., Bedford, Ohio) (1.5-in. o.d., 1.25-in. i.d., 1.5-in. length). The composite system, made up of this unit epoxied to the liquid-filled glass tube, resonates in the 50–60 kHz frequency range, corresponding to the $(1,0,n)$ mode of the system, where 1 refers to the first radial mode, 0 refers to axial symmetry, and n is 2, 3, 4, or 5, corresponding to a spatial pressure distribution along the axis that can be characterized by a wavelength of 2–10 cm. (The axial pressure distribution is not simply sinusoidal because of the irregularities in the shape of the cylinder.) The particular resonance frequency chosen is that for which the acoustic pressure amplitude along the axis of the liquid-filled tube has a maximum in the test region.

The magnitude of the acoustic field can be increased so as to produce a net acoustic force on the test droplet that is equal in magnitude and opposite in direction to the buoyancy force. In other words, the test droplet can be held motionless in the host liquid. By trapping the droplet at a given position, the experimenter can

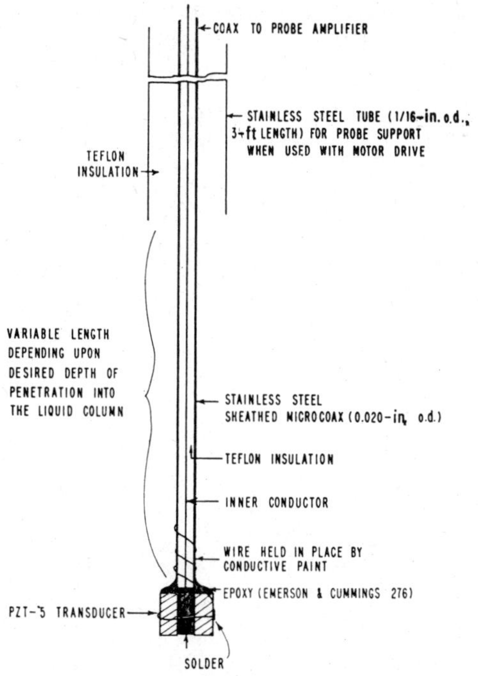

FIG. 2. Acoustic probe construction.

observe its size and can assure that the droplet has time to achieve thermal equilibrium with the immediate surroundings. As explained later, the pressure amplitude can then be increased until the droplet explodes into its vapor.

In addition to this droplet-trapping function, the acoustic unit can also be used to assist the evacuation system in the process of degassing the host liquid.

D. Acoustic and Thermal Probes and Their Motor Drive

1. Thermocouple

The commercial chromel–alumel thermocouple unit (Nanmac Co., Framingham Centre, Massachusetts) consists of a ribbon junction (ribbon width, $\frac{1}{8}$ in.) in the shape of a loop of $\frac{1}{8}$-in. diam that protrudes from a stainless steel tube of $\frac{1}{8}$-in. o.d. and 3.5-ft length. The ribbon thermocouple is designed with some attempt to minimize thermal conduction errors. These errors are expected to be especially small in the region of nearly constant temperature (the test region).

2. Acoustic Probe

The acoustic monitoring system is far more complex than the thermocouple. The complexity arises from the difficulty in constructing an acoustic transducing device that has the following properties:

(a) a sufficiently small size in order that it not disturb or detune the acoustic field and so that it has a fairly flat frequency response in the frequency range of interest;

(b) a construction that can withstand temperatures as high as 170°C continually, acoustic pressure amplitudes of several bars, and rough treatment; and

(c) a sensitivity that is not affected markedly by small changes in temperature.

Such a probe is shown and described in Fig. 2. The basic elements of this probe design are a cylindrical PZT-5 lead zirconate transducer (Clevite Co., Bedford, Ohio) (0.0625-in. o.d., 0.0425-in. i.d., 0.0625-in. length); microcoax (4-ft length, 0.020-in. o.d. of stainless-steel sheath, Teflon insulation); and electrical connections described as follows: The inner conductor of the Microcoax is soldered to the silvered inner surface of the ceramic. One end of a fine wire is soldered to the silvered outer surface of the ceramic. The other end is wrapped around the stainless sheath of Microcoax and is held in place by conductive paint (Dynaloy 375). (The paint is cured by heating to 160°C for about 45 min.)

3. Motor Drive[13]

Both thermocouple and acoustic probe are motor driven. In this experiment, the probes are driven separately, although dual drive is possible. One important feature of the setup is a slotted-support tube which prevents the acoustic probe from buckling. Such support is not required for the more rigid thermocouple. Both up and down drives are possible, and a hand switch allows the experimenter to control vertical excursions of the probes to within 0.4 mm (with practice).

II. MEASUREMENT PROCEDURE

The host liquid, glycerine, meets the following requirements: It is heavier than the test liquid; it is inert, in that the test liquid is insoluble in it; the nominal vaporization temperature is far above the estimated droplet-vaporization temperature; and it is sufficiently viscous so that a small test droplet will rise slowly enough to assure that the droplet is always in thermal equilibrium with its immediate surroundings.

If such conditions are met, the test droplet will be heated as it rises in the host liquid. If no acoustic field is present, the droplet will rise until it reaches the point at which it can no longer maintain the liquid state. The position at which the explosion occurs is observed. Before and after the experiment, the temperature profile of the column is measured with the motor-driven thermocouple. The temperature is measured at millimeter spacings, as observed on a cathetometer, which is mounted about 5 ft from the tube. This cathetometer can resolve vertical displacements of order 0.15 mm. The temperature is measured $\frac{1}{4}$-in. off axis (the acoustic probe is on axis). The difference between temperatures on and off axis has been measured at less than about 0.25°C. Also, the effect of the displacement of the host liquid by the thermocouple has been shown to lead to a negligible correction to the temperature profile. Small-droplet explosions produce very small changes in the temperature profile. If, however, a large droplet (over 2-mm diam) explodes, then, before another droplet is introduced, a short time is allowed so that the temperature profile can restabilize itself.

Glycerine is a desirable liquid for measurements of the tensile strength of ether, because it not only satisfies the properties described above, but it also has a high cavitation threshold. This is desirable if the threshold of the ether is to be measured without the detuning effects of cavitation in the host. After degassing the host liquid by evacuating and agitating acoustically, the pressure above the liquid is returned to atmospheric by letting air into the system through an air filter. In preparing for the measurements, the acoustic probe is placed in the test region, and the frequency of the driver is adjusted to produce a maximum probe output voltage. The axial pressure field is then observed. If the axial pressure field has a strong maximum in the test region, then this mode is used during the test. The axial field is measured as a function of vertical position. The probe is then removed from the liquid. At this point a test droplet is introduced at the bottom of the inner cylinder. When it approaches the test region, the acoustic pressure amplitude is increased until the test droplet comes to a standstill. The position of the droplet is recorded with the aid of a cathetometer. The cathetometer can also be used to measure the vertical (undistorted) diameter of the droplet within 0.2 mm. Then, the acoustic pressure is increased slightly (say 10%) and the droplet moves to a new equilibrium position, which is observed and recorded. This procedure is continued until the droplet explodes into its vapor. In Sec. III it is shown how the acoustic profile, the trapping and explosion positions, and the corresponding driver input voltages can be utilized in determining the trapping and explosion pressures, as well as the probe calibration.

III. CALIBRATION OF THE SYSTEM

A. Thermocouple

The voltage difference between the thermocouple junction in the host liquid and the reference junction in the ice bath is measured by a Ballantine digital voltmeter (model 353) on which changes of 0.01 mV, corresponding to about 0.25°C, can be easily resolved. Voltage readings are converted to degrees centigrade with the aid of a standard conversion chart for chromel-alumel thermocouples.

B. System Calibration

Calibration of an acoustic probe at high temperatures is not a simple task. Therefore, a self-calibration scheme for the system has been adopted. This scheme is based on the knowledge of an analytic expression for the force on a compressible sphere in a liquid in which the shape of the acoustic standing wave pattern is known. Yosioka and Kawasima,[14] Eller,[15] Gould,[16] Crum,[17] and Crum and Eller[18] have presented similar expressions for this force. Crum has also performed droplet-trapping experiments with results within 10% of his theoretical predictions.[17]

Crum gives the following expression for the pressure amplitude required to trap the droplet for the case in which the axial pressure field is sinusoidal:

$$\frac{-P_A{}^2 k_z \sin(2k_z z)}{2} = 2(1-\lambda)g\rho^2c^2 \Big/ \left[\frac{1}{\lambda\delta^2} - \left(\frac{5\lambda-2}{2\lambda+1}\right)\right]. \quad (2)$$

In this expression, P_A is the peak pressure amplitude, k_z is the axial wavenumber (the z direction is along the axis, and up is positive), λ is the ratio of the density of the droplet to the density of the host (ρ^*/ρ), δ is the ratio of the speed of sound in the droplet material to that in the host (c^*/c), and g is the gravitational constant.

The left-hand side of Eq. 2 is just $P(z)\cdot dP(z)/dz$, the product of pressure and pressure gradient at any point z on the axis. It is simple to show that if the axial pressure field is not sinusoidal, we can rewrite Eq. 2 in the general form[19]

$$-P(z)\frac{dP(z)}{dz} = 2(1-\lambda)g\rho^2c^2 \Big/ \left[\frac{1}{\lambda\delta^2} - \left(\frac{5\lambda-2}{2\lambda+1}\right)\right]. \quad (3)$$

It is also a simple matter to show that in the calculation of the pressure required to trap an ether droplet in glycerine: $(5\lambda-2)/(2\lambda+1) \ll 1/\lambda\delta^2$. If we neglect the smaller term in Eq. 3 and use the definitions of λ and δ, we have

$$-P(z)dP(z)/dz = 2\rho^* c^{*2}(\rho-\rho^*)g \equiv K. \quad (4)$$

Two points are noteworthy: (1) Equation 4 is independent of the radius of the droplet. The experimental results in the next chapter complement the results of Crum in confirming this prediction.[17] (2) Equation 4 is also independent of the acoustic frequency. (The above derivation assumes that percentage changes in the acoustic pressure amplitude are small over distances comparable to the droplet radius.)

We now ask how Eq. 4 can be used to calibrate the acoustic probe. We first assume linearity in two respects: (1) The voltage output of the probe amplifier v_0 is proportional to the pressure at the probe position in the liquid. (2) The pressure in the liquid increases linearly with the input voltage to the driving transducer. (This can be assured so long as the electric input impedance to the driver is constant as the voltage is increased. The waveform can also be observed for distortion.)

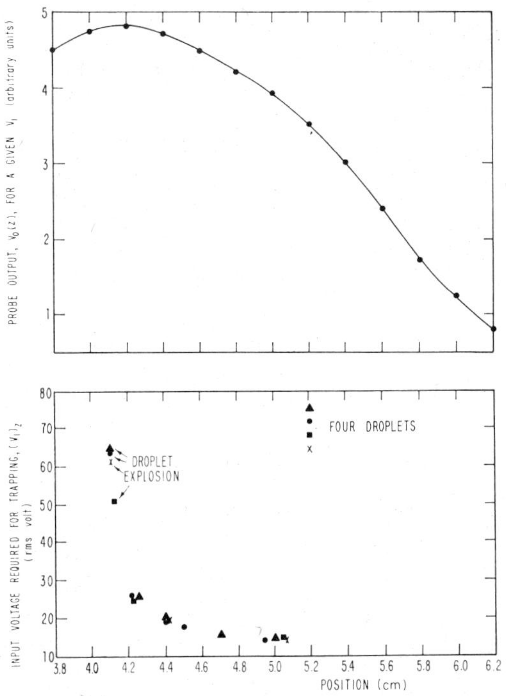

FIG. 3. Measurements of field distribution and trapping conditions (134.4–137°C).

The first condition can be written

$$P(z)=\alpha v_0(z),\quad dP/dz=\alpha dv_0/dz,$$
$$\alpha=\text{PROPORTIONALITY CONST.}\tag{5}$$

If the probe amplifier output v_0 is β times the actual probe output v_P, then the acoustic probe sensitivity is

$$v_P/P=v_0/\beta P=1/\beta\alpha.\tag{6}$$

In logarithmic form, the sensitivity S is just

$$S\equiv 20\log_{10}(v_p/P)=-20\log_{10}\alpha\beta.\tag{7}$$

The experimental procedure for probe calibration is as follows: with the driver input at v_i, the acoustic field is probed along the axis; the output, as a function of z, is designated $v_0(z)$. The probe is then removed from the liquid.

A droplet of ether is introduced into the column of glycerine. When it reaches the test region, the driver input voltage is increased until the droplet is trapped. The input voltage with the droplet trapped at the position z is designated $(v_i)_z$. Assuming linearity, this would correspond to a trapping pressure

$$P(z)=\alpha\left[\frac{(v_i)_z}{v_i}v_0(z)\right].\tag{8}$$

[The bracketed term is just what the probe output voltage would be if the probe were at z with the driver voltage, $(v_i)_z$.]

Substituting Eq. 8 (and the derivative of it) into Eq. 4 yields

$$\alpha^2[(v_i)_z/v_i]^2\cdot v_0(z)\cdot[-dv_0(z)/dz]=K\tag{9}$$

or

$$\alpha=\frac{v_i}{(v_i)_z}\left\{\frac{K}{v_0(z)[-dv_0(z)/dz]}\right\}^{\frac{1}{2}}.\tag{10}$$

We have included the minus sign with dv_0/dz because this term must be negative if a droplet is to be trapped. Since α, v_i, and K are constant, this relation implies

$$(v_i)_z^2 v_0(z)dv_0(z)/dz=\text{CONST},\tag{11}$$

where z equals trapping position, $(v_i)_z$ is input voltage when droplet is trapped at z, and $v_0(z)$ is probe output voltage at z when driver input voltage is v_i. As the input voltage is increased, the droplet's position will change and Eq. 11, which is one test of the theory, can be checked. Crum has indirectly confirmed that this prediction is consistent with experiment. Problems of adequate resolution, however, occur in regions in which the pressure gradient is so small that errors in the measurement of $dv_0(z)/dz$ can be large. For practical purposes, therefore, Eq. 11 should be most accurate in regions of maximum pressure gradient. Once α in Eq. 10 is known, the sensitivity of the probe can be calculated using Eq. 7, and Eq. 8 can be used to give the trapping pressure (provided, of course, that it is less than the pressure that will cause the droplet to explode into its vapor).

We point out, also, that the probe calibration can be circumvented in determining trapping and explosion pressures. This is often desirable because the calculation of α assumes that the acoustic field has been probed precisely where the droplet is trapped. Slight probe misalignment would lead to errors in the probe calibration. The following expression for the trapping pressure does not require this precise alignment. Using Eq. 5, we can rewrite Eq. 4 as follows:

$$-P\frac{dP}{dz}=P\alpha\frac{(v_i)_z}{v_i}\cdot\frac{dv_0(z)}{dz}=K,$$

or

$$P(z)=\frac{K}{\alpha}\frac{v_i}{(v_i)_z}\left[-\frac{1}{dv_0(z)/dz}\right].$$

With Eq. 10 for α, we have

$$P(z)=\left\{Kv_0(z)\Big/\left[-\frac{dv_0(z)}{dz}\right]\right\}^{\frac{1}{2}}.\tag{12}$$

Once again, the equation is most accurate at trapping positions for which $dv_0(z)/dz$ is not difficult to measure accurately. In order to find the trapping pressure at a position z' where $dv_0(z)/dz$ is small, we just use the known trapping pressure at z and Eq. 8 to eliminate the proportionality constant, and we solve for $P(z')$:

$$P(z')=P(z)\frac{v_0(z')}{v_0(z)}\cdot\frac{(v_i)_{z'}}{(v_i)_z}. \quad (13)$$

We can summarize the above as follows:

(1) Equations 12 and 13, the formulas for converting voltage measurements to pressure measurements, are based on the assumption that Eq. 4 for the trapping pressures is accurate. The recent work of Crum[17] and Eller[15] suggests that results within 10% of the theory are to be expected. The fact that the theory predicts (and experiment confirms) that the trapping pressure is independent of droplet radius, provided that the radius is small compared to an acoustic wavelength, is rather comforting.

(2) Since $P(z)$ in Eq. 12 depends on the ratio $[v_0(z)/(dv_0/dz)]$, the actual magnitude of the probe output does not enter into the prediction of the trapping pressure. (For instance, if the probe is slightly off axis, this ratio will have the same value as that along the axis. The probe must, of course, move vertically.) For probe calibration, however, the actual magnitude of $v_0(z)$ is, of course, important.

IV. RAW DATA

The procedure for measuring the conditions that cause droplet trapping and droplet vaporization is illustrated by the following example.

A. Example

In Fig. 3, two curves are plotted. The upper ordinate scale is the output of the acoustic probe amplifier at a given probe position in the sound field when a constant rms voltage is maintained across the acoustic driving unit. This axial distribution depends on the temperature profile and on the particular acoustic mode utilized. For a given test, the temperature varied by less than 1°C, being higher near the maximum of the acoustic pressure, where most of the droplets exploded. For this particular experiment, the acoustic system resonates in the (1,0,3) mode corresponding to a frequency of about 50 kHz.

Having probed the acoustic and thermal fields, we remove both probes from the liquid. A droplet of ether is introduced into the column of liquid and rises into the test region where it is trapped in the glycerine by appropriately adjusting the input voltage to the acoustic driving unit. The position of the trapped droplet above the pressure maximum and the rms voltage to the driving unit are recorded. The input voltage to the driving unit is then changed, causing the droplet to move to a new equilibrium position closer to the pressure maximum. Input voltage and position are again recorded. This procedure is continued until the acoustic amplitude becomes too great, at a given temperature, and the droplet explodes into its vapor. A minimum of 1 sec is allowed at each voltage before the input voltage to the driver is changed.

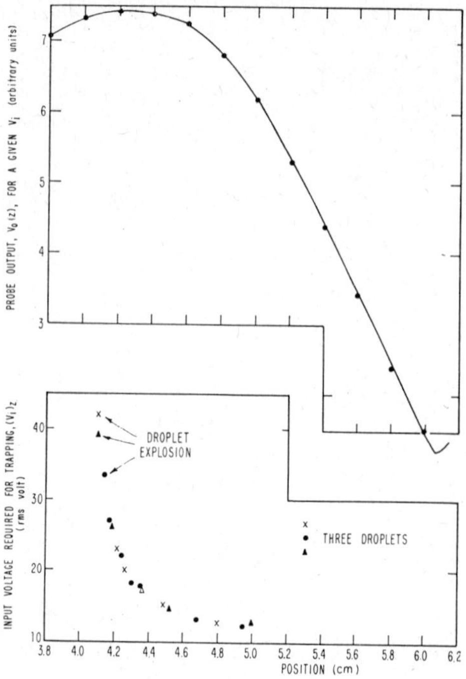

FIG. 4. Measurements of field distribution and trapping conditions (140.2°C).

The bottom ordinate in Fig. 3 indicates the trapping input voltage versus position for four ether droplets. The input voltages at which the droplets exploded are indicated on the curve.

Additional results for two different temperature ranges are shown in Figs. 4 and 5.

B. Observations

One interesting feature of the results is that the data for droplets of different size (ranging from about 0.2- to 2.0-mm diam) appear to lie on the same curve. This is consistent with the "trapping theory," which predicts the independence of trapping pressure on droplet size, provided that the droplet is small compared to the acoustic wavelength.

According to Eq. 11, the product of $(v_i)_z^2$, $v_0(z)$, and dv_0/dz should be a constant if the theory for droplet trapping is to be trusted. For the region $z=4.4$–5.0, in which most of the droplet trapping measurements are made, this product varies by less than 10%. The calculated product shows even less variation if the point at $z=4.4$, where the slope of $v_0(z)$ is more difficult to measure, is not considered.

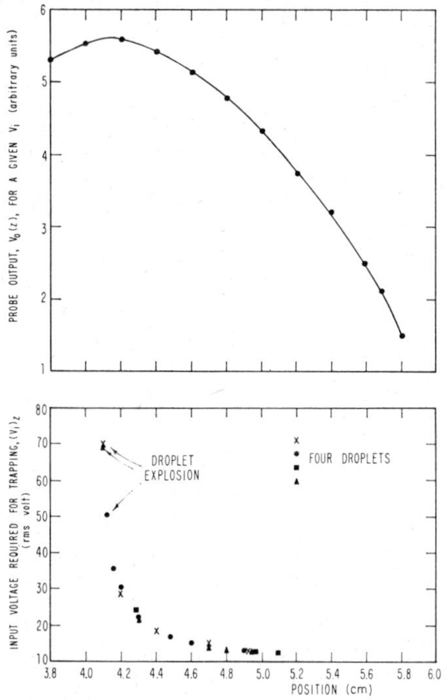

FIG. 5. Measurements of field distribution and trapping conditions (130.5–130.7°C).

The results became progressively less repeatable as the mote content of the glycerine increased. An indication of the mote content was obtained by noting the Tyndal effect when a beam of light was directed through the test region, perpendicular to the viewing direction. The repeatability always improved after the glycerine was filtered and was degassed with the assistance of high-intensity acoustic agitation.

The repeatability of the measurements decreased with increasing droplet size and waiting time, presumably because the likelihood of a collision of a mote and a droplet also increases with these factors. This presumption is supported by the fact that the repeatability always improved after the glycerine was filtered. It can be pointed out, incidentally, that motes in the test region that are more compressible and less dense than the glycerine will tend to be trapped near a pressure maximum and, therefore, will be more likely to collide with an ether droplet than motes that are less compressible and more dense than the glycerine and that will be forced away from the pressure maximum.

V. INTERPRETATION OF RESULTS

In order to convert voltage readings into pressure readings, we use Eqs. 12 and 13.

In order to make the conversion from voltages to pressures, we must be able to approximate ρ^* and c^* at temperatures for which these parameters have not been measured. We have, therefore, been forced to rely on rather extensive extrapolations. The procedure for these extrapolations is described in Ref. 3, pp. 160–164.

Unfortunately, it is difficult to assess the accuracy of these extrapolations. In hopes of remedying this situation, we propose a scheme in Sec. VII for the measurement of the speed of sound in superheated liquids. For the time being, however, all that can be said is that the extrapolations were made *before* the nucleation experiments were performed! Curves of ρ^*, c^*, and $(K)^{\frac{1}{2}}$ are plotted in Fig. 6.

Having approximated $K(T)$, we can now use Eqs. 12 and 13 and the raw data to calculate the trapping and explosion pressures.

Although the formula for the trapping pressure can be used for any position on the curve of v_0 vs z, more accuracy can be obtained at the position for which $v_0(z)/(dv_0/dz)$ can be most accurately measured. This occurs where dv_0/dz has a relatively large value; that is, at low trapping pressures. For instance, at $z=5.0$ in Fig. 3 we have

$$\left.\frac{v_0(z)}{-(dv_0/dz)}\right|_{z=5.0}=\frac{3.95}{1.71}=2.30 \text{ cm}$$

or

$$\underset{\text{TRAPPING}}{P(5.0)}=\left[K\frac{v_0}{(-dv_0/dz)}\right]^{\frac{1}{2}}$$

$$=1.55\times10^6\times1.515=2.35 \text{ bars.}$$

Continuing this example, the droplet exploded at $z=z'=4.1$ cm. Using Eq. 13, the explosion pressure is

$$\underset{\text{EXPLOSION}}{P(4.1)}=\underset{\text{TRAPPING}}{P(5.0)}\frac{(v_i)\text{EXPLOSION}}{(v_i)\text{TRAPPING}}\cdot\frac{v_0(4.1)}{v_0(5.0)}$$

$$=2.35\times\frac{64.5}{14.0}\times\frac{4.8}{3.95}=13.2 \text{ bars.}$$

The instantaneous peak pressure in the liquid during the negative phase of the acoustic cycle is given by $(-P_{\text{EXPLOSION}}+P_{\text{HYDROSTATIC}})$.

The main results of this experiment are now presented. In Table I the peak acoustic pressure amplitude that produces ether vaporization (P_E) and the peak negative pressure in the liquid (P_E-P_H) are listed as a function of temperature for 30 ether droplets. The last five results are measurements of the limit of superheat of ether for $P_E=0$. We note that at 127°C and for the highest pressure attainable in this system (about 17 bars), small droplets would remain trapped for over a minute without vaporizing. The eventual explosion of these droplets was presumably induced by a mote lodging on the surface or by local heating effects resulting from intense acoustic excitation.

We now turn to a consideration of the accuracy of the results presented in Table I. Temperature measurements are thought to be in error by less than ±0.5°C. These errors arise from a slight temperature drift in the column of glycerine, as well as from calibration errors. Errors in probe positioning and voltage readings are estimated at no more than 5%. Neglecting small terms in the formula for the trapping pressure leads to less than 3% change in the predicted trapping pressure. A more serious uncertainty exists in the extrapolations required to obtain values of c^* at temperatures where the speed of sound has not been measured. Fortunately, the trapping pressure depends on the inverse of the *square root* of the compressibility, so that a 2-dB error in the compressibility leads to an uncertainty of only 1 dB in the trapping pressure. The total uncertainty in the formula for the trapping pressure probably does not exceed 15%.

VI. COMPARISON WITH THEORY AND DISCUSSION

In comparing the experimental results with the predictions from Eq. 1, the liquid pressure P_0 is taken to be the peak tensile stress in the liquid during the acoustic cycle; that is, $(P_E - P_H)$, as indicated in the last column of Table I. We justify this by the observation that, at a frequency of 50 kHz, ether is subjected to a pressure amplitude that is at least 98% of the peak tensile stress for about 1 μsec per cycle, a time interval that is very long compared to the time it takes for a "critical" cavity to be formed.

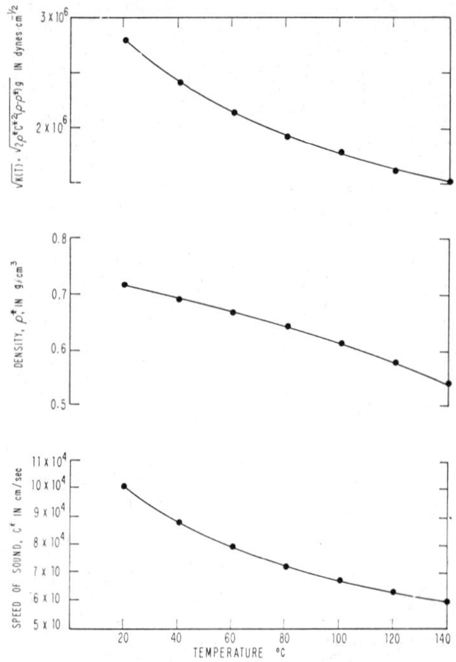

FIG. 6. Temperature dependence of $K^{\frac{1}{2}}(T)$, c^*, ρ^*.

TABLE I. Pressure at which superheated ether droplet vaporizes.

Droplet number	Peak acoustic amplitude at explosion P_E in bars	Temperature °C (±0.5°)	°K	Peak tensile stress $(P_E - P_H)$ bars (±0.5)
1	>16.6[a]	127.0	400.2	>15.6
2	15.5	129.5	402.7	14.6
3	16.2	129.5	402.7	15.2
4	15.7	130.5	403.7	14.7
5	17.0	130.5	403.7	16.0
6	13.1	131.7	404.9	12.1
7	15.3	131.7	404.9	14.3
8	15.5	131.7	404.9	14.5
9	15.5	131.7	404.9	14.5
10	13.2	134.4	407.6	12.2
11	13.2	134.4	407.6	12.2
12	13.2	134.4	407.6	12.2
13	12.2	136.0	409.2	11.2
14	10.3	137.0	410.2	9.3
15	11.0	137.0	410.2	10.0
16	6.6	140.2	413.4	5.6
17	7.7	140.2	413.4	6.7
18	7.8	140.2	413.4	6.8
19	8.4	140.2	413.4	7.4
20	8.4	140.2	413.4	7.4
21	3.2	144.7	417.9	2.2
22	3.9	144.7	417.9	2.9
23	2.6	145.0	418.2	1.6
24	2.8	145.0	418.2	1.8
25	3.5	145.0	418.2	2.5
26	0	145.0	418.2	−1.0
27	0	146.0	419.2	−1.0
28	0	146.0	419.2	−1.0
29	0	147.0	420.2	−1.0
30	0	147.0	420.0	−1.0

[a] Droplet did not explode at the maximum attainable pressure at this temperature.

Using an explicit expression for $C(T)$ (Ref. 3, p. 62), we find that $\ln C(T) \sim 53$. We can now numerically solve Eq. 1 and calculate the P–T states for the onset of nucleation. In Fig. 7 these predictions are compared with the experimental results listed in Table I. The experimental results are in gratifying agreement with theory.

The success of this experimental procedure results from three major considerations—the minimization of the foreign nucleation sites in the liquids (ether and glycerine), the choice of liquids, and the choice of the temperature range in which these nucleation measurements were performed:

(1) Minimization of foreign nucleation sites: By using a small droplet of liquid ($<10^{-3}$ cm³) isolated from solid boundaries by an inert host liquid, we have in one stroke limited the access of motes to the test liquid and eliminated any concern about nucleation at a solid container surface. We can further improve this situation by filtering, which removes the large motes, and by acoustically assisted degassing, which inactivates some of the motes. Moreover, it is presumed that motes are more thoroughly wetted at these elevated tempera-

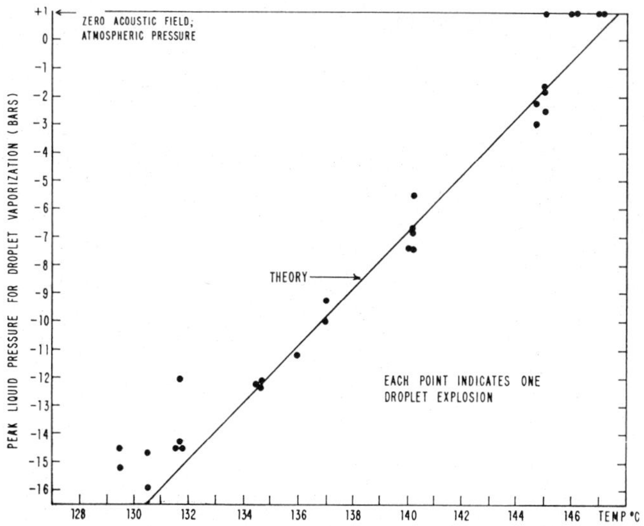

FIG. 7. Pressure–temperature (P–T) states for ether vaporization.

tures than at, say, room temperature, thus making them less likely to act as nucleation sites.

(2) Choice of liquids: The choice of ether as the test liquid has helped in two ways: (a) It is expected that many motes were thoroughly wetted by the ether, and, therefore, were unable to act as nucleation sites. (b) It is also unlikely that the presence of absorbed impurities would greatly reduce the already low value of ether's surface tension (3.6 dyn/cm at 145°C). For liquids of higher surface tension, such as water, the adsorption of impurities can produce substantial changes. This may be one of the reasons for the more favorable comparison of theory and experiment for liquids of low surface tension.[20] Furthermore, because of ether's low surface tension and glycerine's high surface tension, the probability of nucleation occurring at the ether–glycerine interface is far below that for homogeneous nucleation within the droplet.[21]

(3) Choice of temperature range: This experiment is the only one that the author is aware of in which superheat and tensile stress were simultaneously used to produce the liquid-to-vapor phase transition. One difficulty in measuring the tensile strength of a liquid is producing adequate tensile stresses. (The theoretical estimate of ether's room-temperature tensile strength is about 170 bars.) By operating in the superheat range, we were able to bring the theoretical predictions down to where the experiment could get to them! Furthermore, because of ether's relatively low vapor pressure at these high temperatures ($P=16.2$ bars at 145°C), we were able to produce changes of over 100% in $(P-P_0)$, leading to a relatively large slope in the curve plotted in Fig. 7 of the P–T states for homogeneous nucleation.

VII. A NEW TOOL FOR STUDYING LIQUIDS

One major result of this work is the development of a new technique for studying the properties of liquids under conditions that are not ordinarily accessible. This technique involves the use of an acoustic standing-wave field established in a column of one liquid in order to trap an immiscible droplet of another liquid.

We have indicated the possibilities of using this technique for measuring the tensile strength of a liquid. This technique ought now to be used for the study of a wide variety of liquids.

We should be able to use this technique for studying the liquid-to-solid phase transition. For example, a droplet of liquid could be injected at the *top* of a column of a host liquid that is less dense than the droplet and that has been cooled at the bottom. The droplet will then be supercooled as it falls. Once in the test region, the acoustic field can be adjusted to trap the droplet. If the pressure is great enough during the positive phase of the acoustic cycle, then the supercooled droplet can be expected to solidify (in crystalline or glass form). Because of the change in the compressibility that accompanies this change of phase, the droplet would no longer be trapped and would move rapidly from its previous equilibrium position. The pressure and temperature at which the phase change takes place could, therefore, be measured.

This technique should also be able to be used for acoustical measurements in superheated and supercooled liquids. By trapping a droplet in the host liquid, we can get at it acoustically. The droplet is expected to have a resonance frequency that depends on the adiabatic compressibility of the droplet and known properties of the host liquid. By using an independent low amplitude "probe signal" of variable frequency and noting the frequency of resonance, we might be able to calculate the adiabatic compressibility of the droplet. Then, with the knowledge of the droplet's density, we could calculate the speed of sound in the liquid at the given temperature. We might also find the speed of sound by echo-ranging experiments in which the time history of a test sound pulse reflecting off a droplet of known size could be measured.

This projected program of research offers the possibility of substantially increasing our knowledge of superheated and supercooled liquids and thus advancing our understanding of the liquid state.

ACKNOWLEDGMENT

It is a pleasure to acknowledge the assistance of Professor F. V. Hunt during the course of this research, which has been supported by the U. S. Office of Naval Research.

* Presented in part at the 79th Meeting of the Acoustical Society of America, 21–24 April 1970.

[1] M. Volmer, "Kinetic der Phasenbildung," Band IV of *Die Chemische Reaktion* (Theodor Steinkopff, Dresden, 1939); transl. U. S. Intelligence Dep., ATI No. 81935, p. 73.
[2] M. Planck, *Thermodynamics* (Dover, New York, 1926), p. 140.
[3] R. Apfel, Acoust. Res. Lab., Harvard Univ. Tech. Memo. No. 62 (Feb. 1970).
[4] H. Wakeshima and K. Takata, J. Phys. Soc. Japan **13**, 1398–1403 (1958).
[5] F. Kenrick, C. Gilbert, and K. Wismer, J. Phys. Chem. **28**, 1297–1307 (1924).
[6] L. Briggs, J. Appl. Phys. **26**, 1001–1003 (1955).
[7] W. Döring, Z. Phys. Chem. **B36**, 371–386 (1937); **B38**, 292–294 (1938).
[8] J. Fisher, J. Appl. Phys. **19**, 1062–1067 (1948).
[9] M. Berthelot, Ann. Chim. (Ser. 3) **30**, 232–237 (1850).
[10] E. Roedder, Science **155**, 1413–1417 (1967).
[11] L. Briggs, J. Appl. Phys. **21**, 721–722(L) (1950).
[12] M. Greenspan and C. Tschiegg, J. Res. Nat. Bur. Stand. **C71**, 299–312 (1967).
[13] Not shown in Fig. 1. See Ref. 3, p. 128.
[14] K. Yosioka and Y. Kawasima, Acustica **5**, 167–173 (1955).
[15] A. Eller, J. Acoust. Soc. Amer. **43**, 170–171(L) (1968).
[16] R. Gould, J. Acoust. Soc. Amer. **43**, 1185–1086(L) (1968).
[17] L. Crum (private communication); J. Acoust. Soc. Amer. **47**, 82(A) (1970); U. S. Naval Acad. Tech. Rep. No. C-1 (Feb. 1970).
[18] L. Crum and A. Eller, Acoust. Res. Lab., Harvard Univ. Tech. Memo. No. 61 (Jan. 1969).
[19] Ref. 3, pp. 132–134.
[20] Ref. 3, p. 22.
[21] Ref. 3, pp. 63–69; 140–153. Also, R. E. Apfel, "Vapor Nucleation at a Liquid–Liquid Interface," J. Chem. Phys. **54** (Jan. 1971).

Author Index

Subject Index